Methods in Enzymology

Volume 344

G Protein Pathways

Part B
G Proteins and Their Regulators

EDITED BY

Ravi Iyengar

MOUNT SINAI SCHOOL OF MEDICINE
NEW YORK, NEW YORK

John D. Hildebrandt

MEDICAL UNIVERSITY OF SOUTH CAROLINA
CHARLESTON, SOUTH CAROLINA

D1562328

ACADEMIC PRESS

San Diego   London   Boston   New York   Sydney   Tokyo   Toronto

Academic Press
A Harcourt Science and Technology Company
525 B Street, Suite 1900, San Diego, California 92101-4495, USA
http://www.academicpress.com

Academic Press
Harcourt Place, 32 Jamestown Road, London NW1 7BY, UK
http://www.academicpress.com

International Standard Book Number: 0-12-182245-1

PRINTED IN THE UNITED STATES OF AMERICA
01 02 03 04 05 06 07 SB 9 8 7 6 5 4 3 2 1

Table of Contents

Section I. Activation of G Proteins by Receptors or Other Regulators

v

Section III. Functional Analysis of G Protein Subunits

Section V. RGS Proteins and Signal Termination

Contributors to Volume 344

Article numbers are in parentheses following the names of contributors.
Affiliations listed are current.

DONALD M. APANOVITCH (44), *Bayer Corporation, West Haven, Connecticut 06516*

INAKI AZPIAZU (8), *Department of Anesthesiology, Washington University Medical School, St. Louis, Missouri 63110*

CATHERINE H. BERLOT (19, 32), *Department of Cellular and Molecular Physiology, Yale University School of Medicine, New Haven, Connecticut 06520*

MICHAEL L. BERNARD (37), *Department of Pharmacology, Medical University of South Carolina, Charleston, South Carolina 29403*

LAURIE BETTS (49), *Department of Pharmacology, and UNC Biomolecular X-ray Data Collection Facility, The University of North Carolina at Chapel Hill, Chapel Hill, North Carolina 27599*

LUTZ BIRNBAUMER (20), *Department of Anesthesiology, School of Medicine, University of California, Los Angeles, California 90095*

TRILLIUM BLACKMER (30), *Institute for Neuroscience and Department of Molecular Pharmacology and Biological Chemistry, Northwestern University Medical School, Chicago, Illinois 60613*

GUYLAIN BOULAY (20), *Département de Pharmacologie, Faculté de Médecine, Université de Sherbrooke, Sherbrooke, Québec, Canada J1H 5N4*

GREG M. BROTHERS (49, 51), *Ontario Cancer Institute/Amgen Institute, University of Toronto, Toronto, Ontario M5G 2C1, Canada*

GEORGE P. BROWN (25), *Department of Pharmacology, Mount Sinai School of Medicine, New York, New York 10029*

ELIZABETH BUCK (36), *Department of Pharmacology, Mount Sinai School of Medicine, New York, New York 10029*

MARK BUSMAN (41), *Department of Cell and Molecular Pharmacology and Experimental Therapeutics, Medical University of South Carolina, Charleston, South Carolina 29425*

THERESA M. CABRERA-VERA (5), *Department of Molecular Pharmacology and Biochemistry, Institute for Neuroscience, Northwestern University, Chicago, Illinois 60611*

VANESSA CHANG (18, 35), *Department of Anesthesiology, Washington University School of Medicine, St. Louis, Missouri 63110*

PETER CHIDIAC (48), *Department of Pharmacology and Toxicology, Faculty of Medicine and Dentistry, University of Western Ontario, London, Ontario N6A 5C1, Canada*

STEVEN R. CHILDERS (3), *Department of Physiology and Pharmacology, Center for Investigative Neuroscience, Wake Forest University School of Medicine, Winston-Salem, North Carolina 27157*

STEPHEN CHUNG (49), *Ontario Cancer Institute/Amgen Institute, University of Toronto, Toronto, Ontario M5G 2C1, Canada*

MARY J. CISMOWSKI (11), *Neurocrine Biosciences, Incorporated, San Diego, California 92121*

LANA A. COOK (16, 41), *Department of Pharmacology, Medical University of South Carolina, Charleston, South Carolina 29425*

YEHIA DAAKA (30), *Department of Surgery, Duke University Medical Center, Durham, North Carolina 27710*

E. J. DELL (30), *Institute for Neuroscience and Department of Molecular Pharmacology and Biological Chemistry, Northwestern University Medical School, Chicago, Illinois 60613*

KARYN M. DEPREE (5), *Department of Pharmacology and Toxicology, West Virginia University, Morgantown, West Virginia 26506*

JONATHAN M. DERMOTT (21), *Laboratory of Molecular Immunoregulation, National Cancer Institute-Frederick Cancer Research and Development Center, Frederick, Maryland 21702*

LUC DE VRIES (46), *Department of Cellular and Molecular Medicine, University of California, La Jolla, California 92093*

N. DHANASEKARAN (21), *Department of Biochemistry, Fels Institute for Cancer Research and Molecular Biology, Temple University School of Medicine, Philadelphia, Pennsylvania 19140*

JANE DINGUS (13, 15, 16, 34), *Department of Pharmacology, Medical University of South Carolina, Charleston, South Carolina 29425*

MARK D. DISTEFANO (18), *Department of Chemistry, University of Minnesota, Minneapolis, Minnesota 55455*

HENRIK G. DOHLMAN (43, 44), *Department of Biochemistry and Biophysics, University of North Carolina at Chapel Hill, Chapel Hill, North Carolina 27599*

MERCEDES DOSIL (7), *Departamento de Bioquímica y Biologica Molecular, Campus Miguel de Unamuno, University of Salamanca, 37007 Salamanca, Spain*

EMIR DUZIC (11), *Millennium Pharmaceuticals, Incorporated, Cambridge, Massachusetts 02139*

CATHERINE A. DYE (20), *Nina Ireland Laboratory of Developmental Neurobiology, Langley Porter Psychiatric Institute, University of California, San Francisco, California 94143*

MARILYN GIST FARQUHAR (46), *Department of Cellular and Molecular Medicine, University of California, La Jolla, California 92093*

MICHAEL FREISSMUTH (33), *Institute of Pharmacology, University of Vienna, A-1090 Vienna, Austria*

MARTHA E. GADD (48), *Department of Pharmacology, Emory University School of Medicine, Atlanta, Georgia 30322*

TIFFANY RUNYAN GARRISON (43, 44), *Department of Pharmacology, Yale University School of Medicine, New Haven, Connecticut 06536*

N. GAUTAM (8, 18, 35), *Departments of Anesthesiology and Genetics, Washington University School of Medicine, St. Louis, Missouri 63110*

KYLE R. GEE (29), *Molecular Probes, Inc., Eugene, Oregon 97402*

ANNETTE GILCHRIST (4), *Department of Molecular Pharmacology and Biochemistry, Institute for Neuroscience, Northwestern University, Chicago, Illinois 60611*

ALFRED G. GILMAN (47), *Department of Pharmacology, University of Texas Southwestern Medical Center, Dallas, Texas 75390*

STEPHEN G. GRABER (5), *Department of Pharmacology and Toxicology, West Virginia University, Morgantown, West Virginia 26506*

HEIDI E. HAMM (4, 5, 30, 41), *Department of Pharmacology, Vanderbilt University Medical Center, Nashville, Tennessee 37232*

WEI HE (50), *Department of Biochemistry and Molecular Biology, Baylor College of Medicine, Houston, Texas 77030, and Department of Cell Biology, Memorial Sloan-Kettering Cancer Center, New York, New York 10021*

JOHN R. HEPLER (48), *Department of Pharmacology, Emory University School of Medicine, Atlanta, Georgia 30322*

JOHN D. HILDEBRANDT (10, 13, 15, 16, 34, 41), *Department of Pharmacology, Medical University of South Carolina, Charleston, South Carolina 29425*

YEE-KIN HO (9), *Departments of Biochemistry and Molecular Biology, University of Illinois College of Medicine, Chicago, Illinois 60612*

GINGER A. HOFFMAN (43), *Interdepartmental Neuroscience Program, Yale University, New Haven, Connecticut 06536*

YONGMIN HOU (35), *Department of Anesthesiology, Washington University School of Medicine, St. Louis, Missouri 63110*

XI-PING HUANG (22, 23), *Department of Pharmacology, Diabetes and Metabolic Diseases Research Center, University Medical Center, State University of New York, Stony Brook, New York 11794*

RAVI IYENGAR (24, 25, 36), *Department of Pharmacology, Mount Sinai School of Medicine, New York, New York 10029*

GREGOR JANSEN (6), *Eukaryotic Genetics Department, NRC Biotechnology Research Institute, Montreal, Québec, Canada H4P 2R2*

MEISHENG JIANG (20), *Department of Anesthesiology, School of Medicine, University of California, Los Angeles, California 90095*

RICHARD A. KAHN (14), *Department of Biochemistry, Emory University School of Medicine, Atlanta, Georgia 30322*

TAMARA A. KALE (18), *Department of Chemistry, Washington University School of Medicine, St. Louis, Missouri 63110*

HEE C. KANG (29), *Molecular Probes, Inc., Eugene, Oregon 97402*

DANIEL R. KNAPP (34), *Department of Pharmacology, Medical University of South Carolina, Charleston, South Carolina 29425*

JAMES B. KONOPKA (7), *Department of Molecular Genetics and Microbiology, State University of New York, Stony Brook, New York 11794*

ANDREJS M. KRUMINS (47), *Department of Pharmacology, University of Texas Southwestern Medical Center, Dallas, Texas 75390*

OLIVER KUDLACEK (33), *Institute of Pharmacology, University of Vienna, A-1090 Vienna, Austria*

STEPHEN M. LANIER (10, 11, 37), *Department of Pharmacology, Louisiana State University Health Sciences Center, New Orleans, Louisiana 70118*

EKKEHARD LEBERER (6), *Eukaryotic Genetics Department, NRC Biotechnology Research Institute, Montreal, Québec, Canada H4P 2R2; and Department of Experimental Medicine, McGill University, Montreal, Québec, Canada H3A 1B1*

REHWA HO LEE (9), *Department of Anatomy and Cell Biology, University of California Los Angeles, School of Medicine and Developmental Neurology Laboratory, Veteran Administration Medical Center, Sepulveda, California 91343*

ANLI LI (4), *Institute for Neuroscience, Northwestern University, Chicago, Illinois 60611*

OLIVIER LICHTARGE (38), *Department of Molecular and Human Genetics, Program in Structural and Computational Biology and Molecular Biophysics, Baylor Human Genome Sequencing Center, Program in Developmental Biology, Baylor College of Medicine, Houston, Texas 77030*

JIE LIU (27), *Metabolic Diseases Branch, National Institute of Diabetes, Digestive, and Kidney Diseases, National Institutes of Health, Bethesda, Maryland 20892*

CHIENLING MA (11), *OSI Pharmaceuticals, Incorporated, Tarrytown, New York 10591*

CRAIG C. MALBON (22, 23), *Department of Pharmacology, Diabetes and Metabolic Diseases Research Center, University Medical Center, State University of New York, Stony Brook, New York 11794*

DAVID R. MANNING (1), *Department of Pharmacology, University of Pennsylvania School of Medicine, Philadelphia, Pennsylvania 19104*

ANGELES MARTÍN-REQUERO (20), *Centro de Investigaciones Biologicas, Consejo Superior de Investigaciones Cientificas, 28006 Madrid, Spain*

DYKE P. MCEWEN (29), *Department of Pharmacology, University of Michigan, Ann Arbor, Michigan 48109*

WILLIAM E. MCINTIRE (13, 34), *Department of Pharmacology, Medical University of South Carolina, Charleston, South Carolina 29425; and Department of Pharmacology, University of Virginia School of Medicine, Charlottesville, Virginia 22908*

PAMELA E. MENTESANA (7), *Department of Molecular Genetics and Microbiology, State University of New York, Stony Brook, New York 11794*

DAVID MICHAELSON (17), *Departments of Medicine and Cell Biology, New York University School of Medicine, New York, New York 10016*

SUCHETANA MUKHOPADHYAY (26), *Department of Biological Sciences, Purdue University, West Lafayette, Indiana 47907*

SUSANNE M. MUMBY (28), *Department of Pharmacology, University of Texas Southwestern Medical Center, Dallas, Texas 75390*

CHRISTIAN NANOFF (33), *Institute of Pharmacology, University of Vienna, A-1090 Vienna, Austria*

RICHARD R. NEUBIG (29), *Departments of Pharmacology and Internal Medicine, University of Michigan, Ann Arbor, Michigan 48109*

JOHN K. NORTHUP (2), *Laboratory of Cellular Biology, National Institute on Deafness and Other Communication Disorders, National Institutes of Health, Rockville, Maryland 20850*

ANNE PHILIPPI (38), *Department of Molecular and Human Genetics, Baylor College of Medicine, Houston, Texas 77030*

MARK R. PHILIPS (17), *Departments of Medicine and Cell Biology, New York University School of Medicine, New York, New York 10016*

SERGUEI POPOV (45), *Advanced Biosystems, Manassas, Virginia 20110*

VICTOR REBOIS (2), *Laboratory of Molecular and Cellular Neurobiology and Biology, National Institute of Neurological Diseases and Stroke, National Institutes of Health, Bethesda, Maryland 20892*

EITAN REUVENY (30), *Department of Biological Chemistry, Weizmann Institute of Science, Rehovot 76100, Israel*

CATALINA RIBAS (10), *Centro de Biología Molecular "Severo Ochoa" (C51C-UAM), Universidad Autónoma de Madrid, Cantoblanco, 28049 Madrid, Spain*

JANET D. ROBISHAW (31), *Weis Center for Research, Geisinger Clinic, Danville, Pennsylvania 17822*

RAMON ROSAL (30), *Division of Environmental Health Sciences, School of Public Health, Columbia University, New York, New York 10032*

THOMAS A. ROSENQUIST (22, 23), *Department of Pharmacology, Transgenic Mice Facility, University Medical Center, State University of New York, Stony Brook, New York 11794*

ELLIOTT M. ROSS (26, 42), *Department of Pharmacology, University of Texas Southwestern Medical Center, Dallas, Texas 75390*

UWE RUDOLPH (20), *Institute of Pharmacology, University of Zürich, CH-8057, Zürich, Switzerland*

TARA ANN SANTORE (24), *Department of Pharmacology, Mount Sinai School of Medicine, New York, New York 10029*

MOTOHIKO SATO (10), *First Department of Internal Medicine, Asahikawa Medical College, Hokkaido 078-8510, Japan*

KEVIN L. SCHEY (16, 34, 41), *Department of Pharmacology, Medical University of South Carolina, Charleston, South Carolina 29425*

PETER SCHUCK (2), *Molecular Interactions Resource, Division of Bioengineering and Physical Science, Office of Research Services, National Institutes of Health, Bethesda, Maryland 20892*

WILLIAM SCHWINDINGER (31), *Weis Center for Research, Geisinger Clinic, Danville, Pennsylvania 17822*

JAMIE K. SCOTT (39), *Department of Molecular Biology and Biochemistry, Simon Fraser University, Burnaby, British Columbia, Canada V5A 1S6*

LEE R. SHEKTER (30), *Department of Pharmacological and Physiological Sciences, University of Chicago, Chicago, Illinois 60637*

DAVID P. SIDEROVSKI (49, 51), *Department of Pharmacology, UNC Neuroscience Center, and Lineberger Comprehensive Cancer Center, The University of North Carolina at Chapel Hill, Chapel Hill, North Carolina 27599*

LAURA J. SIM-SELLEY (3), *Department of Pharmacology and Toxicology and Institute for Drug and Alcohol Studies, Virginia Commonwealth University Medical College of Virginia, Richmond, Virginia 23298*

NIKOLAI P. SKIBA (30), *Department of Ophthalmology, Harvard Medical School, Howe Laboratories, Boston, Massachusetts 02114*

ALAN V. SMRCKA (39), *Department of Pharmacology and Physiology, University of Rochester, School of Medicine and Dentistry, Rochester, New York 14642*

BRYAN E. SNOW (49, 51), *Ontario Cancer Institute/Amgen Institute, University of Toronto, Toronto, Ontario M5G 2C1, Canada*

JOHN SONDEK (49), *Departments of Biochemistry and Biophysics, and Pharmacology, Lineberger Comprehensive Cancer Center, The University of North Carolina at Chapel Hill, Chapel Hill, North Carolina 27599*

XIAOSONG SONG (22), *Department of Pharmacology, Diabetes and Metabolic Diseases Research Center, University Medical Center, State University of New York, Stony Brook, New York 11794*

MATHEW E. SOWA (38), *Program in Structural and Computational Biology and Molecular Biophysics, Department of Biochemistry and Molecular Biology, W. M. Keck Center for Computational Biology, Baylor College of Medicine, Houston, Texas 77030*

KARSTEN SPICHER (20), *Institut für Pharmakologie, Freie Universität Berlin, 14195 Berlin, Germany*

AYA TAKESONO (11), *National Human Genome Research Institute, National Institutes of Health, Bethesda, Maryland 20892*

WEI-JEN TANG (12), *Ben May Institute for Cancer Research, The University of Chicago, Chicago, Illinois 60637*

JIANGCHUAN TAO (22), *Department of Pharmacology, Diabetes and Metabolic Diseases Research Center, University Medical Center, State University of New York, Stony Brook, New York 11794*

BRONWYN S. TATUM (15), *Department of Pharmacology, Medical University of South Carolina, Charleston, South Carolina 29425*

DAVID Y. THOMAS (6), *Eukaryotic Genetics Department, NRC Biotechnology Research Institute, Montreal, Québec, Canada H4P 2R2; and Departments of Biology, and Anatomy and Cell Biology, McGill University, Montreal, Québec, Canada H3A 1B1*

TARITA O. THOMAS (5), *Department of Molecular Pharmacology and Biochemistry, Institute for Neuroscience, Northwestern University, Chicago, Illinois 60611*

TUOW DANIEL TING (9), *Departments of Biochemistry and Molecular Biology, University of Illinois College of Medicine, Chicago, Illinois 60612*

TAMMY C. TUREK (18), *Department of Chemistry, Washington University School of Medicine, St. Louis, Missouri 63110*

GOVINDAN VAIDYANATHAN (15), *Department of Pharmacology, Medical University of South Carolina, Charleston, South Carolina 29425*

JURGEN VANHAUWE (5), *Department of Molecular Pharmacology and Biochemistry, Institute for Neuroscience, Northwestern University, Chicago, Illinois 60611*

HILLARY A. VAN VALKENBURGH (14), *GlaxoSmithKline, King of Prussia, Pennsylvania 19406*

HSIEN-YU WANG (22, 23), *Department of Physiology and Biophysics, University Medical Center, State University of New York, Stony Brook, New York 11794*

QIN WANG (31), *Department of Pharmacology, University of Michigan Medical School, Ann Arbor, Michigan 48109*

LEE S. WEINSTEIN (27), *Metabolic Diseases Branch, National Institute of Diabetes, Digestive, and Kidney Diseases, National Institutes of Health, Bethesda, Maryland 20892*

GEZHI WENG (40), *Department of Pharmacology, Mount Sinai School of Medicine, New York, New York 10029*

THEODORE G. WENSEL (50), *Department of Biochemistry and Molecular Biology, Baylor College of Medicine, Houston, Texas 77030*

MALCOLM WHITEWAY (6), *Eukaryotic Genetics Department, NRC Biotechnology Research Institute, Montreal, Québec, Canada H4P 2R2; and Department of Biology, McGill University, Montreal, Québec, Canada H3A 1B1*

MICHAEL D. WILCOX (13, 16), *Department of Pharmacology, Medical University of South Carolina, Charleston, South Carolina 29425*

THOMAS M. WILKIE (45), *Department of Pharmacology, University of Texas Southwestern Medical Center, Dallas, Texas 75390*

ROLF T. WINDH (1), *Adlolor Corporation, Malvern, Pennsylvania 19355*

GUANGYU WU (37), *Department of Internal Medicine, University of Cincinnati, Cincinnati, Ohio 45267*

SHUI-ZHONG YAN (12), *Ben May Institute for Cancer Research, The University of Chicago, Chicago, Illinois 60637*

DAVID YOWE (45), *Millennium Pharmaceuticals, Cambridge, Massachusetts 02171*

KAN YU (45), *Department of Pharmacology, University of Texas Southwestern Medical Center, Dallas, Texas 75390*

SHUHUA YU (27), *Laboratory of Biochemical Genetics, National Heart, Lung, and Blood Institute, National Institutes of Health, Bethesda, Maryland 20892*

Preface

The heterotrimeric G proteins are the central component of one of the primary mechanisms used by eukaryotic cells to receive, interpret, and respond to extracellular signals. Many of the basic concepts associated with the entire range of fields collectively referred to in terms such as "signal transduction" originate from the pioneering work of Sutherland, Krebs and Fischer, Rodbell and Gilman, Greengard, and others on systems that are primary examples of G protein signaling pathways. The study of these proteins has been and remains at the forefront of research on cell signaling mechanisms. The G Protein Pathways volumes (343, 344, and 345) of *Methods in Enzymology* have come about as part of a continuing attempt to use the methods developed in studying the G protein signaling pathways as a resource both within this field and throughout the signal transduction field.

Several volumes of this series have been devoted in whole or in part to approaches for studying the heterotrimeric G proteins. Volumes 109 and 195 were the earliest to devote substantial parts to G protein-mediated signaling systems. In 1994 Volumes 237 and 238 comprehensively covered this field. The continued growth, the ever increasing impact of this field, and the continued evolution of approaches and questions generated by research related to G proteins have led inevitably to the need for a new and comprehensive treatment of the approaches used to study these proteins. Each volume of G Protein Pathways brings together varied topics and approaches to this central theme.

Very early in the development of the concepts of G protein signaling mechanisms, Rodbell and Birnbaumer recognized at least three components of these signaling systems. They compared them to a receiver, a transducer, and an amplifier. The receiver was the receptor for an extracellular signal. The transducer referred to those mechanisms and components required for converting an extracellular signal into an intracellular response. The amplifier was synonymous with the effector enzymes that generate the beginning of the intracellular signal. Over the years these ideas have evolved in many ways. We now know an immense amount about the receptors and their great range of diversity. Through the work of Gilman, along with his associates and contemporaries, the transducer component turned out to be nearly synonymous with the heterotrimeric G proteins themselves. Nevertheless, the complexity of this component of the system continues to become more and more apparent with the recognition of the diversity of these proteins, the many ways they interact, and the increasing number of regulatory influences on their function. The key concepts associated with the effector enzymes, such as adenylyl cyclase that produces the intracellular "second messenger" cAMP, have been broadened substantially to include other enzymes, ion channels, and the components of other signaling systems that form an interacting network of systems inside cells.

The organization of Volumes 343, 344, and 345 is still conveniently centered on these three components of the G proteins signaling pathway: receptor, G protein, and effector. The evolution of the field, however, inevitably left a mark on the form of these volumes. So, for example, receptors in Volume 343 and G proteins in Volume 344 no longer stand alone as individual components in these volumes, but share space with other directly interacting proteins that influence their function. In addition, Volume 345 addresses, more generally, effector mechanisms and forms a bridge between the many different cell regulatory mechanisms that cooperate to control cellular function. Thus, there are chapters that include methods related to small GTP binding proteins, ion channels, gene regulation, and novel signaling compounds.

As we learn more about G protein signaling systems, we acquire an ever increasing appreciation of their complexity. In the previous volumes on G proteins it was already evident that there were many different isoforms of each of the three heterotrimeric G protein subunits. Initial analysis of the Human Genome Project suggests some constraints on the number of members of these proteins with 27 α subunit isoforms, 5 β subunits, and 13 γ subunits. It is interesting that in recent years we had nearly accounted for all or most of the β and γ subunit isoforms, but that there are nearly twice as many potential α subunits as the ones we currently understand. These recent discoveries, along with the recognition of the existence of between 600 and 700 G protein-coupled receptors in the human genome, may place limits on the complexity of the G protein signaling system itself. These potential limits, however, are balanced by the immense number of possible combinations of interactions that can be generated from all these components, by the possible variation of all of these proteins at levels after their genomic structure, and by our continual discovery of additional interacting components of the system.

Perhaps one of the really substantial gains in our understanding of this system since the earlier volumes is the increasing recognition of the role of accessory proteins in G protein signaling pathways. Those proteins recognized nearly 20 years ago that work at the level of the receptor continue to grow and have a prominent place in Volume 343. One of the real breakthroughs though has been the rapid development of our knowledge of accessory proteins that interact with the G proteins themselves. Prominent among these are the RGS (regulators of G protein signaling) proteins that act as GTPase activating proteins for selective G protein α subunits. A fairly substantial section of these volumes is devoted to these proteins. One apparent aside related to the RGS proteins though, is that as much as we have rapidly learned about them, there is much more yet to be learned, because there is wide variation in the structure of these proteins outside their G protein interaction sites. These proteins likely have many different stories yet to be developed based on their interactions with the G proteins, perhaps mediating functions that we do not yet know about. The RGS proteins are not the end of the interacting proteins either, however, with the description of additional G protein-interacting proteins such as the AGS (activators of G protein signaling) proteins. These are likely a heterogeneous group of proteins with several different mechanisms of interaction and roles in G protein signaling mechanisms.

The range of topics covered in these volumes turned out to be quite large. This is a result of the wide range of approaches that creative scientists can develop to gain an understanding of a complex and rapidly evolving field. There are several chapters that provide a theoretical basis for the analysis and interpretation of data, several chapters on the application of modeling techniques at several different levels, and chapters on structural biology approaches, classical biochemical techniques juxtaposed to protein engineering, molecular biology, gene targeting strategies addressing physiological questions, and DNA array approaches to evaluating the effects of pathway activation. In all likelihood, the G protein signaling field will continue to be one that moves at the forefront of scientific approaches to studying events at the interface between biochemistry and molecular biology, on the one hand, and physiology and cell biology, on the other. Thus, it is our hope that these volumes will serve a scientific readership beyond those that study G proteins per se, or even those that study cell signaling mechanisms. We would hope that the approaches and techniques described here would hold relevance for those large number of scientists involved in many different kinds of projects that address the interface between the molecular/biochemical world and the cell/tissue/organism world.

We owe a tremendous debt of gratitude to our colleagues who so readily contributed chapters to these volumes. Truly, without their so willing participation this work would not have evolved into as substantial and comprehensive a work as it ultimately became. We would also like to thank Ms. Shirley Light for her support, encouragement, and patience throughout this long process.

JOHN D. HILDEBRANDT
RAVI IYENGAR

METHODS IN ENZYMOLOGY

Section I

Activation of G Proteins by Receptors or Other Regulators

[1] Analysis of G Protein Activation in Sf9 and Mammalian Cells by Agonist-Promoted [^{35}S]GTPγS Binding

By ROLF T. WINDH and DAVID R. MANNING

Introduction

G-protein-coupled receptors (GPCRs) mediate many of their effects through the activation of heterotrimeric G proteins. The activated receptor accelerates an exchange of GTP for GDP on the G protein α subunit. The corresponding change in conformation of the α subunit results in the release of $\beta\gamma$, and both the GTP-bound α subunit and released $\beta\gamma$ dimers interact with a variety of effectors, including adenylyl cyclase, phospholipases, and ion channels. In time, the intrinsic GTP hydrolysis activity of the α subunit converts GTP to GDP, and the GDP-bound α subunit reassociates with $\beta\gamma$ to reform a heterotrimer.

Traditional analysis of GPCR signaling has relied on changes in the activity of downstream effectors as readouts for receptor and G protein function. Amplification at each step in the transduction pathway makes events distal to receptor activation easy to detect biochemically, and a great deal has been deduced about both signal transduction pathways and receptor–ligand interactions from these studies. It is clear, however, that receptor pharmacology is often system-dependent; that is, the relative efficacies of a variety of receptor ligands differ depending on which of various downstream events is measured.[1] Differences in the type of G protein activated or which G protein subunit interacts with the effector, as well as differential regulation of each step of the amplification scheme, can contribute to this system-dependent pharmacology.

By measuring the activation of G proteins themselves, the first step in the traditional signaling cascade, some of these issues can be bypassed. Direct assessment of G protein activation by G-protein-coupled receptors is typically accomplished using either of two conserved features of the G protein cycle. One group of assays measures agonist-induced increases in the rate of GTP hydrolysis by the α subunit.[2] In these GTPase assays, the release of inorganic phosphate from radiolabeled GTP is determined. These assays can be performed either as single turnover studies, by preloading the G proteins with [^{32}P]GTP, or as steady-state production of inorganic [^{32}P]phosphate release in the continued presence of [^{32}P]GTP.

Exchange of GDP for analogs of GTP on the α subunit is the basis for the other major assay of G protein activation by receptor. Hydrolysis-resistant, radiolabeled

---

[1] T. Kenakin, *Pharmacol. Rev.* **48**, 413 (1996).
[2] D. Cassel and Z. Selinger, *Biochem. Biophys. Acta* **752**, 538 (1976).

forms of GTP are used to monitor the exchange. The most widely used GTP analog is guanosine $5'$-(γ-[^{35}S]thio)triphosphate ([^{35}S]GTPγS). Since GTPγS cannot be hydrolyzed to GDP, GDP–GTPγS exchange assays measure the progression of an irreversible activation rather than steady-state activation/deactivation cycles. As GDP release is the rate-limiting step in the activation of G proteins,[3] however, these assays can measure a highly relevant aspect of G protein activation. Furthermore, because the background binding can be controlled more tightly than in GTP hydrolysis studies, the signal strength for GDP–GTPγS exchange assays tends to be greater than for GTPase assays.

Whereas direct assay of G protein activity, rather than changes in effector activity, provides a closely linked measure of receptor–ligand interactions, neither GTPase nor GDP–GTPγS exchange assays, when applied to cell membranes, indicate which G proteins the receptor is engaging. Identifiying the G proteins that couple to a given receptor provides a great deal of information about the downstream pathways to be regulated. It has become clear that receptors can couple to multiple G proteins. A receptor often couples to multiple members of a single family of G proteins; a receptor that activates G_{i2} will probably also activate other members of the G_i family.[4,5] Whereas it is not surprising that structurally related family members will couple to the same receptor, it is not uncommon for a receptor to engage members of two, three, or even all four families of G proteins.[6–9]

We describe here a GDP–GTPγS exchange assay in which the [^{35}S]GTPγS-bound G protein α subunit is immunoprecipitated with subtype-specific antisera, making it possible to confidently identify which subtypes of G proteins are being activated by receptor in membranes expressing multiple G proteins. In addition to providing a means of identifying which G proteins are activated by a given receptor, the immunoprecipitation method also greatly improves the signal-to-noise ratio. In bypassing the binding of [^{35}S]GTPγS to the filtration membrane or other GTP-binding proteins, the immunoprecipitation method greatly reduces background binding without significantly altering binding induced by ligand. Thus the measured GTPγS binding following activation represents a larger increase over basal, as high as 20-fold in some preparations.[10] This increased signal strength

[3] B. R. Conklin and H. R. Bourne, *Cell* **73,** 631 (1993).

[4] T. W. Gettys, T. A. Fields, and J. R. Raymond, *Biochemistry* **33,** 4283 (1994).

[5] P. Butkerait, Y. Zheng, H. Hallak, T. E. Graham, H. A. Miller, K. D. Burris, P. B. Molinoff, and D. R. Manning, *J. Biol. Chem.* **270,** 18691 (1995).

[6] D. T. Hung, Y. H. Wong, T. K. Vu, and S. R. Coughlin, *J. Biol. Chem.* **267,** 20831 (1992).

[7] S. Offermanns, K.-L. Laugwitz, K. Spicher, and G. Schultz, *Proc. Nat. Acad. Sci. U.S.A.* **91,** 504 (1994).

[8] K.-L. Laugwitz, A. Allgeier, S. Offermanns, K. Spicher, J. Van Sande, J. E. Dumont, and G. Schultz, *Proc. Nat. Acad. Sci. U.S.A.* **93,** 116 (1996).

[9] A. J. Barr, L. F. Brass, and D. R. Manning, *J. Biol. Chem.* **272,** 2223 (1997).

[10] R. T. Windh and D. R. Manning, unpublished observations.

allows for more accurate ranking of efficacies for partial agonists and partial inverse agonists as well as a more confident assignment of the G protein activated. We describe these assays using cell expression models that we have employed successfully—Sf9 (*Spodoptera frugiperda*) cells expressing mammalian receptors and G proteins,[9,11,12] and HEK293 (human embryonal kidney) cells[12,13] and CHO (Chinese hamster ovary) cells[10] in which selected receptors have been introduced by transfection. In some instances, we have been able to measure activation of G proteins in mammalian cells through receptors endogenous to these cells,[12] although not in all instances because of issues of sensitivity.

Experimental Procedures

Preparation of Protein A

To 0.5 g of protein A immobilized on Sepharose CL-4B (Sigma, St. Louis, MO), add 10 ml of 10 mM HEPES, pH 7.4, and shake gently for 30 min at 4° to allow the beads to swell and to wash away stabilizers added for storage and shipment. Pellet the protein A-Sepharose beads by centrifuging for 5 min at approximately 4000g. Pour off the supernatant, and add 10 ml of ice-cold wash buffer (50 mM Tris-HCl, pH 7.4, 150 mM NaCl, 20 mM MgCl$_2$) containing 0.5% (v/v) Nonidet P-40 (NP-40) and 4 mg/ml of bovine serum albumin (BSA) to block nonspecific binding sites on the Sepharose beads. Mix to resuspend the slurry, and shake 30 min at 4°. Pellet the protein A-Sepharose beads as before, and wash the pellet twice by resuspending in 10 ml of cold wash buffer with NP-40 but without BSA. Finally, resuspend the pellet in 10 ml of cold wash buffer containing NP-40 and 0.33% (v/v) aprotinin (Sigma, St. Louis, MO; or 5μg protein/ml), and store at 4°.

Preparation of Cell Membranes

HEK293 or CHO cells (or Sf9 cells; see Ref. 14 for growth, infection, and membrane preparation protocols) grown in monolayer and expressing the desired receptors (and G proteins in Sf9 cells) should be washed several times on the plate with cold phosphate-buffered saline and left on ice. Add 0.5 ml/plate of ice-cold HE/PI buffer (20 mM HEPES, pH 8.0, 2 mM EDTA; add protease inhibitors [2 μg/ml aprotinin, 10 μg/ml leupeptin, and 0.1 mM phenylmethylsulfonyl fluoride (PMSF)] immediately before using, scrape cells, and transfer

[11] A. J. Barr and D. R. Manning, *J. Biol. Chem.* **272,** 32979 (1997).
[12] R. T. Windh, M.-J. Lee, T. Hla, S. An, A. J. Barr, and D. R. Manning, *J. Biol. Chem.* **274,** 27351 (1999).
[13] Y. Wang, R. T. Windh, C. A. Chen, and D. R. Manning, *J. Biol. Chem.* **274,** 37435 (1999).
[14] R. T. Windh and D. R. Manning, *Methods Enzymol.* **343,** 417 (2002).

to a microfuge tube. Break the scraped cells by passing gently through a 26-gauge needle 15 times. Pellet the nuclei and unbroken cells by centrifuging for 5 min at 660g, and spin the supernatant for 30–45 min at 20,000g, to pellet the membranes. Resuspend the pellet in HE/PI buffer such that the final protein concentration is 1–3 mg/ml (usually 0.1 or 0.25 ml HE/PI buffer per original confluent 10 cm plate of CHO or HEK293 cells, respectively). Determine protein concentration, aliquot the membranes into lots appropriate for individual assays, freeze in a dry ice/ethanol bath, and store at −70°. Membranes prepared in this manner can be thawed and used only once and keep well for at least 2–3 months.

$[^{35}S]GTP\gamma S$ Assay Overview

A standard assay to demonstrate agonist-induced activation of a given G protein requires a minimum of three conditions (Fig. 1). In the first two conditions, the G protein α subunit is immunoprecipitated with a subunit-selective antiserum following incubation of the membranes with $[^{35}S]GTP\gamma S$ in the absence or

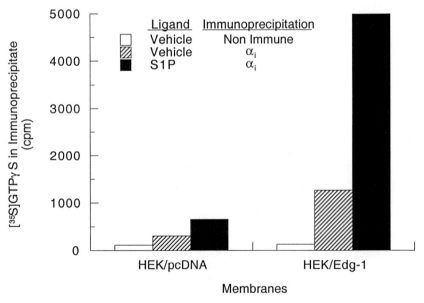

FIG. 1. Typical $[^{35}S]GTP\gamma S$ binding experiment in HEK293 cell membranes. Membranes from HEK293 cells transfected with vector (HEK/pcDNA) or Edg-1 (HEK/Edg-1) were resuspended in TMEN buffer containing GDP (10 μM) and incubated with $[^{35}S]GTP\gamma S$ (1 nM) for 10 min in the absence (hatched bars) or presence (filled bars) of sphingosine 1-phosphate (S1P, 10 μM). Endogenous G_i was immunoprecipitated subsequently, using an antiserum raised against the C terminus of $\alpha_{i1/i2}$. Nonimmune serum (open bars) was used as a control.

presence of agonist. The third condition serves as a control for the immunoprecipitation step; membranes that have been incubated with [^{35}S]GTPγS and vehicle are carried through the immunoprecipitation protocol with a preimmune or nonimmune serum. Each condition is performed in duplicate, so the most basic assay is six individual assay points.

Precoupling Antibodies

Prepare two sets of 1.5 ml microcentrifuge tubes, the number of tubes in each set corresponding to the number of assay points; one set of tubes will be used to "preclear" the samples, and the other will be used for the primary immunoprecipitation. In each preclear tube, add 100 μl of a pansorbin cell suspension (as supplied by Calbiochem, San Diego, CA), and pellet the cells by centrifuging at 4° for 1 min at 10,000g. Aspirate the supernatant, add 2 μl of nonimmune serum and 25 μl of wash buffer (50 mM Tris-HCl, pH 7.4, 150 mM NaCl, and 20 mM MgCl$_2$), and vortex to completely resuspend the pellet. To each primary immunoprecipitation tube, add 10 μl of nonimmune or immune serum and 100 μl of protein A-Sepharose in IP buffer (see above). Mix well, and shake both sets of tubes for 2 hr or more at 4° in a reciprocal or orbital shaker.

Preparing Assay Reagents

Thaw frozen membranes quickly at 30°, and centrifuge at 4° for 30 min at 20,000g to pellet them. During this time, add GDP to the TMEN buffer (50 mM Tris-HCl, pH 7.4, 4.8 mM MgCl$_2$, 2 mM EDTA, and 100 mM NaCl), making enough to resuspend the membranes and to dilute the [^{35}S]GTPγS (see below). GDP concentrations will vary depending on which G protein is being assayed. Although the optimal concentrations will depend on the cell type and the goals of the assay and thus should be determined experimentally, good starting points are 0.1–1 μM for G$_q$, G$_{12}$, G$_{13}$, and G$_z$, 1–10 μM for G$_s$, and 10–30 μM for G$_{i1-3}$ and G$_o$. When the membranes have pelleted, aspirate the supernatant and resuspend the pellet in TMEN/GDP by passing through a 26-gauge needle 15 times such that 55 μl contains 20 μg of membrane protein.

Since 2 μl of ligand is added into a total final volume of 62 μl, ligands must be prepared at 31-fold the desired final concentration. Vehicles should be tested carefully for possible effects on [^{35}S]GTPγS binding. We have found that many commonly used solvents, including water, TMEN, and dilute dimethyl sulfoxide (DMSO), BSA, and ascorbic acid solutions, do not affect [^{35}S]GTPγS binding. High concentrations of ethanol and methanol both significantly reduce total [^{35}S]GTPγS binding, although adding BSA to these solvents helps to buffer their inhibitory effects.

[^{35}S]GTPγS (NEN Life Science Products, Boston, MA; ~1250 Ci/mmol) should be prepared at 12.4-fold the final desired concentration in the TMEN/GDP buffer, since 5 μl of diluted [^{35}S]GTPγS will be added into a total final volume of 62 μl. Concentrations will vary depending on incubation conditions and expression levels of receptor and G protein, but 5 nM (G$_q$, G$_{12}$, G$_{13}$, and G$_z$) and 1 nM (G$_s$, G$_{i1-3}$, and G$_o$) are good initial concentrations. Because the final concentration of [^{35}S]GTPγS is so much lower than the purchased stock solution ($\sim$10 μM), it is more economical to prepare assay-sized aliquots of [^{35}S]GTPγS at 10–20× the final concentration in TMEN buffer (without GDP) and store them at $-70°$ until use.

For each assay point, 3.6 ml of immunoprecipitation (IP) buffer will be needed: 0.6 ml to stop the reaction, and three 1 ml washes. To the amount of wash buffer (50 mM Tris-HCl, pH 7.4, 150 mM NaCl, 20 mM MgCl$_2$) required for about 4 ml per assay point, add Nonidet P-40 (0.5%, v/v) and stir to dissolve. Then add GDP and GTP (100 μM each) and aprotinin (0.3%, v/v), and store in the dark at 4° until ready for use.

[^{35}S]GTPγS Binding and Immunoprecipitation

For each assay point, add 2 μl of ligand or appropriate vehicle and 55 μl of membranes in TMEN/GDP to the bottom of a 1.5 ml microfuge tube. Mix gently, taking care not to spatter the contents, and incubate at 30° for 10 min or more to promote ligand binding to the receptor. Then add 5 μl of the [^{35}S]GTPγS diluted in TMEN/GDP, mix gently, and incubate an additional 10 min at 30°.

Stop the reaction by moving the tubes to ice, adding 600 μl per tube of IP buffer, and vortexing to mix. From this point, all procedures should be performed on ice. Shake tubes at 4° for 30 min to allow solubilization of membrane proteins, then transfer the contents of the reaction tubes into the preclear tubes (containing Pansorbin and nonimmune serum) prepared earlier. Mix well and shake an additional 20 min at 4° to allow Pansorbin cells and the nonimmune serum to interact with membrane proteins. Pellet the Pansorbin cells by spinning 3 min at 20,000g at 4°, and transfer the supernatant to the primary IP tubes, taking care not to transfer any of the pelleted Pansorbin cells. Shake at 4° for 1 hr.

Centrifuge the primary IP tubes for 3 min at 20,000g at 4° to pellet the protein A-Sepharose, and wash the pellet by adding 1 ml of IP buffer, vortexing gently, and repelleting the protein A-Sepharose. Wash two additional times with IP buffer, then once more with wash buffer.

Following the final wash and aspiration, move the tubes to room temperature and add 0.5 ml of 0.5% (w/v) sodium dodecyl sulfate (SDS) per tube, vortex, and incubate 2–3 min at 85–90°. Allow the tubes to cool, then transfer the entire contents of each tube to a scintillation vial containing approximately 5 ml of liquid scintillant. Mix well to disperse the SDS–protein A slurry, and count for 2 min in a scintillation spectrophotometer.

Technical Considerations

Time of Incubation

Two major binding events occur in the G protein activation assay: the ligand binds to the receptor, and the [^{35}S]GTPγS binds to the α subunit. In order to make kinetic interpretation of assay data easier, two different incubation periods are used to isolate these two binding events. The first incubation allows the ligand to bind the receptor in the absence of [^{35}S]GTPγS. When determining the efficacies of a variety of ligands, it is important that differences in ligand binding kinetics, which can be determined using traditional radioligand binding assays, be controlled. Thus the first incubation period should be of sufficient length that ligand–receptor binding approaches equilibrium; this incubation time may vary substantially from the 10-min guideline used in the detailed protocol. The subsequent incubation of [^{35}S]GTPγS with membranes is allowed to occur only after the ligand–receptor binding has reached equilibrium. Although the [^{35}S]GTPγS binding is empirically saturable in most cases (both the basal and activation states reach a plateau; Fig. 2), it is important to remember that [^{35}S]GTPγS binding is a measure of a one-time activation event. Thus, choosing a duration for incubation with [^{35}S]GTPγS

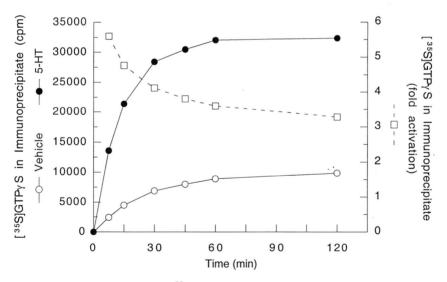

FIG. 2. Time course of binding of [^{35}S]GTPγS to α_{i2} in Sf9 cells. Membranes from Sf9 cells expressing G_{i2} (α_{i2}, β_1, γ_2), and the human 5-HT$_{1A}$ receptor were resuspended in TMEN buffer containing GDP (30 μM) and incubated with [^{35}S]GTPγS (1 nM) in the absence (open circles) or presence (filled circles) of serotonin (5-HT, 1 μM) for the indicated times. α_{i2} was immunoprecipitated subsequently. The fold activation (5-HT-stimulated [^{35}S]GTPγS binding/vehicle-stimulated [^{35}S]GTPγS binding; open squares) is plotted in reference to the right-hand y axis.

is somewhat arbitrary. Short incubation times, those under 10–15 min, result in fewer total counts in the immunoprecipitate than longer incubation times. However, since stimulation with agonist preferentially increases the initial rate of binding, greater stimulation relative to incubation with vehicle is observed with these shorter incubation times.

Guanine Nucleotide Concentrations

Whereas receptor-enhanced GDP–GTPγS exchange on the G protein α subunit is the basis of the this binding assay, considerable GDP–GTPγS exchange, and thus [^{35}S]GTPγS binding, in the absence of ligand can markedly reduce the signal-to-noise ratio and even mask the effects of ligand. Our studies in Sf9 cells, where the G protein can be expressed and characterized in the absence of receptor, indicate that some of this basal GDP–GTPγS exchange is independent of receptor; that is, it represents constitutive activity of the G protein itself.[9,12] In mammalian cells, the basal activity likely reflects a combination of G protein constitutive activity and activation by a variety of receptors present in those cells. In any case, controlling the basal GDP–GTPγS exchange is one of the best ways to improve signal strength. Decreasing the GTPγS : GDP ratio, by including excess GDP in the assay buffer and using modest [^{35}S]GTPγS concentrations, does an excellent job of reducing receptor-independent GDP–GTPγS exchange.

Of course, these methods of controlling receptor-independent GDP–GTPγS exchange influence receptor-mediated (both agonist-dependent and -independent) GDP–GTPγS exchange as well. Thus the optimal GDP and [^{35}S]GTPγS concentrations represent a compromise between decreased receptor-independent [^{35}S]GTPγS binding and maximal agonist-stimulated [^{35}S]GTPγS binding. As the expression levels of both G protein and receptor will affect sensitivity to GDP, the optimal GDP and GTPγS concentrations for assay of any given α subunit are best determined experimentally. We find that testing concentrations of GDP from 0.1 to 100 μM, in half-log units, provides a clear peak or plateau of maximal agonist stimulation relative to vehicle (Fig. 3). Although the optimal GDP concentration is consistent for different receptors within a given cell line, it may vary slightly across cell lines for a given G protein.

The fact that the most favorable conditions for assessing G protein activation by GPCRs varies significantly across different G proteins makes it difficult to draw quantitative conclusions about the relative preference among G proteins of a single receptor. Whereas qualitative assessment of G protein coupling can be made with great certainty, and the relative efficacies of ligands in activating a G protein through a given receptor can determined accurately, it is not possible to use the relative increase in [^{35}S]GTPγS binding in response to agonist as a measure of the preference of a receptor for different G proteins. In these cases, use of several different cell lines and conditions can be used in building an argument for

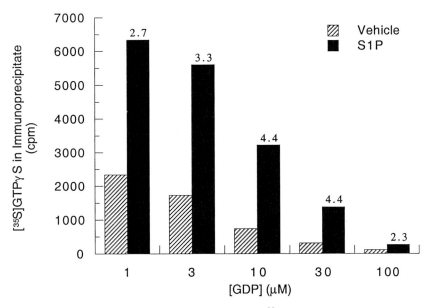

FIG. 3. Effects of GDP concentration on agonist-promoted [^{35}S]GTPγS binding. Membranes from HEK293 cells stably expressing the Edg-1 receptor were resuspended in TMEN buffer containing the indicated concentrations of GDP. Membranes were incubated with [^{35}S]GTPγS (1 nM) for 10 min in the absence (hatched bars) or presence (filled bars) of S1P (10 μM), and endogenous α_i was immunoprecipitated subsequently. The fold stimulation of [^{35}S]GTPγS binding to endogenous G_i by S1P relative to vehicle is indicated above each S1P point.

preferences among G proteins. Sometimes, selective constitutive activity toward a single G protein will make it possible to propose the primary coupling partner (Fig. 4).

Choice of Antibody

The antibody must meet two important requirements: it must be subunit-selective, and it must bind to a site on the α subunit that is accessible when the G protein is properly folded and bound to GTP. Perhaps the best way to satisfy both of these requirements is to use anstisera directed against the C terminus of the α subunit. Generally C-terminal directed antisera recognize their antigens especially well, and although C-terminal directed antisera do not always distinguish between some members of the G_i family (the C terminus sequences of α_{i1} and α_{i2} are identical, and antisera directed against the C terminus of α_{i3} cross-react strongly with α_o), the C termini of most α subunits are quite distinct. Furthermore, the C terminus is accessible when the protein is properly folded and is not thought

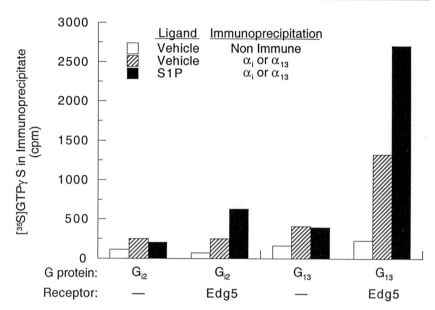

FIG. 4. Agonist-independent stimulation of [^{35}S]GTPγS binding as an indication of G-protein coupling preferences. Membranes from Sf9 cells expressing G_{i2} (α_{i2}, β_1, γ_2) or G_{13} (α_{13}, β_1, γ_2) alone or with the rat Edg5/H218 receptor were resuspended in TMEN buffer containing GDP (30 μM for G_{i2}, 0.1 μM for G_{13}) and incubated for 10 min with [^{35}S]GTPγS (1 nM for G_{i2}, 5 nM for G_{13}) in the absence (open bars and hatched bars) or presence (filled bars) of S1P (10 μM). G_{i2} or G_{13} was immunoprecipitated subsequently. Nonimmune serum was used as a control for the immunoprecipitation step (open bars). Note that in the absence of agonist, expression of Edg5 markedly increased the binding of [^{35}S]GTPγS by G_{13} but not by G_{i2}, suggesting that Edg5 couples more efficiently to G_{13} than to G_{i2}.

to be affected by nucleotide binding.[15] We have used our own antisera as well as commercially obtained antisera with success. In some cases, for example immunoprecipitating α_z, antisera directed against the N terminus have also proved useful.[9] When the identity of the activated G protein cannot be absolutely assumed from the immunoprecipitation (e.g., α_{i3} or α_o), immunoblots using internally directed antisera can help resolve which G proteins are present in a given cell line, often eliminating one of the possibilities.

Because data are usually expressed as fold increase over basal, antibody titer is effectively normalized by the calculation, and thus very high titer is not necessary for satisfactory results. In fact, we find that the extent of activation over basal conditions does not vary significantly with titer (Fig. 5). However, when the absolute amount of G protein is limited, particularly for G proteins with low

[15] H. R. Bourne, *Curr. Opin. Cell Biol.* **9**, 134 (1997).

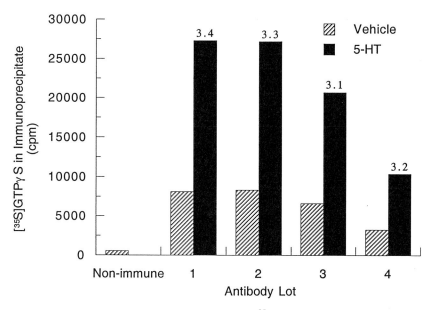

FIG. 5. Effect of antibody titer on immunoprecipitated [^{35}S]GTPγS. Membranes from Sf9 cells expressing G_{i2} and the human 5-HT$_{1A}$ receptor were resuspended in TMEN buffer containing GDP (30 μM) and incubated for 10 min with [^{35}S]GTPγS (1 nM) in the absence (hatched bars) and presence (filled bars) of 5-HT (1 μM). G_{i2} was subsequently immunoprecipitated using several different bleed dates of antiserum. Nonimmune serum (leftmost bar) was used as a control. The stimulation of [^{35}S]GTPγS binding by 5-HT relative to vehicle (fold activation) is indicated above each 5-HT bar.

GDP–GTP exchange rates such as α_z, higher titer may be required to immuno-precipitate enough [^{35}S]GTPγS to allow confident measurement of basal activity. Since different bleed dates or antibody lots will have significantly different titers, consideration of these variables should be made if pooling of unnormalized data is desired.

The use of a nonimmune control for the immunoprecipitation step is strongly advised. The nonimmune control ensures that the immunoprecipitated [^{35}S]GTPγS is antiserum-specific, and it can provide valuble information about whether other variables in the immunoprecipitation procedure, such as the number of washes, are optimized and consistent among assays. Additionally, the nonimmune control indicates the effective minimum number of counts that can be expected; even if no agonist effect can be discerned, the counts immunoprecipitated by α-subunit-specific antisera following incubation with vehicle usually surpass those of the nonimmune control because of constitutive GDP–GTP exchange on the α subunit. When no difference between nonimmune and immune sera is observed, expression levels of appropriately folded G protein should be investigated. Despite its usefulness in these regards, we find that the counts in the nonimmune control do not provide a means of normalizing results across assays. Preimmune sera, those taken from the same animal as the antisera prior to immunogen injection, represent the perfect control for these experiments. However, preimmune sera are precious commodities for other assays, and we find that normal serum derived from the same species as the antiserum serves as a perfectly adequate control in almost all cases.

Duration of Immunoprecipitation

The primary immunoprecipitation step in this assay is considerably shorter than those we use for other purposes.[16] The times described here represent a compromise between ensuring an adequate immunoprecipitation efficiency and preventing continued binding of free $[^{35}S]GTP\gamma S$ to the α subunits or excess degradation of bound $[^{35}S]GTP\gamma S$. Diluting the reaction mixture 10-fold and performing the immunoprecipitation in the presence of excess GDP and GTP means that minimal association of $[^{35}S]GTP\gamma S$ with α subunits during the immunoprecipitation is observed. Although the concentration of excess GDP and GTP in the immunoprecipitation buffer can be tailored to each subunit, we use 100 μM in all cases.

Limitations of Expression

We have rarely come across situations in receptor overexpression paradigms (infected Sf9 cells or transfected HEK293 or CHO cells) where good responses to at least one G protein are not obtained. Our studies in mammalian cells suggest that it is the receptor, rather than the G protein, that is limiting in terms of detecting a signal with ligand. We have not performed extensive evaluation on the lower limits of receptor expression necessary for effective assay, but receptor expression levels of less than 50–100 fmol/mg protein have inconsistently provided adequate signal. This seems to depend on the cell type and the G protein coupled. For example, we were able to measure the activation of G_i and G_{13} by sphingosine 1-phosphate (S1P) in HEK293 cells.[12] Although both polymerase chain reaction and Northern analysis indicated that the HEK293 cells express S1P receptors, no $[^{32}P]S1P$ binding could be detected in them.[17] The $[^{32}P]S1P$ binding assay is not especially sensitive, but we were nonetheless surprised to detect activation by receptors that are present at these relatively low levels.

[16] C. A. Chen and D. R. Manning, *in* "G Proteins: Techniques of Analysis" (D. R. Manning, ed.), p. 99. CRC Press, New York, 1999.
[17] M. J. Lee, J. R. Van Brocklyn, S. Thangada, C. H. Liu, A. R. Hand, R. Menzeleev, S. Spiegel, and T. Hla, *Science* **279**, 1552 (1998).

[2] Elucidating Kinetic and Thermodynamic Constants for Interaction of G Protein Subunits and Receptors by Surface Plasmon Resonance Spectroscopy

By R. VICTOR REBOIS, PETER SCHUCK, and JOHN K. NORTHUP

Introduction

Surface plasmon resonance (SPR) spectroscopy is a useful technique for investigating protein–protein interactions in real time. The technique can produce information about the specificity, kinetics, and affinities of these interactions. In only a few instances have the kinetic or thermodynamic constants associated with G protein subunit interactions been determined quantitatively.[1-4] SPR spectroscopy offers the opportunity to make such quantitative comparisons without some of the complexities of other methods, and SPR spectroscopy is becoming increasingly popular for investigating the interaction of various components of G-protein-mediated signal transduction systems,[5] such as those between G protein subunits, their receptors, and regulatory proteins such as regulators of G protein signaling (RGS) proteins. SPR spectroscopy is a biosensor-based method that observes the interaction of a protein in the mobile phase with specific site(s) immobilized at the sensor surface. The detection principle is based on changes of the optical properties of a surface layer with increasing mass of adsorbed protein. In brief, the angular dependence of surface plasmon resonance excitation from light in total internal reflection is monitored. This allows the measurement of refractive index changes of the solution within the evanescent field close to the sensor surface. If compensated for refractive index changes of the mobile phase flowing across the surface, these data directly reflect changes in the amount of surface-bound protein. Mathematical analysis of the time course of binding allows determination of association rate constants, and, if the mobile binding partner is removed from the mobile phase, dissociation rate constants. Also, from the binding data in steady-state, binding isotherms can be generated for the thermodynamic determination of equilibrium dissociation constants. For a general introduction in the practice of biosensing, see, e.g., Ref. 6.

[1] J. K. Northup, P. C. Sternweis, and A. G. Gilman, *J. Biol. Chem.* **258,** 11361 (1983).

[2] R. E. Kohnken and J. D. Hildbrandt, *J. Biol. Chem.* **264,** 20688 (1989).

[3] H. Heithier, M. Fröhlich, C. Dees, M. Baumann, M. Häring, P. Gierschik, E. Schiltz, W. L. C. Vaz, M. Hekman, and E. J. M. Helmreich, *Eur. J. Biochem.* **204,** 1169 (1992).

[4] N. A. Sarvazyan, A. E. Remmers, and R. R. Neubig, *J. Biol. Chem.* **273,** 7934 (1998).

[5] V. Z. Slepak, *J. Mol. Recog.* **13,** 20 (2000).

[6] P. Schuck, L. F. Boyd, and P. S. Andersen, *in* "Current Protocols in Protein Science" (J. E. Coligan, B. M. Dunn, H. L. Ploegh, D. W. Speicher, and P. T. Wingfield, eds.), Vol. 2, pp. 20.2.1–20.2.21. John Wiley & Sons, New York, 1999.

METHODS IN ENZYMOLOGY, VOL. 344

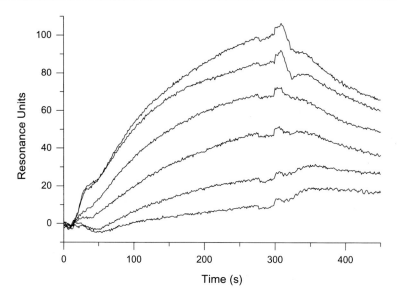

FIG. 1. Sensorgrams of Gβγ binding to immobilized biotinylated G$_i$α, captured on a streptavidin modified surface, at concentrations of 10, 20, 40, 60, 80, and 100 nM.

In our laboratory, we use the commercial SPR instruments Biacore X and Biacore 2000, and therefore most of the following description refers to the use of Biacore instruments. However, all the techniques described can be used or adapted for other surface plasmon resonance or related sensors. In the Biacore instruments, the refractive index near the sensor surface is recorded at a rate of up to 10 Hz, and the time course of the refractive index is referred to as a sensorgram (see Fig. 1). The units of a sensorgram are usually called resonance units (RU) with one RU being equivalent to the binding of approximately 1 pg of protein per mm^2 of SPR chip surface.[7,8] Several different sensor chips allowing different strategies for the immobilization of proteins are commercially available; most of them have a layer of carboxymethylated dextran that forms a flexible hydrogel of thickness 100–200 nm. For the delivery of protein to the sensor surface, a microfluidic system is employed, which can generate a constant flow of the mobile phase across two (Biacore X) or four (Biacore 2000) flow paths in series, or it can address each flow path individually. Pneumatic valves and an HPLC-like injection loop allow switching of the mobile phase from running buffer to a plug of protein solution. Usually, one of the flow paths is specifically modified for the experiment, while a

[7] R. Karlsson, H. Roos, L. Fägerstam, and B. Persson, *Methods: Companion Methods Enzymol.* **6,** 99 (1994).
[8] E. Stenberg, B. Persson, H. Roos, and C. Urbaniczky, *J. Coll. Interface Sci.* **143,** 513 (1991).

second one serves as a reference surface to monitor nonspecific binding and the refractive index of the mobile phase. Additional sensor flow paths in the Biacore 2000 can be used for several experiments (or controls) to be performed in parallel. For greater economy in sample consumption and increased contact times, we have taken advantage of an alternative sample delivery method with oscillating flow.[9]

Investigating G Protein Subunit Interactions by SPR

Preparation of G Protein Subunits

G_i is purified form bovine brain as previously described.[10] Heterotrimeric transducin (G_t) is purified by GTP extraction[11] of rod outer segment disks isolated from bovine retina.[12] The α subunit of G_t ($G_t\alpha$) is separated from its heterodimeric counterpart ($G_t\beta\gamma$) in two column chromatographic steps involving first ω-aminooctyl-agarose (Sigma) and subsequently Blue Sepharose CL-4B (Pharmacia, Piscataway, NJ) as previously described.[13,14] Recombinant $\beta_1\gamma_2$ and $\beta_1\gamma_{8\text{-olf}}$ (also called $\beta_1\gamma_9$) dimers are prepared by expression in Sf9 (*Spodoptera frugiperda* ovary) cells coinfected with viruses encoding $G\beta_1$ and γ_2 or $\gamma_{8\text{-olf}}$. Postnuclear membranes isolated from these Sf9 cells are extracted at 2 mg/ml protein with 1% sodium cholate and purified by ion-exchange chromatography over DEAE-Sephacel followed by size exclusion chromatography over Ultrogel AcA44 (BioSepra) as described.[15] Peak fractions of $G\beta\gamma$ from AcA44 chromatography are pooled, concentrated in an Amicon (Danvers, MA) stirred concentrator with PM30 membranes, and chromatographed over FPLC (fast protein liquid chromatography) Sephadex HR75 in 10 mM MOPS, pH 7.5, 100 mM NaCl, and 8 mM CHAPS for storage at $-80°$.

Strategies for Immobilizing G Protein Subunits

For studies of protein–protein interactions by SPR spectroscopy, one of the proteins involved in the interaction must be immobilized on the surface of an SPR chip either by covalently cross-linking it to the chip or through a high-affinity noncovalent interaction with the surface. A simple, and therefore frequently used,

[9] M. Abrantes, M. T. Magone, L. F. Boyd, and P. Schuck, *Anal. Chem.* **73,** 2828 (2001).
[10] J. D. Roof, J. L. Applebury, and P. C. Sternweis, *J. Biol. Chem.* **260,** 16242 (1985).
[11] H. Kuhn, *Nature* **283,** 587 (1980).
[12] D. S. Papermaster and W. J. Dreyer, *Biochemistry* **13,** 2438 (1974).
[13] A. Yamazaki and M. W. Bitensky, *Methods Enzymol.* **159,** 702 (1988).
[14] A. B. Fawzi, D. S. Fay, E. A. Murphy, H. Tamir, J. J. Erdos, and J. K. Northup, *J. Biol. Chem.* **266,** 12194 (1991).
[15] D. E. Wildman, H. Tamir, E. Leberer, J. K. Northup, and M. Dennis, *Proc. Natl. Acad. Sci. U.S.A.* **90,** 794 (1993).

method for covalent immobilization is to convert the carboxyl moieties of the dextran layer on the surface of an SPR chip to N-hydroxysuccinimide esters that will form amide bonds with primary amines on the protein (the protocol is given below).[6,16] The cross-linking is facilitated by increasing the concentration of the protein within the negatively charged dextran matrix through electrostatic attraction. This is accomplished by dissolving the protein in a solution at a pH below the pI of the protein so that it is positively charged. In order to effectively concentrate G protein subunits within the dextran matrix the pH of the solution must be 4.5 or lower. In our experience, and that of others,[5] this destroys the ability of Gα to associate with G$\beta\gamma$. To avoid this problem, G protein subunits can be covalently coupled on the SPR chip at neutral pH by thioester bonds as has been described for affinity chromatography to purify G protein subunits.[17,18] This procedure is described below.

Another very useful strategy for immobilizing G protein subunits is their capture through high-affinity noncovalent interactions. Most antibodies are tolerant to brief exposure to low pH, and therefore can be easily cross-linked to the sensor surface. Antibodies against the carboxy-terminal decapeptide of the stimulatory G protein α subunit (G$_s\alpha$) might be used to immobilize G$_s\alpha$ for studying its interaction with G$\beta\gamma$.[19] Unfortunately, this strategy will not work for G$_t$ since antibodies against the carboxy terminus of G$_t\alpha$ can interfere with subunit interaction and cause G$_t$ subunit dissociation,[20] and the same may be true for other heterotrimeric G proteins. By using recombinant techniques an epitope tag, such as FLAG or HA, or a hexahistidine (His$_6$) tag can be introduced into G protein subunits to use as a site for immobilization with the appropriate commercially available antibody, or with Ni-nitriloacetic acid-modified chips, respectively. In principle, this approach to immobilizing G protein subunits avoids the requirement for scrupulous purification of the protein and provides a unique, homogeneous attachment of the G protein to the surface of the SPR chip. His$_6$-modified Gγ_2 in combination with Gβ_1 has proven successful for affinity chromatography of Gα,[21] but we have not examined its utility for SPR capture.

The observation that antibody binding to an intrinsic amino acid sequences of G$_t\alpha$ reduces its affinity for G$\beta\gamma$ suggests that potential interference with subunit interactions may be a critical factor for the choice of the immobilization technique when studying the interaction of naturally occurring G protein subunits. Therefore, the method for immobilization employed by us utilizes biotinylated heterotrimeric G protein, which is subsequently separated into subunits, and then captured by a

[16] B. Johnsson, S. Löfås, and G. Lindquist, *Anal. Biochem.* **198,** 268 (1991).
[17] I.-H. Pang and P. C. Sternweis, *Proc. Natl. Acad. Sci. U.S.A.* **86,** 7814 (1989).
[18] I. H. Pang, A. V. Smrcka, and P. C. Sternweis, *Methods Enzymol.* **237,** 164 (1994).
[19] M. Toyoshige, S. Okuya, and R. V. Rebois, *Biochemistry* **33,** 4865 (1994).
[20] M. R. Mazzoni and H. E. Hamm, *Biochemistry* **28,** 9873 (1989).
[21] T. Kozasa and A. G. Gilman, *J. Biol. Chem.* **270,** 1734 (1995).

surface modified with immobilized streptavidin, which has a very high affinity for biotin (K_D of 10^{-15} M). Importantly, by biotinylating the heterotrimer the subunit interface is likely to be protected from covalent modification, thus minimizing interference with subunit interactions due to vitiation of the subunit interface or immobilization in an orientation that would sterically hinder subunit binding interactions.

Covalent Coupling of G Protein Subunits to SPR Chips

Procedures for the immobilization of $G_o\alpha$ and bovine brain $G\beta\gamma$ have been described for affinity chromatography to purify G protein subunits.[17,18] This procedure involves the cross-linking of G protein subunits to dextran matrices via thioester coupling with the heterobifunctional reagent GMBS (N-[γ-maleimido-butyryloxy]succinimide ester, Pierce Chemical Co., Rockford, IL) that contains a short spacer group. The procedure for affinity chromatography can be adapted with essentially identical chemistries. Because the flow coupling in the SPR instrument led to inconsistent results (usually poor coupling), we performed the following operations by pipetting volumes of 50 μl directly to the exposed sensor surface. This seems to have the advantage of allowing for more uniform immobilization and longer contact times. First, a primary amine linkable group is created by reacting the carboxymethyldextran hydrogel with 1 M diaminoethane, 200 mM N-ethyl-N'-(dimethylaminopropyl)carbodiimide (EDC), and 50 mM N-hydroxysuccinimide (NHS) in 100 mM KP_i pH 7. After 20 min at room temperature, unreacted reagents are removed by 10 successive rinsing of the surface with 50 μl distilled water. One molar ethanolamine dissolved in water is then applied and allowed to react for 10 min to inactivate all remaining activated carboxyl residues, followed by 10 rinses with water. The surface is then reacted with 2 mM GMBS in 100 mM KP_i pH 7 for 30 min at room temperature, followed by 10 rinses of the surface with distilled water. Finally, the G protein subunits are coupled by addition to the surface in 100 mM KP_i pH 7 with 0.2% (w/v) Lubrol 12A9 for $G\alpha_o$, $G\alpha_{i1}$, or brain $G\beta\gamma$ (detergent is omitted for retinal $G\alpha_t$ and $G\beta_1\gamma_1$) for 30 min at room temperature. Uncoupled G protein is removed by 10 successive rinses of the surface with the 100 mM KP_i pH 7 and 0.2% Lubrol (omitted for retinal G protein subunits). The G-protein-coupled sensor chip is then inserted into the instrument for measurement of subunit interactions as described below. Typically this procedure yields 600–1000 RU of coupled G protein subunit with a binding capacity of about 400–700 RU for the cognate subunit. This coupling procedure produces less heterogeneity in the orientation of immobilized protein than amine coupling because there are fewer surface-accessible cysteines than primary amines. Furthermore, it does not require the exposure of the protein to low pH. However, this procedure is more demanding than the indirect immobilization method described below and in our studies did not provide clear advantages.

Biotinylation of G Protein Subunits

The biotinylation of G protein subunits at cysteines and primary amines accessible at the protein surface can be achieved with commercially available reagents. Reagents that react with primary amines such as sulfosuccinimidobiotin (EZ-Link sulfo-NHS-biotin), sulfosuccinimidyl-6-(biotinamido)hexanoate (EZ-Link sulfo-NHS-LC-biotin), and sulfosuccinimidyl-12-(biotinamido)-6-hexanamidohexanoate (EZ-Link sulfo-NHS-LC-LC-biotin) can be obtained from Pierce. All of these reagents were found to be equally effective, and biotin is incorporated into both $G\alpha$ and $G\beta$ subunits (although the α subunit is somewhat more extensively modified than the β when the holoprotein is biotinylated). The extent of biotinylation is dependent on the ratio of biotinylating reagent to G protein. With the biotinylation protocol described below, the incorporation of biotin can be easily detected when the mole ratio of biotin to G protein is 2.5 : 1; and the incorporation of biotin increases with a ratio of up to 200 : 1. Reagents that modify sulfhydryl moieties, such as 1-biotinamido-4-[4'-(maleimidomethyl)cyclohexanecarboxamido]butane (EZ-Link biotin-BMCC) were found to biotinylate G protein subunits, but the incorporation of biotin was much less than with reagents that react with primary amines.

The following protocol is suitable for obtaining biotinylated G proteins that can be used to study subunit interactions by SPR spectroscopy. To a 100 μl volume of 50 mM HEPES, pH 7.5, containing 100 μg of heterotrimeric G protein add an equal volume the same buffer containing a 2.5 mM concentration of EZ-Link sulfo-NHS-LC-LC-biotin and incubate at room temperature for 30 min. Under these conditions the mole ratio of biotinylating agent to G protein is approximately 200 : 1. Unreacted biotinylating reagent is removed by desalting with a PD-10 column (Pharmacia) and the sample is concentrated with a Centricon 30 filter unit (Amicon). It is important to remove all unreacted biotinylating agent so that it will not compete with the biotinylated G protein subunit for binding to the streptavidin on the SPR chip.

Although G proteins are usually stored in detergent containing solutions, it is not necessary to add detergents to retinal G protein subunits; neither have we found it to be necessary to have detergents present when working with $G_i\alpha$ subunits from mammalian tissue or recombinant $G\alpha_i$ subunits expressed in *Escherichia coli*. If detergent is used, then the best choices are probably sodium cholate or CHAPS, which will not interfere with biotinylation by reagents described here and are not retained by the Amicon stirred concentrators or Centricon filter units using PM30 membranes, thus avoiding concentration of the detergent together with the protein during ultrafiltration.

The procedure used for subsequent purification of biotinylated G protein subunits will depend on the G protein under study. Biotinylated G_i subunits can be dissociated by incubation in 50 mM HEPES, pH 7.5, with 10 mM NaF, 20 μM

$AlCl_3$, 3 mM $MgCl_2$, 100 mM NaCl, and 0.3% sodium cholate for 1 hr, and re-solved on an ω-aminooctyl-agarose column as previously described.[2] Both the subunit dissociation and the column chromatography are performed at room temperature. The elution of the biotinylated protein has been found to be qualitatively indistinguishable from that of unmodified G_i protein subunits. Biotinylated $G_t\alpha$ prepared in this way has been found to be unstable, as judged by its poor ability to reassociate with $G\beta\gamma$: In rate zonal sedimentation experiments the migration of only a small percentage of the biotinylated $G_t\alpha$ can be shifted to that correspond-ing to the heterotrimeric G_t, even in the presence of a sevenfold molar excess of $G\beta\gamma$. However, this behaviour does not seem to be a consequence of biotinylation. G_α subunits belonging to the G_i subclass of G proteins, which includes G_t, are sub-strates for pertussis toxin when they form an unactivated heterotrimer with $G\beta\gamma$. When nonbiotinylated G_t subunits are separated as described above and subse-quently recombined with $G\beta\gamma$ in an attempt to regenerate the heterotrimer, $G_t\alpha$ is a poor substrate for the toxin when compared with an equimolar amount $G_t\alpha$ in a preparation of holo-G_t that has never been dissociated. On the other hand, when $G_i\alpha$ is prepared as described above and mixed with $G\beta\gamma$, it is just as good a substrate for pertussis toxin, as an equimolar amount of $G_i\alpha$ in a preparation of heterotrimeric G_i, and biotinylated $G_i\alpha$ rapidly and completely reassociate with $G\beta\gamma$ as assessed by rate zonal sedimentation experiments. For studies with G_t we recommend the procedure described in the section on preparation of G protein subunits above.[13] Biotinylated $G_i\alpha$ is stable for at least 1 year if stored at $-20°$ in 20 mM HEPES, pH 8.0, 100 mM NaCl, 1 mM EDTA, 1 mM dithiothreitol (DTT) 0.1% sodium cholate, and 30% glycerol at a protein concentration of approximately 50 μg/ml.

Separated subunits can be precipitated with streptavidin that has been cross-linked to Sepharose (Ultralink immobilized NeutrAvidin, Pierce). This represents a simple assay to predict the coupling efficiency to an SPR chip. When the mole ratio of biotinylating agent to G_i is 20 : 1 only 10% to 15% of the biotinylated $G_i\alpha$ and $G\beta\gamma$ can be precipitated with immobilized NeutrAvidin, whereas essentially all of the biotinylated subunits can be precipitated when the ratio is 200 : 1.

Cross-Linking Streptavidin to SPR Chips

Streptavidin is cross-linked to CM5 sensor chips with standard amine coupling protocols.[16] Briefly, 35 μl of a 50 mM solution of NHS and 200 mM EDC is injected. The surface is rinsed with standard running buffer (10 mM HEPES, pH 7.5, 150 mM NaCl, 3 mM EDTA, and 0.005% Tween 20, Solution A), followed by an injection of approximately 25 μl of 50 μg/ml streptavidin (e.g., Pierce or Sigma) dissolved in sodium acetate pH 5.0 (see Table I). Typically, this resulted in a signal increase of 2000–3000 RU. Unreacted N-hydroxysuccinimide ester are blocked by injecting 35 μl of 1 M ethanolamine, pH 9.0, followed by equilibration

TABLE I
COMPOSITIONS OF SOLUTIONS USED FOR SPR OF G PROTEIN SUBUNITS
AND RHODOPSIN

Solution	Composition
A	10 mM HEPES, pH 7.5, 150 mM NaCl, 3 mM EDTA, and 0.005% Tween 20
B	Solution A containing 4 mM MgSO$_4$ and 1 mM dithiothreitol (DTT)
C	50 mM MOPS, pH 7.5, 150 mM NaCl, 3 mM MgSO$_4$, 10 μM CaCl$_2$, and 10 μM MnCl$_2$
D	10 mM MOPS, pH 7.5, 100 mM NaCl, 3 mM MnCl$_2$, and 3 mM CaCl$_2$

of the chip surface with Solution A. All operations are conducted at a flow rate of 5 μl/min.

Capturing Biotinylated G_i Subunits with Streptavidin-Modified SPR Chips and Verifying Specificity of Interaction

Both biotinylated-G$\beta\gamma$ and biotinylated Gα can be immobilized to the streptavidin-modified surface. Although immobilized biotinylated G$\beta\gamma$ has been used successfully for studying its interaction with a variety of proteins involved in G-protein-mediated signal transduction,[5] in our studies of G_i subunit affinity we have focused on surfaces with immobilized $G_i\alpha$.

For the immobilization of $G_i\alpha$ the biotinylated subunit is diluted to a concentration of 5–10 μg/ml in Solution. A containing 4 mM MgSO$_4$ (approximately 1 mM free Mg^{2+}) and 1 mM DTT (Solution B), and 30 μl is delivered at a flow rate of 1 μl/min to the experimental surface of an SPR chip that has streptavidin cross-linked to it. Under these conditions the binding begins to plateau in about 30 min. Alternatively to the constant-flow immobilization, the sensor surface can be incubated with a 5 to 10 μl volume of biotinylated $G_i\alpha$ using the oscillating flow configuration in the Biacore X in conjunction with a separate syringe pump.[9] Both configurations produced a surface with 600–1000 RU of $G_i\alpha$ bound. Once the biotinylated G-protein subunit has been immobilized, any unoccupied streptavidin binding sites can be blocked with biotin. Routinely, we injected 20 μl of a 0.5 mM solution of biotin in Solution B at a flow rate of 10 μl/min to block binding sites on both the experimental and reference surfaces.

Once $G_i\alpha$ is attached to the sensor surface as described above, it will bind G$\beta\gamma$ dissolved in Solution B. Typically, the injection of 100 nM G$\beta\gamma$ results in a signal increase of 100 to 150 RU in 5 min, while concentrations as low as 10 nM G$\beta\gamma$ give a small, but significant increase in RU (Fig. 1). This association phase can be followed by rinsing the surface with running buffer, which causes dissociation of the surface-bound $G\beta\gamma$. Experimental and analytical strategies for the determination of the binding constants will be described below.

There is ample evidence to indicate that the binding of $G\beta\gamma$ to the immobilized $G_i\alpha$ is specific: (1) Under the condition described above there is very little binding of $G\beta\gamma$ to the reference surface; (2) $G_i\alpha$ if present in the mobile phase will compete with the immobilized $G_i\alpha$ in binding $G\beta\gamma$; (3) $G\beta\gamma$ bound to the immobilized $G_i\alpha$ is washed away by the high Mg^{2+} concentrations of the regeneration buffer; (4) 40 mM Mg^{2+} in the mobile phase prevents $G\beta\gamma$ from binding to the immobilized $G_i\alpha$; (5) other proteins such as antibodies do not bind to the immobilized $G_i\alpha$; and (6) although the immobilized $G_i\alpha$ is surprisingly stable it does lose activity during prolonged (i.e., >12 hr) washing of the chips surface at 25° in Solution B.

Regenerating the Surface by Stripping It of $G\beta\gamma$

Most experimental strategies for biosensor experiments require multiple cycles of binding and dissociation of $G_i\beta\gamma$. Before each cycle the bound $G_i\beta\gamma$ must be completely dissociated from the $G_i\alpha$ with little or no deterioration in its ability to bind $G_i\beta\gamma$ in subsequent cycles. Because of the relatively slow dissociation of $G\beta\gamma$ from $G_i\alpha$, regeneration of the surface by changing the chemical environment is necessary. High concentrations of Mg^{2+} lower the affinity of $G\alpha$ for $G\beta\gamma$, and as a consequence, $G\beta\gamma$ can be rapidly removed from the surface by injecting 20 μl of Solution A containing 120 mM $MgSO_4$, 1 mM DTT, and 1.0% Lubrol PX at a flow rate of 10 μl/min. Including Lubrol PX facilitates the $MgSO_4$-induced dissociation of $G\beta\gamma$ from the immobilized $G_i\alpha$. Stripping the SPR chip of bound $G\beta\gamma$ by this procedure has little effect on the ability of the immobilized $G_i\alpha$ to bind $G\beta\gamma$ in subsequent experimental cycles. Even after as many as 40 cycles of stripping and rebinding $G\beta\gamma$ over a period of 8 hr the loss in binding capacity of the immobilized $G_i\alpha$ is less than 3%.

In solution, high concentrations of Mg^{2+} normally cause the guanine nucleotide to dissociate from $G_i\alpha$, producing an unstable state of the protein that is susceptible to denaturation.[22] Denaturation is accompanied by a loss in the ability of $G\alpha$ to bind $G\beta\gamma$,[19] and this can be prevented by GDP, by GTP or its nonhydrolyzable analogs, or by AlF_4^- at concentrations sufficient to ensure occupancy of the guanine nucleotide binding site. Consequently, 100 μM GDP is routinely included in the regeneration buffer that is used to strip $G\beta\gamma$ from the immobilized $G_i\alpha$. However, it should be noted that this practice may not be necessary. We have observed that exposure of immobilized $G_i\alpha$ to 120 mM Mg^{2+} in the absence of guanine nucleotides for a period of time that would denature the soluble protein does

---

[22] T. Katata, M. Oinuma, and M. Ui, *J. Biol. Chem.* **261**, 8182 (1986).

not prevent the immobilized protein from subsequently binding $G\beta\gamma$. Further, we found the affinity of the immobilized $G_i\alpha$ for $G_i\beta\gamma$ to be independent of the guanine nucleotide present in the mobile phase. These data suggest that the immobilized $G_i\alpha$ subunit is able to retain bound guanine nucleotide under conditions promoting nucleotide dissociation from the protein in solution, thus rendering the immobilized $G_i\alpha$ surprisingly stable.

Effects of Detergents, Salts, and Magnesium on Binding of $G\beta\gamma$ to Immobilize $G_i\alpha$

If NaCl is omitted from Solution B, nonspecific binding of G protein subunits to the SPR chip surface occurs. The concentration of salt in Solution B (150 mM) is sufficient to prevent nonspecific binding. It is possible to substitute $NaSO_4$ for NaCl in Solution B. Magnesium has a marked effect on G_i subunit interaction. As noted above, no binding of $G\beta\gamma$ to the immobilized $G_i\alpha$ is observed in the presence of 40 mM $MgSO_4$, and even the Mg^{2+} concentration present in Solution B reduced the amount of binding, presumably by shifting the equilibrium toward the dissociated subunits. Magnesium is routinely included in the SPR mobile phase to ensure tight binding of guanine nucleotide to $G\alpha$, but the concentration could safely be reduced since micromolar concentrations are all that is required for guanine nucleotide binding to $G\alpha$.

It is not necessary to use detergents when investigating G_i subunit interactions by SPR spectroscopy. However, the maximum binding for a given concentration of $G\beta\gamma$ to the immobilized $G_i\alpha$ is reduced severalfold in the absence of detergent when compared with the binding in Solution B, which contains 0.005% Tween 20. It appears that either the concentration of $G\beta\gamma$ reaching the surface of the SPR chip is lower, or the affinity of $G\beta\gamma$ for the biotinylated-$G_i\alpha$ is lower in the absence of detergent. Although the reason for the reduced binding is unclear, a reduction of $G\beta\gamma$ concentration due to precipitation in the absence of detergent seems unlikely since this should also result in nonspecific and irreversible binding to the surface of the SPR chip, which we have not observed.

Investigating G Protein Subunit Interaction with G-Protein-Coupled Receptors by SPR

Strategies for Immobilizing Rhodopsin

The principal regulation of G protein subunit interaction and guanine nucleotide exchange occurs through cell surface receptor activation. Techniques for the isolation and reconstitution of G-protein-coupled receptors (GPCRs) with G proteins have been available for some years now. The techniques for the introduction of the GPCRs into synthetic lipid vesicles or bilayer have recently been adapted to the SPR measurement of rhodopsin–retinal G protein interactions [for reviews see Refs. 23 and 24]. These approaches have demonstrated light-regulated G protein binding to rhodopsin, as has been known since the pioneering work of Herman Kuhn,[11] and these SPR approaches have enabled measurement of equilibrium dissociation constants for the heterotrimer G protein binding to rhodopsin. The inclusion of lipid surfaces in these procedures represents a significant drawback to the examination of most GPCR–G protein interactions, as the nonretinal G proteins require detergents for solubility, and the detergent conditions necessary for G protein solubility are detrimental to the lipid matrices. In addition, investigations using lipid matrices have measured a significant contribution to the SPR signal by interaction of the retinal G protein subunits with the lipids per se.[25,26] Therefore, we have adopted an alternative strategy for coupling functionally active GPCR to the sensor chip.[27] In this approach, detergent-extracted rhodopsin is immobilized indirectly by noncovalent binding to covalently immobilized concanavalin A (ConA). The reversible binding to ConA is a routine procedure for the purification of rhodopsin from detergent extracts of rod outer segment (ROS) disks.[28] We show that the rhodopsin bound to ConA retains full biochemical activity for GTP exchange on retinal G protein subunits. Further, rhodopsin immobilized in this way remains stable for several days at 25° in the Biacore instrument. These procedures are conceptually appealing in that the extracellular carbohydrate modifications for most GPCRs do not appear to be essential for ligand binding or G protein regulation. Further, this immobilization strategy, as opposed to direct amine linkage, results in a less heterogeneous orientation of the receptor with the intracellular surfaces oriented away from potential steric interferences by the ConA and carboxymethyldextran hydrogel. Lastly, initial experiments utilizing direct amine coupling chemistries resulted in complete inactivation of detergent-extracted rhodopsin. Although rhodopsin immobilized indirectly via ConA binding is stable to the detergents necessary for solubility of G protein subunits, repeated and prolonged exposure to detergents denatures the rhodopsin. Therefore, the running solutions used for these experiments do not contain detergent, and detergents are added only with G protein subunits at minimal concentrations sufficient for protein solubility. The running solution utilized for these experiments is based on reagents used for measuring GTP binding exchange reactions catalyzed by rhodopsin[14] in order to provide for a direct comparison of protein interaction affinity constants determined directly by SPR and indirectly by saturation of G protein activation measurements.

[23] C. Bieri, O. P. Ernst, S. Heyse, K. P. Hofmann, and H. Vogel, *Nature Biotech.* **17,** 1105 (1999).

[24] Z. Salamon, M. F. Brown, and G. Tollin, *Trends Biochem.* **24,** 213 (1999).

[25] Z. Salamon, Y. Wang, J. L. Soulages, M. F. Brown, and G. Tollin, *Biophys. J.* **71,** 283 (1996).

[26] S. Heyse, O. P. Ernst, Z. Dienes, K. P. Hofmann, and H. Vogel, *Biochemistry* **37,** 507 (1998).

[27] W. A. Clark, X. Jian, L. Chen, and J. K. Northup, *Biochem. J.* **358,** 389 (2001).

[28] J. W. Clack and P. J. Stein, *Proc. Natl. Acad. Sci. U.S.A.* **85,** 9806 (1988).

Cross-Linking Concanavalin A to SPR Chips

Concanavalin A is a tetrameric protein with four identical 28 kDa subunits.[29] The minimal structure exhibiting high affinity binding of carbohydrates is the dimeric form. The dimer form of ConA is an active stable structure, and the tetramer–dimer transition can be promoted by low pH or succinylation of the protein.[30] For reasons of economy we use unmodified concanavalin A (Sigma). Because the dissociation of dimer from tetramer leads to a continual decline in the SPR baseline, our coupling procedures first promote the tetramer–dimer dissociation prior to linkage. To accomplish this, ConA is dissolved and stored at 4° in 100 mM sodium acetate, pH 4.8, prior to coupling. This also provides for optimal concentration of the ConA dimer by ionic attraction to the unactivated carboxyl groups. For limiting the ConA coupling the pH can be adjusted to 5.4 and still promote the dimer form. However, extensive washing of the surface to allow complete tetramer dissociation is then necessary. Linkage of the ConA is accomplished by standard NHS/EDC amine coupling[16] as described for streptavidin above, but with a modification in the running solution (Solution C, 50 mM MOPS, pH 7.5, 3 mM MgSO$_4$, 150 mM NaCl with 10 μM CaCl$_2$ and 10 μM MnCl$_2$. First, the surface is activated by injection of a mixture of 0.2 M EDC with 50 mM NHS in water at a flow rate of 5 μl/min, exposure time of 5–7 min. This is followed by injection of ConA at 0.1–0.3 mg/ml in 100 mM sodium acetate, pH 4.8, with an exposure time of 5–7 min. Activated carboxyls are then quenched by reaction with 1 M ethanolamine in water with an exposure time of 4 min. These procedures lead to the immobilization predominantly of dimeric ConA. The immobilized ConA is quite stable, so that rhodopsin coupling does not have to proceed immediately. However, the coupling reactions are sufficiently reproducible and rapid that we find no advantage to prior preparation of ConA-modified surfaces. For experiments defining the independent binding of Gα or G$\beta\gamma$ described below, surfaces modified with 7000–12,000 RU of ConA have been utilized. Under these conditions, the presence of immobilized tetrameric ConA is a significant factor. When high concentrations of ConA and/or aggressive NHS/EDC activation conditions are used, we routinely wash the ConA surface for 30–60 min with Solution C at a flow rate of 5 μl/min to allow for complete tetramer–dimer dissociation, prior to rhodopsin immobilization.

Extraction and Immobilization of Rhodopsin on Concanavalin-A-Modified SPR Chips

Urea-washed ROS disks are prepared from hypotonically washed, GTP-extracted ROS disks to remove phosphodiesterase, rhodopsin kinase, and

[29] A. J. Kalb and A. Lustig, *Biochim. Biophys. Acta* **168**, 366 (1968).

[30] Gr. Gunther, J. L. Wang, I. Yahara, B. A. Cunningham, and G. M. Edelman, *Proc. Natl. Acad. Sci. U.S.A.* **70**, 1012 (1973).

G proteins by repeated incubation and sedimentation with 5 M urea at $4°$ under ambient illumination as described,[31] resulting in the depletion of all peripheral proteins from the ROS disks and a mixture of meta-II and meta-III rhodopsin isomers in the disks. These preparations are snap frozen and stored at $-80°$. Prior to immobilization, rhodopsin is solubilized from a preparation of urea-washed ROS disks by addition of 20–25 mM CHAPS to a suspension of ROS disks containing 3–3.5 mg/ml protein (approximately 100 μM rhodopsin) in 10 mM Tris, pH 7.5, on ice. After 30 min, unextracted ROS disks are sedimented for 10 min at approx. 14,000g at $4°$ in an Eppendorf microfuge, and the supernatant is used for immobilization. Solubilized rhodopsin is diluted between 5- and 10-fold with a solution of chilled 50 mM MES pH 6.0 with 1 mM $CaCl_2$ and 1 mM $MnCl_2$. The two divalent cations are essential for the carbohydrate binding by ConA, and they must be present to retain rhodopsin bound to ConA.[32] The 10 μM concentrations of Ca^{2+} and Mn^{2+} in Solution C are sufficient to prevent detectable dissociation of rhodopsin from ConA measured over a 10 min time period. The diluted rhodopsin is warmed to $25°$ for 1–2 min prior to injection. Because the association of rhodopsin with ConA is a slow reaction and the strategy is to saturate the ConA with rhodopsin, the flow rate is reduced to 2 μl/min to conserve the amount of receptor required for coupling. The exposure time is typically 25 min, by which time saturation of the binding equilibrium for ConA has been achieved. At this time the flow rate is returned to 5 μl/min for washing out unbound rhodopsin. We have designed these procedures to be limited by the amount of ConA immobilized to address issues of nonspecific interactions of G proteins with ConA discussed below. The immobilized rhodopsin has a surprisingly long half-life at $25°$ of approximately 40–60 hr based on the binding capacity for $G\beta\gamma$. However, several factors may decrease the useful lifetime of a rhodopsin surface; most critically, repeated detergent exposures and exposure to denaturing concentrations of detergent are detrimental to the rhodopsin. All of our research has utilized either Sodium-cholate or CHAPS as the detergent for solubilizing G protein subunits, and we routinely limit the concentrations of these to less than 800 μM in the sample injections with exposure times of 5 minutes or less. Because of the variability in expected rhodopsin surface lifetimes, we generally use a surface for only a single day of experimental tests, synthesizing a new surface for each experiment.

A typical high capacity surface coupling is documented in Fig. 2. This shows the sequential activation of carboxyl groups by NHS/EDC; binding and coupling of ConA at pH 4.8; quenching of the activated carboxyls with 1 M ethanolamine; and subsequent binding of CHAPS-extracted bovine rhodopsin to the immobilized ConA.

[31] A. B. Fawzi and J. K. Northup, *Biochemistry* **29**, 3804 (1990).
[32] G. N. Reeke, J. W. Becher, B. A. Cunningham, G. R. Gunther, J. L. Wang, and C. M. Edelman, *Ann. NY Acad. Sci.* **234**, 369 (1974).

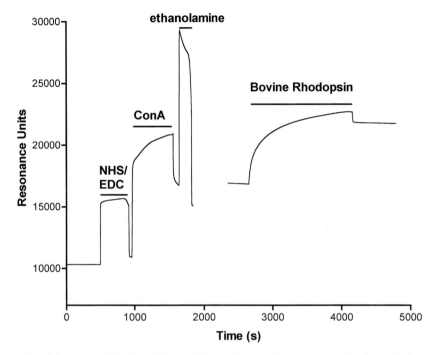

Fig. 2. Sensogram of the immobilation of rhodopsin by binding to a concanavalin-A-modified surface. Concanavalin A was immobilized by NHS/EDC coupling at 0.3 mg/ml as described. Subsequent
to the coupling of ConA and washing of the surface, rhodopsin, which had been extracted at 100 μM
with 25 mM CHAPS, was diluted to 10 μM as described and injected for capture by the immobilized
ConA. Reproduced with permission from W. A. Clark, X. Jian, L. Chen, and J. K. Northup, *Biochem.
J.* **358,** 389 (2001).

Verifying That Immobilized Rhodopsin Has Biological Activity

Although it is difficult to assess the physical structure of the immobilized
rhodopsin produced by the ConA coupling procedures outlined above, it is possible to examine the catalytic activity of this preparation. The carboxymethyl
dextran layer of the SPR surface is chemically identical to beaded carboxymethyl
dextran produced by Pharmacia. The coupling procedures utilized for the immobilization described above are modified for bulk application in the following procedures. Because ionic attraction is not required for the bulk coupling
to be successful, the pH of the rhodopsin immobilization is changed to 7.5, at
which the CHAPS- extracted protein is more stable. For bulk immobilization,
CM Sephadex C-25 (Pharmacia) resin (0.5 g) is hydrated, then washed repeat-
edly with 100 mM MES (pH 5.0). The resin is sedimented by gentle centrifugation and the supernatant removed. Typically, the resin expands to approximately

5 ml on hydration. The resin (1 to 2 ml) is then incubated with an equal volume of EDC [30 to 60 mg/ml in 100 mM MES (pH 5.0)] followed by the same volume of ConA [2 mg/ml in 100 mM MES (pH 5.0)]. After a 1 hr incubation at room temperature, the resin is sedimented and resuspended for 10 min in 6 ml of ethanolamine (0.1 M). After sedimentation, the resin is then resuspended three times, 5 to 10 min each instance, in 5 ml of Solution D [10 mM MOPS (pH 7.5), 100 mM NaCl, 3 mM MnCl$_2$, 3 mM CaCl$_2$] at room temperature followed by two more washes in 5 ml of Solution D at 4°. ROS disks (about 100 μM rhodopsin, 250 to 300 μl) are extracted on ice for 10 min with 0.1 M CHAPS (125 to 150 μl). Four ml of Solution D is added to the extracted disks and incubated an additional 20 min on ice. The extracted ROS solution is then added to the washed resin and the mixture is rocked gently at 4° for 12 to 18 hr. The final resin is then sedimented and washed 5 times with Solution D as described above. We have utilized rhodopsin immobilized as described for affinity chromatography of G$\beta\gamma$. The carboxymethyldextran matrix provides a superior support to commercially available ConA-agarose, as retinal G protein subunits are adsorbed by the latter and not by CM-Sephadex. However, ConA-immobilized rhodopsin has more limited utility for routine use as an affinity chromatography for G$\beta\gamma$ subunits as compared to immobilized G$_o\alpha$ or G$_i\alpha$.

To examine the biochemical activity of immobilized rhodopsin, the CM-dextran beads can be introduced into an *in vitro* assay for catalyzed turnover of GTPγS binding by the procedures of Ref. 14. For reproducible distribution of the beaded dextran, the aperature of a disposable pipette tip needs to be widened by slicing with a razor blade. In addition, it is essential that the timed incubation of the assay be conducted in a shaking water bath, to keep the bead well mixed with the assay reagents. All other aspects of the rhodopsin-catalyzed binding assay are conducted as described for assay of rhodopsin in ROS disk membranes. Even with care, since the smallest particle distributed in the assay is a bead, some quantitative variability of the assay is to be expected. Estimation of the specific activity of immobilized rhodopsin is accomplished by constructing a standard curve with known concentrations of rhodopsin in ROS disks and comparing the activity of varying volumes of dextran-immobilized rhodopsin. The concentration of rhodopsin immobilized on the CM-dextran is assessed by SDS–PAGE analysis. The beads are treated with Laemmli sample buffer, incubated at room temperature for 45 min, and run on 12% acrylamide gels. For optimal visualization of rhodopsin, we find that silver staining[33] provides the most clearly stained bands. Two protein bands of equal staining intensity are visualized from this procedure, the 28 kDa ConA subunit and the approximately 35 kDa rhodopsin band. When calibrated with known amounts of rhodopsin from ROS

---

[33] W. Wray, T. Boulikas, V. P. Wray, and R. Hancock, *Anal. Biochem.* **118,** 197 (1981).

disk membranes, these procedures yield $80 \pm 36\%$ of the specific catalytic activity of rhodopsin in the native membrane environment.[27] Of equal importance, the catalyzed turnover, as found for rhodopsin in ROS disks, requires the presence of $G\beta\gamma$.

Synergistic Binding of Retinal G Protein Subunits to Immobilized Rhodopsin

ConA-immobilized rhodopsin surfaces provide a basis for examining the interactions of G protein subunits with rhodopsin independent of guanidine nucleotide exchange. As a final preparation step prior to SPR analysis, all samples are exchanged into Solution C by chromatography over Sephadex G-50 (Pharmacia). For the hydrophobic $G\beta\gamma$ samples, Solution C is supplemented with 8 mM CHAPS. This procedure eliminates refractive index differences between the protein samples and the running solution, thereby diminishing the bulk refractive index contribution to the SPR signal. Storage of protein samples at $4°$ for periods of more than a few days can lead to higher dissolved gas content than is present in the initial running Solution C, thus altering the refractive index. For long-term storage, samples are snap frozen and kept at $-80°$. It is also important that the samples be desalted in the identical preparation of the running Solution C, as minor differences in solute concentration can lead to significant bulk refractive index changes. The Solution C used for these studies is based on the composition of the reagent used for the analysis of G protein subunit saturation of rhodopsin interaction by [^{35}S]GTPγS binding.[14] The binding reagent differs from the SPR Solution C in the inclusion of BSA in the binding reagent (which has been eliminated from the SPR Solution C), the addition of $MnCl_2$ and $CaCl_2$ to the SPR Solution C for retention to ConA, and the increase of NaCl from 100 mM in the binding reagent to 150 mM in the SPR Solution C to eliminate ionic adsorption to the carboxymethyldextran. None of these changes influences the rates of GTPγS binding catalyzed by rhodopsin, except that the recovery of bound GTPγS is increased by the presence of BSA at low concentrations of $G\alpha$.

In a typical SPR experiment G protein subunits are injected over the ConA–rhodopsin surface at flow rates of 5 μl/min. This flow rate balances the requirements for rapid mixing of the flow paths (60 nl volume equilibrates in about 3 sec) for kinetic studies with the expense of adequate volumes of G proteins at the concentrations required for binding analysis. This trade-off is most dramatically illustrated in the experiments measuring the kinetics of transducin α subunit presented in Figs. 3 and 4A. For the slower binding reactions of the hydrophobic $G\beta\gamma$ dimers 5 μl/min is clearly adequate for kinetic analysis (see Fig. 5). Figure 3 presents the binding interactions of bovine retinal G protein subunits independently and in concert both to the immobilized rhodopsin and to the ConA–dextran used to immobilize the rhodopsin. What is revealed is that rhodopsin displays

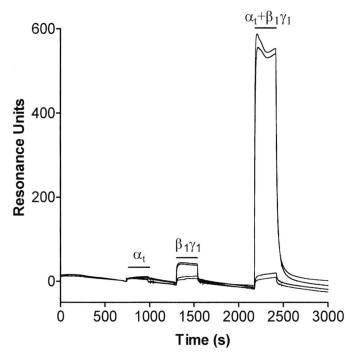

FIG. 3. Simultaneous recordings of sensograms for retinal G protein subunits through four flow paths. All four flow paths were modified with ConA as described. Two paths were additionally bound with rhodopsin. The injections were 0.48 μM α, 0.65 μM $\beta\gamma$ dimer and the combination of 0.48 μM α with 0.65 μM $\beta\gamma$. The resonance signal from the immobilized ConA or ConA-captured rhodopsin has been subtracted from the signals. Reproduced from W. A. Clark, X. Jian, L. Chen, and J. K. Northup, *Biochem. J.* **358**, 389 (2001).

a measurable binding of G$\beta\gamma$ independent of Gα, and that there is a profound synergy of the binding reaction when both Gα and G$\beta\gamma$ subunits are present. Additional data demonstrate that this binding is selective for the Gα_t species and the GDP conformation of Gα_t. All of these results recapitulate data firmly established by *in vitro* biochemical analysis of GTP exchange, except that the binding of G$\beta\gamma$ could not otherwise be measured. It is important to note that the results for the synergistic binding of Gα and G$\beta\gamma$ are dramatically influenced by the temperature of the samples when mixed. If the subunits are mixed and held on ice prior to injection, the synergism is profoundly reduced. Therefore, the samples are diluted into the running Solution C, which has been equilibrated to 25° about 120 sec before injection and the mixed samples are then maintained at 25° prior to injection.

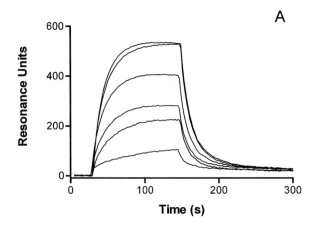

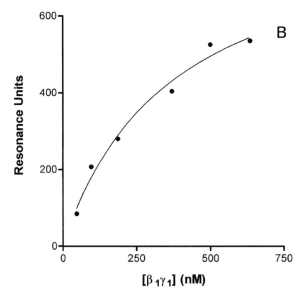

FIG. 4. (A) Sensograms of the synergistic binding of retinal $\beta\gamma$ in the presence of a constant amount of retinal α. Rhodopsin was immobilized to a ConA modified flow path as described. Mixtures of resolved $\beta\gamma$ were made at varying concentrations from 64 to 640 nM in the presence of 1 μM α. The individual sensograms are overlayed here to compare the binding signals. (B) Saturation analysis of the equilibrium binding of retinal $\beta\gamma$ in the presence of retinal α. The plateau values for the binding signals (equilibria) have been analyzed with a single-site model as a Langmuir binding isotherm [see Eq. 2 on p. 37]. Reproduced with permission from W. A. Clark, X. Jian, L. Chen, and J. K. Northup, *Biochem. J.* **358,** 389 (2001).

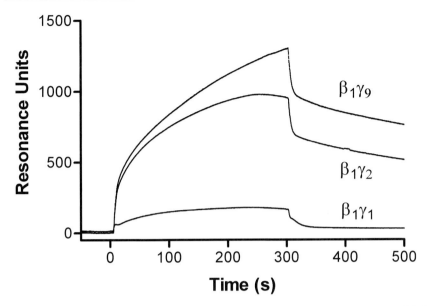

FIG. 5. Sensograms of the independent binding of $G\beta\gamma$ dimers to rhodopsin. A ConA modified flow path was bound with rhodopsin as described. Three preparations of $G\beta\gamma$ dimer with distinct γ subunit composition were injected in sequence. The $\beta_1\gamma_1$ was injected first. After the injection of $\beta_1\gamma_2$ the flow path was washed for 7000 sec to allow complete dissociation prior to the injection of $\beta_1\gamma_9$. Both $\beta_1\gamma_2$ and $\beta_1\gamma_9$ were diluted with running Solution C to a final CHAPS concentration of 400 μM to equalize the refractive index contribution. All $\beta\gamma$ dimers were injected at 200 nM.

Equilibrium Dissociation Constants for Retinal G Protein Subunits

The data shown in Fig. 3 demonstrate the rapidity of the equilibration of binding of retinal G protein subunits to rhodopsin. A number of experiments suggest that both rates of $G\alpha$ and $G\beta\gamma$ association with and dissociation from rhodopsin are too rapid for precise measurement under these conditions. Indeed, for $G\alpha$ it appears that equilibration is completed within the mixing time of the flow path at a 5 μl/min flow rate. Therefore, the kinetics of retinal G protein subunit interactions have not been pursued. However, equilibrium binding is readily obtained, and these values can be determined in experiments varying the concentrations of free G protein subunits in the injection. These experiments utilize the synergistic binding of the combination of $G\alpha$ and $G\beta\gamma$ to examine the binding constants for each. The concentration of one of the subunits is held constant, while that for its interacting partner is varied to assess the saturation of binding to rhodopsin.

An example of this type of experiment is provided in Fig. 4A in which the apparent affinity of rhodopsin for retinal $G\beta\gamma$ has been determined in the presence of a constant 0.5 μM of retinal $G\alpha$. Figure 4B has plotted the plateau values for

SPR for each injection as a function of the concentration of $G\beta\gamma$ added. These data conform reasonably well to a simple bimolecular binding model, which is the basis for the curve drawn. In this instance, the apparent K_d value obtained is about 350 nM, which is in good agreement with data for retinal $G\beta\gamma$ saturation of the rhodopsin-catalyzed binding of GTPγS to $G\alpha$.[14] Ideally, a family of equilibrium binding curves will be generated covarying both $G\alpha$ and $G\beta\gamma$ concentrations, from which the individual equilibrium constants may be obtained. Such an evaluation is inherently model dependent, as the impact of variation of each subunit on the binding interactions of its partner is governed by the molecular mechanism of rhodopsin catalysis of G protein activation. SPR analysis could provide important clues to the precise mechanism.

The limitations on this approach are the binding constants for the G protein subunits. While the empirical determination of saturation by retinal $G\beta\gamma$ was obtainable in these experiments, that for the retinal $G\alpha$ was not. It appears to have too low an affinity for affordable measurement by these methods (we estimate $K_d > 1$ μM). As for the determination of the rate constants for the binding interactions of $G\alpha$ with rhodopsin, the repeated injection of 20–50 μl of G protein subunits at several micromolar concentration is impractical for tissue-derived proteins.

Independent Binding of $G\beta\gamma$ Subunits to Immobilized Rhodopsin

The SPR approach allows the examination of the molecular interaction of G protein signaling components that do not lead to functional changes. Although all of the protein interactions discussed thus far can be assessed indirectly through functional assays—$G_t\alpha$–$G\beta\gamma$ and $G_i\alpha$–$G\beta\gamma$ binding by pertussis toxin ADP-ribosylation assays[14,34] and $G_t\alpha$–$G\beta\gamma$–rhodopsin binding by GTPγS binding catalysis[14,35]—the binding of $G\beta\gamma$ to receptor in the absence of $G\alpha$ has no identified functional sequelae. Figure 3 provides an indication of the independent binding of retinal $G\beta\gamma$. As there are potentially a wide variety of $G\beta\gamma$ dimers of distinct $G\beta$ and $G\gamma$ composition, SPR provides an opportunity to examine the influence of this diversity on receptor interaction. To test the specificity for rhodopsin, $G\beta\gamma$ dimers of defined compositions are expressed by coinfection in Sf9 cells.[15] Figure 5A illustrates the examination of $G\beta\gamma$ selectivity for rhodopsin binding determined by SPR. As seen here, the binding interactions of $G\beta_1\gamma_1$, $G\beta_1\gamma_2$, and $G\beta_1\gamma_9$ all at 200 nM give distinct kinetic and equilibrium values. Most significantly, while the $G\beta_1\gamma_1$ completely dissociates from rhodopsin within seconds of terminating the injection, both $G\beta_1\gamma_2$ and $G\beta_1\gamma_9$ dissociate slowly. Calculated half-times for

[34] P. J. Casey, M. P. Grazioano, and A. G. Gilman, *Biochemistry* **28**, 611 (1989).

[35] A. Yamazaki, M. Tatsumi, D. C. Torney, and M. W. Bitensky, *J. Biol. Chem.* **262**, 9316 (1987).

their dissociations are on the order of 1000 sec. These rates are not facilitated by the addition of detergent to the running Solution C, nor metal ions, nor modest pH changes. Indeed, to date the only conditions that have been identified to enhance the rate of this dissociation reaction lead to denaturation of the rhodopsin. Hence, the examination of the hydrophobic $G\beta\gamma$ dimers is limited in that there are no convenient regeneration procedures, and sequential injections require extensive washing times to allow for complete dissociation of the $G\beta\gamma$. In addition, except for the naturally occurring $G\beta_1\gamma_1$ and $G\beta_3\gamma_{8\text{-cone}}$ dimers, all of the Sf9 produced products we have tested behave as integral membrane proteins and require detergent for maintaining solubility. These samples are prepared and stored in 8 mM CHAPS solutions, and the running Solution C for these dimers is supplemented with 8 mM CHAPS as described above. The introduction of CHAPS with the $G\beta\gamma$ presents a technical difficulty in the SPR signal due to the differences in the refractive index between running Solution C without and with CHAPS. This is illustrated in Figs. 5A and 5B by the steep increases and decreases in the SPR signal at the onset and termination of injections of $G\beta_1\gamma_2$ and $G\beta_1\gamma_9$ samples. As discussed above, the stability of the rhodopsin binding activity for G proteins is drastically reduced in the presence of detergent, so that this refractive index change cannot be avoided by the inclusion of a constant concentration of detergent to the running Solution C. The contribution of CHAPS refractive index to the SPR signal for $G\beta\gamma$ injections can be addressed by the strategy shown in Fig. 3. For the Biacore 2000 and later instruments, the experimental design of Fig. 3 allows for the subtraction of all contributions to the SPR signal that are not due to the protein–protein interactions with rhodopsin. For single flow path monitoring, a somewhat more cumbersome approach can be adopted using the same strategy. For these experiments, reference injections of the hydrophobic $G\beta\gamma$ samples are made using the immobilized ConA surface, prior to binding the rhodopsin. These injections result in nearly square-wave SPR signals from the refractive index changes. The stored reference signals can then be subtracted from the signals obtained for the same $G\beta\gamma$ samples with the rhodopsin-coupled surface.

Sequential Binding of G Protein Subunits to Immobilized Rhodopsin

The long-lived binding of $G\beta_1\gamma_2$ to rhodopsin allows the examination of sequential binding interactions of $G\alpha$ and $G\beta\gamma$. This experiment asks if the G protein must bind to receptor as a heterotrimer, or if the binding interactions can be ordered. In this experimental design, the impact of $G\beta_1\gamma_2$ binding to rhodopsin on the subsequent binding of $G\alpha$ is examined by sequential injection of $G\beta_1\gamma_2$ followed by $G\alpha$. A significant consideration in the design of this experiment is the contribution of CHAPS to the SPR signal for the $G\beta_1\gamma_2$ injection. Sufficient time must be allowed for complete washout of the CHAPS solution from that injection prior to injection of $G\alpha$ to allow for clearly interpretable SPR signals from the $G\alpha$. This of course limits the fractional saturation of rhodopsin by $G\beta_1\gamma_2$ to a value less than unity. However, the dissociation of $G\beta_1\gamma_2$ from rhodopsin is sufficiently slow that this is not an important limitation. Figure 6 presents an example of the sequential injection of G protein subunits by this paradigm. This should be contrasted with the results obtained for coinjection of the retinal G protein subunits. First, it is quite apparent that the binding interaction of $G\alpha$ obtained in the sequential injection paradigm equilibrates more rapidly than the mixing time of the flow cell, and the dissociation is also rapid. Note also that the SPR signal from the $G\alpha$ binding is considerably less than that from the $G\beta_1\gamma_2$ despite the fact that the $G\alpha$ concentration is micromolar while the $G\beta_1\gamma_2$ was 200 nM. As found for the coinjection synergy experiments, it has been technically unfeasible to obtain $G\alpha$ saturation in the sequential injection paradigm, owing to the low affinity

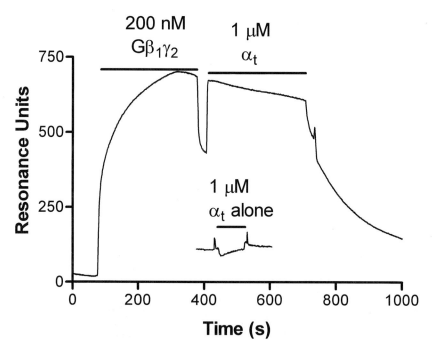

FIG. 6. Sensogram of sequential binding of G protein subunits to immobilized rhodopsin. The design of this experiment is described in the text. The SPR signal for the injection of 1 μM $G\alpha$, which was obtained prior to the binding of $\beta_1\gamma_2$, is presented below the subsequent injection of that same $G\alpha$ after the injection of 240 nM $\beta_1\gamma_2$. Reproduced with permission from W. A. Clark, X. Jian, L. Chen, and J. K. Northup, *Biochem. J.* **358,** 389 (2001).

of this interaction. However, it is possible to examine the influence of $G\beta_1\gamma_2$ saturation on the binding of $G\alpha$ to the rhodopsin–$G\beta_1\gamma_2$ surface. Because the binding of $G\alpha$ equilibrates essentially instantly, the injection times required to determine the binding plateau are relatively short. By spacing several 1- to 2-min injections of $G\alpha$ during the time course of dissociation of $G\beta_1\gamma_2$ from rhodopsin, it is possible to obtain a quantitative relationship between fractional binding of $G\beta_1\gamma_2$ and the binding of $G\alpha$. As found in the coinjection paradigm, this latter experiment also demonstrates an absolute requirement for $G\beta\gamma$ binding for the binding of $G\alpha$ to rhodopsin.

Analytical Methods for Determining Kinetic Rate Constants and Equilibrium Dissociation Constants

In general, there are three main approaches for the determination of binding constants: the analysis of the binding kinetics from the time course of binding, the analysis of the equilibrium surface binding isotherm from plateau signals such as shown above in Fig. 4A, and the analysis of solution competition isotherms. Each of these methods requires different experiments to be performed and has specific advantages and drawbacks. The analysis of the binding kinetics and the equilibrium surface binding isotherm can be done with commercial software, while the automation of the competition analysis can be accomplished with scripts for general purpose modeling programs.

The analysis of the time course of binding is based on the mathematical modeling of sequences of association and dissociation obtained at different concentrations (c) of the protein in the mobile phase, such as shown in Fig. 1 for $G_i\beta\gamma$ binding to and dissociating from $G_i\alpha$. In theory, for a bimolecular reaction, the association proceeds as

$$R(t) = R_{eq}(c)\,[1 - \exp(-(k_a c + k_d)\,t)] \tag{1}$$

with the equilibrium plateau signal

$$R_{eq}(c) = R_{max}\,[1 + k_d/(k_a c)]^{-1} = R_{max}\,[1 + K_D/c]^{-1} \tag{2}$$

(with R_{max} denoting the maximal binding capacity of the surface). The dissociation follows

$$R(t) = R(t_0)\,\exp[-k_d(t - t_0)] \tag{3}$$

From the obtained values of the association rate constant, k_a, and the dissociation rate constant, k_d, one can calculate the equilibrium constant as $K_D = k_d/k_a$. One difficulty that frequently occurs with SPR data is that the sensorgrams, such as those shown in Figs. 1 and 5, do not conform to the single-exponential processes predicted for simple bimolecular reaction kinetics.[36,37] It is obvious that in this case, no single rate constants can be reported. When empirically using data subsets for analysis, we found apparent equilibrium constants that could vary over several orders of magnitude, depending on the data range and numerical analysis method chosen. One possible course of action is the use of more complex kinetic models; however, this can lead to ambiguous results, in particular, as many sensor-related artifacts can mimic complex binding kinetics. Such possible artifacts and corresponding control experiments are described in Refs. 6 and 36. Therefore, we will focus on more robust strategies for the determination of the equilibrium constant K_D, which can be combined with estimates of the apparent dissociation rate constant to give approximate association rate constants.

The analysis of the equilibrium isotherm (see Figs. 4A and 4B) is based on modeling the plateau binding signals with Eq. (2). This is the traditional Langmuir isotherm analysis, and equivalent to the use of Scatchard plots as used in many other techniques. Since no information on the binding progress is used, this approach is very robust, and independent of mass transport limitations or other transient artifacts of surface binding. Also, it can be used over a very wide range of affinities. One possible limitation can be the relatively long contact time that may be required to reach equilibrium. In this case, the use of an oscillatory sample delivery system may be helpful.[9] Alternatively, a solution competition experiment can be performed. Competition experiments have the advantage of providing equilibrium dissociation constants for the unmodified proteins in solution.

The basic strategy for the solution competition experiments is the injection of a constant concentration of mobile reactant, in our case $G_i\beta\gamma$ ($C^{tot}\beta\gamma$), in equilibrium with varying concentrations of the soluble form of the immobilized binding partner, here $G_i\alpha$ ($C^{tot}\alpha$). SPR spectroscopy is used only as a concentration detector. Since the soluble $G_i\alpha$ competes with the surface immobilized $G_i\alpha$, only the remaining free $G_i\beta\gamma$ ($C^{free}\beta\gamma$) will be detected at the sensor surface. It follows the isotherm

$$
\begin{aligned}
C^{free}\beta\gamma = C^{tot}\beta\gamma - 0.5\big[&C^{tot}\beta\gamma + C^{tot}\alpha + K_D{}^{sol} \\
&- \{(C^{tot}\beta\gamma + C^{tot}\alpha + K_D{}^{sol})^2 - 4C^{tot}\beta\gamma\, C^{tot}\alpha\}^{0.5}\big]
\end{aligned}
\tag{4}
$$

Mathematical modeling of the data of $C^{free}\beta\gamma$ as a function of competitor $C^{tot}\alpha$ allows the determination of the solution binding constant $K_D{}^{sol}$ for the G-protein subunits.

[36] P. Schuck, *Annu. Rev. Biophys. Biomol. Struct.* **26**, 541 (1997).
[37] P. Schuck, *Curr. Opin. Biotechnol.* **8**(4), 498 (1997).

Several different strategies for quantifying the concentration of free $G_i\beta\gamma$ from the sensorgrams can be used. We found the following approach both economical in material and numerically robust: First, a series of sensorgrams at increasing concentration of $G_i\beta\gamma$ in the absence of $G_i\alpha$ are generated (e.g., 10, 20, 40, 60, 80, and 100 nM). This serves as a reference to generate a standard curve for calibration of the sensor response in terms of $C^{\text{free}}\beta\gamma$. Then a series of competition experiments is performed with equilibrium mixtures of constant 100 nM $G_i\beta\gamma$ and nonbiotinylated $G_i\alpha$ at different concentrations (ideally starting at concentrations much larger than K_D, in twofold dilutions to concentrations much smaller than K_D). This is followed by a second set of binding experiments in the absence of $G_i\alpha$ in order to verify the stability of the sensor response. From the graphical superposition of the sensorgrams obtained in the calibration experiments, an arbitrary but characteristic concentration-dependent feature can be selected. We chose the average slope of the curves in a predetermined time window of $\sim$100 sec from the association and dissociation phases (other possible choices could be a binding signal at steady state or after a fixed time). For each $G_i\beta\gamma$ concentration, the slopes are calculated, and the negative slope in the dissociation phase can be subtracted from the slope in the association phase in order to increase the signal. This results in a reproducible value for each $G_i\beta\gamma$ concentration that can be used as a standard curve. A fourth-order polynomial is fitted to the standard curve to allow interpolation of the values. The same procedure for calculating the slopes is applied to the competition experiments. As the concentration of $G_i\beta\gamma$ capable of surface binding is reduced because of heterotrimer formation with $G_i\alpha$ in solution, smaller values for the slopes are obtained, and the corresponding concentration $C^{\text{free}}\beta\gamma$ can be determined from the standard curve. Modeling of the competition isotherm with Eq. (4) reveals the value of the equilibrium dissociation constant. This approach is illustrated in Fig. 7A, which shows the standard curve with slopes at different values of $G_i\beta\gamma$ both in the first (triangles) and the second calibration series (circles), as well as the interpolated curve. Figure 7B shows an example for the competition isotherm so obtained.

Although the competition approach requires a larger number of experiments, it is very general and completely independent of most sensor-related artifacts, including immobilization artifacts, as it relies exclusively on the reproducibility of a characteristic concentration-dependent feature of the sensorgrams, and the fact that the soluble competitor does not interfere with surface binding. One drawback of this method is that it does not permit the determination of affinities in a range as wide as the direct equilibrium isotherm analysis. However, it provides for additional experimental flexibility. For example, the ability of Mg^{2+} (e.g., 120 mM) to promote dissociation of the nucleotide from $G_i\alpha$ in solution makes it possible to load the $G_i\alpha$ subunit with various guanine nucleotides, so that their effects on

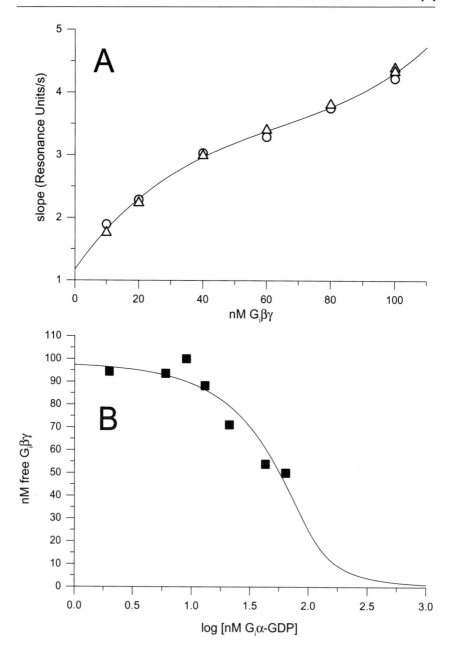

G protein subunit affinity can be investigated. (As noted above, it was very difficult to exchange nucleotides at the surface immobilized $G_i\alpha$.) It is clear that there is a qualitative difference between guanine nucleotides in their effect on G protein subunit affinity because under conditions where nonhydrolyzable GTP analogs are effective in causing subunit dissociation, GDP is not. We have used SPR spectroscopy to make quantitative comparisons of the effects of various guanine nucleotides on G protein subunit affinity, and obtained equilibrium dissociation constants for soluble G_i ranging from approximately 10 nM when GDP is bound to $G_i\alpha$ to approximately 0.6 μM when GTPγS is bound to $G_i\alpha$, with intermediate values for Gpp[CH$_2$]p and Gpp[NH]p.

Because the presence of guanine nucleotides in the mobile phase does not affect the affinity of the immobilized $G_i\alpha$ for $G_i\beta\gamma$ there is no need to generate a different standard curve for each guanine nucleotide whose effect on the equilibrium dissociation constant for G_i subunits is being determined by the competition method. However, this situation makes it difficult to investigate the effects of various guanine nucleotides on the kinetic rate constants for G_i subunits by using an experimental protocol in which $G\alpha$ is immobilized on an SPR chip. It is noteworthy that the nucleotide binding site on the immobilized $G_i\alpha$ is not entirely refractory since AlF_4^-, whose binding to GDP-liganded $G_i\alpha$ does not require nucleotide exchange, was found to alter the affinity of immobilized $G_i\alpha$ for $G\beta\gamma$.

After the analysis of the equilibrium constant using either the direct surface binding analysis or the competition approach, this value can be used to constrain the rate constants in the kinetic analysis of the sensorgrams. For example, using the dissociation rate constant from the sensorgrams, association rate constants can be calculated via $k_a = k_d/K_D$. Using this approach for $G\beta\gamma$ interacting with GDP-liganded $G_i\alpha$, we have estimated the association rate constants to be in the range of $4 \times 10^4\ M^{-1}\ \mathrm{sec}^{-1}$ and the dissociation rate constant to be approximately 2–$3 \times 10^{-3}\ \mathrm{sec}^{-1}$.

FIG. 7. (A) Example of a standard curve from a series of injections of $G\beta\gamma$ binding to immobilized biotinylated $G_i\alpha$, from experiment similar to that shown in Fig. 1. Calibration values were obtained by combining for each sensorgram the slope during $\sim$100 sec in the association phase with that during $\sim$100 sec of the dissociation phase. Two sets of calibration curves were collected, one before (triangles) and one after (circles) the competition experiment. The similarity of the slope values from both sets indicates the stability of the surface and the reproducibility of the calibration method. The solid line represents the fourth-order polynomial used for inverting the calibration curve and for obtaining free $G\beta\gamma$ concentrations from the slope data in the competition experiments. (B) Competition data of free $G\beta\gamma$ as a function of the concentration of nonbiotinylated $G_i\alpha$ with GDP (symbols). The solid line represents the best-fit competition isotherm according to Eq. (4), which results in a K_D of 15 ($\pm$12) nM.

Conclusions

SPR spectroscopy has proven to be a useful tool for investigating protein–protein interactions between G protein subunits and other proteins involved in signal transduction. We have described procedures for the immobilization of $G\alpha$ subunits, $G\beta\gamma$ subunits, and rhodopsin that should be applicable to the entire family of G proteins and GPCRs. We also describe several approaches for the analysis of kinetic and equilibrium constants obtained from SPR data. Because SPR affords activity-independent analysis of the protein–protein interactions among these interacting species it offers a complementary approach to the examination of GPCR-G protein regulatory processes to that of *in vitro* biochemistry. Importantly, protein interactions that do not lead to activity changes become accessible for examination. We expect that complementary investigation of the influence of such factors as metal ions and nucleotide and protein subunit composition on protein binding interactions as well as G protein activity will enable a clearer understanding of the molecular mechanisms involved. Thus, SPR offers a unique additional window of insight to the study of the regulation of G protein signaling.

[3] Neuroanatomical Localization of Receptor-Activated G Proteins in Brain

By LAURA J. SIM-SELLEY and STEVEN R. CHILDERS

Introduction

G-protein-coupled receptors have been traditionally localized in brain by specific radioligand binding assays, including *in vitro* autoradiography of radioligand binding in frozen brain sections.[1,2] Although these techniques provide important information on the localization of receptors in brain, anatomical studies have been limited in their ability to demonstrate functional receptor activity. Conversely, anatomical methods that have shown function in the brain (e.g., 2-[^{14}C]deoxyglucose autoradiography; Fos immunohistochemistry) provide data regarding general neuronal activity rather than that of individual receptor systems. The finding that specific receptor agonists stimulated the binding of a hydrolysis-resistant analog of GTP, [^{35}S]GTPγS (guanosine 5′-O-(γ-thio)triphosphate), in membranes from various tissues and isolated cells,[3] and in particular the discovery that a number of $G_{i/o}$-coupled receptors stimulated [^{35}S]GTPγS binding to

[1] M. Herkenham and C. B. Pert, *Proc. Natl. Acad. Sci. U.S.A.* **77,** 5532 (1980).
[2] W. S. Young and M. J. Kuhar, *Brain Res.* **179,** 255 (1979).
[3] T. Wieland and K. H. Jakobs, *Methods Enzymol.* **237,** 3 (1994).

neuronal membranes,[4,5] provided an opportunity to adapt this technique to frozen brain sections. [^{35}S]GTPγS autoradiography was first reported in 1995 with the demonstration of mu opioid, GABA$_B$, and cannabinoid CB1 receptor-stimulated [^{35}S]GTPγS binding in rat brain,[6] and this technique has now been used to localize a wide variety of receptor-activated G proteins.

[^{35}S]GTPγS autoradiography is used to detect receptor-activated G proteins by visualization of receptor-stimulated [^{35}S]GTPγS binding to the G-protein α subunit. The technique depends on the catalytic exchange of GDP by GTP on Gα in the presence of an appropriate receptor agonist.[7] In particular, receptor agonists increase the affinity of Gα for GTP by several hundredfold, with a parallel decrease in affinity for GDP.[8,9] When [^{35}S]GTPγS is used as a GTP analog in the presence of specific receptor agonists, the result is incorporation of [^{35}S] in membrane-bound Gα, which is detected by film autoradiography or phosphorimaging. [^{35}S]GTPγS is an ideal radioligand for such studies: it is resistant to hydrolysis, has a high affinity for all types of Gα subunits, and is available at high specific activity (>1000 Ci/mmol). The most important step in [^{35}S]GTPγS autoradiography is the reduction of the high level of basal [^{35}S]GTPγS binding normally present in brain membranes and sections; this is accomplished by addition of sodium and millimolar concentrations of GDP (see below).

[^{35}S]GTPγS Autoradiography Assay

Although [^{35}S]GTPγS autoradiography is based on a technique previously used to measure receptor-activated G proteins in membrane homogenates, with adaptation to tissue sections based on standard autoradiographic protocols, the specific conditions used in the [^{35}S]GTPγS autoradiographic assay are uniquely different from either receptor autoradiography or agonist-stimulated [^{35}S]GTPγS binding in membrane preparations and are considered in this article.

Tissue Preparation

Tissue preparation for [^{35}S]GTPγS autoradiography follows standard autoradiographic protocols. Animals are sacrificed and the brain is immediately removed and slowly immersed in isopentane (2-methylbutane) at $-30°$. After immersion for 3–5 min, the brain is placed on dry ice for approximately 5 min to evaporate remaining isopentane, then stored at $-80°$ until sectioning. Brains are sectioned at

[4] A. Lorenzen, M. Fuss, H. Vogt, and U. Schwabe, *Mol. Pharmacol.* **44,** 115 (1993).

[5] J. R. Traynor and S. R. Nahorski, *Mol. Pharmacol.* **47,** 848 (1995).

[6] L. J. Sim, D. E. Selley, and S. R. Childers, *Proc. Natl. Acad. Sci. U.S.A.* **92,** 7242 (1995).

[7] A. G. Gilman, *Ann. Rev. Biochem.* **56,** 615 (1987).

[8] D. E. Selley, L. J. Sim, R. Xiao, Q. Liu, and S. R. Childers, *Mol. Pharmacol.* **51,** 87 (1997).

[9] C. S. Breivogel and S. R. Childers, *J. Biol. Chem.* **273,** 16865 (1998).

20 μm thickness on a cryostat maintained at $-20°$. Sections are thaw mounted onto gelatin subbed slides by quickly running a finger under the slide; the use of a hot plate to thaw-mount sections greatly diminishes agonist-stimulated [^{35}S]GTPγS binding. Sections are collected in duplicate or triplicate, and slides are paired so that basal and agonist-stimulated [^{35}S]GTPγS binding can be evaluated in adjacent sections. Slides are collected on ice in a sealed container, then transferred to a desiccator at $4°$. After overnight storage at $4°$, slides are boxed with desiccant and stored at $-80°$ until assay. Sections used for [^{35}S]GTPγS autoradiography are more temperature-sensitive than those processed for receptor autoradiography. On the day of the assay, slides are removed from the freezer 30–60 min prior to processing and brought to room temperature under cool air. Although some autoradiographic procedures include overnight storage at $-20°$ prior to assay, this procedure reduces [^{35}S]GTPγS binding. Slides should be processed as soon as possible following sectioning. Slides may be stored at $-80°$ for extended periods (up to 1 year), but [^{35}S]GTPγS binding (both basal and agonist-stimulated) decreases with increased storage time.

Assay Components

The assay conditions used for [^{35}S]GTPγS autoradiography were derived from those reported for agonist-stimulated [^{35}S]GTPγS binding in membrane homogenates.[4,5,10,11] In general, assay conditions for the autoradiographic and homogenate assays are similar, so that conditions can be tested in membrane assays for application in autoradiography. The notable exception to this rule is the concentration of GDP. GDP is included in agonist-stimulated [^{35}S]GTPγS binding assays to decrease basal [^{35}S]GTPγS binding by favoring G-protein inactivation. GDP concentrations in homogenate assays range from 10 to 100 μM, whereas 1–2 mM GDP is required to sufficiently reduce basal binding in brain sections. This is illustrated in Fig. 1, which shows the effect of varying GDP concentrations on mu opioid-stimulated [^{35}S]GTPγS binding in rat brain sections. Unlike membranes, where significant agonist-stimulated [^{35}S]GTPγS binding is observed at 10–100 μM GDP, brain sections show very high basal binding and undetectable agonist-stimulated [^{35}S]GTPγS binding at the same GDP concentrations. However, at 1–2 mM GDP, basal binding is reduced and the agonist-stimulated signal is clearly discernible in specific regions of the section (in this case, thalamus and amygdala). The reason for this increased GDP requirement appears related to the relative amounts of protein in sections vs assay tubes: generally 2–20 μg of brain protein is added per assay tube, whereas a typical brain section at the level of the striatum contains approximately 175 μg protein. This higher level of membrane protein provides high levels of nucleotide hydrolases that hydrolyze GDP to

[10] G. Hilf, P. Gierschik, and K. H. Jakobs, *Eur. J. Biochem.* **186,** 725 (1989).
[11] S. Lazareno, T. Farries, and N. J. M. Birdsall, *Life Sci.* **52,** 449 (1993).

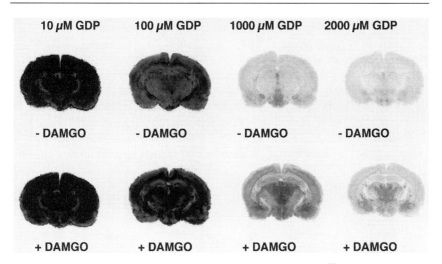

FIG. 1. Effect of GDP on basal and mu opioid (DAMGO)-stimulated [^{35}S]GTPγS in rat brain sections at the level of thalamus and amygdala. At low GDP concentrations, basal binding is too high for significant agonist stimulation to be observed. Agonist stimulation of [^{35}S]GTPγS binding is observed in thalamus and amygdala with 1–2 mM GDP. [Reproduced from L. J. Sim, D. E. Selley, and S. R. Childers, *Proc. Natl. Acad. Sci. U.S.A.* **92,** 7242 (1995).]

guanosine; thus, higher levels of GDP are required to compensate for this increased metabolism.

The standard [^{35}S]GTPγS autoradiographic buffer is 50 mM Tris-HCl containing 3 mM MgCl$_2$, 0.2 mM EGTA, and 100 mM NaCl (pH 7.4) (assay buffer). Magnesium, which is required for G-protein function, has a biphasic effect on G proteins. Low (high nanomolar to low micromolar) concentrations of magnesium are required for hydrolysis of Gα-GTP, whereas high (high micromolar to low millimolar) concentrations promote agonist-stimulated G-protein activation.[7,12] Concentrations of magnesium greater than 10 mM can decrease [^{35}S]GTPγS binding to membranes[13] and should be avoided. Sodium is included to decrease basal [^{35}S]GTPγS binding by inhibiting spontaneously active receptors. Standard conditions use 100 mM NaCl, which appears appropriate for most G-protein-coupled receptors, particularly those coupled to G$_{i/o}$.

The level of basal [^{35}S]GTPγS binding is a major concern in [^{35}S]GTPγS autoradiography. Since basal [^{35}S]GTPγS binding appears to be more highly localized in specific brain regions, it appears that basal activity of specific G-protein-coupled receptors may contribute to this activity. This is supported by studies showing that 5'-p-fluorosulfonylbenzoyl guanosine (an irreversible G-protein inhibitor)

[12] D. R. Brandt and E. M. Ross, *J. Biol. Chem.* **261,** 1656 (1986).
[13] P. Gierschik, R. Moghtader, C. Straub, K. Dieterich, and K. H. Jakobs, *Eur. J. Biochem.* **197,** 725 (1991).

and *N*-ethylmaleimide decrease basal as well as agonist-stimulated [^{35}S]GTPγS binding.[14] One receptor that clearly contributes to basal [^{35}S]GTPγS binding is the adenosine A$_1$ receptor.[15,16] Receptor autoradiography studies have shown that adenosine is present in tissue sections and may be produced in sections by the breakdown of endogenous adenosine precursors.[17] Therefore, the [^{35}S]GTPγS assay procedure has been modified to include 10 mU/ml of adenosine deaminase in the preincubation and incubation solutions to inactivate endogenous adenosine.[18] This technique reduces basal [^{35}S]GTPγS binding in areas high in adenosine A$_1$ receptors, such as hippocampus, thalamus, cortex, and cerebellum, but has little effect on areas with a low density of A$_1$ receptors, including hypothalamus and brain stem.[19] At the present time, it is not known whether endogenous ligands for other G-protein-coupled receptors also contribute to basal [^{35}S]GTPγS binding; however, in most cases this would not likely be a problem since the presence of GDP and sodium favor the low-affinity state of the receptor, which dissociates most ligands.[20]

Certain receptor ligands may require additional assay ingredients. Assays using lipophilic agents, such as cannabinoid ligands, include 0.1–0.5% bovine serum albumin to inhibit adsorption of ligands to the tissue or container. Peptide ligands require the addition of protease inhibitors to the preincubation solution to prevent degradation. A general protease inhibitor cocktail containing 0.2 mg/ml each of bestatin, leupeptin, pepstatin A, and aprotinin can be added to the preincubation solution (10 μl cocktail/ml preincubation solution). Commercially available protease inhibitor solutions may contain EDTA, which could affect the magnesium concentration in the assay and should therefore be avoided.

Since [^{35}S]GTPγS is the radioligand used for all receptors, any agonist for the receptor of interest can be used in the assay. The choice of agonist is generally based on: (1) specificity for the appropriate receptor and (2) sufficient efficacy to produce a high signal relative to basal [^{35}S]GTPγS binding levels. The appropriate concentration of agonist for use in [^{35}S]GTPγS autoradiography can be determined using membrane homogenate assays. Although concentration–effect curves can also be generated using brain sections,[6] most autoradiographic assays are conducted using an E_{max} (i.e., a maximally effective) concentration of agonist, as determined empirically from homogenate binding assays. For most agonists, E_{max} concentrations are in the 1–10 μM range; however, this can vary depending

[14] C. Waeber and M. L. Chiu, *J. Neurochem.* **73**, 1212 (1999).

[15] J. T. Laitinen, *Neuroscience* **90**, 1265 (1999).

[16] R. J. Moore, R. Xiao, L. J. Sim-Selley, and S. R. Childers, *Neuropharmacology* **39**, 282 (2000).

[17] F. Parkinson and B. B. Fredholm, *J. Neurochem.* **58**, 941 (1992).

[18] L. J. Sim, Q. X. Liu, S. R. Childers, and D. E. Selley, *J. Neurochem.* **70**, 1567 (1998).

[19] J. Fastbom, A. Pazos, and J. M. Palacios, *Neuroscience* **22**, 813 (1987).

[20] S. R. Childers and S. H. Snyder, *J. Neurochem.* **34**, 583 (1980).

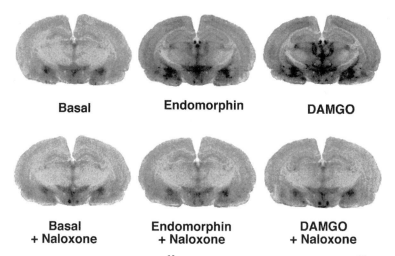

FIG. 2. Pharmacological specificity of [^{35}S]GTPγS autoradiography. Stimulation of [^{35}S]GTPγS binding by mu-opioid agonists DAMGO and endomorphin in rat thalamus and amygdala is blocked by addition of the mu antagonist naloxone. [Reproduced from L. J. Sim, Q. X. Liu, S. R. Childers, and D. E. Selley, *J. Neurochem.* **70,** 1567 (1998).]

on agonist potency. It is important to note that agonist concentrations will be considerably higher for stimulation of [^{35}S]GTPγS binding than receptor binding, since the assay is conducted in the presence of GDP and sodium, favoring the low-affinity state of the receptor.

During development of [^{35}S]GTPγS autoradiography for a new agonist, particularly for a new receptor, it is important to include a positive control in the assay. Appropriate standard controls include 10 μM [D-Ala^2,N-Me-Phe^4, Gly-ol^5]enkephalin (DAMGO), a mu (μ-) opioid agonist, or 1 μM phenylisopropyladenosine, an adenosine A$_1$ agonist, both of which produce high levels of stimulation of [^{35}S]GTPγS binding in numerous brain structures. The pharmacological specificity of agonist-stimulated [^{35}S]GTPγS binding can be verified by using a specific antagonist in the presence of an EC$_{50}$ concentration of agonist to demonstrate antagonist reversibility (Fig. 2). In addition, the distribution of agonist-stimulated [^{35}S]GTPγS binding should correspond to the known receptor localization, although the relative levels of receptors versus activity may differ somewhat (see below).

The storage of [^{35}S]GTPγS and GTPγS is an important consideration, since the compound is somewhat unstable. Immediately on receipt, [^{35}S]GTPγS (New England Nuclear) is diluted 1 : 10 in distilled H$_2$O and stored in aliquots at $-80°$. Unlabeled GTPγS (used both to define nonspecific binding, and in [^{35}S]GTPγS saturation analyses) is also stored in aliquots at $-80°$. Multiple freeze–thaw cycles of GTPγS can lead to degradation of the compound.

Several laboratories have reported minor modifications of the basic [^{35}S]GTPγS autoradiography protocol described above. In some cases, assay conditions were modified to correspond to those used for other techniques in the laboratory, such as receptor autoradiography. Other modifications have been implemented to optimize the signal-to-noise ratio in the experiment. Receptor autoradiography buffers, such HEPES and glycylglycine buffers, have been substituted for Tris buffer. Studies have used magnesium at concentrations from 1 to 10 mM. Most [^{35}S]GTPγS autoradiographic experiments have used 1–2 mM GDP and 100 mM sodium, although slight variations have been reported. Several laboratories have included dithiothreitol (0.2 to 1 mM) in the assay and/or wash buffers. Finally, [^{35}S]GTPγS has been used at concentrations from 0.04 to 0.2 nM, with most studies reporting 0.04–0.05 nM. It is best to test these conditions during assay development, as optimal conditions may vary based on factors such as receptor, species, and treatment paradigm. Since the apparent K_D value for [^{35}S]GTPγS binding to agonist-activated brain G proteins (primarily G$_o$) is 1–3 nM,[8] the occupancy of total G proteins by [^{35}S]GTPγS is relatively low under these conditions, and results must be interpreted accordingly.

Incubation Conditions

All incubations are conducted in a shaking water bath maintained at 25°. Once slides reach room temperature, they are equilibrated in assay buffer for 10 min at 25°. There are several reasons for this wash: (1) to dissociate endogenous ligands; (2) to equilibrate the tissue to assay conditions; and (3) to remove any mounting medium that remains from the sectioning process. After this rinse, slides are immediately placed in the preincubation solution containing 2 mM GDP and 10 mU/ml adenosine deaminase in assay buffer for 15 min at 25°. Slides are placed in slide mailers (5 slide capacity) or Coplin jars (10 slide capacity) for preincubation and incubation. Slides are then transferred to the incubation solution, which contains agonist, 0.04 nM [^{35}S]GTPγS, 2 mM GDP, and 10 mU/ml adenosine deaminase in assay buffer for 2 hr at 25°. Basal binding is measured in the absence of agonist and nonspecific binding is measured with 10 μM GTPγS. After incubation, slides are transferred into slide racks and rinsed twice for 2 min each in 50 mM Tris buffer (pH 7.0 at room temperature) on ice. Slides are washed for 30 sec in distilled H$_2$O on ice and dried completely under cool air.

Several variations on the incubation conditions have been reported. Preincubation in GDP has been increased to 30–60 min in an attempt to decrease basal [^{35}S]GTPγS binding. Incubation times have also been varied from 1 to 4 hr to maximize agonist-stimulated [^{35}S]GTPγS binding. Finally, incubation temperatures from 20° to 37° have been reported. Optimal incubation conditions may vary somewhat based on the ion and nucleotide concentrations used (see above) as well as characteristics of the receptor or tissue of interest. Thus, it is important to conduct test assays to optimize assay conditions when developing [^{35}S]GTPγS autoradiography protocols.

Since relatively large volumes of solutions are used in autoradiography, [^{35}S]GTPγS assays can be quite expensive, mainly because of the high GDP requirement. Preincubation and incubation solutions may be used twice (i.e., sequential runs of slides) in order to maximize use of materials. This is particularly advantageous for studies comparing control and treated groups of animals, where hundreds of slides may be processed during the experiment. A strategy to conserve GDP in smaller assays is to add the agonist and [^{35}S]GTPγS to the preincubation solution for use in the incubation. The main concern in this method is that the solution be completely mixed for incubation. Finally, incubations may be performed directly on slides in a humidified chamber rather than by immersion in solutions.[21]

Visualization and Analysis

[^{35}S]GTPγS binding can be visualized using film autoradiography, phosphorimaging screens, or slide-coated emulsion by following basic autoradiography protocols. A major advantage of [^{35}S]GTPγS autoradiography compared to receptor autoradiography is the use of ^{35}S, which has higher energy for β emission than [^{3}H] (used for most receptor binding assays), and therefore shorter exposure time. Several types of film are appropriate for ^{35}S, including Kodak (Rochester, NY) X-O-Mat or Biomax MR and Amersham Hyperfilm βmax. Film exposures generally range from 2 to 10 days and depend primarily on ligand efficacy and receptor density. Phosphorimaging screens decrease the exposure times considerably, to 3–12 hr. Experimental procedures, such as chronic drug treatment or lesioning, may decrease the levels of [^{35}S]GTPγS binding and thus increase exposure time. Although the exposure for [^{35}S]GTPγS autoradiography is usually short enough to allow technical development, it is possible to perform an immediate analysis using a scintillation counter. In this case, the section is wiped from the slide using a small piece of filter paper and subject to scintillation spectrophotometry.

It is necessary to include appropriate standards during film exposure to allow densitometric analysis of [^{35}S]GTPγS autoradiography. ^{14}C microscales (Amersham) are included in each film cassette and used to calibrate the films. Calibration may be corrected for ^{35}S by using a brain paste standard assay. The correction equation can be calculated once for a particular type of film and applied to subsequent experiments. Brains are homogenized and centrifuged at 1000 rpm for 10 min. The supernatant is discarded and the pellet is saved as paste. Eighteen 250 mg samples are taken and mixed with various concentrations of [^{35}S]GTPγS (serially diluted [^{35}S]GTPγS stock from 1 : 100 to 1 : 64,000). Each

---

[21] K. M. A. Kurkinen, J. Koistinaho, and J. T. Laitinen, *Brain Res.* **769,** 21 (1997).

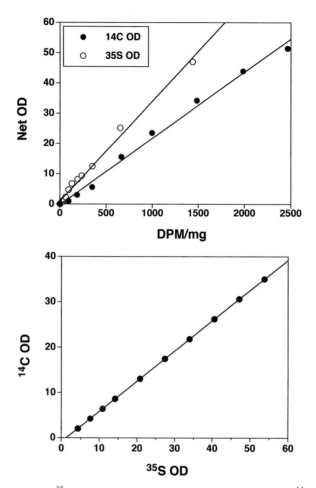

FIG. 3. Calculation of ^{35}S dpm/mg values from optical density (OD) data, using ^{14}C as a standard. *Top*: Graph shows OD readings from [14]C (commercially available) and [35]S (prepared) standards. *Bottom*: For each isotope, OD values are calculated at specific values of dpm/mg and plotted. The equation of this line provides the correction factor of ^{14}C to [35]S.

brain paste/^{35}S]GTPγS sample is loaded into a sealed 1 ml syringe and centrifuged at 1000 rpm for 10 min. The sample is then frozen at $-80°$, and frozen samples are transferred to cold cryovials. Each sample is cut into 20 μm sections and collected as follows: 10 sections for weighing, 10 sections per vial for scintillation counting (triplicate), and 10 sections on a slide for film exposure. Calculations will be based on sample weight, measured radioactivity, and optical density. First, net optical density is plotted vs dpm ^{35}S]GTPγS /mg tissue for both ^{35}S and [14]C. [14]C (dpm) values can be obtained from the microscale data sheet. From this graph, ^{35}S and ^{14}C values are calculated at several points, which are then plotted on a separate graph. The equation derived from this plot is the correction factor for translation of ^{14}C to ^{35}S (Fig. 3).

[^{35}S]GTPγS autoradiography on film can be analyzed using standard imaging equipment; this would include a digital camera, a lightbox with a stable light source, and a computer fitted with an appropriate video capture card. Alternatively, phosphorimaging screens offer a more rapid alternative to film, although neuroanatomical resolution (particularly of small brain structures) often does not match that of standard film autoradiography. For image analysis, a variety of software is commercially available, including the public domain software NIH Image.

There are several strategies used to measure optical density in sections. Sections may be analyzed anatomically by drawing regional boundaries to select the appropriate area of interest. Samples may also be selected based on levels of optical density; in this case, a threshold level of stimulation is set and areas are selected for measurement based on this criterion. Finally, small samples within a region can be selected for analysis; the potential problem with this strategy is that the specific area selected may influence results. Duplicate or triplicate sections are analyzed, and the densitometric results are averaged to calculate the mean nCi/g for that region. A minimum of 3–5 brains is necessary for densitometric analysis. However, under many experimental conditions (e.g., chronic drug administration), 5–10 brains are required to detect statistically significant changes.

There are several ways to express [^{35}S]GTPγS binding data. Net agonist-stimulated [^{35}S]GTPγS binding is calculated by subtracting basal from agonist-stimulated [^{35}S]GTPγS binding. Percent stimulation of [^{35}S]GTPγS binding is calculated by dividing net stimulated [^{35}S]GTPγS binding by basal [^{35}S]GTPγS binding. Data can also be converted to fmol [^{35}S]GTPγS/mg or fmol [^{35}S]GTPγS/g based on the specific activity of the [^{35}S]GTPγS. An adequate level of stimulation is necessary in order to obtain accurate quantitative measures of agonist-stimulated [^{35}S]GTPγS binding. In general, net stimulated [^{35}S]GTPγS binding levels less than 50 nCi/g or representing less than 20% stimulation over basal are difficult to assess quantitatively. This may be the case for low-efficacy partial agonists or low-density receptors.

Limitations

Most laboratories have reported that agonist-stimulated [^{35}S]GTPγS autoradiography in brain sections is detected only for those receptors that couple to $G_{i/o}$ proteins. This may reflect the relative levels of G-proteins in brain, since G_o represents 0.1–1% of brain protein.[22] In addition, the assay conditions favor $G_{i/o}$

[22] P. C. Sternweis and J. D. Robishaw, *J. Biol. Chem.* **259**, 13806 (1984).

TABLE I
RECEPTORS DETECTED BY [^{35}S]GTPγS AUTORADIOGRAPHY IN
BRAIN SECTIONS

Receptor	Reference
Adenosine A$_1$	15,16
Cannabinoid CB1	6,39
Dopamine D$_2$	24
GABA$_B$	6
Histamine H$_3$	42
5HT$_{1A}$/5HT$_{1B}$	28,43,44
Mu, delta, kappa opioid	6,27,45
Muscarinic acetylcholine	21,46
Neuropeptide Y	47
Nociceptin/orphanin FQ	31,45,48
Norepinephrine α_2-adrenergic	25,49
Endothelium-differentiation gene	14

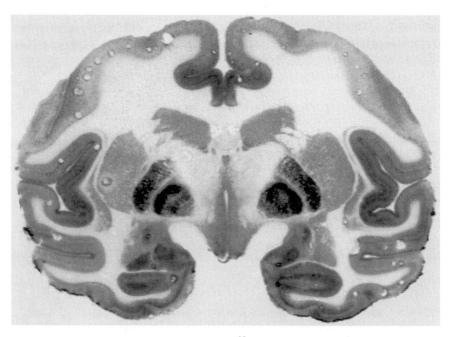

FIG. 4. Cannabinoid CB1 receptor-stimulated [^{35}S]GTPγS autoradiography in cynomolgus monkey brain, using WIN 55212-2 as a cannabinoid agonist. Prominent stimulation of [^{35}S]GTPγS binding is observed in internal and external portions of globus pallidus, as well as hippocampus and cortical layers. [Reproduced from L. J. Sim-Selley, J. B. Daunais, L. J. Porrino, and S. R. Childers, *Neuroscience* **94**, 651 (1999).]

activity, in regard to ion and GDP concentrations as well as kinetics of GDP/GTP exchange. Agonists for G_s- (corticotropin-releasing factor, β-adrenergic) or G_q- (substance P, α_1-adrenergic, 5-serotonin$_2$) coupled receptors do not stimulate [^{35}S]GTPγS binding above basal levels (L. Sim-Selley and S. Childers, 1998, unpublished observations), which may indicate that different assay conditions are necessary to visualize receptors coupled to these G proteins. Moreover, the signals for some $G_{i/o}$-coupled receptors such as dopamine D_2 and α_2-adrenergic are lower than predicted from their receptor number in brain; these difficulties may reflect coupling efficiencies and receptor/G-protein amplification,[23] although reports of both dopamine D_2-[24] and α_2-adrenergic[25]-stimulated [^{35}S]GTPγS autoradiography have appeared.

When [^{35}S]GTPγS autoradiograms are compared to autoradiography of receptor binding using 3H radioligands, the neuroanatomical resolution of the ^{35}S film is not as high as the ^{3}H film. This is a natural consequence of the higher energy of ^{35}S radioactive decay compared to ^{3}H, which is somewhat compensated by the much faster exposure time for ^{35}S.

[^{35}S]GTPγS autoradiography detects the total population of G proteins activated by a particular receptor in a brain section and cannot distinguish between individual types of Gα subunits. Current studies in progress are addressing this question by using the GTP photoaffinity probe [^{32}P]AAGTP (azidoanilido-GTP) binding to sections followed by immunoprecipitation and sodium dodecyl sulfate–polyacrylamide gel electrophoresis (SDS–PAGE) to identify individual Gα subunits activated by receptor agonists in different brain regions. Preliminary data (S. Childers, P. Prather, and M. Rasenick, 2001, unpublished observations) reveal that appreciable [^{32}P]AAGTP is incorporated into 20 μm sections which can then be separated by gel electrophoresis.

Applicability

[^{35}S]GTPγS autoradiography has been used to identify receptor-activated G proteins for several receptor systems (see Table I). [^{35}S]GTPγS autoradiography has been used in several species, including rat, mouse, guinea pig, chick, and monkey. One application has been the use of [^{35}S]GTPγS binding in primate brain (Fig. 4). Opioid, cannabinoid, and 5-HT$_{1A}$ receptor-stimulated [^{35}S]GTPγS binding have been visualized autoradiographically in nonhuman primate brain,[26,27] and 5-HT$_{1A}$ receptor-activated G proteins have been reported in postmortem human brain.[26] The combination of [^{35}S]GTPγS autoradiography with other anatomical

[23] L. J. Sim, D. E. Selley, R. Xiao, and S. R. Childers, *Eur. J. Pharmacol.* **307,** 95 (1996).

[24] S. M. Khan, T. S. Smith, and J. P. Bennett, *J. Neurosci. Res.* **55,** 71 (1999).

[25] H. K. Happe, D. B. Bylund, and L. C. Murrin, *Eur. J. Pharmacol.* **399,** 17 (2000).

[26] D. S. Dupuis, P. J. Pauwels, D. Radu, and H. Hall, *Eur. J. Neurosci.* **11,** 1809 (1999).

[27] L. J. Sim-Selley, J. B. Daunais, L. J. Porrino, and S. R. Childers, *Neuroscience* **94,** 651 (1999).

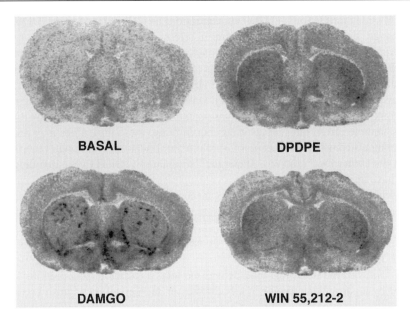

FIG. 5. Comparison of mu opioid (DAMGO)-, delta opioid (DPDPE)-, and cannabinoid (WIN 55212-2)-stimulated [^{35}S]GTPγS binding in rat brain at the level of caudate/putamen. Stimulation of [^{35}S]GTPγS binding by the three agonists is comparable, despite the fact that rat caudate contains more than 10 times as many cannabinoid receptors as mu or delta receptors. This illustrates differences in amplification between receptors and G-protein activation. [Reproduced from L. J. Sim, D. E. Selley, R. Xiao, and S. R. Childers, *Eur. J. Pharmacol.* **307**, 95 (1996).]

techniques further extends the type of information that can be generated regarding receptor function. Lesion experiments, followed by processing for [^{35}S]GTPγS and receptor autoradiography, can provide data not only on receptor localization, but also on receptor function. For example, lesion studies have demonstrated that nociceptin/orphanin FQ (N/OFQ) receptors are located on somatodendritic elements in the anterior cingulate cortex by showing a decrease in receptor binding correlated with elimination of receptor-stimulated G-protein activity following ibotenic acid lesions (L. Sim-Selley, B. Vogt, and S. Childers, submitted). The use of emulsion autoradiography with agonist-stimulated [^{35}S]GTPγS binding further enhances the ability to use [^{35}S]GTPγS autoradiography for detailed anatomical studies.[28]

Agonist-stimulated [^{35}S]GTPγS autoradiography has been utilized to study various aspects of receptor–G protein function. One example is the amplification between receptor binding sites and receptor-activated G proteins. For example, by autoradiography,[6] mu opioid and cannabinoid-stimulated [^{35}S]GTPγS binding

[28] C. Waeber and M. A. Moskowitz, *Mol. Pharmacol.* **52**, 623 (1997).

in adjacent brain sections is similar in magnitude, despite a tenfold excess of cannabinoid vs opioid receptors (Fig. 5). More detailed studies in both brain membranes and sections confirmed that amplification of receptor–G protein activation is greater for opioid than cannabinoid receptors.[23] Comparison of receptor binding with agonist-stimulated [^{35}S]GTPγS binding in alternate brain sections also indicated that the levels of receptor activation of G proteins in a single receptor system may vary by region.[6] This observation has also been confirmed with the demonstration that regional differences in receptor amplification exist for cannabinoid[29] and mu opioid[30] receptors.

Another important application of [^{35}S]GTPγS autoradiography has been the localization of novel receptor systems. Since [^{35}S]GTPγS autoradiography uses an unlabeled ligand, novel endogenous transmitters can be visualized despite the lack of a radiolabeled ligand appropriate for receptor autoradiography. An example is the localization of ORL-1 receptor-activated G-proteins in brain,[31] using [^{35}S]GTPγS binding stimulated by the peptide nociceptin/orphanin-FQ[32,33] (N/OFQ) before the radiolabeled peptide was available. These results demonstrated a different brain distribution of N/OFQ-activated G proteins compared to mu (μ), delta (δ), or kappa (κ) opioid (Fig. 6).

An advantage of [^{35}S]GTPγS binding is the ability to examine agonist efficacy, since efficacy is determined at the level of G-protein activation for G-protein-coupled receptors.[34] Although full agonists are generally selected for [^{35}S]GTPγS assays, it is possible to identity partial agonists, antagonists, and inverse agonists using [^{35}S]GTPγS autoradiography[8] (Fig. 7). Perhaps the most effective application of these studies is the combination of [^{35}S]GTPγS autoradiography with [^{35}S]GTPγS assays in membrane homogenates to biochemically characterize the receptor system.[8] Another advantage of using [^{35}S]GTPγS autoradiography and agonist-stimulated [^{35}S]GTPγS binding in brain tissue is the ability to examine agonist efficacy under native conditions. Receptor number is known to affect apparent agonist efficacy. For example, in overexpressed transfected systems, the mu opioid agonist morphine is a full agonist, whereas in the brain morphine is a partial agonist.[35]

A number of studies have utilized [^{35}S]GTPγS autoradiography to evaluate receptor desensitization following chronic drug administration (Fig. 8). Chronic

[29] C. S. Breivogel, L. J. Sim, and S. R. Childers, *J. Pharmacol. Exp. Ther.* **282**, 1632 (1997).

[30] C. E. Maher, D. E. Selley, and S. R. Childers, *Biochem. Pharmacol.* **59**, 1395 (2000).

[31] L. J. Sim, R. Xiao, and S. R. Childers, *NeuroReport* **7**, 729 (1996).

[32] R. K. Reinscheid, H.-P. Nothacker, A. Bourson, A. Ardati, R. A. Henningsen, J. R. Bunzow, D. K. Grandy, H. Langen, F. J. Monsma, and O. Civelli, *Science* **270**, 792 (1995).

[33] J.-C. Meunier, C. Mollereau, L. Toll, C. Suaudeau, C. Moisand, P. Alvinerie, J.-L. Butour, J.-C. Guillemot, P. Ferrara, B. Monsarrat, H. Mazararguil, G. Vassart, M. Parmentier, and J. Costentin, *Nature* **377**, 532 (1995).

[34] M. Keen, *Trends Pharmacol.* **12**, 371 (1991).

[35] D. Selley, Q. Liu, and S. R. Childers, *J. Pharmacol. Exp. Ther.* **285**, 496 (1998).

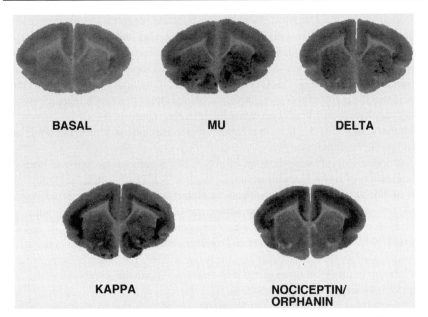

FIG. 6. [^{35}S]GTPγS autoradiography in guinea pig brain, comparing stimulation of [^{35}S]GTPγS binding by mu, delta, and kappa opioid receptor agonists to the ORL-1 receptor peptide agonist nociceptin/orphanin-FQ. [Reproduced from L. J. Sim and S. R. Childers, *J. Comp. Neurol.* **386,** 562 (1997).]

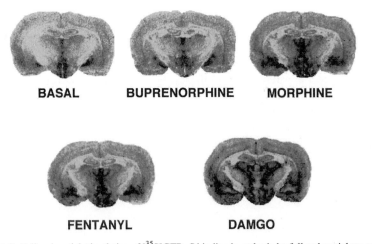

FIG. 7. Full and partial stimulation of [^{35}S]GTPγS binding in rat brain by full and partial mu opioid agonists. The full agonist DAMGO produces the highest level of activated G proteins in thalamus and amygdala, with lower stimulation by the high-efficacy partial agonists morphine and fentanyl, and even lower stimulation by the low-efficacy partial agonist buprenorphine. [Reproduced from D. E. Selley, L. J. Sim, R. Xiao, Q. Liu, and S. R. Childers, *Mol. Pharmacol.* **51,** 87 (1997).]

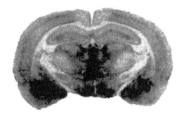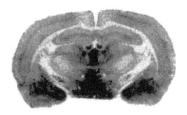

CONTROL   **CHRONIC HEROIN**

Fig. 8. Chronic agonist administration produces desensitization of receptor-activated G proteins in brain. Rats were administered heroin for 40 days, and mu opioid receptor-activated G proteins were examined using [^{35}S]GTPγS autoradiography of brain sections with DAMGO. Decrease in DAMGO-stimulated [^{35}S]GTPγS binding is apparent in thalamus and amygdala of heroin-treated rats, despite the finding of no decrease in mu receptor binding sites in the same animals (not shown). [Reproduced from L. J. Sim-Selley, D. E. Selley, L. J. Vogt, S. R. Childers, and T. J. Martin, *J. Neurosci.* **20,** 4555 (2000).]

administration of opioid,[36,37] cannabinoid,[38–40] and 5-HT$_{1A}$[41] agonists produces desensitization, as defined by a decrease in receptor-activated G proteins. Interestingly, the magnitude and regional extent of desensitization produced varies greatly among receptors. For example, regionally widespread changes are detected in the cannabinoid system following chronic Δ^9-THC, whereas regionally specific changes are found for opioid and 5-HT$_{1A}$ receptors following morphine and heroin or buspirone, respectively. Chronic drug studies have also compared the effects of drug treatment on receptor levels with receptor function. For cannabinoid receptors, decreases in both receptors and agonist-stimulated [^{35}S]GTPγS binding were measured in most regions,[40] whereas in the opioid system increased receptor binding was found in some areas that exhibit desensitization.[37] These observations illustrate the advantage of examining both receptor levels and functional activity in receptor studies.

A number of studies have now demonstrated the applicability of [^{35}S]GTPγS autoradiography to a variety of questions. The ability to study functional receptor activity with anatomical specificity has provided data that significantly extend our understanding of receptor systems. Since manipulations such as chronic drug

[36] L. J. Sim, D. E. Selley, S. I. Dworkin, and S. R. Childers, *J. Neurosci.* **16,** 2684 (1996).

[37] L. J. Sim-Selley, D. E. Selley, L. J. Vogt, S. R. Childers, and T. J. Martin, *J. Neurosci.* **20,** 4555 (2000).

[38] J. Romero, F. Berrendero, L. Garcia-Gil, S. Y. Lin, A. Makriyannis, J. A. Ramos, and J. J. Fernandez-Ruiz, *Neurochem. Int.* **34,** 473 (1999).

[39] L. J. Sim, R. E. Hampson, S. A. Deadwyler, and S. R. Childers, *J. Neurosci.* **16,** 8057 (1996).

[40] C. S. Breivogel, S. R. Childers, S. A. Deadwyler, R. E. Hampson, L. J. Vogt, and L. J. Sim-Selley, *J. Neurochem.* **73,** 2447 (1999).

[41] L. J. Sim-Selley, L. J. Vogt, R. Xiao, S. R. Childers, and D. E. Selley, *Eur. J. Pharmacol.* **389,** 147 (2000).

treatment can produce functional changes in receptor/G protein coupling in brain without affecting receptor number, this approach is crucial in understanding effects on receptor function.

Acknowledgment

Our work has been supported by grants from the National Institute on Drug Abuse, including DA-00287 (L.J.S.) and DA-06634, DA06784, and DA02984 (S.R.C.).

[42] J. T. Laitinen and M. Jokinen, *J. Neurochem.* **71,** 808 (1998).
[43] L. J. Sim, R. Xiao, and S. R. Childers, *Brain Res. Bull.* **44,** 39 (1997).
[44] D. S. Dupuis, M. Perez, S. Halazy, F. C. Colpaert, and P. J. Pauwels, *Mol. Brain Res.* **67,** 107 (1999).
[45] L. J. Sim and S. R. Childers, *J. Comp. Neurol.* **386,** 562 (1997).
[46] M. L. Capece, H. A. Baghdoyan, and R. Lydic, *J. Neurosci.* **18,** 3779 (1998).
[47] R. J. Primus, E. Yevich, and D. W. Gallager, *Mol. Brain Res.* **58,** 74 (1998).
[48] I. Shimohira, S. Tokuyama, A. Himeno, M. Niwa, and H. Ueda, *Neuroscience Letters* **237,** 113 (1997).
[49] U. H. Winzer-Serhan and F. M. Leslie, *J. Neurobiol.* **38,** 259 (1999).

[4] Design and Use of C-Terminal Minigene Vectors for Studying Role of Heterotrimeric G Proteins

By ANNETTE GILCHRIST, ANLI LI, and HEIDI E. HAMM

Introduction

Many biologically active molecules convey their signals via receptors coupled to heterotrimeric guanine nucleotide binding proteins (G proteins). Molecular cloning has resulted in the identification of 16 distinct $G\alpha$ subunits which are commonly divided into four families based on their sequence similarity (see review 1). The G_i family includes G_t, G_{gust}, G_{i1}, G_{i2}, G_{i3}, G_{o1}, G_{o2}, and G_z; the G_s family members are G_s and G_{olf}; the G_q family includes G_q, G_{11}, G_{14}, and $G_{15/16}$; and the G_{12}/G_{13} family members are G_{12} and G_{13}. Similarly, five $G\beta$ and 11 $G\gamma$ subunits have been identified (see review 1).

On activation, G-protein-coupled receptors (GPCRs) interact with their cognate heterotrimeric G protein, inducing GDP release with subsequent GTP binding to the α subunit (see reviews 2, 3). The exchange of GDP for GTP leads to

[1] S. Rens-Domiano and H. E. Hamm, *FASEB J.* **9,** 1059 (1995).

dissociation of the $G\beta\gamma$ dimer from the $G\alpha$ subunit, and both initiate unique intra-cellular signaling responses. In all G proteins studied GTP is bound as a complex with Mg^{2+}, and the GTP- and Mg^{2+}-binding sites are tightly coupled. Dominant negative constructs of the $G\alpha$ subunit have been made in which mutations are introduced at residues known to contact the magnesium ion. For the α subunit of G proteins, this includes mutations of the Gly residue within the invariant sequence (G203T; G204A), as well as mutations of a Ser residue in the effector loop, switch I region (S47C) in either $G\alpha_o$ or $G\alpha_i$ 2.[8,9] Although this approach was quite successful with p21[ras] and other small G proteins,[4,5] dominant negative $G\alpha_i$, $G\alpha_o$, $G\alpha_q$, and $G\alpha_{11}$ have been less effective.[6-10] This is probably due to the degree to which Mg^{2+} is necessary to support GDP binding. p21[ras] forms a tight and nearly irreversible $GDP \cdot Mg^{2+}$ complex, whereas $G\alpha$ subunits bind Mg^{2+} in the $GDP \cdot Mg^{2+}$ complex with lower affinity than in the $GTP \cdot Mg^{2+}$ complex.[11-13]

Thus, we looked to other regions on G protein α subunits that could serve to block receptor G protein interactions, and consequently serve as dominant negatives. Studies utilizing ADP ribosylation by pertussis toxin, site-directed mutagenesis, peptide-specific antibodies, and chimeric proteins indicate that the C terminus of $G\alpha$ is essential for receptor contact. *In vitro* assays, as well as microinjection studies of intact cells, indicate $G\alpha$ C-terminal peptides can competitively block G-protein-coupled downstream events (see reviews 2, 14). We have shown that the carboxyl termini from various G protein α subunits are important sites of receptor binding, and peptides corresponding to the carboxyl terminus can be used as competitive inhibitors of receptor–G protein interactions.[15-17] Using a combinatorial peptide library Martin *et al.*[18] have shown that specific residues within

[2] H. E. Hamm and A. Gilchrist, *Curr. Opin. Cell Biol.* **8,** 189 (1996).

[3] H. Hamm, *J. Biol. Chem.* **273,** 669 (1998).

[4] J. John, H. Rensland, I. Schlichting, I. Vetter, G. Borasio, R. Goody, and A. Wittinghofer, *J. Biol. Chem.* **268,** 923 (1993).

[5] L. Quilliam, K. Kato, K. Rabun, M. Hisaka, S. Huff, S. Campbell-Burk, and C. Der, *Mol. Cell Biol.* **14,** 1113 (1994).

[6] S. Hermouet, J. J. Merendino, J. S. Gutkind, and A. M. Spiegel, *Proc. Natl. Acad. Sci. U.S.A.* **88,** 10455 (1991).

[7] S. Osawa and G. L. Johnson, *J. Biol. Chem.* **266,** 4673 (1991).

[8] V. Z. Slepak, M. W. Quick, A. M. Aragay, N. Davidson, H. A. Lester, and M. I. Simon, *J. Biol. Chem.* **263,** 21889 (1993).

[9] V. Slepak, A. Katz, and M. Simon, *J. Biol. Chem.* **270,** 4037 (1995).

[10] S. Winitz, S. K. Gupta, N.-X. Qian, L. E. Heasley, R. A. Nemenoff, and G. L. Johnson, *J. Biol. Chem.* **269,** 1889 (1994).

[11] T. Higashijima, K. Ferguson, P. Sternweis, M. Smigel, and A. Gilman, *J. Biol. Chem.* **262,** 762 (1987).

[12] T. Higashijima, K. Ferguson, M. Smigel, and A. Gilman, *J. Biol. Chem.* **262,** 757 (1987).

[13] E. Lee, R. Taussig, and A. Gilman, *J. Biol. Chem.* **267,** 1212 (1992).

[14] H. Bourne, *Curr. Opin. Cell Biol.* **9,** 134 (1997).

TABLE I
CARBOXYL TERMINUS SEQUENCES[a]

Subunit	Sequence										
$G\alpha_{i1/2}$	I	K	N	N	L	K	D	C	G	L	F
$G\alpha_q$	L	Q	L	N	L	K	E	Y	N	A	V
$G\alpha_s$	Q	R	M	H	L	R	Q	Y	E	L	L
$G\alpha_iR$	N	G	I	K	C	L	F	N	D	K	L

[a]Alignment of the final 11 amino acid residues of the C termini from human $G\alpha_{i1/2}$, $G\alpha_q$, and $G\alpha_s$ subunits. Also shown is the peptide sequence of $G\alpha_iR$, the $G\alpha_{i1/2}$ sequence in random order, used to construct a control minigene.

the C terminus of $G\alpha_t$ are critical for high-affinity binding of the $G\alpha$ peptide to rhodopsin. This interaction is quite specific as a change of one amino acid can annul the ability of the $G\alpha_{i1/2}$ peptide to bind the A1 adenosine receptor–G protein interface.[19]

To selectively antagonize G protein signal transduction events *in vivo* by expressing peptides that block the receptor–G protein interface, we generated "minigene" plasmid constructs that encode carboxyl terminal peptide sequences from each of the $G\alpha$ subunits.[20] As a control we constructed a minigene vector for the carboxyl terminus of $G\alpha_{i1/2}$ in random order ($G\alpha_iR$, Table I). The minigene plasmid vectors were designed to express the C-terminal peptide sequence of the various $G\alpha$ subunits following their transfection into mammalian cells.

Construction of $G\alpha$ Carboxyl-Terminal Minigenes

Oligonucleotides corresponding to the last 11 amino acids of the C terminus of each of the different $G\alpha$ subunits were designed with new 5' and 3' ends.[20] The 5' end contained a unique restriction enzyme site followed by a ribosome-binding consensus sequence, a methionine for translation initiation, and a glycine to protect the ribosome binding site during translation and the nascent peptide

[15] H. E. Hamm, D. Deretic, A. Arendt, P. A. Hargrave, B. Koenig, and K. P. Hoffman, *Science* **241**, 832 (1988).
[16] H. M. Rarick, N. O. Artemyev, J. S. Mills, N. P. Skiba, and H. E. Hamm, *Methods Enzymol.* **238**, 13 (1994).
[17] M. M. Rasenick, M. Watanabe, M. B. Lazarevic, G. Hatta, and H. E. Hamm, *J. Biol. Chem.* **269**, 21519 (1994).
[18] E. L. Martin, S. Rens-Domanio, P. J. Schatz, and E. H. Hamm, *J. Biol. Chem.* **271**, 361 (1996).
[19] A. Gilchrist, M. Mazzoni, B. Dineen, A. Dice, J. Linden, T. Dunwiddie, and H. E. Hamm, *J. Biol. Chem.* **273**, 14912 (1998).
[20] A. Gilchrist, M. Bünemann, A. Li, M. M. Hosey, and H. E. Hamm, *J. Biol. Chem.* **274**, 6610 (1999).

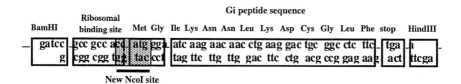

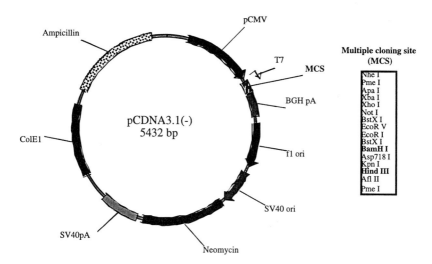

FIG. 1. The cDNA minigene constructs. All Gα carboxyl-terminal peptide minigenes contain a *Bam*HI restriction enzyme site at the 5′ end followed by a ribosomal binding site sequence, a methionine for translation initiation, a glycine for stabilization of the peptide, the peptide sequence, a stop codon, and a *Hind*III restriction enzyme site at the 3′ end. The Gα$_i$R contains the Gα$_{i1/2}$ carboxyl peptide sequence in random order. Following annealing, complementary oligonucleotides were ligated into *Bam*HI/*Hind*III cut pcDNA 3.1(−) plasmid vector.

against proteolytic degradation. A unique restriction enzyme site was synthesized at the 3′ end immediately following a translation stop codon. The oligonucleotides were synthesized, annealed, and ligated into mammalian pcDNA3.1 (−) vector that had previously been cut with the appropriate restriction enzymes (Fig. 1).

Design of Oligonucleotides

1. The cDNA encoding the last 11 amino acids of Gα subunits is synthesized (Great American Gene Company) with newly engineered 5′ and 3′ ends (Fig. 1). The 5′ end contained a *Bam*HI restriction enzyme site followed by the human ribosome-binding consensus sequence (5′-GCCGCCACC-3′), a methionine (ATG) for translation initiation, and a glycine (GGA) to protect

the ribosome binding site during translation and the nascent peptide against proteolytic degradation. A *Hind*III restriction enzyme site is synthesized at the 3′ end immediately following the translational stop codon (TGA). Thus, the full-length 57 bp oligonucleotides ordered for the $G_{i1/2}$ carboxyl-terminal sequence are 5′-gatccgccgccaccatgggaatcaagaacaacctgaaggactgc-ggcctcttctgaa-3′ and the complementary strand is 5′-agctttcagaagaggccg-cagtccttcaggttgttcttgattcccatggtggcggcg-3′. As a control, oligonucleotides encoding the $G\alpha_{i1/2}$ carboxyl terminus in random order ($G\alpha_i$R) with newly engineered 5′ and 3′ ends can also be synthesized.

2. The DNA is brought up in sterile doubly distilled H_2O (stock concentration 100 μM).

3. Complimentary DNA is annealed in 1× NEBuffer 3[50 mM Tris-HCl, 10 mM MgCl$_2$, 100 mM NaCl, 1 mM (dithiothreitol) DTT; New England Biolabs Beverly, MA] at 85° for 10 min, then allowed to cool slowly to room temperature.

4. The DNA is run on a 4% agarose gel and the annealed band is excised and the DNA purified according to the manufacturer's protocol (GeneClean II Kit, Bio 101).

5. After digestion with each restriction enzyme the pcDNA 3.1 (−) plasmid vector is run on an 0.8% agarose gel, the appropriate band cut out, and the DNA purified according to the manufacturer's protocol (GeneClean II Kit, Bio 101).

6. The annealed/cleaned cDNA is ligated for 1 hr at room temperature into the cut/cleaned pcDNA 3.1 plasmid vector (Invitrogen) previously cut with *Bam*HI and *Hind*III. For the ligation reaction several ratios of insert to vector are plated ranging from 25 μM : 25 pM to 250 pM : 25 pM annealed cDNA to vector cDNA.

7. Following the ligation reaction, the samples are heated to 65° for 5 min to deactivate the T4 DNA ligase.

8. The ligation mixture (1 μl) is electroporated into 50 μl competent cells (ARI814; Bio-Rad Hercules, CA, *Escherichia coli* Pulsar; REF) and the cells immediately placed into 1 ml of SOC (Gibco, Gaithersburg, MD).

9. After 1 hr shaking at 37°, 100 μl of the electroporated cells containing the minigene plasmid DNA is spread on LB/ampicillin plates and incubated at 37° for 12–16 hr.

10. To verify that insert is present, colonies are grown overnight in LB/Amp and their plasmid DNA purified (Qiagen SpinKit). The plasmid DNA is digested with *Nco*I (New England Biolabs, Inc.) for 1 hr at 37° and run on a 1.5% (3 : 1) agarose gel. Vector alone produces 3 bands (3.3 kb, 1.35 kb, and 0.74 kb), whereas vector with insert results in 4 bands (3.3 kb, 1.0 kb, 0.74 kb, and 0.38 kb) (Fig. 2). DNA with the correct pattern is sequenced to confirm the appropriate sequence.

Lane 1 2 3 4 5

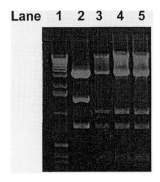

FIG. 2. *Nco*I digest of minigene vector. Following the ligation of the annealed oligonucleotides into the precut vector and electroporation into competent cells, colonies were picked and grown overnight in LB/Amp. Plasmid DNA was purified, digested for 1 hr with *Nco*I at 37°, and then separated on a 1.5% agarose gel to determine if insert was present. Lane 1 is a 1 kb DNA Ladder; Lane 2 is pcDNA3.1; Lane 3 is pcDNA-Gα_i; Lane 4 is pcDNA-Gα_iR; and Lane 5 is pcDNA-Gα_q. When the 56 bp annealed oligonucleotide insert in present there is a new NcoI, site resulting in a shift in the band pattern, such that the digest pattern goes from three bands (3345 bp, 1352 bp, 735 bp) to four bands (3345 bp, 1011 bp, 735 bp, 380 bp).

Cell Culture and Transfection

As the minigene approach depends on competitive inhibition, a key element for success is the expression of adequate amounts of peptides to block intracellular signaling pathways. To confirm the presence of the minigene constructs in transfected cells, total RNA is isolated 48 hr posttransfection, cDNA made using RT-PCR (reverse transcriptase-polymerase chain reaction), and PCR analysis is performed using the cDNA as template with primers specific for the Gα carboxyl-terminal peptide insert. Separation of the PCR products on 1.5% agarose gels indicates the presence of the Gα carboxyl-terminus peptide minigene RNA by a single 434 bp band.[20] Control experiments are done using a T7 forward primer with the vector reverse primer to verify the presence of the pcDNA3.1 vector, and G3DPH primers (Clonetech) to approximate the amount of total RNA.

To verify that the peptide is being produced in the transfected cells, 48 hr posttransfection, cells are lysed and homogenized. Cytosolic extracts are analyzed by HPLC, and peaks (Fig. 3) are analyzed by ion mass spray analysis. The mass spectrometer analysis for peak 1 from the G$\alpha_{i1/2}$ peptide vector (pcDNA-Gα_i) transfected cells and peak 1 from cells, transfected with a vector expressing a random sequence from G$\alpha_{i1/2}$ (pcDNA-Gα_iR) indicates that a 1450 molecular weight peptide is found in both cytosolic extracts. This is the expected molecular weight for both 13 amino acid peptide sequences. The fact that they are the major peptides found in the cytosol from cells transiently transfected with the pcDNA-Gα_i or pcDNA-Gα_iR vectors strongly suggests that the vectors are producing the

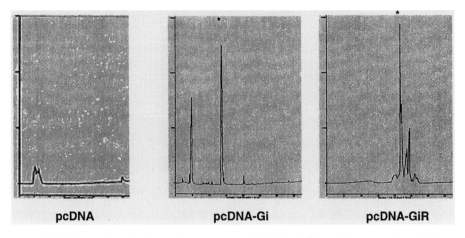

pcDNA   **pcDNA-Gi**   **pcDNA-GiR**

FIG. 3. Transient transfection of minigene vectors. HEK 293 cells that stably express M_2 mAChR were transiently transfected with DNA from GIRK1/4 and pcDNA3.1, pcDNA-Gα_i, or pcDNA-Gα_iR. To verify that the peptide was being produced in the transiently transfected cells, the cells were lysed 48 hr posttransfection and homogenized, and cytosolic extracts analyzed by HPLC. The peaks noted with the asterik (*) were analyzed by ion mass spray analysis.

appropriate peptide sequences. Therefore, analysis of the transiently transfected HEK 293 cells indicates not only that the minigene vectors are present, but also that the peptides are being expressed.[20]

Cell Culture and Transfection

1. Human embryonic kidney (HEK) 293 cells, stably expressing the M_2 mA-ChR ($\sim$400 fmol receptor/mg protein; ref. 21) are grown in Dulbecco's modified Eagle's medium (DMEM, Gibco) supplemented with 10% fetal bovine serum (Gibco), streptomycin/penicillin (100 U each; Gibco) and G418 (500 mg/liter; Gibco). Cells are grown under 10% (v/v) CO_2 at 37°.
2. In all transfections for electrophysiological studies the CD8 reporter gene system is used in order to visualize transfected cells.[22] Dynabeads coated with anti-CD8 antibodies are purchased from Dynal.
3. Typically we use the following amounts of cDNAs for transient trans-fections: pC1-GIRK1 (rat), 1 μg; pcDNA1-GIRK4 (rat), 1μg; πH3-CD8 (human), 1 μg; pcDNA3.1, pcDNA-Gα_i, pcDNA-Gα_iR, pcDNA-Gα_q, or pcDNA-Gα_s, 4 μg. Thus, typically the total amount of cDNA used for trans-fecting one 10 cm dish is 7 μg. The cDNAs for the GIRK1 and GIRK4 were gifts from F. Lesage and M. Lazdunski (Nice, France).

[21] R. Pals-Rylaarsdam, Y. Xu, P. Witt-Enderby, J. Benovic, and M. Hosey, *J. Biol. Chem.* **270,** 29004 (1995).
[22] M. Jurman, L. Boland, Y. Liu, and G. Yellen, *Biotechniques* **17,** 876 (1994).

4. A standard calcium phosphate procedure is used for transient transfection of HEK cells.[23]

5. All assays are performed 48–72 hr posttransfection.

Determining RNA Expression

1. To determine minigene RNA expression transiently transfected cells are washed twice with PBS (1.08 g $Na_2HPO_4 \cdot 7H_2O$, 0.53 g Na_2HPO_4, 0.53 g NaCl, 0.1 g KH_2PO_4, 0.1 g KCl. Bring volume to 500 ml with sterile double distilled H_2O).

2. Lyse a 10 cm culture dish of transiently transfected cells with 350 μl of RLT lysis buffer (Qiagen, Rneasy Mini Kit).

3. Homogenize transiently transfected cells using a QIAshredder column (Qiagen), and process total RNA according to the manufacturer's protocol. Total RNA is eluted in diethyl pyrocarbonate (DEPC)-treated water, quantified, and stored at $-20°$.

4. cDNA is made from total RNA using a reverse transcribed polymerase chain reaction (RT-PCR; Clontech Advantage RT-for-PCR kit) according to the manufacturer's protocol.

5. To verify the presence of insert in cells transfected with pcDNA-Gα_i, or pcDNA-Gα_iR constructs, their cDNA is used as the template for PCR with forward and reverse primers that correspond to Gα insert and vector, respectively (forward: 5′-ATCCGCCGCCACCATGGGA reverse: 5′-GCGAAAGGAGCGGGCGCTA). The primers for the Gα minigenes amplify a 434 bp fragment only if the insert carboxyl termini oligonucleotides are present; no band is observed in cells transfected with the empty pcDNA3.1 vector.[20] As a control, PCR can also performed using T7 forward with the vector reverse primer that amplifies a 486 bp fragment.

Determining Minigene Peptide Expression

1. Forty-eight hours posttransfection, cells are trypsinized and pelleted. The cell pellet is washed twice with phosphate-buffered saline (PBS), and then stored at $-80°$.

2. Cellular extracts are prepared by homogenizing the cell pellets for 15 sec (ESGE Biohomogenizer M133/1281-0) in fractionation buffer (10 mM HEPES, pH 7.3, 11.5% sucrose, 1 mM EDTA, 1 mM EGTA, 1 mM phenylmethylsulfonyl fluoride).

3. The homogenate is centrifuged at 3,000g for 20 min.

4. The supernatant from the previous spin (Step 3) is centrifuged at 100,000g for 30 min.

---

[23] E. Schenborn and V. Goiffon, *Methods Mol. Biol.* **130**, 135 (2000).

5. The resulting supernatant is collected (cytosolic fraction) and stored at $-80°$ until needed.

6. For HPLC analysis 100 μl of cytosolic fraction extract is loaded onto a C_4 column (Vydac) equilibrated with 0.1% trifluoroacetic acid (TFA) in ddH_2O. Elution of the peptide was performed using 0.1% TFA in acetonitrile. The amount of acetonitrile is increased from 0 to 60% over 45 min.

7. Peaks are collected (Fig. 3), lyophilized, and analyzed using ion mass spray analysis (University of Illinois-Urbana Champaign).

Cellular Effects of Minigene Peptide Expression

To test whether minigene constructs encoding the C-terminal 11 amino acid residues from $G\alpha$ subunits could effectively inhibit G-protein-coupled receptor-mediated cellular responses, we have chosen a system in which (1) the importance of the carboxyl terminus and (2) the downstream effector system have been well established. Many studies have shown that the M_2 muscarinic receptor (MAChR) couples exclusively to the G_i/G_o family.[24–27] The M_2 mAChR can efficiently couple to mutant $G\alpha_q$ in which last five amino acids of $G\alpha_q$ are substituted with the corresponding residues from $G\alpha_i$ or $G\alpha_o$,[28] suggesting that this receptor contains domains that are specifically recognized by the carboxyl termini of $G\alpha_{i/o}$ subunits. The effector system we selected was that of the M_2 mAChR activated inwardly rectifying K^+ channel (I_{KACh}). In cardiac cells, the I_{KACh} channel is formed as a heterotetramer of G protein regulated inwardly rectifying K^+ channels (GIRK), with two GIRK1 and two GIRK4 subunits.[29,30] This channel is activated on stimulation of M_2 mAChR in a manner that is completely pertussis toxin-sensitive and is the prototype for a direct $G\beta\gamma$ activated channel.[31–33]

GIRK channels modulate electrical activity in many excitable cells (see reviews 34–36). Because the channel opens as a consequence of a direct interaction with $G\beta\gamma$, whole cell patch clamp recording of I_{KACh} can be used as a readout of

[24] M. Dell'Acqua, R. Carroll, and E. Peralta, *J. Biol. Chem.* **268**, 5676 (1993).

[25] J. Lai, S. Waite, J. Bloom, H. Yamamura, and W. Roeske, *J. Pharmacol. Exp. Ther.* **258**, 938 (1991).

[26] S. Offermanns, T. Wieland, D. Homann, J. Sandmann, E. Bombien, K. Spicher, G. Schultz, and K. Jakobs, *Mol. Pharmacol.* **45**, 890 (1994).

[27] E. Thomas and F. Ehlert, *J. Pharmacol. Exp. Ther.* **271**, 1042 (1994).

[28] J. Liu, B. Conklin, N. Blin, J. Yun, and J. Wess, *Proc. Natl. Acad. Sci. U.S.A.* **92**, 11642 (1995).

[29] S. Corey, G. Krapivinsky, L. Krapivinsky, and D. Clapham, *J. Biol. Chem.* **273**, 5271 (1998).

[30] G. Krapivinsky, E. Gordon, K. Wickman, B. Velimirovic, L. Krapivinsky, and D. Clapham, *Nature* **374**, 135 (1995).

[31] G. Krapivinsky, L. Krapivinsky, K. Wickman, and D. Clapham, *J. Biol. Chem.* **270**, 29059 (1995).

[32] G. Krapivinsky, M. Kennedy, J. Nemec, I. Medina, L. Krapivinsky, and D. Clapham, *J. Biol. Chem.* **273**, 16946 (1998).

[33] M. Sowell, C. Ye, D. Ricupero, S. Hansen, S. Quinn, P. Vassilev, and R. Mortensen, *Proc. Natl. Acad. Sci. U.S.A.* **94**, 7921 (1997).

G protein activity in single intact cells. Thus, we have tested whether the $G\alpha$ C-terminal peptide minigenes could inhibit M_2 mAChR activation of inwardly rectifying K^+ currents.

Measurement of I_{KACh} Currents

1. For the measurement of inwardly rectifying K^+ current whole cell currents are recorded using an extracellular solution containing 120 mM NaCl; 20 mM KCl; 2 mM CaCl$_2$; 1 mM MgCl$_2$; and 10 mM HEPES–NaOH, pH 7.4. The solution for filling the patch pipettes is composed of 100 mM potassium glutamate; 40 mM KCl; 5 mM MgATP; 10 mM HEPES–KOH, pH 7.4; 5 mM NaCl; 2 mM EGTA; 1 mM MgCl$_2$; and 0.01 mM GTP.

2. To minimize variations due to different transfections or culture conditions, control experiments (transfection with pcDNA, or pcDNA-Gα_iR) are done in parallel.

3. Membrane currents are recorded under voltage clamp, using conventional whole cell-patch techniques.[37,38] Patch pipettes are fabricated from borosilicate glass capillaries (GF-150-10, Warner Instrument Corp.) using a horizontal puller (P-95 Fleming & Poulsen) and are filled with the solutions listed above. For our experiments, the DC resistance of the filled pipettes range from 3 to 6 MΩ.

4. Membrane currents are recorded using a patch-clamp amplifier (Axopatch 200, Axon Instruments). Signals are analog filtered using a lowpass Bessel filter (1–3 kHz corner frequency). Data are digitally stored using an IBM-compatible PC equipped with a hardware/software package (ISO2 by MFK, Frankfurt/Main, Germany) for voltage control, data acquisition, and data evaluation.

5. In order to measure K^+ currents in the inward direction, the potassium equilibrium potential is set to about -50 mV and the holding potential is -90 mV.[37,38]

6. Agonist-induced currents are evoked by application of acetylcholine (ACh; 1 μM) using a solenoid-operated superfusion device that allows for solution exchange within 300 msec.

7. Linear voltage ramps (from -120 mV to $+60$ mV within 500 msec) are applied every 10 sec.

8. By subtracting nonagonist-dependent currents we are able to resolve the current–voltage properties of the agonist-induced currents.

[34] R. G. Breitwiese, *J. Membr. Biol.* **152,** 1 (1996).

[35] L. Jan and Y. Jan, *Curr. Opin. Cell Biol.* **9,** 155 (1997).

[36] K. Wickman and D. Clapham, *Curr. Opin. Neurobiol.* **5,** 278 (1995).

[37] M. Bunemann, B. Brandts, D. zu Heringdorf, and C. van Koppen, K. Jakobs, and L. Pott, *J. Physiol.* **489,** 701 (1995).

[38] M. Bunemann and L. Pott, *J. Physiol.* **482,** 81 (1995).

9. For analysis of the data the maximal current densities (peak amplitudes) of ACh-induced inwardly rectifying K^+ currents are measured at -80 mV and compared.

10. To exclude experiments in which we record currents from cells that may not have expressed the functional channel, only those cells that exhibit a basal nonagonist-dependent Ba^{2+} ($200 \ \mu M$) sensitive inwardly rectifying current are used for analysis.

Superfusion of HEK 293 cells transiently transfected with GIRK1/GIRK4 and either pcDNA-Gα_i or pcDNA-Gα_iR DNA with 1 μM ACh reveals that cells transfected with pcDNA-Gα_i DNA have a dramatically impaired response to the M_2 mAchR agonist (Fig. 4). Thus, it appears that the Gα_i minigene construct

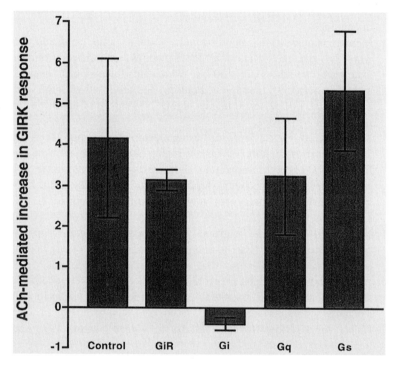

FIG. 4. Minigenes encoding C-terminal Gα_i peptides inhibit M_2 mAChR activated I_{KACh} while Gα_s or Gα_q C-terminal minigenes do not. HEK 293 cells stably expressing the M_2 mAChr were transiently transfected with DNA from GIRK1/4 and control vector (pcDNA), pcDNA–Gα_iR, pcDNA–Gα_i, pcDNA–Gα_s, or pcDNA–Gα_q. The Ach-mediated increase in maximum current was calculated as [(stimulated – basal/basal) – 1]. Cells transfected with pcDNA-Gai had a significantly reduced increase in Ach evoked pA/pF when compared to cells transfected with empty vector (control) or pcDNA-Gα_iR. The pcDNA-Gα_q and pcDNA- Gα_iR vectors had no significant effect on Ach evoked pA/pF when compared to cells transfected with empty vector (control) or pcDNA-Gα_iR.

completely blocks the agonist-mediated M_2 mAChR GIRK 1/4 response while the control minigene constructs (empty vector—pcDNA3.1, and $G\alpha_{i1/2}$ C-terminal peptide in random order—pcDNA-$G\alpha_i$R) has no effect on the agonist-mediated M_2 mAChR GIRK1/4 response. We have also looked to see if the minigene effects are specific by transiently transfected HEK 293 cells stably expressing the M_2 mAChR with GIRK1/GIRK4 and with minigene constructs encoding $G\alpha$ carboxyl termini for $G\alpha_q$ or $G\alpha_s$. ACh-stimulated I_{KACh} currents from cells transfected with pcDNA-$G\alpha_q$ or pcDNA-$G\alpha_s$ (Figure 4) are not significantly different from those of cells transfected with the control vectors. Thus, it appears that our C-terminal minigenes can specifically block agonist-mediated responses. In addition our minigene approach appears to be a promising method for specifically turning off any G–protein-mediated response *in vitro* and *in vivo*. Transfection of different $G\alpha$ carboxyl–terminal peptide allows us to selectively block signal transduction through any G protein. Therefore our approach provides a novel dominant negative strategy to explore the coupling mechanisms of receptors that interact with multiple G proteins, and tease out the downstream responses mediated by each G protein.

Acknowledgments

This work was supported by Grants EY06062 (H.E.H.), EY10291 (H.E.H.) and HL60678 (H.E.H./A.G) from the National Institutes of Health.

[5] Dissecting Receptor–G Protein Specificity Using Gα Chimeras

By Theresa M. Cabrera-Vera, Tarita O. Thomas, Jurgen Vanhauwe, Karyn M. Depree, Stephen G. Graber, and Heidi E. Hamm

Introduction

Heterotrimeric G proteins are composed of four major families defined by the $G\alpha$ subtype, namely $G\alpha_s$, $G\alpha_i$, $G\alpha_q$, and $G\alpha_{12}$.[1] $G\alpha$ subunits all demonstrate the following common features: (1) GDP/GTP binding, (2) GTPase activity, and (3) $G\beta\gamma$ association. Nonetheless, $G\alpha$ subunits uniquely regulate various effectors and demonstrate different profiles of receptor interaction. Strategies used to determine receptor–G protein specificity include the use of antisense technology,[2,3]

[1] M. I. Simon, M. P. Strathmann, and N. Gautam, *Science* **252**, 802 (1991).

[2] C. J. Kleuss, C. Hescheler, W. Ewel, G. Rosenthal, G. Schultz, and B. Wittig, *Nature* **353**, 43 (1991).

[3] C. Kleuss, H. Scherubl, J. Hescheler, G. Schultz, and B. Wittig, *Science* **259**, 832 (1993).

sequence-specific peptide inhibitors,[4,5] G protein antibodies,[6] and chimeric $G\alpha$ subunits.[7,8] Although the primary receptor recognition region is believed to be localized to the carboxyl-terminal domain of $G\alpha$ subunits,[9-14] at least four other regions in $G\alpha$ are involved in receptor interaction: the amino-terminal domain[14-16]; the α_2 helix and $\alpha_2-\beta_4$ loop regions[12,15]; the α_4 helix and $\alpha_4-\beta_6$ loop domain[7,12,17]; and the $\alpha_3-\beta_5$ region.[18] Overall, studies investigating the molecular determinants of receptor G protein specificity suggest that the relative contribution of these potential $G\alpha$ contact sites is dependent on the particular receptor examined.[8,18,19] In addition, receptors may utilize different regions on $G\alpha$ to mediate G protein binding and activation.[20] Do G-protein-coupled receptors (GPCRs) within a particular family utilize the same $G\alpha$ sequence identifiers as contact and/or activation sites? Does the profile of critical contact sites vary among subtypes of the same receptor family or do the contact sites direct more subtle features of specificity such as the efficiency of receptor coupling? Do all GPCRs interact simultaneously with several regions on heterotrimers? These are some of the questions that remain to be answered. Although segments of the β and γ subunits may also contribute to the receptor interacting surface of heterotrimers,[21-26] this chapter will discuss the use of chimeric $G\alpha$ subunits to resolve some of these outstanding questions regarding receptor–G protein specificity.

[4] M. R. Mazzoni, S. Taddei, L. Giusti, P. Rovero, C. Galoppini, A. D'Ursi, S. Albrizio, A. Triolo, E. Novellino, G. Greco, A. Lucacchini, and H. E. Hamm, *Mol. Pharmacol.* **58,** 226 (2000).

[5] E. L. Martin, S. Rens-Domiano, P. J. Schatz, and H. E. Hamm, *J. Biol. Chem.* **271,** 361 (1996).

[6] H. E. Hamm, D. Deretic, K. P. Hofmann, A. Schleicher, and B. Kohl, *J. Biol. Chem.* **262,** 10831 (1987).

[7] H. Bae, K. Anderson, L. A. Flood, N. P. Skiba, H. E. Hamm, and S. G. Graber, *J. Biol. Chem.* **272,** 32071 (1997).

[8] H. Bae, T. M. Cabrera-Vera, K. M. Depree, S. G. Graber, and H. E. Hamm, *J. Biol. Chem.* **274,** 14963 (1999).

[9] B. R. Conklin, Z. Farfel, K. D. Lustig, D. Julius, and H. R. Bourne, *Nature* **363,** 274 (1993).

[10] J. Liu, B. R. Conklin, N. Blin, J. Yun, and J. Wess, *Proc. Natl. Acad. Sci. U.S.A.* **92,** 11642 (1995).

[11] E. Kostenis, B. R. Conklin, and J. Wess, *Biochemistry* **36,** 1487 (1997).

[12] R. Onrust, P. Herzmark, P. Chi, P. D. Garcia, O. Lichtarge, C. Kingsley, and H. R. Bourne, *Science* **275,** 381 (1997).

[13] M. M. Rasenick, M. Watanabe, M. B. Lazarevic, G. Hatta, and H. E. Hamm, *J. Biol. Chem.* **269,** 21519 (1994).

[14] H. E. Hamm, D. Deretic, A. Arendt, P. A. Hargrave, B. Koenig, and K. P. Hofmann, *Science* **241,** 832 (1988).

[15] C. H. Lee, A. Katz, and M. I. Simon, *Mol. Pharmacol.* **47,** 218 (1995).

[16] E. Kostenis, J. Gomeza, C. Lerche, and J. Wess, *J. Biol. Chem.* **272,** 23675 (1997).

[17] M. R. Mazzoni and H. E. Hamm, *J. Biol. Chem.* **271,** 30034 (1996).

[18] G. Grishina and C. H. Berlot, *Mol. Pharmacol.* **57,** 1081 (2000).

[19] S. R. Sprang, *Annu. Rev. Biochem.* **66,** 639 (1997).

[20] S. M. Wade, W. K. Lim, K. L. Lan, D. A. Chung, M. Nanamori, and R. R. Neubig, *Mol. Pharmacol.* **56,** 1005 (1999).

Because of the overall sequence homology[1] and structural similarity of Gα subunits, functional chimeric Gα subunits can be generated and used to assess the unique nature of receptor–G protein interactions without interfering with $\beta\gamma$ association or nucleotide binding and hydrolysis. We have successfully examined the structural determinants of receptor–G protein specificity with the aid of two biochemical approaches: (1) measurement of the ability of the agonist to stimulate receptor-mediated GDP release and subsequent binding of [^{35}S]GTPγS to the Gα subunit; and (2) measurement of the ability of an agonist to promote receptor–G protein coupling thereby "shifting" the receptor to a higher affinity state than is observed for the uncoupled receptor (affinity shift assay). The following paragraphs will outline the methodology used for these assays and will describe the advantages, disadvantages, and points to consider when choosing to use these assays for the study of receptor–G protein interactions.

General Comments

The assays described in the following paragraphs require the generation of the following components: (1) purified native or chimeric Gα subunits, (2) purified native G$\beta\gamma$ subunits, and (3) membrane preparations containing high levels of the receptor of interest. Each of these components is generated separately, and during the course of the assays the receptor–G protein complex is reconstituted *in vitro*. The advantage of using a reconstituted *in vitro* assay is that the receptor: G protein ratio can be strictly controlled and easily manipulated. In addition, the *in vitro* assays limit any confounding influences from cellular proteins that may interfere with the receptor for G protein binding and/or activation. One disadvantage of using a reconstituted system is that one occasionally encounters a chimeric or native protein (either G protein or receptor) that cannot be readily expressed or purified in a functional form. This problem is encountered more often with proteins expressed in *Escherichia coli* but to a lesser extent when expressed in eukaryotic cells such as Sf9 (*spodoptera frugiperda* ovary) insect cells. Whereas *E. coli* expression is relatively easy and only requires generating a construct in an

[21] J. M. Taylor, G. G. Jacob-Mosier, R. G. Lawton, A. E. Remmers, and R. R. Neubig, *J. Biol. Chem.* **269,** 27618 (1994).

[22] J. M. Taylor, G. G. Jacob-Mosier, R. G. Lawton, M. VanDort, and R. R. Neubig, *J. Biol. Chem.* **271,** 3336 (1996).

[23] O. Kisselev, A. Pronin, M. Ermolaeva, and N. Gautam, *Proc. Natl. Acad. Sci. U.S.A.* **92,** 9102 (1995).

[24] O. Kisselev, M. Ermolaeva, and N. Gautam, *J. Biol. Chem.* **270,** 25356 (1995).

[25] O. Kisselev, M. V. Ermolaeva, and N. Gautam, *J. Biol. Chem.* **269,** 21399 (1994).

[26] H. Yasuda, M. A. Lindorfer, K. A. Woodfork, J. E. Fletcher, and J. C. Garrison, *J. Biol. Chem.* **271,** 18588 (1996).

appropriate bacterial expression vector, expression of chimeras and receptors in Sf9 cells requires the generation of recombinant baculovirus constructs. The isolation of a single high-titer virus can take several weeks but is generally worth the extra effort when constructs cannot be expressed in *E. coli*. Two kits that we have used successfully in our laboratory for this purpose include BAC-TO-BAC (Life Technologies, Gaithersburg, MD) and BaculoGold (PharMingen, San Diego, CA). One disadvantage of *E. coli* expression is that constructs expressed in these bacterial cells do not undergo posttranslational modification such as palmitoylation and myristoylation, which may be required to improve membrane targeting and achieve the optimal interaction between $G\alpha$ and the receptor.[27–29] However, if needed, $G\alpha$ subunits can be expressed in *E. coli* cells that have been transformed with the *N*-methyltransferase gene in order to generate myristoylated $G\alpha$ subunits.[30]

Purification of $G\alpha$ Subunits from *E. coli* or Sf9 Cells

The technique used in our laboratory to generate chimeric $G\alpha$ subunits has been reported previously in detail and is carried out in accordance with standard molecular biological approaches.[31] Likewise, the expression and purification of $G\alpha$ subunits from *E. coli* cells has been detailed elsewhere.[8,32] Briefly, $G\alpha$ chimeras are typically N-terminally epitope-tagged with a hexahistidine (His_6) sequence and then purified in the presence of 50 μM GDP using a Ni^{2+}–nitrilotriacetic acid–agarose resin column in accordance with the manufacturer's instructions (Qiagen). The purified protein is then dialyzed in 50 mM Tris-HCl, pH 8.0, 50 mM NaCl, 5 mM $MgCl_2$ in 20% (v/v) glycerol supplemented with 0.1 mM phenylmethylsulfonyl fluoride and 2 mM 2-mercaptoethanol to remove excess salts and further purified by high-performance liquid chromatography (HPLC) (Waters Protein-Pak QHR-15, Waters Chromatography). Based on our laboratories' experience, $G\alpha_i$ and $G\alpha_s$ subunits are readily expressed in *E. coli* cells, whereas $G\alpha_o$, $G\alpha_q$, and $G\alpha_{12}$ are more readily expressed in Sf9 cells.[33–35] $G\alpha_t$ is not easily expressed

[27] J. Morales, C. S. Fishburn, P. T. Wilson, and H. R. Bourne, *Mol. Biol. Cell* **10**, 1 (1998).

[28] P. Wedegaertner, D. Chu, P. Wilson, M. Levis, and H. Bourne, *J. Biol. Chem.* **268**, 25001 (1993).

[29] A. Wise, M. Parenti, and G. Milligan, *FEBS Lett.* **407**, 257 (1997).

[30] M. E. Linder, I. H. Pang, R. J. Duronio, J. I. Gordon, P. C. Sternweis, and A. G. Gilman, *J. Biol. Chem.* **266**, 4654 (1991).

[31] N. P. Skiba, T. O. Thomas, and H. E. Hamm, *Methods Enzymol.* **315**, 502 (2000).

[32] N. P. Skiba, H. Bae, and H. E. Hamm, *J. Biol. Chem.* **271**, 413 (1996).

[33] T. Kozasa and A. G. Gilman, *J. Biol. Chem.* **270**, 1734 (1995).

[34] S. G. Graber, R. A. Figler, and J. C. Garrison, *Methods Enzymol* **237**, 212 (1994).

[35] T. Kozasa, X. Jiang, M. J. Hart, P. M. Sternweis, W. D. Singer, A. G. Gilman, G. Bollag, and P. C. Sternweis, *Science* **280**, 2109 (1998).

in either cell type. However, a $G\alpha_{t/i}$ chimeric protein (Chi6; amino acid residues 216–295 of $G\alpha_t$ have been replaced with $G\alpha_{i1}$ residues 220–299) originally described by Skiba $et\ al.$[32] can be readily expressed in $E.\ coli$ and is used routinely in our laboratory because of its demonstrated $G\alpha_t$-like character.[32] For a comprehensive discussion of the expression and purification of G proteins produced using the baculovirus expression system, the reader is referred to Graber $et\ al.$[34]

Assessing Functional Status of Purified $G\alpha$ Subunits

After purification of $G\alpha$ constructs, each protein is tested for its ability to undergo an AlF_4^- dependent increase in tryptophan fluorescence.[32] This assay is based on the ability of AlF_4^- to induce the active conformation of $G\alpha$ in the presence of GDP resulting in an increase in intrinsic fluorescence of a conserved tryptophan residue located in the switch II region of $G\alpha$ subunits. The details of this assay have been published elsewhere.[7,32]

Purification of $\beta\gamma$ Subunits from Tissue Extracts

$G\beta_1\gamma_1$ subunits are commonly purified from bovine retinal rod outer segment membranes. Bovine eyes can be obtained from an area slaughterhouse for this purpose. A detailed protocol for the extraction of $\beta\gamma$ subunits from bovine retina has been published previously.[36] The most common $\beta\gamma$ subunit present in these retinal preparations is $\beta_1\gamma_1$.[37,38] Although this dimer composition will be suitable for most purposes, one might need to consider either extracting other dimer combinations such as $\beta_1\gamma_2$ (prevalent in brain) or coexpressing $\beta\gamma$ subunits in Sf9 insect cells.[39,40] The functional interaction between $G\beta\gamma$ and $G\alpha$ can be tested either by using the $[^{35}S]GTP\gamma S$ assay as described herein with the wild-type $G\alpha$ subunit of interest or by assessing $G\beta\gamma$-dependent ADP ribosylation of $G\alpha$ by pertussis toxin.[41,42]

[36] M. R. Mazzoni, J. A. Malinski, and H. E. Hamm, $J.\ Biol.\ Chem.$ **266,** 14072 (1991).

[37] N. Ueda, J. A. Iniguez-Lluhi, E. Lee, A. V. Smrcka, J. D. Robishaw, and A. G. Gilman, $J.\ Biol.\ Chem.$ **269,** 4388 (1994).

[38] J. A. Iniguez-Lluhi, M. I. Simon, J. D. Robishaw, and A. G. Gilman, $J.\ Biol.\ Chem.$ **267,** 23409 (1992).

[39] E. J. Dell, T. Blackmer, N. P. Skiba, Y. Daaka, L. R. Shekter, R. Rosal, E. Reuveny, and H. E. Hamm, $Methods\ Enzymol.$ (in press).

[40] S. G. Graber, R. A. Figler, and J. C. Garrison, $J.\ Biol.\ Chem.$ **267,** 1271 (1992).

[41] J. F. Vanhauwe, N. Fraeyman, B. J. Francken, W. H. Luyten, and J. E. Leysen, $J.\ Pharmacol.\ Exp.\ Ther.$ **290,** 908 (1999).

[42] T. Katada, K. Kontani, A. Inanobe, I. Kobayashi, Y. Ohoka, H. Nishina, and K. Takahashi, $Methods\ Enzymol.$ **237,** 131 (1994).

Expression of Receptors in Sf9 Insect Cell vs Mammalian Cell Membranes

Both the $[^{35}S]GTP\gamma S$ binding assay and the affinity shift assay can be carried out using membranes generated either from insect cells (e.g., Sf9) or from one of a number of mammalian cell lines. The advantages of using insect cell membrane preparations include the minimization of background signal from endogenous G proteins, and the ease with which large amounts of membrane preparations can be generated. This large-scale preparation minimizes the number of times that receptor densities must be determined and limits variability introduced by the repeated transient transfection of mammalian cells. Although the use of stably transfected mammalian cell lines would reduce receptor expression variability, the receptor density may be lower in stable cell lines. Expression and purification of the membrane receptors in Sf9 cells is as previously described by Clawges et al.[43] When choosing to use mammalian membranes, one must keep in mind that receptors expressed in mammalian cells are often at a lower density than in Sf9 cells though the expression of receptors in mammalian cells typically results in the expression of sufficient receptors to obtain a reliable signal in $[^{35}S]GTP\gamma S$ binding assays. Background $[^{35}S]GTP\gamma S$ binding (i.e., $[^{35}S]GTP\gamma S$ binding in the absence of agonist) can be particularly troublesome when using mammalian cell membrane preparations because of the presence of endogenous GPCRs and heterotrimeric G proteins in the membrane. Thus, assay conditions should be optimized to reduce basal $[^{35}S]GTP\gamma S$ binding vs the total signal of $[^{35}S]GTP\gamma S$ binding. This can be achieved by changing the concentration of GDP, $MgCl_2$, or NaCl and/or by changing the temperature or assay time. Preincubation of membranes for 1 hr at $4°$ with $1 \mu M$ adenosine $5'-(\beta,\gamma$-imino)triphosphate lowers basal $[^{35}S]GTP\gamma S$ binding by blocking nonselective nucleotide binding sites. For a comprehensive overview of $[^{35}S]GTP\gamma S$ binding assays, we refer to Lazareno.[44] In addition, one might need to denature endogenous G proteins with urea[45–47] or inactivate endogenous G proteins which may interfere with the assays (for example, by pretreating cells or membranes with pertussis toxin for inactivation of $G_{i/o}$ family members).[41,42]

Preparation of Mammalian Membranes for $[^{35}S]GTP\gamma S$ Binding Assays

For a particular study, a large pool of cell membranes can be prepared (40–60 145 mm^2 petri dishes). Each pool should be tested for receptor expression

[43] H. M. Clawges, K. M. Depree, E. M. Parker, and S. G. Graber, *Biochemistry* **36**, 12930 (1997).
[44] S. Lazareno, *Methods Mol. Biol.* **106**, 231 (1999).
[45] J. I. Hartman and J. K. Northup, *J. Biol. Chem.* **271**, 22591 (1996).
[46] M. R. Hellmich, J. F. Battey, and J. K. Northup, *Proc. Natl. Acad. Sci. U.S.A.* **94**, 751 (1997).
[47] M. A. Lindorfer, C. S. Myung, Y. Savino, H. Yasuda, R. Khazan, and J. C. Garrison, *J. Biol. Chem.* **273**, 34429 (1998).

(using a saturating concentration of radiolabeled ligand) and amount of stimulation of [^{35}S]GTPγS binding using different concentrations of protein. For the preparation of membranes, cells are subcultured from 175 cm^2 tissue culture flasks to 145 cm^2 petri dishes. If desired, 5 mM sodium butyrate can be added 24 hr prior to harvesting the cells to increase the receptor expression level.[48] After aspirating the culture medium, the petri dishes are washed once with 5 ml ice-cold phosphate-buffered saline (PBS). If needed, the cells can be stored at –70° until further use. Then, petri dishes are thawed and 5 ml 10 mM Tris-HCl (pH 7.4), 1 mM EDTA, and 1 mM 4-(2-aminoethyl)benzenesulfonylfluoride hydrochloride (buffer B) are added per dish. The cells are gently removed from the petri dishes by pipetting and then homogenized by 10 strokes with a dual homogenizer (motor-driven Teflon pestle and conical glass tube). The homogenate is centrifuged (10 min at 1000g, 4°) and the resulting pellet is resuspended in buffer B and centrifuged again (10 min at 1300g, 4°). The two supernatants are pooled and centrifuged at 50,000g for 1 hr at 4°. The resulting pellet is resuspended in 50 mM Tris-HCl (pH 7.4), containing 10% (v/v) glycerol, and stored in aliquots at –70°.

Determination of Receptor Density in Membrane Preparations

Because of the inherent variation in receptor density that occurs between receptor preparations, the receptor density must be determined for each membrane preparation. This characterization is achieved using a saturation radioligand binding isotherm with an appropriate antagonist and determination of B_{max} using nonlinear regression analysis of the binding data. Although parameters such as incubation time, temperature, buffer composition, and ligand concentration will vary among receptors, the assay conditions can generally be adapted from conditions reported in the literature for a given receptor. The following is a set of conditions that have been found to be appropriate for a number of receptors utilized in our laboratory. An aliquot of frozen membrane is thawed and pelleted by centrifugation in a chilled microfuge at 12,000 rpm for 10 min. The pellet is resuspended in binding buffer (50 mM Tris-HCl, 5 mM MgCl$_2$, 0.5 mM EDTA, pH 7.5) at an appropriate concentration and 20–100 μg of resuspended membrane protein is added to each tube in a binding assay containing increasing concentrations of radioligand. Typically the total volume of this reaction will be 1 ml or less. The tubes are incubated at 25° for 1–2 hr in a shaking water bath and then filtered over Whatman (Clifton, NJ) GF/C filters using a Brandel harvester. Each filter is rinsed three times with 4 ml ice-cold wash buffer (50 mM Tris-HCl, 5 mM MgCl$_2$, 0.5 mM EDTA, 0.01% sodium azide, pH 7.5

[48] D. P. Palermo, M. E. DeGraaf, K. R. Marotti, E. Rehberg, and L. E. Post, *J. Biotechnol.* **19**, 35 (1991).

at 4°) and placed in 4.5 ml CytoScint counted to constant error in a scintillation counter. Ideally the radioligand concentrations will range from below to 10- to 20-fold above the ligand's K_D for the receptor. Because of the presence of endogenous G proteins, antagonists are the preferred ligands: they exhibit the same affinity for receptors that are either coupled or uncoupled to G proteins. If an appropriate antagonist is not readily available, a radiolabeled agonist may be used. However, if an agonist is used the binding data are likely to best fit a two-site model depending on the extent of receptor coupling with the endogenous G proteins and the total receptor number. The significance of the endogenous G proteins can be estimated by the use of a nonhydrolyzable GTP analog, such as GTPγS, which in the presence of low concentrations of Mg^{2+} binds nearly irreversibly to Gα thereby shifting the receptors to their low-affinity state for agonist. Generally 50 μM GTPγS is added to the binding assay and the contribution of the endogenous G proteins can be determined by the degree to which radioligand binding is reduced compared to binding in the absence of GTPγS.

Protocol for [^{35}S]GTPγS Binding Assay using Sf9 Membranes in a 96-Well Plate

The following is a description of the method routinely used in our laboratory to determine the rates of agonist-stimulated [^{35}S]GTPγS binding using a reconstituted *in vitro* assay in a 96-well plate format.[7,8] Using this format, agonist concentration–response curves and time course experiments can be readily completed. Membranes containing the receptor of interest are incubated with 1 mM AMP–PNP [adenosine 5'-(β,γ-imino)triphosphate] at 37° for 1 hr. Receptor coupling to G protein α and $\beta\gamma$ subunits is reconstituted on ice in 70 μl of reaction buffer A (25 mM Na–HEPES, pH 7.5, 100 mM NaCl, 5 mM MgCl$_2$, 1 mM EDTA, 1 mM dithiothreitol) for 30 min and then diluted with 200 μl of reaction buffer A containing 150 nM GDP and 60 nM GTPγS. The addition of 30 μl of [^{35}S]GTPγS ($\sim$7 × 10^6 cpm) initiates the reaction, which is incubated at 25°. The final concentrations of receptor, agonist, and G proteins added during reconstitution will depend on the particular constituents being examined. For serotonin (5-HT) 1B receptors we have previously used the following final concentrations: 1.28 nM 5-HT$_{1B}$ receptor, 40 nM Gα, and 40 nM retinal $\beta\gamma$. For agonist activation, 1 μM of 5-HT was included. As previously mentioned, changing the concentration of GDP, MgCl$_2$, or NaCl and/or changing the temperature or assay time may be required for optimization of the signal-to-noise ratio (Lazareno[44]). Samples (20 μl) are withdrawn at various times and the reaction is terminated by passing through a Millipore (Bedford, MA) Multiscreen-HA 96-well filtration plate followed immediately by five washes with 200 μl ice-cold wash buffer (20 mM Tris-HCl pH 7.4, 100 mM NaCl, 25 mM MgCl$_2$). The filters are air dried for 5 min and punched out using a Millipore Multiscreen Puncher into scintillation vials. After addition of liquid scintillation fluid to the vials, the amount of radioactivity retained on the filters is quantitated using a liquid scintillation counter.

Adaptions to [^{35}S]GTPγS Binding Protocol for Use with Mammalian Membrane Preparations

As mentioned, a drawback to using mammalian cell membranes is the presence of endogenous G proteins. For receptors coupling to the $G_{i/o}$ proteins G_{i1}, G_{i2}, G_{i3}, G_{oA}, G_{oB}, G_t, this problem can be eliminated by treatment with pertussis toxin, which ADP-ribosylates the C-terminal cysteine residue and prevents interaction with the receptor. Chimeric G proteins can then be added to membranes from cells that were pertussis toxin treated. Alternatively, chimeric G proteins can be coexpressed with the receptor, but ADP-ribosylation sites must be removed (e.g., by mutating them to serine residues). The disadvantage of coexpression is that the receptor: G protein ratio cannot be easily controlled.

The [^{35}S]GTPγS binding assay is carried out in a final volume of 500 μl containing 50 mM Tris-HCl (pH 7.4), 100 mM NaCl, 5 mM MgCl$_2$, 1 mM EGTA, 0.1 mM dithiothreitol, 10 μM guanosine diphosphate (GDP), and 0.25 nM [^{35}S]GTPγS. The buffer composition should be optimized for each different receptor to obtain maximal stimulation of [^{35}S]GTPγS binding. The proposed buffer composition can be used as a starting point. Concentrations of the following compounds can be optimized between the suggested range: NaCl (10–200 mM), MgCl$_2$ (1–20 mM) and GDP (0.1–20 μM). Also, addition of saponin (10 mg/ml) can increase the signal by increasing the permeability of the membrane vesicles and therefore the amount of G protein that can be stimulated.

Prior to the start of the assay, membranes are thawed and rehomogenized using an Ultra-Turrax homogenizer. Membranes (10 μg protein/assay) and ligands are preincubated with the ligands (without [^{35}S]GTPγS) for 30 min at 30° to obtain steady-state receptor occupation. After addition of [^{35}S]GTPγS, membranes are incubated for an additional 30 min. The incubation time with [^{35}S]GTPγS can be optimized as well. In our experiments, we have never reached equilibrium for incubation times up to 2 hr; the best stimulation of [^{35}S]GTPγS binding (vs basal [^{35}S]GTPγS binding) was found at 30 min. Basal [^{35}S]GTPγS binding is measured in the absence of compounds. The reactions are terminated by rapid filtration through Whatman GF/B filters and the filters are washed with 5 ml of ice-cold 50 mM Tris-HCl (pH 7.4), 100 mM NaCl, 1 mM EGTA. Note that the filters are soaked in incubation buffer *without* 0.1% polyethyleneimine (which is often used in radioligand binding assays with iodinated ligands). Addition of 0.1% polyethyleneimine will drastically increase (background) [^{35}S]GTPγS binding to the filter. The amount of radioactivity retained on the filters is determined by liquid scintillation counting. The maximal amount of bound [^{35}S]GTPγS should be less than 10% of [^{35}S]GTPγS added.

GTPγs Data Analysis

The data from this assay can be analyzed in one of several ways. The percent increase in [^{35}S]GTPγS binding elicited by the agonist can be determined by the following equation.

$$\% \text{ Increase in GTP}\gamma\text{S binding} = \{([^{35}\text{S}]\text{GTP}\gamma\text{S binding in} $$
$$\text{presence of agonist} - [^{35}\text{S}]\text{GTP}\gamma\text{S binding in absence of} $$
$$\text{agonist}) \div [^{35}\text{S}]\text{GTP}\gamma\text{S binding in absence of agonist}\} \times 100 $$

If the backbones of the chimeric Gα subunits (e.g., Gα_i and Gα_t) interact equally well with a particular receptor (e.g., rhodopsin) that is not the receptor being studied (e.g., 5-HT$_{1B}$), then [^{35}S]GTPγS binding data can alternatively be normalized to the percent of maximal binding which occurs in the presence of a saturating amount of this second receptor (e.g., rhodopsin). This normalization serves to control for differences in active Gα protein content among purified protein preparations. In order to determine the rate of [^{35}S]GTPγS binding, agonist-stimulated and -unstimulated [^{35}S]GTPγS binding must be measured at several intervals over time (typically 1–30 min). A linear regression analysis is then carried out on the data after subtracting the basal spontaneous [^{35}S]GTPγS binding which occurs in the absence of the agonist. The initial rate of [^{35}S]GTPγS binding is given by the slopes of the resulting lines.

Affinity Shift Assay

Assay Theory

The high-affinity agonist binding state of a GPCR is the ternary complex of receptor and guanine nucleotide-free G protein heterotrimer. In membranes where expressed receptors are either in excess of or unable to couple with endogenous G proteins, receptor–G protein coupling may be studied by the addition of purified, exogenous G proteins to the membranes. By measuring the increase in high-affinity agonist binding to the receptor of interest in the presence of exogenous wild-type or chimeric G proteins, one can determine whether the mutations introduced into the G protein affect receptor–G protein coupling. Affinity shift activity is defined as the fold enhancement above nonreconstituted buffer controls of high-affinity agonist binding in membranes reconstituted with exogenous G protein heterotrimers in 40- to 100-fold molar excess over the expressed receptors. Gα subunits with no ability to produce a high-affinity agonist binding state would have an affinity shift activity of 1, whereas those subunits with abilities similar to those of wild-type Gα subunits exhibit affinity shift activities >1, typically from 4 to 15 depending on the receptor. Although this measure is beneficial for the majority of receptors examined, in a few instances agonists may exhibit a similar affinity for both coupled and uncoupled receptors,[41,49] and hence no measurable shift in affinity can be detected.

General Considerations

The affinity shift assay requires that partially purified membranes containing the expressed receptor be reconstituted with exogenous G proteins. A radioligand binding assay using a low concentration of agonist (near the high-affinity K_D for the receptor such that little or no binding occurs to uncoupled receptors; see below) carried out on the reconstituted preparation as discussed above is used to determine the enhancement of binding due to the exogenous G proteins. This assay has several advantages over assays in which receptors and G proteins are cotransfected into cells and their functional interaction determined by measuring the activity of a downstream effector. First, the functional interaction between receptor and G protein is measured directly by radioligand binding, eliminating interference from other proteins that may modulate the activity of the downstream effector or of the activated Gα subunit. Second, in the reconstitution system the stoichiometry of receptors and G proteins can be tightly regulated, which allows for detection of relatively subtle changes in their functional interaction resulting either from differing affinities of a receptor for a particular G protein or from a change in the agonist affinity of a particular ternary complex. Because amino acid substitutions made in the chimeric Gα subunits may give rise to only small changes in receptor–G protein interactions,[7] the affinity shift assay utilizing chimeric G proteins is a particularly sensitive tool in the analysis of selectivity in receptor–G protein interactions.

G Protein Concentration in Assay

The G protein concentration is an important consideration in setting up an affinity shift assay with a particular receptor. The G proteins must be present at saturating concentrations with respect to the receptor in order to ensure maximum binding at the concentration of agonist chosen. Because different receptors have distinct affinities for Gα subunits (Ref. 43 and S. Graber and H. Hamm, unpublished observations), the G protein saturation point needs to be determined for each

[49] M. B. Assie, C. Cosi, and W. Koek, *Eur. J. Pharmacol.* **386,** 97 (1999).

receptor examined. Titration of the G proteins in the reconstitution is used to determine the saturating concentration of individual G proteins, which is used routinely in affinity shift assays for that receptor. This allows for meaningful comparisons among activities resulting from reconstitutions with equivalent concentrations of native or chimeric $G\alpha$ subunits.

Receptor Density in Affinity Shift Assay

In principle, the magnitude of the affinity shift will depend on both receptor number and the difference between the high- and low-affinity K_D values for a given receptor. In order to obtain a measurable affinity shift, the expressed receptors must be in excess of endogenous G proteins capable of coupling with the receptor. As the number of uncoupled receptors increases, the magnitude of the shift will also increase. This phenomenon must be taken into consideration when comparing affinity shift activities from different receptor preparations. If it is difficult to express the receptor in excess of the endogenous G proteins, the endogenous G proteins may be removed by urea treatment of the membrane preparation.[45-47] Treatment with urea will remove proteins not tightly associated with the membranes. Although experiments with $GTP\gamma S$ reveal that a small portion of the endogenous G proteins remain after the urea stripping (unpublished data from S. Graber laboratory and Refs. 45 and 46), this treatment does significantly improve affinity shift activity.

Reconstitution and Affinity Shift Assay

Purified receptor-containing membranes are pelleted at 12,000 rpm in a refrigerated microfuge (4°) for 10 min and resuspended at 7–10 $\mu g/\mu l$ in reconstitution buffer (25 mM Na HEPES, 100 mM NaCl, 5 mM $MgCl_2$, 1 mM EDTA, 0.04% CHAPS, 500 nM GDP, pH 7.5 at 4°). The G protein subunits are premixed in this same buffer with a $G\beta\gamma$ concentration 1.2 times that of $G\alpha$. The resuspended membranes are then reconstituted for 15 min at 25° with the premixed G proteins such that the concentration of α subunit is saturating for receptor to ensure maximal coupling of G protein to receptor. Typically 1–2 μl of G protein will be added to 20 μl membrane suspension. The reconstituted membranes are then diluted 10- to 15-fold with binding buffer and held on ice until the start of the binding assay. Fifty microliters reconstituted membranes is added to the binding tubes containing binding buffer. If the protein concentration in the binding assay is low (for example, when receptors are expressed at very high levels) then 0.1 $\mu g/\mu l$ bovine serum albumin (BSA) may be added to the binding buffer to minimize protein loss due to adsorption. The assay is started by the addition of radiolabeled agonist (used at a concentration near the high-affinity K_D for the receptor to ensure that ligand binding will be almost entirely to coupled receptors). Total binding assay volume will usually be 150–200 μl and each sample is assayed in triplicate. Nonspecific binding is measured with a 1000-fold excess of competing ligand in a set of control tubes. The tubes are incubated for an appropriate time in a shaking water bath after which the assay is terminated by filtration as discribed above.

Affinity Shift Activity Data Analysis

Affinity shift activity is calculated as the ratio of specific binding activity in the presence of exogenous G protein to the specific binding activity in the nonreconstituted control membranes. Thus, nonreconstituted controls and inactive G proteins will have affinity shift activities of 1, while values greater than 1 are indicative of functional interactions between the G protein and receptor. Because the magnitude of the difference between high- and low-affinity states varies among GPCRs, affinity shift activity must be normalized in order to compare activities among different receptors. Normalized affinity shift activity (NASA) is defined by the following relationship:

NASA = (Chimera reconstituted binding − nonreconstituted binding)

÷ native G protein reconstituted binding

− nonreconstituted binding)

When normalized affinity shift activity is calculated, native or chimeric G proteins that do not interact with a given receptor will have normalized activities of 0, while fully active G proteins will have activities of 1. Generally, an appropriate native G protein will produce the largest affinity shift activity with a given receptor, though occasionally chimeric G proteins with enhanced coupling properties with affinity shift activities significantly greater than those of native G proteins have been observed. Such chimeras have normalized affinity shift activities significantly greater than 1.

Summary

In conclusion, by taking advantage of the overall sequence homology and structural similarity of Gα subunits, functional chimeric Gα subunits can be generated and used as tools for the identification of sequence-specific factors that mediate receptor: G protein specificity. The [^{35}S]GTPγS binding assay and the affinity shift activity assay are two sensitive biochemical approaches that can be used to assess receptor: G protein coupling *in vitro*. These *in vitro* assays limit confounding influences from cellular proteins and allow for the strict control of receptor: G protein ratios.

[6] Use of Dominant Negative Mutations in Analysis of G Protein Function in *Saccharomyces cerevisiae*

By GREGOR JANSEN,* EKKEHARD LEBERER,* DAVID Y. THOMAS,*
and MALCOLM WHITEWAY*

Introduction

The use of dominant negative mutations as a means of disrupting genetic function in organisms that were refractory to direct gene disruptions received wide acceptance through the publication of an influential review by Ira Herskowitz.[1] This review pointed out that high-level expression of some mutant proteins may interfere with normal protein activity, and thus cells could be made phenotypically null for a function even in the presence of normal levels of a wild-type protein. This approach has been widely used in mammalian cells to create null mutants of cloned genes without the necessity of creating gene disruptions. Such mutations have been extensively used in the analysis of mammalian signaling pathways; see for example Refs. 2–5.

Although originally developed to allow genetic manipulation of genetically intractable organisms, dominant negative mutations have proven useful in the investigations of organisms such as the yeast *Saccharomyces cerevisiae* where genetic disruption of gene functions is routine. Dominant negative mutations have helped to dissect complex pathways such as messenger RNA splicing,[6,7] ubiquitin-mediated proteolysis,[8,9] and secretion.[10] In the case of yeast dominant negative mutants, simple inactivation of function is not the usual goal, because gene disruptions can readily create null mutations. Rather, the dominant mutations are used to disrupt one or more activities of a multifunctional protein, thus providing an informative separation of function.

---

* National Research Council of Canada. This is NRCC publication #44791.

[1] I. Herskowitz, *Nature* **329**, 219 (1987).

[2] R. G. Watts, C. Huang, M. R. Young, J. J. Li, Z. Dong, W. D. Pennie, and N. H. Colburn, *Oncogene* **17**, 3493 (1998).

[3] S. Roy, R. Luetterforst, A. Harding, A. Apolloni, M. Etheridge, E. Stang, B. Rolls, J. F. Hancock, and R. G. Parton, *Nature Cell Biol.* **1**, 98 (1999).

[4] M. Sakaue, D. Bowtell, and M. Kasuga, *Mol. Cell Biol.* **15**, 379 (1995).

[5] G. Ferrari and L. A. Greene, *EMBO J.* **13**, 5922 (1994).

[6] M. Plumpton, M. McGarvey, and J. D. Beggs, *EMBO J.* **13**, 879 (1994).

[7] B. Schwer and C. Guthrie, *Mol. Cell Biol.* **12**, 3540 (1992).

[8] Q. Wang and A. Chang, *EMBO J.* **18**, 5972 (1999).

[9] A. Banerjee, R. J. Deshaies, and V. Chau, *J. Biol. Chem.* **270**, 26209 (1995).

[10] D. Roth, W. Guo, and P. Novick, *Mol. Biol. Cell* **9**, 1725 (1998).

The tools of yeast molecular biology provide many advantages for the identification and characterization of such dominant negative mutations. Overexpression of the mutant protein is typically necessary for efficient inhibition of function through a dominant negative protein, and several stable yeast vectors containing strong constitutive and regulated promoters have been developed.[11,12] Clearly, if the function being inhibited is important or essential for growth, selection of a regulated promoter is critical. The use of strong promoters allows the expression of mutant versions of cloned genes at high levels. Efficient mutagenesis of cloned sequences can be applied to generate a wide variety of genetic variants, and the ease of handling many independent yeast colonies allows these variants to be screened for dominant negative characteristics. Therefore specific mutations do not have to be constructed to create the interfering phenotype; random mutants can be tested for this behavior. This allows the identification of dominant negative variants through phenotypic characterization, and provides the opportunity to identify dominant negative mutations even in proteins whose specific molecular function is unknown. The identification of dominant negative mutant proteins ensures that a partially functional protein has been selected; this characteristic allows a researcher to avoid the relatively uninformative, completely nonfunctional proteins that are frequently created by frameshifts and nonsense mutations during saturation mutagenesis protocols.

Dominant negative mutations have been used extensively to study the role of G protein subunits on the pheromone response pathway in yeast.[13–17] In this pathway, a heterotrimeric G protein consisting of the α subunit Gpa1p, the β subunit Ste4p, and the γ subunit Ste18p couples the mating pheromone receptors Ste2p and Ste3p to a downstream MAP (mitogen activated protein) kinase cascade.[18] This coupling involves direct physical association between the G$\beta\gamma$ subunit and both the Ste20p kinase[19] and the Ste5p MAP kinase scaffold protein.[20] Activation of the signal requires Ste5p complex dimerization[21,22] and membrane localization of the

[11] S. A. Parent, C. M. Fenimore, and K. A. Bostian, *Yeast* **1,** 83 (1985).

[12] D. Mumberg, R. Muller, and M. Funk, *Gene* **156,** 119 (1995).

[13] E. Leberer, D. Dignard, L. Hougan, D. Y. Thomas, and M. Whiteway, *EMBO J.* **11,** 4805 (1992).

[14] M. Whiteway, D. Dignard, and D. Y. Thomas, *Biochem. Cell Biol.* **70,** 1230 (1992).

[15] A. V. Grishin, J. L. Weiner, and K. J. Blumer, *Mol. Cell Biol.* **14,** 4571 (1994).

[16] A. V. Grishin, J. L. Weiner, and K. J. Blumer, *Genetics* **138,** 1081 (1994).

[17] M. S. Whiteway and D. Y. Thomas, *Genetics* **137,** 967 (1994).

[18] E. Leberer, D. Y. Thomas, and M. Whiteway, *Curr. Opin. Genet. Dev.* **7,** 59 (1997).

[19] T. Leeuw, C. Wu, J. D. Schrag, M. Whiteway, D. Y. Thomas, and E. Leberer, *Nature* **391,** 191 (1998).

[20] M. S. Whiteway, C. Wu, T. Leeuw, K. Clark, A. Fourest, D. Y. Thomas, and E. Leberer, *Science* **269,** 1572 (1995).

[21] C. Inouye, N. Dhillon, and J. Thorner, *Science* **278,** 103 (1997).

[22] D. Yablonski, I. Marbach, and A. Levitzki, *Proc. Natl. Acad. Sci. U.S.A.* **93,** 13864 (1996).

Ste5p/MAP kinase complex.[23] Random and site-directed mutagenesis procedures were applied to cloned G protein β and γ subunits, and the mutants generated were used both to probe aspects of G protein function, as well as to initiate hunts for other components of the pheromone response pathway.[13–17]

Procedures

Mutagenesis

Several mutagenesis strategies have been applied to generate random populations of Gβ and Gγ mutations, and these mutations have been screened for various phenotypes, including dominant negative mutations. Because the strategy of dominant negative selections involves high-level expression of the mutant protein, these selections involved genes encoding the G protein subunits β (Ste4p) and γ (Ste18p) cloned behind strong yeast promoters in yeast/*Escherichia coli* shuttle vectors. The promoter could be constitutive, such as *ADH1* or *PGK1*, or regulated, such as *GAL1, GAL10,* or *MET25*. For the analysis of the Gγ subunit we[14,17] and others[15] have used high copy 2μ plasmids that expressed the *STE18* gene from the *ADH1* promoter. In the case of the Gβ subunit, independent selections were applied to mutants generated in centromere plasmids with the *STE4* gene expressed from the *GAL1* promoter.[13,16] The selection of the vector system can be directed by the choice of the mutagenesis protocol. Strategies that involve either site-directed mutagenesis or the use of randomly doped oligonucleotides required plasmids with f1 origins of replication that allowed the production of single-stranded DNA versions of the plasmids.[24,25] However, mutagenic strategies that rely on PCR[26,27] or on chemical treatment of DNA[28,29] do not place any specific limitations on the vector.

Any efficient mutagenesis protocol is suitable for generating dominant negative versions of cloned yeast G protein subunits. We have had success using doped oligonucleotide mutagenesis while others have used chemical mutagenesis. These two approaches yielded similar spectra of mutations, although in the case of the Gβ subunit, the doped oligonucleotide strategy identified mutations in a region that was not detected by the chemical mutagenesis.[13,16] We have efficiently mutagenized the *STE4* gene using a PCR (polymerase chain reaction) approach.[30]

There are advantages and disadvantages to each method of mutagenesis. The oligonucleotide-directed strategy is highly efficient and gives good control of the

[23] P. M. Pryciak and F. A. Huntress, *Genes Dev.* **12,** 2684 (1998).

[24] T. Vernet, D. Dignard, and D. Y. Thomas, *Gene* **52,** 225 (1987).

[25] R. S. Sikorski and P. Hieter, *Genetics* **122,** 19 (1989).

[26] Y. H. Zhou, X. P. Zhang, and R. H. Ebright, *Nucleic Acids Res.* **19,** 6052 (1991).

[27] V. Svetlov and T. G. Cooper, *Yeast* **14,** 89 (1998).

[28] D. Shortle and D. Botstein, *Methods Enzymol.* **100,** 457 (1983).

[29] J. T. Kadonaga and J. R. Knowles, *Nucleic Acids Res.* **13,** 1733 (1985).

[30] S. J. Dowell, A. L. Bishop, S. L. Dyos, A. J. Brown, and M. S. Whiteway, *Genetics* **150,** 1407 (1998).

region of the mutagenesis. In particular, it may be the method of choice if a limited region of the protein is to be targeted. However, it is relatively expensive and requires access to oligonucleotide synthesizing capacity. In addition, even though the area of mutagenesis is targeted, occasional nontargeted residues can be changed, so it is safest either to sequence the entire gene for each identified mutant, or to subclone a limited region of the mutated gene into a wild-type gene background and sequence the subcloned region. Still, with efficient automated facilities providing access to rapid sequence analysis, sequencing of the mutagenized templates is unlikely to be a limiting factor in the analysis of dominant negative mutations, and sequencing the gene is a requirement regardless of the mutagenic approach applied.

Chemical mutagenesis is much less expensive than oligonucleotide-directed mutagenesis. However, it does require the use of mutagenic chemicals and thus necessitates care in the handling of the mutagenesis and the subsequent cleanup. In addition, whereas the doped oligonucleotide approach is essentially unbiased in the generation of mutations, chemical strategies have strong biases toward certain molecular changes, and thus the chemically induced mutant libraries will be less random. As noted above, in a direct comparison between chemical and doped oligonucleotide approaches in the analysis of the *STE4* gene, the chemical mutagenesis failed to identify a class of mutations found by the oligonucleotide-directed strategy.[13,16]

PCR mutagenesis has become a popular strategy for generating libraries of mutant variants of a cloned gene. Because PCR fragments can be efficiently combined into a linearized vector by *in vivo* recombination in yeast,[31] mutant libraries can be generated and analyzed directly in yeast without the need for an intermediate cloning step in *E. coli*. The mutagenesis can be accomplished with the level of mutations created by *Taq* polymerase under standard conditions, or can involve enhancements caused by biasing the nucleotide pools or replacing magnesium with manganese.[32] In general, the mutagenesis should be adjusted to generate single mutations on average, because the complications of separating the roles of multiple substitutions typically outweigh the advantages of high levels of mutants.

Selection of Dominant Negative Variants

The pheromone response pathway is very convenient for the identification of dominant negative versions of pathway components. Activation of the pathway leads to cell cycle arrest. Dominant negative variants that interfere with this arrest can be identified because they permit inappropriate cell division and thus can be readily selected from a pool of mutagenized variants.

[31] D. Muhlrad, R. Hunter, and R. Parker, *Yeast* **8,** 79 (1992).
[32] J. L. Lin-Goerke, D. J. Robbins, and J. D. Burczak, *Biotechniques* **23,** 409 (1997).

In the studies on the Gγ subunit Ste18p, the mutagenized libraries are directly transformed into yeast strains and transformants are screened for resistance to α-pheromone mediated arrest. In a screen of doped oligonuceotide mutagenized *STE18*, mutagenized plasmids are introduced into strain M200-6C, a strain supersensitive to the mating pheromone α factor because of mutations in the *SST1* and *SST2* genes.[33] Colonies showing enhanced resistance to pheromone-mediated cell cycle arrest are identified by replica plating patched transformants to selective plates spread with 100 μl of a 100 μg/ml solution of α factor dissolved in 90% (v/v) methanol. Dilution of the colonies prior to replica plating to the pheromone-containing plates reduces the background growth and facilitates the identification of resistant colonies. This identification is accomplished by initially replica plating the patched colonies to a nonselective YPD plate, and then using the freshly seeded YPD plate immediately as the source of cells for replica plating to a fresh velvet. The second velvet is used to transfer the cells to the pheromone-containing plate; this two-step replica plating results in the transfer of a limited number of cells per patch, and makes identification of the resistant colonies straightforward. However, the use of supersensitive strains is not essential for these selections. In a screen of hydroxylamine mutagenized *STE18,* the mutagenized plasmids are transformed into a wild-type strain and pheromone-resistant variants identified by direct plating on medium containing 1 mM α-factor.[15]

In the analysis of the dominant negative Gβ subunit, an enrichment of the Ste4p mutants is achieved by preparing sublibraries of nonfunctional proteins by transforming the mutagenized plasmid populations into a *gpa1 ste4* mutant strain (YEL107). Because overexpression of a functional Ste4p leads to cell cycle arrest, and because complementation of the *ste4* mutation in the *ste4 gpa1* double mutant would lead to cell cycle arrest due to the *gpa1* mutation, only transformants carrying a nonfunctional Ste4p will grow.[13] Similar selections have been applied during a Ste4p mutagenesis using PCR as the mutagen.[30] However, such enrichment schemes are not essential to the identification of dominant negative mutations. Direct screening of chemically mutagenized plasmids after transformation is sufficient to identify mutant Ste4 proteins that block pheromone-mediated cell cycle arrest.[16] However, only one of two regions conferring this phenotype is detected by direct screening, whereas two domains are found after the enrichment procedure.

Use of Dominant Negative Mutations in Analysis of Yeast G Protein Function

The logic of dominant negative mutant proteins is that they interfere with the function of the cell by performing some but not all the normal functions

---

[33] M. Whiteway, L. Hougan, and D. Y. Thomas, *Mol. Gen. Genet.* **214,** 85 (1988).

of the wild-type molecule. In signaling pathways such partial function can be disruptive in several ways. For example, if a multiprotein complex has to form for proper signaling, overexpression of a mutant component that makes some but not all the interactions could complex a limiting component into a nonfunctional association. Such dominant mutations would thus define residues necessary for an essential protein–protein association. Alternatively, if proper localization of a protein complex were critical for function, loss of localization signals on one component would be likely to create an interfering mutant protein. If the protein has to receive a signal from an upstream component and transfer it to a downstream target, overexpression of a protein that has lost the ability to either receive or transmit the signal, while still maintaining its other functions, could easily block the signaling pathway.

In addition to defining key residues involved in aspects of function of the signaling process, dominant negative mutants can serve as the starting point for selections to identify further signaling components. For example, if a mutant signaling protein is blocking the pathway by associating a limiting component in a nonfunctional complex, overexpression suppression of the block could identify the limiting protein. Alternatively, if a protein is defective in activating a downstream element, overexpression of this target protein may overcome the block. Thus dominant negative mutations can serve to identify key residues involved in G protein function, and can also be used for the initiation of searches for interacting signaling elements.

Gγ Mutants

A variety of dominant negative derivatives of the Gγ subunit encoded by the *STE18* gene have been generated (Fig. 1). Both site-directed and random mutants have shown that the C terminus of the protein is readily modified to create a dominant negative phenotype. Deletion of the C terminus[14,15] as well as mutation of residues within the CAAX box motif that directs posttranslational prenylation of the γ subunit[15,17] creates dominant negative proteins. Direct biochemical analysis showed that the C107Y substitution changes the mobility of the protein in a manner consistent with loss of the farnesyl modification.[15] Interestingly, C-terminal truncation mutations tested reduced the coupling of the G protein to the pheromone receptor, whereas the C107Y substitution mutation did not affect receptor coupling.[15]

Two-hybrid analysis shows that the dominant negative mutations do not block interaction with Gβ (Fig. 2). However, loss of C-terminal prenylation does influence membrane localization of the protein. A green fluorescent protein (GFP) fusion of Ste18p was created (Jansen *et al.*, in preparation), and the localization of the overexpressed protein containing the C106S substitution mutation was compared to the overproduced wild-type protein. Although the wild-type protein was

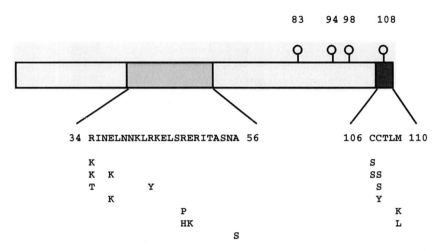

FIG. 1. Dominant negative mutations in Ste18p. Dominant negative mutations have been identified in two regions of Ste18p. These include a central region encompassing amino acids 34 to 56, and the carboxyl terminus. A variety of single and double amino acid substitutions were generated by both doped oligonucleotides and chemical mutagenesis in two different studies. Both strategies also identified C-terminal mutations; these include nonsense mutations at positions 83, 94, and 98 that caused premature termination and eliminated the CAAX box that directs carboxyl-terminal processing necessary for membrane association of the $G\gamma$ subunit. In addition, both random and site-directed mutagenesis protocols identified dominant negative mutations that substituted either the lipid-modified cysteine residues at positions 106 and 107, or amino acids in the last three amino acids that play roles in defining the specificity of the processing machinery.

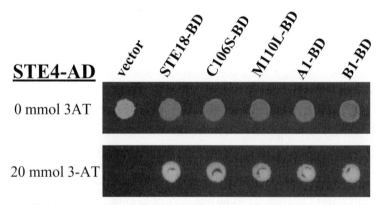

FIG. 2. *STE18* and mutants (*GAL4-BD*) in a two-hybrid assay with *STE4* (*GAL4-AD*). Transformants of strain *YPJ187* containing both plasmids were grown for three days on plates lacking histidine ± 3-AT. Shown are (from left to right): vector (pMBD), *STE18* (pMBDSTE18), *STE18-C106S* (pMBDSTE18-C106S), *STE18-M110L* (pMBDSTE18-M110L), *STE18-R34T R43Y* (pMBDSTE18-A1), *STE18-A56S* (pMBDSTE18-B1).

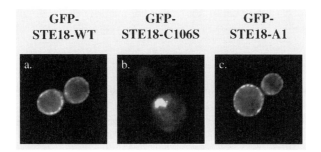

GFP- **STE18-WT**	**GFP-** **STE18-C106S**	**GFP-** **STE18-A1**

FIG. 3. *GALGFPSTE18* mutant localization in W303-1A Δ*ste18::LEU2* strain. Cells were grown for 14h in selective media containing galactose as a carbon source to induce expression of the fusion protein. (a) GFP-STE18 (pGREG576-STE18); (b) GFP-STE18-C106S (pGREG576-C106S); (c) GFP-STE18-R34T R43Y (pGREG576-A1). GFP fluorescence was visualized using a Leitz Aristoplan microscope equiped with a UV light source: images were acquired with a $100 \times$ objective using a Micro Max camera (Princeton Instruments Inc.) and Northern Eclipse imaging software (Empix Imaging Inc.) and were processed using Adobe PhotoShop.

membrane localized, the signal from the C106S variant was found primarily in discrete clusters in the cytoplasm—the significance of this localization pattern is not known. The mislocalization of the protein is due to the CAAX box mutation and not the dominant negative characteristics of the protein, because GFP fusions of dominant negative Ste18p variants that have proper CAAX box structures localize normally to the membrane (Fig. 3).

We speculated that sequestration and mislocalization of Ste4p could explain the dominant negative phenotype of the Gγ subunits lacking the membrane targeting prenylation signal provided by the CAAX box. A simple test of this hypothesis was to determine if overproduction of Ste4p could overcome the block in pheromone responsiveness caused by the mutation in Gγ. Although specific dominant negative versions of Ste18p were suppressed by high-level expression of Ste4p (the C106S CAAX box mutant and the R34T R43Y double mutant), the majority of the dominant negative mutants tested were not suppressed by overproduction of the Gβ subunit when cell cycle arrest blockage was analyzed.[14,17] However, overexpression of Gβ suppressed the mating defects caused by a variety of dominant negative Gγ mutants.[15] Therefore it may be that the ability of the dominant negative mutations to block cell cycle arrest is superior to their ability to block mating. As seen below, a similar behavior is evident for dominant negative alleles of Ste4p. We tested whether overexpression of Ste18p lacking the membrane association signal could mislocalize a Ste4 GFP fusion protein. The localization of the Ste4 GFP fusion protein was essentially the same in the presence of overproduced wild-type or dominant negative Ste18p, which is consistent with the observation that simple overproduction of Ste4p could not rescue the majority of the dominant negative Ste18p mutants.

Gβ Mutants

In the initial search for dominant negative mutants, the entire coding sequence of *STE4* was subjected to doped oligonucleotide mutagenesis, and the 18 pools of mutagenized plasmids were analyzed for galactose-induced interference with pheromone-mediated cell cycle arrest (Fig. 4). Although the mutagenesis of each segment was efficient and led to between 10% and 30% nonfunctional proteins, only two short regions generated proteins that blocked pheromone-mediated cell cycle arrest.[13] An independent study, using hydroxylamine mutagenesis, also identified the class of mutants located near the N terminus of the protein; these mutants were termed "sd" for signaling defective.[16] These results suggested either that these identified regions contained specific functions involved in transmitting the arrest signal from the pheromone receptor to the downstream pheromone response pathway,[13] or that these residues were involved in the mechanism of adaptation to the pheromone signal.[16]

Detailed analysis of the mutants showed that their primary effect was in blocking cell cycle arrest mediated either by pheromone or constitutive activation of the pathway by deletion of the Gα subunit encoded by *GPA1*.[13,16] Morphological

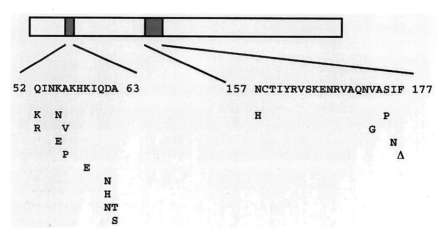

FIG. 4. Dominant negative mutants of Ste4p. Chemical treatment and doped oligonucleotides were used to mutagenize *STE4* and mutant proteins were identified that prevented normal cell cycle arrest in response to mating factor. These mutants mapped to two small regions of the protein; the region encompassing amino acids 52 to 63 contains residues implicated in interaction between Ste4p and both Ste20p and Ste5p. The function of the residues in the second region has not been determined.

changes induced by pheromone treatment, mating ability, and induction of pheromone-mediated gene expression were all relatively normal in the cells over-expressing the dominant negative mutant proteins.[13,16]

The dominant negative Ste4 variants were defective in normal signal trans-mission, because when expressed as the only Ste4 proteins, they were unable to support mating. This behavior was used to identify downstream elements of the pheromone response pathway. Overexpressed proteins that permitted cells con-taining high levels of a specific dominant negative allele of Ste4p to mate were identified after screening a multicopy plasmid library.[34] This approach identi-fied the Ste20 protein kinase as a high-copy suppressor of the D62N allele of Ste4p. Subsequent analysis established that Ste4p directly associated with Ste20p, and that this association was disrupted by mutations in the residues defining the N-terminal cluster of dominant negative alleles.[19] This region mapped to the sec-tion of Ste4p that would be predicted, by alignment of the yeast proteins with the X-ray structure determined for mammalian G proteins,[35,36] to form part of the coiled coil with Ste18p. In addition, residues in this region also are involved in physical association of Ste4p with Ste5p.[30] Thus the dominant negative mutations identified a region of Ste4p that was critical to the protein's function as a signaling molecule. The specific function defined by the second class of dominant negative mutations has not yet been determined.

Conclusions

Dominant negative mutations of G protein β and γ subunits have proven to be useful tools in the analysis of yeast G protein function. These alleles have permitted the identification of residues involved in protein localization and in protein–protein interactions. In addition, suppression of dominant negative alleles of Gβ allowed the detection of the Ste20p kinase as an important downstream component of the pheromone response pathway.

[34] E. Leberer, D. Dignard, D. Harcus, D. Y. Thomas, and M. Whiteway, *EMBO J.* **11,** 4815 (1992).

[35] D. G. Lambright, J. Sondek, A. Bohm, N. P. Skiba, H. E. Hamm, and P. B. Sigler, *Nature* **379,** 311 (1996).

[36] M. A. Wall, D. E. Coleman, E. Lee, J. A. Iniguez-Lluhi, B. A. Posner, A. G. Gilman, and S. R. Sprang, *Cell* **83,** 1047 (1995).

[7] Functional Assays for Mammalian G-Protein-Coupled Receptors in Yeast

By PAMELA E. MENTESANA, MERCEDES DOSIL, and JAMES B. KONOPKA

Introduction

Saccharomyces cerevisiae yeast cells are proving to be a useful expression system for the study of G-protein-coupled receptors (GPCRs) because they offer unique experimental advantages. Yeast grow rapidly and are highly amenable to manipulation by genetic and molecular biological procedures. A particular advantage for the study of GPCRs is that it has been possible to achieve functional expression of mammalian receptors in yeast.[1,2] A wide range of mammalian and other GPCRs have been shown to function in place of the G-protein-coupled mating pheromone receptors that promote conjugation in yeast. The ability of heterologous receptors to function in yeast can be readily assessed by analyzing their ability to activate the mating pheromone signal pathway. This strategy permits the direct analysis of a particular receptor in the absence of other receptor subtypes. In addition, yeast expression systems offer opportunities for high throughput screening assays and genetic selections to identify agonists, antagonists, receptor mutants, etc. Yeast has also been proven to be a useful system to produce high levels of receptor protein for purification. Therefore, in this chapter, we will describe the basic techniques needed to express a GPCR in yeast and monitor its function.

Selecting Expression Vector

The expression of foreign genes in yeast has become fairly routine. However, in view of the wide variety of plasmids that are available, care should be taken to choose an expression vector with features that are appropriate for the goals of the study. Yeast plasmids are divided into three types according to their mode of replication: (i) Yeast episomal plasmids (YEp) are probably the first choice for expression of foreign GPCR genes in yeast because they replicate to high copy number. YEp plasmids are derived from the 2μ circle, an autonomously replicating plasmid found endogenously in most laboratory yeast strains, and are thus also called 2μ plasmids. (ii) Yeast centromeric plasmids (YCp) replicate autonomously but are present in low copy number (1 or 2 copies per cell) because they contain

---

[1] M. H. Pausch, *Trends Biotechnol.* **15,** 487 (1997).
[2] J. R. Broach and J. Thorner, *Nature* **384,** 14 (1996).

a centromeric sequence (CEN) that ensures even partitioning at mitosis. (iii) Yeast integrating plasmids (YIp) are maintained by integration into a chromosome via homologous recombination.

Most currently used yeast plasmids are based on pBluescript or pUC19 plasmids[3,4] and can therefore replicate in both yeast and *Escherichia coli.* In addition to an Amp^R gene for selection in *E. coli,* these shuttle plasmids typically carry one of several different marker genes that provide the selective pressure for plasmid maintenance in yeast by complementing a specific auxotrophy in the host strain. The most commonly used selectable markers are *URA3, HIS3, LEU2,* and *TRP1,* which are involved in the biosynthesis of uracil, histidine, leucine, and tryptophan, respectively.

The shuttle plasmids that serve as the basis for most expression vectors in use today carry a promoter inserted into the multiple cloning site[5,6] and many are available from the American Type Culture Collection (ATCC, Manassus, VA). Constitutive expression is usually driven by the *GPD1, TEF1,* or *ADH1* promoters, with the *GPD* promoter being the strongest.[6] For regulated gene expression, the *GAL1* and *MET25* promoters are commonly used.[5] The *GAL1* promoter is induced by galactose and is one of the strongest inducible promoters. Transcriptional terminators can be placed downstream of the gene being expressed but are often left out as they are not as important in yeast as they are in animal cells.

To illustrate the considerations involved in expressing a foreign receptor in yeast, the methods our laboratory used to express the rat A_{2a}-adenosine receptor will be described. This receptor serves as a good control because it was shown to constitutively activate the yeast pheromone pathway.[7] YEp plasmid, p426GPD,[6] was used because it replicates to high copy number, contains the strong constitutive *GPD1* promoter, has a multiple cloning site, and is readily available from the ATCC. The transcription start site is included on the plasmid vector but not the translation start site. Thus, a receptor cDNA sequence can be easily cloned by using PCR to introduce a restriction site upstream of the ATG start codon. As yeast cells generally start translation at the first ATG codon encountered, it is important to make sure that the cloning procedures do not inadvertently introduce another ATG. As shown in Fig. 1, plasmid pA$_2$aHA was constructed by cloning the adenosine receptor into *SpeI–SmaI* sites of p426GPD. An HA (hemagglutinin) epitope tag was cloned downstream to facilitate detection of the receptor protein.

[3] R. D. Gietz and A. Sugino, *Gene* **74,** 527 (1988).

[4] R. S. Sikorski and P. Hieter, *Genetics* **122,** 19 (1989).

[5] D. Mumberg, R. Muller, and M. Funk, *Nucleic Acids Res.* **22,** 5767 (1994).

[6] D. Mumberg, R. Muller, and M. Funk, *Gene* **156,** 119 (1995).

[7] L. A. Price, J. Strnad, M. Pausch, and J. R. Hadcock, *Mol. Pharmacol.* **50,** 829 (1996).

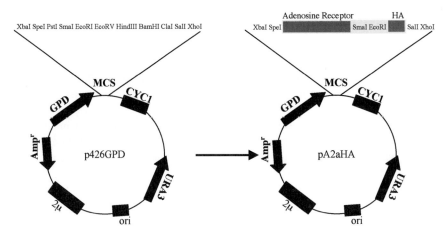

FIG. 1. Expression vector. The shuttle vector p426GPD[6] is depicted on the left. A triple HA tag was cloned into the MCS as an *Eco*RI–*Sal*I fragment and the rat adenosine A_{2a} receptor was inserted under control of the GPD1 promoter as a *Spe*I–*Sma*I fragment to create the expression plasmid, pA₂aHA.

Selecting Yeast Strain

In choosing an appropriate yeast strain for expression of GPCRs, it is important to distinguish whether the goal is high-level production of receptors for purification or functional analysis of G protein activation. For functional studies, yeast strains modified to enhance the ability of the foreign GPCR to couple to the yeast pheromone-responsive G protein are used. For GPCR overproduction, yeast strains that grow well and contain mutations in endogenous protease genes are often selected to enhance the yield of intact protein.[8] A list of genetic modifications that should be considered when choosing or constructing a yeast strain for functional expression is shown in Table I.

To facilitate the discussion of these modified yeast strains, we will first provide a quick review of yeast nomenclature and the endogenous pheromone pathway. Yeast genes are designated by three letters, in italics, followed by a number corresponding to a specific genetic locus (e.g., *STE2*). Dominant alleles are written in capital letters; recessive alleles are written in lowercase letters. Wild-type allele may be designated with a superscript plus sign, but this is optional. Mutant alleles are distinguished by the addition of a dash followed by a specific identifying allele number (e.g., *ste2-3*). Loss of function alleles in which the corresponding gene is deleted, or disrupted with another DNA sequence are indicated either with a Δ sign (e.g., *ste2*Δ) or a double colon followed by the name of the marker gene used to create the deletion allele (e.g., *ste2::LEU2*). Protein nomenclature does

[8] N. E. David, M. Gee, B. Andersen, F. Naider, J. Thorner, and R. C. Stevens, *J. Biol. Chem.* **272,** 15553 (1997).

TABLE I
GENES THAT CAN BE MODIFIED TO IMPROVE HETEROLOGOUS GPCR COUPLING IN YEAST

Gene	Function	Considerations
STE2	α Factor receptor	Delete endogenous GPCR to prevent interference
GPA1	Yeast Gα subunit	Modify to promote better coupling of GPCR to pheromone pathway
FUS1	Pheromone-induced gene	Fuse to reporter gene, lacZ or HIS3
FAR1	Promotes growth arrest in G_1	Delete to allow growth in response to activation (e.g., FUS1-HIS3 reporter)
SST2	RGS protein (GAP)	Deletion enhances signaling

not follow a standard format, but protein designations differ from genes by the absence of italics (e.g., STE2, Ste2, or Ste2p).

The key elements of the pheromone signal pathway are summarized in Fig. 2. This pathway is activated when haploid cells of opposite mating type (**a** and α) stimulate each other with secreted mating pheromones to undergo conjugation and form an **a**/α diploid cell (reviewed in Ref. 9). The MATa and MATα haploid cells primarily differ in the combination of receptor and pheromone they produce. MATa cells produce **a**-factor and the receptors for α-factor (STE2); MATα cells produce α-factor and the receptors for **a**-factor (STE3). A haploid cell must be selected for the functional expression of GPCRs because the **a**/α diploid cells are insensitive to pheromone. In practice, either MATa or MATα cells can be used because they activate the same postreceptor signal pathway. In our laboratory, we use MATa strains because α-factor pheromone, which is commercially available (Sigma, St. Louis, MO; Bachem, King of Prussia, PA), can be added to cells carrying the α-factor receptor gene as a convenient control for pathway function.

Ligand binding stimulates receptors to form an activated complex with a G protein that leads to the exchange of GDP for GTP on the α subunit (Gpa1p) and the subsequent release of the βγ complex (Ste4p, Ste18p) (reviewed in Ref. 9). The free βγ subunits then recruit Ste5p to the membrane. Ste5p, a scaffolding protein, brings with it the components of a MAP kinase cascade including a MAPKKK (Ste11p), a MAPKK (Ste7p) and a MAPK (Fus3p). This ultimately leads to the activation of the pheromone-responsive transcription factor (Ste12p) that induces genes such as FUS1. The FAR1 gene, which acts to promote pheromone-induced cell division arrest in G1, is also induced and then the Far1 protein is activated by MAPK phosphorylation.

There are several modifications that can be made to optimize the system and increase the coupling of heterologous GPCRs to the pheromone pathway. Starting at the beginning of the pathway, the endogenous receptor gene should be deleted from the chromosome. The endogenous pheromone receptors diminish the ability

[9] E. A. Elion, *Curr. Opin. Microbiol.* **3,** 573 (2000).

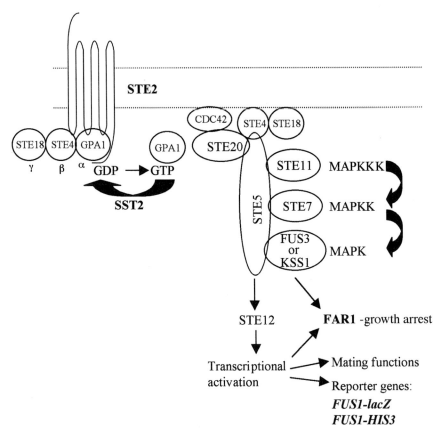

FIG. 2. Yeast pheromone signal transduction pathway. The key proteins known to be involved in the pheromone pathway are indicated as described in the text. Those protein indicated in bold are points of consideration when designing or choosing a yeast strain for these studies.

to detect signaling from foreign GPCRs,[1] apparently by sequestering G proteins into inactive preactivation complexes.[10] Next, the ability of most foreign receptors to signal in yeast can be facilitated by altering the Gα subunit to improve receptor coupling. In particular, substitution of the C terminal sequences of the pheromone-responsive Gα with sequences from mammalian Gα subunits has been successful.[11-13] The degree to which it will be necessary to improve signaling is variable as some mammalian GPCRs are able to activate the endogenous yeast G protein adequately.[7,14] The ability to detect signaling can be further enhanced by deleting the *SST2* adaptation gene, which encodes an RGS protein that promotes

[10] M. Dosil, K. Schandel, E. Gupta, D. D. Jenness, and J. B. Konopka, *Mol. Cell. Biol.* **20,** 5321 (2000).

TABLE II
MAMMALIAN GPCRs EXPRESSED IN YEAST

Receptor	Comments	Reference
Somatostatin (SSTR2)	Coupled to GPA1 and GPA1-$G\alpha_{i2}$	[11]
	Dose-dependent growth responses to SSTR2 subtype-selective agonists	
Adenosine (A_{2a})	Coupled to GPA1	[7]
	Detected agonist binding with cell membranes	
FPRL-1	Coupled to GPA1-$G\alpha_{i2}$	[12]
	Coexpressed plasmid library encoding small peptides to screen for ligands	
Melatonin (Mel_{1a})	Coupled to GPA1	[38]
	Determined order of potency known agonists	
Edg-2/Vzg-1	Coupled to GPA1	[14]
	Detected high specific binding to LPA	
C5a	Coupled to GPA1-$G\alpha_{i2}$	[32]
	Autocrine stimulation by coexpression of α-factor prepro/C5a ligand fusion protein	
Various receptors	Showed that chimeric $G\alpha$ subunits couple a broader range of mammalian GPCRs in yeast	[13]

GTP hydrolysis by the $G\alpha$ subunit.[15] The *FAR1* gene should also be deleted to prevent cell division arrest in response to activation of the pathway.[16] Finally, it is important to introduce sensitive reporter genes to detect signaling. Most reporter gene constructs make use of the promoter from the *FUS1* gene, which has a low basal level of transcription that increases dramatically with pheromone induction.[17] The *FUS1* promoter is commonly fused to reporter genes such as *lacZ* or *HIS3* to provide convenient assays for detection of receptor activation.[17,18]

Yeast strains containing many of these modifications can be obtained from investigators who have published successful expression studies (see Table II). In addition, most labs with an interest in the pheromone signal pathway could

[11] L. A. Price, E. M. Kajkowski, J. R. Hadcock, B. A. Ozenberger, and M. H. Pausch, *Mol. Cell. Biol.* **15,** 6188 (1995).

[12] C. Klein, J. I. Paul, K. Sauve, M. M. Schmidt, L. Arcangeli, J. Ransom, J. Trueheart, J. P. Manfredi, J. R. Broach, and A. J. Murphy, *Nature Biotechnol.* **16,** 1334 (1998).

[13] A. J. Brown, S. L. Dyos, M. S. Whiteway, J. H. White, M. A. Watson, M. Marzioch, J. J. Clare, D. J. Cousens, C. Paddon, C. Plumpton, M. A. Romanos, and S. J. Dowell, *Yeast* **16,** 11 (2000).

[14] J. R. Erickson, J. J. Wu, J. G. Goddard, G. Tigyi, K. Kawanishi, L. D. Tomei, and M. C. Kiefer, *J. Biol. Chem.* **273,** 1506 (1998).

[15] H. G. Dohlman and J. Thorner, *J. Biol. Chem.* **272,** 3871 (1997).

[16] F. Chang and I. Herskowitz, *Cell* **63,** 999 (1990).

[17] J. Trueheart, J. D. Boeke, and G. R. Fink, *Mol. Cell. Biol.* **7,** 2316 (1987).

[18] D. C. Hagen, G. McCaffrey, and G. F. Sprague, Jr., *Mol. Cell. Biol.* **11,** 2952 (1991).

probably provide a strain. In our experience, it is usually a good idea to try more than one strain as different levels of expression or signaling may be obtained. Two strains frequently used in our lab are JKY127-36-1 (*MATa sst2-1 bar1 ste2Δ mfa1::LEU2 mfα2::his5 far1 arg4 ade2 leu2 ura3 his3 mfa2::FUS1-lacZ*)[19] and PMY1. PMY1 is isogenic to JKY127 and differs only in that it is *SST2*⁺. Modified yeast strains can also be constructed using well-established methods including simple PCR strategies to introduce gene deletions and other mutations into the genome.[20]

Growth and Storage of Yeast

The manner in which yeast cells are grown and stored[21] is similar to the procedures used for *E. coli*. A rich medium (YPD) is used for general propagation of strains under nonselective conditions. Plasmids are maintained by growth of cells on chemically defined synthetic medium. Drug resistance markers such as Neo^R have been developed, but are less commonly used. Thus, most yeast laboratories make use of a variety of different plates lacking certain amino acids or other essential components such as uracil. The starting point for synthetic media is "Yeast Nitrogen Base," which consists of salts, vitamins, and ammonium sulfate as nitrogen source.[21] The medium is supplemented with amino acids and other appropriate constituents except for the one that will provide the selective pressure. Dextrose is used as an energy source for optimal growth unless the experiment involves the use of galactose to regulate inducible promoters. For laboratories in which the growth of yeast is not a common course of action, it may be advantageous to purchase media. In our experience, media purchased from Qbiogene (formerly Bio101, Carlsbad, CA) are comparable to our homemade stocks. Recipes for preparing commonly used media are given below.

Nonselective YPD Media

1. Dissolve 10 g yeast extract, 20 g Bacto-peptone, and 0.120 g adenine* in 900 ml distilled water with a magnetic stir bar in a 2 liter flask. If making solid medium, add 18 g Bacto-agar.
2. Autoclave for 30 min.
3. Add 100 ml of sterile 20% dextrose (w/v).

---

[19] P. Dube and J. B. Konopka, *Mol. Cell. Biol.* **18**, 7205 (1998).
[20] M. S. Longtine, A. McKenzie III, D. J. Demarini, N. G. Shah, A. Wach, A. Brachat, P. Philippsen, and J. R. Pringle, *Yeast* **14**, 953 (1998).
[21] F. Sherman, *Methods Enzymol.* **194**, 3 (1991).
 *Adenine is optional but should be added to suppress the formation of a toxic red pigment that accumulates in cells carrying certain mutations in the adenine biosynthetic pathway (e.g., *ade2*).

Selective Media

10× Media Supplement

1. Dissolve 0.2 g each of histidine, arginine, and methionine, 0.3 g each of tyrosine, isoleucine, and lysine, 0.4 g each of adenine sulfate, uracil, and tryptophan, 0.5 g phenylalanine, 0.6 g leucine, and 1.0 g each of glutamic acid and aspartic acid, 1.5 g valine, 2.0 g threonine, and 4.0 g serine in 1 liter distilled water.
2. Omit constituent(s) that will be used for selection.
3. Autoclave 100 ml aliquots for 30 min.

Selective Media (Dropout Media)

1. Dissolve 6.7 g Bacto yeast nitrogen base (e.g., Difco, Detroit, MI) in 800 ml distilled water. If making solid medium add 18 g Bacto-agar.
2. Autoclave 200 ml aliquots for 30 min.
3. Add 25 ml sterile 20% dextrose to cooled aliquots.
4. Add 25 ml of the appropriate 10× Media Supplemental solution (minus the appropriate additive).

Aminotriazole Plates (Selective Plates for FUS1-HIS3 Reporter Gene)

1. Prepare a filter-sterilized stock solution of 2.5 M 3(1,2,4)-aminotriazole (e.g., Sigma, St. Louis, MO).
2. Prepare liquid or solid medium lacking histidine as described above.
3. When liquid medium is cool (just prior to pouring solid medium plates), add 3(1,2,4)-aminotriazole from the stock solution. Aminotriazole is a competitive inhibitor of the *HIS3* gene product imidazoleglycerol-phosphate dehydratase. Concentrations ranging from 1 mM to 10 mM suppress growth due to basal level of *FUS1-HIS3* expression.

X-Gal Plates for β-Galactosidase Reporter Gene Detection

To perform a colorimetric assay for detection of the *FUS1-lacZ* (β-galactosidase) reporter gene, add X-Gal (5-bromo-4-chloro-3-indoyl-β-D-galactoside) (40 µg/ml final concentration) just prior to pouring, when agar is cool. Note that special medium buffered to pH 7 must be used to detect β-galactosidase activity.

1. Prepare 1 liter KPi solution pH 7 [1 M KH$_2$PO$_4$, 0.15 M (NH$_4$)$_2$SO$_4$, 0.75M KOH].
2. Prepare solid medium lacking the appropriate amino acid as described above.
3. When medium is cool (just prior to pouring plates), add 100 ml of the appropriate 10× Media Supplemental solution (minus the appropriate additive),

100 ml 20% dextrose, 100 ml KP$_i$, and 2 ml of a 20 mg/ml stock of X-Gal in dimethylformamide.

Storage of Yeast Cultures

Yeast can usually be stored at 4° for 1–3 months. Most strains will remain viable for even longer periods of time, especially on rich media, but there is the inadvertent danger of selecting for mutant strains on old plates. Yeast go into a deep resting state on long-term storage so it is strongly advised to restreak older cells onto a fresh plate to allow a period of logarithmic growth prior to commencing new experiments. To keep more permanent stocks of yeast for the laboratory, a fresh culture of yeast should be adjusted to contain 15% (v/v) glycerol and then frozen at −70°. Yeast stored in this way should be viable for many years and can be recovered at any time by scraping some of the cells with a sterile toothpick and transferring them to a fresh plate.

Transforming Plasmids into Yeast

The most common method for the introduction of foreign DNA into intact yeast cells is based on the lithium acetate method [22] as modified by Schiestl and Gietz.[23] This relatively simple method can yield up to 1×10^6 transformants/μg of DNA.[24] Note that the key steps that are absolutely required for a successful transformation include the addition of properly prepared single-stranded carrier DNA, the use of the proper-sized polymer of polyethylene glycol (PEG), and the heat shock at 42°. A handy Internet site that describes recent variations on these protocols can be consulted at: www.umanitoba.ca/faculties/medicine/units/human_genetics/gietz/.

Preparation of Single-Stranded Carrier DNA

1. Dissolve salmon testis DNA (Sigma) in TE (10 mM Tris-HCl pH 8.0, 1 mM EDTA) to 2 mg/ml by mixing on a magnetic stirrer overnight at 4° until fully dissolved.
2. Sonicate to break up DNA. Check the size of the resulting DNA on an agarose gel (fragments ranging from 2 to 15 kb are desired). Aliquot DNA into tubes and store at −20°.
3. Prior to use, boil an aliquot for 5 min and then cool in an ice–water bath. If kept frozen or on ice it is not necessary to boil before each use.

---

[22] H. Ito, Y. Kukuda, K. Murate, and A. Kimura, *J. Bacteriol.* **153,** 163 (1983).
[23] R. H. Schiestl and R. D. Gietz, *Curr. Genet.* **16,** 339 (1989).
[24] R. D. Gietz, R. H. Schiestl, A. R. Willems, and R. A. Woods, *Yeast* **11,** 355 (1995).

Transformation Protocol

Special Reagents

LiAc/TE (10 m*M* Tris, pH 7.5; 1 m*M* EDTA; 100 m*M* lithium acetate, pH 7.5) PEG/lithium acetate/TE (10 m*M* Tris, pH 7.5; 1 m*M* EDTA; 100 m*M* lithium acetate, pH 7.5; 40% PEG 3350 (e.g., Sigma)

Procedure

1. Use a fresh overnight culture to inoculate 50 ml YPD to a cell density of 2.5×10^6 cells/ml ($OD_{660} \sim 0.15$).
2. Grow for $\sim$4 hr at 30°, with shaking, until cells reach $5–10 \times 10^6$ cells/ml ($OD_{660} = 0.4–0.7$).
3. Harvest cells by centrifugation at 3000*g* for 5 min.
4. Wash cell pellet once with sterile water, once with LiAc/TE and then resuspend the final cell pellet in 0.5 ml LiAc/TE (1/100 volume).
5. Add 50 μl of the yeast cell suspension to a tube containing 2 μl (200 μg) carrier DNA and 2–4 μl transforming DNA (up to 5 μg) and vortex. Keep the volume of added DNA to a minimum to prevent dilution of the other reagents.
6. Add 0.4 ml PEG/LiAc/TE and vortex again.
7. Incubate at 30° for 30 min with occasional mixing.
8. Heat shock for 15 min at 42°.
9. Add 1 ml selective media or water, mix and centrifuge 10 sec.
10. Resuspend pellet in 0.2 ml selective medium and plate on selective agar plate.
11. Incubate at 30° for 2–3 days to allow colonies to form.

Analysis of Receptor Protein Production in Whole Cell Extracts

Western blot analysis is probably the most convenient method to detect receptor proteins produced in yeast. The preparation of yeast cell extracts requires either enzymatic or mechanical disruption of the cell wall. For most purposes, mechanical disruption by vortexing cells with glass beads is the simplest and will be described here. If whole cell extracts do not give a sufficient signal, added sensitivity can be gained by analyzing a crude membrane fraction that can be prepared quickly as described below. Methods to separate the protein extract by polyacrylamide gel electrophoresis and transfer to nitrocellulose are common practices that are described in detail elsewhere.[25]

---

[25] E. Harlow and D. Lane, "Antibodies, a Laboratory Manual." Cold Spring Harbor Laboratory Press, Cold Spring Harbor, NY, 1988

Special Reagents

1 M NaN$_3$ and 1 M KF

Acid-treated 450 μm glass beads (e.g., Sigma)

PAGE sample buffer (8 M urea, 0.3% SDS, 25 mM Tris pH 6.8, 0.01% bromphenol blue, 0.01% xylene cyanol)

Procedure

1. Place all buffers on ice and cool centrifuges and tubes.
2. Grow a 50 ml culture of yeast to a density of about 1 × 10^7 cells/ml (OD$_{660}$ = 0.5–1). Either YPD or synthetic medium can be used.
3. Transfer to a 50 ml conical tube and harvest cells by centrifugation at about 1000g for 5 min.
4. Wash cell pellet with 10 ml sterile water. Vortex to mix and centrifuge 5 min.
5. Add 1 ml sterile water to pellet and transfer to 1.5 ml microfuge tube.
6. Centrifuge at 14,000g for 30 sec in a microfuge and then remove the supernatant. (Cells may be stored at −70° at this point or placed on ice until needed.)
7. Quickly resuspend cells in 100 μl 2× gel sample buffer per 10^8 cells. Add an equal volume of glass beads. Keep the final volume below 0.5 ml to prevent overcrowding or switch to a larger tube. *Note:* Do not overcrowd the tube or it will not be possible to get good mixing action.
8. Vortex 4 times for 1 min each. Cool on ice between vortexing steps.
9. Warm tubes at 37° for 10 min prior to loading gel. Centrifuge at top speed in a microfuge for 3 min prior to loading.
10. Do not boil GPCR samples as they will aggregate.
11. Perform gel electrophoresis and Western blot according to standard procedures.

Analysis of Receptor Production in Crude Membrane Fractions

1. Grow cells and harvest cell pellet in a microfuge tube as described above.
2. Suspend 5 × 10^8 cells in 250 μl TE/PP (50 mM Tris-HCl pH 7.5, 1 mM EDTA, 100 μg/ml PMSF, 2 μg/ml pepstatin A) and add 250 μl glass beads.
3. Lyse the cells by vortexing at high speed for 1 min followed by incubation on ice for 1 min. Repeat three times.
4. Transfer cell extract to fresh tube (leaving behind the glass beads).
5. Centrifuge at 330g for 5 min at 4° to remove unbroken cells.
6. Transfer supernatant to a new tube. Centrifuge at top speed for 15 min at 4° in a microfuge to pellet membranes. Carefully remove supernatant.
7. The crude membrane pellet can be used immediately or stored at −70°.

8. Suspend crude membrane pellet in 100 μl 2× PAGE loading buffer. Warm tubes at 37° for 10 min prior to loading gel. Centrifuge at top speed in a microfuge for 3 min prior to loading.
9. Do not boil GPCR samples as they will aggregate.
10. Perform gel electrophoresis and Western blot according to standard procedures.

Gel Shift Assays for Posttranslational Modifications

The yeast α-factor receptor undergoes at least three forms of posttranslational modification: N-linked glycosylation, phosphorylation, and ubiquitination. Protocols will be described for the analysis of some of these modifications as they could have significant effects on the function of heterologous GPCRs in yeast. For example, glycoslyation on nonnative sites could have deleterious effects. Phosphorylation of serine and threonine residues located at the C terminus of many GPCRs, including the yeast α-factor receptor,[26] promotes negative regulation of signaling. Furthermore, phosphorylated receptors in yeast are subsequently modified by ubiquitination, which promotes internalization of receptors by endocytosis and their subsequent degradation in the vacuole (yeast lysosome).[27]

The addition of oligosaccharide and phosphate groups to GPCRs can be assessed by a number of means. However, gel mobility assays are the most convenient and will be described in detail here. As shown in the Western blot in Fig. 3, these methods detect the loss of high molecular weight bands from samples that were treated with endoglycosidase H. Note that in contrast to mammalian cells, N-linked glycosylation remains endoglycosidase H sensitive in yeast. Phosphatase treatment can be used in a similar manner to analyze phosphorylation. Other approaches include the use of glycosylation inhibitors such as tunicamycin, or radiolabeling of receptors with $^{32}PO_4$.[26] Ubiquitination can be detected with anti-ubiquitin antibodies or by coexpression of an epitope-tagged ubiquitin construct.

Procedure

1. Harvest cells and prepare crude membrane extracts as described previously. Suspend membrane pellet in 100 μl TE.
2. Treat membranes with either endo H or λ protein phosphatase.
   (a) Incubate 10 μl protein with 3000 units endo H (New England Biolabs, Beverly, MA) in endo H buffer (100 mM NaCl; 50 mM Tris pH 7.5; 1 mM EDTA) at 37°. The amount of time will need to be determined as it may vary for different proteins.

[26] Q. Chen and J. B. Konopka, *Mol. Cell. Biol.* **16**, 247 (1996).
[27] L. Hicke, B. Zanolari, and H. Riezman, *J. Cell Biol.* **141**, 349 (1998).

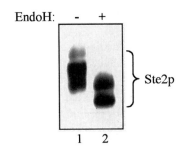

FIG. 3. Analysis of GPCR modification. Western blot analysis showing gel mobility shift of the α-factor receptor after treatment with endoglycosidase H to remove N-linked glycosylation.

      (b) Incubate 10 μl protein with 100–200 units λ protein phosphatase (New England Biolabs) in λPP buffer (50 mM Tris-HCl pH 7.5; 0.1 mM EDTA; 5 mM dithiothreitol (DTT); 0.1% Brij 35; 2 mM MnCl$_2$) at 30° for 1 hr.

  3. Add PAGE loading buffer and run Western blot as previously described.

Subcellular Localization of GFP-Tagged Receptors

The *Aequorea victoria* green fluorescent protein (GFP) is a useful visual marker to monitor gene expression and protein localization.[28] Fusion of GFP to the C terminus of the α-factor receptor results in a receptor protein that displays wild-type functional properties, yet can be detected by fluorescence microscopy. An example of this is shown in Fig. 4. Fluorescence is seen as a prominent ring around the cell, demonstrating that the protein is stable and properly transported to the plasma membrane. Mutants that do not localize properly or receptors that are down-regulated by endocytosis are generally found in punctate-looking intracellular compartments.[29] Thus, fusion of GFP to other GPCRs expressed in yeast should provide a rapid method to assess their production and membrane localization. To obtain the best results, a version of GFP that is enhanced for brightness and whose codons are optimized for expression in yeast codons should be used.[30]

Ligand Binding Assay

The following protocol for the binding of α-factor pheromone to whole yeast cells is adapted from the procedure developed by Jenness and colleagues.[31] It can

[28] R. Y. Tsien, *Annu. Rev. Biochem.* **67,** 509 (1998).

[29] Y. Li, T. Kane, C. Tipper, P. Spatrick, and D. D. Jenness, *Mol. Cell. Biol.* **19,** 3588 (1999).

[30] B. P. Cormack, G. Bertram, M. Egerton, N. A. Gow, S. Falkow, and A. J. Brown, *Microbiology* **143,** 303 (1997).

[31] D. D. Jenness, A. C. Burkholder, and L. H. Hartwell, *Cell* **35,** 521 (1983).

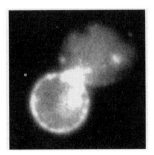

FIG. 4. Receptor localization. A GFP tag fused to the C terminus of the α-factor receptor was used to visualize receptor localization by fluorescence microscopy. Note the fluorescence at the periphery of the cell indicating plasma membrane localization as well as punctate clusters of fluorescence indicative of intracellular pools of receptor.

be modified for testing the ability of other small ligands to bind to cell surface receptors in yeast. However, if the ligand under study is too large and is blocked by the yeast cell wall, or displays a high degree of nonspecific binding to the cell wall, the procedure can be easily modified to assay the binding of ligand to membrane fractions that are collected in Whatman (Clifton, NJ) GF/F filters.[10]

Special Reagents

   IM (inhibitor medium) (YPD medium with 10 mM NaN$_3$ and 10 mM KF and
      filtered through a 0.2 sterile filter)
   IM + 2× TAME (IM containing 20 mM TAME, p-tosyl-L-arginine methyl
      ester) Whatman GF/C filters

Procedure

   1. Inoculate yeast into 200 ml medium and grow at 30° to log phase (0.5–1 ×
      10^7 cells/ml).
   2. Inhibit receptor endocytosis by adjusting medium to 10 mM NaN$_3$ and
      10 mM KF.
   3. Harvest cells in a cold tube already containing by centrifugation at 4° for
      10 min at 1500g and then suspend cell pellet in 40 ml of ice-cold IM. Repeat.
   4. Centrifuge cells and suspend in IM to a final volume of 1 ml.
   5. Vortex and then dilute to 1 × 10^9 cells/ml in IM. Keep on ice. (Cell concen-
      tration can be determined using a hemacytometer.)
   6. Prepare a dilution series for the ligand. For α-factor use a range from at
      least 2 nM to 200 nM (K_d = 5 nM) of ^{35}S-labeled α-factor (50 Ci/mmol)
      in IM + 2× TAME. Make enough to allow for duplicate samples and label
      these tubes as "hot."
   7. Add cold α-factor at 100× the concentration of each hot dilution to the
      tubes labeled "cold."

8. For each α-factor concentration label four 0.5 ml microfuge tubes (e.g., a,b,c,d). Add 50 μl of the cells to each microfuge tube.

9. Start binding reactions by adding 50 μl of the "hot" α-factor to tubes a and b, and add 50 μL of the corresponding "cold" α-factor to tubes c and d.

10. α-Factor has a rapid off rate. Therefore, either use a multiwelled cell harvester that can process many samples quickly, or else stagger each reaction to provide time to wash samples.

11. After 30 min, filter 90 μl of the sample in tube 1a onto a GF/C filter wetted with IM-TAME, wash twice with 2 ml IM-TAME, and then allow to dry. Continue with other samples.

12. Quantify the amount of radioactivity bound to the cells in a scintillation counter.

FUS1-HIS3 Reporter Gene Assay for Receptor Signaling

The *FUS1-HIS3* reporter gene provides a relatively simple way to assay the ability of receptors to activate G protein signaling in yeast. Activation of the pathway enables cells carrying a *FUS1-HIS3* reporter gene to grow on media lacking histidine (provided a *FUS1-HIS3 his3 far1* mutant strain is used). The relative ability of a GPCR to induce signaling is determined by assaying the ability of the cells to grow on media containing different concentrations of 3-aminotriazole, an inhibitor of *HIS3* function. Thus, this convenient assay has been used analyze ligand specificity[7] and to identify novel ligands.[12] Although convenient, this assay does require several days to observe the difference in growth. Therefore, as an alternative, induction of the *FUS1-lacZ* reporter gene can be monitored after a short induction by assaying for the *lacZ*-encoded β-galactosidase as will be described below.

One potential limitation of this assay is that the ligand must be small enough to pass through the cell wall to reach the plasma membrane. In general, ligands smaller than about 20–30 kDa may be able to get through the cell wall. Alternatively, peptide ligands can be coexpressed to stimulate signaling in an autocrine manner as has been achieved for the C5a receptor.[32] Another concern is that it may be necessary to adjust the pH of the medium in order for some ligands to bind efficiently since yeast medium is generally pH 4–5. The medium can be buffered to maintain a higher pH, but yeast do not grow well above pH 7.

Several variations for the analysis of ligand-induced activation of the *FUS1-HIS3* have been described.[1] The general outline of these approaches are as follows:

---

[32] T. J. Baranski, P. Herzmark, O. Lichtarge, B. O. Gerber, J. Trueheart, E. C. Meng, T. Iiri, S. P. Sheikh, and H. R. Bourne, *J. Biol. Chem.* **274,** 15757 (1999).

1. Assay the ability of cells to grow on solid medium agar plates containing different concentrations of ligand and 3-aminotriazole. Prepare a series of plates containing different concentrations of 3-aminotriazole (try a range from 1 mM to 50 mM in initial studies) using recipes described above in the Cell Growth section. Streak the cells on the plate and compare the growth rate of single colonies. For more quantitative results spot different dilutions of cells on the plates.
2. To simplify the number of plates, spread a lawn of about 10^6 cells on an agar plate and then place filter disks containing different concentrations of ligand on the plate. Diffusion of the ligand should result in a zone of growth that is proportional to the sensitivity to ligand.
3. Another variation is to carry out the above assays in liquid medium using a 96-well plate.

FUS1-lacZ Reporter Gene Assay for Receptor Signaling

This protocol uses a colorimetric substrate, ONPG, to assay β-galactosidase activity[33] as modified for use with permeabilized yeast cells. Wild-type *S. cerevisiae* yeast cells ordinarily do not produce β-galactosidase. Thus, reporter genes that make use of the *E. coli* β-galactosidase gene (*lacZ*) provide a sensitive and convenient indicator of gene expression, an example of which is shown in Fig. 5. In addition to assaying β-galactosidase in liquid cultures, another colorimetric substrate (X-Gal) can be incorporated into solid medium plates for detection of signaling using media described in the section on Cell Growth.

Special Reagents

Z buffer (10 mM KCl, 1 mM MgSO$_4$, 60 mM Na$_2$HPO$_4 \cdot$ 7H$_2$O, 40 mM NaH$_2$PO$_4 \cdot$ H$_2$O, 50 mM 2-mercaptoethanol). (*Note:* add 2-mercapto-ethanol just prior to use.)
ONPG (4 mg/ml *o*-nitrophenyl-β-D-galactopyranoside in 0.1 M phosphate buffer pH 7.0)

Procedure

1. Grow cells overnight at 30° so that the cells remain in log phase (i.e., OD$_{660}$ < 0.7; less than 10^7 cells/ml). To ensure that the basal level remains at a consistently low level, cells should be kept growing for 2 days before the assay is performed.

---

[33] J. H. Miller, *in* "Experiments in Molecular Genetics," p. 325. Cold Spring Harbor Laboratory Press, Cold Spring Harbor, NY, 1972.

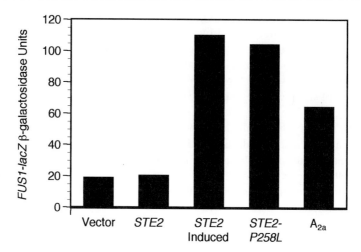

FIG. 5. Analysis of *FUS1-lacZ* reporter gene. β-Galactosidase units indicating the level of *FUS1-lacZ* in cells carrying either the empty vector, wild-type receptor gene (*STE2*), a constitutively active α-factor receptor mutant (*ste2-P258L*), or the adenosine A_{2a} receptor. For comparison of basal and induced states, α-factor was added at 10^{-7} M final concentration to the wild-type *STE2* cells for 2 hr.

2. Adjust the culture to 2.5×10^{6} cells/ml in 2 ml and then add α factor (1×10^{-7} M) and incubate for 2 hr. A longer incubation can be used for greater sensitivity.

3. Place the cells on ice. For careful measurements add cycloheximide to 10 μg/ml.

4. Measure the OD_{600} of the cells.

5. Add 0.1 ml of the cell culture to a 1.5 ml tube containing 0.7 ml Z buffer with 2-mercaptoethanol added.

6. Add 50 μl chloroform and 50 μl 0.1% SDS. Vortex for 30 sec.

7. Add 0.16 ml ONPG and mix by vortexing.

8. Incubate at 37° for 1 hr or until the reactions turn visibly yellow.

9. Quench the reactions by adding 0.4 ml 1 M Na$_2$CO$_3$.

10. Spin tubes for 10 min to remove debris and then read OD_{420}.

11. Calculate the β-galactosidase units using the formula:

$$\text{Units} = (1000\,OD_{420})/(t\,V\,OD_{600})$$

where t is time of incubation in minutes and V is volume of cell culture added to Z buffer in ml.

Genetic Strategies for Identifying Mutant Receptors

Once functional expression of a GPCR in yeast has been achieved, the experimental accessibility of yeast can be used to screen or select for new mutant

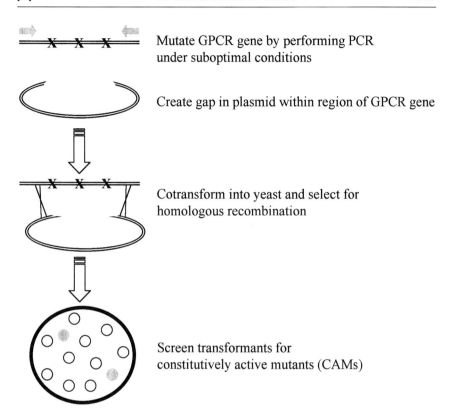

FIG. 6. Gap repair mutagenesis. PCR amplification with *Taq* polymerase under suboptimal conditions is used to introduce random mutations within the GPCR gene. The PCR products are then transformed into yeast along with a GPCR plasmid that has been linearized within the same region. Homologous recombination that occurs between the plasmid and the PCR product repairs the gap and introduces mutations into the targeted region. Yeast carrying a plasmid are selected for on selective media. Constitutive mutants can be identified as blue colonies on an X-Gal indicator plate due to ligand-independent activation of the *FUS1-lacZ* reporter gene.

phenotypes to help identify functional domains of receptors. For example, we have developed rapid screening techniques for constitutively active[34] and dominant-negative mutants[35] in the α-factor receptor. Here, we will describe a "gap repair" method for introducing random mutations into a yeast plasmid that can then be screened for constitutive mutants. The basis of this technique is outlined in Fig. 6. Basically, a region within the gene of interest is mutated by performing PCR under suboptimal conditions using error-prone *Taq* polymerase. This fragment is then cotransformed into yeast along with a gapped plasmid whose ends are homologous

[34] J. B. Konopka, M. Margarit, and P. Dube, *Proc. Natl. Acad. Sci. U.S.A.* **93**, 6764 (1996).
[35] M. Dosil, L. Giot, C. Davis, and J. B. Konopka, *Mol. Cell. Biol.* **18**, 5981 (1998).

to those of the PCR DNA. The gap in the plasmid is then repaired by homologous recombination with the ends of the PCR product.[36]

Procedure

1. PCR amplify the region of interest using *Taq* polymerase lacking proofreading function, but limit one nucleotide by using one-fifth the normal concentration.
2. Amplify using a standard protocol, such as 94°, 1 min; 55°, 1 min; 72°, 1 min per kb, 30 cycles; 72°, 10 min.
3. Create a gap within the region being mutagenized by digesting the plasmid with unique restriction sites. There should be at least 50 bp of homology between the PCR product and the ends of the gap on the plasmid for optimal efficiency.
4. Cotransform the purified PCR fragment and the gapped plasmid, at a molar ratio of at least 3 : 1, by the lithium acetate method.
5. Plate transformations on selective medium.
6. Incubate at 30° for 2–3 days or until colonies are visible. (A control transformation of gapped plasmid alone should give only a few colonies.)
7. Replica-plate colonies to indicator plates containing X-Gal (see Cell Growth section). Incubate at 30° for 3–5 days. Constitutive mutants will be identified as blue colonies.
8. Recover the plasmid using a previously described procedure.[37] Basically, the cell pellet from 1.5 ml of liquid culture is resuspended in 200 μl lysis buffer (2% Triton X-100, 1% SDS, 100 mM NaCl, 10 mM Tris pH 8.0, 1 mM EDTA) and then mixed with 200 μl phenol/chloroform. Add 200 μl glass beads and vortex for 2 min. Spin 5 min in a microfuge. One microliter of the supernatant can be used to transform *E. coli*. The efficiency is low because of an inhibitor in the yeast extract.
9. Retransform yeast to confirm that the observed phenotype is plasmid-dependent.

Future Directions

The techniques described in this section have been successfully used to express a variety of different mammalian GPCRs in yeast and to detect their ability to activate the pheromone-responsive MAP kinase cascade.[7,11,38] Unfortunately, the function of some receptors is not readily detectable in yeast. However, because

---

[36] S. Kunes, H. Ma, K. Overbye, M. S. Fox, and D. Botstein, *Genetics* **115,** 73 (1987).
[37] C. S. Hoffman and F. Winston, *Gene* **57,** 267 (1987).
[38] T. Kokkola, M. A. Watson, J. White, S. Dowell, S. M. Foord, and J. T. Laitinen, *Biochem. Biophys. Res. Commun.* **249,** 531 (1998).

of the genetic advantages of yeast, multiple options are available to circumvent this problem. One strategy to improve coupling is to replace the endogenous yeast $G\alpha$ (GPA1) with a chimeric G_α protein containing sequences from both *GPA1* and various mammalian G_α subunits. These studies have revealed that replacing the C-terminal tail of the yeast G_α with as few as five amino acids of a mammalian G_α greatly improved signaling.[13,39,40] Fusion of the G_α subunit directly to the C terminus of the receptor may help to further improve signaling.[41] Ultimately, it will be interesting to develop yeast strains in which entire signal pathways are expressed to reconstitute a homologous pathway. Other possibilities for improving the function of GPCRs in yeast include the coexpression of a second protein. For example, some GPCRs may require a coreceptor for heterooligomerization, an accessory protein such as a RAMP, or a chaperone to promote proper folding. Strategies could be developed to coexpress cDNA libraries in yeast to identify proteins that enhance receptor function.

Although it is desirable to optimize the detection of signaling by heterologous receptors in yeast, partial ability to activate the pheromone pathway will be sufficient for most purposes. The pheromone-responsive reporter genes are induced about 100-fold during maximum stimulation so it only takes a small degree of activation to provide a sufficient signal to carry out structure–function studies on receptors, and to identify agonists and antagonists. In addition, it is possible to use yeast as an expression system for other applications such as the analysis of orphan receptors or to search for mutant receptor alleles associated with human diseases.

Acknowledgments

We thank Bill Parrish for helpful comments on the manuscript and we also thank the members of our laboratory for contributing protocols and suggestions. P.M. was supported in part by a training grant (T32 CA09176) from the National Cancer Institute. Our research was supported by National Institutes of Health Grant GM55107 awarded to J.B.K.

[39] N. S. Olesnicky, A. J. Brown, S. J. Dowell, and L. A. Casselton, *EMBO J.* **18,** 2756 (1999).
[40] P. Coward, S. D. Chan, H. G. Wada, G. M. Humphries, and B. R. Conklin, *Anal. Biochem.* **270,** 242 (1999).
[41] R. Medici, E. Bianchi, S. G. Di, and G. P. Tocchini-Valentini, *EMBO J.* **16,** 7241 (1997).

[8] Role of G Protein $\beta\gamma$ Complex in Receptor–G Protein Interaction

By INAKI AZPIAZU and N. GAUTAM

Introduction

The activation of a G protein by a receptor is the first step and among the most critical in modulating signal transduction. The molecular mechanisms that underlie this crucial interaction are still little understood. Although it is clear that the $\beta\gamma$ complex is a requirement for this interaction, the role of this subunit complex in receptor activation of a G protein is not known. To study the function of the $\beta\gamma$ complex at the receptor surface two different approaches have been used. (i) Peptides specific to the G protein $\beta\gamma$ complex have been used to identify domains on the complex that contact a receptor.[1,2] (ii) Purified G protein $\beta\gamma$ subunits have been used to examine the effect of different subunit types and mutants on coupling of G protein to receptor.[3,4] Both approaches can provide information about the mechanistic basis of G protein activation by receptors.

Receptors for these studies can be obtained from appropriate mammalian tissue or obtained through heterologous expression. Membranes from tissue cannot be used directly because of the low level of expression of the endogenous receptors. In contrast, membranes from cell lines overexpressing a receptor are better sources of receptors for receptor–G protein interaction assays since the concentration of the receptor is significantly higher in these membranes. If necessary, the receptor can also be purified from these membranes and reconstituted into artificial membranes. The use of purified receptor and G protein subunits in a reconstituted membrane system allows us to perform assays in which the relative concentrations of the proteins can be manipulated carefully and the kinetics of the resultant activity determined precisely in the absence of potentially confounding influence of other proteins endogenous to the cell line used for expression.

Preparation of M2 Receptor-Containing Membranes

CHO Cell Culture and Membrane Purification

CHO (Chinese hamster ovary) cells expressing muscarinic receptors were developed by Dr. E. G. Peralta and have been successfully used in studies of G-protein

---

[1] O. G. Kisselev, M. V. Ermolaeva, and N. Gautam, *J. Biol. Chem.* **269**, 21399 (1994).

[2] I. Azpiazu, H. Cruzblanca, P. Li, M. Linder, M. Zhuo, and N. Gautam, *J. Biol. Chem.* **274**, 35305 (1999).

and receptor interaction.[2,5,6] Cells are thawed from a liquid nitrogen frozen stock and seeded. CHO cells expressing M2 are grown in Dulbecco's Modified Eagle's (DME)/F12 media containing 10% dialyzed fetal bovine serum, 2 mM glutamine, 6 μg/ml penicillin, 10 μg/ml streptomycin, and 50 nM methotrexate in an atmosphere of 5% (v/v) CO_2 in air [DME/F12, glutamine, penicillin, and streptomycin were from Tissue Culture Center (Washington University, St. Louis, MO)]. The cultured cells are grown in monolayer mode in 150 mm tissue culture plates, expanded in four passages to 40–50 plates, and harvested close to confluence. While passaging and for harvesting, cells are lifted off the plates with phosphate-buffered saline (PBS) supplemented with 0.02% EDTA. After harvesting, cells are centrifuged and pellets frozen by immersion in liquid nitrogen. Levels of expressed M2 receptor are about 400,000 receptor per cell.

For membrane preparation, cells are thawed at 37°, resuspended in hypotonic buffer containing 20 mM HEPES pH 7.4, 1 mM EDTA, 2 mM MgCl$_2$, and a cocktail of protease inhibitors containing 10 μg/ml each of leupeptin, aprotinin, antipain, 1 mM benzamidine, and 0.2 mM phenylmethylsulfonyl fluoride (PMSF). Unless otherwise noted, all chemicals are available from Sigma-Aldrich, St. Louis, MO). Cells are freeze-thawed in liquid nitrogen, homogenates are centrifuged at low speed, and the supernatant is centrifuged again at 100,000g. High-speed pellets are resuspended in the same buffer and membranes are frozen at −85° until further use.

Uncoupling of Muscarinic Receptors

Muscarinic receptors expressed in heterologous systems, such as CHO cells, are coupled to endogenous G proteins. In assays of G protein interaction with the receptor, the receptor-containing membranes need to be free of endogenous G proteins. To obtain such G-protein-uncoupled receptors, membranes are treated with the same buffer containing 5 M urea and 100 μM GTPγS at 4° for 60 min (GTPγS is from Calbiochem, San Diego, CA). Membranes are washed twice, centrifuged, and resuspended in buffer. Immunoblot analysis with antibodies specific to the β1 subunit[7] is used to measure the level of contamination with endogenous G proteins. This procedure decreases the concentration of β subunit in membrane significantly. The amount of [^{3}H] N-methyl scopolamine ([^{3}H]NMS) binding to the membranes is measured by mixing serially diluted membrane stocks with a saturating concentration of [^{3}H]NMS (10 nM, final) in the membrane buffer. After

[3] O. Kisselev and N. Gautam, *J. Biol. Chem.* **268**, 24519 (1993).

[4] C. S. Myung, H. Yasuda, W. W. Liu, T. K. Harden, and J. C. Garrison, *J. Biol. Chem.* **274**, 16595 (1999).

[5] E. G. Peralta, J. W. Winslow, G. L. Peterson, D. H. Smith, A. Ashkenazi, J. Ramachandran, M. I. Schimerlik, and D. J. Capon, *Science* **236**, 600 (1987).

[6] M. L. Dell'Acqua, R. C. Carroll, and E. G. Peralta, *J. Biol. Chem.* **268**, 5676 (1993).

[7] A. N. Pronin and N. Gautam, *Methods Enzymol.* **237**, 482 (1994).

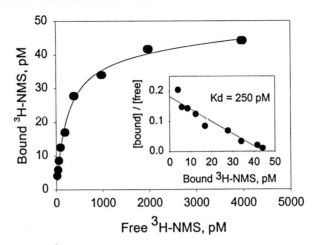

FIG. 1. Saturation of [^{3}H]NMS binding sites. Membrane preparations of M2 receptors were incubated for 60 min at room temperature with [^{3}H]NMS in 500 μl of buffer containing 20 mM sodium phosphate and 5 mM MgCl$_2$ ([^{3}H]NMS is available from Amersham-Pharmacia, Piscataway, NJ). Nonspecific activity was measured in the presence of 10 μM atropine. Samples were transferred onto Whatman glass filters (GF/B) and washed with cold saline phosphate buffer, and radioactivity was measured by scintillation counting.

60 min of incubation at room temperature, the reaction mixtures are filter-washed with ice-cold phosphate-buffered saline (PBS). Filtration is performed through GF/B glass microfiber filters (Whatman, Clifton, NJ) and radioactivity in filters is counted after suspending in scintillation liquid.

The membrane receptor bound radiolabeled [^{3}H]NMS with an affinity (Fig. 1) similar to that published earlier.[8]

We usually get about 0.2–0.4 nmol of M2 ([^{3}H]NMS binding protein) from 50 plates of M2-CHO cells with a concentration of 0.1 nmol M2/mg of membrane protein.

Purification and Reconstitution of M2

Expression of His-m2 Receptor in Sf9 Cells

Sf9 (*Spodoptera frugiperda* ovary) insect cells (Pharmingen, San Diego, CA) are grown in suspension in a shaking incubator (Innova 4000, New Brunswick Scientific, Edison, NJ) in IPL-41 medium (Gibco-Life Technologies, Rockville, MD) supplemented with 10% fetal bovine serum, 0.1% Pluronic F68 (Gibco-Life Technologies), gentamicin (50 μg/ml) and Fungizone (50 μg/ml) (Fungizone and gentamicin were from Gibco or the Tissue Culture Center, Washington University).

[8] E. C. Hulme, N. J. Birdsall, and N. J. Buckley, *Annu. Rev. Pharmacol. Toxicol.* **30**, 633 (1990).

Serum quality is critical for healthy growth of cells. We obtain serum from Atlanta Biologicals (Atlanta, GA). Protocols for Sf9 cell culture and baculovirus expression are available as manuals,[9] or from commercial sources for baculovirus vectors (e.g., Gibco-Life Technologies, Pharmingen, San Diego, CA; Invitrogen, Carlsbad, CA).

Cells are cultured in suspension in Bellco glass conical flasks at 27°. Usually 1–2 liters of culture at densities between 1 and 1.5 million cells/ml are infected with baculovirus. The ratio of virus to cell is 3 : 1. Infected cells are harvested 48–60 hr later and resuspended in 50–100 ml of lysis buffer (50 mM HEPES, pH 7, 100 mM NaCl plus a cocktail of protease inhibitors : 1mM benzamidin, 0.2 mM phenylmethylsulfonyl fluoride, and 10 μg/ml each of antipapain, leupeptin, aprotinin—the first two being added fresh). Cells are broken by nitrogen cavitation in a Parr high-pressure chamber (40 ml) at 4° (Parr Instrument Company, Moline, NJ). Cell disruption is confirmed by optical microscopy. Nuclei and cells are spun down at 500g for 10 min at 4° and the supernatant is spun down at 200,000g which results in a compact microsomal pellet. The latter is resuspended in lysis buffer and spun down again (200,000g) to 45 min at 4°. Resuspended membranes carry the functional receptor. Muscarinic receptor yield is determined by [^{3}H]NMS binding as described before.

We obtain 1–2 nmol [^{3}H]NMS binding sites per liter of cell culture. This translates to 60–120 μg of receptor protein per liter of culture. Receptor containing membranes are stored by freezing in liquid nitrogen.

His-M2 Receptor Purification

This protocol is a modified version of a previously described procedure.[10] All operations are performed at 4°.

Preparation of Metal-Chelate Beads. Cobalt-chelate beads are prepared by mixing 500 μl (for each 100 μg of M2) of iminodiacetic acid beads (Sigma, St. Louis, MO) with 500 μl of 200 mM cobalt chloride (Sigma). The mixture is shaken well for a few seconds and washed at least five times with 10 volumes of water. Solubilization buffer is used for the last wash.

Solubilization. Receptor containing membrane stock from above is diluted down to 5 mg protein/ml in solubilization buffer (50 mM HEPES, pH 7, 50 mM NaCl containing protease inhibitors). A solution of digitonin (Calbiochem)/sodium cholate is added to 1/0.5% final concentration. We use a stock solution of 5/2.5% detergent. The solution is incubated in a rotary shaker at 4° for at least 1 hr and then spun at 200,000g for 30 min. The supernatant is recovered.

Solubilized supernatant is diluted with 4 volumes of the solubilization buffer containing 625 mM NaCl and 0.1/0.05% of digitonin/sodium cholate solution.

[9] D. R. O'Reilly, L. K. Miller, and V. A. Luckow, "Baculovirus Expression Vectors: Laboratory Manuals." W. H. Freeman and Company, New York, 1992.

[10] M. K. Hayashi and T. Haga, *J. Biochem. (Tokyo)* **120**, 1232 (1996).

Cobalt-beads   ABT-beads

M   S   U   B   U   B

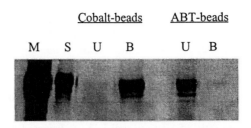

FIG. 2. Immunoblot probed with peptide directed M2 specific antibody (Chemicon International, Temecula, CA). Sf9 cell membranes expressing His-M2 (M); solubilized digitonin/cholate supernatant (S), unbound (U), and bound (B) fractions from cobalt beads. Bound (B) and unbound (U) fractions from ABT beads.

Five hundred microliters of cobalt beads is added and batch incubated on a rotary shaker at least 1 hr at 4°. The mixture is poured into a spin column. Beads are washed with 10 volumes of solubilization buffer containing 500 mM NaCl and 0.1/0.05% digitonin/cholate. His-M2 is eluted with 2–3 volumes of the buffer (solubilized buffer including NaCl and detergent) containing 200 mM imidazole. Solubilization with digitonin is usually complete and the eluted His-M2 retains 50% of the initial [³H]NMS binding activity. Binding is determined by mixing an aliquot of the eluate with [³H]NMS, adding the mixture to a 5 ml Sephadex G-50 (fine) gel that has been equilibrated with solubilization buffer containing 0.1/0.05% digitonin/cholate. [³H]NMS present in the exclusion volume is measured. Purified soluble receptors are stable at 4° for at least 12 hr.[11] However, we do not store the receptors at this point but continue with the reconstitution as described below. An immunoblot with an antibody against the i3 loop of M2 (Chemicon International) is used to examine the integrity of the purified reconstituted M2 receptor. An example of an immunoblot shown in Fig. 2 demonstrates the absence of degraded products (containing this epitope).

We have synthesized and tested an ABT-agarose resin [ABT: 3-(2-amino-benzhydryloxy)tropane] for affinity chromatography applications with the receptor as described before,[12] but the yields obtained were low and the levels of receptor bound to the resin (see Fig. 2) indicated no bound M2 protein. It is unclear why this approach was not successful.

Reconstitution of M2 Receptor into Lipids

Brain lipids (Folch fraction VII, Sigma) are stored in chloroform at −85° and prepared for use by evaporation and freeze drying. Solubilization buffer is added

---

[11] A. Rinken and T. Haga, *Arch. Biochem. Biophys.* **301,** 158 (1993).
[12] K. Haga and T. Haga, *J. Biol. Chem.* **260,** 7927 (1985).

to obtain 10 mg lipid/ml. Complete solubilization requires warming to 37° and agitation. If necessary a bath sonicator can be used. This suspension can be made and aliquots stored at −85°.

Typically, 100 μl (1 mg lipid) of the aqueous lipid suspension is added to 18 μl of 10% sodium deoxycholate and 4 μl of 10% sodium cholate. Suspension clarifies on shaking. Warming may be necessary. Suspension needs to be chilled before mixing with receptors. One hundred to 500 μl of pure receptor is mixed with solubilized lipids (~122 μl from above). The mixture is applied to a 10–12 ml column of Sephadex G-50 (fine), equilibrated with solubilization buffer. The first 4 ml is collected. Vesicles are visible as turbidity after the first 2 ml. The column is washed with solubilization buffer and the procedure repeated. Otherwise the system can be scaled up. Vesicle fractions are pooled and concentrated down 3–5 times with an Amicon stirring cell (chamber of 10 ml or larger, YM30 membrane or similar—Millipore, Bedford, MA) at 4°. We usually freeze membranes in 20 μl aliquots by immersion in liquid nitrogen. The quantity of functional His-M2 is measured as above for M2 in native membranes. Yields are usually 10–15% relative to the original native membrane stock and 20–30% relative to the purified receptor stock. Other groups have obtained similar results.[13]

The composition of the brain lipid mixture is sphingomyelin (20%), phosphatidylethanolamine (30%), phosphatidylserine (20%), and other lipids according to the distributor (Sigma). The choice of this lipid mixture was based after comparing a variety of lipids. We have also tested other lipid compositions including brain lipids, cholesterol, egg, and soy lecithin. The yields of receptors were, however, similar.

Purification of G Protein

Expression and Purification of G Protein α Subunit

The α subunits of G proteins are expressed in Sf9 cells or bacteria (*Escherichia coli*) and purified using protocols that are available.[14,15] α_o subunit protein used in some of the experiments described here is synthesized using an plasmid that expresses rat α-O and yeast myristoyltransferase.[16] Expression and purification are performed based on a previous published procedure with modifications.[15,17] Cells are grown at 20° with shaking at 100 rpm to ensure slow growth. Induction is initiated at an OD_{600} of about 0.2 and cells harvested 12 hr later. Expression of α_o is ~1 mg protein/liter of cell culture. The protein was purified as previously

[13] G. H. Biddlecome, G. Berstein, and E. M. Ross, *J. Biol. Chem.* **271**, 7999 (1996).
[14] T. Kozasa and A. G. Gilman, *J. Biol. Chem.* **270**, 1734 (1995).
[15] E. Lee, M. E. Linder, and A. G. Gilman, *Methods Enzymol.* **237**, 146 (1994).
[16] M. E. Linder, C. Kleuss, and S. M. Mumby, *Methods Enzymol.* **250**, 314 (1995).
[17] S. M. Mumby and M. E. Linder, *Methods Enzymol.* **237**, 254 (1994).

described except for the use of a gel-exclusion column (Superose 12, Pharmacia, Piscataway, NJ) instead of a hydroxyapatite column. In order to ensure α_o protein stability, bacterial extracts were supplemented with 30 μM aluminum chloride, 1 mM sodium fluoride, and 50 μM GDP. The final purity was 90% according to visual observations from of SDS–PAGE gels stained with Coomassie blue staining. More than 80% of the protein was found to be functional according to the levels of GTPγS uptake.

Expression and Purification of G Protein $\beta\gamma$ Complex

The $\beta\gamma$ dimer is produced in Sf9 insect cells, by triple infection of His-α_{i2}, β_1, and γ subunit viruses. With each virus, the virus : cell ratio was 3 : 1. Cells are harvested 48–62 hr postinfection. Membranes are purified as described above. Our protocol was a modification of a previously published procedure.[14] Before elution with buffer E,[14] the protein bound column was equilibrated at 30° in buffer A[14] containing 0.7% sodium cholate. The $\beta\gamma$ complex eluted from the Ni-NTA beads (Qiagen, Valencia, CA) was subjected to dialysis and concentration. The purity was about 90%. When higher purity was desired, the purified untagged $\beta\gamma$ complex was incubated with 200–500 μl of Ni-NTA beads and centrifuged through spin columns. This process significantly reduced the concentration of nonspecific contaminant proteins that had been eluted with the heterotrimer earlier because of their affinity for the Ni beads. The resulting preparation is more than 95% pure. The $\beta\gamma$ was quantitated by laser densitometry using bovine serum albumin (BSA) as a standard.

Concentration of $\beta\gamma$ complex using either Centricon (YM30 or lower molecular weight cutoff) or Amicon pressurized stirred-cell devices can cause an unwanted accumulation of detergent in the final protein mixture (Centricon devices are available from Millipore). We usually determine the presence of CHAPS or sodium cholate using thin-layer chromatography with Silica Gel 60 (Merck, Whitehouse Station, NJ). The solvent is chloroform/methanol/ammonia (1/1/0.1). To calibrate the amount of detergent in the sample, known concentrations of detergents are loaded as standards. After drying the plates, the detergent is visualized with iodine vapor. If accumulation of detergent is found, dialysis or an additional chromatographic step is used to lower detergent concentration.

Assays to Measure G Protein Coupling to Receptor

G Protein–Receptor Complex

The stock lipid reconstituted His-M2 preparation (25–40 nM) and the G proteins of defined composition (25–250 nM, preincubated for 30 min) are mixed in the assay buffer (20 mM HEPES, pH 8, 100 mM NaCl, 5 mM DTT, and 2 mM MgCl$_2$) to a final concentration of 5 nM His-M2. The proteins are allowed to equilibrate under continuous shaking at 4° for 30 min. This protein mixture is diluted five-fold in the final reaction mixture, which includes the radiolabeled nucleotide, 10 μM GDP, and vehicle/agonist/antagonist. The receptor–G protein complex prepared in this manner is used immediately in the assays described below.

Measurement of Receptor-Stimulated GTPγS Binding to G Protein

Receptor activation of a G protein can be determined by measuring the rates of nucleotide incorporation into the G protein α subunit. This is measured directly as nucleotide-bound Gα subunit using a radiolabeled nonhydrolyzable analog of GTP like [^{35}S]GTPγS. Because the bound GTP is not hydrolyzed, the α subunit goes through one cycle of activation by the receptor. This decreases the sensitivity of the reaction and therefore requires an excess of G protein α subunit relative to the receptor. Because the $\beta\gamma$ complex can function catalytically, lower concentrations of the $\beta\gamma$ complex suffice. The ratio of $\alpha : \beta\gamma : R$ is $100 : 10 : 1$ with 1 nM receptor.

M2 receptors (1 nM, all final concentrations) were equilibrated with G protein α_o (100 nM) and G protein $\beta\gamma$ complex (1–10 nM) by mixing using a vortex mixer at 4°. The buffer is 20 mM HEPES, pH 8, 100 mM NaCl, 5 mM DTT, 2 mM MgCl$_2$ and includes 10 μM GDP. The volume of the reaction mixture is 5–10 μl. Agonist, antagonist, or vehicle is added with the radiolabeled 0.2 μM GTPγS (100–300 cpm/fmol nucleotide) and incorporation is allowed to proceed for 0.5–10 min ([^{35}S]GTPγS is available from Amersham-Pharmacia, Piscataway, NJ). Reaction is stopped by mixing aliquots with 20–50 μl ice cold reaction buffer above containing 1 mM atropine and 200 μM cold GTP. Samples are filter-washed on nitrocellulose membranes (HAWP-0250, Millipore, Bedford, MA). Nonspecific binding is measured in the absence of α subunit and is usually negligible. Figure 3 shows the levels of agonist stimulated activity in the presence of the $\beta\gamma$ complex. This activity is increased 3- to 4-fold relative to that in the absence of $\beta\gamma$ (Fig. 3) or in the absence of the agonist (data not shown). This agonist-stimulated activity is observed only in the presence of moderate concentration of GDP and is not seen in its absence.

Steady-State GTPase Assay

In a steady-state GTPase assay the P$_i$ released by the GTPase activity of the α subunit is measured as the heterotrimer transits through multiple cycles of activation by the receptor. The steady-state GTPase rate is limited by the degree of G protein activation and by the endogenous GTPase rate of the α subunit. Our initial studies with the GTPase assay encountered the fact, noticed before,[18,19] that the basal activity of G$_o$ is about 0.25 min^{-1} (Fig. 4).

---

[18] S. E. Senogles, A. M. Spiegel, E. Padrell, R. Iyengar, and M. G. Caron, *J. Biol. Chem.* **265**, 4507 (1990).

[19] H. Kurose, J. W. Regan, M. G. Caron, and R. J. Lefkowitz, *Biochemistry* **30**, 3335 (1991).

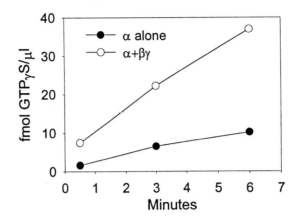

FIG. 3. G protein activation assay : GTPγS binding assay. 1 nM M2, 100 nM α_0 and 10 nM $\beta\gamma$ wild-type (or vehicle) were equilibrated at room temperature with assay buffer containing 20 mM HEPES, pH 8, 100 mM NaCl, 2 mM MgCl$_2$, 5 mM DTT, and 10 μM GDP. At time zero 0.2 μM [^{35}S]GTPγS and 1 mM carbachol were added. Reaction was stopped at the times indicated with stop buffer (described in text). Samples were filter-washed and radioactivity was counted. The α_0 present in 1 μl of the reaction mixture was 100 fmol, indicating that less than 50% of the α_0 had bound GTPγS in 6 min.

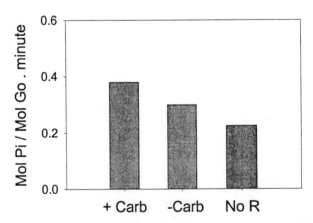

FIG. 4. Measurement of receptor-activated GTPase. Preincubated 10 nM α_0, 10 nM $\beta\gamma_5$ and 1 nM M2 or no receptor (R) were mixed at room temperature in assay buffer containing 20 mM HEPES, pH 8, 100 mM NaCl, 2 mM MgCl$_2$, 5 mM DTT. At time zero 0.2 μM[γ-^{32}P]GTP ([γ-^{32}P]GTP is available from Amersham-Pharmacia) and 1 mM carbachol (Carb) were added and after 15 min the reaction was arrested with ice-cold 50 mM potassium phosphate buffer (pH 7.0) + 5% activated charcoal. Supernatants after centrifugation were measured for radioactivity.

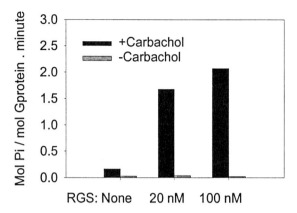

FIG. 5. Measurement of receptor activated GTPase. Preincubated 1 nM M2, 2 nM α_o, 2 nM $\beta\gamma_5$ and the indicated concentration of RGS4 were mixed at room temperature in assay buffer containing 20 mM HEPES pH 8, 100 mM NaCl, 2 mM MgCl$_2$, 5 mM DTT, and 10 μM GDP. At time zero 0.2 μM [γ-^{32}P]GTP and 1 mM carbachol or vehicle was added and after 10 min the reaction was stopped by adding ice-cold 50 mM potassium phosphate buffer (pH 7.0) and 5% activated charcoal. Supernatant after centrifugation was counted for radioactivity.

Agonist-stimulated activity represented only 50% increase over basal levels in the absence of receptor. Addition of 10 μM GDP reduces the basal rate of GTPase to about 0.02–0.05 min^{-1} and the agonist-stimulatory activity to about 0.1–0.35 min^{-1} depending on the concentration of G protein heterotrimer (Fig. 5). In the presence of the agonist-bound receptor, guanine nucleotide exchange is rapid and the k_{cat} for the α subunit GTPase limits G protein activation rate. To increase the sensitivity of the signal from the GTPase assay, we add RGS4 protein to the reaction (Fig. 5).

RGS4 is a GAP (GTPase activating protein) that stimulates the GTPase rate of the α subunit.[20] The inclusion of RGS4 in the GTPase assay results in a considerable increase in the agonist-induced GTPase rate of the reconstituted mixture. RGS4 does not affect the GTPase rate of the G protein in the absence of the agonist. The magnitude of increase in GTPase activity in response to the agonist is 50- to 100-fold over that in the absence of the agonist. This increase confirms that in the absence of RGS4, GTPase rate and not nucleotide exchange of α_o was rate limiting. The concentration of RGS4 needed to attain maximal steady state GTPase rate was 20–100 nM. The rate of the receptor stimulated GTPase reaction is now dependent on other factors, such as receptor interaction with G protein, α–$\beta\gamma$ reassociation, and nucleotide exchange. Steady-state GTPase activity can be titrated with increasing concentrations of carbachol indicating an EC$_{50}$ of about 2 μM (Fig. 6).

[20] D. M. Berman and A. G. Gilman, *J. Biol. Chem.* **273**, 1269 (1998).

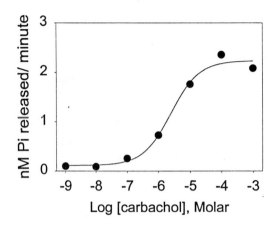

FIG. 6. Agonist dependence of GTPase activity. Preincubated 1 nM M2, 2 nM α_o, 2 nM $\beta\gamma_5$, and 100 nM RGS4 were mixed at room temperature with assay buffer above (Fig. 5, legend). At time zero 0.2 μM [γ-^{32}P]GTP and the indicated concentrations of carbachol were added and after 10 min the reaction was arrested by addition of cold 50 mM potassium phosphate buffer (pH 7.0) and 5% activated charcoal. Supernatant after centrifugation was counted for radioactivity.

The protocol for the GTPase assay follows the same lines as for the GTPγS assay. However, an important distinction is that in the case of the GTPase assays the concentrations of heterotrimer used are between 0.3 and 10 nM and the ration between G protein and receptor close to 1 : 1. At these concentrations of G protein and receptor, subtle differences in the receptor activation of G proteins containing mutant subunits or different subunits are more likely to be detected. The reaction is performed in the presence of the same buffer as in the GTPγS binding assay, including 1 nM His-M2, 0.2 μM[γ-^{32}P]GTP (100–300 cpm/fmol nucleotide), 10 μM GDP, and 2 mM MgCl$_2$. Reactions were performed at room temperature for 5–20 min. Within this period the rate of P$_i$ release was found to be linear with time and the amount of [^{32}P]GTP substrate consumed was kept at 10–20% total GTP added. The ^{32}P is measured by sequestration of organic phosphate with charcoal and radioactivity is measured after mixing the supernatants of samples with scintillation liquid. Nonspecific binding is measured by performing the reaction in the presence of 200 μM GTP.

Testing Effect of Peptides Specific to G Protein $\beta\gamma$ Complex on Receptor-G Protein Interaction

We have examined the effect of peptides specific to the C terminus of the γ subunit on receptor coupling of a G protein.[1,2] In principle this approach can be used to examine the effect of any peptide specific to the $\beta\gamma$ complex. Since the

C termini of the γ subunits are prenylated—an isoprenoid, either farnesyl (C_{15}) or geranylgeranyl (C_{20}) is attached to the cysteine—peptides specific to this region need to be chemically prenylated. This is especially important since the prenyl group influences $\beta\gamma$ complex activity at the receptor surface.[21,22]

Chemical Synthesis of Prenylated Peptides

Peptides containing appropriate amino acid sequences are synthesized and then prenylated. The geranylgeranyl bromide is commercially available from American Radiolabeled Chemicals (St. Louis, MO). Farnesyl and geranyl bromide are available from Aldrich (Milwaukee, WI).

Peptide (2 μmol) and prenyl bromide (4 μmol) are mixed in a solution (1 ml) of butanol : methanol : water (1 : 1 : 1, volume) previously purged under nitrogen atmosphere. Since the prenylated peptides will need to be purified anyhow (see below), we start with peptides that have not been purified. Butylhydroxytoluene (10 μg) is provided as antioxidant. The reaction is started with the addition of drops (5–10 μl) of 0.5 M sodium carbonate until the pH reaches between pH 8.5 and 9 (checked by pH paper indicator Baker-pHIX pH 2-9, Fisher Scientific, Pittsburgh, PA). The container (with a small magnetic flea for agitation) is purged under nitrogen atmosphere, sealed and placed in the dark, and continuously agitated for 18 and 24 hr at room temperature. The reaction is stopped with a few drops of acetic acid to bring the pH to 7.0. Samples are frozen at $-85°$.

Purification of Prenylated Peptides

Prenylated peptides are purified by reverse chromatography on a PepRPC-FPLC column HR 10/16 (Pharmacia Biotech Inc.) using a linear (0–100%) gradient of acetonitrile in water containing 0.1% trifluoroacetic acid. The prenylated compounds elute at a position of the gradient corresponding to around 50–60% acetonitrile content. Figure 7A shows chromatograms of samples analyzed 6 hr and 24 hr after the start of the prenylation reaction (described above). Elution peak(s) at 20–25 min corresponds to peptide species and the peak at 35–37 min corresponds to the prenylated peptide whose intensity increases in the 24-hour samples as expected from the progression of the reaction. The peak at 35–37 min is absent when unprenylated peptide preparations are run through the same column. Later peaks may arise from unincorporated geranylgeranyl bromide and derivatives. Farnesylated peptides elute 2–4 min (about 33–35 min) earlier than their geranylgeranylated counterparts. Elution times may differ for each peptide depending on its hydrophobicity. Peptides are usually converted to prenyl peptide

---

[21] O. Kisselev, M. Ermolaeva, and N. Gautam, *J. Biol. Chem.* **270,** 25356 (1995).

[22] H. Yasuda, M. A. Lindorfer, K. A. Woodfork, J. E. Fletcher, and J. C. Garrison, *J. Biol. Chem.* **271,** 18588 (1996).

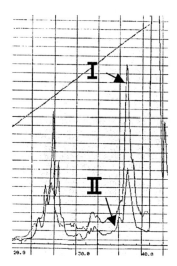
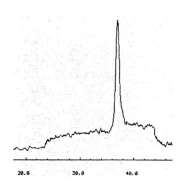

FIG. 7. (A) Chromatogram of geranylgeranylated Gγ C-terminal peptides. A PepRPC HR 10/16 at 4° was used to purify prenylated peptides. Elution was monitored at 205 nM using a Uvicord SII detector. Samples were injected from a 500 μl loop on to the column, equilibrated with 0.1% trifluoroacetic acid in deionized H$_2$O at a flow rate of 5 ml/min. The column was then washed for at least 5 min, and the gradient was started at time 10 min and ended at 60 min at 100% acetonitrile. Interval between minutes 18 and 34 is shown. Peaks corresponding to the geranylgeranylated peptide obtained at different time points from the reaction mixture are labeled. I : 6 hr; II : 24 hr (B) Chromatogram of prenylated γ_5 peptide. Purified prenylated γ_5 peptide was run through a PepRPC HR 10/16 column again. Elution was monitored using a Uvicord SII with interference filter at 205 nM. Gradient started at 10 min and ended at 60 min at 100% acetonitrile. Interval between minutes 18 and 47 is shown.

with a 30–50% yield, which after purification and other operations results in 15 to 25% neat prenylated peptide. Prenyl peptides are stored in butanol : methanol : water (1 : 1 : 1, volume) or dimethyl sulfoxide at −85°.

The molecular weight of the prenylated peptide is checked with mass spectrometric methods. The concentration of the modified peptide is determined by amino acid analysis. The modified peptide is acid digested and the content of acid-resistant amino acids such as glutamic and aspartic acid estimated. The integrity of the modified peptides in stocks is checked regularly by chromatography in a PepRPC column by FPLC. Figure 7B shows a chromatogram resulting from running 10 μmol of a purified geranylgeranylated peptide through a PepRPC column. The sharpness of the peak and its elution timing indicate its purity and integrity.

Effect of γ Subunit Peptides on M$_2$ Receptor Activation of a G Protein

G protein activation by a muscarinic receptor is inhibited by a prenylated γ_5 subunit specific peptide as shown in Fig. 8. The peptide is specific to the

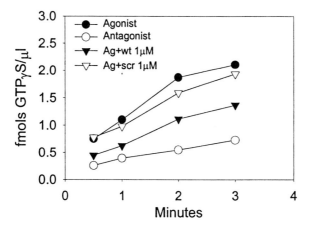

FIG. 8. Prenylated peptides inhibit M2 stimulated GTPγS uptake into G proteins. CHO cell membrane suspensions of M2 (5 nM) were reconstituted with G proteins (100 nM) (Gα_{i2} and brain $\beta\gamma$), in the presence or absence of 1 μM geranylgeranylated Gγ_5 wild-type (wt) or Gγ_5 scrambled (scr) in ice. The buffer was made up of 20 mM HEPES, pH 8, 5 mM MgCl$_2$, 100 mM NaCl, 1 μM GDP, and 0.02% sodium cholate. Reactions were started by adding (final concentrations) 0.2 μM [^{35}S]GTPγS and 1 mM agonist-carbachol or antagonist-atropine. Aliquots were taken at the indicated times. GTPγS uptake was arrested by mixing aliquots in ice-cold buffer containing 500 μM GTP and 1 mM atropine. Samples were filtered through nitrocellulose membranes and washed with cold phosphate-buffered saline, and radioactivity was counted in scintillation liquid.

14 C-terminal residues (inclusive of and upstream of the prenylated cysteine). The peptide inhibits GTPγS incorporation into the G protein α_{i2} subunit. If such inhibition by a peptide is due to sequence-specific interaction with a receptor, a peptide with the same amino acid sequence scrambled should not be active as shown in Fig. 8. The effective concentration at which inhibition occurs can be determined by plotting concentration of peptide and extent of inhibition. The role of the prenyl moiety can be addressed by testing the unprenylated peptide and peptide that is modified with a different isoprenoid such as geranylgeranyl in the same assays.

Acknowledgments

Some of the reagents used in the experiments described here were kind gifts from Drs. T. Kozasa, M. Linder, E. Peralta, and E. Ross.

[9] Phosducin Down-Regulation of G-Protein Coupling: Reconstitution of Phosducin and Transducin of cGMP Cascade in Bovine Rod Photoreceptor Cells

By YEE-KIN HO, TUOW DANIEL TING, and REHWA HO LEE

Introduction

Photoexcitation in vertebrate rod photoreceptor cells occurs via a light-activated cGMP enzyme cascade. Absorption of a photon by the receptor molecule, rhodopsin, leads to the activation of a latent cGMP phosphodiesterase (PDE) which rapidly hydrolyzes cytosolic cGMP. The transient decrease of cGMP concentration causes the closure of the cGMP-sensitive cation channels on the plasma membrane and results in hyperpolarization of the cell.[1] Signal coupling between photolyzed rhodopsin (R*) and PDE is mediated by the retinal G protein, transducin (T).[2] Transducin is composed of three polypeptides: T_α, T_β, and T_γ. The GDP-bound form of T_α has a high affinity for $T_{\beta\gamma}$ and rhodopsin. R* catalyzes the exchange of GTP for bound GDP on T_α. The T_α–GTP complex then dissociates from $T_{\beta\gamma}$/R* and activates the latent PDE. After the hydrolysis of the tightly bound GTP, T_α-GDP recombines with $T_{\beta\gamma}$ for another cycle of activation. The apparent role $T_{\beta\gamma}$ is to assist the binding of T_α to rhodopsin.[3] Phosducin (Pdc), a retinal phosphoprotein, has been isolated and shown to form a specific complex with $T_{\beta\gamma}$.[4] The soluble phosducin/$T_{\beta\gamma}$ complex sequesters $T_{\beta\gamma}$ in cytosol, thus limiting the availability of $T_{\beta\gamma}$ needed for the recycling of T_α and the continuous activation of the cGMP cascade.

Phosducin is phosphorylated by protein kinase A at serine residue 73[5,6] and dephosphorylated by 2A phosphoprotein phosphatase.[7] Its phosphorylation state is light-dependent with high phosphorylated level in the dark and dephosphorylated in light when the cGMP cascade is activated. Consequently, the phosphorylated phosducin exhibited reduced affinity toward $T_{\beta\gamma}$. The formation of phosducin/$T_{\beta\gamma}$ complex and the phosphorylation state of phosducin represent a major regulatory site of the visual excitation processes in rod outer segments. In this chapter, we

[1] P. A. Liebman, K. R. Park, and E. A. Dratz, *Ann. Rev. Physiol.* **49,** 765 (1987).
[2] L. Styer, *J. Biol. Chem.* **266,** 10711 (1991).
[3] Y.-K. Ho, V. N. Hingorani, S. E. Navon, and B. K.-K. Fung, *Curr. Top. Cell. Reg. Vol.* **30,** 171 (1989).
[4] R. H. Lee, B. S. Lieberman, and R. N. Lolley, *Biochemistry* **26,** 3983 (1987).
[5] R. H. Lee, B. M. Brown, and R. N. Lolley, *J. Biol. Chem.* **265,** 15860 (1990).
[6] R. H. Lee, B. M. Brown, and R. N. Lolley, *Biochemistry* **20,** 7532 (1981).
[7] R. H. Lee and B. M. Brown, *Invest. Ophthal. Visual Sci.* **32,** 1054 (1991).

describe the reconstitution assays in studying the phosducin down-regulation of the cGMP cascade in photoreceptor cell.

Preparation of Rod Outer Segment Membrane and Purification of Proteins

All proteins involved in the retinal cGMP cascade including rhodopsin, transducin, phosducin, and the cGMP phosphodiesterase (PDE) are purified from bovine retinas. The detailed procedures of their purification have been reported in previous volumes.[8,9] They will be briefly described in this chapter. Bovine retinas are obtained from Brown Packing Co., South Holland, IL. The retinas are dissected in the dark and stored at $-70°$. Rod outer segment (ROS) disk membrane is isolated from bovine retina by the sucrose flotation (38% w/v) method in MOPS buffers at pH 7.5 with 0.1 mM PMSF (phenylmethylsulfonyl fluoride), 1 mM DTT (dithiothreitol), 2 mM $MgCl_2$, and 50 mM NaCl. Rhodopsin (R) in reconstituted membrane is prepared by the detergent dialysis method.[10] Stripped ROS membrane is prepared by washing the purified ROS membranes 6 times with buffer containing 2 mM MOPS, 4 M urea, 1 mM DTT, and 1 mM EDTA (ethylenediaminetetraacetic acid), pH 7.5. Stripped ROS membrane is void of transducin and PDE activity and is used as a source of rhodopsin in the reconstituted assays. Transducin is extracted from photolyzed ROS membrane in MOPS buffer with GTP and purified by hexylagarose column chromatography.[11] T_α-GDP and $T_{\beta\gamma}$ are separated from transducin by ω-aminooctylagarose column eluted with a salt gradient from 0.075 to 0.4 M.[12] The T_α-Gpp(NH)p is obtained by incubating transducin with Gpp(NH)p [guanosine $5'$-(β,γ-imidotriphosphate)] in the presence of photolyzed rhodopsin (R^*) prior to the ω-aminooctylagarose column separation. PDE is purified by a procedure of using DEAE (diethylaminoethyl)-Sephadex, S-300 gel filtration and ω-aminooctylagarose chromatographies.[9] Purified PDE is a latent enzyme that can be activated by trypsin treatment or T_α-Gpp(NH)p complex. The phosducin/$T_{\beta\gamma}$ complex is purified from bovine retinas with a hydroxyapatite column.[4] The complex is dissociated to phosducin and $T_{\beta\gamma}$ by chromatography on a Q-sepharose column with a 300–800 mM Tris-HCl gradient.[13] All purified proteins are stored at $-20°$ in Tris or MOPS buffers at pH 7.5 with 0.1 mM PMSF, 1 mM DTT, 2 mM $MgCl_2$ and 40% glycerol. Sodium dodecyl sulfate–polyacrylamide gel electrophoresis (SDS–PAGE) with subsequent Coomassie blue staining reveals

[8] T. D. Ting, S. B. Goldin, and Y.-K. Ho, *Methods Neurosci.* **15**, 180 (1993).

[9] A. Tar, T. D. Ting, and Y.-K. Ho, *Methods Enzymol.* "G-proteins," **161**, 3 (1994).

[10] K. Hong and W. L. Hubbell, *Biochemistry* **12**, 4517 (1973).

[11] B. K.-K. Fung, *J. Biol. Chem.* **258**, 10495 (1983).

[12] Y.-K. Ho and B. K.-K. Fung, *J. Biol. Chem.* **259**, 6694 (1984).

[13] R. H. Lee and R. H. Lolley, *Methods Neurosci.* **15**, 196 (1992).

that the purified phosducin contains a single polypeptide of 33 kDa (molecular mass of phosducin is calculated to be 28 kDa from its amino acid sequence),[14] T_α contains a single polypeptide of 40 kDa, and $T_{\beta\gamma}$ contains two bands of 36 kDa and 8 kDa. There is no cross-contamination of subunits in the purified T_α and $T_{\beta\gamma}$ samples. Purified PDE contains three polypeptides of 88 kDa (P_α), 84 kDa (P_β), and 14 kDa (P_γ). $P_{\alpha\beta}$ exhibits the catalytic activity of cGMP hydrolysis and P_γ is the inhibitory peptide of the latent enzyme complex. Protein concentrations are determined by the Coomassie blue binding method[15] using γ-globulin from Bio-Rad as the standard. Molar ratio of proteins in the reconstituted is calculated from protein concentration as mg/ml without further correction for difference in dye binding property. The rhodopsin content is determined from the absorbance at 498 nm with a molar extinction coefficient of 42,700 $cm^{-1} M^{-1}$.

Phosducin Inhibition of Retinal cGMP Cascade

Phosducin Inhibition of Activation of cGMP Phosphodiesterase

To examine the possible regulatory role of phosducin on the retinal cGMP cascade, the effect of phosducin on the light-activated PDE activity is examined in a reconstituted system containing stripped ROS membrane, purified transducin, and PDE. The PDE activity is monitored by the decrease of medium pH due to the hydrolysis of cGMP.[16] The reaction mixture of 200 μl contains 3 μM photolyzed rhodopsin (R*, in Meta II photo-intermediate stage), 2 mM cGMP, 3 μg transducin, and 3 μg PDE in 1 mM Tris, 200 mM NaCl, and 2 mM MgCl$_2$, pH 7.5. Reactions are initiated by the addition of 100 μM GTP to activated the cascade. The change of pH in the reaction medium is monitored by a pH microelectrode (Microelectrodes Inc., Londonderry, NH) and a Radiometer PHM 82 pH meter. The results are recorded on a Soltex strip chart recorder. The change in pH is then converted to the amount of cGMP hydrolyzed. The results are shown in Fig. 1. In the control experiment (lane 1, Fig. 1), containing R*, transducin, and PDE, the obtained PDE activity is set as 100% for comparison. Purified PDE is an inactive latent enzyme in the absence of R* and transducin (lane 2, Fig. 1). Purified phosducin has no cGMP hydrolysis activity (lane 3, Fig. 1). The addition of 5 and 10 μg of phosducin to the control sample reduces the PDE activity by approximately 30% and 40%, respectively (lanes 4 and 5, Fig. 1). The inhibition by phosducin is overcome by the addition of exogenous $T_{\beta\gamma}$. By adding 5, 10, and 30 μg of $T_{\beta\gamma}$ to the 10 μg phosducin-inhibited sample, the PDE activity is recovered to 70, 80, and 100%, respectively

[14] R. H. Lee, A. Fowler, J. F. McGinnis, R. N. Lolley, and C. M. Craft, *J. Biol. Chem.* **265**, 15867 (1990).
[15] M. M. Bradford, *Anal. Biochem.* **72**, 248 (1976).
[16] R. Yee and P. A. Liebman, *J. Biol. Chem.* **253**, 8902 (1978).

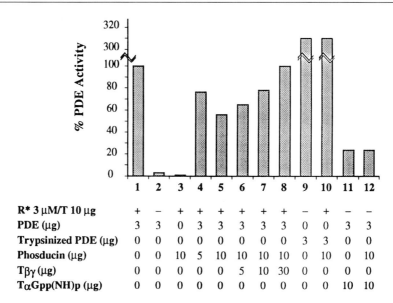

	1	2	3	4	5	6	7	8	9	10	11	12
R* 3 μM/T 10 μg	+	−	+	+	+	+	+	+	−	+	−	−
PDE (μg)	3	3	0	3	3	3	3	3	0	0	3	3
Trypsinized PDE (μg)	0	0	0	0	0	0	0	0	3	3	0	0
Phosducin (μg)	0	0	10	5	10	10	10	10	0	10	0	10
$T_{\beta\gamma}$ (μg)	0	0	0	0	0	5	10	30	0	0	0	0
T_{α}Gpp(NH)p (μg)	0	0	0	0	0	0	0	0	0	0	10	10

FIG. 1. Effect of phosducin on the activation of cGMP PDE in ROS membrane. The PDE activity was assayed by monitoring the change of pH in the reaction mixture due to cGMP hydrolysis. Samples that contained or did not contain 3 μM R* and 10 μg transducin are denoted by + or −, respectively. The amount (μg) of PDE, trypsin-activated PDE, phosducin, the $T_{\beta\gamma}$ subunit, and purified T_{α}-Gpp(NH)p complex present in each sample is shown under the graph. All samples contain 2 mM cGMP and the reactions were initiated by the addition of 100 μM of GTP to activate the cascade. The cGMP hydrolysis rate of the control sample (lane 1) containing 3 μM R*, 10 μg T, and 3 μg PDE is set at 100% for comparison. Activities of trypsin and T_{α}-Gpp(NH)p activated PDE activities were measured in the absence of added GTP. The reaction was initiated by the addition of cGMP. Reproduced with permission from R. H. Lee, T. D. Ting, B. S. Lieberman, D. E. Tobias, R. N. Lolley, and Y. K. Ho, *J. Biol. Chem.* **267**, 25104 (1992).

(lanes 6–8, Fig. 1). These observations suggest that the inhibition of phosducin may be due to its binding to $T_{\beta\gamma}$. To eliminate the possibility of direct inhibition to the enzymatic activity of PDE by phosducin, the effect of phosducin on the activity of trypsin-activated PDE is examined. The latent PDE is treated with TPCK [L-1-(tosylamido)-2-phenylethyl chloromethyl ketone]trypsin for 3 min at a PDE to trypsin ratio of 100 : 1 (w/w) to remove the inhibitory P_{γ} subunits. Proteolysis is terminated by the addition of soybean trypsin inhibitor. The trypsin activated PDE demonstrates a cGMP hydrolysis activity approximately threefold higher (lane 9, Fig. 1) than the control sample (lane 1, Fig. 1). The addition of 10 μg of phosducin has no effect on the trypsin-activated PDE (lane 10, Fig. 1). In a similar experiment, phosducin shows no inhibitory effect on the activity of PDE activated by purified T_{α}-Gpp(NH)p (lanes 11 and 12, Fig. 1). One may conclude that the regulatory effect of phosducin on the cGMP cascade is related to the activation of transducin and not on PDE enzymatic activity.

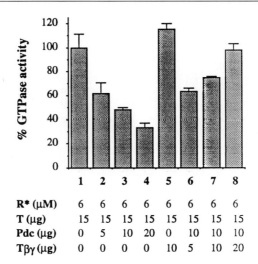

R* (µM)	6	6	6	6	6	6	6	6
T (µg)	15	15	15	15	15	15	15	15
Pdc (µg)	0	5	10	20	0	10	10	10
T$\beta\gamma$ (µg)	0	0	0	0	10	5	10	20

FIG. 2. Effect of phosducin on transducin GTPase activity. Each reaction mixture contained 30 µl of 6 µM R*, 15 µg transducin, and various amounts of phosducin or $T_{\beta\gamma}$ as shown under the graph. The reactions were initiated by the addition of 30 µl of 60 µM [γ-^{32}P]GTP. The control sample (lane 1) contained 6 µM R* and 15 µg T and is set at 100% for comparison. The samples (lanes 1–4) demonstrate the inhibition of transducin GTPase by phosducin; the samples (lanes 5–8) demonstrate the reversal of this inhibition by the addition of $T_{\beta\gamma}$. Reproduced with permission from R. H. Lee, T. D. Ting, B. S. Lieberman, D. E. Tobias, R. N. Lolley, and Y. K. Ho, *J. Biol. Chem.* **267**, 25104 (1992).

Phosducin Inhibition of Transducin GTP Hydrolysis Cycle

The inhibitory effect of phosducin on PDE activation suggests that phosducin interacts with transducin by binding to $T_{\beta\gamma}$. Since $T_{\beta\gamma}$ is essential for the activation of transducin by presenting T_α for binding of R*,[11] the $T_{\beta\gamma}$-phosducin interaction may reduce the level of activated T_α and attenuate PDE activation. The effect of phosducin on the R*-catalyzed GTPase activity of transducin is shown in Fig. 2. The GTP hydrolysis activity of transducin catalyzed by R* is assayed by measuring the release of [^{32}P]P$_i$ from [γ-^{32}P]GTP (10 Ci/mmol from ICN Radiochemicals (Costa Mosa, CA).[17] The reaction mixture (30 µl) contains 6 µg transducin and 6 µM R*. Reactions are initiated by the addition of 30 µl of 100 µM [γ-^{32}P]GTP. The reaction is stopped after 10 min by the addition of 0.2 M perchloric acid and followed by ammonium molybdate precipitation. The inorganic phosphate precipitates are filtered onto Whatman (Clifton, NJ) glass fiber filters, and the radioactivity is counted. Under the experimental conditions, the GTP hydrolysis rate is linear up to 20 min. Therefore, the assayed activity at the 10-min time point represents the steady-state rate of GTP hydrolysis due to continuous turnover of transducin. In

[17] B. K.-K. Fung and L. Stryer, *Proc. Natl. Acad. Sci. U.S.A.* **77**, 2500 (1980).

the control experiment (lane 1, Fig. 2), the steady-state rate of GTP hydrolysis in a reconstituted R* and transducin is set at 100% for comparison. Phosducin inhibits GTPase activation in a dose-dependent manner. The addition of 5, 10, and 20 μg of phosducin to the reconstituted sample reduces the GTPase activity 30, 40, and 60%, respectively (lanes 2–4, Fig. 2). The inhibition is recovered by the addition of exogenous $T_{\beta\gamma}$ in a dose-dependent manner back to 100% of the control level (lanes 6–8, Fig. 2). Exogenous $T_{\beta\gamma}$ in the absence of phosducin shows little activating effect on the GTPase activity (lane 5, Fig. 2). These results indicate that phosducin interacts with soluble $T_{\beta\gamma}$ to prevent its reassociation with T_{α}-GDP and subsequently interrupts the transducin coupling cycle.

Dissociation of T_{α} and $T_{\beta\gamma}$ Subunits by Phosducin

Both phosducin and T_{α}-GDP are capable of forming a soluble complex with $T_{\beta\gamma}$. The relative affinity between phosducin and T_{α}-GDP competing for $T_{\beta\gamma}$ is essential to determine the significance of the phosducin regulation. If phosducin binding to $T_{\beta\gamma}$ is stronger than that of T_{α}-GDP, the cascade will be shut off. Otherwise, the T_{α}–$T_{\beta\gamma}$ complex will continue the signal coupling between R* and PDE. The subunit interaction of transducin can be monitored in solution directly by gel filtration column chromatography or indirectly by pertussis toxin-catalyzed ADP-ribosylation of T_{α}. Both assays are conducted in the absence of ROS disk membrane.

Phosducin Inhibition of Pertussis Toxin ADP-Ribosylation of Transducin

Purified T_{α} is a poor substrate for pertussis toxin-catalyzed ADP-ribosylation; however, in the presence of $T_{\beta\gamma}$, the reaction can be enhanced severalhundred fold.[18] This enhancement can be used as an indirect measurement of the subunit interaction of transducin. Pertussis toxin catalyzed ADP-ribosylation is carried out by incubating a reaction mixture containing purified transducin (1.0 mg/ml), 10 μg/ml pertussis toxin (List Biological, Campbell, CA) activated with 20 mM DTT at 30° for 10 min, 1 mM [*adenylate*-^{32}P] NAD (0.25 Ci/mmol), and 1 mM ATP in buffer containing 10 mM MOPS, 200 mM NaCl, 2 mM DTT, and 2 mM MgCl$_2$, pH 7.5 for up to 25 min at 30°. The radioactive labeling was stopped by the addition of 10-fold excess of nonradioactive NAD. The sample was separated on 13% SDS–polyacrylamide gel electrophoresis, the band corresponding to T_{α} was excised, and the labeling was quantitated by scintillation counting. Under the experimental conditions, the rate of ADP-ribosylation remained linear for 50 min, and the incorporation of [^{32}P] ADP-ribose in the T_{α} subunit of transducin reached a molar ratio of 1 : 1 in approximately 2 hr. It is assumed that the

[18] P. A. Watkins, D. L. Burns, Y. Kanaho, T.-Y. Liu, E. L. Hewlett, and J. Moss, *J. Biol. Chem.* **260**, 13478 (1985).

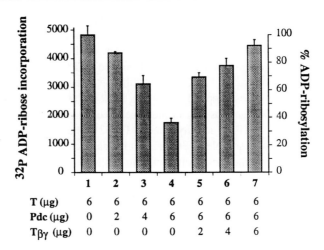

	1	2	3	4	5	6	7
T (μg)	6	6	6	6	6	6	6
Pdc (μg)	0	2	4	6	6	6	6
T$_{\beta\gamma}$ (μg)	0	0	0	0	2	4	6

FIG. 3. Effect of phosducin on pertussis toxin catalyzed ADP-ribosylation of transducin. The reaction mixture contained 6 μg transducin, 10 μg/ml pertussis toxin, 1 mM [adenylate-^{32}P]NAD, 1 mM ATP, and various amounts of phosducin or T$_{\beta\gamma}$ as shown under the graph. Subunits of the ADP-ribosylated transducin were separated on SDS–polyacrylamide gel and the incorporation of radioactive ADP-ribose to T$_{\alpha}$ was quantitated by excising the protein band for scintillation counting. The control sample (lane 1) did not contain phosducin or T$_{\beta\gamma}$ and is set at 100% for comparison. Reproduced with permission from R. H. Lee, T. D. Ting, B. S. Lieberman, D. E. Tobias, R. N. Lolley, and Y. K. Ho, J. Biol. Chem. **267,** 25104 (1992).

25-min samples represented the initial rate of ADP-ribosylation. The effect of phosducin on the pertussis-toxin catalyzed ADP-ribosylation of transducin is shown in Fig. 3. ADP-ribosylation of transducin, a 1 : 1 complex of T$_{\alpha}$/T$_{\beta\gamma}$, is inhibited by the addition of phosducin. The degree of inhibition increases proportionally to the concentration of phosducin (lanes 2–4, Fig. 3). Similar to the inhibition of the cGMP cascade, the addition of exogenous T$_{\beta\gamma}$ reverses the phosducin inhibition of the ADP-ribosylation reaction (lanes 5–7, Fig. 3). Phosducin is not a substrate for pertussis toxin-catalyzed ADP-ribosylation and does not inhibited the action of pertussis toxin directly. The level of ADP-ribosylation of purified T$_{\alpha}$ in the absence of T$_{\beta\gamma}$, though at a lower level, is not affected by phosducin. In the absence of phosducin, the addition of excess T$_{\beta\gamma}$ to transducin has no effect on the level of ADP-ribosylation (data not shown). Since the phosducininhibited ADP-ribosylation of transducin can be fully recovered with exogenously added T$_{\beta\gamma}$, the initial inhibition can only be due to the direct blockage of the T$_{\beta\gamma}$ enhancement of ADP-ribosylation via the formation of the T$_{\beta\gamma}$–phosducin complexes. This disruption of the T$_{\alpha}$T$_{\beta\gamma}$ complex reduces the efficiency of the ADP-ribosylation of T$_{\alpha}$.

Superose-12 Gel Filtration Chromatography

Protein complex formation and dissociation among phosducin and the subunits of transducin were examined by gel filtration chromatography using a

Superose-12 column (Pharmacia, Piscataway, NJ HR 10/30) connected to a Waters HPLC (high performance liquid chromatography) system. Samples in 200 μl aliquots containing 50 μg of either individual proteins or mixtures of purified proteins were injected and eluted with isotonic buffer.[19] The elution profiles of phosducin, T_α, and $T_{\beta\gamma}$ were visualized by the analyses of the protein contents of each fractions by SDS–polyacrylamide gel electrophoresis with subsequent Coomassie staining. The Superose-12 column was calibrated with the following protein standards: catalase (52,000), yeast alcohol dehydrogenase (46,000), lactate dehydrogenase (42,000), bovine serum albumin (35,000), ovalbumin (28,000), and myoglobin (18,000). The void volume and total volume of the column were determined by the elution of blue dextran and adenosine, respectively. Figure 4 A–C shows the elution profiles of purified T_α, $T_{\beta\gamma}$, and phosducin. Based on the calibration the phosducin/$T_{\beta\gamma}$ complex elutes with a Stokes radius of 39.5 Å, the T_α subunit of 28.5 Å, and $T_{\beta\gamma}$ of 21 Å. The elution position of $T_{\beta\gamma}$ is retarded as compared to the predicted position based on its molecular mass, which may be due to the hydrophobic interaction between the protein and the gel matrix. Figure 4D shows the elution profiles of phosducin/$T_{\beta\gamma}$ and transducin ($T_\alpha T_{\beta\gamma}$) complexes. Phosducin/$T_{\beta\gamma}$ complexes directly isolated from retinal extract or reconstituted using purified phosducin and $T_{\beta\gamma}$ are eluted at fraction 23. Figure 4E shows the chromatographic result of transducin complex ($T_\alpha T_{\beta\gamma}$). There is a minor dissociation of the T_α and $T_{\beta\gamma}$ subunits by the Superose-12 column presumably via the hydrophobic interaction with the column matrix. The distributions of the T_α and $T_{\beta\gamma}$ subunits in the fractions are identical to that of purified subunit as shown in Fig. 4A and B. Similar results on separating transducin subunits by TSK column has been reported.[20] Figure 4F shows the elution profile of a mixture of the transducin complex and phosducin. $T_{\beta\gamma}$ is absent at fraction 26; instead it coeluted with phosducin at fraction 23, which indicates the formation of a phosducin/ $T_{\beta\gamma}$ complex. T_α is eluted with a peak at fraction 24 in a manner similar to that of purified T_α, which indicates the dissociation of T_α from the phosducin/$T_{\beta\gamma}$ complex. As can be seen, the phosducin/$T_{\beta\gamma}$ elution profile of panel F in the presence of T_α is identical to that of Fig. 4D in the absence of T_α. This observation indicates that phosducin could dissociate $T_{\beta\gamma}$ from T_α. Moreover, unlike the $T_\alpha T_{\beta\gamma}$ complex, the phosducin/$T_{\beta\gamma}$ complex does not dissociate during the HPLC separation, which suggests stronger interaction between phosducin and $T_{\beta\gamma}$ than between T_α and $T_{\beta\gamma}$.

Role of ROS Disk Membranes in Phosducin/ Transducin Interaction

Phosducin being a soluble protein only interacts with transducin molecules that are released from ROS membrane and does not directly interrupt rhodopsin-bound

[19] R. H. Lee, B. S. Lieberman, and R. N. Lolley, *Exp. Eye Res.* **51**, 325 (1990).
[20] B. K.-K. Fung, J. Hurley, and L. Stryer, *Proc. Natl. Acad. Sci. U.S.A.* **78**, 152 (1981).

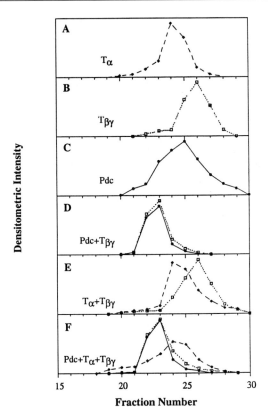

FIG. 4. Column separation of T_α from phosducin-$T_{\beta\gamma}$. Separation of transducin subunits and phosducin was conducted by HPLC using a Superose-12 (Pharmacia) column. After chromatographic separation, the protein composition of the fractions was analyzed by SDS–polyacrylamide gel electrophoresis. (A) Elution profile of purified T_α (◆). (B) Elution profile of purified $T_{\beta\gamma}$ (□). (C) Elution profile of purified phosducin (●). (D) Elution profile of the phosducin/$T_{\beta\gamma}$ complex isolated either directly from retinal extract or reconstituted using purified phosducin and $T_{\beta\gamma}$. (E) Elution profile of the transducin complex ($T_\alpha/T_{\beta\gamma}$). (F) Elution profile of the phosducin/transducin ($T_\alpha/T_{\beta\gamma}$) complex. Reproduced with permission from R. H. Lee, T. D. Ting, B. S. Lieberman, D. E. Tobias, R. N. Lolley, and Y. K. Ho, *J. Biol. Chem.* **267,** 25104 (1992).

transducin. The initial activation of the visual response in photoreceptor cell is mediated by ROS disk membrane associated transducin. Thus, phosducin does not affect the initial cycle of the transducin activation. However, after the initial activation, $T_{\beta\gamma}$ becomes soluble and is sequestered by phosducin. The sequestration of $T_{\beta\gamma}$ in cytosol stops the recycling of the T_α for continuous activation by R*. To demonstrate this feature, the fast kinetics of transducin coupled GTP hydrolysis is measured to show that the initial phase is not, but the recycling phase is affected by phosducin. In addition, the presence of ROS disk membrane protects transducin complex from phosducin-induced dissociation and solubilization.

Fast Kinetics of GTP Hydrolysis by Transducin

If phosducin has no effect on membrane-bound transducin, it should not be involved in regulating the initial activation of transducin. However, phosducin may inhibit the recycling of activated transducin that is dissociated from ROS membrane. This implies that phosducin may function as a negative regulator for the turning-over of the cGMP cascade, and may play a role in the light/dark adaptation process of the photoreceptor cells. To illustrate this point, the effect of phosducin on the pre-steady-state kinetics of GTP hydrolysis by transducin is examined. Using a rapid acid quenching method, the formation of inorganic P_i from GTP hydrolysis in the initial activation of membrane-associated transducin and the recycling of transducin under steady-state turnover can be monitored.[21] In a reconstituted system containing R^* and transducin, $[\gamma\text{-}^{32}P]GTP$ was mixed rapidly with the sample to initiate the reaction. After a predetermined incubation time from 2 to 60 sec, perchloric acid was added to quench the reaction and the kinetics of the formation of $[^{32}P]Pi$ was followed. The rate of P_i formation due to GTP hydrolysis exhibited biphasic characteristics with an initial burst of P_i formation occurring between 1 and 4 sec, which was followed by a slow steady-state rate. Kinetic analyses indicate that the burst represents the rapid activation of transducin by R^*-catalyzed GTP/GDP exchange followed by fast hydrolysis of GTP at the binding site of transducin which results in the accumulation of T_α-GDP $\cdot$ P_i complexes. The steady-state rate represents the slow release of P_i from T_α-GDP $\cdot$ P_i and the recycling of T_α-GDP for additional activation by R^*. Under these conditions, the initial burst of P_i formation represents the initial activation of R^*-associated transducin and the steady-state rate of P_i formation represents the turnover of solubilized transducin. The reaction mixtures containing 2 μM transducin (0.06 nmol.) and 20 μM R^* in the presence or absence of phosducin are incubated at room temperature for 5 min. Phosducin is added after brief incubation of R^* and transducin to ensure the binding of transducin to the R^* membrane. The reaction was initiated by the addition of 30 μl of 60 μM $[\gamma\text{-}^{32}P]GTP$ and was stopped at the appropriate times ranging from 2 to 60 sec by the addition of 0.2 M perchloric acid. The amount of inorganic phosphate in the samples was assayed by molybdate precipitation, followed by filtration onto Whatman glass fiber filters and radioactivity counting. As shown in Fig. 5, the presence of 10 μg phosducin in a sample containing 20 μM R^*, 20 μg transducin, and 60 μM $[\gamma\text{-}^{32}P]GTP$ has little effect on the rate of the initial burst of P_i formation. Furthermore, the sizes of the initial burst, which are proportional to the amount of R^*-associated transducin that is available for the initial activation, are similar between the control and phosducin samples. However, the steady-state rate of GTP hydrolysis of the phosducin-containing sample is decreased approximately 40% as compared with the control sample. These results further corroborate the idea that phosducin only modulates the turnover of soluble transducin but not its initial activation by R^*.

[21] T. D. Ting and Y.-K. Ho, *Biochemistry* **30**, 8996 (1991).

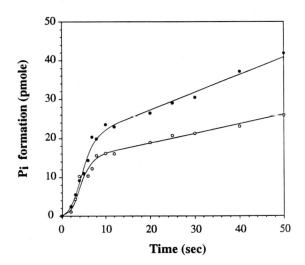

Fig. 5. The effect of phosducin on the fast kinetics of transducin GTPase activity. The pre-steady kinetic analysis of GTP hydrolysis by transducin was carried out according to Ting and Ho.[21] Using a rapid acid quenching procedure the formation of P_i due to GTP hydrolysis was monitored. The assays were conducted at $22°$. The reaction mixture contained 30 μl solution of 6 μM R* and 20 μg T. The reaction was initiated by the addition of 30 μl of 60 mM $[\gamma\text{-}^{32}P]GTP$. Time courses of Pi formation due to GTP hydrolysis by transducin in the absence of phosducin (●) and in the presence of 10 μg phosducin (○) are shown. The initial burst indicates the rapid hydrolysis of transducin-bound and the steady rate of P_i formation represents the steady rate of GTP hydrolysis which relates to the turnover of the transducin coupling cycle. Reproduced with permission from R. H. Lee, T. D. Ting, B. S. Lieberman, D. E. Tobias, R. N. Lolley, and Y. K. Ho, *J. Biol. Chem.* **267**, 25104 (1992).

ROS Membrane Protection of Transducin Subunit Dissociation by Phosducin

The binding of transducin to ROS membrane was assayed by a centrifugation method.[22] In general, transducin was mixed with urea–EDTA stripped ROS membrane (150 μM R*). After a 10 min incubation at $25°$, the solubilized transducin was separated from the membrane-bound transducin by centrifugation in a Beckman Airfuge (5 min at 20 psi at room temperature). The soluble transducin in the supernatant was electrophoretically separated on a 13% SDS–polyacrylamide gel. The amount of transducin bound or released is quantitated from the Coomassie blue stained bands. The results are shown in Fig. 6. Lane 1 (Fig. 6) is a control sample of photolyzed stripped ROS membrane that contains little soluble protein. Lane 2 (Fig. 6) contains 20 μg each of purified transducin and phosducin, the amount of protein used in all the experiments. In the absence of ROS membrane,

---

[22] R. H. Lee, T. D. Ting, B. S. Lieberman, D. E. Tobias, R. N. Lolley, and Y.-K. Ho, *J. Biol. Chem.* **267**, 25104 (1992).

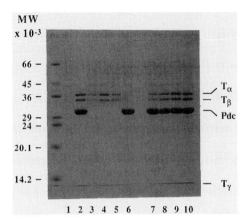

FIG. 6. Effect of phosducin on the binding and release of transducin from photolyzed ROS membrane. The binding and release of transducin from the ROS membrane in the absence or presence or GTP were carried using the centrifiguation method. Soluble transducin was separated from ROS membrane by centrifugation in a Beckman. The amounts of transducin in the supernatants were quantitated from the corresponding protein bands separated by SDS–polyacrylamide gel electrophoresis. Where appropriate, each sample contained 100 μM stripped ROS, 20 μg transducin, 20 μg phosducin, or 200 μM GTP in a volume of 50 μl. The incubation time for binding of the various components is 10 min for all samples. The samples are as follows: lane 1, supernatant from the centrifugation of stripped ROS membrane; lane 2, control amounts of transducin and phosducin in the absence of ROS membrane; lane 3, transducin incubated with stripped ROS membrane in the absence of GTP; lane 4, addition of GTP to the sample in lane 3 with the soluble and membrane-bound transducin separated immediately; lane 5, same sample content as lane 4 except the sample was incubated for 30 min to allow complete hydrolysis of the added GTP; lane 6, phosducin incubated with stripped ROS membrane; lane 7, transducin incubated with stripped ROS membrane for 10 min prior to the addition of phosducin; lane 8, transducin incubated with phosducin for 10 min prior to the addition of stripped ROS membrane; lane 9, GTP added to the sample in lane 7; lane 10, sample of lane 9 with 30 min incubation allowing the complete hydrolysis of the GTP. Reproduced with permission from R. H. Lee, T. D. Ting, B. S. Lieberman, D. E. Tobias, R. N. Lolley, and Y. K. Ho, *J. Biol. Chem.* **267,** 25104 (1992).

all the transducin and phosducin are in soluble form. Therefore, lane 2 (Fig. 6) represents the maximum amount of transducin and phosducin in all other samples. After transducin is incubated with photolyzed ROS membrane, the majority (>90%) of transducin binds to the membrane and low amounts remain in the supernatant (lane 3, Fig. 6). On the addition of 200 μM GTP, ~90% of the membrane-bound transducin dissociates from ROS membrane to form the soluble T_α-GTP and $T_{\beta\gamma}$. As shown in lane 4 (Fig. 6), the amount of transducin in the supernatant increases compared to lane 3 (Fig. 6). After incubating the same sample in lane 4 for 30 min to allow the complete hydrolysis of the added GTP, T_α-GDP reassociates with $T_{\beta\gamma}$ and transducin rebinds to ROS membrane. As shown in lane 5 (Fig. 6), transducin in the soluble fraction is decreased. Lanes 6–10 (Fig. 6) are samples containing phosducin. When phosducin alone is mixed with stripped ROS membrane, all the phosducin molecules remain in the soluble fraction as shown in lane 6 (Fig. 6), which indicates that phosducin does not interact with rhodopsin strongly. To test whether phosducin can dissociate membrane-bound transducin, transducin is first incubated with ROS membrane to allow for binding. After a 10 min incubation period, phosducin is added and further incubated for an additional 10 min before centrifugation. As shown in lane 7 (Fig. 6), $\sim$90% of transducin remains bound on the membrane, which indicates that phosducin is not capable of dissociating ROS membrane-bound transducin. This result can be interpreted in terms that the phosducin-binding site of $T_{\beta\gamma}$ may be blocked when transducin is bound to ROS membrane or that phosducin does not interact with ROS membrane at all. However, when transducin is first incubated with phosducin for 10 min and then ROS membranes is introduced, $\sim$65% of transducin remains in the soluble fraction (lane 8, Fig. 6). This observation supports the suggestion that phosducin competes with soluble T_{α} to form a complex with $T_{\beta\gamma}$. The phosducin/$T_{\beta\gamma}$ is a soluble complex and the dissociated T_{α} without the assistance of $T_{\beta\gamma}$ is unable to bind to ROS membrane. When GTP is added to the sample of lane 7 (Fig. 6) in which transducin is incubated with ROS membrane prior to the addition of phosducin, $>$90% of the membrane-bound transducin is released into solution (lane 9, Fig. 6). However, with phosducin in the system, $\sim$80% of the soluble transducin is blocked from rebinding to membrane after a 30 min incubation period to allow the complete hydrolysis of GTP (lane 10, Fig. 6). Since the addition of GTP leads to the dissociation of T_{α}-GTP and $T_{\beta\gamma}$ from ROS membranes, phosducin can form a complex with $T_{\beta\gamma}$. The depletion of free $T_{\beta\gamma}$ inhibits the rebinding of T_{α} to ROS membrane. These results clearly indicate that phosducin does not dissociate transducin from ROS membrane, but does prevent its rebinding to membrane after it has been solubilized upon activation.

Concluding Remarks

Visual excitation in rod outer segment provides a unique system to study the G-protein coupled second messenger cascade. All the proteins from receptor to effector have been purified and reconstituted assays established for each step of the coupling cycle. The investigator can control the ratio of individual components and their sequence of addition to elucidate their role in the cascade. The phosducin regulation of the cascade has been revealed. Phosducin does not interfere with the initial activation of transducin but negatively regulates the continuous turnover of transducin. By forming a tight complex with $T_{\beta\gamma}$, phosducin blocks the binding of T_{α} to ROS membrane and the cGMP cascade is down-regulated.[22] The phosducin–$T_{\beta\gamma}$ interaction represents a novel mechanism for the regulation of G-protein coupling. Phosducin is phosphorylated by protein kinase A at Ser-73 and its level of phosphorylation is modulated by light. In the dark-adapted state, phosducin is phosphorylated and on illumination, it is rapidly dephosphorylated.

Both the phosphorylated and dephosphorylated forms of phosducin are capable of forming complexes with $T_{\beta\gamma}$. In competitive binding studies, the dephosphorylated form of phosducin seems to form a tighter phosducin/$T_{\beta\gamma}$ complex.[23] It is likely that phosphorylation of phosducin reduces the affinity to $T_{\beta\gamma}$ sufficiently for T_α to compete for $T_{\beta\gamma}$ reassociation. The $T_\alpha T_{\beta\gamma}$ complex will bind to ROS membrane, avoiding phosducin. As a result, the cGMP cascade is reset to its starting point.

The crystal structures of phosducin-$T_{\beta\gamma}$ have been solved in the native complex from bovine retinas and the recombinant complexes (lacking the T_γ C-terminal farnesylation site) of the dephosphorylated and phosphorylated forms.[24–26] The T_β forms a seven-bladed β propeller structure with $T_{\beta\gamma}$ forming a coil–coil structure in the N terminus. Phosducin consists of two domains, an N-terminal helical domain containing the phosphorylation site and a C-terminal thioredoxin-like mixed $\alpha\beta$ structure. The phosducin N terminus forms a cap on top of the T_β propeller structure blocking part of the T_α interacting site. The phosducin C-terminal domain interacts on the side of the propeller, which opens the β propeller structure internalizing the T_γ farnesyl group, which in turn controls the membrane dissociation/dissociation of $T_{\beta\gamma}$. The crystal structure provides a molecular interpretation of the phosducin regulation. Phosducin is found in tissues other than retinas and phosducin-like proteins (PHLP) are more widely expressed in other organisms.[27–29] In general, it is thought to play a similar role in regulating other $G\beta\gamma$. In the search for potential phosducin-interacting proteins other than $G\beta\gamma$, a yeast two-hybrid system has been developed to screen the yeast genome. The SUG1 gene product, a subunit of the 26S proteosome, has been identified as interacting with phosducin, raising the possibility that phosducin may target $G\beta\gamma$ for degradation in cell as a long-term regulatory mechanism.[30]

[23] F. Chen and R. H. Lee, *Biochem. Biophys. Res. Commun.* **233,** 370 (1997).

[24] A. Loew, Y.-K. Ho, T. Blundell, and B. Bax, *Structure* **6,** 1007 (1998).

[25] R. Gaudet, A. Bohm, and P. B. Sigler, *Cell* **87,** 577 (1996).

[26] R. Gaudet, J. R. Savage, J. N. McLaughlin, D. M. Willardson, and P. B. Sigler, *Mol. Cell* **3,** 649 (1999).

[27] C. M. Craft, J. Xu, V. Z. Slepak, X. Zhan-Poe, X. Zhu, B. Brown, and R. N. Lolley, *Biochemistry* **37,** 15758 (1991).

[28] P. H. Bauer, S. Muller, M. Puzicha, S. Pippig, B. Obermaier, F. J. M. Helmreich, and M. J. Lohse, *Nature* **358,** 73 (1992).

[29] S. Danner and M. J. Lohse, *Proc. Natl. Acad. Sci. U.S.A.* **93,** 10145 (1996).

[30] X. Zhu and C. M. Craft, *Mol. Vis.* **4,** 13 (1998).

[10] Analysis of Signal Transfer from Receptor to G_o/G_i in Different Membrane Environments and Receptor-Independent Activators of Brain G Protein

By CATALINA RIBAS, MOTOHIKO SATO, JOHN D. HILDEBRANDT, and STEPHEN M. LANIER

Introduction

A major determinant of signaling specificity for G-protein-coupled receptors is the cell-specific expression of the subtypes of the primary signaling entities, R, G, and effector (E). The regulation of specific cell signaling pathways is also influenced by cell architecture and the stoichiometry of signaling components. The variable efficiency/specificity of coupling observed for many receptors of this class in different cells actually suggests that there are additional unidentified, cell-specific proteins/lipids influencing receptor coupling and basal activity of G-protein signaling systems. One working hypothesis that encompasses several recent observations related to signaling events is that signaling efficiency/specificity is determined in part by "accessory" proteins, distinct from R, G, and E, found in the microenvironment of the receptor that, together with R, G, and E, contribute to the formation of a signal transduction complex at the cytoplasmic face of the receptor. Such accessory proteins may segregate the receptor to microdomains of the cell and regulate the efficiency and/or specificity of signal transfer from R to G and G to E. As one approach to this issue, we focused on the regulation of signal transfer from R to G and developed a "signal restoration" assay in which we could evaluate receptor coupling to the same population of G proteins in different cellular environments.[1] We then further extended this approach to develop a solution-phase assay for evaluation of entities that influence the activation state of G protein independent of receptor.[1,2]

Signal Restoration Assay

General Considerations

The interaction of G-protein-coupled receptors with G proteins in membrane preparations is monitored with three major readouts: Gpp(NH)p-sensitive, high-affinity binding of agonists; GTPase activity; and [^{35}S]GTPγS binding. Although clearly related, the interaction of receptor with G proteins [i.e., Gpp(NH)p-

---

[1] M. Sato, R. Kataoka, J. Dingus, M. Wilcox, J. Hildebrandt, and S. M. Lanier, *J. Biol. Chem.* **170**, 15269 (1995).

[2] M. Sato, C. Ribas, J. D. Hildebrandt, and S. M. Lanier, *J. Biol. Chem.* **271**, 30052 (1996).

sensitive, high affinity binding of agonists] and the final transfer of signal from R to G (i.e., GTPγS binding) are distinct events. The latter is most commonly measured in GTPγS binding assays. Receptor-mediated activation of specific G proteins can be detected by photolabeling of activated G proteins with [^{32}P]azidoanilido-GTP followed by immunoprecipitation with specific Gα antisera.[3,4]

Agonist-induced increases in GTPγS binding in membrane preparations is of particular use for receptor coupling to G_i/G_o proteins and less so for evaluation of receptor coupling to G_q or G_s, primarily due to the differences in nucleotide exchange/subunit interaction properties for the specific G proteins. The signal-to-noise ratio in such assays is optimized for G_i/G_o activation by using small amounts of membrane protein (10 μg) and nonsaturating amounts of GTPγS (0.2–1 nM), and by the inclusion of GDP (1–10 μM), which lowers background nucleotide binding and optimizes the detection of receptor activation. Basal specific binding (without agonist) progressively decreases with increasing concentrations of GDP. This assay system was optimized for analysis of α_2-AR receptor mediated activation of G proteins in different cell types[1,5,6] and subsequently modified to develop a "signal restoration assay."[1] The latter assay involves the elimination of receptor coupling to endogenous G_i/G_o proteins by cell pretreatment with pertussis toxin and the restoration of agonist-mediated signal by addition of purified G_i/G_o protein to the membrane preparation. This approach allowed the evaluation of receptor coupling to the same population of G proteins in different cellular environments. Of course this assay works only if the receptor of interest in a specific cell type couples primarily to G_i/G_o proteins. Receptor coupling to endogenous G proteins can also be eliminated by removal of G proteins and associated peripheral membrane proteins by urea washing of membranes.[7,8] Subsequent addition of G protein to these urea-stripped membranes also restores agonist-induced activation of G proteins. The urea stripping may remove unknown regulatory components from the membrane environment as well. Other variations on this theme include the expression of the receptor of interest in Sf9 (*spodoptera frugiperda* ovary) cells followed by reconstitution of exogenous G proteins with the Sf9 membranes and subsequent analysis of high-affinity binding of agonist or agonist-induced increases in GTPγS binding.[9–11]

[3] K. L. Laugwitz, K. Spicher, G. Schultz, and S. Offermanns, *Methods Enzymol.* **237**, 283 (1994).

[4] T. W. Gettys, T. A. Fields, and J. R. Raymond, *Biochemistry* **33**, 4283 (1994).

[5] W.-N. Tian, S. M. Lanier, E. Duzic, and R. C. Deth, *Mol. Pharmacol.* **45**, 524 (1994).

[6] Q. Yang and S. M. Lanier, *Mol. Pharmacol.* **56**, 651 (1999).

[7] H. Shichi and R. L. Somers, *J. Biol. Chem.* **253**, 7040 (1978).

[8] M. Glass and J. K. Northup, *Mol. Pharmacol.* **56**, 1362 (1999).

[9] M. A. Kazmi, L. A. Snyder, A. M. Cypess, S. G. Graber, and T. P. Sakmar, *Biochemistry* **39**, 3734 (2000).

[10] R. A. Figler, S. G. Graber, M. A. Lindorfer, H. Yasuda, J. Linden, and J. C. Garrison, *Mol. Pharmacol.* **50**, 1587 (1996).

[11] S. G. Graber, A. Figler, and J. C. Garrison, *Methods Enzymol.* **237**, 212 (1994).

Materials

Membrane Preparations. The rat RG-20 α_2-AR was stably expressed in NIH 3T3 fibroblasts, PC-12 pheochromocytoma cells, or DDT-MF2 smooth muscle cells by cotransfection with the receptor gene in the expression vector pMSV and pNEO, a plasmid that confers G418 resistance.[12,13] None of the cell lines used for gene transfection expressed the endogenous α_2-AR gene as determined by radioligand binding experiments and RNA blot analysis. Confluent 100 mm plates of cells are harvested by scraping with a rubber policeman into 3 ml/plate of cell washing solution at 4° (137 mM NaCl, 2.6 mM KCl, 1.8 mM KH$_2$PO$_4$, 10 mM Na$_2$HPO$_4$) and pelleted by centrifugation (500g). Generally, 2–4 confluent plates of cells are sufficient for one assay. Pelleted cells are lysed in 1 ml/plate cell lysis buffer (10 mM Tris-HCl, pH 7.5, 5 mM EDTA, 5 mM EGTA) using a 26-gauge needle and centrifuged in a microfuge at 4° to generate a crude membrane pellet. The membrane pellet is washed twice in buffer A (50 mM Tris-HCl, pH 7.4, 5 mM MgCl$_2$, 0.6 mM EDTA) again using the 26-gauge needle to resuspend the pellet. The final membrane pellet is resuspended in the same buffer at a membrane protein concentration of ~2 mg/ml. All buffers contain a cocktail of protease inhibitors (phenylmethylsulfonyl fluoride 0.1 mM, aprotinin 10 μg/ml, leupeptin 10 μg/ml).

Brain G Protein. G proteins are purified from bovine brain as described.[14] The G-protein preparation consists of approximately 63% G_{oA}, 4% G_{oB}, 16% G_{oC}, 16% G_{i1}, 1% G_{i2}. The heterotrimeric G-protein preparation is isolated in its GDP–ligand form, and greater than 80% of the heterotrimer is functional based on the amount of GTPγ^{35}S binding expected from protein determinations.

Nucleotides. [^{35}S]GTPγS (1250 Ci/mmol) (Dupont/NEN, Boston, MA); guanosine diphosphate, and guanosine 5′-O-3-thiotriphosphate (Boehringer-Mannheim, Indianapolis, IN).

Additional Reagents/Materials. Pertussis toxin (Research Biochemicals Inc., Natick, MA); BA85 nitrocellulose membranes and #32 glass fiber membranes (Schleicher & Schuell, Keene, NH); Thesit (polyoxyethylene-9-lauryl ether; Boehringer-Mannheim, Indianapolis, IN); [^{3}H]RX821002 (Amersham-Pharmacia Biotech, Piscataway, NJ). Thesit is deionized on a AG501-X8 (20-50 mesh, Bio-Rad, Hercules, CA) resin prior to use.[15]

Method of Assay

1. Pretreat cells with 100 ng/ml pertussis toxin for 18 hr prior to membrane preparation. All experiments utilize freshly prepared membranes without

[12] E. Duzic, I. Coupry, S. Downing, and S. M. Lanier, *J. Biol. Chem.* **267,** 9844 (1992).

[13] E. Duzic and S. M. Lanier, *J. Biol. Chem.* **267,** 24045 (1992).

[14] J. Dingus, M. Wilcox, R. Kohnken, and J. Hildebrandt, *Methods Enzymol.* **237,** 457 (1994).

[15] J. Codina, W. Rosenthal, J. D. Hildebrandt, L. Birnbaumer, and R. D. Sekura, *Methods Enzymol.* **109,** 446 (1985).

intervening freeze/thaw. Determine protein concentration of membrane preparations. One confluent 100 mM plate will yield ~600 μg of membrane protein.

2. Pellet the membrane suspension in a microfuge at 4° and resuspend this pellet in buffer A (50 mM Tris-HCl, pH 7.4, 5 mM MgCl$_2$, 0.6 mM EDTA) at a membrane protein concentration of 2 mg/ml.

3. A preincubation mixture is prepared for each assay point. A single assay point consists of eight test tubes: four tubes without agonist—two for total binding, two for nonspecific binding; four tubes with agonist—two for total binding, two for nonspecific binding. The assay may also be done triplicate. Nonspecific binding is defined with 100 μM GTPγS or Gpp(NH)p. The final concentration of GDP is 1–10 μM. In the standard assay system, the preincubation mixture consists of brain G protein (125 nM) and 60 μg of membranes in 120 μl of buffer A containing 0.005% Thesit and 5–50 μM GDP. The concentrations of brain G protein and GDP in the preincubation mixture are five times the final concentration desired in the assay tube.

4. Continue the preincubation for 1 hr at 4° and then add 20 μl of the mixture to assay tubes containing 80 μl of buffer A plus (final concentrations) 150 mM NaCl, 1 mM dithiothreitol, 50,000 cpm (~0.2 nM) [^{35}S]GTPγS, and agonist, vehicle, or 100 μM GTPγS (total final volume, 100 μl).

5. Continue incubation with shaking at 24° for various times (generally 30 min) and terminate the reaction by rapid filtration through nitrocellulose filters (BA85, Schleicher & Schuell) with 4 × 4-ml washes (50 mM Tris-HCl, 5 mM MgCl, pH 7.4, 4°). Radioactivity bound to the filters is determined by liquid scintillation counting.

6. Determine receptor density in each membrane preparation (25 μg of membrane protein per binding tube) using saturating concentrations of [^{3}H]RX821002 (20 nM), a selective α_2-AR antagonist. The glass fiber membranes are used for receptor binding assays. The objective here is to define the level of receptor expression for each experiment. Depending on the receptor gene transfected, the vector used, and the cell type, the stability of receptor expression may vary and it is important to monitor receptor expression on a weekly basis and to have a receptor density value for each membrane preparation used in the GTPγS binding experiments. Whereas the use of the membranes for receptor coupling to G proteins in the signal restoration assay is compromised by freeze/thaw, the receptor density measurements are not markedly altered and they can be frozen and assayed later for receptor binding.

Properties of Assay System

Based on earlier studies involving signal reconstitution,[16–19] several experimental paradigms were evaluated to achieve optimal conditions for reconstitution of the agonist-induced signal. Using the agonist-induced increase in GTPγS binding as a readout, we evaluated different preincubation times of membranes and G protein, different detergent concentrations, and different buffer conditions (Mg^{2+}/GDP). The time course for G-protein activation should be determined for each experimental system as it may vary in different cell types. The preincubation and subsequent dilution is a commonly used strategy to facilitate G-protein association within the membrane environment. This strategy involves a preincubation at a detergent concentration just above the critical micellar concentration with a dilution into the final assay tubes such that the detergent is present at a final concentration below its critical micellar concentration. This system can also be used to evaluate receptor interaction with G protein by evaluating Gpp(NH)p-sensitive binding of the selective α_2-AR agonist [^{3}H]UK14304 (1 nM) with the exclusion of GDP/sodium and the use of higher amounts of G protein (50–300 nM).[1]

In the intact PC-12 cell, the α_2-AR couples to both PT-sensitive and PT-insensitive pathways to regulate cellular cAMP levels.[13] Under the incubation conditions used in this assay system, the receptor-mediated increases in GTPγS binding in α_2-AR transfectants were pertussis toxin-sensitive, and the maximal agonist-induced signal was observed at 30-min incubation time points. Detection of strong receptor-mediated activation of pertussis toxin-insensitive G proteins in the GTPγS binding assay may require different incubation conditions or factors that are lost upon membrane preparation.

Applications

This assay system was used to address cell-specific aspects of signal transfer from R to G following stable expression of the α_2-AR in different membrane environments (NIH 3T3 fibroblasts, DDT1-MF2 cells, and the pheochromocytoma cell line PC12). The activation of endogenous G proteins in response to stimulation of α_2-AR is indicated in Fig. 1A. Receptor coupling to endogenous G proteins in each cell type was eliminated by pertussis toxin pretreatment (Fig. 1B) and R-G signal transfer restored by reconstitution of cell membranes with purified brain G protein (Fig. 2A,B). The strength of the agonist-induced signal was dependent on the concentration of agonist, magnesium, and G protein (Figs. 2B and 3A,B). As receptor-mediated increases in GTPγS binding were completely blocked by pertussis toxin pretreatment, any signals observed in the signal restoration system must involve the added G protein. Thus, the receptor has access to the same population of G proteins in the different cellular environments. In this signal restoration

[16] P. C. Sternweis, J. K. Northup, M. D. Smigel, and A. G. Gilman, *J. Biol. Chem.* **256**, 11517 (1981).

[17] R. E. Kohnken and J. D. Hildebrandt, *J. Biol. Chem.* **169**, 12508 (1984).

[18] A. Kikuchi, O. Kozawa, T. Katada Kaibuchi, M. Ui, and Y. Takai, *J. Biol. Chem.* **261**, 11558 (1986).

[19] H. M. Kim and R. R. Neubig, *Biochemistry* **26**, 3664 (1987).

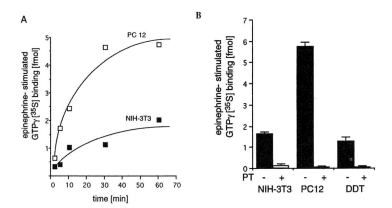

FIG. 1. Increase in GTPγ^{35}S binding elicited by agonist in NIH 3T3 and PC-12 $\alpha_{2A/D}$-AR transfectants. Clonal receptor transfectants were generated and receptor densities determined as described.[1] In each experiment the level of [^{35}S]GTPγS binding was determined in the presence or absence of epinephrine (10 μM) for various incubation times (30 min in B) at 24°. (A) Receptor density (fmol/mg): NIH 3T3, ~1400; PC-12, ~1100. Data presented in (A) are representative of 3–5 separate experiments with different clonal cell lines. (A) Basal [^{35}S]GTPγS binding at 2, 5, 10, 30, and 60 min: NIH 3T3, 0.57, 0.88, 1.37, 1.66, 2.50; PC-12, 0.98, 1.20, 1.91, 3.29, 4.27. (B) Membranes were prepared from control or pertussis toxin (PT)-treated cells as described.[1] Aliquots of the membrane preparations were then evaluated for agonist-mediated effects on [^{35}S]GTPγS binding. Results are presented as the mean ± SEM of four experiments. Receptor density (fmol/mg): NIH 3T3, ~5200; PC-12, ~4550; DDT1-MF2, ~4000. Epinephrine = 10 μM. Basal [^{35}S]GTPγS [fmol]: NIH-3T3, 1.61 ± 0.105; PT+ 0.27 ± 0.045; PC12, 2.12 ± 0.14, PT+ 0.47 ± 0.08; DDT1-MF2, 1.56 ± 0.1, PT+ 0.19 ± 0.04. In (B), DDT refers to DDT1-MF2 cells. (Figure and legend adapted from Ref. 1).

assay, agonist-induced activation of G was 3- to 9-fold greater in PC-12 as compared to NIH 3T3 α_2-AR transfectants (Fig. 3), and this difference was observed over a range of receptor densities.[1]

Receptor-Independent Regulators of G-Protein Activation State

General Considerations

Cell-type specific differences in the GTPγS binding assay described above led us to hypothesize the existence of a cell-specific factor that influenced the transfer of signal from R to G or the basal activation state of G independent of receptor. To pursue the latter, we modified the membrane assay system described above to allow the evaluation of G-protein regulators in a solution phase assay. Specifically, we prepared membrane extracts by detergent solubilization and designed a solution phase assay to detect factors in the extract that influenced the activation state of brain G protein using GTPγS binding as a readout. GTPγS binding to G protein

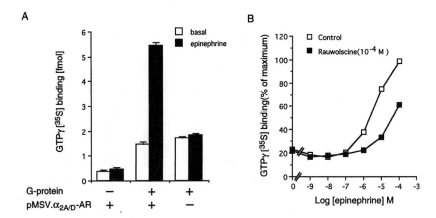

FIG. 2. Agonist-induced activation of G protein in the signal restoration system. Membranes were prepared from cells pretreated with pertussis toxin and reconstituted with 25 nM bovine brain G protein. (A) Agonist-induced activation of G-protein in membrane preparations from PC12 cells transfected with pMSV. $\alpha_{2A/D}$-AR or resistance plasmid alone in the presence and absence of added G protein. Data are presented as the mean $\pm$ SEM of four experiments using different membrane preparations. (B) Effect of increasing epinephrine concentrations on [35S]GTPγS binding in the presence and absence of increasing epinephrine concentrations on [35S]GTPγS binding in the presence and absence of the α_2-AR antagonist rauwolscine following reconstitution with G protein as described in the text. Receptor density, $\sim$5400 fmol/mg. The results are representative of three experiments using different clonal cell lines and are expressed as the percent of the epinephrine-induced increase in [35S]GTPγS binding at 100 μM agonist [35S]GTPγS binding: basal $= 0.81$ fmol, epinephrine (100 μM) $= 3.45$ fmol). (Figure and legend adapted from Ref. 1).

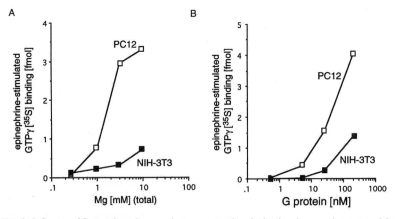

FIG. 3. Influence of G-protein and magnesium concentrations in the signal restoration system. Membranes were prepared from pertussis toxin-treated cells and reconstituted with 25 nM brain G protein. (A) Basal [35S]GTPγS binding [fmol] ranged from 0.75 to 1.24 in NIH 3T3 transfectants and from 1.19 to 2.14 in PC12 transfectants. Data are presented as the mean of duplicate determinations using different clonal cell lines. Experiments were repeated twice in each transfectant with similar results. Epinephrine, 10 μM. (Figure and legend adapted from Ref. 1).

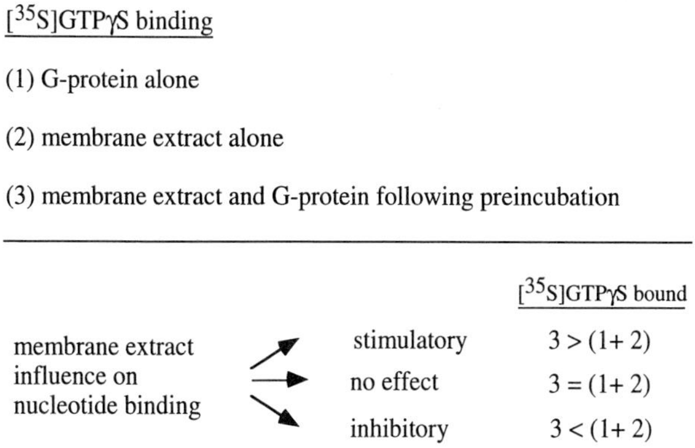

[^{35}S]GTPγS binding

(1) G-protein alone

(2) membrane extract alone

(3) membrane extract and G-protein following preincubation

			[^{35}S]GTPγS bound
membrane extract influence on nucleotide binding	↗	stimulatory	$3 > (1 + 2)$
	→	no effect	$3 = (1 + 2)$
	↘	inhibitory	$3 < (1 + 2)$

FIG. 4. Solution phase assay for G-protein regulators.

was evaluated in three sets of tubes for each experiment: (1) G-protein alone, (2) membrane extract alone, and (3) sample extract and G protein following preincubation (Fig. 4). Thus, greater or less than additive amounts of nucleotide binding in set 3 compared with the sum of sets 1 and 2 would indicate the presence of a factor that influence nucleotide binding in a stimulatory or inhibitory manner, respectively.

Materials

Membrane Extracts

Confluent 100 mM plates of cells are harvested by scraping with a rubber policeman into 3 ml/plate of cell washing solution at 4° and pelleted by centrifugation (500g). Pelleted cells are homogenized in 1 ml/plate cell lysis buffer with a Dounce homogenizer and centrifuged at 35,000g to generate a crude membrane pellet. The membrane pellet is washed twice by homogenization/centrifugation in buffer A (50 mM Tris HCl, pH 7.4, 5 mM MgCl$_2$, 0.6 mM EDTA) with resuspension in the same buffer at a membrane protein concentration of ∼10 mg/ml. An equal volume of solubilization buffer (2% detergent in buffer A containing protease inhibitors, phenylmethylsulfonyl fluoride 0.1 mM, leupeptin 10 μg/ml, aprotinin 10 μg/ml) is then added to the resuspended membranes and the mixture is homogenized twice with a Dounce homogenizer followed by a 1 hr rotating incubation at 4°. The detergent : protein ratio is thus 2 : 1. The solubilized membrane extract is isolated by centrifugation at 100,000g for 1 hr at 4°. We have evaluated several detergents for solubilization including deoxycholate (Sigma, St. Louis, MO), CHAPS (Boehringer Mannheim), digitonin (Garard Schlesinger), and Thesit (polyoxyethylene-9-lauryl ether, Boehringer Mannheim). Each of these detergents solubilizes 40–60% of membrane protein. The nonionic detergent Thesit has proved most effective for our purposes.[1]

The NG108-15 G-protein activator is partially purified as described.[2] The purification scheme involves DEAE ion-exchange chromatography (Bio-Rad DEAE 20 (MacroPrep DEAE 10 μM support), ultrafiltration (100,000 molecular weight filter, Amicon, Danvers, MA), and gel filtration (Pharmacia, Piscataway, NJ Superose 12). The elution profile of the G-protein activator and recoveries are described elsewhere.[2] The overall degree of purification achieved is ~600-fold with a recovery of ~28% based on the specific activity of the crude solubilized preparation.

Additional Reagents/Materials. As described above for the membrane assay.

Method of Assay

1. Prepare a preincubation mixture for each set of assay tubes. The preincubation mixture consists of brain G protein, GDP, and other components in buffer A (50 mM Tris-HCl, pH 7.4, 5 mM MgCl$_2$, 0.6 mM EDTA). The concentrations of brain G protein and GDP in the preincubation mixture were five times the final concentration desired in the assay tube. A typical preincubation cocktail consists of 5 μM GDP, 6–25 nM brain G protein, 30 μl crude solubilized extract or partially purified NG108-15 G-protein activator in a total volume of 50 μl (50 mM Tris-HCl pH. 7.4, 5 mM MgCl$_2$, 0.6 mM EDTA). The detergent concentration during preincubation is 0.6% if a crude extract is used or 0.06% if a fractionated extract is used as during purification, proteins are eluted from resins using buffer containing 0.1% Thesit.

2. Incubate the preincubation mixture at 4° for 1 hr.

3. Add 10 μl of the preincubation mixture to duplicate assay tubes containing buffer A plus (final concentrations) 150 mM NaCl, 1 mM dithiothreitol, ~230,000 cpm (~2 nM) [^{35}S]GTPγS, vehicle, or 100 μM GTPγS for nonspecific binding (total final volume, 50 μl). The detergent concentration in the assays tubes is now 0.12% if a crude extract is used or 0.012% if a fractionated extract is being assayed.

4. Incubate samples at 24° for 10 to 60 min.

5. Terminate reaction by rapid filtration through nitrocellulose filters (Schleicher and Schuell, Keene, NH BA85) with 4 × 4-ml washes with buffer B (50 mM Tris, 5 mM MgCl$_2$, 1 mM EDTA, pH 7.4, 4°). Radioactivity bound to the filters is determined by liquid scintillation counting.

Properties of the Assay System

The conditions for the solution phase assay are based upon the optimal conditions required to detect receptor-mediated activation of G$_i$/G$_o$ in membrane preparations as described above. To slow the rate of association of GTPγS to G proteins, an excess of GDP was added; thus the system is perhaps biased to detect entities that accelerate nucleotide exchange or perhaps increase the affinity of G protein for GTPγS. Relatively low concentrations of G protein (0.5–25 nM) and membrane extract (1–10 μg protein) are used in the assay system. The readout in this system increases with increasing amounts of G protein. The use of low amounts of brain G protein, added GDP, and nonsaturating concentrations of GTPγS all work to decrease basal GTPγS binding and increase the signal-to-noise ratio for entities that increase GTPγS binding to brain G protein. Under these incubation conditions [^{35}S]GTPγS binding to G protein alone ranges from 2000 to 10,000 dpm. The level of [^{35}S]GTPγS binding in tubes containing extract alone depends upon the tissue used to prepare the extract. A higher background is obtained in brain extracts because of the high amounts of G_o in brain, and this complicates detection of activators for the small amount of added G protein. Our highest signal-to-noise ratios are obtained using extracts from neuronal cell lines.

Although the binding of GTPγS reaches a plateau in this assay after 30–60 min, in some ways it is a complicated assay that may not be truly at equilibrium. One consequence of the assay conditions used (i.e., nonsaturating concentrations of GTPγS and added GDP) is that positive readouts likely reflect increased GTPγS binding to a portion of the added G protein. This is in contrast to other assay conditions commonly used in reconstitution assays,[20] where saturating concentrations of GTPγS (1 μM) are used and G-protein activation is reflected as an increase in the rate of GTPγS binding, rather than the total amount of GTPγS bound as in the assay system described here.

Applications

The solution phase assay for G-protein activators was used to demonstrate the existence of a factor in PC-12 cell membrane extracts that increased the amount of GTPγS bound to G protein. This bioactivity was absent in a detergent-solubilized extract of membranes prepared from NIH 3T3 fibroblasts.[1] Screening of additional cell lines and tissues indicated that this bioactivity was most abundant in membrane extracts prepared from the neuroblastoma × glioma cell hybrid NG108-15 (Fig. 5). Preincubation of membrane extracts from NG108-15 increased the binding of nucleotide to bovine brain G protein by 460% (Fig. 5). The stimulatory effect of the membrane extract from NG108-15 cells was heat-sensitive and was retained in the concentrate following centrifugation in a YM30 microconcentrator (fractionation size ~30,000 daltons) indicating that this factor is likely a protein rather than a small organic molecule.[1,2] Several series of experiments were

[20] M. Ross and T. Higashijima, *Methods Enzymol.* **237**, 26 (1994).

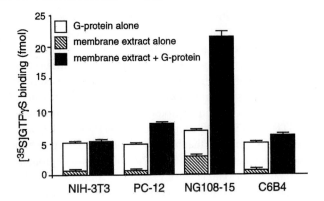

FIG. 5. Effect of cell membrane extracts on guanine nucleotide binding behavior of brain G protein. Membranes from NG108-15 and C6B4 glioma, PC-12, and NIH-3T3 cell lines were solubilized with 1% deoxycholate (detergent : protein, 2 : 1) and used in the solution phase assay to evaluate effects on the binding of [^{35}S]GTPγS to purified G protein. For each set of experiments, the binding of [^{35}S]GTPγS (2 nM) was determined in membrane extract (11 μg protein) alone, G protein alone (1.25 nM), and following preincubation of membrane extract (11 μg protein) with G protein. Data are presented as the mean ± SEM of four experiments. (Figure and legend adapted from Ref. 2).

performed to determine optimal conditions for solubilization of the bioactivity from NG108-15 cells. The nonionic detergent thesit proved most effective in terms of the total amount of bioactivity extracted and the specific activity of the extract.[1] The NG108-15 neuroblastoma–glioma hybrid cell line was used as a source for partial purification and characterization of the bioactivity.

The partially purified material markedly increased GTPγS binding to brain G protein in a manner that was dependent on the time and amount of fractionated membrane extract added to the assay system (Fig. 6).[21] Although the purified bovine brain G-protein preparation is a mixture of G proteins, ~65–80% represents GoA. The partially purified NG108-15 G-protein activator also increased GTPγS to GoA that had been isolated from the bovine brain G-protein preparation by further fractionation[2] (Fig. 7). Interestingly, the NG108-15 G-protein activator also increased GTPγS to Gα which was free of Gβγ (Fig. 7), indicating that this assay system could be set up with different Gα expressed in and purified from recombinant systems to define potential activators of other G-proteins.

Based on these series of studies, an expression cloning system was developed in *Saccharomyces cerevisiae* to rapidly screen mammalian cDNAs for

---

[21] C. Ribas, M. Sato, J. D. Hildebrandt, and S. M. Lanier, manuscript in preparation.

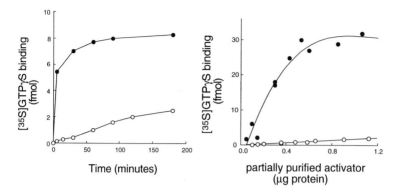

FIG. 6. Properties of the partially purified NG108-15 G-protein activator. The NG108-15 G-protein activator was partially purified by DEAE ion-exchange chromatography, ultrafiltration (100,000 molecular weight filter-Amicon), and gel filtration and assayed as described in the text. The final concentration of G protein in the assay tubes was 1.25 nM. The concentration of GTPγS in (A) and (B) was 2 nM; 0.6 μg of partially purified extract was used in (A). The differences in maximal effect of the partially purified factor in (A) and (B) are due to loss of bioactivity over time. Data are representative of two experiments with duplicate determinations. (Figure adapted from Ref. 21). (A) Open circles—G protein alone; closed circles—G protein plus partially purified extract. (B) Open circles—partially purified extract; closed circles—G protein plus partially purified extract.

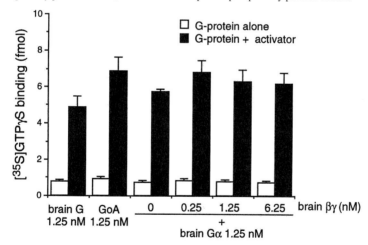

FIG. 7. Effect of G-protein activator on G_{oA} and free $G\alpha$. Fractions of NG108-15 membrane extracts eluted from the DEAE ion-exchange resin containing peak bioactivity were pooled and used to determine the effect of the G-protein activator on free $G\alpha$ vs intact heterotrimers.[1] G_{oA} was purified from the bovine brain G protein preparation by fractionation on a Mono Q resin.[17,29] Bovine brain $G\alpha$ was resolved from $G\beta\gamma$ as previously described.[14,17,28] Approximately 90% of the resolved $G\alpha$ was capable of [^{35}S]GTPγS. The effect of the G-protein activator (0.3 μg extract protein) on [^{35}S]GTPγS (2 nM) binding to $G\alpha$ or heterotrimeric G protein was determined as described in the text. Data are presented as the mean $\pm$ SEM of four experiments. (Figure and legend adapted from Ref. 2.)

receptor-independent regulators of G proteins (see elsewhere in this volume).[22,23] The latter strategy resulted in the identification of three proteins (AGS1-3, activators of G-protein signaling) that activated G proteins in the absence of a receptor and did so by distinct mechanisms.[22-28] The properties of the NG108-15 G-protein activator differ from those of the three AGS proteins, suggesting the existence of additional, as yet undefined, postreceptor regulators of G-protein function. Such proteins may provide alternative modes of input to G-protein-regulated signaling pathways independent of a classical G-protein-coupled receptor and could potentially serve as binding partners for $G\alpha$ or $G\beta\gamma$ independent of heterotrimer formation.

Acknowledgments

This work was supported by grants (NS24821, MH5993—S.M.L.) (DK37219—J.D.H.) from the National Institutes of Health. Dr. Ribas was the recipient of a MUSC Health Sciences Foundation Research Fellowship and a Ministerio de Educacion y Cultura Postdoctoral Fellowship (Spain). Dr. Sato was a visiting scientist from Asahikawa Medical College, Asahikawa, Japan.

[22] M. Cismowski, A. Takesono, C. Ma, J. S. Lizano, S. Xie, H. Fuernkranz, S. M. Lanier, and E. Duzic, *Nature Biotech.* **17,** 878 (1999).

[23] M. J. Cismowski, A. Takesono, C. Ma, S. M. Lanier, and E. Duzic, *Methods Enzymol.* **344,** [11] 2002 (this volume).

[24] A. Takesono, M. J. Cismowski, C. Ribas, M. Bernard, P. Chung, S. Hazard III, E. Duzic, and S. M. Lanier, *J. Biol. Chem.* **274,** 33202 (1999).

[25] M. Cismowski, C. Ma, C. Ribas, X. Xie, M. Spruyt, J. S. Lizano, S. M. Lanier, and E. Duzic, *J. Biol. Chem.* **275,** 23421 (2000).

[26] M. Bernard, Y. K. Peterson, P. Chung, J. Jourdan, and S. M. Lanier, *J. Biol. Chem.* **276,** 1585 (2001).

[27] Y. K. Peterson, M. L. Bernard, H. Ma, S. Hazard, S. G. Graber, and S. M. Lanier, *J. Biol. Chem.* **275,** 33193 (2000).

[28] M. Natochin, B. Lester, Y. K. Peterson, M. L. Bernard, S. M. Lanier, and N. O. Artemyev, *J. Biol. Chem.* **275,** 40981 (2000).

[29] M. D. Wilcox, J. Dingus, E. A. Balcueva, W. E. McIntire, N. D. Mehta, K. L. Schey, J. D. Robishaw, and J. D. Hildebrandt, *J. Biol Chem.* **270,** 4189 (1995).

[11] Identification of Modulators of Mammalian G-Protein Signaling by Functional Screens in the Yeast *Saccharomyces cerevisiae*

By Mary J. Cismowski, Aya Takesono, Chienling Ma, Stephen M. Lanier, and Emir Duzic

Introduction

G-protein-coupled receptor (GPCR) signaling pathways are one of the most widely used mechanisms in nature for transducing signals from the extracellular to the intracellular environment. Signals generated by activated GPCRs are rarely transduced through a linear pathway. Rather, they often modulate, and are modulated by, other signaling pathways within the cell.[1–3] The complexity of intracellular signaling networks generated by activated GPCRs necessitates multiple levels of regulation. The loss of this regulation, leading to inappropriate activation or inactivation of GPCR signaling cascades, is strongly implicated in a variety of human diseases. Somatic mutations of genes coding for key signaling components, such as receptor or heterotrimeric $G\alpha$, have been implicated as causative factors for some diseases.[4,5] For others, the mechanisms underlying aberrant G-protein mediated signaling are less clear.[6–9] Though a variety of proteins modulating GPCR signaling have been identified (e.g., RGS proteins, RAMPs, Homer proteins),[10–21] many of the detailed mechanisms required for regulating and properly directing the input signals remain to be elucidated. The identification of additional modulators of these signaling pathways will greatly aid our understanding of GPCR signaling and its relationship to human disease, as well as provide potentially novel targets for therapeutic intervention.

A variety of approaches have been taken to identify modulators of GPCR signaling pathways, including protein–protein interaction assays,[17–19] expression cloning in *Xenopus* oocytes,[20] biochemical purification,[12,14] and differential

[1] J. S. Gutkind, *J. Biol. Chem.* **273,** 1839 (1998).

[2] L. A. Selbie and S. J. Hill, *Trends Pharmacol. Sci.* **19,** 87 (1998).

[3] L. M. Luttrell, Y. Daaka, and R. J. Lefkowitz, *Curr. Opin. Cell Biol.* **11,** 177 (1999).

[4] A. M. Spiegel, *Annu. Rev. Physiol.* **58,** 143 (1996).

[5] A. M. Spiegel, *J. Inherit. Metab. Dis.* **20,** 113 (1997).

[6] T. W. Gettys, V. Ramkumar, R. S. Surwit, and I. L. Taylor, *Metabolism* **44,** 771 (1995).

[7] J. T. Meji, *Mol. Cell. Biochem.* **157,** 31 (1996).

[8] M. McLaughlin, B. M. Ross, G. Milligan, J. McCulloch, and J. T. Knowler, *J. Neurochemistry* **57,** 9 (1991).

[9] C. O'Neill, B. Wiehager, C. J. Fowler, R. Ravid, B. Winblad, and R. F. Cowburn, *Brain Res.* **636,** 193 (1994).

expression.[21] Each of these techniques, though valuable, is limited either by the lack of functional information or by the throughput. In any search for novel signaling modulators it is advantageous to have a high-throughput, inexpensive system that incorporates functionality in the primary screen. For GPCR signaling pathways the yeast *Saccharomyces cerevisiae* presents just such a system. Yeast cells are inexpensive and easy to propagate. The transformation efficiencies attainable in yeast can easily accommodate high throughput screening of cDNA expression libraries.[22,23] Further, many of the proteins involved in the yeast G-protein-coupled receptor pathway required for response to pheromone are structurally and functionally redundant with corresponding orthologs from higher eukaryotes.[24–27] Indeed, this redundancy has been successfully exploited in the identification of mammalian proteins that negatively regulate the native yeast pheromone pathway.[28,29]

The identification of positive regulators of G-protein signaling in yeast is problematic, however, as the normal consequence of activation of the yeast pheromone response pathway is growth arrest (see below). To avoid this, yeast can be engineered to give a readout of growth on activation of this pathway.[27,30] In addition, the functional redundancy between the pheromone response pathway and

[10] T. Higashijima, J. Burnier, and E. M. Ross, *J. Biol. Chem.* **265,** 14176 (1990).

[11] S. M. Strittmatter, S. C. Cannon, E. M. Riss, T. Higashijima, and M. C. Fishman, *Proc. Natl. Acad. Sci. U.S.A.* **90,** 5327 (1993).

[12] C. Nanoff, T. Mitterauer, F. Roka, M. Hohenegger, and M. Friessmuth, *Mol. Pharm.* **46,** 806 (1995).

[13] T. Okamoto, S. Takeda, Y. Murayama, E. Ogata, and I. Nishimoto, *J. Biol. Chem.* **270,** 4205 (1995).

[14] M. Sato, C. Ribas, J. D. Hildebrandt, and S. M. Lanier, *J. Biol. Chem.* **271,** 30052 (1996).

[15] H. G. Dohlman and J. Thorner, *J. Biol. Chem.* **272,** 3871 (1997).

[16] Y. Odagaki, N. Nishi, and T. Koyama, *Life Sci.* **62,** 1537 (1998).

[17] Y. Luo and B. M. Denker, *J. Biol. Chem.* **274,** 10685 (1999).

[18] N. Mochizuki, G. Cho, B. Wen, and P. A. Insel, *Gene* **181,** 39 (1996).

[19] R. T. Premont, A. Claing, N. Vitale, J. L. R. Freeman, J. A. Pitcher, W. A. Patton, M. Moss, M. Vaughan, and R. J. Lefkowitz, *Proc. Natl. Acad. Sci. U.S.A.* **95,** 14082 (1998).

[20] L. M. McLatchie, N. J. Fraser, M. J. Main, A. Wise, J. Brown, N. Thompson, R. Solari, M. G. Lee, and S. M. Foord, *Nature* **393,** 333 (1998).

[21] P. R. Brakeman, A. A. Lanahan, R. O'Brien, K. Roche, C. A. Barnes, R. L. Huganir, and P. F. Worley, *Nature* **386,** 284 (1997).

[22] D. M. Becker and L. Guarente, *Methods Enzymol.* **194,** 182 (1991).

[23] R. Elble, *BioTechniques* **13,** 18 (1992).

[24] M. Raymond, P. Gros, M. Whiteway, and D. Y. Thomas, *Science* **256,** 232 (1992).

[25] J. L. Brown, L. Stowers, M. Baer, J. Trejo, S. Coughlin, and J. Chant, *Curr. Biol.* **6,** 598 (1996).

[26] M. Hutchinson, K. S. Berman, and M. H. Cobb, *J. Biol. Chem.* **273,** 28625 (1998).

[27] M. J. Cismowski, A. Takesono, C. Ma, J. S. Lizano, X. Xie, H. Fuernkranz, S. M. Lanier, and E. Duzic, *Nature Biotechnol.* **17,** 878 (1999).

[28] K. M. Druey, K. J. Blumer, V. R. Kang, and J. H. Kehrl, *Nature* **379,** 742 (1996).

[29] B. H. Spain, K. S. Bowdish, A. R. Pacal, S. F. Staub, D. Koo, C.-Y. R. Chang, W. Xie, and J. Colicelli, *Mol. Cell. Biol.* **16,** 6698 (1996).

[30] A. Takesono, M. J. Cismowski, C. Ribas, M. Bernard, P. Chung, S. Hazard III, E. Duzic, and S. M. Lanier, *J. Biol. Chem.* **274,** 33202 (1999).

mammalian G-protein signaling pathways allows one to target individual signaling components in the yeast pathway by replacing them with their mammalian counterparts. Here we show that by replacing the yeast Gα protein with a human–yeast chimeric Gα subunit, mammalian cDNA libraries can be screened to identify proteins that activate G-protein pathway signaling at the level of the heterotrimeric Gα, with different mechanisms of action and different Gα specificities. We also show that these identified activating proteins can be used directly to screen yeast for G-protein pathway inhibitors.

General Considerations in Working with Yeast

Although a detailed description of methodologies required for the propagation and genetic manipulation of the yeast *Saccharomyces cerevisiae* is beyond the scope of this chapter, excellent general references are available to researchers.[31] All laboratory strains of yeast carry somatic mutations in genes required for biosynthesis of amino acids and/or nucleosides and, therefore, are propagated in supplemented "rich" medium. Episomal and integrating plasmids typically carry one or more complementary wild-type biosynthetic genes and are selected and/or maintained by propagating yeast in medium lacking the appropriate amino acids or nucleosides. Growth media are generally supplemented with a carbon source at 2%; glucose is the preferred carbon source for most applications. Specific applications, such as induction of galactose-inducible promoters, require the use of different carbon sources (see below). Genetic manipulations are performed using standard gene disruption/deletion techniques and appropriate integrating plasmids.[31] In general gene disruptions are made here using the *URA3* selectable marker followed by growth on 5-fluoroorotic acid to restore uracil auxotrophy. Transformation of yeast with episomal plasmids is performed as described.[23]

Creation of Screening Strains and Libraries

Genetic Alterations of Yeast Pheromone Response Pathway

Haploid *Saccharomyces cerevisiae* exists in two mating types, designated **a** and α (see J. Kurjan[32] and L. Bardwell *et al.*[33] for general reviews of the yeast pheromone response pathway). Each haploid constitutively secretes mating pheromone, **a**-factor, or α-factor respectively, that activates G-protein-coupled receptors on the surface of haploids of the opposite mating type. On receptor activation by pheromone, its associated heterotrimeric G-protein undergoes subunit dissociation into GTP-bound activated Gα and G$\beta\gamma$ dimer (Fig. 1). Free G$\beta\gamma$ then

[31] C. Guthrie and G. R. Fink (eds.), *Methods Enzymol.* **194** (1991).
[32] J. Kurjan, *Annu. Rev. Genet.* **27,** 147 (1993).
[33] L. Bardwell, J. G. Cook, C. J. Inouye, and J. Thorner, *Dev. Biol.* **166,** 363 (1994).

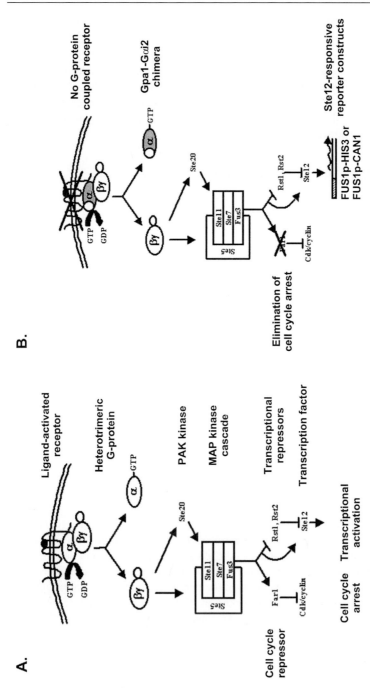

FIG. 1. Engineering of the yeast pheromone response pathway for functional screening. (A) Schematic of the yeast pheromone response pathway. Major signaling components, determined either by functional analysis or by homology to signaling components of higher eukaryotes, are indicated. α, Gpa1; β, Ste4; γ, Ste18. (B) Modifications to the pheromone response pathway in the screening strains. Screening for G-protein activators is performed in strains carrying the $FUS1p\text{-}CAN1$ construct. Screening for G-protein repressors is performed in strains carrying the $FUS1p\text{-}HIS3$ construct. See text for details.

transduces a signal through a p21-activated kinase to a mitogen-activated protein (MAP) kinase cascade, leading to activation of the transcription factor Ste12 as well as Far1-mediated growth arrest in the G_1 phase of the cell cycle. Activated Ste12 then binds to specific pheromone-responsive promoters to induce of a variety of genes required for the cytoskeletal rearrangements associated with mating.

In screening yeast for cDNAs that express G-protein pathway activators, the cell cycle arrest normally associated with pheromone pathway activation can be circumvented by deletion of *FAR1*,[34] thereby uncoupling pathway activation from growth arrest. The introduction of an essential biosynthetic gene whose expression is driven by a pheromone-responsive promoter provides the means for identifying pheromone pathway activators through a growth-based screen. For the screens described here, the promoter for the pheromone-responsive gene *FUS1* (designated *FUS1p*) was ligated to *HIS3,* a gene encoding imidazoleglycerolphosphate dehydratase and required for histidine biosynthesis in yeast. Growth of *far1his3* yeast strains carrying an integrated *FUS1p-HIS3* construct is therefore inhibited in medium lacking histidine unless the pheromone response pathway is activated.

In searching for intracellular modulators of this G-protein-coupled signaling pathway it can also be advantageous to delete the native GPCR. This eliminates possible interference from modulators acting at the receptor–G-protein interface or acting directly on the GPCR. Finally, to target the mammalian cDNA screens toward different components of the signaling pathway, individual yeast genes can be replaced by their mammalian counterparts. For example, to target the heterotrimeric G-protein we replaced the native yeast Gα, Gpa1, with a human Gα_{i2} chimera containing the first 41 amino acids of Gpa1 and introduced it into a *his3 far1 FUS1p-HIS3* strain (CY1316).[27,30] This chimeric Gα functionally couples to the yeast G$\beta\gamma$ as evidenced by its ability to suppress pheromone pathway activity in the absence of Gpa1 (see Table I, below).

cDNA Library Construction

General Considerations. The source material used in generating cDNA libraries may significantly influence the type of modulators identified. As a nonspecific source of activators we used mRNA derived from adult human liver. For a more directed activator screen, we used mRNA derived from the neuroblastoma–glioma hybrid cell line NG108-15, as there is strong biochemical evidence for the presence of a potent G-protein pathway activator in these cells.[35] In addition, as there is evidence of inappropriate up-regulation of G-protein activity in many disease states, it is quite possible that cDNA libraries derived from diseased tissue

---

[34] F. Chang and I. Herskowitz, *Cell* **63,** 999 (1990).

[35] C. Ribas, M. Sato, J. D. Hildebrandt, and S. M. Lanier, *Methods Enzymol.* **344,** [10] 2002 (this volume).

TABLE I
Gα Selectivity of Mammalian G-Protein Activators

Yeast strain[a]	Plasmid	β-Galactosidase activity	
		Glucose	Galactose
No Gα	pYES2	103.4 ± 5.7	76.0 ± 6.9
Gpa1-Gαi2		11.1 ± 0.9	8.7 ± 0.5
Gpa1-Gαi3		10.0 ± 0.6	9.8 ± 0.6
Gαs		13.3 ± 1.0	11.9 ± 0.8
Gpa1-Gα16		23.7 ± 0.7	24.3 ± 1.6
Gpa1		5.6 ± 0.3	5.4 ± 0.3
No Gα	pYES2-AGS1	145.0 ± 5.9	140.1 ± 4.5
Gpa1-Gαi2		5.6 ± 0.6	124.8 ± 4.1
Gpa1-Gαi3		5.6 ± 0.6	63.3 ± 7.8
Gαs		10.1 ± 0.5	6.4 ± 1.1
Gpa1-Gα16		22.5 ± 1.2	22.5 ± 1.6
Gpa1		3.6 ± 0.5	7.6 ± 0.8
No Gα	pYES2-AGS2	89.9 ± 7.1	118.4 ± 5.7
Gpa1-Gαi2		6.2 ± 0.5	17.1 ± 2.0
Gpa1-Gαi3		6.1 ± 0.5	17.0 ± 1.2
Gαs		11.3 ± 1.3	26.0 ± 1.1
Gpa1-Gα16		34.1 ± 2.8	86.9 ± 8.1
Gpa1		5.1 ± 0.8	5.0 ± 0.3
No Gα	pYES2-AGS3	116.2 ± 10.8	87.6 ± 8.4
Gpa1-Gαi2		8.7 ± 0.8	44.0 ± 9.4
Gpa1-Gαi3		8.9 ± 0.8	28.0 ± 2.9
Gαs		11.7 ± 0.7	6.9 ± 0.6
Gpa1-Gα16		37.8 ± 3.1	16.7 ± 2.1
Gpa1		6.0 ± 0.5	4.4 ± 0.8

[a] Yeast strains are derivatives of CY1316 expressing either no Gα or the indicated mammalian Gα or Gα chimera. See Cismowski et al.[27] and text for details.

as well as their normal counterparts can be used in differential screens to identify disease-specific activators of heterotrimeric G-protein signaling pathways.

When choosing an appropriate vector for cDNA library expression, one can consider the level of expression desired as well as the potential advantages or limitations of constitutive vs inducible systems. In general a high copy number vector and a strong promoter allow the identification of cDNAs expressing weak activators, and an inducible system both minimizes the effects of expression of toxic proteins and allows for a simple and rapid assessment of plasmid-dependent growth. Several commercially available yeast vectors are suitable for cDNA library expression as epitope-tagged or non-epitope-tagged proteins [e.g., pYES vectors (Invitrogen, Carlsbad, CA), pYEX vectors (Clontech, Palo Alto, CA), and pESC vectors (Stratagene, La Jolla, CA)]. For our screens we chose pYES2, a high copy vector carrying an inducible *GAL1* promoter. The expression of proteins from this promoter is repressed in the presence of glucose and highly induced in the presence of galactose.

Preparation of cDNA Library from Neuroblastoma–Glioma NG108-15 Cells. NG108-15 cells are grown at $37°$ under 5% CO_2 in Dulbecco's modified Eagle's medium (DMEM) supplemented with 10% (v/v) fetal bovine serum, 100 μM hypoxanthine, and 0.4 μM aminopterin. Twenty confluent 100 mm plates are used to prepare mRNA utilizing one of the standard mRNA preparation kits available (e.g., Qiagen Oligotex, Qiagen, Valencia, CA; Stratagene Messenger RNA Isolation Kit, Stratagene, La Jolla, CA; Ambion Poly (A) Pure Kit, Ambion, Austin, TX). cDNA libraries are generated using the Stratagene ZAP-cDNA synthesis kit with minor modifications. Readers are referred to the instruction manual for this kit for complete details.

First strand cDNAs are synthesized from 5 μg mRNA using an oligo(dT) linker primer carrying a *Xho*I restriction site and Moloney murine leukemia virus reverse transcriptase. Methylated dCTP is used during the first strand synthesis to protect synthesized cDNAs from subsequent restriction digestion with *Xho*I. Second strand reactions are then generated using DNA polymerase I in the presence of RNase H. These reactions include trace amounts of $[\alpha\text{-}^{32}P]$dATP to facilitate size fractionation (see below). Double-stranded DNAs are made blunt ended using cloned *Pfu* DNA polymerase, then purified by phenol : chloroform : isoamyl alcohol extraction and ethanol precipitation. Double-stranded oligonucleotide adaptors carrying an *Eco*RI restriction site are ligated to the double-stranded DNAs, followed by restriction digestion with *Xho*I. DNAs are then loaded onto a 1 ml Sepharose CL-2B column and 100 μl fractions collected. Aliquots of each fraction are analyzed by autoradiography following separation on a 5% nondenaturing acrylamide gel. Fractions containing cDNAs greater than 1 kb are pooled and ligated to *Eco*RI and *Xho*I digested pYES2. In our hands, optimal ligation efficiency is obtained using 40 ng total DNA in 10 μl at a molar ratio of 1 : 5 vector : insert and ligating for 48 hr at $4°$. Under these conditions, approximately 6.5×10^5 colonies/μg DNA can be obtained. Multiple 2 μl aliquots of ligation mixtures are then used to transform 50 μl aliquots of XL-1 Gold *Escherichia coli* (Stratagene), followed by amplification for 2 hr at $37°$ in liquid Luria broth containing 50 μg/ml ampicillin (LB+amp). Plasmid DNA is isolated by column fractionation using Qiagen columns and reagents. Typically, 70–80% of the resulting bacterial transformants contain cDNA inserts and insert sizes range from 400 to 3000 base pairs.

Yeast Screens for Pheromone Pathway Activators

Screening Strategies

Although *his3 far1* yeast strains carrying a *FUS1p-HIS3* construct are, in principle, absolutely conditional for growth on pheromone pathway activation in medium lacking histidine, they often show strain-specific differences in background growth.

This background growth is likely due to variations in basal *FUS1* promoter activity[36] and can be influenced in these screens by the effectiveness of engineered mammalian Gα constructs in sequestering yeast G$\beta\gamma$ (see Table I). To reduce background growth in *FUS1p-HIS3* strains low levels of aminotriazole (AT), a histidine analog that competitively inhibits imidazoleglycerol phosphate dehydratase activity, can be added to the culture medium. The amount of AT required to adequately suppress background growth is empirically determined and is usually in the low millimolar range.

The genetic tractability of yeast, while making functional screens for cDNA activators of the pheromone response pathway feasible, also may create an additional source of background. Spontaneous somatic mutations can confer a growth advantage to cells under selective pressure, and the need to perform these functional screens in haploid yeast cells increases the frequency of phenotypic expression of these mutations. Though false positives such as these are unavoidable, the expression of cDNAs under control of an inducible promoter such as that of galactokinase (*GAL1*) allows for a rapid and simple secondary screen for their identification and elimination. However, the transformation efficiency of yeast plated directly onto medium containing galactose is low relative to efficiencies obtained when plating onto medium containing either glucose or sucrose (M. J. C., unpublished observations). To avoid this problem, yeast transformed with cDNA libraries can be plated onto either sucrose or glucose medium, then replica-plated onto medium lacking histidine and containing both AT and galactose. This allows one to efficiently screen up to 20,000 yeast colonies per 100 mm plate. Because glucose actively represses transcription from the *GAL1* promoter[37] and sucrose neither activates nor represses this transcription, we typically plate initial transformants onto medium containing sucrose.

Yeast Activator Screen

A diagram of the yeast screens used to identify pheromone pathway modulators is shown in Fig. 2. To screen for activators, a 100 ml culture of CY1316 expressing a Gpa1-Gα_{i2} chimera is grown in glucose medium at 30° to a cell density of approximately 1×10^7/ml prior to transformation. Cells are removed to sterile 50 ml conical tubes and centrifuged at 2000 rpm in a Sorvall RC-7 centrifuge for 5–10 min at room temperature. Medium is aspirated, and cells are resuspended in 1/10 volume sterile water, pooled into one tube, and recentrifuged. Cells are resuspended by thorough vortexing at a density of approximately 3.3×10^8/ml in a freshly prepared solution of 40% (w/v) polyethylene glycol (average molecular weight 3350; Sigma, St. Louis, MO), 0.1 *M* lithium acetate, 10 m*M* Tris-HCl,

[36] J. P. Manfredi, C. Klein, J. J. Herrero, D. R. Byrd, J. Trueheart, W. T. Wiesler, D. M. Fowlkes, and J. R. Broach, *Mol. Cell. Biol.* **16,** 4700 (1996).
[37] J. C. Schneider and L. Guarente, *Methods Enzymol.* **194,** 373 (1991).

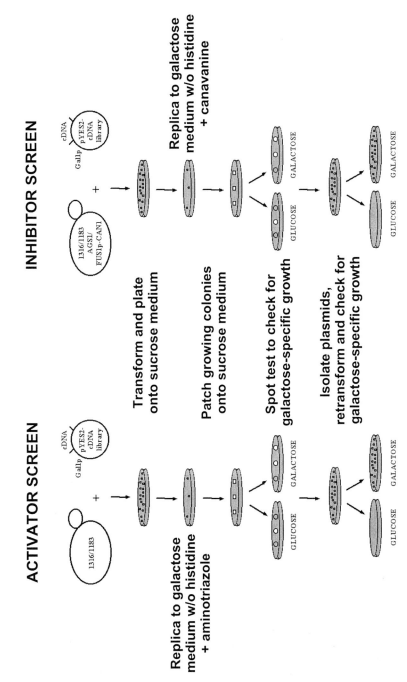

FIG. 2. Schematic of the yeast G-protein pathway modulator screens. See text for details (w/o, without).

pH 7.5, 1 mM EDTA. Carrier DNA (Clontech, Palo Alto, CA; approximately 30 μg) is added with 20 μg pYES2-cDNA library. Cells are vortexed again and left undisturbed at 25° for 24 hr. Cells are then incubated undisturbed at 42° for 20 min. The transformation solution is carefully removed by aspiration, and cells are resuspended in 5 ml sterile water and centrifuged at 2000 rpm for 2 min at room temperature. Cells are resuspended in 6 ml sterile water, 200 μl aliquots are plated onto 100 mm sucrose plates, and plates are incubated at 30°. Typically, $0.5–1.0 \times 10^6$ total transformants can be screened on 30 plates.

Transformed colonies first become visible after 20–22 hr. At this time, colonies are replica plated with sterile velvets onto galactose-plates lacking histidine and containing 1 mM AT. Background growth on galactose plates can be significantly reduced by replica plating transformants when they first become visible as "pinpoint" colonies. Replicas are incubated at 30° for approximately 72 hr until growing colonies are > 1–2 mm in diameter. Colonies are picked with sterile toothpicks and streaked as small patches onto sucrose plates for recovery. Plates are grown at 30° for 24 hr and aliquots of each isolate are resuspended in 1 ml sterile water. Optical densities at 600 nm (OD_{600}) are read, and cultures are diluted to an OD_{600} of 0.05. Four microliter aliquots (approximately 2000 cells) are spotted onto 1 mM AT plates lacking histidine and containing either glucose or galactose (1 OD_{600} of haploid yeast is equivalent to approximately 1×10^7 cells/ml); plates are grown 48 hr at 30° to identify isolates with galactose-dependent growth.

Plasmids are recovered from the selected isolates using the following protocol:

1. Approximately 2×10^6 cells from parent plates are seeded into 5 ml liquid cultures of glucose medium lacking uracil (to select for the pYES2 library plasmid) and grown overnight at 30°, 200 rpm.
2. Of each culture, 1.5 ml is centrifuged at top speed at 25° in a microfuge tube for 1 min. Pellets are resuspended in 1 ml water and recentrifuged.
3. Supernatants are removed and pellets resuspended in 200 μl 10 mM Tris-HCl, pH 8.0, 100 mM NaCl, 1 mM EDTA, 0.1% (w/v) sodium dodecyl sulfate (SDS). An equal volume phenol : chloroform : isoamyl alcohol is added along with 300 μl acid-washed 0.5 mm glass beads (Biospec Products, Inc., Bartlesville, OK).
4. Tubes are capped tightly and vortexed at maximum speed at 25° for 3 min; 200 μl 10 mM Tris-HCl, pH 8.0, 1 mM EDTA is added; and samples are vortexed for an additional 30 sec.
5. Samples are centrifuged at top speed in a microfuge at 25° for 5 min and approximately 300 μl of each aqueous layer is removed to a fresh microfuge tube.
6. Sodium acetate, pH 5.0, is added to a final concentration of 0.3 M, samples are briefly mixed, and 3 volumes 100% ethanol are added.
7. Samples are thoroughly vortexed, incubated 10 min at −20°, and centrifuged 15 min at maximum speed in a microfuge.

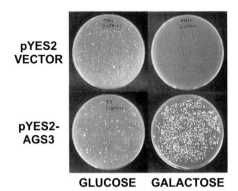

FIG. 3. Growth phenotype of a representative yeast pheromone pathway activator. Transformants of yeast strain CY1316 expressing a Gpa1-Gαi2 chimera and carrying either pYES2 or the G-protein activator AGS3 on plasmid pYES2 are replica plated onto glucose and galactose medium lacking histidine and containing 1 m*M* AT. Plates are grown at 30° for 2 days prior to photography.

8. Supernatants are removed, and pellets are washed with 75% ethanol and air-dried. Pellets are resuspended in 20 μl 10 m*M* Tris-HCl, pH 8.0, 1 m*M* EDTA, and 1 μl each is used to transform electrocompetent *E. coli*. Typically, plating of 1/20 of each transformation onto LB+amp plates yields >1000 bacterial colonies.

9. Plasmids are isolated from bacterial colonies by column purification using Qiagen kits and protocols.

Once isolated, plasmids are digested with *Eco*RI and *Xho*I and electrophoresed on 1% agarose gels for analysis of insert size. Fresh cultures of CY1316 expressing a Gpa1-Gα_{i2} chimera are transformed with each isolated plasmid, plated onto sucrose medium, and grown at 30° for 22 hr. Replicas are made as described above to glucose and galactose medium lacking histidine and containing AT to identify plasmids that confer universal galactose-dependent growth (Fig. 3). The sequence of cDNA inserts of interest is then determined by automated dideoxy sequencing using primers homologous to pYES2 vector sequences in the T7 promoter region and the *CYC1* terminator region, as well as primers internal to the insert sequence.

Using Yeast to Evaluate Site of Action and Mechanism of Isolated G-Protein Activators

General Principles

Epistasis Analysis. The relative ease of genetic manipulation of yeast provides a means for the rapid identification of the site of action of isolated G-protein pathway activators through epistasis analysis. Because most of the proteins in the yeast pheromone response pathway are not required for yeast viability,[32,33] their corresponding genes can be deleted. If an identified mammalian protein continues to activate the yeast pheromone response pathway in a strain carrying a deletion of one of the pheromone pathway genes, it is likely that this protein functions either downstream of the deleted component or in a parallel pathway. Similarly, if the mammalian protein fails to activate the pheromone response pathway in a deletion strain, it is likely that this protein functions upstream of, or at the same level as, the deleted component. In addition to deleting components of the pheromone response pathway, one can also express mutated forms of individual components or express additional proteins that impinge upon the signaling pathway. For example, by expressing a mutant $G\alpha_{i2}$ chimera unable to stably bind GTP ($G\alpha_{i2}$ with a glycine to alanine mutation at amino acid 204),[27] one can determine whether an identified mammalian activator requires GTP exchange on the heterotrimeric $G\alpha$ for its function.

Analysis of $G\alpha$ Selectivity. The cDNA screens described in this chapter are designed to identify mammalian modulators of heterotrimeric $G\alpha$ by replacing the yeast $G\alpha$ with a human $G\alpha_{i2}$ chimera. Several other mammalian $G\alpha$ subunits have been engineered to couple to the yeast pheromone response pathway (see Cismowski *et al.*[27] for details). Therefore the function of identified mammalian G-protein pathway activators can be tested in yeast in a variety of $G\alpha$ backgrounds. Variations in the observed potency of activators expressed in these different $G\alpha$ backgrounds are then presumably due to differences in mechanisms of action and/or differences in $G\alpha$ selectivity.

Quantitative Measurement of Pheromone Pathway Activation

In designing epistasis experiments for pheromone response pathway activators, one can utilize either a growth readout with a *FUS1p-HIS3* reporter construct or a colorimetric readout with a *FUS1p-lacZ* reporter construct. Growth readouts by "spot-testing" as described above provide a relatively quick and simple means to analyze large numbers of putative activators, while colorimetric readouts provide a more quantitative assessment of the degree of pheromone pathway activation. Relatively large-scale colorimetric readouts can be obtained from strains grown in liquid culture by modifying standard β-galactosidase assay protocols for yeast[38] to accomodate a 96-well format. Under conditions that require galactose-mediated protein induction, the optimal induction time is influenced by the stability of both the galactose induced activator and the induced β-galactosidase reporter. Though optimal induction conditions should be empirically determined, we have found

---

[38] F. M. Ausubel, R. Brent, R. E. Kingston, D. D. Moore, J. G. Seidman, J. A. Smith, and K. Struhl, "Current Protocols in Molecular Biology," p. 13.6.1. John Wiley & Sons, Inc., New York, 1989.

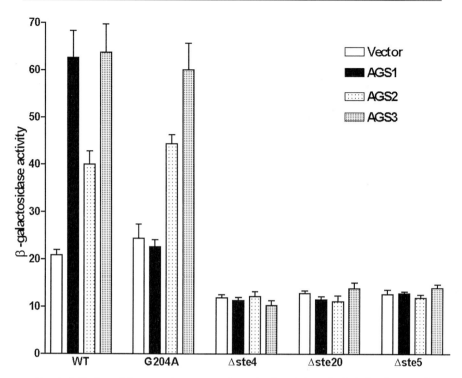

FIG. 4. Activation of the yeast pheromone pathway by three independently isolated mammalian proteins (AGS1-3). Transformants of yeast strains expressing a Gpa1-Gαi2 chimera and carrying a *FUS1p-lacZ* reporter are transformed with either pYES2 or each of the three G-protein activators on plasmid pYES2. Three individual transformants are assayed in triplicate for β-galactosidase activity as described in the text. WT, strain CY1316; G204A, strain CY1316 expressing a glycine-204 to alanine mutated form of Gpa1-Gαi2; Δste4, strain CY1316 carrying a chromosomal disruption of the gene coding for the yeast Gβ *STE4*; Δste20, strain CY1316 carrying a chromosomal disruption of the gene coding for the yeast PAK homolog *STE20*; Δste5, strain CY1316 carrying a chromosomal disruption of the gene coding for the MAP kinase scaffolding protein *STE5*.

that a convenient "overnight" induction protocol provides robust, consistent, and quantitative β-galactosidase activity measurements (see Fig. 4 and Table I).

96-Well β-Galactosidase Assay for FUS1p-lacZ Activity. Yeast strains to be analyzed are maintained on sucrose plates. In general three independent transformants of each isolate are analyzed, and the β-galactosidase activity of each transformant is analyzed is triplicate. Approximately 2×10^6 cells of each transformant are inoculated into 2 ml each liquid medium containing glucose or galactose and incubated overnight (14–16 hr) at 250 rpm, 30°. Cultures are then diluted into 2 ml fresh medium at a density of approximately 2×10^6 cells/ml and incubation continued at 250 rpm, 30°, until a final density of 1×10^7 cells/ml is reached.

Aliquots (100 μl) of each culture are introduced in triplicate into 96-well flat-bottomed microtiter plates (e.g., NUNC microwell 96F); remaining cultures are kept on ice until their OD_{600} is read. Twenty microliters of freshly prepared assay buffer [360 mM Na_2HPO_4, 240 mM NaH_2PO_4, 60 mM KCl, 6 mM $MgSO_4$, 2.5% (v/v) Triton X-100, 16 μl/ml 2-mercaptoethanol, and 10 mM chlorophenol red-β-D-galactopyranoside] is then added to each sample and plates are incubated at 37° with gentle mixing until a reddish color develops. Reactions are terminated by the addition of 60 μl 1 M Na_2CO_3 and absorbance of each well at 575 nm is read using a 96-well plate reader (e.g., Biomek, Beckman Coulter, Fullerton, CA or 3550 Microplate Reader, Bio-Rad, Hercules, CA). β-Galactosidase activities are then normalized for each sample by the following equation:

$$\text{Activity} = A_{575} \times 1000 \,/\, (\text{incubation time in min})(OD_{600} \text{ of culture})$$

Establishing Novel Readouts for Negative Regulators of G-Protein Signaling

Although mammalian proteins that negatively regulate the pheromone response pathway have been previously identified using cDNA screens,[28,29] the isolation of potent G-protein activators presents a unique opportunity to screen for new negative regulators that directly or indirectly affect activator function. By expressing an identified mammalian activator in yeast from a constitutive promoter, such as the promoter for phosphoglycerate kinase (PGK1), the pheromone response pathway can be maintained in an activated state. Mammalian cDNA libraries on plasmid pYES2 can then be screened for galactose-inducible proteins that disrupt this activation as described above (see Fig. 2). A growth readout for pheromone pathway suppression can be obtained in these screens by replacing the FUS1-HIS3 construct with a FUS1p-CAN1 construct. In this system, expression of CAN1, a yeast gene encoding arginine permease,[39] is driven by activation of the pheromone response pathway. Yeast carrying a somatic mutation of the CAN1 gene, such as CY1316, and a FUS1p-CAN1 construct can be grown in the absence of arginine and in the presence of the toxic arginine analog canavanine. On pheromone pathway activation, expressed Can1 protein facilitates uptake of canavanine into yeast, leading to loss of viability.[40] Therefore, cDNAs that encode proteins disrupting pheromone pathway activation will confer growth to yeast under canavanine selection. Isolated cDNAs can then be examined for their function in the presence and absence of activator plasmid, as well as for their site of activity, as described above. As with AT, the amount of canavanine required for adequate selection is empirically determined for each yeast strain and is in the range of 50–250 μg/ml.

[39] W. Hoffman, J. Biol. Chem. 260, 11831 (1985).
[40] R. S. Sikorski and J. D. Boeke, Methods Enzymol. 194, 302 (1991).

Concluding Remarks

Using the strains and protocols described above we identified three different mammalian proteins, two from a neuroblastoma–glioma NG108-15 cDNA library and one from a human liver cDNA library, that function as potent activators of G-protein signaling in yeast.[27,30] The expression of these proteins does not alter G-protein expression levels[27]; therefore, we termed them AGS proteins for activators of G-protein signaling. Epistasis analysis (Fig. 4) indicates that AGS2 and AGS3 (from the neuroblastoma–glioma cDNA library) activate the pheromone response pathway in yeast expressing either wild-type $G\alpha_{i2}$ or the $G\alpha_{i2}$ glycine-204 to alanine mutant, but not in yeast deleted for $G\beta$ (*STE4*) or any downstream component. This is consistent with AGS2 and AGS3 functioning at the level of the heterotrimeric G-protein and functioning independently of nucleotide exchange on $G\alpha$. Interestingly, the mechanisms of action of AGS2 and AGS3 appear to be different, as AGS3 functions selectively in yeast cells expressing $G\alpha_i$ proteins, while AGS2 functions in all mammalian $G\alpha$ backgrounds tested (Table I). The human liver cDNA isolate, AGS1, activates the pheromone response pathway in yeast expressing wild-type $G\alpha_{i2}$, but not in yeast expressing the $G\alpha_{i2}$ glycine-204 to alanine mutant, nor in yeast deleted for $G\beta$ or downstream components (Fig. 4). This strongly implies that AGS1 activates G-protein signaling by facilitating GTP binding on $G\alpha$. Like AGS3, AGS1 shows a strong selectivity for mammalian $G\alpha_i$ proteins (Table I). Biochemical analysis of each of the AGS proteins has verified their distinct mechanisms of action.[30,41] Finally, by constitutively expressing one of these proteins (AGS1) in a *FUS1p-CAN1* and Gpa1-$G\alpha_{i2}$ yeast background, we identified human RGS5 as a protein that potently counteracted AGS1 function by inactivating the chimeric $G\alpha_{i2}$.[27]

One of the goals of these screens was to identify modulators specific for individual proteins in the G-protein signaling cascade by replacing a yeast protein with its functional ortholog from higher eukaryotes. This approach proved quite successful, as two of the three AGS proteins identified show strong selectivity for, and physically interact with, the introduced human $G\alpha_{i2}$.[27,30] In addition RGS5, identified in the screen for negative regulators, has specificity for $G\alpha_i/G\alpha_o$.[42] The screens described in this chapter are also designed to identify intracellular modulators of G-protein signaling; therefore, no GPCR was expressed in the yeast screening strains. The identification of potent activators of G-protein signaling that function at the level of the heterotrimer raises intriguing possibilities about the relative roles of GPCRs and these activators in signaling cascades. AGS proteins may compete with GPCRs for the same pool of heterotrimeric G-proteins, may act independently of GPCR signaling pathways, or may work together with

---

[41] M. J. Cismowski, C. Ma, C. Ribas, X. Xie, M. Spruyt, J. S. Lizano, S. M. Lanier, and E. Duzic, *J. Biol. Chem.* **275**, 23421 (2000).

[42] C. Chen, B. Zheng, J. Han, and S. C. Lin, *J. Biol. Chem.* **272**, 8679 (1997).

activated GPCRs to enhance or prolong signaling. We are currently addressing these possibilities using yeast, mammalian, and *Xenopus* systems.

Though the screens described here are designed to target heterotrimeric $G\alpha_{i2}$, similar screens using other mammalian $G\alpha$ proteins, or screens in which other components of the yeast pheromone responsive signal transduction cascade are replaced by mammalian homologs, should provide the means to identify additional pathway-specific activators and inhibitors. Further, by using cDNA libraries generated from different normal or disease-specific tissue sources, it should be possible to identify novel pathway activators or inhibitors. Finally, analogous screens can be developed that take advantage of the functional redundancy between yeast and higher eukaryotes in other regulatory pathways. Indeed, this type of screen should be readily adaptable to any signaling pathway whose output alters promoter activity. By using transcriptional fusions between promoters that are regulated within the pathway of interest and reporter genes whose expression confers growth under selective conditions, one can conceivably isolate key players in numerous regulatory pathways.

Acknowledgments

We thank Drs. Elliott Ross, Hans Fuernkranz, and Ben Benton for their helpful discussion and comments. We also thank Jeffrey S. Lizano, Michael Spruyt, Xiaobing Xie, Jin Xie, and Ralph Vaccaro for their technical assistance. S.M.L. was supported by the National Institutes of Health Grant RO1-NS24821. A.T. was supported in part by a University of Toyko scholarship.

Section II

Isolation or Production of Native or Modified G Protein Subunits

[12] Expression of α Subunit of G$_s$ in *Escherichia coli*

By SHUI-ZHONG YAN and WEI-JEN TANG

Introduction

Heterotrimeric G proteins amplify and transduce signals from heptahelical receptors to downstream effectors.[1,2] This signaling pathway is widely used in nature to control diverse events ranging from mating in yeast to vision and olfaction in vertebrates. G proteins consist of α and βγ subunits. In the resting state, α subunit binds GDP and is tightly associated with Gβγ. On stimulation by a ligand-bound receptor, GDP is exchanged to GTP, leading to the dissociation of α-GTP from Gβγ. Both α-GTP and βγ serve to modulate the activities of their downstream effectors.

Based on sequence homology, α subunits belong to four subfamilies: G_s, $G_{i/t}$, G_q, and $G_{12/13}$.[2] The G_s family includes $G_{s\alpha}$, which is ubiquitously expressed in all tissues, and $G_{olf}\alpha$, which is predominantly expressed in olfactory epithelium. The best-known effector of the $G_{s\alpha}$ family is adenylyl cyclase, the enzyme that synthesizes cAMP. To date nine isoforms of membrane-bound adenylyl cyclases have been cloned and $G_{s\alpha}$ can potently activate all nine isoforms.[3] The structural basis in $G_{s\alpha}$ activation of adenylyl cyclase has been elucidated.[4,5] $G_{s\alpha}$ has also been shown to modulate the activities of cardiac L-type calcium channel and members of the src kinase family.[6,7] In addition, multiple splicing variants of $G_{s\alpha}$ exist, resulting in extra-large (94 kDa), long (52 kDa), and short (45 kDa) forms of $G_{s\alpha}$.[8–12] Such $G_{s\alpha}$ variants are diverse in their tissue distribution and biochemical properties.

Expression and purification of recombinant G proteins from heterologous expression systems are crucial tasks in analyzing G protein signaling. Here, we

[1] A. G. Gilman, *Ann. Rev. Biochem.* **56**, 615 (1987).

[2] E. J. Neer, *Cell* **80**, 249 (1995).

[3] W.-J. Tang and J. H. Hurley, *Mol. Pharm.* **54**, 231 (1998).

[4] J. J. G. Tesmer, R. K. Sunahara, A. G. Gilman, and S. R. Sprang, *Science* **278**, 1907 (1997).

[5] S. Z. Yan, Z. H. Huang, V. D. Rao, J. H. Hurley, and W. J. Tang, *J. Biol. Chem.* **272**, 18849 (1997).

[6] Y. C. Ma, J. Huang, S. Ali, W. Lowry, and X. Y. Huang, *Cell* **102**, 635 (2000).

[7] A. Yatani, J. Codina, Y. Imoto, J. P. Reeves, L. Birnbaumer, and A. M. Brown, *Science* **238**, 1288 (1987).

[8] B. A. Harris, J. D. Robishaw, S. M. Mumby, and A. G. Gilman, *Science* **229**, 1274 (1985).

[9] M. P. Graziano, P. J. Casey, and A. G. Gilman, *J. Biol. Chem.* **262**, 11375 (1987).

[10] R. H. Kehlenbach, J. Matthey, and W. B. Huttner, *Nature* **372**, 804 (1994).

[11] M. Klemke, H. A. Pasolli, R. H. Kehlenbach, S. Offermanns, S. Gunter, and W. B. Huttner, *J. Biol. Chem.* **275**, 33633 (2000).

[12] H. A. Pasolli, M. Klemke, R. H. Kehlenbach, Y. Wang, and W. B. Huttner, *J. Biol. Chem.* **275**, 33622 (2000).

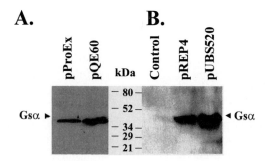

FIG. 1. The optimization of $G_s\alpha$ expression by altering plasmid vector. (A) $G_s\alpha$ expression in *E. coli* BL21(DE3) (pREP4) that harbored the indicated expression vector (pProEx-$G_{s\alpha}$ or pQE60-$G_{s\alpha}$). (B) $G_s\alpha$ expression in *E. coli* BL21 (DE3) (pQE60-$G_{s\alpha}$) that harbored the indicated expression-assistant plasmid (pREP4 or pUBS520). High-speed supernatant (100,000g) of bacterial lysates (10 μl) from *E. coli* BL21 (DE3) cells was loaded onto 13% SDS–PAGE and immunoblotted with anti-$G_s\alpha$ antisera.

describe our effort to improve the expression of the short form of bovine $G_s\alpha$ in *Escherichia coli* (*E. coli*).

Expression of $G_s\alpha$

Functional recombinant $G_s\alpha$ can be expressed and purified from *E. coli*.[13] Lee *et al.* have reported previously that the expression of $G_{s\alpha}$ can be enhanced about 10-fold by lowering the concentration of the inducer, isopropyl-1-thio-β-D-galactopyranoside (IPTG), and using T5 promoter.[14] To extend the analysis of $G_s\alpha$ expression, we compared the expression of $G_s\alpha$ in *E. coli* BL21(DE3) using *E. coli trc* promoters (pProEx-1, Gibco-BRL, Daithersburg, MD) and T5 promoter (pQE60, Qiagen). Both *trc* and T5 promoters use *E. coli* RNA polymerase to transcribe their downstream genes, and *trc* promoter is stronger than T5 promoter. We found that pQE60 provided threefold better expression of $G_{s\alpha}$ than pProEx-1 (Fig. 1A). This is consistent with the result found by Lee *et al.* that the weaker T5 promoter is significantly better than the stronger T7 promoter for $G_{s\alpha}$ expression.[14]

Codon bias can affect protein production in *E. coli*. AGA and AGG codons are commonly used in eukaryotic genes but are rare in gram-negative bacteria such as *E. coli*. dnaY, a gene essential for *E. coli* DNA replication, encodes tRNA to decode the AGA and AGG codons. pUBS520, a plasmid that overexpresses dnaY gene product, enhances the expression of several eukaryotic genes that have high AGA/AGG content.[15] cDNA encoding bovine $G_{s\alpha}$ contains relatively high AGA/AGG codon usage. We thus examined $G_{s\alpha}$ production in BL21(DE3) cells that carried either pUBS520 or a control plasmid, pREP4. Both pUBS520 and

[13] M. P. Graziano, M. Freissmuth, and A. G. Gilman, *J. Biol. Chem.* **264,** 409 (1989).

[14] E. Lee, M. E. Linder, and A. G. Gilman, *Methods Enzymol.* **237,** 146 (1994).

[15] U. Brinkmann, R. E. Mattes, and P. Buckel, *Gene* **85,** 109 (1989).

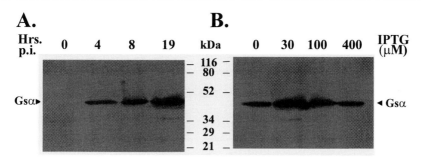

FIG. 2. The optimization for G$_{s}\alpha$ expression by altering time of induction and inducer concentration. (A) G$_{s\alpha}$ expression at the indicated time post-IPTG induction. p.i., post-IPTG induction. (B) G$_{s\alpha}$ expression induced by indicated IPTG concentration. High-speed supernatant (100,000g) of bacterial lysates (10 μl) of *E. coli* BL21 (DE3) containing pQE60-G$_{s\alpha}$ and pUBS520 from (A) different time of post-IPTG induction (30μM IPTG and 1 μg/ml chloramphenicol), (B) induction of the indicated IPTG concentration and 1 μg/ml chloramphenicol were loaded onto 13% SDS–PAGE and immunoblotted with anti-G$_{s}\alpha$ antisera.

pREP4 have p15a origin of DNA replication; thus they are compatible with colE1-based plasmid, pQE60. Both also encode lacI^{q} for the higher-level expression of *lacI* suppressor. We found that the expression of dnaY provided by pUBS520 increased the expression of G$_{s}\alpha$ 3- to 4-fold (Fig. 1B).

Adjusting the time of induction and inducer concentration has been shown to drastically improve G$_{s}\alpha$ expression in BL21(DE3) cells. In the presence of pUBS520 plasmid, we reexamined the optimal condition for both parameters. We found that long induction time (19 hr) and relatively low concentration of IPTG (30 μM) provided best G$_{s}\alpha$ expression (Fig. 2). This is in agreement with the observation from Lee *et al.*[14]

Purification of G$_{s\alpha}$

The following solutions and reagents are needed for the purification of G$_{s\alpha}$:

1. Solutions for *E. coli* culture: T7 medium (20g/liter tryptone peptone (Difco), 10 g/liter yeast extract (Difco), 5 g/liter NaCl, and 50 mM potassium phosphate [pH 7.2]), 100 mg/ml ampicillin, 100 mg/ml kanamycin, 1 mg/ml chloramphenicol, and 100 mM IPTG.
2. Solutions for protein purification: 1 M Tris-HCl (pH 7.7, pH 8.0 at 4°), 0.1 M EDTA (pH 7.0), 4 M NaCl, 1 M imidazole (pH 7.0), 1 M dithiothreitol (DTT), 14.3 M 2-mercaptoethanol, 50 mM guanosine diphosphate (GDP), 1 M MgCl$_{2}$, 100 mM phenylmethylsulfonyl fluoride (PMSF).
3. Columns needed for purification: 5–10 ml Ni-NTA (Qiagen) and 30 ml Q-Sepharose columns (Pharmacia, Piscataway, NJ). Both column media are regenerated as described based on the manufacturer's instructions.

kDa

FIG. 3. Coomassie blue stain of purified $G_s\alpha$ protein (2 μg).

We usually purify recombinant $G_{s\alpha}$ from a 4-liter culture that yields about 16–20 mg of $G_{s\alpha}$ greater than 90% pure (Fig. 3). To express $G_{s\alpha}$, pQE60-$G_{s\alpha}$ and pUBS520 are cotransformed into *E. coli* BL21(DE3) cells. Less than 14 hr incubation at 30° is allowed for the transformants to grow on LB medium plate containing both 50 μg/ml ampicillin and 25 μg/ml kanamycin. Freshly grown BL21(DE3) cells (4 ml) that harbor both pQE60-$G_{s\alpha}$ and pUBS520 are grown in 4 liters of T7 medium containing 50 μg/ml ampicillin and 25 μg/ml kanamycin at 30° in a shaker with agitation set at 250 rev/min. IPTG to 30 μM and chloramphenicol to 1 μg/ml are added when the culture reaches A_{600} of 0.4. Cells are harvested 19 hr after induction and centrifuged at 6000g for 15 min at 4°. After removal of culture medium, the cell pellet is frozen at −80°.

Frozen cells are thawed in 200 ml solution A [20 mM Tris-HCl (pH 8.0), 5 mM 2-mercaptoethanol, 0.1 mM PMSF, 1 mM MgCl$_2$, and 0.1 mM (GDP)]. The cells are lysed by adding 0.1 mg/ml lysozyme and applying 4 min of sonication on ice with a cycle of 1 sec on and 3 sec off. The lysate is spun at 150,000g for 30 min at 4° and the supernatant is collected. NaCl is added to a final concentration of 100 mM.

All the following purification steps are performed in a 4° cold room. The supernatant is loaded onto 5 ml Ni-NTA column equilibrated with solution A containing 100 mM NaCl. The Ni-NTA column is washed with 50 ml solution A containing 500 mM NaCl followed by 100 ml solution A containing 100 mM NaCl and 20 mM imidazole (pH 7.0). The column is then eluted with 100 ml of solution A containing 100 mM NaCl and 150 mM imidazole (pH = 7.0) at 2 ml/min flow rate. Peak fraction (60 ml) is combined and then diluted with 300 ml solution B [20 mM Tris (pH = 8.0), 1 mM dithiothreitol (DTT), 0.5 mM EDTA, 0.1 mM PMSF, 2 mM MgCl$_2$, and 0.1 mM GDP]. The diluted eluate is loaded onto a 50 ml Q-Sepharose column that has been equilibrated with solution B. The absorbed proteins are eluted at 2 ml/min with 240 ml linear gradient of NaCl (100–500 mM) in solution B; 6 ml fractions (2 ml/min) are collected. The peak fractions of $G_{s\alpha}$ are then separated by electrophoresis on 13% SDS–PAGE and analyzed for purity by Coomassie blue staining. The purified $G_{s\alpha}$ is concentrated by ultrafiltration (Amicon positive pressure ultrafiltration device with PM10 membrane) followed by the Centricon

10 microconcentrator (Amicon, Danvers, MA). The concentrated G$_{s\alpha}$ is aliquoted (typically 100 μl) and stored at $-80°$, at a concentration greater than 2 mg/ml.

To assay the ability of G$_{s\alpha}$ to activate adenylyl cyclase, recombinant G$_{s\alpha}$ is activated by 50 μM AlCl$_3$ and 10 mM NaF or by 100 μM GTPγS. Excess GTPγS can be removed by gel filtration chromatography. Adenylyl cyclase can be heterologously expressed and purified from *E. coli* or Sf9 (*spodoptera frugiperda* ovary) cells.[16–19] Adenylyl cyclase assays are performed in the presence of 10 mM MgCl$_2$ and 0.5 mM ATP at 30° for 20 min as described.[20]

Conclusion

In this paper, we show that overexpression of dnaY gene product, the tRNA for AGA/AGG codon, significantly improves G$_{s\alpha}$ expression in *E. coli*. *E. coli* codonplus strains from Stratagene (La Jolla, CA) apply the same principle to enhance the expression of eukaryotic genes that have the eukaryotic arginine, proline, isoleucine, or leucine codon bias. We also found that the weaker T5 promoter, relatively low inducer concentration (30 μM IPTG), and a longer time of induction (19 hr) provide optimal conditions for expression of G$_{s\alpha}$. This is in sharp contrast to protocols commonly used for the expression of recombinant proteins in *E. coli* that apply a stronger promoter (i.e., T7 or *trc*), high inducer concentration (1 mM IPTG), and short time of induction (2–4 hr). In addition, we found that the *E. coli* cultures used need to be as fresh as possible. We obtain the best result using freshly transformed, nonrefrigerated colony/culture. We emphasize that the protocol optimized for G$_{s\alpha}$ expression might not be best suited for other recombinant proteins.[19] We highly recommend that researchers empirically determine the optimal condition for each recombinant protein by altering both the genetic background of the cells and growth and incubation conditions. This simple task could significantly reduce the cost and labor in the subsequent purification steps.

Acknowledgments

This research was supported by NIH Grant R01GM53459, American Heart Association Established Investigator Award, and Brain Research Foundation to W.-J. Tang and Fellowship from U. of Chicago Committee of Cancer Biology and the Ralph S. Zitnik, M.D. Clinical Research Investigatorship from the American Heart Association of Metropolitan Chicago Affiliate to S.-Z. Yan.

[16] W.-J. Tang, J. Krupinski, and A. G. Gilman, *J. Biol. Chem.* **266,** 8595 (1991).

[17] R. Taussig, W. J. Tang, and A. G. Gilman, *Methods Enzymol.* **238,** 95 (1994).

[18] S. Z. Yan, D. Hahn, Z. H. Huang, and W. J. Tang, *J. Biol. Chem.* **271,** 10941 (1996).

[19] S.-Z. Yan and W.-J. Tang, *Methods Enzymol.* (2001).

[20] Y. Salomon, C. Londos, and M. Rodbell, *Anal. Biochem.* **58,** 541 (1976).

[13] Purification of G Protein Isoforms G_{OA} and G_{OC} from Bovine Brain

By Jane Dingus, William E. McIntire, Michael D. Wilcox, and John D. Hildebrandt

Introduction

The heterotrimeric GTP-binding proteins, known as G proteins couple cell surface receptors with a variety of intracellular effectors and ion channels.[1–7] G proteins are composed of α, β, and γ subunits; there are at least 16 α isoforms organized into four families. The β and γ subunits are very tightly associated as a $\beta\gamma$ dimer, and there are at least 5 β and 12 γ subunit genes.[7] Given the large number of subunits, almost 1000 different G protein heterotrimer combinations are possible, and it is likely that the precise subunit composition is important for heterotrimer function.[4,8]

The most abundant G protein by far in mammalian brain is the G_O isoform.[9,10] A given G protein heterotrimer is traditionally named for its α subunit, and the G_O heterotrimer contains the α subunit isoform α_O. The exact function of G_O is not well understood, though it is concentrated at neuronal growth cones[11] and it may mediate effects on multiple downstream targets.[12–14]

There are multiple isoforms of the α subunit of G_O (Fig. 1), which are thought to originate from a single α_O gene.[7] Alternative splicing produces at least four α_O mRNAs,[15–18] but these give rise to only two distinct coding sequences. We

[1] A. G. Gilman, *Biosci. Rep.* **15**, 65 (1995).

[2] H. E. Hamm, *J. Biol. Chem.* **273**, 669 (1998).

[3] L. Birnbaumer, J. Abramowitz, and A. M. Brown, *Biochim. Biophys. Acta.* **1031**, 163 (1990).

[4] T. Gudermann, F. Kalkbrenner, and G. Schultz, *Ann. Rev. Phamacol. Toxicol.* **36**, 429 (1996).

[5] E. J. Neer, *Cell* **80**, 249 (1995).

[6] M. Vaughan, *J. Biol. Chem.* **273**, 667 (1998).

[7] E. H. Hurowitz, J. M. Melnyk, Y.-J. Chen, H. Kouros-Meir, M. I. Simon, and H. Shizuya, *DNA Res.* **7**, 111 (2000).

[8] J. D. Hildebrandt, *Biochem. Pharmacol.* **54**, 325 (1997).

[9] P. C. Sternweis and J. D. Robishaw, *J. Biol. Chem.* **259**, 13806 (1984).

[10] E. J. Neer, J. M. Lok, and L. G. Wolf, *J. Biol. Chem.* **259**, 14222 (1984).

[11] S. M. Strittmatter, D. Valenzuela, T. E. Kennedy, E. J. Neer, and M. C. Fishman, *Nature* **344**, 836 (1990).

[12] G. Ahnert-Hilger, B. Nurnberg, T. Exner, T. Schafer, and R. Jahn, *EMBO. J.* **17**, 406 (1998).

[13] L. T. Chen, A. G. Gilman, and T. Kozasa, *J. Biol. Chem.* **274**, 26931 (1999).

[14] J. D. Jordan, K. D. Carey, P. J. Stork, and R. Iyengar, *J. Biol. Chem.* **274**, 21507 (1999).

[15] J. J. Murtagh, R. Eddy, T. B. Shows, J. Moss, and M. Vaughan, *Mol. Cell. Biol.* **11**, 1146 (1991).

[16] T. Tsukamoto, R. Toyama, H. Itoh, T. Kozasa, M. Matsuoka, and Y. Kaziro, *Proc. Natl. Acad. Sci. U.S.A.* **88**, 2974 (1991).

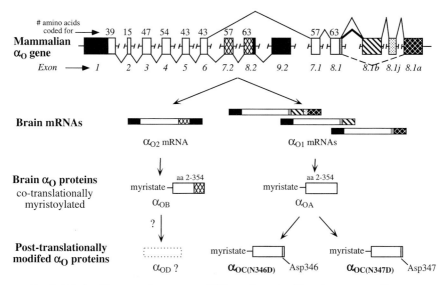

FIG. 1. Relationship among the α_O gene, mRNAs, and proteins. There is one mammalian α_O gene giving rise to at least four mRNAs, but there are only two known coding sequences, referred to as mRNAs O1 and O2. The O1 mRNA actually has three variants differing at their 3' UTR. UTRs are in black or hatched markings. Numbers above exons refer to number of amino acids coded for by the exon. The α_{O1} mRNAs have identical coding regions but different UTRs. The α_O proteins are myristylated and palmitoylated but the palmitate is apparently lost in isolation. The α_{OA} protein is presumably coverted through deamidation into the two forms of α_{OC}.

designate the two mRNAs with different coding sequences as α_{O1} (which is contained in three known mRNAs) and α_{O2} (contained in a single known mRNA). These two mRNA classes differ in sequence near the 3' untranslated region, which results in different C-terminal protein sequences (Fig. 1).[19–21] At the protein level, there are at least three,[22,23] and possibly four,[24] α_O isoforms, designated α_{OA}, α_{OB},

[17] J. J. Murtagh, J. Moss, and M. Vaughan, *Nucleic Acids Res.* **5,** 842 (1994).

[18] P. Bertrand, J. Sanford, U. Rudolph, J. Codina, and L. Birnbaumer, *J. Biol. Chem.* **265,** 18576 (1990).

[19] W. H. Hsu, U. Rudolph, J. Sanford, P. Bertrand, J. Olate, C. Nelson, L. G. Moss, A. E. Boyd, J. Codina, L. Birnbaumer, *J. Biol. Chem.* **265,** 11220 (1990).

[20] M. Strathmann, T. M. Wilkie, and M. I. Simon, *Proc. Natl. Acad. Sci. U.S.A.* **87,** 6477 (1990).

[21] J. J. Murtagh, Jr., J. Moss, and M. Vaughan, *Nucleic Acids Res.* **22,** 842 (1994).

[22] K. Spicher, F. J. Klinz, U. Rudolph, J. Codina, L. Birnbaumer, G. Schultz, and W. Rosenthal, *Biochem. Biophys. Res. Commun.* **175,** 473 (1991).

[23] M. D. Wilcox, J. Dingus, E. A. Balcueva, W. E. McIntire, N. Mehta, K. L. Schey, J. D. Robishaw, and J. D. Hildebrandt, *J. Biol. Chem.* **270,** 4189 (1995).

[24] H. Shibasaki, T. Kozasa, K. Takahashi, A. Inanobe, Y. Kaziro, M. Ui, and T. Katada, *FEBS Lett.* **285,** 268 (1991).

α_{OC}, and α_{OD}. The relationship of these proteins to the four known mRNAs is still somewhat unclear. The α_{OA} protein is transcribed from the α_{O1} mRNA, whereas the α_{OB} protein is transcribed from the α_{O2} mRNA. The third form of the protein, α_{OC}, appears to be a product of the α_{O1} mRNA,[22–24] but differs from α_{OA} at one of two positions near the C terminus (amino acids 346 or 347). At one or the other of these positions, the Asn present in α_{OA} (and coded for in the gene) is an Asp in α_{OC}. This change is most likely the result of posttranslational deamidation of the corresponding Asn residue in α_{OA}.[25,26] This implies that α_{OC} is generated from α_{OA}. Such a mechanism might be important in the regulation of G_O function in neurons. The reason for the existence of multiple α_{O1} mRNAs, and their relationship to the various forms of the protein, is unknown. An α_{OD} protein has also been observed, which may be deamidated α_{OB}; in other words, it is the α_{O2} equivalent of α_{OC} (unpublished observation 1994, and Ref. 24).

The functional significance of these three isoforms of the G_O heterotrimer, G_{OA}, G_{OB}, and G_{OC}, is not understood. The distribution of the α subunits across the brain is very similar for α_{OA} and α_{OC} but that of the α_{OB} splice variant is different.[27] The three G_O heterotrimers contain different populations of $\beta\gamma$ dimers[23,26]; however, after separation of α_O subunits from their $\beta\gamma$ dimers, the purified α_O subunits do not have an inherent preference for a particular $\beta\gamma$ dimer.[27] The difference in $\beta\gamma$ dimer composition between heterotrimers containing α_{OA} and α_{OC} may be related to the deamidation of α_{OA} and its functional consequence, since all three subunits of a heterotrimer contribute to determination of receptor specificity.[28–30]

The G_{OC} protein was first recognized by us as a novel G_O isoform found in bovine brain G protein preparations.[23] It constitutes as much as one-third of all the G_O protein found in brain.[27] Here, we describe methods for purifying the G_O isoforms from bovine brain, including the separation of G_{OA} from G_{OC} and from the other G protein isoforms present in G protein as purified from brain.

General Methods and Materials

The following stock solutions and buffers are required for the G protein purification and/or heterotrimer separation:

1. 1 M Tris-HCl, pH 8.0   (Store at 4°)
2. 1 M Tris-HCl, pH 7.5   (Store at 4°)
3. 1 M Na–HEPES, pH 8.0   (Store at 4°)

[25] W. E. McIntire, K. L. Schey, D. R. Knapp, and J. D. Hildebrandt, *Biochemistry* **37**, 14651 (1998).
[26] T. Exner, O. N. Jensen, M. Mann, C. Kleuss, and B. Nurnberg, *Proc. Natl. Acad. Sci. U.S.A.* **96**, 1327 (1999).
[27] W. E. McIntire, J. Dingus, M. D. Wilcox, and J. D. Hildebrandt, *J. Neurochem.* **73**, 633 (1999).
[28] C. Kleuss, J. Hescheler, C. Ewel, W. Rosenthal, G. Schultz, and B. Wittig, *Nature* **353**, 43 (1991).
[29] C. Kleuss, H. Scherubl, J. Herscheler, G. Schultz, and B. Wittig, *Nature* **358**, 424 (1992).
[30] C. Kleuss, H. Scherubl, J. Hescheler, G. Schultz, and B. Wittig, *Science* **259**, 832 (1993).

 4. 0.4 M NaEDTA, pH 8.0   (Store at 4°)
 5. 1 M Dithiothreitol (DTT)   (Store at −80°)
 6. 10% Sodium cholate (w/v)   (Store at 4°)
 7. 5 M NaCl   (Store at room temperature)
 8. 1 M Phenylmethylsulfonyl fluoride (PMSF)
    in dimethylsulfoxide (DMSO)   (Store at 4°)
 9. 10% Thesit (w/v)   (Store at 4°)
10. 1 M MgCl$_2$ (Store at 4°)
11. 10% CHAPS   (Store at 4°)

Most chemicals are obtained from Sigma (St. Louis, MO), with the exceptions of CHAPS, which is obtained from Pierce (Rockford, IL), and DTT and Thesit (a trademark name for dodecylpolyethylene glycol ether), which are obtained from Boehringer Mannheim (Indianapolis, IN). The sodium cholate must be highly purified, or can be recrystallized.[30a] All stock solutions and buffers, particularly those to be used for FPLC, are filtered through a 0.22 mM filter prior to use. Assay methods are generally the same as given elsewhere in this volume.[30a] For analysis of the heterotrimers, 13% SDS–PAGE gels were prepared as the method of Laemmli,[31] using 30% (w/v) acrylamide with 0.4% (w/v) bisacrylamide, which improves the resolution of the individual subunits.

Preparation of G Protein from Bovine Brain

The G_O isoforms are the major components of mixed bovine brain G protein preparations, which are purified from bovine brain using the procedure from Sternweis and Robishaw,[9] with modifications as previously described.[32,33] Protein is kept at 4° during the purification. Brains from freshly killed young animals (preferably calves) are obtained, and 200–300 g of cortex is dissected out. Tissue is homogenized in a Waring blender in 600 ml of homogenization buffer (300 mM sucrose, 10 mM Tris-HCl, pH 8.0, and 0.5 mM PMSF). After filtration through cheesecloth, the homogenate is centrifuged at 20,000g for 30 min, the supernatant discarded, and membrane pellet resuspended in wash buffer (10 mM Tris-HCl, pH 8.0, 300 mM sucrose, 50 mM NaCl, and 0.1 mM PMSF). Membranes are again collected by centrifugation at 20,000g; after this wash step is repeated, the pellets are stored at −80°.

Membranes are thawed and homogenized in extraction buffer (20 mM Tris-HCl, pH 8.0, 1 mM EDTA, 1 mM DTT, and 2% sodium cholate). Membrane

[30a] J. Dingus, B. S. Tatum, G. Vaidyanathan, and J. D. Hildebrandt, *Methods Enzymol.* **344**, [15] 2002 (this volume).

[31] U. K. Laemmli, *Nature* **227**, 680 (1970).

[32] R. E. Kohnken and J. D. Hildebrandt, *J. Biol. Chem.* **264**, 20688 (1989).

[33] J. Dingus, M. D. Wilcox, R. E. Kohnken, and J. D. Hildebrandt, *Methods Enzymol.* **237**, 457 (1994).

homogenate is centrifuged at 100,000g and extracts collected. The detergent extract from the membranes is then chromatographed on a 5 × 30 cm anion-exchange column (DEAE Sephacel, Amersham Pharmacia, Piscataway, NJ). After loading, the protein is eluted using a linear gradient of 0–225 mM NaCl in 20 mM Tris-HCl, pH 8.0, 1 mM EDTA, 1 mM DTT, and 1% sodium cholate, followed by a final wash with 500 mM NaCl in the same buffer. Fractions are collected and assayed for absorbance at 280 nm, for GTPγS binding activity, and by SDS–PAGE.[31] Fractions containing G protein are pooled and concentrated with a PM30 filter (Amicon, Danvers, MA).

The G protein-containing pool from the DEAE column is further purified using a 5 × 60 cm Ultrogel AcA34 size exclusion column, using 20 mM Tris-HCl, pH 8.0, 1 mM EDTA, 1 mM DTT, 1% sodium cholate, and 100 mM NaCl. Fractions are collected and assayed as described above, and G protein is pooled and concentrated to about 5 ml.

The crude G protein from the size exclusion column is separated using a 1.5 × 20 cm octyl-agarose hydrophobic interaction column (Sigma). The protein is diluted and the buffer composition adjusted so that the final buffer is 20 mM Tris-HCl, pH 8.0, 0.125% sodium cholate, 100 mM NaCl, 1 mM EDTA, 1 mM DTT, and 10 μm GDP. After loading, the column is washed with the same buffer containing 300 mM NaCl. A gradient is then used to elute the G protein, increasing the cholate concentration linearly from 0.125% to 1.3%, while simultaneously decreasing the NaCl concentration from 200 mM to 50 mM. All fractions are assayed as before, and the purified G protein is pooled, concentrated, and exchanged into 20 mM Tris-HCl, pH 8.0, 1 mM EDTA, 1 mM DTT, 100 mM NaCl, and 0.05% Thesit, and stored at −80°. This heterogeneous G protein preparation is a mixture of G$_O$ and G$_i$ isoforms. The protein at various steps in the purification is shown in Fig. 2A, and the recovery of G protein through each step is documented in Fig. 2B.

Purification of G Protein Isoforms

G protein preparations from bovine brain contain predominantly G$_O$, the three isoforms G$_{OA}$, G$_{OB}$, and G$_{OC}$, and the two G$_i$ isoforms, G$_{i1}$ and G$_{i2}$. The separation of these five proteins requires a Pharmacia FPLC (Amersham Pharmacia) system using anion exchange. The column used is a 1 ml HR 5/5 Mono Q column; additional necessary components of the system include two pumps, a UV flow cell and monitor, a fraction collector, the necessary injection and selection valves, a Superloop with peristaltic pump, and a programmable controller. The Mono Q resin is a strong anion-exchange resin with bead size 10 μm, which allows for very good resolution of proteins. The programmable system allows for careful control of the gradient and the flow rate, which is essential for separation of proteins as similar as the G$_O$ and G$_i$ isoforms. The method is essentially that of Wilcox et al.[23]

The initial separation of isoforms is performed on the Mono Q column using a buffer containing 20 mM Tris-HCl, pH 8.0, 1 mM EDTA, 1 mM DTT, 0.7%

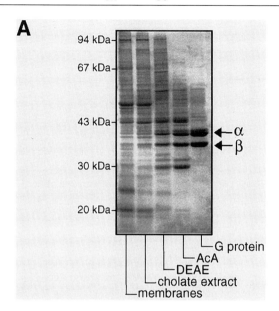

FIG. 2. G protein summary gel. For each step of the G protein purification, 20 pmol of GTPγS-binding activity was analyzed, and the gel stained with Coomassie blue. Lane 1, protein standards: phosphorylase b, 94 kDa; bovine serum albumin, 67 kDa; ovalbumin, 43 kDa; carbonic anhydrase, 30 kDa; soybean trypsin inhibitor, 20 kDa; lane 2, bovine brain membranes; lane 3, cholate extract; lane 4, DEAE column pool; lane 5, AcA34 column pool; lane 6, purified G protein.

CHAPS, and 0.1% Thesit (Mono Q Buffer). Both Thesit and CHAPS were found to be necessary for successful separation of the isoforms with reasonable protein recovery (see Table I). When Thesit was omitted, we obtained poor recovery of protein from the Mono Q column, but good separation of the isoforms was obtained as long as CHAPS was present. If CHAPS was omitted, recovery was good, but separation was poor. Octylglucoside did not substitute satisfactorily for either Thesit or CHAPS. The inclusion of both detergents (0.7% CHAPS and 0.1% Thesit) in the buffer gave separation as good as that seen with CHAPS alone, and recovery as high as seen with Thesit alone (Table I).

Initial column washes and loading steps are done at 0.5 ml/min in Mono Q Buffer with 100 mM NaCl, because the G proteins will bind well in 100 mM salt. The absorbance scale is set at 0.2 or 0.5, depending on how much protein is being loaded; a typical heterotrimer separation starts with 5 to 10 mg of bovine brain G protein. The G protein is diluted to 1 mg/ml and loaded onto the Mono Q column (0.5ml/min) using a Superloop. With the Pharmacia FPLC system, this and all steps are controlled with a program written for the controller. After the sample is loaded, the column is washed with Mono Q Buffer A until the original baseline reading is reached. Then the buffer salt concentration is increased to 120 mM with a short (5 min), steep gradient. The separation portion of the program is a 2 hr

TABLE I
G PROTEIN RECOVERY WITH DIFFERENT DETERGENT CONDITIONS[a]

Detergent	Recovery (%)	Resolution of proteins
0.7% CHAPS	55	+++
2% CHAPS	41	+++
0.05% Thesit	67	+
0.5% Thesit	91	+
40 mM Octylglucoside	80	++
0.7% CHAPS, 0.1% Thesit	91	+++

[a] For each condition 2 mg of mixed G protein was separated on Mono Q as described in the text. The recovery of protein was based on the total amount of GTPγS binding activity that was recovered in all collected fractions. The protein resolution was empirically assigned based on how well the isoforms were separated from each other.

gradient (0.5 ml/min) with the NaCl concentration increasing linearly from 120 mM to 250 mM. The Pharmacia system provides an absorbance chromatogram of the entire run; the separation gradient portion is shown in Fig. 3A. Fractions of 0.5 or 1.0 ml are collected and assayed by SDS–PAGE, as is shown in Fig. 3B, demonstrating the distribution of the G protein isoforms across the chromatogram. This gel is also scanned to better demonstrate the distribution of the various α isoforms across the gradient (Fig. 3C).

The α subunits of the G_O isoforms are resolved by SDS–PAGE at 39 kDa, while the G_{i2} and G_{i1} isoforms are seen at 40 kDa and 41 kDa, respectively. The G_O α subunit protein is resolved into three peaks by this separation (Fig. 3A). The first, at approximately 130 mM NaCl, corresponds to G_O protein containing α_{OB} and precedes the G_{i1} peak of protein, as previously found by others.[34,35] The second peak of G_O α protein (at approximately 170 mM NaCl) corresponds to the major G_{OA} protein originally identified in brain.[9,10] A third peak of G_O α protein not eluting until about 200 mM NaCl, and of variable abundance, corresponds to the G_{OC} heterotrimer. By careful examination of the analytical gels (and gel scans, if necessary), the appropriate fractions from the separation can be combined into pools that contain predominately one isoform. This is shown with the gray boxes in Fig. 3A.

The G protein heterotrimer pools from the initial fractionation are significantly contaminated with other isoforms, and further purification is required. This is accomplished by additional chromatography on the Pharmacia FPLC Mono Q column. Slightly different gradients are used for each heterotrimer pool (see Table II). G_{OA} is the most abundant of the heterotrimers, and the G_{OA} pool is the

[34] P. Goldsmith, P. S. Backlund, Jr., K. Rossiter et al., Biochemistry **27,** 7085 (1988).
[35] I. Kobayashi, H. Shibasaki, K. Takahashi, S. Kikkawa, M. Ui, and T. Katada, FEBS. Lett. **257,** 177 (1989).

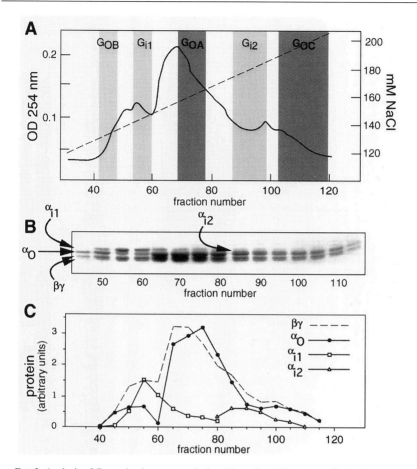

FIG. 3. Analysis of G protein chromatographed on Mono Q. (A) Protein profile for the separation of 10 mg bovine brain G protein on Mono Q with the fractions pooled indicated by shadowing. (B) SDS–PAGE analysis of the fractions from (A). (C) Graph of the distribution of subunits across the chromatogram in (A).

largest both in volume and amount of protein. This protein is rechromatographed on the Mono Q column using a separation gradient of 140 to 180 mM NaCl in 20 mM Tris-HCl pH 8.0, 1 mM EDTA, 1 mM DTT, 0.7% CHAPS, and 0.1% Thesit. After this chromatography step, most of the G_{OA} is sufficiently well separated from contaminating G_{i1} and G_{i2}, and the fractions that contain only G_{OA} are pooled and concentrated.

The G_{OC} pool is similarly rechromatographed on the Mono Q column, but using a separation gradient of 160 to 200 mM NaCl in 20 mM Tris-HCl pH 8.0, 1 mM

TABLE II
RECOVERY AND ACTIVITY OF PURIFIED G PROTEIN HETEROTRIMERS[a]

Isoform	NaCl gradient (mM)	Number of separations	Protein recovered	pmol GTPγS/μg	Percent active (%)
G_{OA}	140–180	2	3.5 mg	11.9	95
G_{OB}	120–140	2	222 μg	12.8	102
G_{OC}	160–200	3	997 μg	13.3	106
G_{i1}	125–150	3	891 μg	11.5	92
G_{i2}	150–170	3	55 μg	9.5	76

[a] Mixed G protein, as prepared in Fig. 2, was used to isolate G_O and G_i isoforms from 10 mg of starting protein. After an initial separation as described in the text, and analogous to the data in Fig. 3, each fraction was chromatographed using a gradient as indicated in the table for the number of separations indicated. Recovery of protein is variable between the isoforms. The specific activity of a theoretical G protein heterotrimer with $\alpha : \beta : \gamma$ stoichiometry of $1 : 1 : 1$ is 12.5 pmol GTPγS/μg protein.

EDTA, 1 mM DTT, 0.7% CHAPS, and 0.1% Thesit. Most of the G_{OC} is sufficiently well separated from the contaminating G_{i2}, and the fractions that contain only G_{OC} are pooled and concentrated. An additional chromatography step on Mono Q can be done if protein of higher purity is desired. Fractions between peaks can also be rechromatographed to increase yields and purification of each isoform, if desired, but the point of diminishing returns is usually reached at three chromatographic steps. The pooled G_{OC} protein is fairly dilute and can be concentrated on the Mono Q by loading in low NaCl buffer and eluting stepwise in buffer containing 300 mM NaCl. Table I illustrates the recovery of the individual G protein isoforms purified from bovine brain preparations of G protein, and the specific activity of the G_O isoforms. SDS–PAGE and immunoblot analysis can be seen elsewhere in this volume.[35a]

Notes and Discussion

The methods and protocols described herein for the separation and purification of the isoforms of G_O require a Pharmacia FPLC system. We have no experience with other types of high-performance protein chromatography systems, but is it likely that these methods can be modified for use with other systems and media, as long as a high-resolution anion exchanger is used.

Successful separation of useable amounts of protein depend on several variables. One of these is the combined use of both CHAPS and Thesit in the chromatography buffers. Another is the use of reasonable (mg amounts) of mixed

[35a] W. E. McIntire, K. L. Schey, D. R. Knapp, J. Dingus, and J. D. Hildebrandt, *Methods Enzymol.* **344,** [34] 2002 (this volume).

G protein in the separation. Even then, the recovery of the different isoforms is variable. G_{i2} is particularly problematic. In addition, the amounts of G_{OC} in these preparations appear to be underestimated.[27] When we compare the amounts of G_{OA} and G_{OC} in different brain regions after immediate tissue extraction (and without purification), we have found that as much as 30% of total G_O is G_{OC} rather than G_{OA}. Recovery of the isoforms by purification suggested that G_{OC} was only about 20% of total G_O. This appears to be due to selective loss of G_{OC} during pooling of fractions during the mixed G preparation.

The purity of the isoforms also varies with the elution profile from the Mono Q column. Purification of the later-eluting proteins is more difficult to achieve than for the early-eluting proteins, even for corresponding levels of abundance (compare G_{OB} and G_{i2}). The most difficult proteins to purify are G_{OC} and, especially, G_{i2}. These proteins tend to contaminate one another even after repeated chromatographic runs.[23]

Subsequent characterization of G_{OC} revealed that it is actually a mixture of two proteins.[25] Both of these proteins represent deamidated forms of G_{OA}, but at different residues. One form is deamidated at Asn-346, while the other is deamidated at Asn-347. These are two adjacent Asn residues near the C terminus of the protein, and at a site that may be a switch for receptor activation. From X-ray crystallography studies, these residues reside at a site where the terminal α helix of α subunits breaks into a disordered structure.[36,37] One model for receptor activation, based on rhodopsin interaction with the C-terminal peptide of $G_t\alpha$,[38] is that activated receptor induces the "disordered" extreme C terminus to form an α helix. This extended α helix would include the deamidated residues of α_{OC} subunits. This might be part of the mechanism by which the receptor promotes GDP exchange on the α subunit, since the nucleotide binding site sits at the other end of this extended α helix. What effect the deamidation of these sites might have on this process is still not clear. But the existence of these two isoforms and the deamidation that differs between the two are likely to be important for the role of α_O in the brain.

Acknowledgment

This work was supported in part by NIH Grant NS38534.

[36] M. A. Wall, D. E. Coleman, E. Lee, J. A. Iniguez, B. A. Posner, A. G. Gilman, and S. R. Sprang, *Cell* **83**, 1047 (1995).

[37] D. G. Lambright, J. Sondek, A. Bohm, N. P. Skiba, H. E. Hamm, and P. B. Sigler, *Nature* **379**, 311 (1996).

[38] O. G. Kisselev, J. Kao, Y. C. Fann, N. Gautam, and G. R. Marshall, *Proc. Natl. Acad. Sci. U.S.A.* **95**, 4270 (1998).

[14] Coexpression of Proteins with Methionine Aminopeptidase and/or N-Myristoyltransferase in Escherichia coli to Increase Acylation and Homogeneity of Protein Preparations

By HILLARY A. VAN VALKENBURGH and RICHARD A. KAHN

Introduction

N-Myristoyltransferase (NMT; EC 2.3.1.97) catalyzes the transfer of the saturated 14-carbon fatty acid myristic acid from myristoyl-CoA to the N-terminal glycine residue of the target protein (reviewed in Ref. 1). N-Myristoylation is essential for the proper function and localization of several signaling proteins, including most members of the ADP-ribosylation factor (Arf) family of 20 kDa GTP binding proteins, some of the α subunits of heterotrimeric G proteins (α_i and α_o), the calcium binder recoverin, and the catalytic subunit of cAMP-dependent protein kinase (PKA).[2-6] The covalently attached myristic acid can play either a structural (e.g., PKA) or regulatory/functional role (e.g., the myristoyl switch in Arfs and recoverin). *N*-Myristoylation occurs cotranslationally, and is thus thought to be a complete and stable modification so that no unacylated protein is found in eukaryotic cells.[7] Thus, the half-life of the bound myristate is the same as that of the protein to which it is attached.

Several factors contribute to the efficient and complete acylation of target proteins. First, the initiating methionine must be cleaved by methionine aminopeptidase to reveal the N-terminal glycine residue to which the myristate will be attached in an amide linkage.[8] Sequences downstream of glycine, most notably residues 6 and 7, can have dramatic effects on K_m and V_{max} values for NMT in *in vitro* assays.[9-13] The concentration of the other substrate, myristoyl-CoA, may

[1] J. A. Boutin, *Cell. Signal.* **9**, 15 (1997).

[2] C. H. Clegg, W. Ran, M. D. Uhler, and G. S. McKnight, *J. Biol. Chem.* **264**, 20140 (1989).

[3] S. M. Mumby, R. O. Heukeroth, J. I. Gordon, and A. G. Gilman, *Proc. Natl. Acad. Sci. U.S.A.* **87**, 728 (1990).

[4] S. M. Mumby and M. E. Linder, *Methods Enzymol.* **237**, 254 (1994).

[5] R. A. Kahn, C. Goddard, and M. Newkirk, *J. Biol. Chem.* **263**, 8282 (1988).

[6] S. Zozulya and L. Stryer, *Proc. Natl. Acad. Sci. U.S.A.* **89**, 11569 (1992).

[7] C. Wilcox, J. S. Hu, and E. N. Olson, *Science* **238**, 1275 (1987).

[8] A. Ben-Basset, *Bioprocess Technol.* 12 (1991).

[9] D. A. Towler, S. P. Adams, S. R. Eubanks, D. S. Towery, E. Jackson-Machelski, L. Glaser, and J. I. Gordon, *Proc. Natl. Acad. Sci. U.S.A.* **84**, 2708 (1987).

[10] D. A. Towler, S. R. Eubanks, D. S. Towery, S. P. Adams, and L. Glaser, *J. Biol. Chem.* **262**, 1030 (1987).

be limiting. Finally, the folding of the nascent polypeptide chain can interfere with acylation, by preventing access of the NMT to the N-terminal glycine. Because N-myristoylation is thought to be complete for endogenous target proteins, these issues are probably not relevant to eukaryotic cell acylation. However, each has been invoked to explain the incomplete N-myristoylation of mammalian target proteins expressed in *Escherichia coli*.

The ability to produce homogeneous protein preparations for structural and biochemical studies has been a challenge to laboratories that work with some of these myristoylated proteins. Because N-myristoylation is a cotranslational event, and bacteria lack endogenous NMT activity, a coexpression system was first employed by Duronio *et al.*[14] that allows expression of both NMT and the target protein in *E. coli*. This system, though quite efficient and useful for several proteins, results in only partial N-myristoylation of many proteins, e.g., $G_o\alpha$, recoverin and mammalian Arfs. This problem is compounded in some cases by the fact that the acyl chain is not solvent exposed so that resolution of the acylated and nonacylated forms of the protein cannot always be readily achieved. In other cases, the acyl chain is likely exposed and can make the protein much less soluble in bacteria. Growth of the bacteria at lower temperatures or including shorter chain or acyl chain analogs in the growth medium have proven helpful in some cases.

Each member of the Arf family of GTPases [including both Arf and Arf-like (Arl) proteins] has a glycine at position 2 and several have been shown to be N-myristoylated in cells. Although the yeast Arf1 and Arf2 proteins are acylated to greater than 95% in bacteria coexpressing yeast or human NMT1, the human Arf proteins under the same conditions are acylated to a much lesser extent, typically 5–10%. This can be increased to about 30% by lowering the growth temperature to 20–30° and adding excess myristic acid to the medium.

Discoveries in the NMT field have sparked hopes for improved, even complete, acylation of recombinant Arf proteins. Cloning of human NMTs revealed that there are two structurally related proteins with NMT activity.[15] Also, the original sequence of human NMT1 was found to have been truncated at the N terminus by 81 residues. Included in this deleted N-terminal region is a positively charged patch that is hypothesized to be involved in ribosome binding (see Fig. 1A).[15] Thus, coexpression of target protein in bacteria with full-length NMT1 or NMT2 may increase the extent of N-myristoylation.

[11] R. J. Duronio, D. A. Rudnick, S. P. Adams, D. A. Towler, and J. I. Gordon, *J. Biol. Chem.* **266,** 10498 (1991).

[12] R. C. Wiegand, C. Carr, J. C. Minnerly, A. M. Pauley, C. P. Carron, C. A. Langner, R. J. Duronio, and J. I. Gordon, *J. Biol. Chem.* **267,** 8591 (1992).

[13] W. J. Rocque, C. A. McWherter, D. C. Wood, and J. I. Gordon, *J. Biol. Chem.* **268,** 9964 (1993).

[14] R. J. Duronio, E. Jackson-Machelski, R. O. Heuckeroth, P. O. Olins, C. S. Devine, W. Yonemoto, L. W. Slice, S. S. Taylor, and J. I. Gordon, *Proc. Natl. Acad. Sci. U.S.A.* **87,** 1506 (1990).

[15] D. K. Giang and B. F. Cravatt, *J. Biol. Chem.* **273,** 6595 (1998).

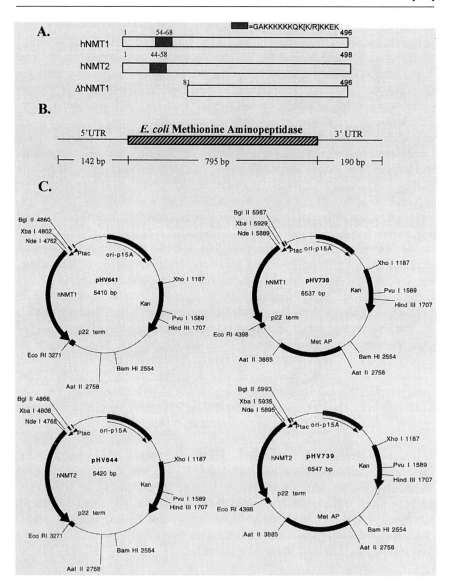

FIG. 1. Vectors described in this paper. (A) Graphical depiction of the different human NMT cDNAs used to generate the various vectors. Note that the full-length hNMT1 and hNMT2 both contain a highly charged positive patch that is absent in the original hNMT1 cloned in 1992.[15] This positive patch has been hypothesized to help the NMTs bind to ribosomes.[15] (B) The E. coli. Met-AP gene used. Note that Met-AP is placed under the control of its own promoter.[16,19] (C) Restriction enzyme sites and features of the various NMT and NMT/Met-AP constructs used in this paper. Note that all four of these constructs contain a p15A origin of replication and kanamycin resistance. This allows them to be coexpressed in

During the course of these investigations we also noted the prevalence in many recombinant protein preparations of another species that was later identified as the full-length target protein, including the initiating methionine. The lack of cleavage of this residue prevents this pool of protein from serving as a substrate for NMT. Thus, the increased expression of bacterial methionine aminopeptidase (Met-AP) (Fig. 1B) was also explored as a means of increasing the covalent processing of N-myristoylated proteins in the Arf family.[16]

[16] D. D. Hwang, L. F. Liu, I. C. Kuan, L. Y. Lin, T. C. Tam, and M. F. Tam, *Biochem. J.* **338,** 335 (1999).

bacteria that harbor various expression vectors (e.g., pET) that contain a ColE1 origin of replication and ampicillin resistance.

hNMT1 and hNMT2 plasmid construction: Both hNMT1 and hNMT2[15] were amplified using the polymerase chain reaction (PCR) from respective cDNAs in pCDNA3 expression vectors. Using oligonucleotides 819 and 820 for hNMT1 and oligonucleotides 817 and 818 for hNMT2 the cDNA was amplified via PCR. Each sense oligo contains a unique *Nde*I site, that includes the starting methionine codon for the NMT protein. The antisense oligonucleotide contains the appropriate stop codon followed by a unique *Eco*RI restriction site. Products were ligated into the pNMT1 vector (Table I) and sequenced.[18] They are designated pHV641 (hNMT1) and pHV644 (hNMT2) (Fig. 1C).

Addition of E. coli methionine aminopeptidase to NMT expression plasmids: The bacterial methionine aminopeptidase (Met-AP) gene was amplified by PCR using BL21 (*DE3*) genomic DNA as a template. The oligonucleotides used for PCR were 907 and 908. These oligonucleotides correspond to *E. coli* Met-AP 5′ and 3′ UTR, respectively, as described in Ben-Bassat *et al.*[19] An *Aat*II site was included in oligonucleotide 908 to facilitate subcloning with a naturally occuring *Aat*II site 5′ of the Met-AP gene. The *Aat*II fragment was ligated into the unique *Aat*II site of pHV641 or pHV644. The resulting plasmids were called pHV738 (hNMT1 and Met-AP) and pHV739 (hNMT2 and Met-AP).

Construction of hArl1 expression plasmid: Human Arl1 was amplified using PCR from cDNA using oligonucleotide 446, which contains a unique *Nde*I restriction site that encodes the starting methionine residue, and oligonucleotide 696, which removes the stop codon and adds a unique a *Not*I restriction site. The PCR product was then placed into pET20b(+) (Novagen), which provides a C-terminal hexahistidine (His$_6$) tag, using the unique *Nde*I and *Not*I restriction sites. The resulting plasmid was sequenced and subsequently named pHV630.

OLIGONUCLEOTIDES USED IN PCR

Number	Sequence
446	5′ GCA GCG CCC GGG ATC CAT ATG GGT GGC TTT TTC TCA AGT 3′
696	5′ TGC GGC CGC CTG TCT GCT TTT TAA TG 3′
817	5′ GAA TTC ATA TGG CGG AGG ACA GTG AG 3′
818	5′ GGA TCC GAA TTC AAT ATC CAT CTA TTG TAG 3′
819	5′ GGA ATT CCA TAT GGC GGA CGA GAG TGA GAC A 3′
820	5′ GGA ATT CTT ATT GTA GCA CCA GTC CAA C 3′
907	5′ GATCGGAAGTCCGGCGCGCT 3′
908	5′ GCTGAGGACGTCGCTTTTATCCCACCGACGGT 3′

We report here four new expression vectors (see Fig. 1C) that contain either the full-length hNMT1 or hNMT2 with and without bacterial Met-AP. These plasmids allowed a doubling in the extent of N-myristoylation of Arfs and should assist workers in the production of more homogeneous protein preparations by promoting the complete removal of the initiating methionine and allowing greater likelihood of any other modifications required at the N terminus, including but not limited to N-myristoylation.

Myristoylation of Proteins in Bacteria Expressing Full-Length hNMT1 or hNMT2

We have subcloned the two cDNAs encoding human NMT1 and 2 into expression vectors containing p15A origins of replication and the gene conferring kanamycin resistance (Table I). These constructs allow coexpression of various NMT plasmids with various Arf family members that are in (pET) expression vectors containing a ColE1 origin of replication and using ampicillin resistance markers.

To determine if either full-length human NMT could increase myristoylation of hArf1 or the Arf-like protein, hArl1, in *E. coli* we cotransformed BL21 (*DE3*) cells with each of the NMT and Arf family member constructs depicted in Table I. Protein expression was induced by the addition of 1 mM isopropyl-β-D-thiogalactoside (IPTG) and acylation was monitored by the incorporation of [^{3}H]myristic acid, as described in Ref. 17.

TABLE I
VECTORS USED IN THIS STUDY

Vector name	Insert(s)	Amino acids	Antibiotic resistance	Origin of replication	Promoter	Ref.
N-Myristoyltransferase vectors						
pHV641	hNMT1	1–496	Kanamycin	P15A	Ptac	This paper
pHV644	hNMT2	1–498	Kanamycin	P15A	Ptac	This paper
pNMT1	hNMT1	81–496	Kanamycin	P15A	Ptac	18
pBB131	yNMT1	1–455	Kanamycin	P15A	Ptac	14
pHV738	hNMT1	1–496	Kanamycin	P15A	Ptac (NMT)	This paper
	MetAP	1–264			Endogenous (MetAP)	
pHV739	hNMT2	1–498	Kanamycin	P15A	Ptac (NMT)	This paper
	MetAP	1–264			Endogenous (MetAP)	
Arf/Arl vectors						
pJCY1-74	yArf1	1–181	Ampicillin	ColE1	T7	20
pOW12	hArf1	1–181	Ampicillin	ColE1	T7	21
pHV631	hArl1-(His)$_6$	1–192	Ampicillin	ColE1	T7	This paper
pJCY1-75	yArf1 G2A	1–181	Ampicillin	ColE1	T7	20

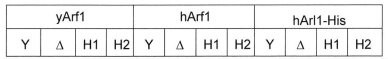

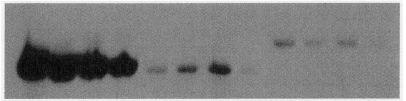

FIG. 2. [³H]Myristic acid incorporation into Arf family members using the various NMT constructs. Incorporation of [³H]myristic acid is shown when bacteria are grown at 37°. Detection is by fluorography. Y, yNMT1 (pBB131); Δ, the deleted version of hNMT1 (pNMT1; see Fig. 1A and Table I); H1, full-length hNMT1 (pHV641); and H2, full-length hNMT2 (pHV644).

As previously reported, yArf1 is myristoylated to a much greater extent than hArf1 or hAr11 (Fig. 2). Previous HPLC analyses of purified Arf proteins have shown that yArf1 is myristoylated to >90% by yNMT1, while the corresponding hArf1 protein is only myristoylated to approximately 5% by the deleted version of hNMT1.[18] All four NMT constructs, yeast NMT1, ΔN-hNMT1, hNMT1, and hNMT2, yield similar levels of modified yeast Arf1. Similar levels of yArf1 and hArf1 protein were expressed in this system (data not shown), so it is evident that hArf1 is not fully myristoylated in *E. coli* with either of the two new hNMT constructs. Full-length hNMT1 myristoylates hArf1 to a greater extent than the truncated version while hNMT2 coexpression results in lesser modification of human Arf1. The results were very similar when the substrate was hAr11.

The mutant protein, yArf1 G2A, in which the site for myristoylation is absent, is not acylated in this system (data not shown).

Myristoylation of hArf1 in Bacteria Expressing hNMT1 and Met-AP

In bacteria, protein synthesis is initiated with *N*-formylmethionine. This formyl group on the initiating methionine residue must be removed by deformylase so that

[17] L. J. Knoll, R. Johnson, M. L. Bryant, and J. I. Gordon, *Methods Enzymol.* **250,** 405 (1995).

[18] P. A. Randazzo and R. A. Kahn, *Methods Enzymol.* **250,** 394 (1995).

[19] A. Ben-Bassat, K. Bauer, S. Y. Chang, K. Myambo, A. Boosman, and S. Chang, *J. Bacteriol.* **169,** 751 (1987).

[20] R. A. Kahn, J. Clark, C. Rulka, T. Stearns, C. J. Zhang, P. A. Randazzo, T. Terui, and M. Cavenagh, *J. Biol. Chem.* **270,** 143 (1995).

[21] O. Weiss, J. Holden, C. Rulka, and R. A. Kahn, *J. Biol. Chem.* **264,** 21066 (1989).

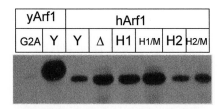

FIG. 3. [³H]Myristic acid incorporation using hNMT1 or 2 in conjunction with *E. coli* Met-AP. Incorporation of [³H]myristic acid is shown when bacteria are grown at 37°. Detection is by fluorography. Y, yNMT1; Δ, the deleted version of hNMT1 (see Fig. 1A); H1, full-length hNMT1 (pHV641); H1/M, pHV738; H2, full-length hNMT2 (pHV644); and H2/M, pHV739.

methionine aminopeptidase can cleave the initiating methionine residue. NMT can then catalyze the acylation of the glycine at the N terminus.

Mass spectroscopic analysis of purified hArf1, produced in the hNMT1 co-expression system, revealed a large amount still contained the initiating methionine residue, while virtually no formylmethionine-free protein could be detected (data not shown). To determine if the removal of the methionine residue was the rate-limiting step for complete myristoylation of hArf, new plasmids were constructed that included the addition of the *E. coli* methionine aminopeptidase (Met-AP) to the hNMT1 or hNMT2 vectors (Table I and Fig. 1C).

Incorporation of myristic acid onto hArf1 increased approximately twofold using this new NMT/Met-AP coexpression system (Fig. 3). This test was performed at 25° and 37°. The increase in the extent of myristoylation was the same at both temperatures. It has been determined previously that more myristoylated Arf protein is soluble when the bacteria are incubated at 25° (data not shown).

To verify that the Met-AP expression is increased we performed immunoblot analysis of total cellular lysates using an antibody against *E. coli* Met-AP (gift from Nathan Brot, Hospital for Special Surgery, NY). There is a dramatic increase in Met-AP expression in cellular extracts that contain the NMT/Met-AP plasmid (Fig. 4). This increase in Met-AP is seen both in the presence and absence of the

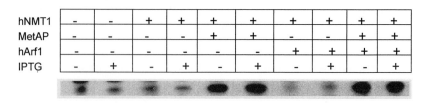

FIG. 4. Western blot of Met-AP expression levels in bacteria expressing different NMT constructs. Western blots were performed using a 1 : 10,000 dilution of rabbit α Met-AP.

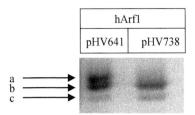

FIG. 5. 12% SDS–PAGE gel of hArfl coexpressed with hNMT1 with and without Met-AP. The upper band (a) corresponds to Arf protein that still contains the initiating methionine residue. The middle band (b) corresponds to Arf protein in which the initiating residue has been removed. The lower band (c) corresponds to myristoylated Arf protein.

Arf plasmid and does not depend on IPTG induction, as Met-AP protein expression is under the control of its own promoter (Fig. 1). When the Met-AP is coexpressed with human Arf1 we can no longer detect any Arf1 still containing the initiating methionine, as seen in Fig. 5 by electrophoretic mobility and confirmed by mass spectroscopic analysis (data not shown).

Summary

New plasmid constructs described in this article allow the coexpression in bacteria of any protein with several different NMT proteins, including the recently cloned full-length human NMT1 and 2,[15] and with increased expression of bacterial Met-AP. Through the use of these plasmids in different combinations it should be possible to improve the homogeneity of a large number of recombinant protein preparations by the complete removal of the initiating methionine and increased extent of N-myristoylation. The new reagents described in this article are available upon request.

Acknowledgments

We thank Dan K. Giang and Benjamin F. Cravatt (Scripps Research Institute, La Jolla, CA) for their gift of human NMT1 and NMT2 cDNA. We also thank Nathan Brot (Hospital for Special Surgery, New York, NY) for the gift of rabbit anti-*E. coli* methionine aminopeptidase.

[15] Purification of G Protein βγ from Bovine Brain

By JANE DINGUS, BRONWYN S. TATUM, GOVINDAN VAIDYANATHAN,
and JOHN D. HILDEBRANDT

Introduction

The heterotrimeric G proteins are one branch of a large family of proteins that bind guanine nucleotides and whose function is regulated by the binding and hydrolysis of GTP. This superfamily is involved in multiple cellular processes, including protein synthesis, vesicle movement inside cells, cell growth regulation, hormone effects on cells, and protein targeting to subcellular compartments. The heterotrimeric G proteins are those GTP-binding proteins involved in sensing extracellular signals and processing them into intracellular responses. The signals themselves are sensed by receptor proteins referred to as seven-transmembrane receptors, or G-protein-coupled receptors.[1-3] There are many different G proteins that respond to a very large number of receptors. These receptors activate G proteins by promoting the exchange of GTP for GDP, and the activated G proteins in turn regulate intracellular enzymes or ion channels.

G protein heterotrimers consist of three nonidentical subunits, α, β, and γ.[4-6] On activation, GTP is exchanged for GDP, and the heterotrimer dissociates into an α subunit and a $\beta\gamma$ dimer. The activated α subunit, which contains the GTP binding site, can activate an effector; it then hydrolyzes the GTP to GDP, which induces reassociation with $\beta\gamma$. A given heterotrimeric G protein is named according to the α subunit contained within the heterotrimer. The β and γ subunits are bound very tightly together in the $\beta\gamma$ dimer and cannot easily be separated from each other.[7] The $\beta\gamma$ dimers also regulate effectors, including some adenylyl cyclase isoforms and calcium channels, and other regulatory processes, including receptor kinases.[8-11] There are multiple genes coding for each of the subunits, as many as

[1] A. G. Gilman, *Annu. Rev. Biochem.* **56,** 615 (1987).
[2] L. Birnbaumer, J. Abramowitz, and A. M. Brown, *Biochim. Biophys. Acta.* **1031,** 163 (1990).
[3] H. R. Bourne, *Curr. Opin. Cell. Biol.* **9,** 134 (1997).
[4] D. E. Clapham and E. J. Neer, *Annu. Rev. Pharmacol. Toxicol.* **37,** 167 (1997).
[5] E. J. Neer, *Cell* **80,** 249 (1995).
[6] J. R. Hepler and A. G. Gilman, *Trends Biochem. Sci.* **17,** 383 (1992).
[7] J. D. Hildebrandt, J. Codina, R. Risinger, and L. Birnbaumer, *J. Biol. Chem.* **259,** 2039 (1984).
[8] K. Yan and N. Gautam, *J. Biol. Chem.* **271,** 17597 (1996).
[9] Y. Chen, G. Weng, J. Li, A. Harry, J. Pieroni, J. Dingus, J. D. Hildebrandt, F. Guarnieri, H. Weinstein, and R. Iyengar, *Proc. Natl. Acad. Sci. U.S.A.* **94,** 2711 (1997).
[10] K. Yan and N. Gautam, *J. Biol. Chem.* **272,** 2056 (1997).
[11] N. Gautam, G. B. Downes, K. Yan, and O. Kisselev, *Cell. Signalling* **10,** 447 (1998).

16 α, 5 β, and 12 γ genes,[12] the products of which are also variably processed to generate a vast number of isoforms.

Bovine brain provides an unusually good source of G proteins, because as much as 1–2% of total particulate protein from brain homogenate is G protein.[13,14] The G protein purified from brain membranes consists largely of G_{OA} and G_{OC}, with lesser amounts of G_{OB}, G_{i1}, G_{i2}, and other isoforms.[15] In addition to heterotrimeric G protein, $\beta\gamma$ dimer free of α is also present in the detergent extract of bovine brain membranes, and its purification is described here. This $\beta\gamma$ separates from the heterotrimer during the initial chromatography step of the heterotrimer purification, and it can be purified separately from and in parallel to the G protein heterotrimer. This free $\beta\gamma$ is herein referred to as bt$\beta\gamma$ (for bovine trailing $\beta\gamma$).

In the purification scheme described here the bt$\beta\gamma$ is purified through three traditional chromatographic steps: (1) anion-exchange chromatography on DEAE Sephacel, (2) gel filtration (size-exclusion) chromatography on UltraGel AcA 34, and (3) hydrophobic interaction chromatography on octyl-agarose. The bt$\beta\gamma$ is then further purified by ion-exchange chromatography using a Pharmacia (Piscataway, NJ) FPLC (fast protein liquid chromatography) system. This method for purification of bt$\beta\gamma$ is generally performed in parallel to the isolation of G protein heterotrimer based on the methods of Sternweis and Robishaw,[13] with modifications as previously described.[16,17]

Solutions and Assay Methods

Stock Solutions Required

1. 1 M Tris-HCl, pH 8.0   (Store at 4°)
2. 1 M Tris-HCl, pH 7.5   (Store at 4°)
3. 1 M Na–HEPES, pH 8.0   (Store at 4°)
4. 0.4 M NaEDTA, pH 8.0   (Store at 4°)
5. 1 M Dithiothreitol (DTT)   (Store at −80°)
6. 10% Sodium cholate (w/v)   (Store at 4°)
7. 5 M NaCl   (Store at room temperature)
8. 1 M Phenylmethylsulfonyl fluoride
   (PMSF) in dimethyl sulfoxide (DMSO)   (Store at 4°)

[12] E. H. Hurowitz, J. M. Melnyk, Y.-J. Chen, H. Kouros-Meir, M. I. Simon, and H. Shizuya, *DNA Res.* **7**, 111 (2000).

[13] P. C. Sternweis and J. D. Robishaw, *J. Biol. Chem.* **259**, 13806 (1984).

[14] E. J. Neer, J. M. Lok, and L. G. Wolf, *J. Biol. Chem.* **259**, 14222 (1984).

[15] M. D. Wilcox, K. L. Schey, J. Dingus, N. D. Mehta, B. S. Tatum, M. Halushka, J. W. Finch, and J. D. Hildebrandt, *J. Biol. Chem.* **269**, 12508 (1994).

[16] R. E. Kohnken and J. D. Hildebrandt, *J. Biol. Chem.* **264**, 20688 (1989).

[17] J. Dingus, M. D. Wilcox, R. E. Kohnken, and J. D. Hildebrandt, *Methods Enzymol.* **237**, 457 (1994).

 9. 10% Thesit (w/v)   (Store at 4°)
10. 1 M MgCl$_2$ (Store at 4°)
11. 10% CHAPS   (Store at 4°)
12. 10 mM guanosine 5'-diphosphate (GDP) (Store at −80°)
13. 10 mM GTPγS   (Store at −80°)

Most chemicals are obtained from Sigma (St. Louis, MO). Thesit (which is a trademark name for dodecylpolyethylene glycol ether), guanosine 5''-O-(3-thiotriphosphate) (GTPγS), and DTT are obtained from Boehringer Mannheim (Indianapolis, IN), and CHAPS is obtained from Pierce (Rockford, IL). All solutions are filtered through a 0.22 μm filter prior to use. All protein-containing samples and fractions should be collected and stored in either polypropylene plastic tubes or siliconized glass tubes. The 10% sodium cholate can be prepared from highly purified sodium cholate (available from Calbiochem, La Jolla, CA). A less expensive alternative involves purifying it from crude cholic acid (Sigma) by recrystallization six times from 95% (v/v) ethanol.[18] It should have an absorbance at 280 nm of 0.25 or less. The cholic acid is dissolved by boiling in 12.5 volumes ethanol, and recrystallized at 4° by the addition of 4 volumes of water. This is done a total of six times, followed by drying in a vacuum oven; a 10% solution prepared from this cholic acid should have an absorbance of less than 0.6 at 280 nm.

Assays

 1. GTPγS binding assays are performed according to the method of Sternweis and Robishaw.[13] Samples are diluted with 10 mM Na–HEPES, pH 8.0, 1 mM EDTA, 1 mM DTT, 0.1% Thesit, so that the concentration is about 0.5 mg/ml. One to 5 μg of protein is included in each sample when assaying DEAE fractions, less if more purified protein is being assayed. The binding assay is started by mixing samples 1 : 1 with 50 mM Na-HEPES, pH 8.0, 40 mM MgCl$_2$, 1 mM EDTA, 1 mM DTT, 200 mM NaCl, 2 μM GTPγS, containing 0.1 μCi [^{35}S]GTPγS (Dupont NEN, Boston, MA) per sample. Samples are started at 30-sec intervals and incubated for 40 min at 32° in a shaking water bath. The binding reaction is stopped by the addition of 3 ml of 20 mM Tris-HCl, pH 8.0, 100 mM NaCl, 25 mM MgCl$_2$ at 4° to each sample. All samples are filtered immediately on nitrocellulose filers using a filtration manifold, and washed four times with 3 ml of the previous buffer each time. The samples are counted with 4 ml Filter Count (Packard, Meriden, CT) in a scintillation counter.

---

[18] J. Codina, W. Rosenthal, J. D. Hildebrandt, L. Birnbaumer, and R. D. Sekura, *Methods Enzymol.* **109,** 446 (1985).

2. SDS–PAGE methods are done by the method of Laemmli,[19] using 11% acrylamide gels, and stained with Coomassie blue. The amounts of each fraction or sample to be run on the gel is given at each step.

3. Electrophoretic transfer and immunoblotting is done using the method of Towbin[20] with modifications. The transfer buffer is 25 mM Tris, 192 mM glycine without methanol, and the protein is transferred using a Bio-Rad (Hercules, CA) Trans-Blot semidry transfer apparatus for 30 min at 20 volts with a limit of 3 mA/cm^2. The buffers used during immunoblotting are 20 mM Tris, pH 7.4, 137 mM NaCl, 0.05% Tween 20 (Sigma), and 20 mM Tris, pH 7.4, 500 mM NaCl, 0.05% Tween 20; 1% bovine serum albumin (BSA, Sigma) and 5% nonfat dry milk are included during the blocking step. The primary antibody used for detection of $\beta\gamma$ is BC1, a polyclonal antibody raised against a synthetic peptide from the β subunit. The peptide sequence is SWDSFLKIWN and is homologous to the last 10 residues at the C terminus of bovine β_1 and β_2. This antibody recognizes β_1, β_2, β_3, and β_4, but not the β_5 isoform; it is routinely used for immunoblotting at a dilution of 1:5000. The transfer is incubated with antibody at room temperature for 3 hr, or at 4° overnight, followed by several washings in the above buffer, then incubated with secondary antibody (HRP-conjugated donkey anti-rabbit, Amersham Pharmacia, Piscataway, NJ), washed, and visualized.

4. Protein assays are done using the BCA assay reagents from Pierce. The protein sample is diluted as much as possible, and heated at 60° for 2 hr prior to addition of the assay buffer, in order to drive off the DTT (which will interfere with the assay, giving falsely high values).

Membrane Preparation and Detergent Extraction

Solutions

1. Transport buffer: 10 mM Tris-HCl, pH 7.5
2. Homogenization buffer: 10 mM Tris-HCl, pH 7.5, 0.3 M sucrose, 1 mM PMSF
3. Membrane buffer: 10 mM Tris-HCl, pH 7.5, 0.3 M sucrose, 50 mM NaCl, 0.1 mM PMSF
4. Wash buffer: 20 mM Tris-HCl, pH 8.0, 1 mM EDTA, 1 mM DTT
5. Extraction buffer: 20 mM Tris-HCl, pH 8.0, 1 mM EDTA, 2% sodium cholate, 1 mM DTT.

[19] U. K. Laemmli, *Nature* **227,** 680 (1970).
[20] H. Iowbin, T. Staehelin, and J. Gordon, *Proc. Natl. Acad. Sci. U.S.A.* **76,** 4350 (1979).

Procedure

Two bovine brains are obtained from freshly slaughtered animals, using calves if possible, and placed in ice-cold transport buffer. From this point, all manipulations, procedures, and assays should be done at 4° unless otherwise stated. Approximately 300 g of cerebral cortex is dissected from the brain and cut into small pieces avoiding white matter, meninges, and blood clots as much as possible. Then, 600 ml homogenization buffer is added, and the tissue is homogenized in a Waring blender at medium speed for 30 seconds. This crude homogenate is filtered through two, and then four layers of cheesecloth, then diluted to 1.5 liters with Homogenization Buffer. The homogenate is distributed to several centrifuge bottles (such as six 280 ml Sorvall Dry-Spin bottles), and centrifuged at 20,000g for 30 min (11,000 rpm in a Sorvall GSA rotor). The pellets are resuspended to 1200 ml (total) with Membrane Buffer and homogenized in six aliquots, five strokes each, using a 200 ml Potter–Elvehjem Teflon-in-glass homogenizer. The homogenate is centrifuged to collect the crude membranes at 20,000g for 60 min. The supernatants are discarded, and the homogenization and centrifugation of the crude membrane pellets is repeated. The final pellets are resuspended in Membrane Buffer to 160 ml per 100 g of starting material, and homogenized with at least three strokes of the homogenizer. The homogenate is divided into three aliquots of approximately 160 ml, and stored at −80°. This freezing step is necessary for the complete lysis of the membranes, so it should not be omitted. An aliquot of this should be saved (and at each of the subsequent steps) for use in the summary analysis of the purification; these membrane aliquots are stored at −80°. Membranes can be stored at −80° for several months.

For the detergent extraction, frozen membranes equivalent to 200 g of cerebral cortex starting material are used. The frozen membranes are broken into small pieces, and stirred in 1 liter of Wash Buffer until fully thawed. The membrane slurry should not be allowed to warm up above 4°. NaCl (5 M) is added to 50 mM final concentration, and the membrane suspension is distributed to six centrifuge bottles and centrifuged at 20,000g for 30 min at 4°.

One centrifuge bottle at a time, the supernatant is discarded, and 100 ml Wash Buffer is added to each pellet. The pellets are transferred to a Potter–Elvehjem homogenizer (200 ml volume, as above) and resuspended with three strokes. All resuspended pellets are combined, and the volume adjusted to 1 liter with Wash Buffer. At this point, an aliquot of these washed membranes is saved as the starting membrane sample. One liter of Extraction Buffer is added, and this suspension stirred for 5 to 10 minutes at 4°. The suspension is homogenized again, in 200 ml aliquots, with two strokes, then stirred for 1 hr at 4°. The homogenate is clarified by centrifugation in a large volume ultracentrifuge rotor (such as a Beckman Type 35) at 100,000g for 60 min. Multiple centrifugations will be required to spin down all of the crude extract, depending on the rotor volume. All the supernatants are collected and pooled, and the total volume of extract obtained is recorded. An aliquot of this cholate extract is saved for summary analysis.

DEAE Column Chromatography

Solutions

   1. Low Salt Gradient Buffer (4.0 liters): 20 m*M* Tris-HCl, pH 8.0, 1 m*M* EDTA, pH 8.0, 1% cholate, 1 m*M* DTT
   2. High Salt Gradient Buffer (1 liter): 20 m*M* Tris-HCl, pH 8.0, 1 m*M* EDTA, pH 8.0, 225 m*M* NaCl, 1% cholate, 1 m*M* DTT
   3. High Salt Wash Buffer (2 liters): 20 m*M* Tris-HCl, pH 8.0, 1 m*M* EDTA, pH 8.0, 500 m*M* NaCl, 1% cholate, 1 m*M* DTT

Procedure

   Routinely, the purification of bt$\beta\gamma$ is done in parallel with a G protein heterotrimer purification. The first separation step for this is DEAE anion exchange chromatography. It is in this first chromatography step that the heterotrimer is separated from the trailing $\beta\gamma$ dimer. The ion exchange column needed here is a very large one, approximately 600 ml (5 × 30 cm) of DEAE Sephacel (Amersham Pharmacia), to accommodate the large amount of protein in the detergent extract. Another contributing factor to the requirement for a large column is the fact that the optimal detergent for isolation of these proteins, cholic acid, is also negatively charged; the presence of cholate decreases the protein binding capacity of the column.

   Before using the column, it should be washed with four volumes of Low Salt Gradient Buffer. The column is then loaded with the entire volume of the detergent extract (about 2 liters) at a flow rate of 100 to 150 ml per hour. The flow-through fraction from the loading step can be collected in bulk. After the column is loaded, it is eluted with a linear gradient of 0–225 m*M* NaCl (Low Salt Gradient Buffer to High Salt Gradient Buffer), total volume of 2 liters, at a flow rate of 100 ml per hour. Fractions (20 ml) are collected during the gradient elution step. When the gradient is finished, the column is then additionally eluted with 2 liters of High Salt Wash Buffer. Fractions are continuously collected from the start of the gradient until the point in this final wash where no more protein elutes from the column.

   As both G protein and bt$\beta\gamma$ are present in the eluant from this column, protein-containing fractions from the column are assayed for GTPγS binding activity (Fig. 1A). After the GTPγS binding is determined, fractions are assayed by SDS–PAGE and are immunoblotted with anti-β antibody BC1, beginning at the point where protein starts eluting and continuing across the peak of GTPγS-binding activity through the high salt protein elution peak. Approximately 5 μl of every fifth or sixth fraction is analyzed. The SDS–PAGE analysis is shown in Fig. 1B, and the immunoblot analysis with BC1 is shown in Fig. 1C. The peak of GTPγS-binding activity corresponds to the heterotrimeric G protein, which contains the guanine nucleotide binding α subunit, and the bt$\beta\gamma$ will be found in the fractions immediately following. The α and β subunits can be seen in the gel analysis

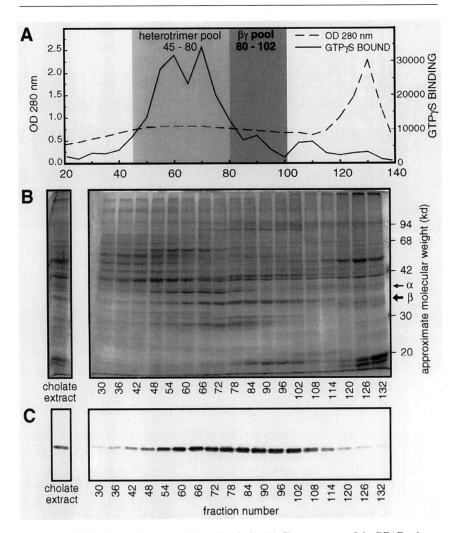

FIG. 1. DEAE column chromatography and analysis. (A) Chromatogram of the DEAE column, displaying the protein profile and the GTPγS-binding activity across the chromatogram, and the fractions pooled for subsequent purification of both heterotrimer and btβγ. (B) SDS–PAGE analysis of every sixth fraction from the separation shown in (A). The large arrow indicates the positions of the β subunits; the small arrow indicates that of the α subunits. (C) Immunoblot with BC1 anti-β antibody of the same fractions analyzed in (B).

(Fig. 1B, arrows). Staining of β is apparent across and beyond the peak of GTPγS-binding activity (compare the distribution of the GTPγS-binding activity and the α subunit in Fig. 1A and 1B with that of the $\beta\gamma$ in Fig. 1B and 1C). The heterotrimer-containing fractions are pooled separately, and the heterotrimer is further purified separately. The bt$\beta\gamma$ is pooled, as shown in Fig. 1A, and concentrated to 30–40 ml using ultrafiltration on an Amicon (Danvers, MA) concentrator with a PM30 membrane. As at every step, an aliquot of the concentrated DEAE pool is saved for summary analysis.

AcA 34 Gel Filtration Chromatography

Solution

1. AcA Column Buffer (4 liters): 20 mM Tris-HCl, pH 8.0, 1 mM EDTA, 1% sodium cholate, 100 mM NaCl, 1 mM DTT

Procedure

The next step in the purification is gel filtration on an Ultrogel AcA 34 column (BioSepra, Marlborough, MA) of approximately 1.1 liter in a 5 × 60 cm column. It should be washed with 2 column volumes of AcA buffer before the concentrated bt$\beta\gamma$ DEAE pool is loaded onto the column. The column is eluted at a flow rate of about 60 ml/hr, with collection of 10 ml fractions. All fractions from the column are assayed for protein (Fig. 2A), then approximately 5 μl of every third or fourth of the protein-containing fractions is analyzed by SDS–PAGE (Fig. 2B; the position of the β subunit is shown with the large arrow), and by immunoblotting with BC1. In the immunoblot (Fig. 2C), the bulk of the β subunit (large arrow) elutes late, as expected. Routinely, some $\beta\gamma$ appears near the void volume of the column, and it is not clear whether this protein is aggregated or specifically bound. The β subunit can be clearly seen on the stained gel (Fig. 2B), as can the α from the small amount of heterotrimer which was included in the $\beta\gamma$ pool from the DEAE (small arrow). The peak of bt$\beta\gamma$ is pooled as shown (Fig. 2A) and is concentrated to approximately 10–15 ml as above. An aliquot of the concentrated AcA pool of bt$\beta\gamma$ is saved for summary analysis.

Octyl-Agarose Hydrophobic Chromatography

Solutions

1. S1 Equilibration Buffer (300 ml): 20 mM Tris-HCl, pH 8.0, 1 mM EDTA, 1 mM DTT, 100 mM NaCl, 0.125% sodium cholate
2. S2. Dilution Buffer (500 ml): 20 mM Tris-HCl, pH 8.0, 1 mM EDTA, 1 mM DTT, 100 mM NaCl

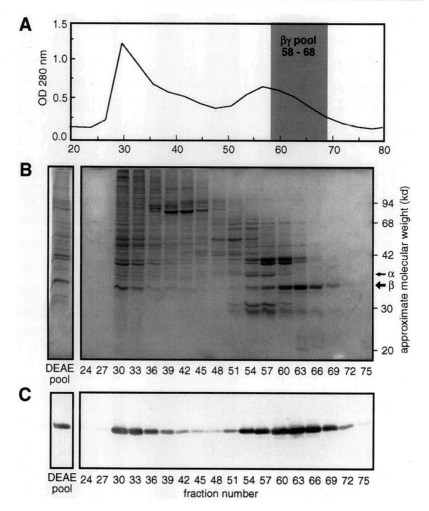

FIG. 2. AcA 34 column analysis. (A) Chromatogram of the AcA column displaying the protein profile, and the bt$\beta\gamma$-containing fractions that were pooled. (B) SDS–PAGE analysis of every third fraction of the separation shown in (A). The positions of the α and β subunits are indicated with arrows. (C) Immunoblot with BC1 anti-β antibody of the same fractions analyzed in (B).

   3. S3 Wash Buffer (100 ml): 20 mM Tris-HCl, pH 8.0, 1 mM EDTA, 1 mM DTT, 300 mM NaCl, 0.125% sodium cholate

   4. S4. Start Gradient Buffer (250 ml): 20 mM Tris-HCl, pH 8.0, 1 mM EDTA, 1 mM DTT, 200 mM NaCl, 0.125% sodium cholate

   5. S5. End Gradient Buffer (250 ml): 20 mM Tris-HCl, pH 8.0, 1 mM EDTA, 1 mM DTT, 50 mM NaCl, 1.3% sodium cholate

6. S6. Final Wash Buffer (50 ml): 20 mM Tris-HCl, pH 8.0, 1 mM EDTA, 1 mM DTT, 500 mM NaCl, 1.3% sodium cholate

7. S7. Urea Wash Buffer (50 ml): 20 mM Tris-HCl, pH 8.0, 1 mM EDTA, 1 mM DTT, 7 M Urea, 2.0% sodium cholate

Procedure

The third step in the purification of bt$\beta\gamma$ is hydrophobic chromatography using a 50 ml column of octyl-agarose (Sigma), which is equilibrated with three column volumes of Solution S1 before loading. The concentrated AcA 34 pool is diluted with seven volumes of Solution S2 (to reduce the cholate concentration to 0.125%). This diluted protein is loaded on to the column at about 30 to 40 ml per hour, and 7 ml fractions are collected throughout all washes and elutions. After loading, the column is washed with 100 ml Solution S3, then eluted with a gradient of 200 ml Solution S4 to 200 ml Solution S5, followed by a wash with 50 ml of Solution S6. This gradient increases hydrophobicity of the eluant by both increasing the detergent and decreasing the salt (as is shown in Fig. 3A). Fractions are collected until all the Solution S6 has been loaded. Fractions from the column are assayed for protein content, and 10 μl of every second or third fraction is run on duplicate SDS–PAGE gels, one for staining and one for immunoblotting with BC1. (An immunoblot of these fractions may not be necessary, as it is easy to see the $\beta\gamma$ on the gel, but is shown in Fig. 3C for illustrative purposes.) $\beta\gamma$ is the bulk of the protein eluting during the gradient (Figs. 3B and C). The bt$\beta\gamma$-containing fractions are pooled as shown in Fig. 3A, and concentrated to approximately 5 ml, saving an aliquot for summary analysis. (The column is regenerated by washing with 50 ml Solution S7 and then 100 ml water.)

FPLC Mono Q Ion Exchange Chromatography

Solutions

1. Mono Q Buffer A (300 ml): 20 mM Tris-HCl, pH 8.0, 1 mM EDTA, 1 mM DTT, 0.1% Thesit, 0.7% CHAPS

2. Mono Q Buffer B (125 ml): 20 mM Tris-HCl, pH 8.0, 1 mM EDTA, 1 mM DTT, 500 mM NaCl, 0.1% Thesit, 0.7% CHAPS

Procedure

After the hydrophobic chromatography step, the bt$\beta\gamma$ is usually about 90% pure, but does contain some contaminating heterotrimer and other proteins. For further purification, the protein is subjected to additional ion-exchange chromatography using a Pharmacia FPLC system (Amersham Pharmacia). The column used is

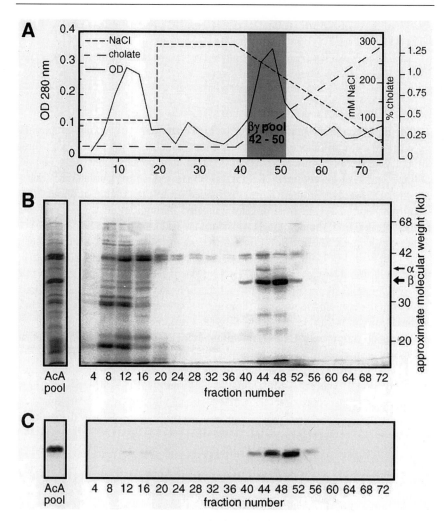

FIG. 3. Octyl-agarose column analysis. (A) Chromatogram of the octyl-agarose column displaying the protein profile, the detergent and salt gradients, and the bt$\beta\gamma$-containing fractions that were pooled. (B) SDS–PAGE analysis of every fourth fraction of the separation in (A). The positions of the α and β subunits are indicated with the arrows. (C) Immunoblot with BC1 anti-β antibody of the same fractions analyzed in (B).

a 1 ml Mono Q column; additional necessary components of the system include two pumps, a UV flow cell and monitor, a fraction collector, the necessary injection and selection valves, a Superloop with peristaltic pump, and a programmable controller.

The pooled and concentrated bt$\beta\gamma$ from the octyl-agarose column must be exchanged into Mono Q Buffer A, as $\beta\gamma$ will not bind well to the Mono Q column unless the salt concentration is reduced to less than 20 mM. This can usually be done by simply diluting the concentrated octyl agarose pool with Mono Q Buffer A. Alternatively, a desalting column can be used, or a concentrator with repeated dilution and reconcentration of the protein. The protein is then diluted to approximately 0.2 mg/ml with the same buffer, for loading on the Mono Q column.

The chromatography on the Mono Q column is performed using a program to control all steps on the chromatography, beginning with buffer exchange. The flow rate is maintained at 0.5 ml/min throughout, and the absorbance scale is set at 0.2 or 0.5. The column is equilibrated with Mono Q buffer A, then the $\beta\gamma$ is loaded onto the column using a SuperLoop. After the sample is loaded, the column is washed with Mono Q Buffer A until the original baseline reading is reached. Then a 100 minute gradient is run, from 0 to 250 mM NaCl [0% Mono Q Buffer B (100% Buffer A) to 50% Mono Q Buffer B], then a rapid 10-min gradient from 250 to 500 mM NaCl [50% Mono Q Buffer B to 100% Mono Q Buffer B (0% Buffer A)]. This is followed with a 10-min wash with 100% Mono Q Buffer B. Fractions (1.0 ml) are collected of all loading and gradient fractions and are analyzed on SDS–PAGE. These fractions are not routinely immunoblotted. This analysis is shown in Fig. 4B, and the protein that is pooled is shown on the chromatogram in Fig. 4A. The $\beta\gamma$ protein usually elutes from the Mono Q column as several sharp peaks, and the pooled protein is often 0.5 mg/ml or greater without concentration, but if necessary, the pooled $\beta\gamma$-containing fractions are concentrated. The protein is aliquotted and stored at $-80°$. *Note:* The Mono Q buffers can be made without Thesit if that detergent is a problem for subsequent uses of the purified protein, but the recovery will be less if CHAPS is the only detergent.

Summary Gel Analysis and Discussion

Following isolation of bt$\beta\gamma$ a summary gel and purification table are prepared comparing the membranes, the cholate extract, the concentrated DEAE pool, the AcA 34 pool, the concentrated octyl-agarose pool, and the final purified bt$\beta\gamma$ from the Mono Q column. A protein determination is done on each sample, and then SDS–PAGE analysis is performed to demonstrate the purification of the bt$\beta\gamma$. From each of the above summary aliquots, 5.0 μg of protein is evaluated by SDS–PAGE (Fig. 5A). At the end of this purification scheme, the bt$\beta\gamma$ should be greater than 95% pure, based upon Coomassie blue staining of an SDS–PAGE gel (Fig. 5A). A second SDS–PAGE gel is run using 0.5 μg total protein from each step, and this second gel is transferred, and immunoblotted with the anti-β antibody BC1. A representative example is shown is Fig. 5B.

The recovery of bt$\beta\gamma$ is shown in Table I. Recovery is expressed as percent, where 100% is defined as the total amount of $\beta\gamma$ present in membranes, the starting material used for the purification. The total $\beta\gamma$ includes the $\beta\gamma$ dimer present both

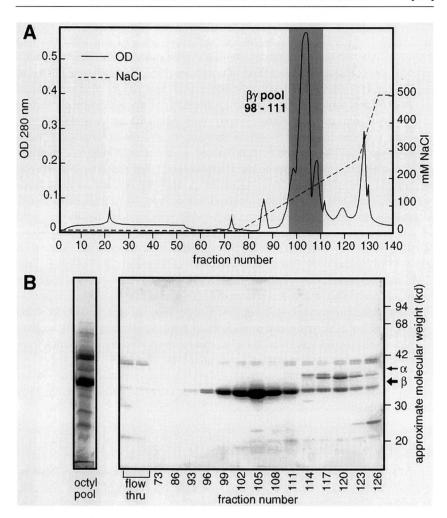

FIG. 4. Mono Q column analysis. (A) Chromatogram of the Mono Q column showing the UV trace for protein and the salt gradient across the separation, and the bt$\beta\gamma$-containing fractions that were pooled. (B) SDS–PAGE analysis of fractions from the separation in (A), with the positions of the α and β subunits indicated.

in heterotrimer and as bt$\beta\gamma$. Results in Table I suggest that as much as 2% of particulate brain protein is $\beta\gamma$. One-third to one-half of the total $\beta\gamma$ from bovine brain is present as bt$\beta\gamma$. Based on this estimate, only about 5% of the total amount of bt$\beta\gamma$ present in the bovine brain membranes is recovered in the purification. This is, however, roughly comparable to the recovery of G protein from bovine brain preparations.

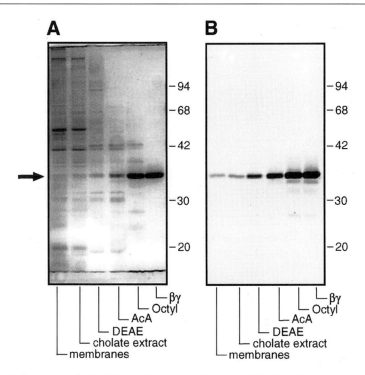

FIG. 5. Summary analysis. (A) 5 μg of total protein from each of the six purification steps analyzed on SDS–PAGE; (B) 0.5 μg of each of the fractions as in (A) is analyzed on a gel identical to that in (A), transferred to nitrocellulose, and immunoblotted with anti-β antibody BC1.

It remains unclear whether the bt$\beta\gamma$ represents $\beta\gamma$ derived from dissociated heterotrimer, or if its presence indicates that $\beta\gamma$ is in excess over α in intact brain. Preliminary analysis by mass spectrometry of the γ subunits associated with bt$\beta\gamma$ suggest that there are subtle differences between it and intact heterotrimer isolated from brain. This could suggest selective loss of $\beta\gamma$ dimers with particular γ subunit isoforms. Nevertheless, the bulk of the protein associated with bt$\beta\gamma$ appears to be essentially the same as the $\beta\gamma$ associated with G protein heterotrimer isolated from bovine brain. In addition, in functional assays, bt$\beta\gamma$ appears to be equivalent to $\beta\gamma$ isolated by the activation and dissociation of mixed G protein heterotrimer isolated from brain.[16,21,22]

[21] J. D. Hildebrandt and R. E. Kohnken, *J. Biol. Chem.* **265,** 9825 (1990).

[22] J. Chen, M. DeVivo, J. Dingus, A. Harry, J. Li, J. Sui, D. J. Carty, J. L. Blank, J. H. Exton, R. H. Stoffel, J. Inglese, R. J. Lefkowitz, D. E. Logothetis, J. D. Hildebrandt, and R. Iyengar, *Science* **268,** 1166 (1995).

TABLE I

SUMMARY OF $\beta\gamma$ PURIFICATION

Purification[a] step	Volume[b] (ml)	Total protein[c] (mg)	$\beta\gamma$ protein[d] (mg)	Percent (%) recovery[e]	Relative specific activity[f]	Purification[g] (-fold)
Combined G protein and $\beta\gamma$ components						
1. Membranes	1000	9900.0	212.4	100.0	0.021	1.0
2. Cholate extract	1708	3484.3	95.3	44.9	0.027	1.3
Free $\beta\gamma$ only component						
3. DEAE	22	281.6	42.6	20.1	0.151	7.2
4. AcA	11	30.8	10.9	5.1	0.354	16.9
5. Octyl-agarose	70	7.7	4.7	2.2	0.610	29.1
6. bt$\beta\gamma$ (FPLC)	10	6.4	5.3	2.5	0.828	39.4

[a] Purification steps relate to those in Fig. 5. In steps 1 and 2, samples contain both free $\beta\gamma$ and $\beta\gamma$ as part of G protein heterotrimer. In steps 3–6, the samples contain primarily free $\beta\gamma$.

[b] Volume of pool during purification.

[c] Total protein in pool from each step, based on a BCA protein assay.

[d] Amount of $\beta\gamma$ protein in the pool based on an immunoblot for the β subunit and using purified $\beta\gamma$ dimer as standard. The BC1 antibody was used in immunoblotting.

[e] Percent recovery is based on the total $\beta\gamma$ in the brain membranes. The 2.5% recovery in step 6 is relative to total $\beta\gamma$. Relative to bt$\beta\gamma$, this represents about 5% recovery, since nearly half of brain $\beta\gamma$ behaves as free $\beta\gamma$.

[f] Relative specific activity (mg $\beta\gamma$/mg total protein) is defined as the fraction of protein that can be accounted for as $\beta\gamma$ based upon the results with the β subunit immunoblot. In this preparation, around 2% of particulate brain protein appears to be $\beta\gamma$. In 6, the relative specific activity should be close to 1.0 since purified $\beta\gamma$ is used as standard in the immunoblot assay.

[g] The purification (fold) is calculated as the ratio of the specific activity in each step to the specific activity of the membrane preparation.

Acknowledgment

This work was supported in part by NIH Grant DK37219.

[16] Separation and Analysis of G Protein γ Subunits

By LANA A. COOK, MICHAEL D. WILCOX, JANE DINGUS, KEVIN L. SCHEY, and JOHN D. HILDEBRANDT

Introduction

G proteins are heterotrimers of α, β, and γ subunits. These proteins bind GTP and are involved in various signal transduction pathways from membrane-bound receptors to intracellular effectors. Activated receptors catalyze the dissociation of GDP from the inactive G protein heterotrimer and facilitate its activation by the binding of GTP to the α subunit. A subsequent step in the activation of the protein is thought to be the dissociation of α from the stable complex made up of the β and γ subunits ($\beta\gamma$ dimer).[1-3] G protein-coupled receptors themselves traverse the plasma membrane with seven helical segments. The G proteins, although usually tightly associated with membranes, are not transmembrane proteins, but are pro-teins tethered to the membrane by their covalent modification with lipids.[4,5] Both the α and the γ subunits are subject to modifications which provide membrane anchors for both α and the $\beta\gamma$ dimer.

The γ subunits are the smallest of the three subunits ranging in molecular mass from 7000 to 9000 kDa. To date 12 γ isoforms have been cloned, but further heterogeneity of the Gγ subunits is generated by their variable posttranslational modifications.[6] The γ subunits are modified at the N and C termini. The N terminus can be acetylated with or without removal of the N-terminal methionine, whereas the C terminus demonstrates a more complex pattern of processing. G protein γ subunits contain a C-terminal CAAX motif and are modified by prenylation. This involves a series of modifications including prenylation of the Cys in the CAAX motif with either farnesyl or geranylgeranyl moieties, proteolytic removal of the three C-terminal amino acids, and carboxymethylation of the new C terminus.[7] Car-boxymethylation is a potentially reversible reaction[8,9] that represents one possible

---

[1] A. G. Gilman, *Biosci. Rep.* **15**, 65 (1995).

[2] L. Birnbaumer, J. Abramowitz, and A. M. Brown, *Biochim. Biophys. Acta.* **1031**, 163 (1990).

[3] H. R. Bourne, *Curr. Opin. Cell. Biol.* **9**, 134 (1997).

[4] H. K. Yamane and B. K. Fung, *Ann. Rev. Phamacol. Toxicol.* **32**, 201 (1993).

[5] J. B. Higgins and P. J. Casey, *Cell. Signal.* **8**, 433 (1996).

[6] L. A. Cook, K. L. Schey, M. D. Wilcox, J. Dingus, and J. D. Hildebrandt, *Biochemistry* **37**, 12280 (1998).

[7] S. Clarke, *Annu. Rev. Biochem.* **61**, 355 (1992).

[8] D. Perez-Sala, E. W. Tan, F. J. Canada, and R. R. Rando, *Proc. Natl. Acad. Sci. U.S.A.* **88**, 3043 (1991).

[9] M. R. Philips, R. Staud, M. Pillinger, A. Feoktistov, C. Volker, J. B. Stock, and G. Weissmann, *Proc. Natl. Acad. Sci. U.S.A.* **92**, 2283 (1995).

variation in the structure of the G protein γ subunit. Carboxymethylation of γ has been shown to alter the membrane association properties of the G protein $\beta\gamma$ dimer.[10] We have also shown that there is variation in the proteolytic processing of the G protein γ subunit after prenylation. For γ_5, there exists both a fully processed form of the protein and a much more prevalent form that retains its last three amino acids after prenylation.[6]

The γ subunits are the most variable of the three subunits of the heterotrimeric G protein. This, and their extensive and variable processing, makes their isolation and analysis important. Described here is the resolution of the γ subunits from each other, and from α and β subunits, by HPLC, and the characterization of both isolated γ subunits and γ subunits that are components of G protein preparations. These methods have been used to isolate and characterize G protein γ_2[11] and γ_5.[6] Methods described for the characterization of γ subunits include using mass spectrometry, gel electrophoresis, differential digestion with acid or proteases, and immunoblotting techniques. These methods are likely to be useful in the characterization of the processing and heterogeneity of the G protein γ subunits, as well as the large number of other small prenylated proteins often involved in cell signaling processes.

Materials and Reagents

Organic solvents: Fisher (Atlanta, GA); Burdick and Jackson (Muskegon, MI)
Donkey anti-rabbit horseradish peroxidase (HRP) secondary antibody: Amersham
Keyhole limpet hemagglutinin (KLH), glutaraldehyde, GDP, GTP, GTPγS and Freund's complete and incomplete adjuvants: Sigma (St. Louis, MO)
[^{35}S]GTPγS: Dupont NEN (Boston, MA)
Coomassie blue and SDS–PAGE reagents: Bio-Rad (Hercules, CA)
Thesit: Boehringer Mannheim (Indianapolis, IN)
MALDI matrices: α-Cyano-4-hydroxycinnamic acid (molecular weight 189.17) and 3,5-dimethoxy-4-hydroxycinnamic acid (molecular weight 224.21): Aldrich (Milwaukee, WI)
MALDI standards (horse heart cytochrome c and bovine insulin): Sigma
Silver nitrate: (Aldrich, Milwaukee, WI)
BCA protein assay reagents: Pierce (Rockford, IL)
Sequencing grade endoproteinase Asp-N: Boehringer Mannheim
Sequencing grade modified trypsin (porcine): Promega (Madison, WI)
Other analytical reagents: Sigma or Fisher, unless otherwise mentioned

---

[10] Y. Fukada, T. Matsuda, K. Kokame, T. Takao, Y. Shimonishi, T. Akino, and T. Yoshizawa, *J. Biol. Chem.* **269**, 5163 (1994).
[11] M. D. Wilcox, K. L. Schey, M. Busman, and J. D. Hildebrandt, *Biochem. Biophys. Res. Commun.* **212**, 367 (1995).

Isolation and Analysis of γ Subunits of G Protein Heterotrimers

Purification of Bovine Brain Cortex G Proteins

Bovine brain G proteins are isolated from membranes as described previously.[12–14] Briefly, membranes are prepared from 200–300 g of gray matter and proteins are solubilized with the detergent cholate (2% final concentration). This procedure is described more fully elsewhere in this volume.[14a] The G proteins are purified from the detergent extract (2–8 g of protein) by sequential separation by three chromatography techniques. After each separation, fractions are analyzed by protein assay, SDS–polyacrylamide gel electrophoresis, and GTPγS binding. G proteins are pooled after the final separation and exchanged into Tris-HCl buffer containing 20 mM Tris-HCl, pH 8.0, 1 mM EDTA, 1 mM DTT, 100 mM NaCl, and 0.1% Thesit on a 40 ml Ultrogel AcA 202 column. Purified G proteins, containing a mixture of primarily G_O and G_i heterotrimers, but also some G_s and G_q isoforms, are stored at a concentration of 2–4 mg/ml at $-80°$ until use. Typically, 30–60 mg of a mixture of G_O and G_i isoforms are obtained, as determined by BCA protein assay. Based on GTPγS binding activity, protein estimates, and staining on Coomassie blue or silver-stained polyacrylamide gels (Fig. 1A), this material contains approximately stoichiometric amounts of α and β (and presumably γ) subunits.

MALDI Mass Spectrometry of Purified G Protein

Matrix-assisted laser desorption ionization mass spectrometry (MALDI-MS) is an analytical technique for determining the mass of proteins and peptides with great enough accuracy and precision to identify Gγ subunit isoforms and to make structural predictions about how these proteins are modified. In comparison to other ionization techniques, MALDI is relatively insensitive to buffer, salts, and detergents. This technique has proved useful for analyzing mixed G protein to characterize Gγ subunit composition.[15] MALDI-MS is performed by adding protein or peptide to an excess of UV-absorbing matrix and irradiating the dried mixture with a nitrogen laser (wavelength 337 nm).[16] The matrix absorbs the energy of the UV light and promotes desorption of the associated proteins (analytes) into the gas phase either as preformed ions or as neutral species that are ionized in the gas phase by excited matrix ions. In the usual configuration of this instrument, mass is determined by time-of-flight (TOF) mass spectrometry, where the time it takes

[12] P. C. Sternweis and J. D. Robishaw, *J. Biol. Chem.* **259**, 13806 (1984).

[13] R. E. Kohnken and J. D. Hildebrandt, *J. Biol. Chem.* **264**, 20688 (1989).

[14] J. Dingus, M. D. Wilcox, R. E. Kohnken, and J. D. Hildebrandt, *Methods Enzymol.* **237**, 457 (1994).

[14a] *Methods Enzymol.* **344**, [13, 15] 2002 (this volume).

[15] M. D. Wilcox, K. L. Schey, J. Dingus, N. D. Mehta, B. S. Tatum, M. Halushka, J. W. Finch, and J. D. Hildebrandt, *J. Biol. Chem.* **269**, 12508 (1994).

[16] K. Biemann, *Annu. Rev. Biochem.* **61**, 977 (1992).

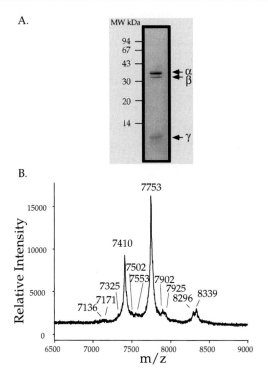

FIG. 1. Analysis of purified heterotrimeric G protein and isolation of γ subunits. (A) Silver-stained gel of purified G protein heterotrimers separated on a 10–20% SDS–polyacrylamide gradient gel, 1 μg. (B) MALDI (Voyager-DE PerSeptive Biosystems Instrument) mass spectrum of mixed G protein heterotrimer showing the γ subunit m/z range, average of 110 scans, 0.16 μg.

an ion to travel from the point of ionization to a fixed detector is inversely related to the square of its mass. This technique usually generates protein ions that are singly charged $[M + H]^+$ ions, with lower abundance of multiply charged ions.[17] The typical nomenclature for ions denotes the charged ion as $[M + nH]^{n+}$, where M is the uncharged mass, n is the number of protons associated, and $n+$ is the charge of the complex. In general, the predominance of singly charged molecular ions simplifies the analysis of the data. This technique requires very little sample (as low as femtomoles) and is relatively simple to perform. We have used both a custom made MALDI-TOF MS instrument and a PerSeptive Biosystems (Framingham, MA) Voyager-DE MS instrument. Optimal conditions for sample analysis are similar for both of these instruments.

[17] M. Schar, K. O. Bornsen, and E. Gassmann, *Rapid. Commun. Mass. Spectrum.* **5,** 319 (1991).

TABLE I
STANDARDS USED FOR CHARACTERIZATION OF G PROTEIN γ SUBUNITS BY
MALDI MASS SPECTROMETRY

Standard	$[M+H]^+$	Working concentration[b]	Typical assay amount
Horse cytochrome c	12361	10–50 μg/ml	0.4–4.0 pmol
Bovine insulin	5735	2.5–5 μg/ml	0.2–1.0 pmol
G2C peptide[a]	1701	2.5–5 μg/ml	0.7–3.0 pmol

[a] Sequence of the G2C peptide is P-A-S-E-N-P-F-R-E-K-K-F-F-C. This peptide was synthesized at the MUSC Protein Sequencing and Peptide Synthesis Facility.

[b] Stock solutions containing 1 mg/ml standard are diluted in matrix solution to these working concentrations. Typical matrix solution contains 50 mM α-cyano-4-hydroxycinnamic acid in 70% (v/v) acetonitrile, 30% (v/v) water, 0.1% (v/v) TFA.

Procedure

1. Heterotrimeric G protein in Tris-HCl buffer (see above) is diluted for use with the matrix α-cyano-4-hydroxycinnamic acid (αC) [50 mM in 70% (v/v) acetonitrile, 30% (v/v) water, 0.1% (v/v) trifluoroacetic acid (TFA)]. The matrix solution is usable for up to 1–2 weeks and should be protected from light as it is light sensitive.

2. Diluted G protein is mixed 1 : 3 with the matrix solution and 1 μl spotted onto a gold MALDI plate (final concentrations should be in the low micromolar range) with internal standards (horse cytochrome c and G2C peptide, see below), air dried and analyzed by MALDI on a PerSeptive Biosystems Voyager-DE mass spectrometer.

3. Mass measurements are based on two-point calibration using the standards horse cytochrome c; bovine insulin; and G2C, a γ_2 peptide (Table I). Cytochrome c and insulin are used as standards for intact γ subunits. Cytochrome c and G2C are used as standards for acid hydrolysis fragments (see below). Mass estimates can be based upon either internal or external calibration procedures using these same standards. Briefly, 1 mg/ml stocks are diluted in matrix solution before use (Table I).

   (a) *Internal calibration* means that the sample mass and standard mass are measured at the same time in the same spectrum. This is done by first spotting standards (0.5–1 μl) on the MALDI plate, air drying, and then spotting and air drying 1 μl of sample over the standards.

   (b) *External calibration* means that a separate spectrum containing the standards is used to create a calibration file to determine the mass of other spectra obtained subsequently on the same day. This is performed by spotting 0.5–1 μl of standards separately from the sample for independent analysis to obtain a high-quality record for that sample.

4. MALDI spectra are usually an average of 75–150 spectra, each based upon one laser shot. This enhances the signal-to-noise ratio in the resulting spectrum and increases the accuracy and precision of the mass estimates.

5. When analyzing γ subunits, data are usually recorded in the range of m/z 500–12,500. The γ isoforms are observed in the m/z range of 6500–9000.

Notes and Discussion. MALDI-MS is an extremely sensitive technique, with attomole sensitivity under optimal conditions. Although the technique is somewhat insensitive to buffers, salts, and detergents, these agents do degrade the extremely high sensitivity of the technique. In order to dilute these components to a level at which they will not interfere with MALDI analysis, mixed G protein (in Tris-HCl, NaCl, and Thesit) must be analyzed at low (micromolar) protein concentrations, which corresponds to low (1–10) picomole amounts in MALDI samples. In addition, water soluble contaminants (i.e., salts) can be removed by gently touching the wet sample well with a KimWipe; this will increase signal quality and decrease drying time.[18] The best MALDI spectra for illustrating γ subunit heterogeneity and diversity are obtained in the absence of standards. On the other hand, the most reliable mass measurements require internal standards. Often we analyze samples in both the presence and absence of internal standards to have high-quality spectra along with good mass measurements. Figure 1B shows the major γ subunits observed in a purified heterotrimeric G protein preparation. At least four strong signals are evident in the γ m/z range and they represent four of the isoforms previously observed and characterized.[15] In addition, numerous minor signals can often be observed in this range.

One potential complication of MALDI analysis is that adducts can form between the sample and the matrix.[19] Such adducts are generally observed at 10–20% of the unchanged ion's peak intensity. One reason for using αC is that it is less likely to form adducts. The αC matrix also works well for small proteins and peptides. Other matrices can be and have been used, such as 50 mM sinapinic acid (3,5-dimethoxy-4-hydroxycinnamic acid). In general, more adducts are seen with sinapinic acid, but more multiply charged ions are observed with αC. The use of multiple matrices provides a test for adduct formation. If the measured mass of sample changes in different matrices, this would support the idea of adduct formation.

Another complication with the use of MALDI is difficulty in the quantitative interpretation of signals. First, different proteins ionize differently depending on their sequence and modifications; therefore, equivalent signals do not necessarily imply equivalent amounts of ion. Second, there is generally a complex or nonlinear relationship between amount of sample present and signal intensity. Above optimal levels, amount of sample causes signal broadening, decreasing the accuracy and reproducibility of mass estimates. Another complication is the effect of mass on

---

[18] G. B. Downes and N. Gautam, *Genomics* **62**, 544 (1999).
[19] R. C. Beavis and B. T. Chait, *Rap. Commun. Mass Spectrom.* **3**, 237 (1989).

detection. Larger masses are detected with less efficiency using microchannel plate detectors commonly employed in TOF instruments. To improve resolution, a reflector can be used.

The precision and accuracy of mass measurements depends on the amount of protein present in the sample, the ability of the protein to ionize, the contaminants in the sample, and matrix adduct formation. The accuracy of commercially available linear MALDI instruments for masses in the range of γ subunits is generally about 0.05% with external calibration and improves with internal calibration. This is equivalent to 1 Da in 2000 mass units or $\sim$4 Da for the γ subunits. Standard errors for multiple γ subunit mass estimates generally range from 2 to 6 Da depending on mass intensity and replicate numbers. These errors are sufficient to detect modifications as small as carboxymethylation (Δmass = +14 Da), but are not generally sufficient to identify changes in the intact protein such as deamidation (Δmass = +1 Da). The ability to resolve multiple masses in a spectrum depends on their relative intensity, the uniformity of peaks, and the difference in mass. Generally, multiple masses should be resolved by differences greater than the associated errors discussed above for the mass estimates.

Resolution of γ Subunits from α and β Subunits

Reversed phase HPLC is used to separate the γ subunits from the α and β subunits in a modified method of Morishita *et al.*[20] Initial attempts by us and others to use C_8 and C_{18} columns were not successful. Although γ separated from α and β, the γ isoforms eluted without resolution. In contrast, a phenyl column eluted with a shallow gradient resolves the Gγ isoforms well.

Procedure

1. Columns: 220 × 4.6 mm Aquapore 7 μm phenyl column and 3 cm phenyl guard column (Applied Biosystems, Foster City, CA)
2. Solvents: see Table II
3. Sample: purified G protein heterotrimer (2–4 mg) in Tris-HCl buffer (see above)
4. Run conditions: equilibration with Solvent A is followed by injection of G protein into 2 ml loop (approximately 1 ml; 2 mg protein) at time 0; the gradient is performed as shown in Table III

A flow rate of 1 ml/min is maintained throughout the run and the elution is monitored by absorbance at 214 nm. Fractions of 1ml are collected and stored at $-20°$ under argon until analysis. After the initial separation of γ from α and β, the column is washed with a gradient of Solvent C to remove remaining α and

[20] R. Morishita, K. Masuda, M. Niwa, K. Kato, and T. Asano, *Biochem. Biophys. Res. Commun.* **194**, 1221 (1993).

TABLE II
COMPOSITION OF HPLC SOLVENTS FOR THE ISOLATION OF G PROTEIN
γ SUBUNITS

Component	Solvent A[a]	Solvent B	Solvent C
Acetonitrile	10	75	10
Water	90	0	0
2-Propanol	0	25	0
n-Propanol	0	0	90
TFA	0.1	0.095	0.1

[a] Values are given as relative proportions in the final solution.

β subunits (Table III). At the conclusion of the n-propanol wash gradient the column is washed in 50% (v/v) methanol for 20 min.

In experiments using an in-line LC/MS instrument (Finnigan LCQ), the effluent from the same separation is split so that 50–300 μl/min enters the electrospray source for continuous mass spectrometry (MS and MS/MS) analysis, while the remainder is collected as fractions for future analysis. Figure 2 shows the separation analysis by HPLC. The γ subunits elute between 20 and 40 min as seen by both the absorbance and ion current profiles. The α and β subunits elute later in the gradient as shown. The fractions containing the γ subunits are analyzed by various methods as follows to determine the identity and structure of isoforms which are found in sufficient abundance for characterization.

TABLE III
TIME AND SOLVENT PROPORTIONS FOR ISOLATION OF G PROTEIN γ SUBUNITS BY HPLC

Time (min)	Solvent A (%)	Solvent B (%)	Solvent C (%)
Separation of γ subunits			
0	95	5	0
5	95	5	0
15	60	40	0
80	54	46	0
85	0	100	0
95	0	100	0
105	100	0	0
Elution of α and β subunits			
0	95	0	5
5	80	0	20
35	40	0	60
45	0	0	100
55	0	0	100

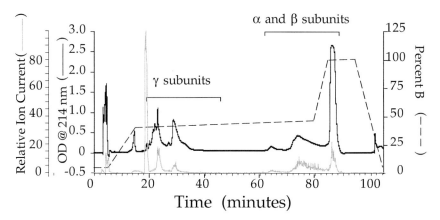

FIG. 2. Resolution of γ subunits from α and β subunits. Reversed-phase HPLC separation of γ subunit isoforms from α and β subunits on a phenyl column performed inline with the Finnigan LCQ for ESI mass spectrometry, 2.6 mg of G protein.

Notes and Discussion. Solvent solutions should be made, degassed, and used the same day since TFA degradation occurs over time. Fractions should be stored at $-20°$ under argon to prevent oxidation of protein.

Analysis of Isolated γ Subunits

MALDI Mass Spectrometry of Isolated γ Subunits

One of the initial steps useful in the characterization of isolated γ subunits in HPLC fractions is MALDI mass spectrometry analysis. This procedure is modified from the one described above for analysis of mixed G protein.

Procedure

1. Aliquots of 25–100 μl of HPLC fractions containing γ subunits are dried in 0.6 ml microfuge tubes under vacuum and resuspended in 1–2 μl of 47.6% (v/v) *n*-propanol, 47.6% (v/v) water, 4.76% (v/v) acetonitrile, and 0.095% (v/v) TFA.
2. A 0.5 μl sample is usually mixed with 1.5 μl (1 : 3 ratio) of the matrix solution containing αC for MALDI analysis.
3. For internal calibration, 0.5–1 μl standards are spotted first and air dried, followed by spotting 0.5–1 μl of sample diluted in matrix. Samples are air dried and analyzed on the MALDI mass spectrometer. For external calibration samples and standards are spotted separately.

Notes and Discussion. Table IV shows the sequences of the 12 human γ isoforms cloned to date, and Table V shows their predicted modifications, masses, and

TABLE IV

SEQUENCES OF HUMAN G PROTEIN γ SUBUNIT ISOFORMS

Gamma subunit[a] number	D/P Site[b]	aa[c]
1	MPVINIEDLTEKDKLKMEVDQLKKEVTLERMLVSKCCEEVRDYVEERSGED / PLVKGIPEDKNPFKELKGG**C**VIS	74
2	MASNNTASIAQARKLVEQLKMEANIDRIKVSKAAADLMAYCEAHAKED / PLLTPVPASENPFREKKFF**C**AIL	71
3	MKGETPVNSTMSIGQARKMVEQLKIEASLCRIKVSKAAADLMTYCDAHACED / PLITPVPTSENPFREKKFF**C**ALL	75
4	MKEGMSNNSTTSISQARKAVEQLKMEACMDRVKVSQAAADLLAYCEAHVRED / PLIIPVPASENPFREKKFF**C**TIL	75
5	MSGSSSVAAMKKVVQQLRLEAGLNRVKVSQAAADLKQFCLQNAQHD / PLLTGVSSSTNPFRPQKV.**C**SFL	68
7	MSATNNIAQARKLVEQLRIEAGIERIKVSKAASDLMSYCEQHARND / PLLVGVPASENPFKDKKP.**C**IIL	68
8 (8 olf)	MSNNMAKIAEARKTVEQLKLEVNIDRMKVSQAAAELLAFCETHAKDD / PLVTPVPAAENPFRDKRLF**C**VLL	70
9 (8 cone)	MAQDLSEKDLLKMEVEQLKKEVKNTRIPISKAGKEIKEYVEAQAGND / PFLKGIPEDKNPFKEKGG.**C**LIS	69
10	MSSGASASALQRLVEQLKLEAGVERIKVSQAAAELQQYCMQNACKD . ALLVGVPAGSNPFREPRS.**C**ALL	68
11	MPALHIEDLPEKEKLKMEVEQLRKEVKLQRQQVSKCSEEIKNYIEERSGED / PLVKGIPEDKNPFKEKGS.**C**VIS	73
12	MSSKTASTNNIAQARRTVQQLRLEASIERIKVSKASADLMSYCEEHARSD / PLLIGIPTSENPFKDKKT.**C**IIL	72
13	MEEWDVPQMKKEVESLKYQLAFQREMASKTIPELLKWIEDGIPKD / PFLNPDLMKNNPWVEKGK.**C**TIL	67

[a] Human G protein γ subunit sequences are from Ref. 26 except γ_{13}, which is from Ref. 27. Sequences are aligned according to an internal Asp/Pro acidic cleavage site, and the Cys residue 4 amino acids from the C terminus, which is the site of prenylation. This Cys is indicated as underlined bold type.

[b] The site of an Asp/Pro bond that is sensitive to acid treatment. This site is present in all γ subunits except γ_{10}. The sequences were aligned relative to this site.

[c] Indicates the total number of amino acids predicted to be in the unprocessed protein. The dots above the sequence indicate each 10th amino acid in the protein sequence.

fragmentation patterns after acid treatment. This series of subunit isoforms can be used as an example of all known γ subunits that would be found in cells or tissues from a given species. The predicted masses of all of these isoforms are unique (Table V) and can be used for the initial identification of γ isoforms based on their intact mass (m/z). Masses in a γ subunit preparation not compatible with any of these predicted masses could be unique (or new) γ subunit isoforms, γ subunits processed in unique or alternative ways, or artifacts of preparation

TABLE V

PREDICTED SINGLY CHARGED IONS AND MODIFICATIONS OF G PROTEIN γ SUBUNITS

Human gamma	Unmodified translation products[a]			Modified translation products[b]			Asp/Pro cleavage products[c]		
	N terminus	[M+H]+	C terminus	N terminus	[M+H]+	C terminus	N [M+H]+	Sequence	C [M+H]+
1	MFVINIE..	8496.9	..KGGCVIS	FVINIE..	8284.7	..KGGC(far)-OCH$_3$	5914.8	GED/PLV	2389.0
2	MASNNTA..	7851.2	..KFFCAIL	Ac-ASNNTA..	7751.1	..KFFC(gg)-OCH$_3$	5161.9	KED/PLL	2608.2
3	MKGETPV..	8305.8	..KFFCALL	MKGETPV..	8294.9	..KFFC(gg)-OCH$_3$	5675.7	GED/PLI	2638.3
4	MKEFMSN..	8389.8	..KFFCTIL	MKEFMSN..	8348.9	..KFFC(gg)-OCH$_3$	5747.6	FED/PLI	2620.3
5	MSGSSSV..	7319.5	..QKVCSFL	Ac-SGSSSV..	7169.4	..QKVC(gg)-OCH$_3$	4869.6	QHD/PLL	2318.9
7	MSATNNI..	7522.8	..KKPCIIL	Ac-SATNNI..	7380.6	..KKPC(gg)-OCH$_3$	5072.7	RND/PLL	2326.9
8	MSNNMAK..	7842.2	..RLFCVLL	Ac-SNNMAK..	7714.1	..RLFC(gg)-OCH$_3$	5174.9	KDD/PLV	2558.2
9 (8c)	MAQDLSE..	7748.0	..KGGCLIS	Ac-AQDLSE..	7563.8	..KGGC(far)-OCH$_3$	5259.0	GND/PFL	2323.8
10	MSSGASA..	7206.4	..FRSCALL	Ac-SSGASA..	7106.3	..FRSC(gg)-OCH$_3$	none	CKDALL	none
11	MPALHIE..	8481.8	..KGSCVIS	PALHIE..	8269.6	..KGSC(far)-OCH$_3$	5982.8	GED/PLV	2305.8
12	MSSKTAS..	8007.3	..KKTCIIL	Ac-SSKTAS..	7865.1	..KKTC(gg)-OCH$_3$	5495.1	FSD/PLL	2389.0
13	MEEMDVP..	7950.3	..KGKCTIL	Ac-MEEMDVP..	7951.5	..KGKC(gg)-OCH$_3$	5452.3	HKD/PFL	2518.1

[a] Characterization of the unmodified translation products (proteins) for each known human γ subunit isoform. The N-terminal and C-terminal sequences of each γ subunit are given, along with the predicted mass of the singly charged ion of the unprocessed protein. All predicted ion masses were generated using the Sherpa 3.3.1 program.[28]

[b] Characterization of the N-terminal and C-terminal sequence and mass of singly charged form of the γ subunit following predicted posttranslational processing patterns. These include acetylation of the N terminus without or with removal of the N-terminal Met, dependent on the residue in position 2, as well as the complex series of modifications at the C terminus associated with prenylation.

[c] Predicted masses of the N-terminal and C-terminal fragments of the processed γ subunits after acid cleavage at an internal Asp/Pro site common to all γ subunits except γ$_{10}$. Also given is the sequence around the Asp/Pro site.

or measurement. Further characterization by MS/MS sequencing, enzymatic digestion, and acid hydrolysis can confirm these assignments or be used to identify new γ isoforms and modifications. This has been done for a γ_5 isoform that was found to be modified differently at the C terminus than is predicted by its sequence. This isoform retains the C-terminal 3 amino acids and; therefore, is not proteolyzed or carboxymethylated.[6]

Greater resolution and sensitivity is obtained with the MALDI analysis of HPLC fractions than for purified G protein (Fig. 1B). The presence of detergent, salt, and the large α and β subunits in the G protein MALDI analysis can suppress ionization and lower signal intensity. In addition, resolution may be affected by sodium, potassium, or detergent adducts. MALDI analyses of γ isoforms from HPLC isolation have typically given best results with 1–10 pmol of γ subunit protein.[11] Interpretable signals can be obtained, however, with femtomole amounts of protein. Using high picomole amounts (greater than 100 picomoles) can make it difficult to obtain high resolution spectra typically seen with low picomole amounts. Under these conditions it can be difficult to reproducibly assign mass estimates because of peak broadening.

SDS–PAGE and Silver Staining

Since the γ subunits do not stain well with Coomassie blue, γ subunits identified in HPLC fractions are visualized by silver staining 10–20% gradient SDS–polyacrylamide gels (Fig. 3A).

Procedure

1. SDS–PAGE is performed as described by Laemmli using 10–20% gradient gels.[21] Gels are stained with Coomassie blue [0.05% Coomassie blue in 45% methanol, 10% (v/v) acetic acid] for at least 1 hr and destained with 10% acetic acid in water. Analysis of proteins by silver staining was performed by the method of Wray *et al.*[22]
2. Gel samples are prepared in 2X Sample Buffer [125 mM Tris-HCl, pH 6.8, 2.0% SDS, 20% glycerol, 0.002% (w/v) Pyronine Y] with 10% 2-mercaptoethanol and boiled at 95° for 5–10 min before loading on the gel.

Notes and Discussion. It is also possible to use 15% polyacrylamide (ratio of 37.5 acrylamide to bisacrylamide) gels, with some loss of separation of γ subunits as compared to 10–20% polyacrylamide gradient gels. Commercial gel systems from Novex (Invitrogen, Carlsbad, CA) can also be used. The 4–12% Bis–Tris NuPAGE gels with MES [2-(N-morpholino)ethanesulfonic acid] running buffer (Invitrogen) provide very good resolution and work well for Western blotting.

---

[21] U. K. Laemmli, *Nature* **227,** 680 (1970).
[22] W. Wray, T. Boulikas, V. P. Wray, and R. Hancock, *Anal. Biochem.* **118,** 197 (1961).

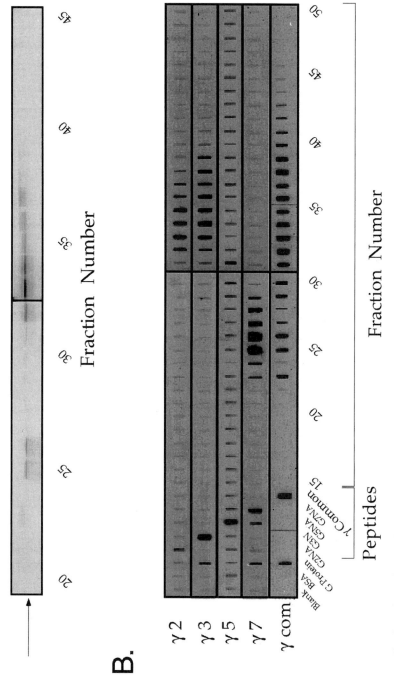

FIG. 3. Analysis of HPLC fractions containing γ subunit isoforms. (A) A representative silver stained 10–20% SDS–polyacrylamide gel of isolated γ isoforms present in indicated HPLC fractions. (B) Representative immunoblots of γ subunit containing HPLC fractions with γ subunit-specific antisera indicating the presence of multiple γ isoforms. Antisera were made to the control peptides used in the immunoblots. Antisera dilutions used for immunoblotting are shown in Table VI.

Tank transfer system conditions using these gels and optimized for γ subunits are 30 min at 30 V, 0.2 A limit. Immunoblotting using a tank transfer system has generally provided better results than transfers using a semidry apparatus.

Slot Blots of HPLC Fractions

The γ isoforms present in the HPLC fractions can be determined by immunoblotting with γ-specific antisera. The use of a slot-blot or dot-blot apparatus greatly facilitates analyzing large numbers of HPLC fractions.

Procedure

1. HPLC fractions containing the isolated γ isoforms are diluted 1 : 40 with TEDA (20 mM Tris-HCl pH 8.0, 1 mM EDTA, 1 mM DTT, and 1% acetonitrile) and are spotted immediately after dilution onto a nitrocellulose membrane (Schleicher and Schuell, Keene, NH, Potran BA 85, 0.45 μm) using a Schleicher and Schuell minifold II slot-blot apparatus under vacuum.
2. Blots are baked under vacuum for at least 90 min at 70–80° and blocked with 1% (w/v) bovine serum albumin (BSA), 5% (w/v) nonfat dry milk in TBS-T (20 mM Tris-HCl, pH 7.4, 137 mM NaCl, 0.05% Tween 20) for at least 30 min at room temperature.
3. Nitrocellulose blots are incubated with the antisera diluted in high-salt TBS-T (20 mM Tris-HCl, pH 7.4, 500 mM NaCl, 0.05% Tween 20).
4. Blots are subsequently washed alternately with TBS-T and high salt TBS-T, four washes total, for 10 min each time.
5. After washing, blots are incubated with donkey anti-rabbit horseradish peroxidase secondary antibody (Amersham, Piscataway, NJ) at 1 : 5000 in high-salt TBS-T and developed with enhanced chemiluminescence reagents (Dupont NEN, Boston, MA) for approximately 1 min followed by exposure to film and development on a Kodak (Rochester, NY) film processor. Alternatively, sample blots can be processed with a Bio-Rad (Hercules, CA) Imager or equivalent instrument.

Notes and Discussion. It is necessary to dilute the organic solvent concentration of the samples to 2% or less to prevent interference by the solvent with binding of the samples to the nitrocellulose. Samples should be filtered immediately after dilution to minimize potential aggregation of protein and loss of the sample. Baking the blots under vacuum is critical for the γ proteins to bind well to the nitrocellulose.

All anti-peptide polyclonal antisera were generated according to the method of Green using female New Zealand rabbits.[23] The peptides used as immunogens are shown in Table VI, along with the dilution of the antibody used for the blots.

[23] N. Green, H. Alexander, A. Olson, S. Alexander, T. M. Shinnick, J. G. Sutcliffe, and R. A. Lerner, *Cell* **28**, 477 (1982).

TABLE VI
GENERATION AND PROPERTIES OF ANTISERA AGAINST G PROTEIN γ SUBUNITS

Name	Specificity	Peptide sequence	Working antibody dilution
G2NA	γ_2	Acetyl-A-S-N-N-T-A-S-I-A-Q-A-R-K-C	1 : 400
G5NA	γ_5	Acetyl-S-G-S-S-S-V-A-A-M-K-K-C	1 : 400
G7NA	γ_7	Acetyl-S-A-T-N-N-I-A-Q-A-R-K-C	1 : 1000
G3N	γ_3	M-K-G-E-T-P-V-N-S-T-M-S-I-G-C-	1 : 500
GBIC	$\gamma_2, \gamma_3, \gamma_7, \gamma_{12}$	P-A-S-E-N-P-F-R-E-K-K-C	1 : 1–5,000

An example immunoblot using this procedure is shown in Fig. 3B. Immunore-activity is seen with each of the subunit-specific antisera, as well as the common antiserum, GBIC, which recognizes γ_2, γ_3, γ_7, and γ_{12} on these blots. The controls spotted on the membrane include each of the immunogen peptides (200 picomoles), purified G protein, bovine serum albumin, and a blank (buffer only). Interestingly, two peaks of γ_3 immunoreactivity are observed: one in fractions 33–34 and the other in fractions 36–37.

Edman Degradation Protein Sequencing and Amino Acid Analysis

HPLC fractions containing γ subunits can be analyzed by Edman degradation and by amino acid analysis. These two analytical techniques allow estimation of the amount of protein that can be sequenced (i.e., unblocked) compared to the total protein present. In general, the samples collected here are compatible with standard techniques implemented in protein sequencing facilities. Amino acid analysis can be performed using a Waters Millipore (Bedford, MA) PicoTag System, as in the MUSC Protein Chemistry Facility. Under standard conditions, 50 μl of an HPLC fraction contains adequate protein for quantitation. For Edman degradation, a 50 μl aliquot of each HPLC fraction is spotted onto glass fiber filters preincubated with Polybrene and sequenced on an automated protein sequencer. In this case, sequencing was on a Procise 494 Applied Biosystems (division of PerkinElmer) instrument in the Protein Chemistry Facility at MUSC.

Notes and Discussion. Comparison of amino acid analysis and Edman degra-dation indicates that the G protein samples analyzed here do not contain significant amounts of proteolytically digested protein, nor do such fragments arise during the isolation of the γ subunits, or during their analysis. For example, the results in Fig. 4 show that only a small portion (approximately 2%) of the total protein in the fractions can be sequenced by Edman degradation. This suggests that the remainder of the proteins have blocked amino termini. The low level of sequenced protein is not, however, the result of nonspecific amino terminal blocking of all protein present during the separation procedures used here. A γ_3 isoform predicted

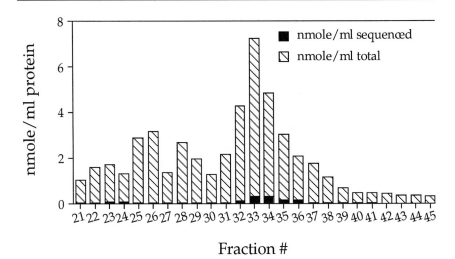

FIG. 4. γ Subunit quantitation and sequencing. Protein quantitation by amino acid analysis (hatched bars) and Edman sequencing (black bars) of HPLC fractions containing γ subunits.

to be unblocked was found in fractions 32–35 by MALDI-MS and was readily sequenced (up to 31 cycles) in fractions 31–37 at levels of 2–16 picomoles. This provided additional evidence for the identity of a γ_3 isoform in the first peak of γ_3 immunoreactivity (Fig. 3B). The second peak of γ_3 immunoreactivity could not be sequenced by Edman degradation. Based upon the results with γ_3, any proteins not N-terminally blocked in these samples should have been sequenced.

The fact that so little protein can be sequenced in the fractions analyzed here argues against the occurrence of nonspecific proteolysis associated with our isolation and characterization procedures because at least half of all proteolytically generated peptides should have an N terminus that could be sequenced. These results are, in fact, compatible with known processing patterns predicted to dictate the modifications of the γ subunits. As shown in Table V, most γ subunits are predicted to be N-terminally blocked by acetylation.

Aspartate–Proline Bond Hydrolysis and MALDI Analysis

Acid hydrolysis of proteins the size of γ subunits can be used for their identification based on the predicted sizes of the fragments generated. A characteristic of most γ subunits cloned to date is the presence of a single Asp-Pro peptide bond (Table IV). The only exception to this is γ_{10}. Asp/Pro bonds are susceptible to acid hydrolysis, which, for the γ subunits, generates an N-terminal fragment in the 4500–6000 mass range and a C-terminal fragment in the 2000–3000 mass range. The exact masses of these fragments can be predicted for a specific γ subunit if its cDNA or protein sequence is known, as well as the pattern of modifications of the protein (Table V).

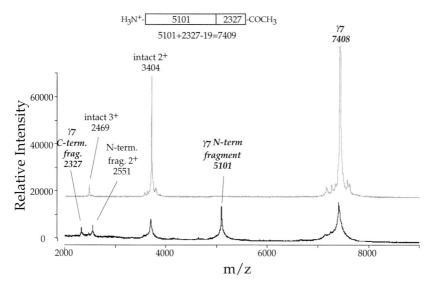

FIG. 5. Asp-Pro bond hydrolysis of γ_7. MALDI mass spectra before (gray) and after (black) acid hydrolysis of the γ_7 subunit, average of 135 and 103 scans, respectively. The inset shows the observed N- and C-terminal fragment ions and how the intact $[M+H]^+$ is calculated from them.

Procedure

1. Aliquots (50–100 μl) of HPLC fractions containing γ subunits are dried under vacuum in 0.6 ml microfuge tubes.
2. The dried sample is resuspended in 2–3 μl of 50% (v/v) acetonitrile in water with 1–3% (v/v) TFA and digested for at least 12 hr at room temperature.
3. Digested samples are assayed by MALDI mass spectrometry as described above.

Notes and Discussion. Data analysis of a typical digest is shown in Fig. 5 for γ_7. The intact $[M+H]^+$ mass observed for γ_7 is seen at m/z 7408. After acid hydrolysis, an N-terminal fragment appears at m/z 5101 and a C-terminal fragment at m/z 2327. These two masses when added with the subtraction of water, account for the intact γ_7 mass seen at m/z 7408 within a 1 Da error (5101 + 2327 − 18 − 1 = 7409).

Some difficulties may be encountered with the Asp-Pro hydrolysis of the intact γ subunits. In general, acid hydrolysis of Asp-Pro bonds is an inefficient reaction that does not go to completion. It is often necessary to continue hydrolysis for up to 48 hr. There are always significant amounts of the intact protein remaining after the reaction. This can, however, be useful in the analysis of the resultant spectrum since both the intact protein and the products are present.

The N- and C-terminal fragment ions of the Asp-Pro hydrolysis of the γ subunit isoforms are shown in Table V. These masses, seen in a MALDI spectrum after acid hydrolysis, can help to identify the γ isoform present in the sample. The existence of an intact mass of predicted size, along with predicted acid cleavage products, is strong presumptive evidence for the identification of a γ subunit isoform. In analyzing the data, it should be recognized that multiply charged intact γ protein or N-terminal fragment ions can fall into the range of C terminal fragment ions. In addition, all combinations of the N- and C-terminal ions, and not just those combinations predicted, should be evaluated to determine if any unknown intact γ possesses those N- and C-terminal ions.

ESI-MS/MS Analysis

ESI-tandem mass spectrometry is a powerful technique which can be used to determine which γ isoforms are present in the HPLC fractions. Electrospray ionization (ESI) is performed by spraying a protein sample from the tip of a needle carrying a high electrical potential, ~4 kV. Charged droplets travel across a voltage gradient to a heated capillary, set at a lower voltage, where they enter a vacuum inside the mass spectrometer. Multiprotonated ions of the intact protein species are generated (MS mode). This generates multiple mass measurements of the sample, which can be used to obtain highly accurate and precise masses. In MS/MS mode, protein ions, selected according to m/z ratio, collide with He gas, which induces fragmentation (called collision-induced dissociation or CID) into b and y ions, which are N-terminal and C-terminal ions generated by cleavage at peptide bonds, respectively.[24] Analysis of these fragmentation patterns can be used to generate sequence information about the protein of interest. The MS/MS analysis can be performed in three different ways: by in-line injection from an HPLC to an ESI mass spectrometer, by direct injection into the mass spectrometer, or by nanospray analysis. These methods were described previously and are in detail in the following section.[6]

Procedure
In-line ESI analysis. The effluent from the HPLC column can be directly connected to the ESI source of the Finnigan LCQ instrument for analysis as described above.

Direct injection ESI analysis. An aliquot of an HPLC fraction is dried under vacuum and resuspended in 5–25 μl of either 47.6% n-propanol/47.6% water/4.76% acetonitrile/0.095% TFA or 50% methanol/26% water/20% n-propanol/4% acetic acid. The sample is delivered to the Finnigan LCQ ion trap mass spectrometer by a direct injection of 5 μl of sample into a flowing stream (50 μl/min) of 50% water/46% methanol/4% acetic acid (v/v/v) directed to the standard ESI source or by nanospray.

[24] K. Biemann and H. A. Scoble, *Science* **237**, 992 (1987).

Nanospray analysis. Nanospray (continuous flow of nl/min) is performed using a custom built nanospray source with a gold sputter-coated pulled glass capillary filled with ~ 1 μl of sample. Voltage (1500 V) is applied to the capillary positioned ~ 1 mm from the heated metal capillary of the electrospray source. Nanospray is also performed using a glass capillary filled with ~ 1 μl of sample with a wire inserted into the liquid sample and the voltage applied. The latter configuration is much easier and therefore has become the method of choice for nanospray.

MS/MS is accomplished by selecting the *m/z* ratio of the precursor ion of interest, that is subsequently fragmented. Figure 6 shows a representative MS/MS spectrum for γ_7, *m/z* 1482.61, $[M+5H]^{5+}$ selected, indicating the b and y ions obtained. These fragment ions can be further analyzed in an MS/MS/MS experiment if there is sufficient sample and signal. This is accomplished by two sequential stages of *m/z* selection fragmentation and analysis. A window of 1 or 2 *m/z* units is used in each stage of ion selection. This procedure and data analysis is fully described elsewhere and in this volume.[6,24a]

Notes and Discussion. The presence of an isoprenyl group could be shown to be associated with many of the γ subunit masses by ESI-ion trap mass spectrometry. This thioether bond is preferentially cleaved by CID (collision induced dissociation) and can be seen by the loss of the mass for the isoprenyl group (272 for geranylgeranyl, 204 for farnesyl) from the intact protein. Many γ subunit masses are of sufficient abundance to allow a mass loss consistent with loss of an isoprenyl group to be seen.[24a]

Of consideration in the use of nanospray capability is its availability. This technique allows analysis of very small quantities of sample (femtomole amounts) for long periods of time (up to 30 min or more). This approach is limited only by sample availability. The long time allowed for sample analysis using this technique has an advantage in that it allows multiple rounds of MS/MS fragmentation of the protein. The nanospray experiments described here were performed on a custom-built external nanospray source that may not be available at (or whose configuration will vary among) other institutions. Commercial nanospray sources are available from Protana (Denmark) and New Objectives, Inc. (Cambridge, MA).

Enzymatic Digestion and LCQ-MS (MS/MS) Analysis

The γ subunits can also be identified by enzymatic digestion (i.e., with trypsin or Asp-N). Masses of the peptide fragments can be observed in MALDI spectra and/or sequenced using ESI-MS/MS. For example, to create a proteolytic map of γ_2, the protein is digested with Asp-N, which cleaves at the amino side of

[24a] K. L. Schey, M. Busmon, L. A. Cook, H. Hamm, and J. D. Hildebrandt, *Methods Enzymol.* **344,** [41] 2002 (this volume).

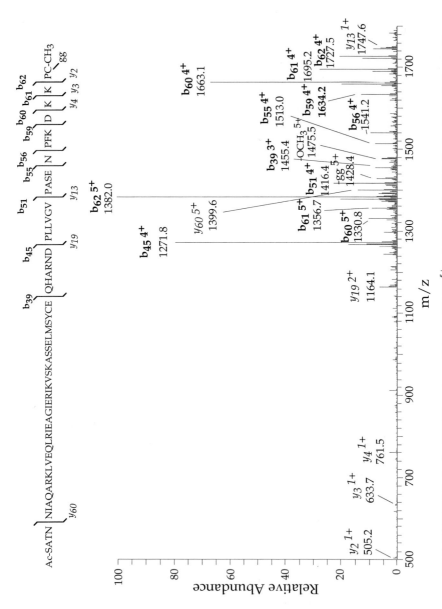

FIG. 6. MS/MS spectrum of intact γ_7. MS/MS spectrum of intact γ_7 $[M+5H]^{5+}$, m/z 1482.61 selected, indicating the b and y fragment ions, average of four scans. Data were collected during the separation of γ subunits from α and β by HPLC inline with ESI-LCQ. *Inset:* Sequence shows the γ_7 sequence and the b and y ions generated during the experiment.

TABLE VII
PREDICTED ASP-N FRAGMENTS OF HUMAN γ SUBUNIT ISOFORMS[a]

1		2		3		4		5		7	
(8-12)	605.3	(26-35)	1058.6	(46-51)	645.2	(30-39)	1044.6	(34-45)	1444.7	(61-65)	876.6
(2-7)	684.4	(36-47)	1380.6	(40-45)	745.3	(40-51)	1418.7	(46-65)	2434.0	(34-45)	1466.6
(13-19)	862.5	(2-25)	2644.0	(52-72)	2753.4	(52-72)	2735.4	(2-33)	3327.9	(46-60)	1583.8
(51-59)	967.5	(48-68)	2723.3	(1-39)	4207.1	(1-29)	3205.8			(2-33)	3509.0
(42-50)	1083.5										
(60-71)	1554.9										
(20-41)	2638.2										

8(olf)		9(8cone)		10		11		12		13	
(46)	134.0	(2-3)	260.1	(46-65)	2372.9	(2-7)	679.4	(65-69)	880.6	(40-44)	529.3
(62-67)	1067.7	(4-8)	591.3	(2-45)	4752.5	(51-59)	967.5	(38-49)	1440.6	(1-4)	636.2
(47-61)	1623.8	(47-55)	1015.5			(60-70)	1470.8	(50-64)	1641.9	(45-50)	702.3
(25-45)	2320.7	(56-66)	1440.8			(8-50)	5207.0	(2-37)	3957.5	(51-64)	1949.5
(2-24)	2643.1	(9-46)	4330.0							(5-39)	4209.0

[a] For each of the 12 known human γ subunit isoforms (with predicted modifications), the first column contains a list of residues contained in their Asp-N fragments. The second column contains the predicted $[M+H]^+$ ion for each fragment. In boxes are fragments unique to one γ subunit isoform.

Asp residues. The bovine and human γ_2 isoforms have three proteolytic sites and therefore should generate four Asp-N fragments (Table VII). Digestion with other proteases such as trypsin can also be done; however, the number of fragments generated increases the complexity of the HPLC separation (Table VIII). Some very small fragments of as little as one amino acid are generated and it is difficult to account for the entire protein. Nevertheless, numerous unique fragments can be generated using these proteases that potentially allow identification of G protein γ subunits even in mixtures of proteins.

Procedure

1. A 200 μl aliquot (approximately 0.4–1.4 nmol of total γ) of an HPLC fraction containing γ_2 as the major component is dried under vacuum and resuspended in 9.2 μl of 50 mM NaP buffer pH 7.4.
2. Endoproteinase Asp-N (0.04 μg/μl in water) is added for a final ratio of approximately 1 : 100–400 w/w of enzyme (0.8 μl) to γ_2. Reconstituted Asp-N should be stored at 4° and used within 1 week.
3. The protein is digested overnight at 32°.
4. The digested protein is dried under vacuum, resuspended in 10 μl of 0.1 M acetic acid (solvent A), and stored at −20° until analysis.
5. HPLC separation of peptides is performed using a MicroTech Scientific (Saratoga, CA) 15 cm × 0.32 mm i.d. Inertsil C_{18} column ODS 5 μm inline

TABLE VIII
PREDICTED TRYPSIN FRAGMENTS OF HUMAN γ SUBUNIT ISOFORMS[a]

1		2		3		4		5		7	
(24)	147.1	(14)	147.1	(18)	147.1	(18)	147.1	(12)	147.1	(12)	147.1
(15-16)	260.2	(65)	147.1	(69)	147.1	(69)	147.1	(26-27)	246.2	(26-27)	260.2
(13-14)	262.1	(28-29)	260.2	(32-33)	260.2	(32-33)	246.2	(64-65)	**507.4**	(61-62)	262.1
(66-68)	389.2	(63-64)	276.2	(67-68)	276.2	(67-68)	276.2	(13-18)	**742.5**	(28-30)	333.2
(69-71)	454.3	(30-32)	333.2	(1-2)	278.2	(1-2)	278.2	(19-25)	**772.4**	(63-65)	**633.4**
(62-65)	505.3	(66-68)	702.4	(34-36)	333.2	(19-24)	**687.4**	(28-36)	**902.5**	(13-18)	**757.5**
(31-35)	**577.3**	(15-20)	729.5	(70-72)	702.4	(70-72)	702.4	(2-11)	**966.5**	(19-25)	**787.4**
(56-61)	658.3	(21-27)	**848.4**	(19-24)	**747.4**	(25-31)	**855.3**	(37-63)	**3015.4**	(2-11)	**1087.5**
(36-41)	**738.3**	(2-13)	**1245.6**	(25-31)	791.4	(3-17)	1583.7			(31-44)	**1582.8**
(25-30)	**746.4**	(33-46)	**1464.7**	(3-17)	1548.7	(51-66)	**1795.0**			(45-60)	**1697.9**
(42-47)	**810.4**	(47-62)	**1783.0**	(37-66)	**3250.7**	(34-50)	**1818.1**				
(48-55)	844.4										
(17-23)	**862.4**										
(2-12)	**1270.7**										

8(olf)		9(8cone)		10		11		12		13	
(13)	147.1	(20)	147.1	(26-27)	260.2	(24)	147.1	(67)	147.1	(11)	147.1
(64)	175.1	(32-34)	**275.2**	(61-63)	**401.2**	(15-16)	260.2	(16)	175.1	(62-63)	**204.1**
(62-63)	262.1	(62-63)	276.2	(64-65)	**495.3**	(13-14)	276.2	(30-31)	260.2	(64)	**408.3**
(27-28)	278.2	(21-23)	375.2	(13-18)	729.5	(66-67)	276.2	(65-66)	262.1	(25-29)	**565.3**
(8-12)	**559.3**	(35-37)	389.2	(19-25)	773.4	(25-27)	375.2	(32-34)	333.2	(12-17)	**704.4**
(65-67)	**668.4**	(24-26)	**390.2**	(2-12)	1076.5	(28-30)	**416.3**	(2-4)	**363.2**	(30-36)	**813.5**
(2-7)	**706.3**	(64-66)	454.3	(46-60)	**1513.7**	(68-70)	**484.3**	(68-69)	**509.3**	(55-61)	**886.4**
(14-19)	**717.4**	(9-12)	**488.3**	(28-45)	**1987.3**	(62-65)	505.3	(17-22)	**744.4**	(18-24)	**925.5**
(20-26)	**858.5**	(58-61)	505.3			(31-35)	**589.3**	(23-29)	817.4	(37-44)	**957.5**
(46-61)	**1738.9**	(27-31)	**557.4**			(56-61)	658.3	(5-15)	**1146.6**	(45-54)	**1189.6**
(29-45)	**1790.1**	(52-57)	658.3			(36-41)	**708.3**	(35-48)	1583.7	(1-10)	**1334.6**
		(2-8)	**832.4**			(42-47)	**823.4**	(49-64)	**1729.0**		
		(13-19)	**876.5**			(48-55)	844.4				
		(38-51)	**1581.7**			(17-23)	**904.5**				
						(2-12)	**1261.7**				

[a] For each of the 12 known human γ subunit isoforms (with predicted modifications), the first column contains a list of residues contained in their tryptic fragments. The second column contains the predicted $[M+H]^+$ ion for each fragment. In boxes are fragments unique to one γ subunit isoform.

with a Finnigan LCQ instrument. The flow (0.5 ml/min) is split before the column to obtain a column flow rate of 5–8 μl/min. The peptides are injected into a 10 μl loop and eluted with a gradient of 3–90% solvent B (100% acetonitrile) for 43 min. The eluate is sent directly into the electrospray source for ionization. Spectra (MS and MS/MS) are collected during the HPLC run for the analysis and sequencing of peptides.

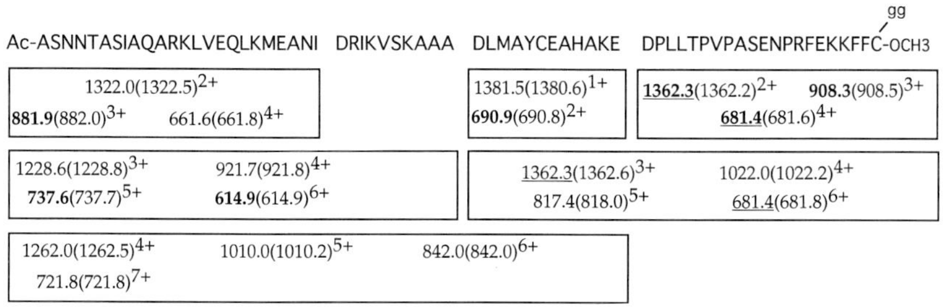

Ac-ASNNTASIAQARKLVEQLKMEANI   DRIKVSKAAA   DLMAYCEAHAKE   DPLLTPVPASENPRFEKKFFC-OCH3

1322.0(1322.5)$^{2+}$

881.9(882.0)$^{3+}$ 661.6(661.8)$^{4+}$

1381.5(1380.6)$^{1+}$ **1362.3**(1362.2)$^{2+}$ 908.3(908.5)$^{3+}$

690.9(690.8)$^{2+}$ <u>**681.4**</u>(681.6)$^{4+}$

1228.6(1228.8)$^{3+}$ 921.7(921.8)$^{4+}$

737.6(737.7)$^{5+}$ **614.9**(614.9)$^{6+}$

<u>1362.3</u>(1362.6)$^{3+}$ 1022.0(1022.2)$^{4+}$

817.4(818.0)$^{5+}$ <u>**681.4**</u>(681.8)$^{6+}$

1262.0(1262.5)$^{4+}$ 1010.0(1010.2)$^{5+}$ 842.0(842.0)$^{6+}$

721.8(721.8)$^{7+}$

FIG. 7. Proteolytic mapping of γ_2. Endoproteinase Asp-N proteolytic map of the γ_2 subunit after ESI-MS and MS/MS indicating the predicted and observed peptide ions. The numbers in parentheses are the predicted m/z ions for the indicated peptides. The underlined m/z values have more than one Asp-N peptide assignment possible and are also in the same fraction of time. The bold m/z values have MS/MS data available.

Notes and Discussion. Analysis of HPLC separated peptides generated MS (and MS/MS) spectra which produced the proteolytic map of γ_2 seen in Fig. 7. As shown in Fig. 7, all fragments in the protein digest can be accounted for using this analysis. In addition, some MS/MS ions provide sufficient signal to sequence at least a portion of the peptide. This technique can furnish similar information on other γ subunit sequences and provide evidence for modifications of a γ subunit.

In general, proteolytic digests do not go to completion. As a result, fully digested peptide fragments are not always obtained. Together with the fact that ESI usually generates multiply charged ions, this underscores the complexity of MS and MS/MS data interpretation. Analysis is greatly facilitated by deconvolution algorithms of the software that help identify the singly charged ion masses. The advantage of ESI is the great accuracy (within 1 Da) and resolution (1 Da resolutions) of the estimates associated with multiple mass measurements.

The existence of partially digested fragments can be used to great advantage during the analysis of the results. This produces overlapping fragment masses that can be used to evaluate not just the size of proteolytic fragments, but also their presumed order (to determine the sequence of unknown γ subunits). This greatly increases the confidence with which a γ subunit structure or isoform can be identified. Additionally, MALDI analysis of peptides can be performed for γ subunit characterization (data not shown). Unique peptide ions can be observed in MALDI mass spectra to identify the presence of particular γ isoforms, termed peptide mass fingerprinting (PMF).[25]

Masses corresponding to trypsin or Asp-N fragments for the human γ subunits are shown in Tables VII and VIII. Unique masses are shown in bold. Some γ isoforms generate fragments whose masses are identical to those of other γ isoforms. Within a γ family (i.e., γ_2, γ_3, γ_4 and γ_1, γ_9, γ_{11}) many trypsin fragments of the same mass represent identical or similar sequences. For γ subunits, peptide mass

[25] J. S. Cottrell, *Peptide Res.* **7**, 115 (1994).

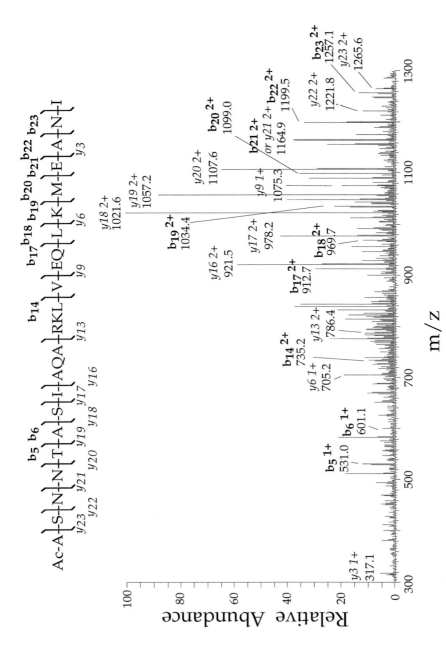

FIG. 8. MS/MS spectrum of γ_2 Asp-N peptide. MS/MS spectrum of the first γ_2 Asp-N peptide $[M+3H]^{3+}$, m/z 881.75, average of 17 scans. Data were collected during the separation of digested peptides by HPLC inline with ESI-LCQ. *Inset*: Sequence shows the γ_2 peptide sequence and the b and y ions generated during the experiment.

fingerprinting is very useful. There must be at least two unique peptide masses observed after digestion for a particular isoform to be identified. For example, for trypsin fragment masses above 500 Da, the presence of two unique peptides would correctly identify the γ isoform. Asp-N digestion generates almost all unique fragments masses (exception m/z 967.5 for γ_1 and γ_{11}). In this case two peptides at any mass (except m/z 967.5) would correctly identify the γ isoform. The one caveat to these considerations is that modifications to peptides cannot always be identified, particularly if uncharacterized.

Figure 8 shows an MS/MS spectrum of the first γ_2 ASP-N peptide $[M+3H]^{3+}$ at m/z 881.75 with the b and y ions labeled. As indicated in Fig. 8, most of the peptide is accounted for by these ions. The sequence information obtained from this spectrum can accurately identify the sequence of this peptide fragment and its modifications.

Discussion

The analysis of γ subunits described here has broad applicability to the characterization of modified proteins. Gel electrophoresis, immunoblotting, and Edman sequencing can elucidate only limited information about a protein (such as crude or even deceiving molecular weight estimates and whether or not the protein is N-terminally blocked), but they cannot predict structure or identify modifications to the protein. Advances in mass spectrometry are occurring at a rapid pace and new methods and instruments are currently developing with even greater capabilities. Digestion of protein, peptide isolation by HPLC, and subsequent sequencing by mass spectrometry can elucidate the sequence and modifications of proteins with great facility. Already we have used this technology to generate evidence for heterogeneous processing of G protein γ subunits.[6] However, the degree to which processing heterogeneity influences cellular function is still far from understood. The techniques described here are likely to play a critical role in evaluating the relationship between the complement of proteins expressed in cells (the proteome) and generation of diversity of biological function.

Acknowledgments

This work was supported in part by NIH Grant DK37219, by NSF Grant EPS9630167, and by institutional support of the Mass Spectrometry, Protein Sequencing and Peptide Synthesis, and Protein Chemistry Facilities of the Medical University of South Carolina. The authors thank Rebecca Ettling for protein sequencing, Dr. Chris Schwabe and Mr. Bob Bracey for amino acid analysis, and Dr. Dan Knapp for use of the Mass Spectrometry Facility.

[26] E. H. Hurowitz, J. M. Melnyk, Y.-J. Chen, H. Kouros-Mehr, M. I. Simon, and H. Shizuya, *DNA Res.* **7,** 111 (2000).

[27] L. Huang, Y. G. Shanker, J. Dubauskaite, J. Z. Zheng, W. Yan, S. Rosenzweig, A. I. Spielman, M. Max, and R. F. Margolskee, *Nature Neurosci.* **2,** 1055 (1999).

[28] J. A. Taylor, K. A. Walsh, and R. S. Johnson, *Rapid Commun. Mass Spectrom.* **10,** 679 (1996).

[17] Activity of Gγ Prenylcysteine Carboxyl Methyltransferase

By David Michaelson and Mark R. Philips

Introduction

Prenylcysteine-directed carboxyl methyltransferase (pcCMT) is the third of three enzymes that sequentially modify the C terminus of CAAX proteins such as Ras and the γ subunits of G proteins. Nascent CAAX proteins are localized in the cytosol where they encounter one of several prenyltransferases[1] that modify the CAAX cysteine with a farnesyl (C_{15}) or geranylgeranyl (C_{20}) lipid. Whereas $G\gamma_t$ is a substrate for farnesyltransferase, the other γ subunits are modified by geranylgeranyltransferase II. Prenylated CAAX proteins are directed to the endomembrane where they encounter a prenyl-CAAX specific protease, Rce1,[2,3] that cleaves the AAX amino acids. The newly C-terminal prenylcysteine residue then becomes a substrate for pcCMT, also localized in the endomembrane,[4] that methylesterifies the α-carboxyl group of the prenylcysteine. The net effect of this sequence of post-translational modifications is to create a hydrophobic domain at the C terminus of otherwise hydrophilic proteins and to thereby engage pathways for specific targeting of CAAX proteins to cellular membranes.[5] The precise nature of these pathways and how specificity for one membrane over another is accomplished remain unresolved.

Although prenylcysteine carboxyl methylation contributes significantly to the hydrophobicity of farnesylated proteins, this is not the case with geranylgeranylated proteins,[6] suggesting a more specific role for this reversible modification in protein–protein interactions, perhaps analogous to protein phosphorylation. Indeed, carboxyl methylation of $G\gamma$ has been shown to influence the interaction of $G\beta\gamma$ with both $G\alpha$ and various effectors.[7,8] Elucidation of the precise biological

---

[1] P. J. Casey and M. C. Seabra, *J. Biol. Chem.* **271,** 5289 (1996).

[2] V. L. Boyartchuk, M. N. Ashby, and J. Rine, *Science* **275,** 1796 (1997).

[3] W. K. Schmidt, A. Tam, K. Fujimura-Kamada, and S. Michaelis, *Proc. Natl. Acad. Sci. U.S.A.* **95,** 11175 (1998).

[4] Q. Dai, E. Choy, V. Chiu, J. Romano, S. Slivka, S. Steitz, S. Michaelis, and M. Philips, *J. Biol. Chem.* **273,** 15030 (1998).

[5] E. Choy, V. K. Chiu, J. Silletti, M. Feoktistov, T. Morimoto, D. Michaelson, I. Ivanov, and M. Philips, *Cell* **98,** 69 (1999).

[6] J. R. Silvius and F. l'Heureux, *Biochemistry* **33,** 3014 (1994).

[7] Y. Fukada, T. Matsuda, K. Kokame, T. Takao, Y. Shimonishi, T. Akino, and T. Yoshizawa, *J. Biol. Chem.* **269,** 5163 (1994).

[8] C. A. Parish, A. V. Smrcka, and R. R. Rando, *Biochemistry* **34,** 7722 (1995).

role of prenylcysteine carboxyl methylation has been confounded by the observation that, whereas yeast require Gγ and Ras homologs for growth, strains of yeast that lack pcCMT (*Saccharomyces cerevisiae* Ste14) are not defective in growth. Nevertheless, pcCMT has been highly conserved through evolution, suggesting an important, if not essential, biological role, and recent observations on Ras trafficking in mammalian cells suggest that pcCMT is required for the proper membrane targeting of Ras.[5]

Although pcCMT activities have been described in mammalian tissues and cells for a decade, until recently, the only pcCMT characterized at the molecular level was Ste14p of *S. cerevisiae*. We have now cloned and expressed human pcCMT from an HL-60 cell cDNA library.[4] In this chapter we describe assays of pcCMT utilizing membranes derived either from myeloid cells or tissue culture cells that overexpress human pcCMT.

In Vitro Methyltransferase Assay

Human pcCMT is a polytopic integral membrane protein localized in the endoplasmic reticulum (ER) and Golgi.[4] The enzymatic activity of pcCMT is sensitive to detergent and therefore the protein cannot be extracted from membranes in active form.[9] Consequently, preparations of membranes from cells relatively rich in endogenous pcCMT or enriched by overexpression of recombinant pcCMT serve as the source of enzyme for methylation assays.

Membranes derived from brain tissue and myeloid cells have been found to possess high levels of endogenous pcCMT activity.[9,10] Neutrophils and HL-60 cells are convenient sources of membranes rich in pcCMT. Human neutrophils isolated from the blood of normal volunteers, when available, offer several advantages over cultured cells as a source of pcCMT-rich membranes. These include rapid isolation of large numbers of cells ($1-3 \times 10^8$ per 150 ml of whole blood) and reduced tissue culture costs. The disadvantages of using freshly isolated human neutrophils include the requirement for a human subjects protocol (expired blood bank donations are not suitable for the isolation of neutrophils because of their very short *ex vivo* shelf-life), the precautions needed when working with human blood products, and the requirement for a relatively expensive nitrogen bomb (see below) for cell disruption to avoid lysis of protease-laden granules. If tissue culture is to be used, the human promyelocytic leukemia cell line, HL-60, offers the advantage of relatively rapid growth to high density in suspension cultures, but like neutrophils, HL-60 cells require a nitrogen bomb for high-quality membrane preparations.

[9] M. H. Pillinger, C. Volker, J. B. Stock, G. Weissmann, and M. R. Philips, *J. Biol. Chem.* **269,** 1486 (1994).
[10] C. Volker, R. A. Miller, W. R. McCleary, A. Rao, M. Poenie, J. Backer, and J. Stock, *J. Biol. Chem.* **266,** 21515 (1991).

TABLE I
ACTIVITY (AFC CARBOXYL METHYLATION) OF ECTOPICALLY
EXPRESSED HUMAN pcCMT

Source of membranes	Specific activity[a] (pmol/mg · min)	Increase (fold)
Neutrophils	3.8	(3)
COS-1 untransfected	1.1	1
COS-1 + vector	1.1	1
COS-1 + human pcCMT	49.6	45

[a] Assayed by AFC methylation as described in text.

Although COS-1 cells express relatively low levels of endogenous pcCMT, they are the cell-of-choice for preparation of membranes expressing recombinant pcCMT because of their ease of transfection and the high levels of expression that can be achieved with vectors that contain robust promoters and an SV40 (simian virus 40) origin of replication (e.g., pcDNA3.1). Although it is more difficult to harvest COS-1 cells in numbers approaching those readily achieved with human neutrophils or even HL-60 cells, overexpression of recombinant pcCMT can yield membranes with a specific activity 10 to 20-fold that obtained from myeloid cells (Table I). Homogenization by physical shearing (e.g., Dounce) is adequate for preparing membranes from COS-1 cells because they lack the protease-rich granules of myeloid cells.

Human Neutrophils

Isolation of human neutrophils from whole blood of healthy volunteers has been described in detail elsewhere.[11] Although isolated neutrophils have a bench-life measured in hours, pcCMT activity is well-preserved for up to 1 year in membrane preparations stored at $-80°$. One hundred and fifty ml of blood from a single human donor can yield $1–3 \times 10^8$ cells and $100–300 \ \mu$g of membranes.

HL-60 Cells

HL-60 cells (obtained from ATCC, Manassas, VA) are maintained in Dulbecco's modified Eagle's medium (DMEM) supplemented with 10% fetal calf serum (FCS, Life Technologies, Rockville, MD) at $37°$ in 5% (v/v) CO_2. Cells are split 1 : 20 twice a week. Because HL-60 are nonadherent cells that settle on the bottom of culture flasks, maximal yields of cells can be obtained with rotating or spinner flasks, although incubating T75 flasks on rockers achieves similar results. Under these conditions, HL-60 cells will grow, doubling approximately once every 24 hours, to a density of 5×10^4 cells per ml of medium. Thus, 500 ml of medium can yield up to 2.5×10^7 cells.

Preparation of Myeloid Cell Membranes by Nitrogen Cavitation

Because neutrophils and HL-60 cells contain numerous granules rich in potent proteases, nitrogen cavitation is the optimal method of cell disruption.[12] This method yields vesiculated light membranes (plasma membrane and endomembrane) without disrupting the granules.

Neutrophils or HL-60 cells to be fractionated are washed twice with cell buffer (150 mM NaCl, 5 mM KOH, 1.3 mM CaCl$_2$, 1.2 mM MgCl$_2$, 10 mM HEPES, pH 7.4) and then resuspended in 18 ml of ice-cold relaxation buffer [100 mM KCl, 3 mM NaCl, 3.5 mM MgCl$_2$, 1 mM ATP, 2 mM phenylmethylsulfonyl fluoride (PMSF), 10 μg/ml leupeptin, 10 μg/ml chymostatin, 10 μg/ml pepstatin A, 27 μg/ml aprotinin, 10 mM HEPES, pH 7.3]. As previously described,[11] cell suspensions are stirred at 350 psi in a nitrogen bomb (Parr Instruments, Moline, IL) for 20 min at 4° and then cavitated by dropwise collection into a tube containing a volume of 20 mM EGTA calculated to yield a final concentration of 1.25 mM. The cavitate is centrifuged at 3000 rpm to remove nuclei and unbroken cells. The postnuclear supernatant is layered on enough ice-cold 40% sucrose ($\eta_D = 1.389$ at 20°, $\rho = 1.150$ gm/cm^3) in 20 mM Tris-HCl pH 7.4 containing 27 μg/ml aprotinin to fill a 12 ml Ultra-Clear (Beckman Coulter, Inc., Fullerton, CA) polycarbonate ultracentrifuge tube. The tube is spun in an SW-41 rotor (Beckman) for 2 hr at 4° at 35,000 rpm. Cytosol is collected from the layer above the sucrose. This cytosol, dialyzed against 20 mM Tris-HCl pH 7.4 containing 3 μg/ml aprotinin and then concentrated by Centricon (Millipore, Bedford, MA), can be used as a source for p21 methylation substrates (see below). Light membranes (plasma membrane and endomembranes) are collected as an opalescent band just below the cytosol/sucrose interface. The granular pellet is discarded. The membrane fraction is suspended in several volumes of ice-cold 20 mM Tris-HCl pH 7.4 containing 27 μg/ml aprotinin, freeze/thawed (ethanol in dry ice) in the same buffer 5–7 times to release cytosol contaminating the reclosed membrane vesicles, pelleted for 30 min at 4° at 150,000g, resuspended in 100–200 μl of the same buffer, and stored in aliquots at −80°. Membrane protein is assayed with the BCA assay (Pierce, Rockford, IL) using 0.1% sodium dodecyl sulfate (SDS) in 150 mM NaCl as the diluent and bovine serum albumin (BSA) as the standard. Approximately 150 μg of membrane proteins (enough for 30–50 methylation reactions) can be obtained from 10^8 cells.

Ectopic Expression of Human pcCMT in COS-1 Cells

COS-1 cells (obtained from ATCC) are grown in 5% CO$_2$ at 37° in DMEM containing 10% FCS and antibiotics. The day before transfection, cells must be

[11] M. R. Philips and M. H. Pillinger, *Methods Enzymol.* **256,** 49 (1995).

[12] M. R. Philips, S. B. Abramson, S. L. Kolasinski, K. A. Haines, G. Weissmann, and M. Rosenfeld, *J. Biol. Chem.* **266,** 1289 (1991).

split and seeded at 10^6 per 10 cm plate in 6 ml of DMEM containing antibiotics and 10% FCS. The cells should be grown to 30% confluence on the day of transfection, which will yield 5×10^6 cells per 10 cm plate at the time of harvesting.

The pcCMT cDNA cloned into a mammalian expression vector, e.g., pcDNA3.1, can be transfected into COS-1 cells with high efficiency using SuperFect reagent (Qiagen, Valencia, CA). Although lipid-based transfection methods such as LipofectAMINE (Life Technologies) give transfection and expression efficiencies similar to those of SuperFect, the lipid can perturb the membrane and is best avoided when membrane preparations are desired. Dilute 10 μg of vector DNA in 300 μl of DMEM without serum and antibiotics and mix gently. Add 40 μl of SuperFect reagent to the DNA solution and mix by pipetting 5 times. Incubate the sample at room temperature for 10–15 min. While incubating, gently wash the cells with phosphate-buffered saline (PBS) prewarmed to 37°. Add 3 ml of DMEM (containing serum and antibiotics) to the transfection mixture. Mix by pipetting and immediately transfer onto the cells in a 10 cm dish. Incubate cells with the transfection mixture for 3 hr at 37° in 5% CO_2. Remove the medium with the transfection complex and wash 3 times with prewarmed PBS. Add fresh DMEM containing serum and antibiotics and place in incubator. Seventy-two hours after initiation of transfection, cells should be 95% confluent. These cells can be harvested by scraping for membrane preparation.

Preparation of COS-1 Membranes by Homogenization

Scrape cells into 1 ml of ice-cold Homogenizing Buffer [HB: 10 mM Tris-HCl pH 7.4, 10 mM KCl, 1 mM dithiothreitol (DTT), 2 mM PMSF, 27 μg/ml aprotinin, 10 μg/ml antipain, 10 μg/ml pepstatin, 10 μg/ml chymostatin]. Keeping cells cold on ice or working in a cold room at 4°, homogenize cells in a tight-fitting Dounce homogenizer with 30 strokes. Immediately transfer the homogenate to a clean tube containing 1/10 volume of 2.5 M sucrose to make the homogenate isotonic.

Centrifuge homogenate at 600g for 5 min at 4°. Collect and save the first postnuclear supernatant (PNS). Resuspend the pellet in 0.5 ml of HB, rehomogenize, adjust the tonicity with 1/10 volume 2.5 M sucrose, and centrifuge at 600g for 5 min at 4°. Collect the second PNS and combine it with the first.

Centrifuge the combined PNS at 8000g for 10 min at 4° to remove mitochondria. Collect the postmitochondrial supernatant (PMS), transfer it to clean tubes, and centrifuge at 150,000g for 90 min at 4°. Resuspend the pellet (crude membranes) in 100–500 μl 25 mM Tris-HCl pH 7.4 containing 27 μg/ml aprotinin. Measure membrane protein concentration with the BCA assay (Pierce) using 0.1% SDS in 150 mM NaCl as diluent. Ten 10 cm plates of confluent COS-1 cells yields up to 3 mg of membrane protein, enough for up to 3000 methylation assays. Store membrane suspensions in aliquots at −80°.

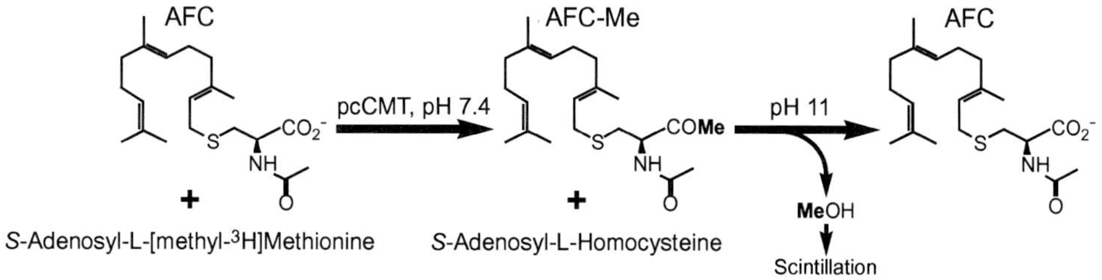

FIG. 1. Carboxyl methylation of N-acetyl-S-trans,trans-farnesylcysteine (AFC) by pcCMT. The two reactions shown form the basis of a rapid, quantitative, and reproducible assay for pcCMT activity. In the first reaction the methyl acceptor, AFC, and the methyl donor, S-adenosyl-L-[methyl-³H]methionine ([³H]AdoMet), are incubated with membranes containing pcCMT at pH 7.4 resulting in the formation of N-acetyl-S-trans,trans-farnesylcysteine methyl ester (AFC-Me) that can be extracted from unreacted [³H]AdoMet in n-heptane. In the second reaction the pH is raised to 11 resulting in hydrolysis of AFC-Me and liberation of volatile [³H]methanol that can be quantitated by partition into scintillation fluid. The radiolabeled methyl group is shown in bold.

In Vitro AFC Methylation Assay

Although a pcCMT assay has been described that utilizes as a methyl acceptor synthetic peptides that terminate in a farnesylcysteine and thereby mimic the C-terminal domains of partially processed CAAX proteins,[13] this assay requires custom synthesis and, unless the peptide is epitope-tagged for affinity separation, relatively expensive S-adenosyl-L-[methyl-¹⁴C]methionine as the methyl donor. A convenient alternative is a rapid, reproducible and quantitative assay for pcCMT activity that utilizes a small farnesylcysteine analog such as N-acetyl-S-trans,trans-farnesyl-L-cysteine (AFC) as a substrate (methyl acceptor) (Fig. 1) that can be easily separated from the unreacted methyl donor. This assay has been previously described in detail.[11] AFC from BioMol (Plymouth Meeting, PA) is prepared as a 100 mM stock solution in DMSO and stored at −20°. Prepare a 1 mM working solution by diluting with distilled H₂O. Combine in a 1.5 ml Eppendorf tube: (a) membrane or fraction to be tested for pcCMT activity (e.g., 1 μg of COS-1 membranes) or a control without membrane; (b) 5 μl 1 mM AFC; (c) 3 μl ³H-labeled S-adenosyl-L-methionine ([³H]AdoMet, 60 Ci/mmol, 0.55 mCi/ml, NEN Life Science Products, Boston, MA; (d) 12.5 μl 4× TE buffer (4×, 200 mM Tris-HCl pH 8.0, 4 mM NaEDTA); and (e) distilled H₂O to 50 μl. Incubate the reaction for 30 min at 37°, tapping occasionally to mix. Terminate the reaction by adding an equal volume (50 μl) 20% trichloroacetic acid (TCA) and vortex for 10 sec.

To the terminated reaction add 400 μl of n-heptane and vortex for 10 sec. Centrifuge the reaction mixture at 16,000 rpm for 2 min at room temperature. The pellet will contain proteins, the bottom aqueous layer will contain unreacted

[13] R. C. Stephenson and S. Clarke, J. Biol. Chem. 265, 16248 (1990).

[^{3}H]AdoMet, and the top organic layer will contain [^{3}H]AFC methyl ester. The organic n-heptane fraction should contain few radiolabeled species other than [^{3}H]AFC methyl ester and therefore, in principle, could be counted directly without alkaline hydrolysis and quantitation of volatilized [^{3}H]methanol. However, the possibility of methylation of small molecules in the membrane preparation that might partition into n-heptane warrants the added step described below which makes the assay specific for carboxyl methylation.

Remove 300 μl of the organic (n-heptane) layer from the top of the tube, being careful not to allow mixing or entry of the pipette tip into the aqueous phase. Transfer the organic layer to a fresh 1.5 ml Eppendorf tube and dry by rotary evaporation. Add 100 μl 1 N NaOH, which will promote alkaline hydrolysis of the [^{3}H]AFC methyl ester to produce volatile [^{3}H]methanol. Promptly transfer the uncapped Eppendorf tube into a scintillation vial containing approximately 7 ml of scintillation fluid (e.g., Ecoscint, National Diagnostics, Atlanta, GA). The cap of the open Eppendorf tube can be used to keep the tube upright in the scintillation vial. Do not allow mixing of the Eppendorf contents with the scintillation fluid. Quickly close scintillation vial and set aside for 24–72 hr in order to allow the volatilized [^{3}H]methanol to partition into the organic scintillation fluid. Preliminary readings can be obtained after 24 hr but $\geq$48 hr should be allowed for definitive results. While incubating, do not disturb the tubes in order to ensure that the scintillation fluid does not splash into the Eppendorf tubes. Determine counts per minute (cpm) in the tritium channel of a scintillation counter.

The assay described above typically yields cpm levels 1000-fold higher than background (no enzyme control) and is therefore very sensitive. However, the concentration of [^{3}H]AdoMet is far below its K_m and therefore the reaction rate is below the maximal value. Accordingly, for kinetic studies of pcCMT the final concentration of [^{3}H]AdoMet should be increased with unlabeled AdoMet (Sigma, St. Louis, MO) to at least 15 μM. The counts per minute detected can be converted into specific activity (pmol AFC methylated/mg membrane protein · min) with the following formula:

Specific activity $=$

$$\frac{\text{cpm}[^3\text{H}]\text{methanol}/(\text{cpm}[^3\text{H}]\text{Adomet}/\text{pmol AdoMet})}{(\text{fraction heptane recovered}=0.75)(\text{efficiency of MeOH partition}=0.8)(\text{mg membrane})(\text{min reaction})}$$

Carboxyl Methylation of $G\gamma$ and $p21s$

Endogenous CAAX proteins such as $G\gamma$ (p6s) and rho GTPases (p21s) derived from myeloid cells or brain homogenates can be used as substrates in an *in vitro* pcCMT assay instead of small-molecule methyl acceptors such as AFC. Although such an assay is less quantitative than AFC methylation, the CAAX protein methylation assay has the advantage of allowing direct visualization of methylated substrates in a polyacrylamide gel (Fig. 2a). Bacterially expressed recombinant

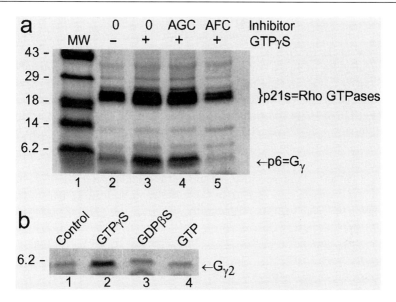

FIG. 2. *In vitro* carboxyl methylation of endogenous Gγ and Ras-related GTPases in subcellular fractions of human neutrophils. (a) Neutrophil light membranes containing pcCMT and G proteins were incubated with neutrophil cytosol containing a mixture of rho GTPases complexed with rhoGDI and with the methyl donor *S*-adenosyl-L-[*methyl*-^{3}H]methionine. In some reactions a competitive pcCMT inhibitor, AFC (lane 5), or an inactive analog, AGC (lane 4), was included. In lane 2 GTPγS was omitted. Reaction products were analyzed by SDS–PAGE and fluorography. (b) Analysis as in (a) showing GTP-sensitivity (±10 μM GTP analog) of carboxyl methylation of Gγ.

CAAX proteins cannot serve directly as substrates unless a system is available for stoichiometric *in vitro* prenylation and AAX proteolysis. Recombinant CAAX proteins produced in insect cells, when harvested from the membrane fraction, can serve as substrates but care must be taken to avoid detergents in the final preparation because of the sensitivity of pcCMT to detergents. We have found that the light membrane fraction of neutrophils contains sufficient Gγ$_2$ to serve as a substrate for *in vitro* methylation.[14] Similarly, we have found that cytosol derived from neutrophils contains several rho family GTPases that are prenylated but remain soluble by forming a complex with GDI and can serve as substrates for pcCMT.[4] Using such substrates, the reaction is accelerated considerably in the presence of a nonhydrolyzable GTP analog (Fig. 2b).[15] In the case of endogenous p21s derived from neutrophil cytosol, the nonhydrolyzable GTP analog acts by

[14] M. R. Philips, R. Staud, M. Pillinger, A. Feoktistov, C. Volker, J. Stock, and G. Weissmann, *Proc. Natl. Acad. Sci. U.S.A.* **92**, 2283 (1995).

[15] M. R. Philips, M. H. Pillinger, R. Staud, C. Volker, M. G. Rosenfield, G. Weissmann, and J. Stock, *Science* **259**, 977 (1993).

promoting dissociation of the rho GTPases from GDI making them accessible to pcCMT. In the case of $G\gamma$, the nonhydrolyzable GTP analog promotes dissociation of $G\beta\gamma$ from $G\alpha$ which has a similar effect.[14]

Combine 1.5 μCi of [^{3}H]AdoMet with 10–25 μg of neutrophil membranes (source of both $G\gamma$ and pcCMT) in 5 μl of 4× TE buffer containing 400 μM GTPγS and bring the total volume to 20 μl with distilled H_2O. To visualize methylated p21s in the same reaction add 100 μg of dialyzed, concentrated neutrophil cytosol. Incubate at 37° for 30 min. In some reactions 100 μM AFC can be included in the methylation mixture as a competitive inhibitor to increase specificity (a cytosolic protein carboxyl methyltransferase that methylates the C-terminal leucine in PP2a is AFC-insensitive and appears as a 36kDa band[4]). Performing the reaction in the presence or absence of GTPγS also increases specificity since the methylation of $G\gamma$ and p21s is GTP-sensitive. The reaction is terminated by adding 20 μl SDS sample buffer. The reaction products are analyzed by 16% Tricine SDS–polyacrylamide gel electrophoresis (e.g., Novex precast Tricine gels available from Invitrogen) and fluorography. The pH of the Tricine gel (pH 8.7) is sufficient to hydrolyzed all aspartyl carboxyl methylesters, leaving the α-carboxyl methyl esters of C-terminal prenylcysteine residues as the only hydrolyzable methyl groups. Carboxyl methylation of radiolabeled proteins can be verified and quantitated by excision of the bands of interest from gels dried onto filter paper (1 mm strips can be excised with a razor). The excised bands are placed in an Eppendorf tube and rehydrated with 100 μl 1 N NaOH to hydrolyze α-carboxyl methyl esters, and the volatilized [^{3}H]methanol that is produced is quantitated by partition into organic scintillation fluid as described above for the dried n-heptane residue containing [^{3}H]AFC-methyl ester.

Methylation of CAAX Proteins in Intact Cells

Assays of prenylcysteine carboxyl methylation on proteins in intact cells offer the advantage of analyzing the reaction under physiologic conditions in resting, stimulated, and genetically manipulated cells and, when combined with immunoprecipitation, add substrate specificity to the system. The disadvantages in relation to the *in vitro* assay include a relatively weak signal and the high cost of L-[*methyl*-^{3}H]methionine, required as a precursor of the methyl donor [^{3}H]AdoMet in metabolic labeling since AdoMet cannot enter cells. Although using this method we have detected prenylcysteine carboxyl methylation of endogenous p21s in neutrophils,[15] methylation of endogenous $G\gamma$ was not reliably detected owing to a high background in the region of the gel below 14 kDa. However, we have found that overexpressing by transfection CAAX substrates, including $G\gamma$, particularly when epitope tagged to facilitate immunoprecipitation, gives quantitative and reproducible results (Fig. 3). The use of EGFP as the epitope tag used for immunoprecipitation of CAAX proteins affords the additional advantage of allowing one to directly visualize by epifluorescence the transfection efficiency and the level of

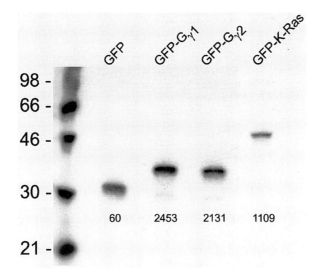

FIG. 3. Carboxyl methylation of ectopically expressed epitope-tagged CAAX constructs in metabolically labeled COS-1 cells. COS-1 cells were transfected with GFP alone or GFP extended at the C terminus with Gγ1, Gγ2, or K-ras, as indicated. Twenty-four hr after transfection the cells were incubated with L-[*methyl*-³H]methionine to label intracellular pools of AdoMet. The cells were lysed and GFP-tagged proteins immunoprecipitated and analyzed by SDS–PAGE and fluorography. Most of the radioactivity that gives rise to the bands on the fluorogram represents L-[*methyl*-³H]methionine incorporated into the primary sequence of the proteins. To analyze carboxyl methylation, the bands were excised from the gel and subjected to alkaline hydrolysis and quantitation of volatile [³H]methanol, and the results (cpm) are given as numbers below each band.

expression. Epitope tags more commonly used for immunoprecipitation (e.g., HA, FLAG, and Myc) can also be used. With a molecular mass of 27 kDa, green fluorescent protein (GFP) offers the additional advantage of increasing the size of Gγ fusions such that they are easily resolved in conventional Tris–glycine gels.

Construction of pEGFP Gγ Constructs

Using the cDNA of choice (Gγ or other CAAX proteins) the coding sequence is PCR (polymerase chain reaction) amplified with 5′ *Eco*RI and 3′ XbaI linkers designed into the primers (Table II). The amplification product is subcloned into the eukaryotic expression vector pEGFP-C3 (Clontech, Palo Alto, CA). The CAAX cysteine can be mutated to serine, when a negative control for methylation is desired, by the design of the 3′ PCR primer.

Transfection and Metabolic Labeling

COS-1 cells are seeded into 6-well plates at a density of 1.5×10^5 cells per well. After 24 hr, the cells are transfected with the various EGFP–Gγ fusion constructs

TABLE II
PCR Primers for Amplification from Gγ cDNAs of Inserts for
pEGFP-C3[a]

Primer	Sequence
Forward Gγ$_1$	5'-ATA AGC TTA TGC CAG TGA TCA ATA TTG-3'
Reverse Gγ$_1$	5'-ATT ATC TAG ATT ATG AAA TCA GAC AG-3'
Forward Gγ$_2$	5'-ATA AGC TTA TGG CCA GCA ACA ACA CC-3'
Reverse Gγ$_2$	5'-TCC TCC TCT AGA CTT AAA GG-3'

[a] To produce EGFP-tagged Gγ constructs.

using SuperFect (Qiagen) as described above for expression of pcCMT, except the volumes can be reduced to give a 0.5 ml final volume of transfection mixture.

LipofectAMINE (Life Technologies) used according to the manufacture's instructions gives results similar to those using SuperFect. Control cells should be transfected with EGFP alone to establish a background. Twenty-four hr after transfection COS-1 cells are ready for metabolic labeling. Expression of EGFP–Gγ fusion constructs can be verified by direct visualization in an epifluorescence microscope with filters optimized for GFP or fluorescein isothiocyanate (FITC). Cells are washed twice with PBS (prewarmed to 37°) and methionine-starved by incubating in 1 ml methionine-free DMEM containing 10% dialyzed FCS at 37° in 5% CO_2 for 30 min. Next, 200 μCi L-[*methyl*-^{3}H]methionine (NEN Life Science Products) is added directly to the medium and the cells are incubated for 30 min. The cells are then washed 3× in ice-cold PBS and can be lysed directly in the 6-well plates with 1 ml RIPA buffer [20 mM Tris-HCl, pH 7.5, 150 mM NaCl, 1% nonidet p-40 (NP-40), 0.1% SDS, 0.1% Na-deoxycholate (DOC), 0.5 mM EDTA, 10 μg/ml leupeptin, 1 mM PMSF, 27 μg/ml aprotinin, 1 mM DTT]. The samples are now ready for immunoprecipitation but can be stored for several days at $-20°$.

Immunoprecipitation, SDS–PAGE, and Analysis of Carboxyl Methylation

Immediately prior to immunoprecipitation the RIPA lysates should be centrifuged at 15,000 rpm for 15 min at 4° to remove detergent-insoluble material. One microliter of rabbit polyclonal anti-GFP antiserum (Molecular Probes, Eugene, OR) is sufficient to immunoprecipitate GFP-tagged proteins from 1 ml of RIPA lysate. The antiserum is added and the mixture is rotated at 4° for 1 hr. Next, 20 μl of a 50% slurry (in PBS) of protein A-conjugated agarose beads (Life Technologies) is added to the lysate/antiserum mixture and mixed at 4° for 1 hr. The agarose beads are then washed 5× with PBS, pelleted by centrifugation at 15,000 rpm for 5 min at 4°, and eluted by boiling in 20 μl of 4× SDS sample buffer. Eluates are loaded onto 10% or 14% Tris–glycine gels (e.g., Novex) and subjected to electrophoresis. Immunoprecipitated proteins labeled by L-[*methyl*-^{3}H]methionine are identified by fluorography. Because the L-[*methyl-*^{3}H]methionine that is not converted to [^{3}H]AdoMet will metabolically label all nascent proteins, GFP-tagged proteins will be visible by fluorography whether they are carboxyl methylated or not. Therefore, carboxyl methylation of labeled proteins must be determined by alkaline hydrolysis of bands excised from the dried gel as described above for *in vitro* methylation of p21s.

Densitometric analysis of the fluorogram gives an accurate determination of protein expression (D). The relative methylation of different EGFP-CAAX constructs (M) can be determined by normalizing counts per minute (cpm) to protein expression. Accordingly, $M = (\text{cpm}/D)[(6 + N)/6]$, where N is the number of methionine residues in the protein or peptide fused to the EGFP that contains six methionines.

Acknowledgment

This work was supported by the National Institutes of Health Research Grants GM55279 and AI36224.

[18] Preparation and Application of G Protein γ Subunit-Derived Peptides Incorporating a Photoactive Isoprenoid

By TAMARA A. KALE, TAMMY C. TUREK, VANESSA CHANG, N. GAUTAM, and MARK D. DISTEFANO

Introduction

Protein prenylation involves the attachment of either C_{15} (farnesyl) or C_{20} (geranylgeranyl) isoprenoids to proteins via thioether linkages. This posttranslational modification to various proteins is catalyzed by a class of enzymes called prenyltransferases that have been the focus of intensive study as possible anticancer drug targets.[1] A large number of prenylated proteins have been identified including small GTP-binding proteins of the Ras superfamily and heterotrimeric G proteins.[2] Geranylgeranylation is the most common modification, occurring on the small monomeric GTP binding proteins and on the γ subunits of heterotrimeric G proteins[3]; farnesylation is much less common, being found on Ras proteins

[1] D. M. Leonard, *J. Med. Chem.* **40,** 2971 (1997).
[2] W. R. Schafer and J. Rine, *Ann. Rev. Genet.* **30,** 209 (1992).
[3] P. J. Casey, *Science* **268,** 221 (1995).

(H-Ras, N-Ras, K-Ras) and on the γ subunit of transducin.[4] To date, the function of protein prenylation is not clearly understood. Prenylation is the first step of a process that involves modification of a specific cysteine residue (or pairs of cysteines) with a C_{15} or C_{20} isoprenoid, followed by endoproteolysis to produce a C-terminal prenyl group and finally carboxymethylation of the C-terminal prenylcysteine to produce a methyl ester.[5] It is clear that this series of enzyme-catalyzed reactions greatly increases the hydrophobicity of the modified proteins and results in localization of the proteins out of the pool of cytosolic proteins and into membranes.[6] However, the question of whether this is the sole function of prenylation remains unclear. Several investigators have proposed that protein prenylation may be a structural feature involved in mediating protein–protein interactions.[7] Work with peptides derived from the C termini of G protein γ subunits has shown that prenylation of these peptides is essential for their ability to bind to receptors.[8,9] Unprenylated G protein $\beta\gamma$ complex fails to stimulate PLC-$\beta2$ *in vitro* or *in vivo*[10,11]; such a complex also does not stimulate adenylyl cyclase II or inhibit adenylyl cyclase I *in vitro*.[12] Earlier structural work with Rho protein GDP-dissociation inhibitor (RhoGDI) suggested that this protein contained a binding site for the isoprenoid moiety of geranylgeranylated RhoA[13]; this has been confirmed by an X-ray crystal structure of the complex between RhoGDI and Cdc42 where a clearly defined prenyl group binding pocket occupied by a geranylgeranyl group was observed.[14]

Given that specific interactions between prenyl groups and proteins may be of general importance, we are interested in developing biochemical probes that will enable us to identify whether there are specific interactions between isoprenoid groups attached to G proteins and their cognate receptors. To accomplish this, we set out to prepare a series of peptides that would contain a photoactive isoprenoid in place of the normal geranylgeranyl group. This idea is based on an early observation that the photoactive geranylgeranyl diphosphate

[4] P. J. Casey, P. A. Solski, C. J. Der, and J. E. Buss, *Proc. Natl. Acad. Sci. U.S.A.* **86,** 8323 (1989).

[5] F. L. Zhang and P. J. Casey, *Ann. Rev. Biochem.* **65,** 241 (1996).

[6] J. A. Glomset and C. C. Farnsworth, *Ann. Rev. Cell Biol.* **10,** 181 (1994).

[7] A. D. Cox and C. J. Der, *Curr. Opin. Cell Biol.* **4,** 2008 (1992).

[8] O. G. Kisselev, M. V. Ermolaeva, and N. Gautam, *J. Biol. Chem.* **269,** 21399 (1994).

[9] I. Azpiazu, H. Cruzblanca, L. Peng, M. Linder, M. Zhuo, and N. Gautam, *J. Biol. Chem.* **274,** 35305 (1999).

[10] A. Katz, D. Wu, and M. I. Simon, *Nature* **360,** 686 (1992).

[11] A. Dietrich, D. Brazil, O. N. Jensen, M. Meister, M. Schrader, J. F. Moomaw, M. Mann, D. Illenberger, and P. Gierschik, *Biochemistry* **35,** 15174 (1996).

[12] J. A. Iniguez-Lluhi, M. I. Simon, J. D. Robishaw, and A. G. Gilman, *J. Biol. Chem.* **267,** 23409 (1992).

[13] Y. Q. Gosser, T. K. Nomanbhoy, B. Aghazadeh, D. Manor, C. Combs, R. A. Cerione, and M. K. Rosen, *Nature* **387,** 814 (1997).

[14] G. R. Hoffman, N. Nassar, and R. A. Cerione, *Cell* **100,** 345 (2000).

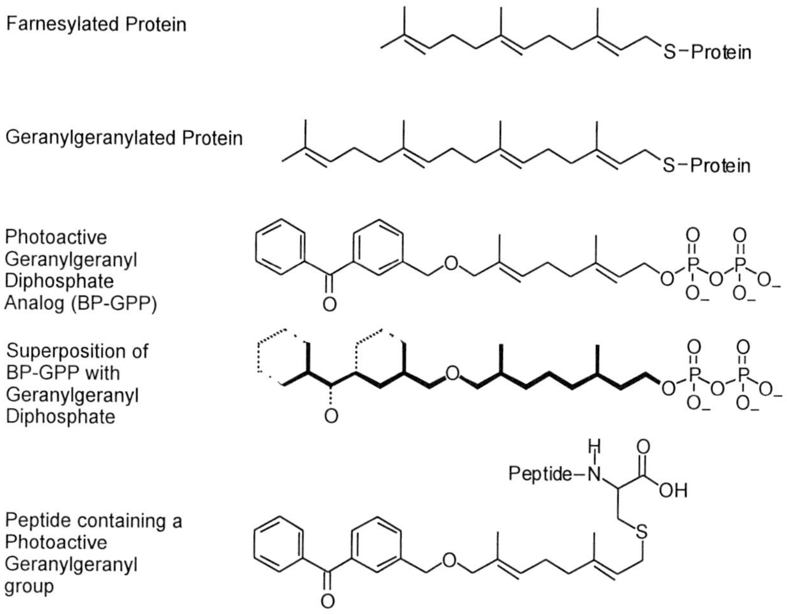

Farnesylated Protein

Geranylgeranylated Protein

Photoactive
Geranylgeranyl
Diphosphate
Analog (BP-GPP)

Superposition of
BP-GPP with
Geranylgeranyl
Diphosphate

Peptide containing a
Photoactive
Geranylgeranyl
group

FIG. 1. Structures of prenylated proteins and photoactive isoprenoid mimics.

analog (BP-GPP), shown in Fig. 1, is a good mimic of a farnesyl or geranylgeranyl group. Work with BP-GPP has shown that it is a competitive inhibitor of yeast protein farnesyltransferase (PFTase) with a K_I of 49 nM.[15] Photocross-linking experiments using a BP-GPP analog and either yeast PFTase or human protein geranylgeranyl transferase type I (PGG Tase-I) show that the probe cross-links to the β subunit of these enzymes, consistent with the known structure of PFTase.[16] Finally, a co-crystal structure of human PFTase and BP-GPP shows that this iso-prenoid mimic binds in the farnesyl diphosphate (FPP) binding site in the same fashion as the natural substrate.[17] Thus, the benzophenone-containing isoprenoid portion of BP-GPP appears to be a good mimic of a farnesyl or geranylgeranyl group. In this article, we describe the preparation of a series of biotinylated pep-tides that incorporate the isoprenoid moiety present in BP-GPP as shown in Fig. 1. This description includes detailed procedures on the synthesis, purification, and characterization via electrospray ionization–mass spectrometry (ESI-MS) of these peptides. As a test case for the utility of these probes, their ability to cross-link to RhoGDI was evaluated. As noted above, RhoGDI is known to interact with

[15] I. Gaon, T. C. Turek, and M. D. Distefano, *Tetrahedron Lett.* **37**, 8833 (1996).

[16] T. C. Turek, I. Gaon, D. Gamache, and M. D. Distefano, *Bioorg. Med. Chem. Lett.* **7**, 2125 (1997).

[17] T. C. Turek, I. Gaon, M. D. Distefano, and C. L. Strickland, *J. Org. Chem.* **66**, 3253 (2001).

peptides and proteins through specific interactions with the geranylgeranyl group. Thus, the positive results obtained here, in which these photoactive peptides cross-link to RhoGDI, suggest that these probes should be useful for studying the role of prenyl groups in the interactions between G proteins and other proteins.

Materials and Methods

Materials

All synthetic reactions (see Scheme 1, structures **1–6**) are conducted under a nitrogen atmosphere, stirred magnetically, and carried out at room temperature unless otherwise noted. Analytical thin layer chromatography (TLC) is performed on precoated (250 μm) silica gel 60 F-254 plates from E. Merck (Darmstadt, Germany). Visualization of plates is done under UV irradiation or by staining with an ethanolic phosphomolybdic acid solution followed by heating. Flash chromatography silica gel (60-120 mesh) is obtained from E. M. Science (Cincinnati, OH) or Scientific Adsorbents, Inc. (Atlanta, GA). CH_2Cl_2 is distilled from CaH_2. Deuterated NMR solvents are used as obtained from Cambridge Isotope Laboratories, Inc. (Andover, MA). Nuclear magnetic resonance (NMR) spectra (^1H) are obtained at 300 MHz or 500 MHz on Varian instruments. Chemical shifts are reported in parts per million (ppm) and J values are given in hertz (Hz). Mass spectrometry analyses are performed using a VG 7070E-HF instrument [fast atom bombardment (FAB)-MS], Finnigan MAT 95 [chemical ionization (CI)-MS], or a Finnigan LCQ ESI-MS. HPLC analyses are carried out using a Beckman (Fullerton, CA) Model 127/166 instrument equipped with a diode array

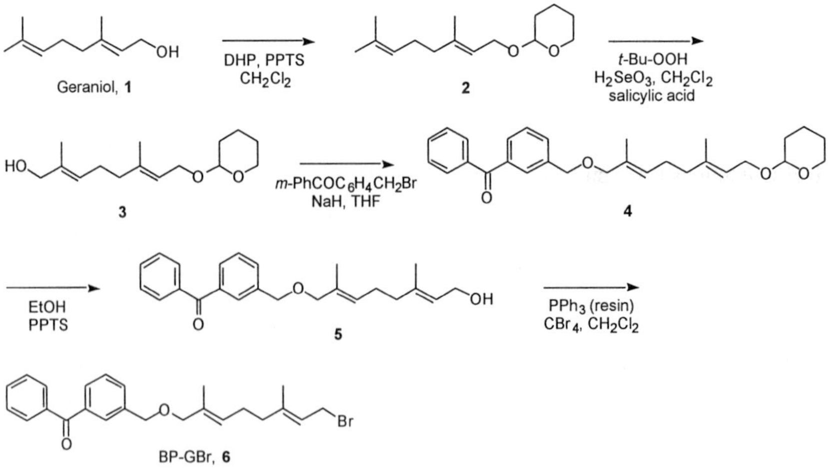

SCHEME 1. Synthesis of photoactive BP-GBr (**6**) from geraniol (**1**).

UV detector and a Varian (Rainin Instruments, Woburn, MA) Dynamax C_{18} column (8.0 μm, 4.6 × 250 mm) equipped with a 5 cm guard column [flow rate: 1.0 ml/min, 500 μl injection loop, 5% to 100% B in 40 min. Solvent A: 95% H_2O, 5% CH_3CN, 0.2% trifluoroacetic acid (TFA); solvent B: 100% CH_3CN, 0.2% TFA] with monitoring at 220 nm and 260 nm. Biotin was appended to the peptides via a pentaamide linker.

3,7-Dimethyl-1-O-THP-2,6-octadiene (2)

A solution of geraniol **1** (6.2 g, 40 mmol) and dihydropyran (5.0 g, 60 mmol) in dry CH_2Cl_2 (60 ml) containing pyridinium-p-toluenesulfonate (PPTS) (1.0 g, 4.0 mmol) is reacted for 4 h at room temperature. The solution is partially evaporated, diluted with ether, and washed with half-saturated brine. The organic layer is dried with anhydrous Na_2SO_4 and concentrated to yield a colorless oil (9.1 g, 98%). R_f 0.60 (silica gel, toluene/ethyl acetate, 5 : 2, v/v); ^{1}H NMR (300 MHz, $CDCl_3$): δ 1.52-1.87 (m, 6H), 1.63 (s, 3H), 1.71 (s, 6H), 2.05-2.15 (m, 4H), 3.51-3.56 (m, 1H), 3.89-3.93 (m, 1H), 4.06 (dd, 1H, $J = 12.0, 6.0$), 4.26 (dd, 1H, $J = 12.0, 6.0$), 4.66 (t, 1H, $J = 6.0$), 5.12 (dd, 1H, $J = 6.0, 1.0$), 5.39 (dd, 1H, $J = 6.0, 1.0$); HR-FAB-MS: $C_{15}H_{27}O_2$ $[M+H]^+$, calculated 239.2004, found 239.2010.

3,7-Dimethyl-1-O-THP-2,6-octadien-8-ol (3)

Compound **2** (5.8 g, 20 mmol) and *tert*-butyl hydroperoxide (8.0 ml, 72 mmol) are stirred in the presence of H_2SeO_3 (52 mg, 0.40 mmol) and salicylic acid (280 mg, 2.0 mmol) in CH_2Cl_2 (20 ml) for 20 h at room temperature. The CH_2Cl_2 is removed under reduced pressure and *tert*-butyl hydroperoxide is removed by repeated washing and evaporation with toluene. The residue is dissolved in ether, washed with 1 *M* $NaHCO_3$ to remove H_2SeO_3, dried with anhydrous Na_2SO_4, and filtered. The organic layer is concentrated and the crude product purified by flash chromatography (silica gel, toluene/ethyl acetate, 5 : 2, v/v), which affords compound **3** (2.8 g, 56%) as a colorless oil. R_f 0.20 (silica gel, toluene/ethyl acetate, 5 : 2, v/v); ^{1}H NMR (500 MHz, $CDCl_3$): δ 1.47-1.83 (m, 6H), 1.63 (s, 3H), 1.65 (s, 3H), 2.05 (t, 2H, $J = 7.5$), 2.14 (t, 2H, $J = 7.0$), 3.46-3.51 (m, 1H), 3.83-3.88 (m, 1H), 3.94 (s, 2H), 3.99 (dd, 1H, $J = 12.0, 7.5$), 4.21 (dd, 1H, $J = 12.0, 6.5$), 4.60 (t, 1H, $J = 3.0$), 5.34 (ddd, 2H, $J = 7.5, 7.5, 1.0$); HR-FAB-MS: $C_{15}H_{25}O_3$ $[M-H]^+$, calculated 253.1797, found 253.1794.

(E,E)-8-O-(3-Benzoylbenzyl)-1-O-THP-3,7-dimethyl-2,6-octadiene (4)

Compound **3** (250 mg, 1.0 mmol) is dissolved in tetrahydrofuran (THF, 2.0 ml) under N_2 and cooled to 0°, and NaH (90 mg of 60% dispersion in oil, 2.3 mmol) is added. After 1 h, 3-benzoylbenzylbromide (450 mg, 1.6 mmol) is added and stirred 1 h at 0° and then 24 h at room temperature. The reaction mixture is poured over water and extracted with ethyl acetate. The combined extracts are washed with water, dried over anhydrous Na_2SO_4, and filtered. The filtrate is evaporated and the residue purified by flash chromatography on silica gel (toluene/ethyl acetate, 5:2, v/v). Evaporation of the solvent gives **4** as a light yellow oil (360 mg, 80%). R_f 0.53 (silica gel, toluene/ethyl acetate, 5:2, v/v); ^{1}H NMR (300 MHz, CDCl$_3$): δ 1.53-1.86 (m, 6H), 1.70 (s, 6H), 2.08-2.23 (m, 4H), 3.44-3.54 (m, 1H), 3.87-3.91 (m, 1H), 3.94 (s, 2H), 4.04 (dd, 1H, $J = 12.0, 6.0$), 4.26 (dd, 1H, $J = 12.0, 6.0$), 4.52 (s, 2H), 4.64 (t, 1H, $J = 6.0$), 5.42 (t, 1H, $J = 12.0$), 5.44 (t, 1H, $J = 12.0$), 7.48 (t, 1H, $J = 6.0$), 7.50 (t, 2H, $J = 6.0$), 7.59-7.62 (m, 2H), 7.73 (d, 1H, $J = 9.0$), 7.81 (s, 1H), 7.83 (d, 2H, $J = 6.0$); HR-FAB-MS: $C_{29}H_{37}O_4$ [M+H]$^+$, calculated 449.2682, found 449.2664.

(E,E)-8-O-(3-Benzoylbenzyl)-3,7-dimethyl-2,6-octadien-1-ol **(5)**

To a solution of **4** (220 mg, 0.50 mmol) in ethanol (4.0 ml) is added PPTS (13 mg, 0.050 mmol) and the reaction mixture is stirred at 55° for 8 h. The solvent is evaporated *in vacuo,* and the resulting residue purified by silica gel flash chromatography (toluene/ethyl acetate, 10:1, v/v) which affords compound **7a** in 100% yield (180 mg). R_f 0.29 (silica gel, toluene/ethyl acetate, 10:1, v/v); ^{1}H NMR (300 MHz, CDCl$_3$): δ 2.10 (s, 6H), 2.07-2.23 (m, 4H), 3.94 (s, 2H), 4.17 (d, 2H, $J = 6.0$), 4.53 (s, 2H), 5.44 (t, 2H, $J = 6.0$), 7.48 (t, 1H, $J = 6.0$), 7.51 (t, 2H, $J = 6.0$), 7.59-7.65 (m, 2H), 7.73 (d, 1H, $J = 9.0$), 7.81 (s, 1H), 7.83 (d, 2H, $J = 6.0$); HR-FAB-MS: $C_{24}H_{27}O_3$ [M-H]$^+$, calculated 363.1953, found 363.1974; $C_{24}H_{27}O_2$ [M-OH]$^+$, calculated 347.2004, found 347.2011.

(E,E)-8-O-(3-Benzoylbenzyl)-3,7-dimethyl-2,6-octadien-1-bromide **(6)**

Compound **5** (178 mg, 0.49 mmol, 1 equivalent), polymer-supported triphenylphosphine (PPh)$_3$ beads (327 mg, 0.98 mmol, 2 equivalents), and 3 ml dry CH_2Cl_2 are combined and the mix is allowed to sit for 30 min under nitrogen to allow the beads to swell. CBr$_4$ (195 mg, 0.59 mmol, 1.2 equivalent) is then added. The mixture is stirred for 5 h under nitrogen at room temperature. TLC analysis (silica gel, toluene/ethyl acetate, 5:2, v/v; R_f 0.88) shows the reaction is complete at this time, and the mixture is poured through a fine fritted filter. The filtrate, containing product, is evaporated and not purified before further use. ^{1}H NMR (300 MHz, CDCl$_3$): δ 1.65 (s, 3H), 1.71 (s, 3H), 2.07-2.21 (m, 4H), 3.91 (s, 2H), 3.95 (d, 2H, $J = 8.4$), 4.48 (s, 2H), 5.38 (t, 1H, $J = 6.5$), 5.51 (t, 1H, $J = 8.4$), 7.25-7.80 (m, 9H). LR-CI-MS: calculated for $C_{24}H_{27}O_2Br$ [M+H]$^+$, 427.1, found 427.2; calculated [M+NH$_4$]$^+$, 444.2, found 444.2.

General Peptide Alkylation Procedure

This procedure is outlined in Scheme 2 (structures **7a–8e**).

A peptide (1 equivalent) is dissolved in a minimal amount of 2:1:1 dimethylformamide (DMF)/butanol or CH_3CN/0.025% TFA with stirring under nitrogen at

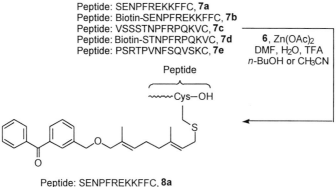

Peptide: SENPFREKKFFC, **7a**
Peptide: Biotin-SENPFREKKFFC, **7b**
Peptide: VSSSTNPFRPQKVC, **7c**
Peptide: Biotin-STNPFRPQKVC, **7d**
Peptide: PSRTPVNFSQVSKC, **7e**

6, Zn(OAc)$_2$
DMF, H$_2$O, TFA
n-BuOH or CH$_3$CN

Peptide: SENPFREKKFFC, **8a**
Peptide: Biotin-SENPFREKKFFC, **8b**
Peptide: VSSSTNPFRPQKVC, **8c**
Peptide: Biotin-STNPFRPQKVC, **8d**
Peptide: PSRTPVNFSQVSKC, **8e**

SCHEME 2. Thiol alkylation of starting peptides **7a–7e**.

room temperature. Bromide **6** (4 equivalents) is dissolved separately in the 2 : 1 : 1 system and then added to the peptide solution. Additional solvent, of 2 : 1 : 1 or of individual component(s), may need to be added at this point to ensure a homogeneous solution. Zinc acetate (5 equivalents of a 1 M solution) is then added dropwise and stirring is continued for 1 h. After 15 min, starting peptide has often disappeared and a product peak is detectable by HPLC. 2-Mercaptoethanol (20 equivalents) is then added after 1 h of reaction time and the solution is allowed to stir for an additional 3 h. Solvents do not necessarily need to be removed before HPLC purification is performed and no workup is needed.

Photolysis Reaction Conditions

All photolysis reactions (50 μl final volume) are performed in silinized quartz test tubes covered with Parafilm at 4° in a UV Rayonet minireactor equipped with 8 RPR-3500° lamps and a circulating platform that allows for irradiation of up to eight samples simultaneously. Buffer consists of 20 mM HEPES, pH 8.0, 5 mM MgCl$_2$, 1 mM NaN$_3$, 100 mM NaCl.

Detection of Biotinylated RhoGDI on Gels

Photolysis reactions consist of 150 nM RhoGDI with either 5 μM, 10 μM, or 20 μM **8b** and are irradiated for 30 min. All reagents are mixed gently and immediately photolyzed. Following photolysis, loading buffer (50 mM Tris-HCl, pH 6.8, 100 mM 2-mercaptoethanol, 2% SDS, 0.1% bromophenol blue, 10% glycerol) is added and samples are heated for 2 min at 100°. Samples are then frozen at −20° until gel analysis, at which time they are thawed and electrophoresed through a 12% acrylamide gel, then transferred to polyvinylidene difluoride (PVDF) membranes (Immobilon-P from Millipore, Bedford, MA). Membranes are blocked overnight at 4° in 5% bovine serum albumin (BSA) in Tris-buffered saline with 0.1% Tween 20 (TBS/T). After the blocking step, membranes are incubated for 1 h at room temperature in streptavidin–horseradish peroxidase (HRP) solution (0.2 μg/ml streptavidin-HRP, diluted in TBS/T). Streptavidin-HRP is obtained from Pierce (Rockford, IL). The membranes are then washed thoroughly with TBS/T and streptavidin binding is detected using the enhanced chemiluminescence light detection system (ECL) from Amersham (Buckinghamshire, England).

Dot-Blot Analyses

Preparations of concentration profile photolysis reactions (three) are as described above for gel analysis. Time course analysis conditions are as follows: 150 nM RhoGDI and 20 μM **8b;** irradiation times of 1, 5, and 30 min. Following photolysis, 50 μl acetone is added to each reaction mixture and white particles can be seen upon addition. The samples are mixed and then centrifuged (13,800g, 5 min, room temperature). Supernatant is removed, 50 μl of a 50% acetone solution is added and the solution mixed to wash the pellet, and the sample is centrifuged again. After removal of supernatant, the washing/pelleting step is repeated and supernatant is once again removed. Samples are placed under vacuum to remove solvents, then frozen at $-20°$ until dot-blot analysis, at which time samples are dotted onto PVDF membranes and allowed to dry thoroughly. Membranes are then blocked overnight in 5% BSA in TBS/T and processed as above.

Negative controls are generated by setting up three 50 μl samples containing buffer and either 5, 10, or 20 μM **8b** (no protein present). These samples are treated with acetone and subjected to the washing/centrifugation steps described above. The three supernatants are frozen at $-20°$ until dot-blot analysis, at which time samples are thawed at treated as above. Values acquired from quantification of these samples are subtracted from values corresponding to their related protein-containing reactions to account for the presence of unincorporated probe. Quantification software used for analyses is Quantity One, version 4.1, from Bio-Rad (Hercules, CA).

Attachment of a Photoactive Isoprenoid to Peptides

Synthesis and Purification of Prenylated Peptide Products

BP-GBr (**6**) can easily be prepared from geraniol (**1**) in five steps in 43% overall yield as shown in Scheme 1. Several different peptides, **7a–7e,** have been successfully modified by thiol alkylation using BP-GBr and their products (**8a–8e,** Scheme 2) subsequently isolated via reverse-phase high performance liquid chromatography (HPLC) using UV monitoring. The zinc acetate peptide

A

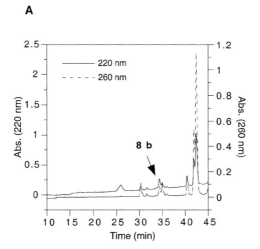

B

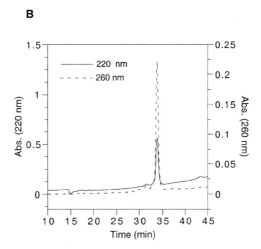

FIG. 2. (A) HPLC chromatogram of **8b** crude reaction mixture. (B) HPLC chromatogram of purified **8b** following HPLC purification.

alkylation reaction conditions described by Xue and co-workers are used with slight modification for this attachment.[18] Because the bromide decomposes when subjected to the HPLC conditions, multiple peaks are seen when the crude reaction mixture is analyzed for product peak development (Fig. 2A). These bromide-associated peaks are also noticeable because there is a vast excess of this reagent

[18] C.-B. Xue, J. M. Becker, and F. Naider, *Tet. Lett.* **33,** 1435 (1992).

TABLE I
CHARACTERIZATION DATA FOR PRODUCTS **8a–8e**

Product	Retention time (min)	Yield (%)	$[M+H]^+$ calculated, found	$[M+H]^+$ calculated, found	$[M+H]^+$ calculated, found	$[M+H]^+$ calculated, found
8a	33.1	>100[a]	1878.2, 1878.2	939.6, 939.7	b	b
8b	34.3	57	2218.4, b	1109.7, 1109.5	740.1, 740.0	b
8c	37.5	76	1896.2, 1896.8	948.6, 948.8	632.7, 632.8	474.8, 475.0
8d	32.1	70	1964.3, 1962.0	982.6, 982.5	b	b
8e	30.1	95	1896.2, 1896.1	948.6, 948.2	b	b

[a] Contaminated by C_{18} resin from HPLC column.
[b] Not found.

used in these reactions, coupled with the intense UV absorbancy of the benzophenone moiety at 260 nm. Comparisons of the HPLC chromatograms of crude reaction mixture vs that of the bromide alone can allow for identification of the product peak. Isolation of peptide products **8a–8e** is successful in each case, as demonstrated with **8b** (Figure 2B). Every desired prenylated peptide elutes with a longer retention time than the starting peptide (9–11 min) and displays a greater increase in absorbance at 260 nm due to the benzophenone chromophore. The prenylated peptides are obtained in good yield (57–95%, Table I) and purity as evidenced by the chromatogram shown in Fig. 2B.

Characterization of Prenylated Peptides

Electrospray ionization-mass spectrometry is used to characterize prenylated peptide products. This technique provides information about the parent molecular ion (MS) as well as sequencing information via subsequent MS-MS analysis. Normally, sequencing via MS-MS is possible for half of a peptide product and occasionally, the entire peptide can be sequenced. C-terminal fragmentation was most common, though N-terminal fragmentation is also seen in some cases. Representative mass spectra are given in Fig. 3 where the parent ion of **8c** and its multiply charged counterparts ($[M+H]^+$, $[M+2H]^{2+}$, $[M+3H]^{3+}$, $[M+4H]^{4+}$) are seen in the initial mass spectrum (Fig. 3A) and the MS-MS sequencing data is seen in Fig. 3B, showing C-terminal sequencing including loss of the benzophenone group at the thioether linkage. A summary of each product generated, HPLC retention time, yield of alkylation, and electrospray MS results is presented in Table I.

Photolysis of a Prenylated Peptide with RhoGDI

RhoGDI is a cytosolic protein responsible for the extraction of prenylated Rho proteins from cell membranes.[19] Rho proteins—normally geranylgeranylated—are

A

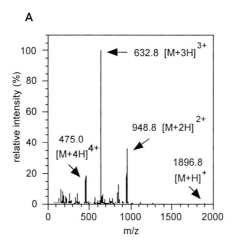

B

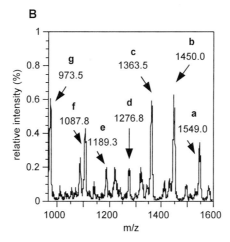

FIG. 3. Electrospray ionization mass spectral analysis of **8c**. (A) Parent and multiply charged ion identification. (B) MS-MS analysis: peaks a–g represent the sequencing of C-terminal peptides with the loss of the BP unit represented by peak a, followed by sequential loss of Cys, Val, Lys, Gln, Pro, and Arg (peaks b–g).

thought to be involved in the oncogenic transformation of cells and are members of the Ras superfamily of proteins.[20] RhoGDI is believed to remove embedded

[19] For a recent review of RhoGDIs see: B. Olofsson, *Cell. Signal.* **11,** 545 (1999).

[20] (a) W. Du, P. F. Lebowtiz, and G. C. Prendergast, *Mol. Cell. Biol.* **19,** 1831 (1991). (b) R. Lin, R. A. Cerione, and D. Manor, *J. Biol. Chem.* **274,** 23633 (1999). (c) J.-P. Mira, V. Benard, J. Groffen, L. C. Sanders, and U. G. Knaus, *Proc. Natl. Acad. Sci. U.S.A.* **97,** 185 (2000).

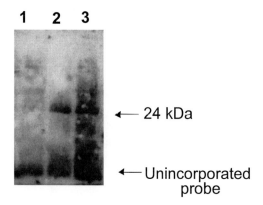

FIG. 4. Western blot of purified RhoGDI following photolysis with varying concentrations of **8b**. [RhoGDI], 150 nM. Concentrations of **8d** in separate lanes: (1) 5 μM, (2) 10 μM, (3) 20 μM. Irradiation time was 30 min.

geranylgeranylated Rho proteins by allowing the protein's prenyl tail, anchored in the membrane, to migrate into a known isoprenoid binding site within RhoGDI, thus solubilizing the Rho protein.[14,21] We use peptide **8b** in photolysis experiments with RhoGDI to explore the ability of using this multifaceted molecule for studying isoprenoid binding sites within proteins. Purified RhoGDI[22] is photolyzed in the presence of increasing concentrations of **8b** and reaction mixtures are analyzed by sodium dodecyl sulfate–polyacrylamide gel electrophoresis (SDS–PAGE). Detection is accomplished using streptavidin–HRP. Evidence for cross-linking is seen in Fig. 4, where increasing amounts of **8b** result in an increased amount of detection. Concentrations of 10 μM and higher provide adequate signal, and 20 μM **8b** has been used in irradiation time course studies.

Another verification of the cross-linking ability of our biotinylated, benzophenone-containing peptides is achieved by analyzing samples of RhoGDI with varying concentrations of **8b** not by electrophoresis but by dot blotting. Samples are prepared and photolyzed in the same manner as described for Fig. 4, but an attempt to remove unincorporated probe is performed by precipitating protein products while leaving unreacted **8b** in the supernatant. Although unincorporated probe is not entirely removed for each of the 5 μM, 10 μM, or 20 μM **8b** cases,

[21] T. K. Nomanbhoy, J. W. Erickson, and R. A. Cerione, *Biochem.* **38**, 1744 (1999).

[22] We used a truncated RhoGDI whose first 25 N-terminal residues had been removed. This modification increases the stability of the protein and the binding affinity is comparable to the wild type [J. V. Platko, D. A. Leonard, C. N. Adra, R. J. Shaw, R. A. Cerione, and G. R. Hoffman, personal communication]. This construct has a mass of approximately 20.5 kDa as determined from sequence analysis, but migrates at roughly the 24 kDa position as seen by SDS–PAGE. For expression and purification of GST-RhoGDI see: D. Leonard, J. M. Hart, J. V. Platko, A. Eva, W. Henzel, T. Evans, and R. A. Cerione, *J. Biol. Chem.* **267**, 22860 (1992).

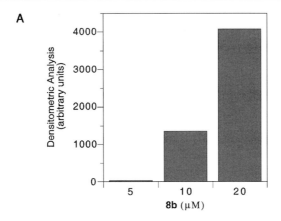

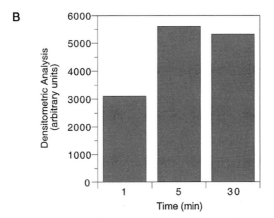

FIG. 5. (A) Quantification of streptavidin–HRP detected products adjusted for the presence of corresponding unincorporated probe [**8b**]. (B) Quantification of streptavidin–HRP detected products related to increasing irradiation times. [RhoGDI], 150 nM; [**8b**], 20 μM.

this can be corrected for using a corresponding negative control (see Materials and Methods) and quantification of the adjusted data is shown in Fig. 5A. Once again, an increase in the amount of **8b** correlates to an increase in the amount of detected cross-linking.

Finally, an irradiation time-course study is carried out using 150 nM RhoGDI and 20 μM **8b.** Three separate photolysis reactions are irradiated for different times: 1 min, 5 min, and 30 min. Tabulation of the dot-blot results is seen in Fig. 5B. Maximum cross-linking is achieved in roughly 5 min of irradiation, and further irradiation does not increase the amount of cross-linking. RhoGDI does not decompose under these conditions, as detected by SDS–PAGE (data not shown).

The photoaffinity labeled peptides synthesized and characterized for this study (**8a–8e**) have shown to be effective for cross-linking to a protein with a known isoprenoid binding site (RhoGDI). Proteins containing putative isoprenoid binding sites for which little structural data is known can also be studied using the techniques presented here. The role of the isoprenoid in prenylated protein–protein interactions, including those of monomeric and heterotrimeric prenylated G-proteins, can be investigated with these and other similarly modified peptides.

Section III

Functional Analysis of G Protein Subunits

[19] Expression and Functional Analysis of G Protein α Subunits in S49 Lymphoma Cells

By CATHERINE H. BERLOT

Introduction

S49 lymphoma cells are useful for studying G protein signaling because their signaling pathways are well characterized and cell lines deficient in α_s and/or protein kinase A (PKA) have been developed from them. These cells are derived from a T cell lymphoma in a BALB/c mouse.[1] Among the G-protein-coupled receptors that they express are the β_2-adrenergic receptor,[2] a prostaglandin (PG) receptor that stimulates adenylyl cyclase in response to PGE_1,[3] and a somatostatin receptor that inhibits adenylyl cyclase via G_i.[4] The adenylyl cyclase activity in these cells is due to type VI adenylyl cyclase,[5] which can be efficiently inhibited by α_i when activated by α_s,[6] and type VII adenylyl cyclase, which is directly activated by protein kinase C.[7]

cAMP is cytocidal to S49 lymphoma cells and the cyc^- cell line, which lacks α_s,[8] was obtained by selecting for cells that survived in the presence of isoproterenol and RO 20-1724, a cAMP phosphodiesterase inhibitor.[9] These selected cells were named cyc^- based on the premise that they lacked adenylyl cyclase. Subsequently, cyc^- cells were found to be deficient in a regulatory protein that when added back to cell membranes could reconstitute hormone-sensitive adenylyl cyclase activity.[10] This reconstitution assay then provided the basis for the identification, purification, and characterization of α_s.[11] Since α_s is ubiquitously expressed, the cyc^- cell serves as an important null cell for expressing and studying the properties of α_s mutants.[12-17] These cells can also be used to determine whether a particular

[1] K. Horibata and A. W. Harris, *Exp. Cell Res.* **60**, 61 (1970).

[2] S. R. Post, O. Aguila-Buhain, and P. A. Insel, *J. Biol. Chem.* **271**, 895 (1996).

[3] L. L. Brunton, M. E. Maguire, H. J. Anderson, and A. G. Gilman, *J. Biol. Chem.* **252**, 1293 (1977).

[4] K. Aktories, G. Schultz, and K. H. Jakobs, *Mol. Pharmacol.* **24**, 183 (1983).

[5] R. T. Premont, O. Jacobowitz, and R. Iyengar, *Endocrinology* **131**, 2774 (1992).

[6] R. Taussig, W.-J. Tang, J. R. Hepler, and A. G. Gilman, *J. Biol. Chem.* **269**, 6093 (1994).

[7] P. A. Watson, J. Krupinski, A. M. Kempinski, and C. D. Frankenfield, *J. Biol. Chem.* **269**, 28893 (1994).

[8] B. A. Harris, J. D. Robishaw, S. M. Mumby, and A. G. Gilman, *Science* **229**, 1274 (1985).

[9] H. R. Bourne, P. Coffino, and G. M. Tomkins, *Science* **187**, 750 (1975).

[10] E. M. Ross, A. C. Howlett, K. M. Ferguson, and A. G. Gilman, *J. Biol. Chem.* **253**, 6401 (1978).

[11] J. K. Northup, P. C. Sternweis, M. D. Smigel, L. S. Schleifer, E. M. Ross, and A. G. Gilman, *Proc. Natl. Acad. Sci. U.S.A.* **77**, 6516 (1980).

[12] G. Grishina and C. H. Berlot, *Mol. Pharmacol.* **57**, 1081 (2000).

[13] G. Grishina and C. H. Berlot, *J. Biol. Chem.* **273**, 15053 (1998).

signal transduction pathway is dependent on α_s. For instance, they were used to show that α_s is required for β-adrenergic receptor-mediated MAPK activation.[18]

Other S49 lymphoma lines with α_s defects have also been isolated. In the *unc* cell line,[19] which was isolated by selecting for cells that grew in the presence of terbutaline, a β-adrenergic agonist, and the phosphodiesterase inhibitors RO 20-1724 and 1-methyl-3-isobutylxanthine, α_s carries a mutation in the carboxyl terminus, R389P, that uncouples it from receptors.[17] In the H21a cell line, which was isolated by selecting for S49 cells that grew in the presence of cholera toxin,[20] α_s carries a mutation, G226A, which prevents an activating conformational change necessary for dissociation from $\beta\gamma$ and activation of adenylyl cyclase.[16,21]

The growth-inhibitory effect of cAMP in S49 lymphoma cells is due to the activity of protein kinase A (PKA). This enabled the isolation of the *kin*⁻ cell line,[22] which is deficient in cAMP binding activity and PKA activity, by selecting for cells that grew in the presence of N^6-2′-O-dibutyryl-cAMP and theophylline, a cAMP phosphodiesterase inhibitor. *kin*⁻ cells can be used to determine the importance of PKA for a particular signaling pathway. For instance, these cells were used to show that PKA plays a key role in homologous desensitization of the β_2-adrenergic receptor.[2] In addition, studies comparing the effects of β_2-adrenergic receptor stimulation in *cyc*⁻ and *kin*⁻ cells demonstrated an α_s-dependent, PKA-independent apoptotic signaling pathway.[23] *cyc*⁻*kin*⁻ cells lacking both α_s and PKA have also been isolated by selecting for *cyc*⁻ cells that can grow in the presence of N^6-2′-O-dibutyryl-cAMP.[17]

Despite the usefulness of S49 lymphoma cells and the cell lines derived from them, they are difficult to transfect using conventional methods such as calcium phosphate precipitation. This chapter provides methods for both transient and stable expression of α subunit constructs in these cells as well as assays for characterizing these constructs. First, a simple transient expression procedure utilizing electroporation is described. A screening assay for determining the abilities

[14] S. R. Marsh, G. Grishina, P. T. Wilson, and C. H. Berlot, *Mol. Pharmacol.* **53,** 981 (1998).

[15] S. Osawa, L. E. Heasley, N. Dhanasekaran, S. K. Gupta, C. W. Woon, C. Berlot, and G. L. Johnson, *Mol. Cell. Biol.* **10,** 2931 (1990).

[16] R. T. Miller, S. B. Masters, K. A. Sullivan, B. Beiderman, and H. R. Bourne, *Nature* **334,** 712 (1988).

[17] K. A. Sullivan, R. T. Miller, S. B. Masters, B. Beiderman, W. Heideman, and H. R. Bourne, *Nature* **330,** 758 (1987).

[18] Y. Wan and X. Y. Huang, *J. Biol. Chem.* **273,** 14533 (1998).

[19] T. Haga, E. M. Ross, H. J. Anderson, and A. G. Gilman, *Proc. Natl. Acad. Sci. U.S.A.* **74,** 2016 (1977).

[20] M. R. Salomon and H. R. Bourne, *Mol. Pharmacol.* **19,** 109 (1981).

[21] E. Lee, R. Taussig, and A. G. Gilman, *J. Biol. Chem.* **267,** 1212 (1992).

[22] P. A. Insel, H. R. Bourne, P. Coffino, and G. M. Tomkins, *Science* **190,** 896 (1975).

[23] C. Gu, Y. C. Ma, J. Benjamin, D. Littman, M. V. Chao, and X. Y. Huang, *J. Biol. Chem.* **275,** 20726 (2000).

of transiently expressed α_s constructs to stimulate cAMP accumulation in response to receptor stimulation is then described. Next, a step-by-step protocol involving retroviral infection that reliably produces stable cell lines is provided. Finally, assays that can be used in these stable cell lines to measure the effects of α_s mutations on stimulation of adenylyl cyclase activity, receptor-mediated activation, and relative receptor affinity are described.

Cell Lines

S49 lymphoma cells, wild-type and the *cyc⁻, kin⁻, cyc⁻kin⁻, cyc⁻* TAg,[14] *unc,* and H21a variants, can be obtained from the Cell Culture Facility at the University of California, San Francisco [phone: (415) 476-1450, fax: (415) 476-2086]. *cyc⁻kin⁻* cells are referred to as CK7. Ψ2 cells[24] and PA12 cells[25] can also be obtained from this source. Wild-type S49 lymphoma cells can also be obtained from the ATCC (Manassas, VA). All of these cell lines are maintained in high glucose Dulbecco's modified Eagle's medium (GIBCO-BRL, Gaithersburg, MD). The media for each cell line is supplemented with a specific type of serum, as described below. S49 lymphoma cells and their derivatives grow in suspension and are optimally maintained in the presence of 8% (v/v) CO_2, although 5% (v/v) CO_2 can be used. The higher concentration of CO_2 is especially helpful during the dilution cloning step in the stable expression procedure (see below). These cells are particularly susceptible to excessively high or low cell densities and should be maintained between 10^5/ml and 10^6/ml. Cells seeded at 10^5/ml generally reach 10^6/ml in 48 hr. Ψ2 and PA12 cells are maintained in 5% CO_2.

Transient Expression of α_s Constructs in *cyc⁻* S49 Lymphoma Cells

Transient transfections are performed using the *cyc⁻* Tag cell line,[14] a subclone of *cyc⁻* cells[9] that stably expresses simian virus 40 (SV40) large T antigen (TAg). These cells are maintained in media containing 10% heat-inactivated horse serum as well as 0.6 mg/ml of geneticin (GIBCO-BRL) to maintain expression of TAg. Transient transfection of cells expressing TAg with vectors containing a simian virus 40 origin of replication has been shown to maximize expression levels.[26] Therefore, vectors such as pcDNA I/Amp (Invitrogen, Carlsbad, CA), which contains a simian virus 40 origin of replication as well as the cytomegalovirus promoter, should be used to electroporate the cells.

[24] R. Mann, R. C. Mulligan, and D. Baltimore, *Cell* **33,** 153 (1983).
[25] A. D. Miller, M.-F. Law, and I. M. Verma, *Mol. Cell. Biol.* **5,** 431 (1985).
[26] N. A. Clipstone and G. R. Crabtree, *Nature* **357,** 695 (1992).

α subunit constructs are introduced into cyc^- TAg cells (2×10^7 cells in 1.0 ml of 20 mM HEPES-buffered Minimal Essential Medium with Earle's salts, without bicarbonate) by electroporation at room temperature. When using a GIBCO-BRL Cell-Porator, in which the electrodes of the electroporation chambers are separated by 0.4 cm, optimal expression in my laboratory is obtained by using a capacitance setting of 1600 μF and a voltage setting of 250 V. Under these conditions many of the cells are killed, which can be seen as cell clumping. After electroporation, the cells are added to 4.0 ml of Dulbecco's modified Eagle's medium containing 10% heat-inactivated horse serum in 60 mm tissue culture dishes and incubated overnight at 37° in a CO_2 incubator.

An alternate method for transient expression in S49 lymphoma cells that uses vaccinia viruses has been described.[27] The vaccinia virus technique is more laborious than this electroporation method, but it yields higher expression levels.

cAMP Accumulation Assay for Receptor-Mediated Activation of G_s

cyc^- TAg cells are labeled with 48 μCi/ml of [2-^{3}H]adenine (Amersham Pharmacia Biotech, Little Chalfont, UK) 24 hr after electroporation. Twenty-four hrs later, the cells are first washed in assay medium (20 mM HEPES-buffered Dulbecco's modified Eagle's medium without bicarbonate). The cells are then transferred to 24-well plates and incubated at 37° for 30 min in the same medium containing 1 mM of the phosphodiesterase inhibitor, 1-methyl-3-isobutylxanthine, with or without the addition of 0.1 mM isoproterenol. During this incubation without serum, the cells attach to the wells. Aspiration and the immediate addition of 5% trichloroacetic acid plus 1 mM each of ATP and cAMP lyse the cells and terminate the reactions. The cell lysates are then applied to ion-exchange columns to separate cAMP from ATP.[28,29] cAMP accumulation is expressed as $1000 \times [^3H]cAMP/([^3H]ATP + [^3H]cAMP)$.

In this assay, basal cAMP levels in cells transfected with 10–90 μg of vector containing α_s vary linearly in proportion to the plasmid dose (Fig. 1A).[14] Stimulation of these α_s-transfected cells with the β-adrenergic agonist, isoproterenol, increases cAMP levels by ~20-fold. Thus, the cAMP levels from these stimulated cells also exhibit a linear relationship to the amount of transfected plasmid (Fig. 1A).

Controls for Specificity

For α_s mutations that decrease receptor-mediated activation, it is important to test for the potential effects of these mutations on expression level and on

[27] F. Quan and M. Forte, *Methods Enzymol.* **237**, 436 (1994).
[28] Y. Salomon, C. Londos, and M. Rodbell, *Anal. Biochem.* **58**, 541 (1974).
[29] C. H. Berlot, in "G Proteins: Techniques of Analysis" (D. R. Manning, ed.), p. 39. CRC Press LLC, Boca Raton, FL, 1999.

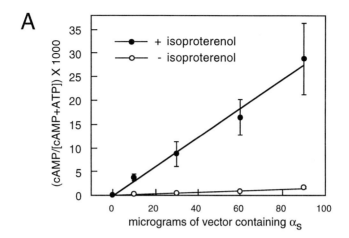

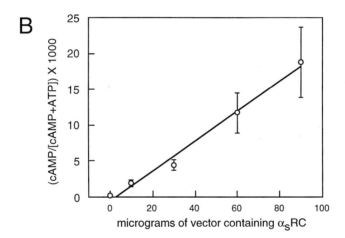

FIG. 1. Transient transfection assay for receptor-mediated activation of α_s. (A) cAMP accumulation in *cyc*⁻ TAg cells electroporated with the indicated doses of vector containing α_s. cAMP levels were measured in the presence and absence of 0.1 mM isoproterenol. (B) Receptor-independent cAMP accumulation in *cyc*⁻ TAg cells electroporated with the indicated doses of vector containing α_sRC. For the 0 μg points, 30 μg of the vector, pcDNAI/Amp, was used. cAMP levels in [³H]adenine-labeled cells were determined as described in the text. All values represent the mean ± standard error of three independent experiments. Reproduced with permission from S. R. Marsh, G. Grishina, P. T. Wilson, and C. H. Berlot, *Mol. Pharmacol.* **53,** 981 (1998).

receptor-independent activation of adenylyl cyclase. The expression levels of transiently expressed α_s proteins in cyc^- TAg cells are not high enough to be detected using an immunoblot. However, both the expression levels and receptor-independent activities of α_s constructs containing the Arg-201 to cysteine (RC) mutation, which causes constitutive activation by inhibiting GTPase activity,[30] can be determined in transiently transfected HEK-293 cells.[29,31] For an α_s construct containing mutations that affect only expression level, but not intrinsic ability to activate adenylyl cyclase, it is possible to normalize the construct's expression level in cyc^- TAg cells to that of α_s. As with α_s-transfected cells, basal cAMP levels in cyc^- TAg cells transfected with 10–90 μg of vector containing α_sRC vary linearly in proportion to the plasmid dose (Fig. 1B). Therefore, to compare receptor-dependent activation of α_s mutants to that of α_s in cyc^- TAg cells, plasmid doses should be identified for which the activities of the α_sRC mutants are similar to that of a given amount, i.e., 30 μg, of the α_sRC-containing plasmid. At these plasmid doses, receptor-dependent cAMP accumulation due to the corresponding α_s mutants can be compared to that of 30 μg of the α_s-containing plasmid. If an α_s construct contains mutations that disrupt receptor-independent adenylyl cyclase activation, then its ability to be activated by receptors cannot be tested using transiently transfected cyc^- TAg cells. However, assays in stable lines of cyc^-kin^- cells can be used to test the receptor interactions of these mutants (see below).

Stable Expression

This method is a modification of a previously described method.[17] It entails retroviral infection using two packaging cell lines, PA12 cells[25] and ψ2 cells.[24] In addition to being a reliable method for expressing constructs in S49 lymphoma cells, retroviral infection has the advantage that the phenotype obtained is likely to be highly stable. This is because retroviruses integrate only once in the host genome,[32] whereas cells transfected with plasmids often harbor multiple copies of the heterologous DNA sequence.

PA12 cells are amphotrophic (broad host range) and are derived from NIH3T3 TK$^-$ cells that were co-transfected with the packaging vector, pPAM, and the herpes simplex *TK* gene as the selectable marker. pPAM is derived from a helper virus, pAM, from which the packaging signal for viral RNA was deleted. ψ2 cells are ecotropic (limited host range). They are derived from NIH3T3 cells that were co-transfected with pMOV-ψ^-, a mutant of Moloney murine leukemia virus with a

---

[30] C. A. Landis, S. B. Masters, A. Spada, A. M. Pace, H. R. Bourne, and L. Vallar, *Nature* **340**, 692 (1989).

[31] R. Medina, G. Grishina, E. G. Meloni, T. R. Muth, and C. H. Berlot, *J. Biol. Chem.* **271**, 24720 (1996).

[32] D. M. Glover, "DNA Cloning: A Practical Approach." IRL Press, Oxford, 1987.

cis-active deficiency for packaging genomic RNA, and pSV2gpt, an SV40 hybrid vector that carries the dominant selectable marker, *gpt*. Thus, both PA12 and ψ2 cells provide retroviral functions in trans that enable production of helper virus-free stocks of infectious retroviral particles that contain retroviral expression vectors. These viruses can introduce α subunit genes into S49 lymphoma cells without turning these cells into retrovirus producers. Virus titers from cells infected with a retroviral vector are at least 10-fold higher than the titers from cells transfected with the same vector.[33] Therefore, the procedure described here involves transfection of PA12 cells followed by infection of ψ2 cells with virus-containing supernatants from the PA12 cells. Virus from the ψ2 cells is used to infect the S49 lymphoma cells.

Step-by-Step Protocol for Retroviral Infection of S49 Lymphoma Cells

The α subunit cDNA is subcloned into a retroviral expression vector such as pMV7,[34] pZIP-NeoSV(X)1,[35] or pLNCX,[36] each of which carries the dominant selectable marker, *neo*, which allows for selection of transduced cells in geneticin. If the goal of producing a stable cell line is to characterize the properties of an α_s mutant construct, especially one with potentially elevated activity, cyc^-kin^- cells should be used to prevent the cytocidal effects of cAMP. Figure 2 illustrates the steps involved in preparation of stable cell lines using this protocol.

Day 1. For each construct, 10^6 PA12 cells are plated onto a 100 mm tissue culture dish. PA12 cells are cultured in medium supplemented with 10% (v/v) fetal bovine serum.

Day 2. The medium is changed on the dish of PA12 cells 2–3 hr prior to transfecting. For each construct, one each of the following is prepared:

Tube A: 93 μl of 2 M CaCl$_2$, 40 μg of plasmid, H$_2$O to 750 μl.
Tube B: 750 μl of 2× HBS (280 mM NaCl, 50 mM HEPES, 2.8 mM Na$_2$HPO$_4$, pH 7.1).

Tube A is added dropwise to Tube B, which is shaken after each addition. The solution should turn cloudy, but not have a large precipitate. The CaPO$_4$/DNA mixture is added dropwise onto the media on the PA12 plate. The plate is then incubated approximately 18 hr at 37° in 5% CO$_2$.

---

[33] L. H. Hwang and E. Gilboa, *J. Virol.* **50**, 417 (1984).
[34] P. T. Kirschmeier, G. M. Housey, M. D. Johnson, A. S. Perkins, and I. B. Weinstein, *DNA* **7**, 219 (1988).
[35] C. L. Cepko, B. E. Roberts, and R. C. Mulligan, *Cell* **37**, 1053 (1984).
[36] A. D. Miller and G. J. Rosman, *Biotechniques* **7**, 980 (1989).

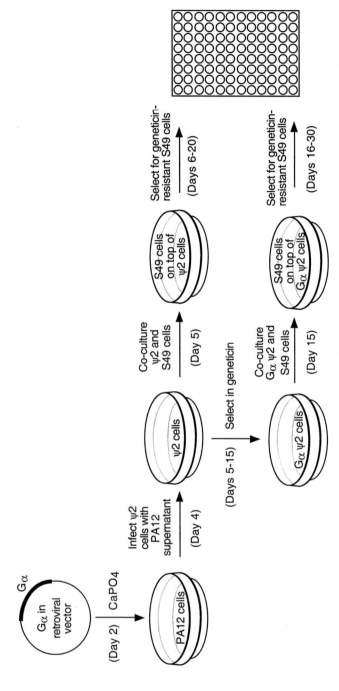

FIG. 2. Diagram of retroviral infection procedure to produce stable lines of S49 lymphoma cells. See text for detailed protocol. *Top:* Diagram illustrates infection of S49 lymphoma cells using transiently produced virus from unselected ψ2 cells. *Bottom:* Same procedure using selected ψ2 cells that have incorporated the retroviral vector. The indicated day for infecting S49 lymphoma cells with virus from the selected ψ2 cells reflects the earliest day that this can be done, which is before the results of the infection with transiently produced virus are known.

Day 3

1. The medium on the PA12 plate is aspirated. The cells are washed twice with 10 ml of prewarmed medium. Ten ml of warm medium is then added and the plate is incubated overnight at 37° in 5% CO_2.
2. ψ2 cells (10^6) are plated on each of two 100 mm tissue culture dishes in medium supplemented with 10% calf serum. Transiently produced virus from one of these plates will be used to infect S49 lymphoma cells on Day 5. The other plate will be used to select for a stable virus-producing cell line that can be used as a backup source of virus. Stable lines of ψ2 cells produce higher titers of virus, but the titers from transiently producing cells are often sufficient, speeding up the process by around 10 days.

Day 4

1. The supernatant from the PA12 plate is collected and centrifuged at 2,000g for 5 min at room temperature in a tabletop centrifuge to remove any cells. The supernatant is then passed through a 0.2 μm filter.
2. The medium on the plates of the ψ2 cells is removed and replaced with the following: 5 ml of the PA12 supernatant, 5 ml of medium supplemented with 10% calf serum, and 10 μl of 8 mg/ml Polybrene, also known as hexadimethrine bromide (Aldrich, Madison, WI).
3. S49 lymphoma cells are seeded at 10^5/ml in 300 ml of medium supplemented with 10% horse serum. Conditioned media from these cells will be used on Day 6 for dilution cloning of the infected S49 lymphoma cells.

Day 5

1. The medium on the plates of the ψ2 cells is removed and discarded. The cells are rinsed twice with medium.
2. The following is added to one of the dishes of ψ2 cells: 10^6 S49 lymphoma cells in 5 ml of medium supplemented with 10% horse serum, 5 ml of medium supplemented with 10% calf serum, and 7.5 μl of 8 mg/ml Polybrene.
3. The other dish of ψ2 cells is split 1 : 5 into selective medium containing geneticin (0.4 mg/ml). The medium is changed every other day for around 10 days until geneticin-resistant cells are obtained. To infect S49 lymphoma cells, 10^6 of the selected ψ2 cells are plated out in medium without geneticin on the day before the infection. Then the procedure continues as described in the previous step.

Day 6. The S49 lymphoma cells are removed from the ψ2 plate. Dilutions are made up in conditioned media from the S49 lymphoma cells seeded on Day 4. Conditioned medium is prepared as described above for the supernatants from the

PA12 cells. If virus from transiently producing $\psi 2$ cells is used for the infection, the recommended dilutions of S49 lymphoma cells are $1:5$, $1:10$, $1:20$, and $1:30$. If virus from selected $\psi 2$ cells is used, the dilutions should be $1:200$, $1:500$, and $1:1000$ because of the higher virus titers. Diluted cells are distributed to 96-well plates (100 μl/well) and 100 μl of 2 mg/ml geneticin is added to each well. Selection is allowed to proceed for $\sim$2 weeks, during which time it is not necessary to change the medium. Cells are considered to be clonal when <20% of the wells are positive. Cells are moved successively to 24-well plates, 60 mm dishes, and then T75 flasks. Selected S49 lymphoma cells are maintained in 0.6 mg/ml of geneticin.

Screening for α Subunit-Expressing Clones

Small-scale membrane preparations from 10^8 cells can be used for immunoblots to determine the expression levels of the clones. Cells are lysed in a hypotonic buffer by passage through a syringe. Nuclei are removed by low-speed centrifugation, and membranes are then isolated by centrifugation in a microcentrifuge, as described.[29,31] Figure 3 shows that the retroviral infection procedure produces S49 lymphoma cell clones with a range of expression levels. When studying a mutant α subunit-expressing clone, it is advisable to compare it to a wild-type α subunit-expressing clone with a similar expression level.

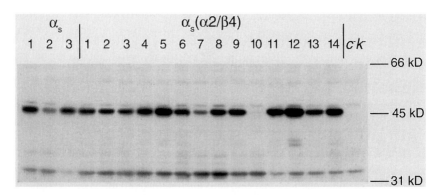

FIG. 3. A range of α subunit expression levels is obtained in stable lines of S49 lymphoma cells. Immunoblot showing expression levels in cyc^-kin^- clones expressing α_s or $\alpha_s(\alpha 2/\beta 4)$, which contains 3 substitutions of α_{i2} homologs for α_s residues (Q236H/N239E/D240G) in the $\alpha 2$ helix and $\alpha 2/\beta 4$ loop [G. Grishina and C. H. Berlot, *Mol. Pharmacol.* **57**, 1081 (2000)]. Both cDNAs contain an epitope referred to as the EE epitope [T. Grussenmeyer, K. H. Scheidtmann, M. A. Hutchinson, W. Eckhart, and G. Walter, *Proc. Natl. Acad. Sci. U.S.A.* **82**, 7952 (1985)], which is described in more detail in [32] in this volume.[36a] Membranes were prepared as described in the text and immunoblotting using an anti-EE monoclonal antibody was performed as described previously [G. Grishina and C. H. Berlot, *J. Biol. Chem.* **272**, 20619 (1997)]. c^-k^- refers to the cyc^-kin^- cell line.

Large-Scale Membrane Preparations

For each large-scale membrane preparation, S49 lymphoma cells are grown in four roller bottles (2.2 liter capacity). Cells are initially seeded at 2×10^5 cells/ml in 500 ml of medium. A stream of 5% CO_2 is directed into the roller bottles for $\sim$1 min before they are sealed. They are rotated at $\sim$1 rpm at 37°. After 2 days, an additional 500 ml of medium is added to each roller bottle and the bottles are again exposed to a stream of 5% CO_2 before sealing. Membranes are prepared 1 day later. At this time the cells are at a density of $\sim$10^6 cells/ml.

Cells are harvested by centrifugation (750g for 5 min, at 4°). All subsequent steps are performed at 4° or on ice. The cells are resuspended in 250 ml of ice-cold Ca^{2+}- and Mg^{2+}-free Dulbecco's phosphate-buffered saline (GIBCO-BRL) and centrifuged again as above. The cells are then resuspended in 100 ml of Bomb Buffer [20 mM HEPES, pH 8.0, 150 mM NaCl, 2 mM MgCl$_2$, 1 mM EDTA, 1 mM 2-mercaptoethanol, 1 mM phenylmethylsulfonyl fluoride (PMSF), 1 mM benzamidine].

The cell suspension is transferred to a beaker on ice inside a Parr cell disruption bomb (Parr Instrument Co., Moline, IL). The chamber is pressurized with N_2 to 500 psi over a period of $\sim$1 min and the cells are maintained at this pressure for 20 min. The cells are lysed by exposure to atmospheric pressure when the exit valve is opened and they pass through the discharge tube. Cell nuclei are pelleted by centrifugation at 750g for 5 min. The supernatant is transferred to a new centrifuge tube and centrifuged at 27,000g for 20 min. The supernatant is aspirated to waste and the pellet is resuspended in 10 ml of membrane buffer (20 mM HEPES, pH 8.0, 2 mM MgCl$_2$, 1 mM EDTA, 1 mM 2-mercaptoethanol, 1mM PMSF, 1 mM benzamidine). The membranes are then centrifuged again as in the previous step.

The pellet is resuspended in 0.5 ml of membrane freezing buffer (membrane buffer with 10% glycerol) and transferred to a 7 ml Kontes Dounce tissue grinder (Fisher Scientific, Pittsburgh, PA). The centrifuge tube is rinsed with 1 ml of membrane freezing buffer, which is also added to the tissue grinder. The membranes are gently dounced until they are homogeneous (5 to 10 strokes). The volume is brought to a total of 3 ml. Protein concentration is determined using the Lowry assay[37] and adjusted to 5 mg/ml. The membrane preparations are aliquotted into microcentrifuge tubes, snap frozen in liquid nitrogen, and stored at −80°. The yield is typically 30–60 mg.

Characterization of α_s Constructs Expressed in Stable Cell Lines

Adenylyl Cyclase Assay

Adenylyl cyclase activity is measured as the conversion of [α-^{32}P] ATP (NEN, Beverly, MA) to [^{32}P]cAMP. [^{3}H]cAMP (NEN) is included in the reactions as an internal standard to correct for the recovery of cAMP in the columns used to separate cAMP from ATP.[28] An $[\alpha\text{-}^{32}\text{P}]\text{ATP}/[^{3}\text{H}]\text{cAMP}$ mixture is prepared allowing for 10 μl per reaction such that each reaction contains 2–3 μCi of $[\alpha\text{-}^{32}\text{P}]\text{ATP}$ and $\sim$15,000 cpm of $[^{3}\text{H}]\text{cAMP}$. Membranes are diluted to 1 mg/ml in TME buffer (10 mM Tris, pH 8.0, 2.5 mM MgCl$_2$, 1 mM EDTA). A 4$\times$ reaction buffer is prepared that consists of 160 mM Tris, pH 8.0, 5 mM MgCl$_2$, 2 mM EDTA, 8 mM 2-mercaptoethanol, 0.4 mg/ml bovine serum albumin, 40 mM phosphocreatine (Sigma, St. Louis, MO), 40 U/ml creatine phosphokinase (Sigma), 1.6 mM ATP, and 4 mM cAMP. Activators such as GTP, GTPγS, GTP plus isoproterenol, and AlF$_4^-$ are prepared at 6.67$\times$ final concentration. Assays are set up in 12 $\times$ 75 glass test tubes on ice as follows: 25 μl of 4$\times$ reaction buffer, 15 μl of activator, and 50 μl of membranes. Reactions are carried out in duplicate or triplicate. Blanks containing 50 μl of H$_2$O instead of membranes are included to control for ^{32}P derived from the $[\alpha\text{-}^{32}\text{P}]\text{ATP}$ that co-elutes with $[^{32}\text{P}]\text{cAMP}$, but does not represent enzymatic conversion of $[\alpha\text{-}^{32}\text{P}]\text{ATP}$ to $[^{32}\text{P}]\text{cAMP}$. The tubes are transferred to a 30$°$ heating block and preincubated for 2–5 min. Production of $[^{32}\text{P}]\text{cAMP}$ is initiated by the addition of 10 μl of the $[\alpha\text{-}^{32}\text{P}]\text{ATP}/[^{3}\text{H}]\text{cAMP}$ mixture. The reactions are incubated for 30 min at 30$°$ and then stopped by the addition of 1 ml of room temperature 0.5% SDS. cAMP and ATP are purified by sequential chromatography on Dowex and alumina columns as described.[28] Adenylyl cyclase activity is expressed as pmol of cAMP/mg of membranes/min.

The adenylyl cyclase assay can be used to assess receptor-mediated activation of α_s mutants even when the mutations decrease the intrinsic ability of α_s to activate this effector. This is because stimulation of G$_s$ by receptors increases the apparent affinity of α_s for GTPγS and this can be measured as a hormone-dependent decrease in the half-maximal effective concentration (EC$_{50}$) for GTPγS stimulation of adenylyl cyclase.

To determine EC$_{50}$ values for stimulation of adenylyl cyclase by GTPγS, the observed adenylyl cyclase activity is fitted to:

$$Y = b + (a - b)/[1 + (X/c)^d] \tag{1}$$

where Y is the observed adenylyl cyclase activity, b is the maximum observed adenylyl cyclase activity, a is the adenylyl cyclase activity observed in the absence of GTPγS, X is the concentration of GTPγS, c is the half-maximal effective concentration (EC$_{50}$) of GTPγS, and d is the slope factor.

As an example, this assay was used to show that substitutions of α_{i2} homologs for α_s residues in the $\alpha3/\beta5$ loop, but not the $\alpha2/\beta4$ loop, decrease receptor-mediated activation of α_s. The mutations in both of these regions decrease activation of adenylyl cyclase.[38] However, the residual activity of these constructs in the

[36a] C. H. Berlot, *Methods Enzymol.* **344,** [32] 2002 (this volume).

[37] O. H. Lowry, N. S. Rosebrough, A. L. Farr, and R. J. Randall, *J. Biol. Chem.* **193,** 265 (1951).

[38] C. H. Berlot and H. R. Bourne, *Cell* **68,** 911 (1992).

adenylyl cyclase assay ($\sim$10% of the activity of wild-type α_s in the presence of 100 μM GTPγS) permitted the use of this assay to test for receptor-mediated activation. The assay showed that $\alpha_s(\alpha3/\beta5)$ exhibits a substantially smaller isoproterenol-dependent decrease in the EC$_{50}$ for GTPγS stimulation of adenylyl cyclase (approximately 2-fold) than α_s and $\alpha_s(\alpha2/\beta4)$ do, which exhibit approximately 5-fold and 10-fold decreases, respectively (Fig. 4).[12] Thus, the mutations in $\alpha_s(\alpha3/\beta5)$ impair activation of α_s by the β_2-adrenergic receptor as well as activation of adenylyl cyclase by α_s. Additional experiments using $\alpha_s(\alpha3/\beta5)$ expressed in and purified from *Escherichia coli* showed that nucleotide handling, activation by A1F$_4^-$, and the ability to undergo GTP-induced conformational changes are normal.[12]

Competitive Receptor Binding Assay

This assay measures an α_s-dependent increase in the affinity of the β-adrenergic receptor for the agonist, isoproterenol, that reflects receptor-G$_s$ interaction.[39,40] The high-affinity isoproterenol-binding state of the receptor requires the presence of G$_s$ in the nucleotide-free state. In the presence of GTP, receptors in membranes of α_s-expressing cells are predominantly in the low-affinity state. Isoproterenol binding is measured in competition with an ^{125}I-labeled β-adrenergic receptor antagonist, iodocyanopindolol (ICYP), which binds to the receptor with the same affinity in the presence and absence of G$_s$.

Membranes are incubated with 75 pM [^{125}I]ICYP in competition with a range of concentrations of isoproterenol (10^{-11} to 10^{-3} M, serial 3-fold dilutions) in the presence or absence of 300 μM GTP or 30 μM GTPγS. [^{125}I]ICYP (2200 Ci/mmol) is obtained from NEN (NEX-174). On the day of the experiment, a 10$\times$ stock (750 pM) of [^{125}I]ICYP is made up in ICYP buffer (50 mM Tris, pH 7.4, 1.76 mg/ml ascorbic acid, 4 mM MgCl$_2$).

For each reaction, the following is assembled in a 12 $\times$ 75 glass test tube on ice: 25 μl of 3mM GTP or 0.3 mM GTPγS or H$_2$O, 25 μl of an isoproterenol dilution, and 15 μg of membranes in 175 μl of buffer consisting of 20 mM HEPES, pH 8.0, 4 mM MgCl$_2$, 1 mM EDTA, and 2 mM 2-mercaptoethanol. Tubes are loaded into a 30° heating block and reactions are started by adding 25 μl of 750 pM [^{125}I]ICYP.

After 1 hr, during which binding reaches equilibrium, the reactions are stopped by adding 5 ml of room temperature Stop Buffer (50 mM Tris, pH 7.4, 4 mM MgCl$_2$). The stopped reactions are immediately filtered through Whatman (Clifton, NJ) GF/C filters held by a filtration apparatus such as the Millipore 1225 Sampling

[39] A. De Lean, M. Stadel, and R. J. Lefkowitz, *J. Biol. Chem.* **255,** 7108 (1980).

[40] H. R. Bourne, D. Kaslow, H. R. Kaslow, M. R. Salomon, and V. Licko, *Mol. Pharmacol.* **20,** 435 (1981).

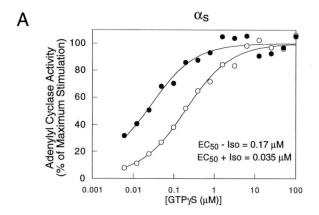

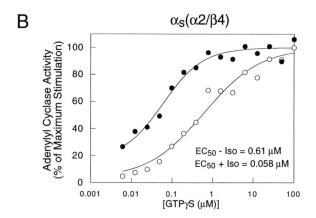

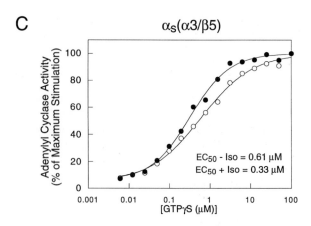

Manifold. The reaction tubes are rinsed once with 5 ml of Stop Buffer, which is also filtered, and then the filters are rinsed twice with 5 ml of Stop Buffer. The filters are then placed in tubes for gamma counting and bound [^{125}I]ICYP is measured. The experimental data are analyzed for competition at two sites by nonlinear least-squares curve fitting as described.[40] The best fits to the data are obtained when K_L and K_H, the low- and high-affinity dissociation constants, respectively, are allowed to vary in the two conditions ($\pm$ guanine nucleotide).

An example of data obtained from this binding assay is shown in Fig. 5. The experiment demonstrates that mutations in the $\alpha3/\beta5$ loop that decrease receptor-mediated increases in the apparent affinity of α_s for GTPγS (Fig. 4) also increase the affinity of G_s for the β_2-adrenergic receptor. The increased receptor affinity is reflected by decreases in K_L and K_H in both the presence and absence of GTP, as well as by an increase in the percentage of high-affinity receptors (R_H) in the presence of GTP. Thus, the decreased receptor-mediated activation caused by the mutations may be due to impaired receptor–G protein dissociation.

Discussion

S49 lymphoma cells offer unique advantages for tracing G protein signaling pathways, in particular those involving G_s. For example, a continuing use for cyc^- cells will be to characterize the defects of loss of function α_s mutants associated with diseases such as Albright's hereditary osteodystrophy. In addition, the genome sequencing projects are likely to reveal novel α_s-like proteins that will be best tested for α_s functions in these cells. Transient expression using electroporation allows large numbers of mutant α_s constructs to be screened for their abilities to interact productively with receptors and effectors. Retroviral infection provides a reliable method to produce stable lines of these cells, overcoming the obstacle that they are difficult to transfect. Although this procedure may appear more involved than transfection, the actual amount of hands-on time is similar. The functional assays that can be performed using these stable lines enable a rigorous characterization of the receptor and effector interactions of α_s constructs in a mammalian cell membrane environment.

FIG. 4. Use of the adenylyl cyclase assay to measure receptor-dependent changes in the apparent affinity of α_s for GTPγS. Adenylyl cyclase activities in membranes of cyc^-kin^- cells stably expressing α_s (A), $\alpha_s(\alpha2/\beta4)$ (B), or $\alpha_s(\alpha3/\beta5)$ (C), which contains 5 α_{i2} homolog substitutions (N271K/K274D/R280K/T284D/I285T) in the $\alpha3$ helix and $\alpha3/\beta5$ loop, were determined in the presence of the indicated concentrations of GTPγS, in the presence (filled symbols) or absence (open symbols) of 100 μM isoproterenol (Iso). Data points represent the means from three independent experiments and are expressed as the percentage of the maximum observed adenylyl cyclase activity. Reproduced with permission from G. Grishina and C. H. Berlot, *Mol. Pharmacol.* **57**, 1081 (2000).

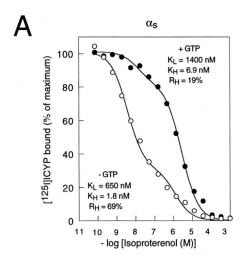

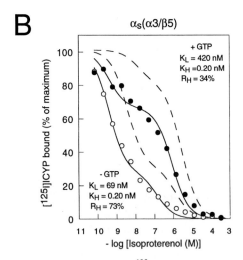

FIG. 5. Competition between isoproterenol and [^{125}I]ICYP for binding to the β_2-adrenergic receptor. Membranes of cyc^-kin^- cells stably expressing α_s (A) or $\alpha_s(\alpha 3/\beta 5)$ (B) were incubated with [^{125}I]ICYP (75 pM) and the indicated concentrations of isoproterenol, in the presence (filled symbols) or absence (open symbols) of 300 μM GTP. Values represent the means of two independent experiments. The solid lines represent a nonlinear least-squares fit to the data, as described [H. R. Bourne, D. Kaslow, H. R. Kaslow, M. R. Salomon, and V. Licko, *Mol. Pharmacol.* **20,** 435 (1981)]. K_L and K_H are the low- and high-affinity dissociation constants, respectively, and % R_H is the percentage of receptors in the high affinity form. In (B), the binding curves for membranes from α_s-expressing cells, from (A), are redrawn as dotted lines. Reproduced and modified with permission from G. Grishina and C. H. Berlot, *Mol. Pharmacol.* **57,** 1081 (2000).

Acknowledgments

I thank Thomas Hynes for helpful discussions and critical reading of the text. Work from the author's laboratory was supported by the National Institutes of Health Grant GM50369. The author is an Established Investigator of the American Heart Association.

[20] Mouse Gene Knockout and Knockin Strategies in Application to α Subunits of G_i/G_o Family of G Proteins

By MEISHENG JIANG, KARSTEN SPICHER, GUYLAIN BOULAY, ANGELES MARTÍN-REQUERO, CATHERINE A. DYE, UWE RUDOLPH, and LUTZ BIRNBAUMER

Introduction

Activation of trimeric G proteins involves two sequential reactions, GDP : GTP exchange and subunit dissociation. Ligand-activated receptors catalyze the exchange reaction and control the kinetic path leading from inactive trimeric G protein occupied by GDP to a free α subunit with GTP bound to it plus a dissociated $\beta\gamma$ dimer. Each, α-GTP and $\beta\gamma$, is competent to signal, i.e., to modulate the activity of effector functions. Because of its intrinsic GTPase activity, the α subunit determines the life span of the activated state of both G protein signaling arms. α-GDP both loses its ability to modulate effector functions and reassociates with $\beta\gamma$. Through the latter action, α-GDP sequesters and deactivates the second arm of the signaling machine.

At the molecular level, each of the trimeric G protein subunits is encoded in a group (family) of separate, nonallelic genes of which a few yield in addition primary transcripts that are alternatively spliced, e.g., $G_o\alpha$ and $G_{i2}\alpha$. To date we know of 16 α, 5 β, and 11 γ genes giving the theoretical possibility of there being 880 distinct G proteins, 935 if the two main variants that result from alternative splicing of the $G_o\alpha$ transcripts ($G_{o1}\alpha$ and $G_{o2}\alpha$) are included. In reality, the molecular diversity of trimeric G proteins in any given cell is much lower because not all combinations are structurally compatible, especially $\beta\gamma$ combinations, and, even when compatible, their expression in cells is severely restricted. It is a safe assumption that a typical cell probably expresses no more than three β and four γ subunits, all able to interact with any one of the expressed α subunits, and that on the average cells express α_s, two α_i, two to three of the α_q family of α subunits (α_q, α_{11} and α_{14} or α_{16}), and α_{12} and α_{13}. $G\alpha_o$ is expressed only in neuronal cell lineages, heart, and endocrine cells, $G\alpha_{i1}$ is mostly neuronal, and $G\alpha_z$ is found predominantly in blood-borne cells. $G\alpha_{14}$ is found in epithelial cells and $G\alpha_{16}$ has variable expression. The remaining $G\alpha$ subunits, $G\alpha_t$ rod and cone, $G\alpha_{olf}$, and $G\alpha_{gust}$, are restricted to the respective sensory neurons: rod and cone photoreceptor cells, olfactory sensory neurons, and taste-sensing neurons.

A few noncanonical γ subunit domains have been found in some of the multi-functional members of the regulators of G protein signaling or RGS family of proteins, adding complexity to the molecular structure of G proteins. RGSs have as their primary biochemical role the activation of the GTPase activity of G protein α subunits. The kinetic consequences of having a GAP activity structurally associated with the inactivated state of the G protein ($\alpha\beta\gamma$ only forms when α is in the GDP-bound form) still needs to be explored.

G proteins are named according to their α subunits without regard to the identity of the $\beta\gamma$ complex they associate with. One thus refers to G_s, G_{olf}, G_i, etc. Based on their susceptibility to be ADP-ribosylated by pertussis toxin (PTX), which causes G protein α subunits to become uncoupled from receptors, often (erroneously) referred to as inactivated, G proteins can be subdivided into two major classes: PTX-sensitive and PTX-insensitive. PTX sensitive are of two types: sensory, having a very restricted set of cells in which they are expressed, and non-sensory. Sensory PTX-sensitive G proteins are the two transducins and gustducin. The nonsensory PTX-sensitive G proteins encompass the three G_i's, G_{o1} and G_{o2}, and G_z. G_z is structurally and functionally a G_i protein, but is PTX-insensitive because of an evolutionary quirk that appears to have led to reverse transcriptase-mediated reincorporation of a $G_i\alpha$ transcript into the genome, its mutation from PTX sensitive to PTX insensitive—changing C352 to I352—acquisition of a promoter, and reexpression as a functionally active pseudogene. Figure 1 presents both the probable evolution of molecular diversity of G protein α subunits and the intron/exon structure of some of the α subunit genes.

At the functional level, heterotrimeric G proteins mediate the effects of a large family of cell surface receptors that range from neurotransmitters, sensory signals, and autacoids to peptide and glycoprotein hormones. Activation of G proteins is responsible for effects as diverse as increase in cAMP production by adenylyl cyclases and inositol trisphosphate plus diacylglycerol by $C\beta$-type phospholipases to stimulation of phosphatidylinositol 3-kinases, activation of potassium channels

---

FIG. 1. Model of chromosomal evolution of G protein α subunit genes based on chromosomal location and similarities in intron–exon structure (adapted from Ref. 16). Intron–exon structure of GNAS, GNAZ, the three GNAI genes and GNAO, highlighting location of regions encoding the (R)KLLLL... identity box, the RXR sequence of linker 1 where the second R is critical for GTPase activity, the switch II DVGGQR sequence, with mutation of Q to L inactivating the subunit's GTPase activity, and the FLNKXD motif important for proper guanine nucleotide binding. Note similar numbering of these amino acids in these proteins. GNAZ has been postulated to be an active pseudogene formed by reverse transcription of α_i precursor gene (RT in boxed scheme[16]). *, amino acids of which spontaneous mutations have been found (adapted from Refs. 17–20).

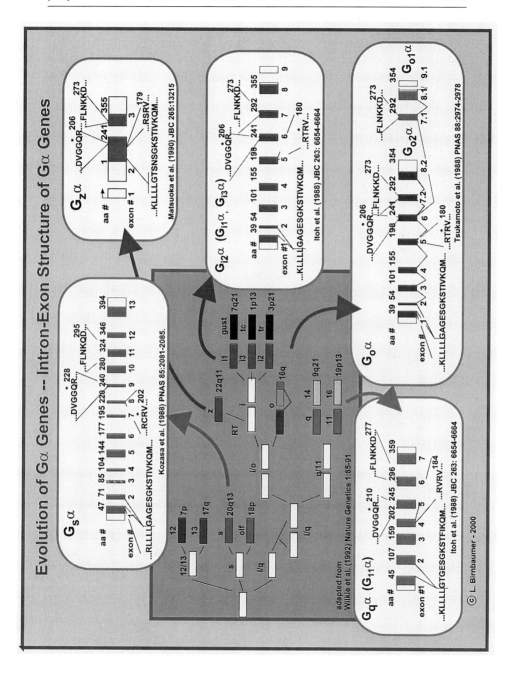

Evolution of Gα Genes -- Intron-Exon Structure of Gα Genes

© L. Birnbaumer - 2000

and *src*-type tyrosine kinases, and inhibition of voltage-gated calcium channels. This has made them central players in the coordination of cellular activities required for normal body homeostasis. Although most of the biochemical functions of trimeric G proteins are probably known, many of their roles as they relate to cellular, organ, and whole-body homeostasis are still poorly understood and in need of further study. One broad approach to the question of G protein involvement in homeostatic or developmental processes is to inactivate and/or alter genes that encode their subunits.

In this article we describe strategies used to inactivate the nonsensory, pertussis toxin sensitive class of heterotrimeric G proteins and present some of the more critical protocols for doing so. The reader is referred to standard texts for basic recombinant DNA techniques.[1,2]

Inactivation of Nonsensory PTX-Sensitive G_i/G_o Class of G Proteins

General Considerations in Gene Targeting

By gene targeting we mean introduction of a mutation into a gene by homologous recombination. The mutation can be the insertion of a foreign gene with or without simultaneous deletion of a segment of the gene, thus causing a gene disruption, or introduction of as little as a single base change, thus changing an amino acid or creating a stop codon.

Genes are targeted *in vitro* in totipotent (also pluripotent) embryonic stem (ES) cells,[3] and genetically engineered mice are obtained by injecting the targeted ES cells into blastocysts that are implanted into the uteri of pseudopregnant foster mothers, where the ES cell-injected blastocysts implant and give rise to live chimeric mice.[4,5] Chimeric mice are mosaic mice formed of cells derived from both the recipient blastocyst and the injected ES cells. The expectation is that ES cell-derived cells in the chimeric mouse (F_o mouse) differentiate into germ cells, allowing the immortalization of the targeted genome. If ES cells differentiate into germ cells, some of the descendants of the F_o chimeras, F_1 mice, will have a genome of which one half is derived from the ES cells and the other half from the mouse to which the chimeric mouse has been mated. If the ES half of the F_1 mouse genome carries the mutation introduced into the ES cell, the F1 mouse is a +/− heterozygote. Interbreeding allows for generation of mice homozygous −/− at the targeted locus where "−" denotes the modification introduced into one of the alleles of the ES cell.

---

[1] T. Maniatis, E. F. Fritsch, and J. Sambrook, "Molecular Cloning: A Laboratory Manual." Cold Spring Harbor Laboratory Press, Cold Spring Harbor, NY, 1982.

[2] J. Sambrook, E. F. Fritsch, and T. Maniatis, "Molecular Cloning: A Laboratory Manual," 2nd Ed. Cold Spring Harbor Laboratory Press, Cold Spring Harbor, NY, 1989.

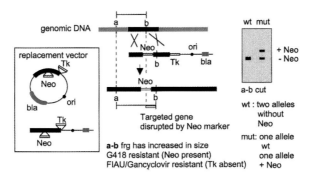

FIG. 2. Basic strategy for gene targeting by homologous recombination using a replacement vector. *Inset:* Replacement vector before and after linearization outside the homology (wide black line = fragment of gene cloned in the vector). The homology is shown disrupted by the presence of the neomycin selection marker. An optional negative selection marker has also been placed into the vector in case the frequency of homologous recombination is very low to allow for further selection against all those cells that have taken up the Neo gene together with the TK gene by nonspecific integration. Thick line: β-lactamase (bla) gene. Note that double reciprocal recombination on both sides of the Neo gene increases the length of one of the two ES cell alleles by the length of the Neo (plus promoter) gene. Right-hand side; Southern analysis of ES cell DNA after successful targeting of one of the alleles. Targeted cells should be G418 and FIAU resistant. Naive ES cells are G418 sensitive and FIAU resistant. Nonspecifically targeted cells are G418 resistant and FIAU sensitive.

Gene targeting thus requires construction of a targeting vector and ES cells. Although disruption by insertion of a selection marker using a replacement vector may be the simplest method (Fig. 2), the promoter driving the inserted selection marker can change expression of nearby genes, creating phenotypic change that is unrelated to the disruption of the gene of interest.[6] It is therefore preferable to modify genes in a manner that does not leave behind foreign selection markers or the promoters that drive them. This is accomplished in the "Hit and Run" (also "insertion/excision" or "In and Out") strategy of gene modification. In this strategy (Fig. 3) a portion of the native gene is first duplicated in the ES cell through the use of an insertion vector, instead of a replacement vector. Insertion is promoted by a double-strand break within the left or right arm of the homology that flanks the mutation one wishes to introduce into the ES genome. In contrast to the replacement vector that carries only a positive selection marker, insertion vectors carry two selection markers: the neomycin resistance gene for positive selection of G418-resistant cells that have incorporated the vector, and thymidine kinase

[3] K. R. Thomas and M. R. Capecchi, *Cell* **51,** 503 (1987).

[4] A. Bradley, M. Evans, M. H. Kaufman, and E. Robertson, *Nature* **309,** 255 (1984).

[5] A. Bradley, *in* "Teratocarcinomas and Embryonic Stem Cells: A Practical Approach" (E. J. Robertson, ed.), pp. 113–151. IRL Press, Oxford, 1987.

[6] E. N. Olson, H. H. Arnold, P. W. Rigby, and B. J. Wold, *Cell* **85,** 1 (1996).

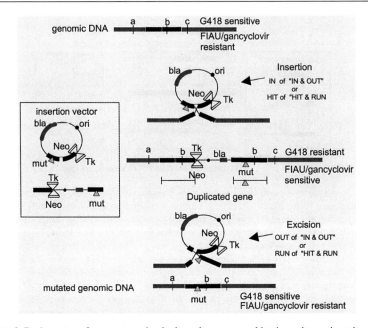

FIG. 3. Basic strategy for gene targeting by homologous recombination using an insertion vector. *Inset:* Insertion vector before and after linearization by creation of a gap within the homology (wide black line = fragment(s) of gene to be targeted used in vector construction). Note that the complete vector is inserted into the gene at the site of the gap, that the gap is repaired, and that the sequences used in the vector have been duplicated. (Middle of main panel: duplicated gene.) After vector insertion the cells are G418 resistant and FIAU sensitive. Selection for FIAU resistant cells selects for cells that have undergone spontaneous intrachromosomal recombination and lost the TK gene. Cells are selected in which the loss of TK is associated with loss of the wild-type duplicate.

for negative counterselection of gancyclovir- or FIAU-resistant cells that have lost the selection marker when incubated with either of these "suicide" substrates. We prefer using FIAU. Insertion of the vector duplicates the region of the gene that is in the vector so that one of the duplicates is wild-type, while the other has the mutation. In the second step, targeted cells with the duplicate are subjected to negative selection for loss of the thymidine kinase gene so as to clone cells that have lost the unmodified duplicate together with the plasmid sequences and selection markers, leaving behind only that portion of the duplicate that bears the mutation (Fig. 3).[7] Details of the construction of one such insertion vector, used to disrupt $G_{i2}\alpha$, can be found in Rudolph *et al.*,[8] as well as in an article by Rudolph *et al.* published in a previous volume of this series.[9]

[7] P. Hasty, J. Ramírez-Solis, R. Krumlauf, and A. Bradley, *Nature* **350**, 243 (1991).

[8] U. Rudolph, P. Brabet, P. Hasty, A. Bradley, and L. Birnbaumer, *Transgen. Res.* **2**, 345 (1993).

[9] U. Rudolph, A. Bradley, and L. Birnbaumer, *Methods Enzymol.* **237**, 366 (1994).

As shown by Anton Berns and colleagues,[10] next to having low passage totipotent ES cells, possibly the most critical step in gene targeting by homologous recombination is the use of isogenic DNA, i.e., the genomic DNA used to construct the vector should be the same as that of the ES cell in which one wishes to promote the homologous recombination process between the ES cell DNA and the homologous DNA of the vector. Each base difference (polymorphism) between the gene to be targeted in the ES cell and the gene used to construct the vector reduces the frequency of homologous recombination. Through inbreeding, mouse strains have accumulated silent—and sometimes not so silent—mutations. Therefore, the natural polymorphisms that exist between a 129 and a C57Black mouse are such that they reduce homologous recombination to frequencies that make it nearly impossible to experimentally isolate the cells that have incorporated the mutated region of the vector by homologous recombination from among those that have done so by random insertion.

129Sv ES cells are highly invasive when injected into C57Black blastocysts. Although the reasons for this are not known, the result is that 12 to 15 ES cells are able to invade the inner cell mass of a 3.5-day-old C57Black blastocyst and contribute to the developed mouse to the extent that often they are almost completely derived from the injected ES cells instead of the cells of the recipient blastocyst. Because of this, essentially all murine gene targeting strategies are based on using 129Sv-derived ES cells injected into C57Black blastocysts.

Vector Constructions

Vector construction requires choosing the 129Sv ES cell in which the gene is to be targeted and a suitable genomic library from which to clone the homologies needed to promote homologous recombination, i.e., the genomic fragments to be used to flank the region that carries the mutation to be introduced. In addition, a plasmid needs to be chosen to carry the homologies, the selection markers, and other features needed to promote the type of homologous recombination that is to be used. These include replacement vs insertion strategies, with or without introduction of loxP sites for creation of conditional knockout mice. As a rule, the final vector has the modifying segment flanked by the 5' and 3' regions of homology along which the homologous recombination will occur, plus auxiliary sequences as dictated by the strategy (e.g., insets in Figs. 2 and 3).

Choice of ES Cells. ES cells are available both from commercial sources and individual investigators. Thus, the exact 129Sv substrain of the particular ES cell available to an investigator will vary. Genomic lineages of commonly used ES cells have been summarized by Simpson *et al.*[11] Although ideally the recombination

[10] H. te Riele, E. R. Maandag, and A. Berns, *Proc. Natl. Acad. Sci. U.S.A.* **89,** 5128 (1992).

[11] E. M. Simpson, C. C. Linder, E. E. Sargent, M. T. Davisson, L. E. Mobraaten, and J. J. Sharp, *Nature Genet.* **16,** 19 (1997).

vector should be constructed with genomic DNA derived from the very ES cell that is to be targeted, this is not always possible, especially if ES cells are provided by colleagues without a matching library. The proven totipotential of the cell is of more concern than the particular 129Sv substrain. We have found that the frequency of homologous recombination in 129SvEv cells is acceptable with homologies derived from 129Sv/Ola DNA (1 : 20–1 : 50, or better).

In G protein disruptions, we have used ES cells established by Alan Bradley (AB1 and AB2.2) and Elizabeth Robertson (EK.CCE) that can all be grown on LIF producing feeder cells and belong to the 129SvEs substrain. In other experiments, we successfully used CMTI-1 cells. Like the commercially available R1 cells, CMTI-1 cells are derived from 129Sv/J mice and need to be grown on primary mouse embryonic fibroblasts (MEFs) in the presence of exogenously supplied LIF.

Choice of Genomic Library. We have found that genomic P1 and bacterial artificial chromosome (BAC) libraries from Incyte Genomics (former Genome Systems, St. Louis, MO, www.incyte.com) to be good sources of DNA for vector construction. To obtain the P1 clones with the desired genomic DNA, PCR oligonu-cleotides can be sent to Incyte Genomics together with an image of the PCR (poly-merase chain reaction) product after agarose gel electrophoresis. Incyte Genomics then prepares three clones yielding a PCR product of the desired size. The clones may or may not be different. To obtain BAC clones, a set of nylon membranes can be purchased from Incyte Genomics onto which a complete genomic library of BAC clones has been spotted. Southern probing these membranes allows the investigator to locate clones positive for the gene of interest. The membranes onto which the sample library has been spotted can be reprobed many times. Clones are sent in live bacteria that are expanded following protocols provided with the clones.

Vector Design. Vector design is tailored to each gene and takes into consid-eration the intron–exon structure of the gene to be targeted, its restriction map, and the targeting method that is to be used: replacement of a disrupted gene by double reciprocal recombination as outlined in Fig. 2, or insertion/excision as out-lined in Fig. 3. The theoretical (and practical) advantages of the insertion/excision approach has been discussed above. Replacement vectors are still used, however, especially in the case of loxP (or other recombinase specific site) introduction as illustrated below for $G_z\alpha$. LoxP sites are sites recognized by the bacteriophage *cre* recombinase, which excises sequences lying between two loxP sites leaving behind only one of the two sites. The loxP/*cre* technique is used to promote *in vivo* excision (and therefore inactivation) of a given gene by mating mice with exons flanked by loxP sites (floxed exons) with mice that express a phage recombinase in selected cells or tissues, leading to tissue specific and conditional gene inactiva-tion. We have introduced three loxP sites (Fig. 4) which, after transient expression of the *cre* recombinase in ES cells, allowed us to generate both ES cells with a $G_z\alpha$ gene lacking exon 2 and ES cells in which exon 2 was simply floxed, and hence prepared for tissue specific inactivation. Insertion vectors are needed to introduce

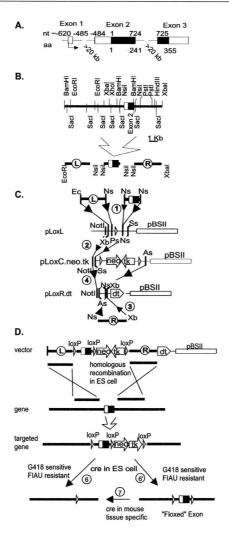

FIG. 4. Strategy for introduction of three loxP sites into the GNAZ (G$_z\alpha$) gene for simultaneous production of ES cells targeted for conventional and conditional G$_z$ inactivation. (A) Intron–exon structure of G$_z\alpha$. (B) Restriction map of the murine GNAZ gene, and segments used to construct a vector to target by replacement. (C) Cloning strategy using pLoxL, pLoxC, and pLoxR. (D) The vector is assembled in pLoxR and carries diphtheria toxin as an additional automatic negative selection marker.

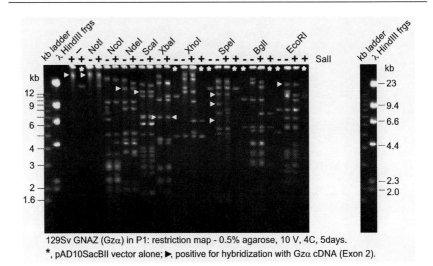

129Sv GNAZ (Gzα) in P1: restriction map - 0.5% agarose, 10 V, 4C, 5days.
*, pAD10SacBII vector alone; ▶, positive for hybridization with Gzα cDNA (Exon 2).

FIG. 5. Typical appearance of an agarose gel after electrophoresis of genomic P1 (or BAC) DNA digested with several restriction enzymes to obtain initial information on the sizes of fragments that carry the gene of interest. Southern analysis showed that this P1 clone had the GNAZ gene in, for example, a 7 kb XbaI fragment. Fragments positive for hybridization to $G_z\alpha$ cDNA have been highlighted (arrow heads).

specific mutations into a given gene, such as is illustrated below for the creation of mice expressing only one of the two G_o splice variants.

Genomic DNA from P1 or BAC clones is analyzed by restriction analysis and hybridization with cDNA or oligonucleotides to identify fragments with the region(s) needed to construct the targeting vector. A typical restriction analysis and identification of bands positive for GNAZ sequences is shown in Fig. 5. The GNAZ ($G_z\alpha$) targeting vector was constructed with sequences derived from the 7 kb XbaI fragment that hybridized with the ^{32}P-labeled $G_z\alpha$ cDNA probe.

Inactivation of Genes Encoding Gα Subunits. Genes encoding the G_{i1}, G_{i2}, G_{i3}, and $G_o\alpha$ subunits were all inactivated by simple insertion of the neomycin (Neo) selection cassette (Figs. 6 and 7). Replacement vectors were used for all, except $G_{i2}\alpha$. Neo was inserted into exon 3 of $G_{i1}\alpha$ and $G_{i2}\alpha$ and exon 6 of $G_o\alpha$. For $G_{i3}\alpha$, Neo was used to replace exon 6 plus flanking sequence. Mice homozygous for loss of G_{i2}, G_o, G_{i3}, and G_{i1} have been obtained.[8,12–14] Figure 8 shows a Southern analysis of tail DNA from mice that are heterozygous and homozygous for the loss

[12] M. Jiang, M. S. Gold, G. Boulay, M. Peyton, K. Spicher, P. Brabet, Y. Srinivasan, U. Rudolph, G. Ellison, and L. Birnbaumer, *Proc. Natl. Acad. Sci. U.S.A.* **95**, 3269 (1998).

[13] K. Spicher and L. Birnbaumer, unpublished.

[14] G. Boulay and L. Birnbaumer, unpublished.

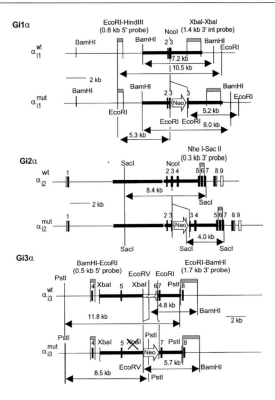

FIG. 6. Diagram of restriction digest patterns for GNAI-1, -2, and -3 before and after targeting with targeting vectors designed to disrupt these genes by insertion of a neomycin selection marker. Thick line, segments used in the targeting vector; vertical bars, exons; gray bars, hybridization probes for Southern analysis.

of these genes. Table I summarizes the sources of genomic DNA used to construct the targeting vectors and the ES cells in which the genes were disrupted and that gave germline transmission.

Table II lists plasmids from which selection markers were derived for vector construction, as well as the plasmids used to construct the replacement vector used to flank exon 2 of Gzα with loxP sites (floxed exon) and drive transient expression of bacteriophage *cre* recombinase in ES cells.

Figure 9 presents the results from applying the strategy outlined in Fig. 4 for simultaneous targeting of the GNAZ ($G_z\alpha$) gene in ES cells for conditional and conventional gene inactivation. Figure 10 illustrates the strategy followed to introduce a Cys-255 to Stop255 mutation (TGC to TGA) and an ATTC tetranucleotide insertion to create an *Eco*RI restriction site (TGA ATTC) in exon 7.2 of the gene encoding the G_{o1} and $G_{o2}\alpha$ subunits. The strategy used an insertion vector which

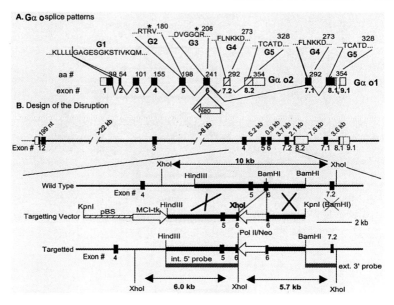

FIG. 7. Diagram of restriction digest patterns. Same as Fig. 6 but for GNAO (Gα_{o1} and Gα_{o2}).

on linearization gave rise to a plasmidless DNA with a double-stranded break that also created a gap. On insertion, the gap was repaired while the duplicated portion of the gene carried the positive and negative selection markers. After the excision step the duplicate was lost, leaving behind the Stop codon together with the new EcoRI restriction site used for diagnostic purposes. Thus, this inactivated one of the G$_{o2}\alpha$ alleles, leaving the G$_{o1}$ allele unaltered.

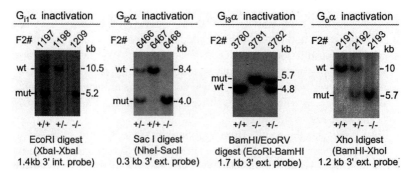

FIG. 8. Southern analysis of DNA extracted from tail biopsies of mice obtained from crossing G$_{i1}$, G$_{i2}$, G$_{i3}$ and G$_o$ +/− mice.

TABLE I
ES CELLS AND GENOMIC LIBRARIES USED TO INACTIVATE G PROTEIN α SUBUNITS
BY HOMOLOGOUS RECOMBINATION

G protein α subunit	Genomic library vector	Strain	ES cell[a]
$G_{i2}\alpha$	λgt10	BALB/c	AB1 (129SvEv)
$G_{i1}\alpha$	λgt10	129Sv	EK.CCE (129SvEv)
$G_{i3}\alpha$	λgt10	129Sv	EK.CCE (129SvEv)
$G_z\alpha$	P1	129SvOLA	EK.CCE (129SvEv)
G_o	P1	129SvOLA	AB2.2 (129SvEv)
G_{o1}	P1	129SvOLA	AB2.2.2 (129SvEv)
G_{o2}	P1	129SvOLA	AB2.2.2 (129SvEv)

[a] AB2.2.2 is a subclone of passage 16 AB2.2.

Southern Blots of P1 and BAC Clones Probed with ^{32}P-Labeled Oligonucleotides

1. Digestion and transfer of BAC DNA for Southern blotting. BAC DNA (2–5 μg) is digested with restriction enzyme(s) and electrophoresed through 0.5–0.6% agarose (0.5 V/cm, overnight at room temperature). After electrophoresis the gel is soaked in 0.5 N NaOH and 1.5 M NaCl for 30 min and neutralized by immersion for 1 hr into 1.5 M NaCl and 0.5 M Tris-HCl, pH 7.5, with agitation on an orbital shaker at ca. 20 cycles per min. The DNA is transferred to GeneScreen Plus (NEN Life Science Product, Boston, MA) (or equivalent) by forced capillary flow with 10× SSC overnight. The membrane with the DNA is then rinsed with 2× SSC and baked at 80° under vacuum for 2 hr.

2. Labeling of oligonucleotide. Ten pmol of an appropriate oligonucleotide (18- to 25-mer) is labeled in a final volume of 20 μl with $[\gamma\text{-}^{32}P]$ATP (50 μCi, ca. 5000 Ci/mmol), 10 U T4 polynucleotide kinase (added last) in 10 mM MgCl$_2$, 5 mM dithiothreitol (DTT), and 70 mM Tris-HCl, pH 7.6. After 45 min at 37°, the reaction is stopped by addition of 75 μl H$_2$O followed by 5 μl of 2 mg/ml salmon sperm DNA in 0.5 M EDTA. Unincorporated $[\gamma\text{-}^{32}P]$ATP and free $[^{32}P]P_i$ are separated either by precipitation with 25 μl 10 M ammonium acetate plus 250 μl 100% ethanol, or by gel exclusion chromatography over Sephadex G-25 (PD10 column, Amersham-Pharmacia, Piscataway, NJ) using 1 mM EDTA and 10 mM Tris-HCl, pH 7.5 as eluent (fraction size: 0.4 ml). Typically ca. 50% of the added $[\gamma\text{-}^{32}P]$ATP is incorporated.

3. Hybridization conditions. Hybridization buffer: 6× SSC, 5× Denhardt's solution, 0.5 mM EDTA, and 0.3 M sodium phosphate, pH 7.0, containing 0.5 mg/ml salmon sperm DNA, and 1×10^6 cpm/ml of labeled

TABLE II
PLASMIDS USED IN GENE TARGETING BY HOMOLOGOUS RECOMBINATION

| Plasmid | Promoter | Size (kb) | | Selection | | |
		Plasmid	Insert	Marker	Drug	Type
Selection markers[a]						
pPolII-Neo	pol II	4.8	1.8	Neomycin	G1418	Positive
pPGK-Neo	PGK[b]	4.6	1.6	Neomycin	G418	Positive
pMC1-TK	HSV-TK[b]	5.0	2.0	Thymidine kinase	FIAU	Negative
pPGK-TK	PGK	5.3	2.3	Thymidine kinase	FIAU	Negative
Conditional knockout plasmids[a]						
ploxL	—	3.0	0.1	—		
ploxC(TK-Neo)	—	6.7	3.8	Neomycin and thymidine kinase	G418/FIAU	
ploxR(DT)	—	4.6	1.6	Diphtheria toxin	—	Negative
pFlox	—	6.3	3.6	Neomycin and thymidine kinase	G418/FIAU	—
pMC-cre	HSV-TK[b]	6.1	1.7 (cre)	Hygromycin	Hygromycin	Positive

[a] Except for the *cre* plasmid, which is based on pUC18, all have pBlueScript II as backbone. pMC-cre was made by replacing the open reading frame of the HSV-TK gene in MC1-TK with the open reading frame of *cre*. Plasmids are available upon request for a nominal handling fee. Nucleotide sequence of plasmids can be found at http//:www.anes.ucla.edu/~lutzb/vectors.html

[b] PGK, Phosphoglycerate kinase; HSV-TK, herpes simplex virus–thymidine kinase.

oligonucleotide. Prehybridization is carried out in hybridization buffer without the labeled oligonucleotide for 2–4 hr at 42°. Hybridization is also carried out at 42° but overnight.

The hybridized blots are then washed with 6× SSC, starting at room temperature and progressively increasing the temperature until the background radioactivity drops to an acceptable level as monitored with a handheld Geiger counter.

4. Autoradiography. The membrane is wrapped in poly(vinyl chloride) film (Saran wrap or similar) and autoradiographed for 2–15 hr.

Development of Genomic Hybridization Probe Useful for Genotyping Targeted Gene

One frequent complication in Southern analysis of genomic DNA is the presence of repetitive sequences in the hybridizing DNA fragments. Therefore, genomic probes should, for the most part, be short. Figure 11 illustrates a typical cleanup process by which an unusable probe became usable by sequential reduction in its size.

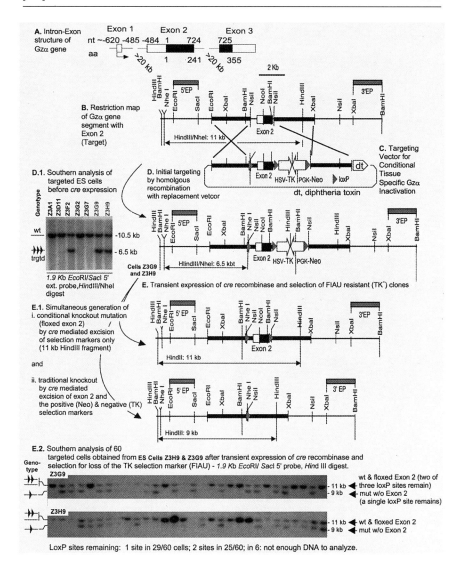

FIG. 9. History of experiments in which ES cells were isolated harboring a floxed or deleted GNAZ exon 2. Note that transient expression of bacteriophage *cre* recombinase under the control of the HSV-TK promoter (MC1) led to an approximate 50:50 (25:29) split of TK-minus and Neo-minus cells (i.e., G418-sensitive and FIAU-resistant) with the floxed exon vs cells that had lost both the selection markers and the exon (reactions 6, 6′, and 7 in Fig. 4D).

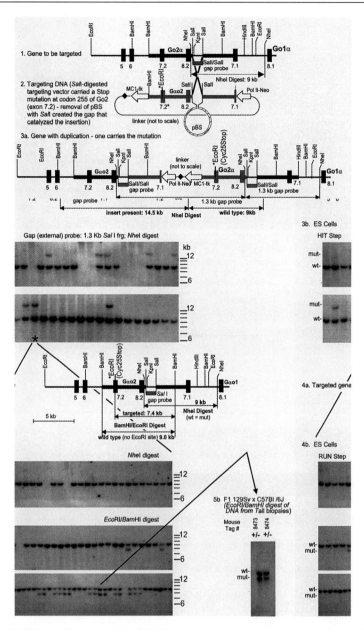

FIG. 10. History of experiments in which ES cells were isolated carrying a Stop codon within exon 7.2 of one of the alleles of the GNAO gene. Southern analysis of DNA from F_1 mice shows germline transmission.

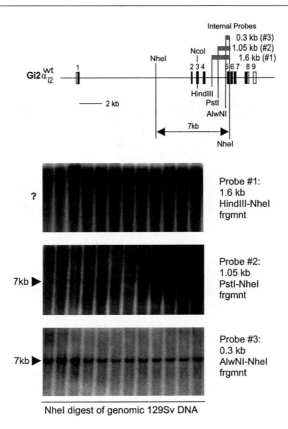

Nhel digest of genomic 129Sv DNA

FIG. 11. Strategy followed for development of a genomic hybridization probe. *Top:* Diagram of $G_{i2}\alpha$ gene and restriction fragments used as hybridization probes. *Bottom:* Southern analysis of 129SV genomic DNA digested with *Nhe*I. Hybridization conditions were adapted from Church and Gilbert[21]: Filters onto which the digested DNA was transferred were baked at 80° for 2 hr in a vacuum oven, prewetted with 2×SSC at 65° for 2 hr, and prehybridized in hybridization buffer without labeled probe for 4 hr at 65°. Hybridization was in 7% SDS, 1% bovine serum albumin, 0.5 mM EDTA, and 0.5 M sodium phosphate, pH 7.2, containing $1–2 \times 10^6$ cpm/ml of denatured ^{32}P-labeled. For labeling by random priming of 25 ng #1, #2, or #3 DNA mixed with 50 μCi [α-^{32}P]dCTP (3000 Ci/mmol), random hexamers, and HiPrime reagents from Roche Molecular Biochemicals. Unincorporated dCTP was removed using a Bio-Rad spin column. Prior to use the probe was denatured by heating to 100° for 10 min and quick-chilling to 4°. After overnight hybridization the filters were washed in 0.5×SSC, 0.1% SDS at room temperature with several changes, followed by washing at 65° (20–30 min) until background became acceptable.

Insertion of Adaptors with Desired Sequences/Restriction Sites

Plasmid preparation: Purified plasmid DNA into which the new sequence is to be introduced is digested with appropriate restriction enzyme(s) and electrophoresed through a 1% agarose (TAE or TBE gel)[2] after addition of 1/5 volume loading dye buffer (60 mM EDTA, 60% (v/v) glycerol, and bromphenol blue).

Selection of restriction enzyme: It is most convenient to insert the new sequence between two noncompatible restriction sites, as it avoids the need of dephosphorylating the vector for the subsequent ligation step. If this is not possible, calf intestinal (alkaline) phosphatase (CIP) is added at the end of the incubation with the restriction enzyme(s) (10U/5 μg DNA) and the incubation is continued for an additional 30 min prior to electrophoresis. After electrophoresis the digested plasmid is excised from the agarose gel and extracted by centrifugation through siliconized glass wool, followed by phenol, chloroform/phenol, and chloroform extraction and precipitation with sodium acetate/ethanol (1/10 volume 3 M sodium acetate, 3 volume 100% ethanol). The DNA pellet is washed with 70% (v/v) ethanol, dried, and resuspended in TE (50 mM Tris-HCl, pH 7.4, 1 mM EDTA). Complementary oligonucleotides, encoding the desired restriction site(s), are synthesized with overhangs that in turn are complementary to the ends of the digested vector.

Annealing: Adaptor concentrations (sense and antisense) are adjusted to 10 μM in TE. Fifty μl of each is pipetted into a screw-cap tube and placed into the heating block at 75°. The block is then allowed to cool to room temperature over a 30-min period and the annealed oligonucleotides are diluted 1 : 5 with TE.

Ligation with vector: one μl of annealed oligonucleotides is added to 20–50 ng of vector prepared as described above in a final volume of 10 μl containing 1 μl T4 DNA ligase (1 U/μl, Roche Molecular Biochemicals, Germany) and 1 μl 10× ligation buffer (10 mM ATP, 50 mM MgCl$_2$, 10 mM DTT, and 600 mM Tris-HCl, pH 7.5). The ligation is allowed to proceed overnight at 16°, and the resulting mixture is used to transform *Escherichia coli* competent cells.

Handling of ES Cells

Low-passage ES cell stocks (1–2 million cells per vial) are kept frozen in liquid N$_2$ in freezing medium (F medium): 20% fetal calf serum (FCS) 10% dimethyl sulfoxide (DMSO), 70% high glucose Dulbecco's modified Eagle's medium (DMEM, Gibco Grand Island, NY), thawed and seeded onto feeder cells at a density of 1–2,000,000 cells/60 mm dish in ES Medium: 15% FCS, 1% penicillin (10,000 U/ml)/streptomycin (10 mg/ml) solution (Gibco) and 1% L-glutamine (200 mM) solution (Gibco), 0.1 mM β-mercaptoethanol, and 1% nonessential amino acids (Gibco) in high-glucose DMEM. They are grown at 37° in a water-jacketed tissue culture incubator with an atmosphere of air supplemented with 5% (v/v) CO$_2$.

Feeder Cells. Depending on the type of ES cell used, ES cells are grown on STO feeder cells carrying a neomycin resistance gene and engineered to secrete

LIF (leukocyte inhibitory factor), or on feeder cells that are primary embryonic fibroblasts (MEFs) isolated from 14-day-old embryos that carry the neomycin resistance gene (usually from readily available knockout mice, but also available from Jackson Labs). The mitotic activity of feeder cells is inactivated the day before either by γ-ray irradiation (3000 rad) or addition of mitomycin C to the culture medium. LIF (ESGRO from Gibco) is added at 1000 units/ml and is used to prevent totipotent cells from differentiating.

Typically, feeder cells are thawed and seeded at a density of 500,000 cells/100 mm dish and grown to confluence in 10% FCS, 1% penicillin/streptomycin (Gibco), 1% L-glutamine in high-glucose DMEM (MEF medium). After reaching confluence, they are split 1 : 5 and grown once more to 80% confluence. If they are to be irradiated, they are lifted from the plate with 0.025% trypsin (5 min) diluted 10-fold in MEF medium, collected by centrifugation in a 50 ml plastic centrifuge tube, resuspended in fresh MEF medium (10 ml per plate), and irradiated. After irradiation, cells are plated evenly onto a new 100 mm plate (or equivalent) that has been precoated for 2 hr at 37° with gelatin (0.1% gelatin in water). If MEFs are to be inactivated with mitomycin C, 80% confluent feeder cells are treated for 2 hr with 10 μg/ml mitomycin C (Gibco) in MEF medium, rinsed twice with Dulbecco's phosphate-buffered saline (DPBS, Gibco) lifted off the plate by trypsinization as above, and plated onto a 100 mm gelatin-coated plate or equivalent (approx. density: 3.5×10^6 cells per 100 mm plate). Alternatively, instead of plating onto gelatin-coated plates, treated MEFs can be frozen in 1.0 ml of the F medium and frozen.

Growth of ES Cells. ES cells are expanded on feeder cell layers prepared in the manner described above in the presence of LIF if necessary. Typically, ES cells are seeded in ES medium at a density of 1–2 million cells per 60 mm plate, then grown for 3 days with daily changes of medium until they reach approximately 80% confluence (final density around 10 million cells/100 mm dish).

Electroporation

Ten million freshly fed ES cells (1 hr before electroporation) are lifted from feeder plates by trypsinization, followed by a 10-fold dilution into ES medium. They are then washed once with Dulbecco's PBS and suspended in 0.8 ml of Dulbecco's PBS plus 25 μg of linearized vector DNA and electroporated in a cuvette with a 4-mm gap at room temperature. We use a Bio-Rad (Hercules, CA) Gene Pulser set at 230 V and 500 μF (time constant, 6–8 msec).

ES Cell Selection and Picking of Colonies

Electroporated cells are resuspended in 24 ml ES medium and placed into four 60 mm plates with feeder cells that had been plated the day before. Selection of neomycin-resistant ES cell clones is begun 24 hr later by addition of 180–200 μg G418 (active ingredient) per ml. Media are changed daily and

G418-resistant colonies are picked on days 8–10 after electroporation. Typically, between 200 and 2000 G418-resistant colonies are obtained, depending on the vector construct. To recover the survivors, the colony-containing plates are rinsed once with DPBS. DPBS is then added to cover the colonies and colonies are picked under a dissecting microscope and transferred with a micropipette tip into wells of 96-well plates containing 25 μl of trypsinization medium [0.05 mg/ml trypsin and 0.53 mM EDTA in Hanks' balanced salt solution (HBSS, Gibco)]. After filling 4 rows of 8 wells, the clones collected in this way are placed in an incubator at 37° for 5 min followed by dilution with 180 μl ES medium containing G418 (and LIF if needed) and vigorous dispersion of the cells by pipetting the content of each well up and down 20 times using an 8-channel pipette. The suspended ES cells are then transferred into the wells of new 96-well plates coated with feeder cells the day before. This cycle of colony picking, trypsinization, dilution, dispersion and plating onto feeder cells is then repeated until 192 or 288 colonies are collected in 2 (or 3) 96-well plates. A skilled operator can collect 2 × 96 clones in a little less than 2 hr. ES cells in these plates are then grown in ES medium with G418 (and LIF) for 3–5 days with daily medium changes until the majority of the clones reach 80% confluence. After removal of ES medium, cells are then rinsed once with DPBS, trypsinized (25 μl), diluted to 150 μl with ES medium, and split into three new plates without feeder cells. Cells of plate #1 ("master plate") are frozen to −70° after receiving 50 μl 2× medium F and 50 μl mineral oil (Sigma). They are then slow-frozen by affixing the plate top with paper adhesive tape, thermal-sealing in a food preservation pouch, and placing into a −70° freezer within a covered Styrofoam container.

Cells in plates #2 and #3 (replicas of the master plate) are diluted to 200 μl with 150 μl of ES medium without G418 or LIF and are expanded by incubation at 37° for 3–5 days until the cells in the majority of the wells are fully confluent. DNA from the expanded cell clones is then isolated and analyzed by mini-Southern,[15,9] to determine which clones have been successfully targeted by homologous recombination.

Thawing Master Plate: Expansion of Targeted ES Clones for Blastocyst Injection or Further Selection [Hit (In) and Run (Out) Strategy]

To recover targeted ES cells from the master plate, the plate is removed from the −70° freezer and placed in a tissue culture incubator at 37° for approximately 5 min, so as to melt the medium in the wells. Immediately after thawing, each of the clones to be used is transferred into a well of 24-well plates that has been seeded with fresh feeder cells the day before and containing 0.5 ml ES medium

[15] J. Ramírez-Solis, J. Rivera-Pérez, J. D. Wallace, M. Wims, H. Zheng, and A. Bradley, *Anal. Biochem.* **201**, 331 (1992).

(with or without LIF, as necessary). The ES cells are then grown to 80% confluence (2–3 days with daily medium changes), trypsinized, and seeded into a well of a 6-well plate (with feeder cells).

After once more growing to ca. 80% confluence the cells are trypsinized and resuspended in 5 ml of ES medium. Some cells (1 ml) are used for injection and the remaining cells are split into four aliquots, spun, resuspended in freezing medium, and frozen for storage until further use. ES cells destined for injection (1 ml) are diluted to 5 ml in ES medium, plated onto a gelatin-coated 60 mm plate, and incubated for 1 hr in a tissue culture incubator. ES cells that have attached to the plate are collected.

Collection of ES Cells for Injection. The supernatant is removed and the attached ES cells are first gently rinsed with ES medium. Then the ES cells to be injected are washed off the plate, leaving behind only the most firmly attached ES cells and feeder cells on which they had grown.

If ES cells from the master plate are to be subjected to a counterselection, as is done in the Hit and Run (or In and Out) strategy (see above), the trypsinized cells from the 6-well plate are diluted with ES medium and seeded at 1 million cells per plate in four 60-mm plates coated with feeder cells. After 24 hr, counterselection is initiated at the first medium change, usually by addition of gancyclovir (2 μg/ml) or fluoroiodoarabinouridine (FIAU 0.2 μM, Moravek Biochemicals). Incubation is continued for 7 to 10 days with daily medium changes until discrete colonies can be picked and transferred into the wells of a 96-well plate for trypsinization, dilution, dispersion, and plating onto feeder cells as described above for isolation of G418-resistant clones.

Comments

The growing times of recombinant clones vary somewhat with the ES cell line used. For example, CMTI-1 cells derived from a male 129Sv/J mouse embryo (Specialtymedia, Phillipsburg, New Jersey) grow faster than AB2.2 or EK.CCE cells, which were both derived from male 129SvEv mouse embryos.

The trypsinization step associated with the initial picking of clones is critical in that no cell clumps should remain. Cell–cell contacts facilitate differentiation processes and decrease totipotentiality.

---

[16] T. M. Wilkie, D. J. Gilbert, A. S. Olsen, X. N. Chen, T. T. Amatruda, J. R. Korenberg, B. J. Trask, P. de Jong, R. R. Reed, M. I. Simon, N. A. Jenkins, and N. G. Copeland, *Nature Genet.* **1,** 85 (1992).

[17] H. Itoh, R. Toyama, T. Kozasa, T. Tsukamoto, M. Matsuoka, and Y. Kaziro, *J. Biol. Chem.* **263,** 6656 (1988).

[18] T. Kozasa, H. Itoh, T. Tsukamoto, and Y. Kaziro, *Proc. Natl. Acad. Sci. U.S.A.* **85,** 2081 (1988).

[19] T. Tsukamoto, R. Toyama, H. Itoh, T. Kozasa, M. Matsuoka, and Y. Kaziro, *Proc. Natl. Acad. Sci. U.S.A.* **88,** 2974 (1991).

[20] M. Matsuoka, H. Itoh, T. Kozasa, and Y. Kaziro, *Proc. Natl. Acad. Sci. U.S.A.* **85,** 5384 (1988).

[21] G. M. Church and W. Gilbert, *Proc. Natl. Acad. Sci. U.S.A.* **81,** 1991 (1984).

Transient Expression of cre *Recombinase in ES Cells prior to Counterselection:*
Generation of Two Types of Targeted ES Cells

Frozen ES cells that have been subjected to the Hit phase of a Hit and Run procedure so as to have three loxP sites (see Figs. 4 and 9) are thawed and expanded to 4–5 million cells on a 60 mm dish coated with feeder cells. They are then prepared as described above for a second electroporation, by resuspending them in 0.8 ml Dulbecco's PBS and electroporating with 1 μg pMC-cre plasmid DNA that carries the bacterial phage *cre* recombinase gene under the control of the HSV thymidine kinase promoter. After electroporation, cells are seeded into three 60 mm dishes with feeder cells and counterselection with FIAU is started after 24 hr, as described.

Acknowledgment

Supported in part by NIH Grant AM-19318 to L.B.

[21] Determining Cellular Role of Gα12

By Jonathan M. Dermott *and* N. Dhanasekaran

Introduction

Heterotrimeric G proteins, consisting of the α, β, and γ subunits, transduce signals from seven transmembrane receptors to a variety of intracellular effector molecules.[1,2] On the basis of the sequence homology of the α subunits, they are classified into four families, namely G_s, G_i, G_q, and G_{12}.[3] Recent studies have indicated that the G_{12} family of G proteins, defined by $G\alpha_{12}$ and $G\alpha_{13}$, is critically involved in the regulation of cell proliferation, differentiation, and apoptosis.[4] Indirect evidence also links $G\alpha_{12}$ to the etiology of soft tissue sarcoma.[5] Studies from different laboratories including ours have demonstrated the ability of GTPase-deficient $G\alpha_{12}$ to activate mitogenic pathways leading to the transformed phenotypes of different fibroblast cell lines.[6–9] More recently, it has been demonstrated

[1] G. L. Johnson and N. Dhanasekaran, *Endo. Rev.* **10**, 317 (1989).
[2] J. R. Hepler and A. G. Gilman, *Trends Biochem. Sci.* **17**, 383 (1992).
[3] M. P. Strathman and M. I. Simon, *Proc. Natl. Acad. Sci. U.S.A.* **88**, 7407 (1991).
[4] N. Dhanasekaran, S-T. Tsim, J. M. Dermott, and D. Onesime, *Oncogene* **17**, 1383 (1998).
[5] A. N-L. Chan, T. P. Fleming, E. S. McGovern, M. Chedid, T. Miki, and S. A. Aaronson, *Mol. Cell. Biol.* **13**, 762 (1993).
[6] N. Xu, L. Bradley, L. I. Ambudkar, and J. S. Gutkind, *Proc. Natl. Acad. Sci. U.S.A.* **90**, 11354 (1993).
[7] H. Jiang, D. Wu, and M. I. Simon, *FEBS Lett.* **3**, 319 (1993).

that these proteins can activate diverse signaling pathways through the activation of the Ras as well as the Rho family of small GTPases.[4,10–12] In this chapter we describe the expression strategies and establishment of a model system to investigate the sequence of signaling events involved in $G\alpha_{12}$-mediated cellular transformation in NIH 3T3 cells.

Strategies

To date, the approaches used to define the role of $G\alpha_{12}$ have relied primarily on the expression of the GTPase-deficient, constitutively activated mutant form of $G\alpha_{12}$ ($G\alpha_{12}$Q229L) in different cell types.[13–16] This was followed by analyses of the biochemical and phenotypic changes brought about by the expression of this activated mutant $G\alpha_{12}$. In some instances, wild-type $G\alpha_{12}$ ($G\alpha_{12}$WT) has been overexpressed to identify the receptors that couple to $G\alpha_{12}$. This has enabled investigators to assess the functional consequence(s) of stimulating $G\alpha_{12}$ through such candidate receptors.[17,18] In many of these studies, either the transient or stable expression strategy has been utilized to define the role of $G\alpha_{12}$ in cell proliferation, differentiation, apoptosis, and cellular homeostasis. In some context, a combination of transient- and stable-expression strategies has been used effectively to elucidate the mechanism of signaling pathways regulated by $G\alpha_{12}$.

Transient Expression Strategies

A comparison of various transfection procedures including electroporation, cationic lipid reagents, DEAE-dextran, and calcium phosphate coprecipitation methods reveals that there is no single procedure that is ideally suited for all types of cells. For example, DEAE-dextran mediated transfection routinely provides high efficiency transfections in COS-1/7 and HEK293 cells, whereas calcium

[8] M. V. V. S. Vara Prasad, S. K. Shore, and N. Dhanasekaran, *Oncogene* **9**, 2425 (1994).

[9] T. Voyno-Yasenetskaya, A. M. Pace, and H. R. Bourne, *Oncogene* **9**, 2559 (1994).

[10] A. M. Buhl, N-L. Johnson, N. Dhanasekaran, and G. L. Johnson, *J. Biol. Chem.* **270**, 24631 (1996).

[11] C. Fromm, O. A. Coso, S. Montaner, N. Xu, and J. S. Gutkind, *Proc. Natl. Acad. Sci. U.S.A.* **94**, 10098 (1997).

[12] T. Kozasa, X. Jiang, M. J. Hart, P. M. Sternweis, W. D. Singer, A. G. Gilman, G. Bollag, and P. C. Sternweis, *Science* **280**, 2109 (1998).

[13] M. V. V. S. Vara Prasad, J. M. Dermott, L. E. Heasley, G. L. Johnson, and N. Dhanasekaran, *J. Biol. Chem.* **270**, 18655 (1995).

[14] L. R. Collins, A. Minden, M. Karin, and J. H. Brown, *J. Biol. Chem.* **271**, 17349 (1996).

[15] T. A. Voyno-Yasenetskaya, M. P. Faur, N. G. Ahn, and H. R. Bourne, *J. Biol. Chem.* **271**, 21081 (1996).

[16] H. Mitsui, N. Takuwa, K. Kurokawa, J. H. Exton, and T. Takuwa, *J. Biol. Chem.* **272**, 4904 (1997).

[17] S. Offermanns, K-L. Laughwitz, K. Spicher, and G. Schultz, *Proc. Natl. Acad. Sci. U.S.A.* **91**, 504 (1994).

[18] A. J. Barr, L. F. Brass, and D. R. Manning, *J. Biol. Chem.* **272**, 2223 (1997).

phosphate coprecipitation is the method of choice for several adherent cell lines including NIH 3T3, Swiss 3T3, and HEK293 cells. The procedure described below consistently yields higher transfection efficiencies as well as optimal transient expressions in COS 1/7, HEK293, and NIH 3T3 cells. The methods involving cationic lipid reagents are not discussed here since these protocols vary to a large extent based on the differences in the lipid formulations and the commercial vendors.

Day 0. Split the cells at an appropriate cell density depending on the transfection method to be used. For electroporation (COS 1/7, NIH 3T3, HEK293) and calcium phosphate methods, seed the actively proliferating cells at a density of 7×10^5 cells per 100-mm culture dish. For the DEAE-dextran method, plate the cells at a density of 3×10^5 cells per 60-mm culture dish.

Day 1. Transfect the cells by the DEAE-dextran, electroporation, or calcium phosphate coprecipitation method.

Transfection by DEAE-dextran (COS 1/7 and HEK 293 cells). DNA/DEAE-dextran solution is prepared as follows. To 5 ml of Dulbecco's modified Eagle's medium (DMEM) containing 10% NuSerun (Collaborative Research, Bedford, MA) sequentially add 0.5–20 μg of DNA contained in 200 μl of sterile Milli-Q water and 200 μl of DEAE-dextran/chloroquine (10 mg/ml DEAE-dextran containing 2.5 mM chloroquine; Sigma, St. Louis, MO).

Prepare the cells for transfection by removing the medium by aspiration and washing the cells twice with 1.5 ml of PBS. Incubate the cells with DNA/DEAE-dextran solution for 2 hr. Aspirate the DNA/DEAE-dextran solution at the end of 2 hr. Shock the cell with 2 ml of PBS containing 10% dimethyl sulfoxide (DMSO; v/v) and incubate for 2 min. Wash the cells twice with PBS and replenish the cells with complete medium consisting of DMEM + 10% fetal bovine serum (FBS).

Transfection by electroporation (COS 1/7, and HEK 293 cells). (See Ref. 19.) Trypsinize and wash the cells twice with PBS. Resuspend the cells in PBS at a concentration of 5×10^6 cells/ml. Transfer 800 μl of cell suspension in 1.5-ml microtube, add 5–10 μg of DNA, and incubate on ice for 10 min with occasional mixing. Transfer the cells into a prechilled electroporation cuvette with 0.4-cm gap (Bio-Rad, Hercules, CA) and place it on ice. Set the electroporator (Bio-Rad) to the following electroporation conditions: voltage: 0.25 kV; capacitance, 500 μF. Electroporate the cells and incubate the cuvette containing the electroporated cells on ice for 10 min, than add 0.5 ml DMEM containing 10% FBS. Mix gently by pipetting and transfer the cells into a 100-mm culture dish containing 10 ml of medium (DMEM + 10% FBS).

Transfection by calcium phosphate coprecipitation (NIH 3T3, Swiss 3T3, and HEK 293 cells). (See Ref. 20.) Prepare 100 ml of 2× HEPES buffered saline (HBS), pH 7.1 (50 mM HEPES, 280 mM NaCl, and 1.5 mM Na$_2$HPO$_4 \cdot$ 2H$_2$O).

---

[19] N. Dhanasekaran, M. V. V. S. Vara Prasad, S. J. Wadsworth, J. M. Dermott, and G. van Rossum, *J. Biol. Chem.* **269,** 11802 (1994).

[20] M. V. V. S. Vara Prasad and G. Shanmugam, *Biochem. Mol. Biol. Int.* **29,** 57 (1993).

Note that the pH is a critical factor for an efficient transfection. Filter-sterilize and store as 5-ml aliquots at $-20°$. Prepare 20 ml of $2\,M$ $CaCl_2 \cdot 2H_2O$. Filter-sterilize and store as 1-ml aliquots at $-20°$.

Two hours prior to transfection, prepare the cells for transfection by replacing the medium in the culture dish with fresh medium (DMEM + 10% FBS). Prepare the calcium phosphate–DNA precipitation solution as follows. Take 5–10 μg of DNA in a 1.5 ml microtube, add 31 μl of $2\,M$ $CaCl_2$ (Mallinckrodt, Paris, KY), and make up the total volume to 250 μl with sterile Milli-Q water and mix well. Alquot 250 μl of $2\times$ HBS in a 1.5-ml microtube and add the DNA–$CaCl_2$ solution dropwise (1 drop/2 sec) into the tube containing $2\times$ HBS with gentle, uniform mixing. Allow the precipitate to stand for 20 min at room temperature. Add the calcium phosphate–DNA solution slowly over the medium covering the cells to be transfected. Mix the solution and the medium by gently rocking the dishes. The medium will become turbid. Incubate the dishes at $37°$ for 24 hr.

Day 2. Change the medium to remove any dead cells.

Day 3 or 4. Harvest the cells 48–72 hr posttransfection and carry out the analyses of interest.

Comments. Care should be taken to normalize the transfection efficiencies between different experimental groups. This can be accomplished by including a reporter plasmid such as pSVβGal (which encodes the enzyme β-galactosidase) along with the DNA of interest.[19] The expression levels of β-galactosidase in the population of the transfected cells can be monitored by biochemical or cytochemical assays. Normalize the amount of cell lysate by comparing equivalent amounts of enzymatic activity rather than total protein.

Stable Expression Strategies

Some physiological responses may require stable expression of the signaling protein of interest. Such stable expression studies have identified the critical role of $G\alpha_{12}$ in cell proliferation and differentiation in many different cell types.[4,21,22] The experimental protocol that has been successfully used for establishing NIH 3T3 and Swiss 3T3 cell lines that stably expresses $G\alpha_{12}$ is described below.

Day 0: Plate the cells at 1.5×10^6 cells per 100 mm dish in DMEM containing 10% FBS.

Day 1: Transfect the cells either by electroporation (the electroporation conditions are 0.2 kV voltage and 960 μF capacitance) or by calcium phosphate coprecipitation method using the methods outlined in the previous section.

Day 2: After 24 hr replace the medium with fresh DMEM containing 10% FBS (without G418).

[21] A. M. Aragay, L. R. Collins, G. R. Post, A. J. Watson, J. R. Feramisco, J. H. Brown, and M. I. Simon, *J. Biol. Chem.* **270**, 20073 (1995).

[22] E-H. Jho, R. J. Davis, and C. C. Malbon, *J. Biol. Chem.* **272**, 24468 (1997).

Day 3: Split the transfected cells 1 : 10 into 100 mm dishes in DMEM containing 10% FBS plus 400 μg/ml of active G418. Feed cells twice every week.

Day 10: By now, G418-resistant clones should be visible; decrease the G418 concentration to 200 μg/ml. Feed cells twice every week.

Day 14: The clones should be ready for isolation. Isolate the individual colonies with cloning rings and grow them in a 24-well plate containing DMEM + 200 μg/ml of G418.

Day 16: When the cells of the 24-well plate become confluent, trypsinize the cells and expand the cultures in 100 mm culture dishes.

Expression of the transfected gene product should be verified by Northern and Western blot analyses. Once a clonal cell line has been established, it is essential that several aliquots be made from a common stock and frozen in liquid nitrogen for future use. Multiple clones should be isolated, analyzed for expression, and stored so that the possible effects of clonal variation can be examined. When preparing cells for freezing, care should be taken that only a limited number of passages is used. Likewise, when the cell are thawed and cultured, care should be taken that the cells have undergone as few passages as possible so that the functional drifting as well as the secondary adaptation of the cells to the expressed protein can be minimized. In the case of NIH 3T3 and Swiss 3T3 cells, cells that have undergone approximately 15–20 passages are generally acceptable.

Establishing Model System in Which $G\alpha_{12}$-Mediated Cellular Transformation Can Be Reversibly Induced

Using the expression protocols presented here, we have established a model system in which $G\alpha_{12}$-mediated transformation can be reversibly induced. It has been observed that the expression of $G\alpha_{12}$WT confers transforming potential to NIH 3T3 cells in a serum-dependent manner.[5] This finding suggests that a "growth factor" present in the serum stimulates a putative $G\alpha_{12}$-coupled seven transmembrane receptor, which results in the activation of $G\alpha_{12}$-mediated signaling pathway(s) leading to the oncogenic transformation NIH 3T3 cells. Following this rationale, the activation of $G\alpha_{12}$, and hence $G\alpha_{12}$-mediated cell transformation, can be either induced or repressed by the addition or removal of serum. Such a model system can be employed to investigate the temporal sequence of signaling events regulated by $G\alpha_{12}$ as follows.

Day 1: A frozen aliquot of $G\alpha_{12}$ WT–NIH 3T3 cells (1×10^7 cells/ml) is thawed and plated in 100 mm dishes containing DMEM + 5% calf serum (CS).

Day 2: On the following day, $G\alpha_{12}$WT-NIH3T3 cells (5×10^4 per plate) are plated in 60 mm culture dishes. Vector control (pcDNA3-NIH3T3) cells

pcDNA3   α₁₂ Wild-type
NIH3T3 Cells NIH3T3 Cells

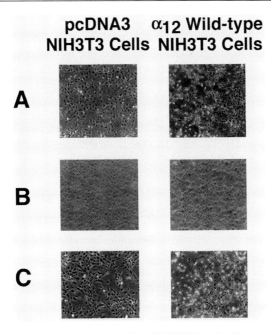

FIG. 1. Serum-dependent transformation of Gα₁₂WT-NIH3T3 cells. The serum-dependent transformations of Gα₁₂WT-NIH3T3 cells were analyzed as described in the text. Control (pcDNA3-
NIH3T3) and Gα₁₂WT-NIH3T3 cells were plated and allowed to recover for 24 hr in DMEM
supplemented with 5% calf serum. Photographs at 4× magnification were taken of the cells that were
actively proliferating (A), serum-deprived for 24 hr (B), and then stimulated with 5% calf serum for
24 hr (C).

are also plated similarly. The actively proliferating Gα₁₂WT-NIH3T3 cells
will exhibit a transformed phenotype, which includes rounded morphology
and loss of contact inhibition.

Day 3: Remove the serum containing medium by replacing the growth medium
with DMEM containing 20 mM HEPES, pH 7.4, and 0.2% bovine serum
albumin (BSA) for 24 hr.

Day 4: Twenty-four hr following serum withdrawal, observe the cells under
the microscope. The serum withdrawal should reverse the transformed
phenotype. Thus, Gα₁₂WT-NIH3T3 cells will exhibit similar growth and
morphological characteristics as those of control NIH3T3 cells. Add 5%
serum to induce cell transformation again.

Day 5: Observe the cells under microscope. Twenty-four hr following serum
stimulation, the cells once again assume the transformed phenotype.

The results from such an experimental protocol are presented in Fig. 1, in
which Gα₁₂WT-NIH3T3 cells stimulated with serum exhibit a transformed phenotype (Fig. 1A) whereas serum withdrawal reverses the transformed phenotype

(Fig. 1B). Subsequent addition of serum again induces the transformed phenotype (Fig. 1C). The results demonstrate a unique model system in which $G\alpha_{12}$-mediated cell transformation can be regulated reversibly. The following representative experimental examples are provided here to illustrate the utility of this model system.

Demonstration of Cell-Growth Stimulation by $G\alpha_{12}$

The growth-promoting properties of $G\alpha_{12}$WT-NIH3T3 cells were compared with the vector control (pcDNA3-NIH3T3) along with the NIH3T3 cells expressing the activated mutant of $G\alpha_{12}$ ($G\alpha_{12}$Q229L-NIH3T3) as follows.

Day -2: $G\alpha_{12}$WT-NIH 3T3 cells (5×10^4) are plated in 20 60-mm culture dishes. $G\alpha_{12}$ QL-NIH3T3 and pcDNA3-NIH3T3 cells are plated similarly.

Day -1: Cells are synchronized by serum-starvation through replacing the growth medium with DMEM containing 20 mM HEPES, pH 7.4, and 0.2% BSA for 24 hr.

Day 0: DMEM supplemented with 5% calf serum is added to initiate cell growth. At this point, cells are counted in triplicate to establish the number of cells present at day 0.

Days 1–5: Cells are counted (three individual dishes for each group) each day for 5 days and the values are plotted against time.

As shown in Fig. 2, $G\alpha_{12}$WT-NIH3T3 cell exhibit an increased growth rate compared to the vector-transfected control cells. This increased growth rate is comparable to that of the cells expressing activated mutant of $G\alpha_{12}$.

Activation of ERK- and JNK-Modules by $G\alpha_{12}$

The $G\alpha_{12}$WT-NIH3T3 cells described here can be used to characterize the regulation of ERK- and JNK-signaling modules during $G\alpha_{12}$-mediated transformation of NIH3T3 cells. Following the procedures outlined in the previous section, cellular transformation is induced in NIH3T3 cells by the addition of 5% CS. To identify the earliest signaling event that distinguishes the transformed phenotype of $G\alpha_{12}$WT cells from that of the control cells, cell lysates are prepared from these cells at varying time points up to 60 min following serum stimulation. Of the several parameters that have been analyzed, the representative examples involving the activation of ERKs and JNKs are presented here.

Day 1: Vector control (pcDNA3-NIH3T3) and $G\alpha_{12}$WT-NIH3T3 cells are plated at the concentration of 4×10^5 cells per 60-mm culture dish and allowed to grow for 24 hr.

Day 2: Cells are serum-starved for 24 hr and the reversal of transformed phenotype is confirmed by microscopy.

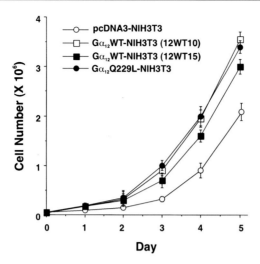

FIG. 2. Stimulation of cell growth by $G\alpha_{12}$. The growth rate of $G\alpha_{12}$WT-NIH3T3 cells (clones 12WT10 and 12WT15) along with pcDNA3 control and $G\alpha_{12}$ Q229L-NIH3T3 were monitored in the presence of DMEM supplemented with 5% calf serum. The cell number was determined from triplicate dishes each day for 5 days. Mean $\pm$ S.E.M. values were determined from three separate experiments.

Day 3: At 24 hr following serum starvation, cell transformation is induced in $G\alpha_{12}$WT cells by serum stimulation (5% CS) for 0, 5, 15, 30, 45, and 60 min. The control, pcDNA3-NIH3T3 cells are subjected to similar treatment.

At the prescribed time points following serum stimulation, cell lysates are prepared from pcDNA3-NIH3T3 and $G\alpha_{12}$WT-NIH3T3 cells. Cells are washed three times with cold PBS (pH 7.4) and lysed in 100–200 μl of RIPA buffer containing PBS (pH 7.4), 1% Nonidet P-40, 0.5% sodium deoxycholate, 0.1% sodium dodecyl sulfate, 2 mM EDTA, 50 mM sodium fluoride, 1 mM phenylmethylsulfonyl fluoride (PMSF), 1 mM 4-(2-amino-ethyl)benzene sulfonyl fluoride (AEBSF), 2 mM benzamidine, 0.2 mM sodium vanadate, 2 $\mu g/ml$ leupeptin, 4 $\mu g/ml$ aprotinin, and 0.1% 2-mer-captoethanol (Sigma-Aldrich, St. Louis, MO).

Equal quantities of protein lysate (25–100 μg) are separated on 9% SDS–PAGE (7 × 8 cm). The separated proteins are transferred onto polyvinylidene difluoride (PVDF) membrane (Millipore, Bedford, MA) using a mini-PROTEAN blotting apparatus (Bio-Rad, Hercules, CA) in 10 mM CAPS (pH 11) buffer containing 20% methanol.[23] The blotting is carried out at 500 mA for 30 min. Protein transfer is checked by Ponceau S [0.1% Ponceau S (w/v) in 5% acetic acid (v/v)] staining of

[23] P. Matsudaira, *J. Biol Chem.* **262**, 10035 (1987).

the PVDF membrane. After confirming protein transfer, the membrane is destained in doubly distilled H_2O after which it is blocked in TBST (20 mM Tris, pH 7.5; 150 mM NaCl; and 0.1% Tween 20) containing 5% milk + 1% BSA. After blocking the membrane for 1 hr at room temperature, it is quickly rinsed and washed three times with TBST (5 min each). The blocked membrane is incubated with the primary antibody in TBST containing 5% BSA at 4° overnight.[24] Rabbit antibodies raised against the phosphorylated forms of ERK1/2 and JNK (antibodies #9101 and #9251, respectively) are from New England Biolabs (Beverly, MA), whereas the rabbit antibodies against the C termini of ERK1, ERK2, JNK1, and N terminus of JNK2 (antibodies sc-93, sc-154, sc-474, and sc-827, respectively) are from Santa Cruz Biotechnology, Inc.

After the overnight incubation, the immunoblot is removed from the primary antibody, quickly rinsed, and washed three times (5 min each) with TBST. The blot is then incubated with the appropriate secondary antibody (horseradish peroxidase conjugated anti-rabbit and antibodies, (Promega, Madison, WI) in TBST containing 5% milk for 1 hr at room temperature. The blot is quickly rinsed with TBST and washed three times (5 min each) with TBST. The immunoblots are developed using Renaissance Western Blot Chemiluminescence Reagent plus (NEN, Boston, MA) according to the manufacturer's protocol. Chemiluminescence is detected by exposing the blots to Kodak (Rochester, NY) scientific imaging film (X-Omat) for times ranging from 5 sec to 15 min. For repeated probing of the blot, the PVDF membrane is stripped by incubating it in 0.2 M glycine hydrochloride, pH 2.5 containing 0.05% Tween 20 at 80° for 2 hr. The membrane is reblocked and continued with the immunoblot procedures as described above.

Interpretation of Results

Western blots developed by antibodies specific for the phosphorylated, and hence activated, forms of ERK1/2 indicate that ERK-activities are potently stimulated in the control as well as $G\alpha_{12}$WT-NIH3T3 cells with little or no difference (Fig. 3). In contrast, the Western blot developed by the phospho-JNK antibodies shows striking differences between the control and $G\alpha_{12}$WT-NIH 3T3 cells (Fig. 4). Upon serum stimulation there is a rapid increase in the phosphorylation of 46 and 54 kDa isoforms of JNKs in pcDNA3-NIH3T3 as well as $G\alpha_{12}$WT-NIH3T3 cells. Whereas this level drops significantly from 15 min onward in the control pcDNA3-NIH3T3 cells, the robust phosphorylation of the 46- and 54-kDa isoforms of JNKs persists throughout the 60 min of serum stimulation in $G\alpha_{12}$WT-NIH3T3 cells (Fig. 4A). In addition, a novel 60 kDa protein, identified by the anti-phospho JNK antibodies, is specifically and continuously stimulated in $G\alpha_{12}$WT-NIH3T3 cells.

The identity of the 46- and 54-kDa JNK isoforms can be deduced from the expression profiles of JNK1 and JNK2 isoforms in these cells. The JNK1 and

[24] G. Gebauer, A. T. Peter, D. Onesime, and N. Dhanasekaran, *J. Cell. Biochem.* **75,** 547 (1999).

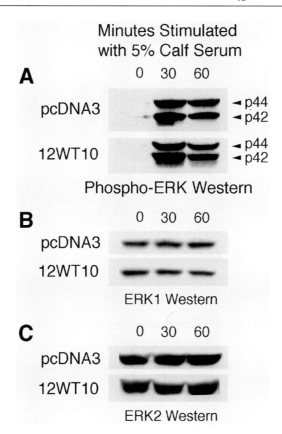

FIG. 3. Phosphorylation of ERK is not potentiated in Gα₁₂WT-NIH3T3 cells. Immunoblot analysis was performed as described. Control (pcDNA3-NIH3T3) and Gα₁₂WT-NIH3T3 (clone 12WT10) cells were serum-deprived for 24 hr and then stimulated with 5% calf serum for 0, 30, and 60 min. The cells were lysed and the lysates were subjected to Western blot analysis using antibodies specific for phospho-ERK (A), ERK1 (B), and ERK2 (C). Identical results were observed in four separate experiments. A typical result is shown.

JNK2 genes can produce two identical isoforms of 46 and 54 kDa each[25] of which only the 54 kDa of JNK2 and both the 46 and 54 kDa isoforms of JNK1 are expressed in NIH 3T3 cells (Figs. 4B and 4C). Therefore, the p46-kDa isoform stimulated in Gα₁₂WT-NIH3T3 cells is likely to be an isoform of JNK1 whereas the stimulated p54-kDa isoform can be JNK1, JNK2, or both. Although the identity of the 60-kDa protein remains to be deduced, since it is recognized by the antibodies directed against the phosphorylated T-P-Y residues of JNKs, it is likely to be an alternate form of JNK or a novel kinase containing the similar "T-P-Y motif." Of

[25] S. Gupta, T. Barrett, A. J. Whitmarsh, J. Cavanaugh, H. K. Sluss, B. Derijard, and R. J. Davis, *EMBO J.* **15,** 2760 (1996).

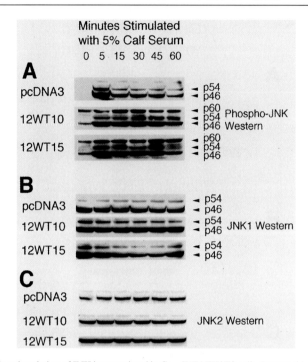

FIG. 4. Phosphorylation of JNK is potentiated in $G\alpha_{12}$WT-NIH3T3 cells. Immunoblot analysis was performed as described in the text. Control (pcDNA3-NIH3T3) and $G\alpha_{12}$WT-NIH3T3 (clones 12WT10 and 12WT15) cells were serum-deprived for 24 hr and then stimulated with 5% calf serum for 0, 5, 15, 30, 45, and 60 min. The cells were lysed and the lysates were subjected to Western blot analysis using antibodies specific for phospho-JNK (A), JNK1 (B), and JNK2 (C). Identical results were observed in four separate experiments. A typical result is shown.

the known JNK isoforms, only the 58-kDa JNK3 α2 isoform[25] is similar in size to the T-P-Y phosphoprotein identified here. However, the expression of JNK3 isoforms has not been observed in NIH3T3 cells. Therefore, it is possible that the 60 kDa T-P-Y kinase activated by $G\alpha_{12}$ is a yet-to-be-identified JNK isoform. Notwithstanding the identity of this kinase, the results indicate that the model system can be used to identify novel proteins involved in a specific response regulated by $G\alpha_{12}$. Moreover, the results presented here point out that the sustained activation of JNK family of kinases is one of the earliest signaling events involved in the oncogenic transformation of NIH 3T3 cells mediated by $G\alpha_{12}$.

Summary

Using the expression strategies described here, we have demonstrated a model system whereby the sequential signaling events involved in cell proliferation and subsequent transformation regulated by $G\alpha_{12}$ can be investigated. The model system presented here can also be used to study the temporal interrelationships between small GTPases, kinases, and other signaling proteins involved in $G\alpha_{12}$-signaling pathways. Further analyses using this model system and the strategies presented here should provide valuable clues in defining the signaling network regulated by $G\alpha_{12}$ in stimulating cell proliferation and oncogenic transformation.

Acknowledgments

Work from the authors' laboratory was supported by United States Public Health Service Grant GM 49897 and a predoctoral fellowship from American Heart Association, Southeastern Pennsylvania Affiliate (J.M.D.). Critical reading of the manuscript by Drs. R. Tantravahi and C. H. Lee is gratefully acknowledged.

[22] Targeted, Regulatable Expression of Activated Heterotrimeric G Protein α Subunits in Transgenic Mice

By XIAOSONG SONG, JIANGCHUAN TAO, XI-PING HUANG,
THOMAS A. ROSENQUIST, CRAIG C. MALBON, and HSIEN-YU WANG

Introduction

Our understanding of G-protein biology has drawn heavily from the creation and detailed analysis of cell lines in which specific G-protein subunits have been overexpressed.[1] Because many G protein α subunits display some level of intrinsic activity, overexpression of wild-type α-subunits can produce a weak gain-of-function mutant from which insights into signaling linkages may be drawn. Elucidation of the mechanism by which the bacterial toxin of *Vibrio cholera* leads to activation of adenylyl cyclase, increasing intracellular cyclic AMP levels and transluminal movement of water and electrolytes characteristic of the pathology of cholera revealed a GTPase activity intrinsic to the $G_s\alpha$ subunit. This activity of $G_s\alpha$ is abolished via ADP-ribosylation by the A 1 subunit of cholera toxin. Cholera toxin ADP-ribosylates the Arg-201 residue, inhibits the GTPase, and renders the activator of adenylylcyclase in the "On" position. Further structure–function studies of G protein α subunits succeeded in identifying a number of mutations that reduced intrinsic GTPase and constitutively activated the molecule in the absence of a ligand-activated G-protein-linked receptor. Examples of the GTPase-deficient mutations of G protein α subunits that are constitutively activated include the

---

[1] A. J. Morris and C. C. Malbon, *Physiol. Rev.* **79,** 1373 (1999).

following: Q227L and G225T versions of $G_s\alpha$; Q205L and G203T versions of $G_{i2}\alpha$; Q209L $G_q\alpha$; Q212L $G_{16}\alpha$; and Q205L $G_o\alpha$. Several human diseases, such as McCune-Albright syndrome (MAS), can be ascribed to activating mutations of G protein α subunits that render the molecule GTPase-deficient and constitutively activated.[1]

Expression of constitutively active (CA) G protein α subunits *in vivo* offers an additional dimension to the analysis of G protein signaling. The fundamental roles that heterotrimeric G proteins play in cell signaling pose a number of formidable challenges to their study *in vivo*. For G-protein α subunits displaying a wide distribution of expression (e.g., $G_s\alpha$), the creation of transgenic mice expressing CA α subunits in virtually all tissues would be expected to yield complex phenotypes, assuming the expression is not lethal. McCune-Albright syndrome, in which patients express an CA form of $G_s\alpha$, is associated with a host of developmental defects, including early death.[2] Expression of a CA α subunit of any one of several G proteins in the pituitary, adrenal, or other endocrine tissues, for example, would likely yield pleiotropic signaling effects that influence cell differentiation, growth, and development and would be difficult to dissect experimentally.

The α subunits of G proteins are ideal targets for manipulation *in vivo*. Although the cell-surface receptors that couple G proteins would seem to be the best targets for manipulation, G-protein-linked receptors display well-described agonist-induced desensitization and down-regulation mechanisms that can readily rectify a signaling pathway with a constitutively activated receptor. A CA receptor would be rapidly nullified by protein phosphorylation and desensitization. Heterotrimeric G proteins, in contrast, are downstream from the element that typically undergoes the rectification and are not subject to desensitization/downregulation. To make optimal use of CA α subunits *in vivo*, inducible and tissue-specific character must be engineered into a construct. We have made use of a construct, based on the phosphoenolpyruvate carboxykinase (PEPCK) gene, that offers inducibility and tissue-specific expression. The PEPCK gene was selected to overcome the most formidable obstacle to expression of G proteins *in vivo*, inducibility. The PEPCK gene is silent *in utero*, its mRNA undetectable by RT-PCR (reverse transcription–polymerase chain reaction). At birth, the PEPCK gene is robustly transcribed, yielding 2–3% of total mRNA. Microinjection of single-cell embryos with an expression vector harboring a CA G protein α-subunit driven by the PEPCK promoter yields an embryo with no dectable α subunit mRNA *in utero*. At birth, the transgene is activated and CA α subunit mRNA accumulates.

The second goal is tissue-specific expression. In our case, interest was focused on tissues active in metabolic regulation, particularly insulin action. For these purposes, the PEPCK tissue-specific elements offered great flexibility. Starting

[2] L. S. Weinstein, A. Shenker, P. V. Gejman, M. J. Merino, E. Friedman, and A. M. Spiegel, *N. Eng. J. Med.* **325,** 1688 (1991).

with a parental genomic fragment of the PEPCK gene, we were able to make use of its tissue-specific elements that enabled robust expression in skeletal muscle, liver, and adipose tissues, while minimizing expression in the proximal tubule of the kidney. As noted below, the construct was able to confer to transgenic mice a robust expression of CA α subunits in the target tissues, while nontarget tissues (e.g., spleen, lung, brain) displayed no expression of the transgene. These studies provided a proof-of-concept that inducible, tissue-specific expression of CA G protein α subunits could be achieved *in vivo*. Tissue-specific expression of α subunits for a number of G proteins has now been reported, primarily in heart. Clearly, sufficient information on tissue-specific gene expression exists to enable the creation of vectors that will express CA G protein α subunits in virtually any specialized tissue or cell type of the mouse.

Methods

Design of Inducible, Targeted Expression of G Protein Subunit

The design of an expression vector for CA G protein α subunits in transgenic mice is quite flexible. Consideration need only be given to the type of induction sought and spectrum of tissue-specific expression desired. The PEPCK system, as indicated above, will provoke expression of a transgene only upon birth and, when engineered as described, only in skeletal muscle, adipose, and liver. The α-myosin heavy chain (α-MHC) promoter, in contrast, directs expression of a transgene specifically in defined regions of the heart throughout development, but can be engineered to be inducible by the Tet promoter.[3] Additionally, the promoter element of the mouse mammary tumor virus-long terminal repeat (MMTV-LTR) has been employed in conjunction with the ecdysone promoter to yield expression in the breast epithelium that can be induced only by pronasterone A. Thus, the proper selection of the proper promoter and tissue-specific elements is an early and important decision.

Construction of PEPCK-Based Expression Vectors

The parental vector is pPCK which is described in Fig. 1 and is readily available from our laboratory. A *Hin*dIII restriction site has been created for optimal cloning. Additional cloning sites can be inserted into the *Hin*dIII site to enable directional insertion of a cDNA for a CA G protein α subunit. Every attempt should be made to eliminate any 3′-untranslated region from the cDNA harboring a G protein α subunit. These 3′-UTR often contain inhibitory sequences that can compromise the level of expression obtainable both *in vitro* and *in vivo*. The construct includes

---

[3] R. S. Passman and G. I. Fishman, *J. Clin. Invest.* **94,** 2421 (1994).

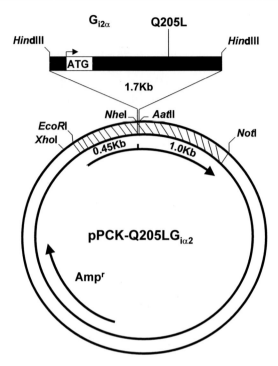

FIG. 1. Schematic of the pPCK expression construct providing regulatable and tissue-specific expression for heterotrimeric G protein subunits. The pPCK vector for the expression of the constitutively active mutant form of $G_{i2}\alpha$ (Q205L) is diagrammed. See the text for details.

an ampicillin resistance gene (Amp^r) but requires cotransfection with a plasmid harboring a neomycin resistance gene (Neo^r).

Transfection and Screening of CA G protein α Subunits in Vitro

We routinely test the functionality of any new construct in rat FTO-2B hepatoma cells in culture, prior to investing the resources and time in generation of transgenic mice. Stably transfected FTO-2B clones display the ability to induce PEPCK gene expression. The response can be induced by cyclic AMP or by an analogue such as the metabolic-resistant, water-soluble 8-(4-chlorphenylthiol, CPT)-cyclic AMP, affording an ideal screen of pPEPCK constructs. Transfection of the FTO-2B hepatoma cells can be achieved using the standard calcium phosphate precipitation method. These cells normally grow well in Dulbecco's modified Eagle's medium (DMEM) containing 5% fetal bovine serum (FBS) at 37° under humidified 95% (v/v) O_2 and 5% (v/v) CO_2. FTO-2B cells were subcultured into a 100 mm plate and incubated overnight to 50% confluence. For expression of the Q205L-G_{i2} α subunit, plasmid pPCKQ205LGi2α DNA (16 μg) and pCW1 DNA (Neo$^+$ plasmid, 4 μg) first are added to 100 μl of 0.25 M CaCl$_2$, and then mixed with 100 μl of

$2\times$ BBS buffer [50 mM N,N-bis(2-hydroxyethyl)-2-aminoethanesulfonic acid (BES, pH 6.95), 1.5 mM Na$_2$HPO$_4$, 280 mM NaCl]. After incubation for 15 min at room temperature, the mixtures are added dropwise to the cells. The cells are incubated overnight at reduced level of CO$_2$ (3%). The cells then are washed with phosphate-buffered saline (PBS) and fed with fresh DMEM containing 5% FBS and grown under 5% (v/v) CO$_2$. Clonal selection is performed in the presence of 200 μg/ml of gentamicin, the neomycin analog often referred to as G418. This concentration of gentamicin is sufficient to kill all of the cells not harboring the neomycin-resistance gene. Single-cell-derived clones were picked and transferred to 24-well plates.

Clonal selection is determined following DNA purification and PCR screening. After growth in 24-well plates to confluence, FTO-2B clones are subcultured into 6-well plates in the continued presence of 100 μg/ml of G418. Genomic DNA from a single well is sufficient for initial screening. The DNA is isolated and purified using DNeasy Tissue Kit (QIAGEN, Inc., Valencia, CA), following manufacturer's instructions. DNA samples are then amplified by polymerase chain reaction (PCR). Water and/or genomic DNA from nontransfected FTO-2B clones serves as a negative control, whereas 15 ng of pPCKQ205LGi2α DNA serves as a positive control. Identification of clones harboring the pPCKQ205LGi2α plasmid, for example, can be identified by PCR using the following primers that will signal uniquely from the construct: the 5′ primer is 5′-GTA CCA ATT CCT CCA GCC TA-3′; the 3′ primer is 5′-CAT CGG GAT TAC ATC TGG CT-3′. The expected PCR product is 410 bp in size. The PCR reaction is performed at 95° for 1 min, 60° for 1 min, 72° for 2 min, with 35 cycles on a GeneAmp PCR System 2400 (PerkinElmer, Norwalk, CT).

Inducing Expression of pPCKCA-Gα in FTO-2B Cells

FTO-2B clones suitable for studies of inducibility of PEPCK gene elements were selected based upon assay of vector mRNA expression using RT-PCR. Stably transfected FTO-2B clones are subcultured into 6-well plates (1×10^5 cells per well) and incubated under standard conditions overnight. To induce expression of a pPCKCA-Gα construct in the FTO-2B clones, 25 μM of 8-(4-chlorphenylthio)adenosine 3′:5′-monophosphate (Roche, Indianapolis, IN) is added to a sample of the clones grown in DMEM. After 3-day treatment, total cellular RNA is isolated and purified using STAT-60 kit (Tel-Test "B", Inc., Friendswood, TX), following the instructions of the manufacturer. Typically we perform reverse transcription according to the commercial protocol for the reverse transcriptase system available from Promega (Madison, WI). This system provides optimal performance. The reverse transcription reaction is conducted in a water bath set at 42° and for a period of 60 min. A 4 μl aliquot of reverse transcription product is subjected to amplification by PCR using the protocol indicated above. Induction with CPT-cyclic AMP should yield a robust transcription of the pPCKCA-Gα construct that can be readily compared to that of clones lacking challenge with cyclic AMP as well as that of wild-type clones induced with cyclic AMP. Assays performed at 6 to 12 days postinduction with cyclic AMP often yield greater responses.

Induction of CA G protein α subunit vector in the stable transfectants can be approached by measuring the level of expression of the targeted G protein subunit with immunoblotting. We specifically use the word "approach," since analysis of CA G protein α subunits is a formidable problem. Tagging the mutant G protein α subunits generally has not been successful, suggesting that the structure cannot tolerate the introduction of a protein tag, such as a polyhistidine, glutathione S-transferase (GST), hemagglutinin (HA), or other such tags. In the absence of a unique tag, the analysis at the level of immunoblotting remains a formidable task since most antibodies generated to date are raised against the C-terminal region of G protein α-subunit, a site distant from the site of the mutation. No antibodies capable of discriminating between point mutants and wild-type G protein α subunits have been described. Consequently, one must seek evidence that the CA G protein α subunits are being expressed over the level of the endogenous wild-type subunits. This can prove very difficult for the simple reason that we and others have observed that expression of a CA G protein α subunit can lead to suppression of the expression of its wild-type counterpart. In general, however, we often observe a 0.3- to 1.5-fold rise in the expression of the target G protein subunit. Validation of this change is most often obtained by measuring a downstream readout (e.g., elevated cyclic AMP levels for CA-$G_s\alpha$, persistently activated phospholipase Cβ for CA-Gqα, or ion channel activation, if present in these screens).

Immunoblotting is performed on protein samples subjected to electrophoresis on 10% polyacrylamide gels in the presence of SDS (SDS–PAGE). To isolate cell membranes, clones are collected in PBS/EDTA and suspended in ice-cold 20 mM HEPES, pH 7.4, 2 mM MgCl$_2$, 1 mM EDTA (HME) buffer containing proteinase inhibitors (5 μg/ml aprotinin, 5 μg/ml leupeptin, and 200 μM phenylmethylsulfonyl fluoride). The suspension is homogenized for 10 sec at setting #3 on a Fisher Scientific 550 Sonic Dismembrator. Intact cells and nuclei are removed by low-speed of centrifugation (1500g) at 4° for 5 min and the supernatant is recovered. Crude membranes are collected by higher-speed centrifugation (15,000g) at 4° for 30 min. Membrane pellets are resuspended in ice-cold HME buffer containing proteinase inhibitors and the amount of protein is determined by the method of Lowry. After separation by SDS–PAGE and transfer to nitrocellulose, the membrane protein blots are blocked with 5% bovine albumin, the blots air-dried, and then probed with the first antibody (1 : 200), typically a rabbit IgG generated against a synthetic dodecapeptide corresponding to the C-terminal region of the Gα. For blotting purposes, we rely on high-quality antipeptide antibodies prepared in the laboratory nearly a decade ago, but have had very good performance with anti-G protein subunit antibodies from Santa Cruz Biotechnology (Santa Cruz, CA) and BD Transduction Laboratories (Los Angeles, CA). The washed blot is probed with a second, goat anti-rabbit immunoglobulin G (IgG) (1 : 5000) coupled with calf alkaline phosphatase. Immune complexes are made visible by assay of the phosphatase product.

Generation of Transgenic Mice

Once a pPCKCA-Gα construct is shown to be inducible, to express the CA-G protein α subunit mRNA, and to express the CA G protein α subunit (or an amplified downstream signal) in response to induction with cyclic AMP in the rat FTO-2B cell screen, the generation of transgenic mice is justified. Large-scale pPCKCA-Gα DNA is prepared and then digested with Xho I and Not I (GIBCO). For the pPCK-Q227LGsα, for example, a 3.2 kb fragment of DNA is generated. The digests are subjected to electrophoresis in a 0.7% low melting temperature agarose gel (Boehringer Mannheim). The DNA is purified by standard silica gel matrix adsorption (Concert Matrix Gel Extraction kit, GIBCO-BRL, Life Technologies). To optimize this step, we have modified the manufacturer's protocol by the inclusion of an additional wash with the binding buffer and with the wash buffer prior to elution. The DNA is eluted into a sterile, injection buffer (10 mM Tris-HCl, 0.2 mM EDTA, pH 7.5). The DNA concentration is determined by electrophoresis of a dilution series with comparison to known standards. Prior to injection, the DNA fragment is diluted to a concentration of 2.5 μg/ml in injection buffer. As an added precaution, the injection solution is centrifuged at 14,000g at 4° in a tabletop microfuge for 10 min to pellet any insoluble material that may impede the flow of the solution during microinjection of the embryos.

Embryo Production. The selection of the proper mouse strains and conditions for generating transgenic mice described below have proven ideal for our purposes over the past decade of study. Female C57BL/6 or FVB mice are used routinely and can be purchased from Charles River Laboratories (Wilmington, MA). The mice are acquired at 10–12 g body weight and are used within 3 days of delivery. Ovulation is induced by injecting the mice i.p. with 5 IU of pregnant mare's serum gonadotropin (Sigma, St. Louis, MO) and again 48 hr later with 5 IU of human chorionic gonadotropin (HCG, Sigma), before housing overnight with stud C57BL/6 males. Mated donor females are euthanized and fertilized eggs isolated from their oviducts. The fertilized eggs are released from cumulus masses by a 5-min incubation in 1 mg/ml hyaluronidase (Sigma) in M2 medium (Sigma). Eggs are rinsed free of debris in M2 medium and then incubated at 37° in 5% (v/v) CO_2 in M16 medium (Sigma). All steps with the eggs outside of the incubator are performed in M2. Both M2 and M16 media must be supplemented with penicillin G (100 U/ml), streptomycin sulfate (100 μg/ml), and lactic acid (23.3 mM). The penicillin–streptomycin combination is obtained from Life Technologies and lactic acid is from Sigma.

DNA Injection. Eggs are injected between 2 and 4 P.M. at 22–24 hr post-HCG injection. Standard manipulations are as published for surgical procedures,[4] for

[4] J. R. Mann, *Methods Enzymol.* **225,** 782 (1993).

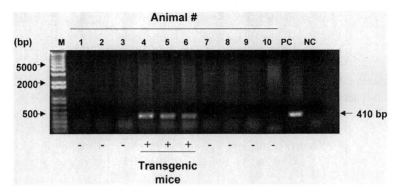

FIG. 2. Identification of mice harboring the pPCK expression construct for the regulatable and tissue-specific expression of the constitutively active mutant form of $G_{i2}\alpha$ (Q205L) using polymerase chain reaction amplification of tail DNA. See the text for details.

production of transgenic mice,[5] and for information on factor influencing the frequency production of transgenic mice.[6] DNA solution is back-loaded by capillary action into self-filling microcapillary tubes, pulled on a Sutter pipette forge to a terminal diameter of roughly 0.5 μm. DNA injection is controlled by a Nikon microspritz controller pressurized with N_2. The injection pipette is kept continuously under a 1 psi positive pressure and DNA is delivered by a 0.5- to 2-sec pulse of 20–30 psi. Male pronuclei are injected with roughly 10 pl of DNA. Injection volume is controlled subjectively by watching expansion of the injected pronuclei to roughly 150% of the original diameter. Following injection the eggs are incubated at 37° and 5% CO_2 for at least 1 hr.

Embryo Transfer. Foster mothers are obtained by mating C57BL/6 × DBA/2 F1 females with Swiss Webster vasectomized males (purchased from Taconic Farms, Germantown, NY). The foster mothers are housed continuously with the vasectomized males. Females with mating plugs on the day of injection are used as pseudo-pregnant egg recipients. Typically, 13 viable injected eggs are transferred to each oviduct of foster mothers. Pups are born 19 days post-transfer.

Detection of Transgenic Founder Mice. Routinely transgenic mice are distinguished from their littermates by analysis of tail DNA (Fig. 2). Tail segments (0.5 to 1.0 cm) are digested and the DNA is isolated and purified using DNeasy Tissue Kit, according to instructions included in the kit (Qiagen, Valencia, CA). To achieve a complete digestion, tail samples must be digested overnight in the presence of protease K (1.0 μg/mg of protein) at 55°. In order to prevent evaporation during overnight digestion, 1.5 ml screw-cap microtubes should be used

[5] J. W. Gordon, *Methods Enzymol.* **225**, 747 (1993).
[6] J. R. Mann and A. P. McMahon, *Methods Enzymol.* **225**, 771 (1993).

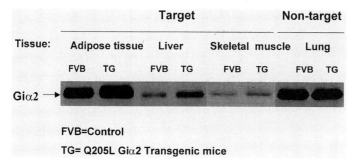

FIG. 3. Identification of mice harboring the pPCK expression construct for the regulatable and tissue-specific expression of the constitutively active mutant form of $G_{i2}\alpha$ (Q205L) using immunoblotting. Samples of membrane protein were subjected to SDS–PAGE and transferred to nitrocellulose, and the blots were probed with anti-$G_{i2}\alpha$ subunit antibodies. The sampling was performed in mice carrying the pPCK-Q205LCA-$G_{i2}\alpha$ transgene and their nontransgenic littermate controls or age/sex-matched FVB controls using both target and nontarget tissues to measure the expression. See the text for details.

for the incubation. The DNA is extracted from the digest and the DNA samples are subjected to amplification by PCR. Water and/or samples of genomic DNA from naive C57B6 mice serve as negative controls; samples of pPCKCA-Gα serve as a positive control. The confirmation of the initial selection based in PCR is provided by immunoblotting of the membrane proteins obtained from target and nontarget tissues samples of transgenic mice as well as samples obtained from either nontransgenic littermates or age/sex-matched control mice (Fig. 3). Note that the transgenic mice harboring the transgene for the expression of the CA-$G_{i2}\alpha$ display elevated levels of immunoreactive $G_{i2}\alpha$ in the target tissues, but not in the nontarget tissues.

Measurement of G Protein Function in Vitro and in Vivo

The measurement of G-protein function *in vivo* is the same as that noted for the *in vitro* screening of the FTO-2B clones harboring the inducible constructs. The primary goal of the expression of CA G protein α subunits *in vivo* is to explore the downstream signaling pathways, taking advantage of the constitutively activated α subunits to identify persistently activated effectors and effector pathways. Mice harboring gain-of-function mutations in a tissue-specific, inducible manner can provide powerful insights into novel effectors and downstream pathways. The insulinomimetic character of the transgenic mice with the Q205LG$_{i2}\alpha$ mutant, for example, provided a model in which an important and novel role of heterotrimeric G proteins in insulin signaling was discovered.[7] Assay of downstream elements for

[7] J. F. Chen, J. H. Guo, C. M. Moxham, H. Y. Wang, and C. C. Malbon, *J. Mol. Med.* **75,** 283 (1997).

the signaling by heterotrimeric G protein α subunits must commence with a functional readout anticipated by study of the known effectors of the CA G protein α subunit introduced into the cells or mice, followed by detailed analysis of the more complex signaling relationships in which heterotrimeric G proteins are believed to play novel roles, including the processes of cellular differentiation, growth, development, and apoptosis.

[23] Inducible, Tissue-Specific Suppression of Heterotrimeric G Protein α Subunits in Vivo

By XI-PING HUANG, THOMAS A. ROSENQUIST, HSIEN-YU WANG, and CRAIG C. MALBON

Introduction

Analysis of the function of heterotrimeric G proteins has benefited by study of cell lines in which gain-of-function and/or loss-of-function of a specific G-protein subunit has been achieved. Loss-of-function mutants for G-protein α subunits were first reported nearly a decade ago.[1] Through the use of oligodeoxynucleotides (ODN) antisense, but not sense, to $G_s\alpha$ it was shown that $G_s\alpha$ suppresses the adipogenic conversion of mouse NIH 3T3-L1 cells in culture. ODN are now used routinely in creation of loss-of-function cells as well as in clinical trials designed to suppress the products of unwanted or aberrant genes. Not all cells are amenable to the actions of ODN, and the use of expression vectors capable of producing antisense RNA emerged as another milestone in facile creation of loss-of-function mutants.[2] In the mid-1990s, the evolution of antisense strategies led to the creation of transgenic mice in which inducible, tissue-specific expression of antisense RNA (in the context of a larger mRNA) was achieved.[3,4] The basic principles that enabled early work using ODN as well as the more recent application of antisense RNA to the study of the loss of function of G protein subunits are simple. With attention paid to several experimental details, investigators can avoid the many pitfalls amply documented in the literature. In addition to gene inactivation by recombination, there exist other approaches, including Cre recombinase-mediated alterations of the mouse genome.[5] The purpose of this article is to highlight the basic principles

[1] H. Y. Wang, D. C. Watkins, and C. C. Malbon, *Nature* **358,** 334 (1992).
[2] D. C. Watkins, G. L. Johnson, and C. C. Malbon, *Science* **258,** 1373 (1992).
[3] C. M. Moxham, Y. Hod, and C. C. Malbon, *Science* **260,** 991 (1993).
[4] C. M. Moxham and C. C. Malbon, *Nature* **379,** 840 (1996).
[5] S. Fiering, M. A. Bender, and M. Groudine, *Methods Enzymol.* **306,** 42 (1999).

and methodology required to make facile use of inducible, tissue-specific constructs that express RNA antisense to G proteins in transgenic mice. Unlike use
of homologous recombination to inactivate genes, which is labor-intensive and
expensive and often takes several years to succeed, the very nature of the introduction of a transgene into a single cell embryo that can effectively suppress the
product of a targeted G-protein subunit gene is comparatively fast and inexpensive.
To facilitate greater understanding of the issues, generic issues will be expanded
with recent examples aimed at creating an inducible, tissue-specific construct for
the suppression of the $G_s\alpha$ subunit *in vivo*.

Methods

Design of Antisense RNA

One of the critical points to the overall success of the strategy is selection of
the proper region to target for antisense RNA. Mathematical modeling of ODN
has revealed that antisense sequences should be more than 7–10 bases and fewer
than 30 in length. The greater the length of the antisense product, the greater the
probability of unintentional suppression of other, nonspecified targets. Potential
sequences should be compared with the GeneBank database for homology search.
Adjustments can be made to regions of the target sequence that maximize specificity and minimize the length of the antisense RNA. A 39-nucleotide antisense
RNA target sequence selected to suppress $G_s\alpha$, for example, is located within
the 5′-untranslated region including the translation initiation codon, 5′-CAT GGC
GGC GGC GGG GCG CGG CCG GGC TGC GGG GCG GCG-3′. Sequences
targeting the same region of mRNAs for Gi2α and for Gqα have been used successfully to suppress the expression of $G_{i2}\alpha$[3] and $G_q\alpha$[6] *in vivo*, respectively.

Construction of Expression Vector

The inducibility and tissue-specific character of this approach is determined
by the goals of the investigator. In the case of heterotrimeric G protein ablation
by homologous recombination, we suspected early on that the ablation of a Gα
subunit might prove lethal to a developing embryo. This concern proved true
in several cases. To obviate the lethality problem, induction at birth seemed a
useful approach to ensure viable pups. Tissue-specific expression was desired in
several tissues, particularly fat, liver, and skeletal muscle, in order to facilitate
our interests in study of insulin signaling.[4] To achieve these goals we employed
elements of the phosphoenolpyruvate carboxykinase (PEPCK) gene that conferred
silence *in utero,* robust inducibility at birth, and tissue-specific expression in these
three target tissues (Fig. 1). The 3′-UTR of the PEPCK gene offered stability

---

[6] P. A. Galvin-Parton, X. Chen, C. M. Moxham, and C. C. Malbon, *J. Biol. Chem.* **272,** 4335 (1997).

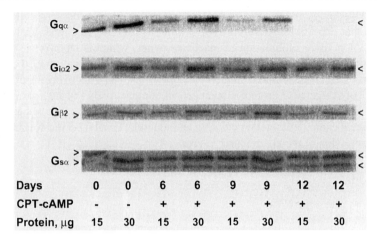

Days	0	0	6	6	9	9	12	12
CPT-cAMP	-	-	+	+	+	+	+	+
Protein, μg	15	30	15	30	15	30	15	30

FIG. 1. The expression of $G\alpha_q$ is suppressed in FTO-2B hepatoma cell stably transfected with pPCK-ASGα_q only upon induction of the antisense expression with a nonhydrolyzable analog of cyclic AMP. The 5′-untranslated region from $G\alpha_q$ (nucleotides 33 to +3) was compared against $G\alpha 14$ and $G\alpha 11$, additional members of the G_q family to identify a target for antisense unique for $G\alpha_q$. The pPCK-AS $G\alpha_q$ construct for inducible expression of RNA antisense was engineered to accomplish tissue-specific expression as well as inducibility at birth (to avoid lethality *in utero*). The 36-nucleotide sequence upstream of and including the translation initiation codon was inserted into the first exon of the rat phosphoenolpyruvate carboxykinase gene (*PCK*) to provide a 2.8-kb hybrid pPCK-ASGα_q antisense RNA, driven by a promoter which is silent *in utero* and activated at birth. Crude membranes (0.2 mg of protein/SDS–polyacrylamide gel electrophoresis lane) were prepared from rat hepatoma FTO-2B cells that were stably transfected with the pPCK-ASGα_q construct and induced with CPT-cyclic AMP for 0, 6, 9, and 12 days, subjected to SDS–polyacrylamide gel electrophoresis, transferred to nitrocellulose blots, and probed with rabbit polyclonal antisera specific for the G-protein subunits indicated. Immune complexes were made visible with goat anti-rabbit IgG coupled to calf alkaline phosphatase and colorimetric development.

elements that ensured an extended half-life of the mRNA harboring the antisense sequence. Finally, the positioning of the antisense region within the PEPCK gene was selected in an effort to keep the antisense region exposed and the insertion was made in the first exon of the mouse PEPCK gene, which also precludes expression of a protein. Target antisense of 39 nucleotides was inserted into multiple cloning sites of the pLNCX vector to obtain pLNCX-ASGα plasmid, which shuttles into the pPEPCK-AS Gα construct employed for further analysis and for microinjection of single-cell embryos.[3]

It should be noted that the inducibility and tissue specificity can be tailored to accommodate other needs. Promoters turned "on" earlier/later in development and tissues other than fat, liver, and skeletal muscle can be targeted through engineering the pPEPCK-ASGα construct, thereby retaining the obvious value of a tried-and-tested construct with proper secondary structure and 3′-UTR stability features of interest while replacing the PEPCK promoter and tissue-specific elements with others.

Transfection and Screening of Constructs in Cells

Although not essential, we prefer to test the functionality of any new construct in cells in culture prior to investing the resources and time in generation of transgenic mice. For the pPEPCKASGα constructs, we employ the rat FTO-2B hepatoma cell line. FTO-2B cells display the ability to induce PEPCK gene expression in response to cyclic AMP, which affords an ideal screen of pPEPCK constructs. The calcium phosphate precipitation method has proven ideal for the stable transfection of FTO-2B cells. These cells normally are grown in Dulbecco's modified Eagle's medium (DMEM) containing 5% fetal bovine serum (FBS) at 37° under humidified 95% (v/v) O_2 and 5% (v/v) CO_2. FTO-2B cells are subcultured into a 100-mm plate and incubated overnight to 50% confluence. pPEPCK-ASGα DNA (16 μg) and pCW1 DNA (Neo$^+$ plasmid, 4 μg) are first added to 100 μl of 0.25 M CaCl$_2$, and then mixed with 100 μl of 2× BBS buffer [50 mM BES (pH6.95), 1.5 mM Na$_2$HPO$_4$, 280 mM NaCl, BES for N,N-bis(2-hydroxyethyl)-2-aminoethanesulfonic acid)]. After incubation for 15 min at room temperature, the mixtures are added dropwise to the cells. The cells are incubated overnight at reduced level of CO_2 (3%). The cells are washed with phosphate-buffered saline (PBS) and fed with fresh DMEM containing 5% FBS and grown under 5% CO_2. Clonal selection is performed in the presence of 200 μg/ml of the neomycin analog G418. This concentration of neomycin analog is sufficient to kill all of the cells not harboring the neomycin resistance gene. Single-cell derived clones are picked and transferred to 24-well plates.

Clone selection requires DNA purification and PCR screening. After being cloned in 24-well plates to confluence, FTO-2B cells are subcultured into 6-well plates and grown in the presence of 100 μg/ml of G418. Genomic DNA is isolated and purified using DNeasy Tissue Kit (Qiagen, Inc., Valencia, CA), according to manufacturer's instructions. DNA samples are then amplified by polymerase chain reaction (PCR). Water and/or genomic DNA from nontransfected FTO-2B clones serves as a negative control, whereas 15 ng of pPEPCK-ASGα DNA serves as a positive control. Identification of clones harboring the pPEPCKASGsα plasmid, for example, can be identified by PCR using the following primers that will signal only from the antisense: the 5' primer (P1) is 5'-CGA GAC GCT CTG AGA CGT TTA GTG-3'; the 3' primer (P2) is 5'-GCA GAG AAG TCC AGA CCA TTA TGC-3' (Fig. 1). The expected PCR product is 513 base pairs in size. The PCR reaction is performed at 95° for 1 min, 60° for 1 min, 72° for 2 min with 35 cycles on a GeneAmp PCR System 2400 (PerkinElmer, Norwalk, CT).

Inducing Expression of pPEPCKASGα in FTO-2B Cells

The clones suitable for studies of inducibility of PEPCK gene elements were selected by reverse transcription (RT)-PCR. Stably transfected FTO-2B clones are subcultured into 6-well plates (1 × 10^5 cells per well) and incubated under standard conditions overnight. To induce expression of a pPEPCKASGα construct in the

FTO-2B clones, 25 μM of the cyclic AMP analog, 8-(4-chlorphenylthio)-adenosine 3′,5′-monophosphate (CPT-cAMP) (Boehringer Mannheim, Indianapolis, IN), is added to a sample of the clones grown in DMEM. After 3-day treatment, total RNA is isolated and purified using STAT-60 kit (TEL-TEST "B", Inc., Friendswood, TX), according to manufacturer's instructions. Reverse transcription is performed according to the commercial protocol for the RT system available from Promega (Madison, WI). This system provides optimal perfromance. The RT reaction is conducted in a water bath set at 42° and for a period of 60 min. A 4 μl aliquot of RT product is subjected to amplification by PCR using the protocol indicated above. Induction with cyclic AMP yields robust transcription of the pPEPCKASGα construct that can be readily compared to that of clones lacking cyclic AMP and wild-type clones also induced with cyclic AMP.

Successful induction of the antisense vector in the stable transfectants can be validated by measuring the level of expression of the targeted G protein subunit with immunoblotting. Immunoblotting is performed on protein samples subject to electrophoresis on 10% polyacrylamide gels in the presence of SDS (SDS–PAGE). To isolate cell membranes, clones are collected in PBS/EDTA and suspended in ice-cold 20 mM HEPES, pH 7.4, 2 mM MgCl$_2$, 1 mM EDTA (HME) buffer containing proteinase inhibitors (5 μg/ml aprotinin, 5 μg/ml leupeptin, and 200 μM phenylmethylsulfonyl fluoride). The suspension is homogenized for 10 sec at setting #3 on a Fisher Scientific's 550 Sonic Dismembrator. Intact cells and nuclei are removed by low speed of centrifugation (1,500g) at 4° for 5 min and the supernatant is recovered. Crude membranes are collected by high-speed centrifugation (15,000g) at 4° for 30 min. Membrane pellets are resuspended in ice-cold HME buffer containing proteinase inhibitors and the amount of protein is determined by the method of Lowry. After separation by SDS–PAGE and transfer to nitrocellulose, the membrane proteins are blotted in 5% bovine albumin and incubated with the first antibody (1 : 200), typically a rabbit IgG generated against a synthetic dodecapepetide corresponding to the C-terminal region of the Gα. The washed blot is probed with a second, goat anti-rabbit IgG (1 : 5000) coupled with calf alkaline phosphatase. Immune complexes are made visible by assay of the phosphatase product.

To illustrate the utility of the FTO-2B clones as a screen for a pPEPCKASGα construct, immunoblotting is performed using crude membrane prepared from FTO-2B clones stably transfected with pPEPCKASGαq (Fig. 1). FTO-2B clones harboring the antisense RNA vector targeting Gαq are challenged with or without CPT-cAMP for 6, 9, or 12 days. Aliquots (15 and 30 μg protein/SDS–PAGE lane) of crude cell membranes are subjected to SDS–PAGE and the resolved proteins transferred to nitocellulose blots. The blots are stained with antibodies to one of the following G-protein subunits (G$_q\alpha$, G$_i\alpha_2$, Gβ_2, or G$_s\alpha$) and the immune complexes made visible using the calf alkaline phosphatase second antibody. The results demonstrate a reduction (day 6 with CPT-cAMP) and then disappearance (day 12 with CPT-cAMP) of G$_q\alpha$. The expression of the other subunits measured, in stark contrast, was virtually unaffected by the treatment with the CPT-cAMP, which activates the PEPCK promoter in these clones.[6]

Generation of Transgenic Mice

Once the pPEPCKAS Gα is shown to be inducible, to express the antisense RNA, and to suppress the expression of the targeted subunit in response to induction with cyclic AMP in the FTO-2B screen, the generation of transgenic mice is warranted. pPEPCKASGα DNA are digested with *Xho*I and *Not*I (GIBCO). For the pPEPCKASGsα, for example, 2.2 kb fragment of DNA is generated. The digests are subjected to electrophoresis in a 0.7% low melting temperature agarose gel (Boehringer Mannheim). The DNA is purified by standard silica gel matrix adsorption (using the Concert Matrix Gel Extraction kit, GIBCO BRL, Life Technologies). To optimize this step, the manufacturer's protocol has been modified by the incorporation of an additional wash with the binding buffer and wash buffer prior to elution. The DNA is eluted into a sterile, injection buffer (10 mM Tris-HCl, 0.2 mM EDTA, pH 7.5). The DNA concentration is determined by electrophoresis of a dilution series with comparison to known standards. Prior to injection the DNA is diluted to a concentration of 2.5 μg/ml in injection buffer. As an added precaution, the injection solution is centrifuged at 14,000g in a tabletop microcentrifuge for 10 min to pellet any insoluble material that may impede the flow of the solution during microinjection of the embryo.

Embryo Production. The selection of the proper mouse strains and conditions for generating transgenic mice described below have been ideal for our purposes over the past decade of study. Female C57BL/6 mice are used routinely and can be purchased from Charles River Laboratories (Wilmington, MA). The mice are acquired at 10–12 g body weight and are used within 3 days of delivery. Ovulation is induced by injecting the mice i.p. with 5 IU of pregnant mare's serum gonadotropin (Sigma, St. Louis, MO) and again 48 hr later with 5 IU of human chorionic gonadotropin (HCG, Sigma) before housing overnight with stud C57BL/6 males. Mated donor females are euthanized and fertilized eggs isolated from their oviducts. The fertilized eggs are released from cumulus masses by a 5-min incubation in 1 mg/ml hyaluronidase (Sigma) in M2 medium (Sigma). Eggs are rinsed free of debris in M2 medium and then incubated at 37° in 5% CO_2 in M16 medium (Sigma). All steps with the eggs outside of the incubator are performed in M2. Both M2 and M16 media must be supplemented with penicillin G (100 U/ml), streptomycin sulfate (100 μg/mL), and lactic acid (23 mM). Penicillin–streptomycin is from Life Technologies and lactic acid is from Sigma.

DNA Injection. Eggs are injected between 2 and 4 P.M. at 22–24 hr post-HCG injection. Standard manipulations are as published for surgical procedures[7] and for production of transgenic mice.[8,9] DNA solution is back-loaded by capillary

[7] J. R. Mann, *Methods Enzymol.* **225,** 782 (1993).

action into self-filling microcapillary tubes, pulled on a Sutter pipette forge to a terminal diameter of roughly 0.5 μm. DNA injection is controlled by a Nikon microspritz controller pressurized with N_2. The injection pipette is kept continuously under a 1 psi positive pressure and DNA is delivered by a 0.5- to 2-sec pulse of 20–30 psi. Male pronuclei are injected with roughly 10 pl of DNA. Injection volume is controlled subjectively by watching expansion of the injected pronuclei to roughly 150% of the original diameter. Following injection the eggs are incubated at 37° and 5% CO_2 for at least 1 hr.

Embryo Transfer. Foster mothers are obtained by mating C57BL/6 × DBA/2 F1 females with Swiss Webster vasectomized males (purchased from Taconic Farms, Germantown, NY). The foster mothers are housed continuously with the vasectomized males. Females with mating plugs on the day of injection are used as pseudopregnant egg recipients. Thirteen viable injected eggs are transferred to each oviduct of foster mothers. Pups are born 19 days posttransfer.

Detection of Transgenic Founder Mice. Tail samples (0.5 to 1.0 cm) are digested and tail DNA is isolated and purified using the DNeasy Tissue Kit from Qiagen, according to manufacturer's instructions. For complete digestion, tail samples must be digested overnight in the presence of protease K at 55°. In order to prevent evaporation during overnight digestion, 1.5 ml screw-cap microtubes should be used. DNA samples are subjected to amplification by PCR. Water and/or samples of genomic DNA from naïve C57B6 mice serve as negative controls; samples of pPEPCKASGα serve as a positive control.

Measurement of G-Protein Function in Vitro and in Vivo

Creating a transgenic mouse in which a loss-of-function mutation occurs in an inducible, tissue-specific manner provides a wealth of information on the role of the targeted G-protein in neonatal growth, development, and signaling.[10] Since only the targeted tissues will sustain the loss of function, it is easy to isolate the function of interest (e.g., hepatic signaling and metabolic regulation) and perform analyses at the biochemical and cellular levels. The fact that other, nontargeted tissues fail to express the transgene facilitates the analysis. One can examine the effects of the loss of a Gα subunit in liver, while the spleen, pancreas, and lung have no direct involvement. As an example, immunoblots of samples (0.1 mg protein/SDS–PAGE lane) from two target (fat, liver) and two nontarget (brain, spleen) tissues of a transgenic mouse harboring the pPEPCKASGαq transgene were prepared. The blots are stained with antibodies to one of the following G-protein subunits (G$_q\alpha$, G$_{i2}\alpha$ Gβ_2, or G$_s\alpha$) and the immunocomplexes made visible using the calf alkaline phosphatase second antibody. The blots were prepared from the transgenic mouse (T) as well as a littermate control (C) of the same sex (Fig. 2). Again the immunoblotting

[8] J. W. Gordon, *Methods Enzymol.* **225,** 747 (1993).

[9] J. R. Mann and A. P. McMahon, *Methods Enzymol.* **225,** 771 (1993).

[10] A. J. Morris and C. C. Malbon, *Physiol. Rev.* **79,** 1373 (1999).

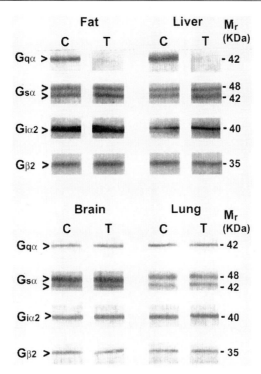

FIG. 2. The expression of Gα_q is suppressed in mice harboring the pPCK-ASGα_q transgene upon induction at birth. Crude membranes were prepared from epididymal and epoophoron white fat and liver, brain, and lung tissues obtained from 24-week-old control (C) and transgenic (T, carrying the pPCK-ASGα_q transgene) mice. Samples (15 μg of protein/lane) were subjected to SDS–polyacrylamide gel electrophoresis on a minigel apparatus and transferred to nitrocellulose for immunoblot analysis of various G protein α and β subunits. For immunoblotting, the sample loading was limited to 15 μg/lane, within the range established for linearity between sample loading and quantification of immunostaining (not shown). Quantification of the blots revealed no significant change in the G-protein subunits tested between control and transgenic mouse tissues, with the exception of the loss of Gα_q in liver and fat tissues. Expression of Gα_s was normal in liver and fat, although reduced (<15%) occasionally in fat, but not liver, of some transgenic mice (not shown). Scanning densitometry values for immunoblots of fat tissue from control and transgenic mice, respectively, were as follows: Gα_q, 0.26, 0.01; Gα_s, 0.92, 0.88; Gβ_2, 0.31, 0.33; and Gα_{i2}, 0.72, 0.75 arbitrary OD units. Scanning densitometry values for immunoblots of liver tissue samples from control and transgenic mice, respectively, were as follows: Gα_q, 0.51; 0.02; Gα_s, 0.68, 0.64; Gβ_2, 0.27, 0.25; and Gα_{i2}, 0.22, 0.25 arbitrary OD units. Scanning densitometry of immunoblots from brain and lung revealed no significant differences in the values obtained with tissues from transgenic as compared to control mice (not shown). The antibodies employed for staining of immunoblots for specific G-proteins subunits were as follows: E973 for Gα_q; CM112 for Gα_{i2}; CM129 for Gα_s; E976 for Gα_{11}; and CM162 for Gβ_2.

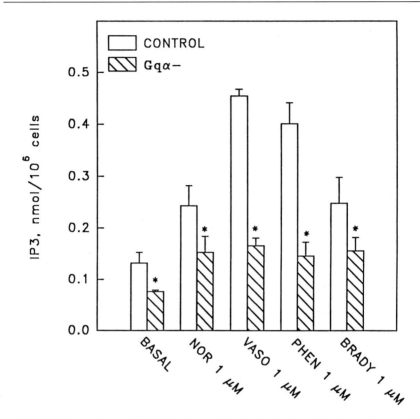

FIG. 3. Adipocytes from pPCK-ASGα$_q$ transgenic mice display loss of function with respect to activation of PLC and α$_1$-adrenergic regulation of lipolysis. White adipocytes were isolated from epididymal and epoophoron fat of transgenic mice and control littermates by collagenase digestion. Accumulation of IP$_3$ at 30 s following stimulation by 1 μM agonist (NOR, norepinephrine; VASO, vasopressin; PHEN, phenylephrine; BRADY, bradykinin) were measured in cells from mice 18–24 weeks of age. Intracellular IP$_3$ accumulation was measured by the mass assay employing the rabbit cerebellar IP$_3$-binding protein. DAG was assayed also using a DAG kinase assay followed by thin-layer chromatographic separation of radiolabeled phosphate generated by the reaction. The DAG mass was calculated from a standard curve using authentic DAG. The results from the DAG analysis were in close agreement with the changes in IP$_3$. The data presented are mean values ± S.E. from at least three separate experiments, each performed on separate occasions. An asterisk denotes statistical significance with $p \leq 0.05$ for the difference between the mean values for transgenic (Gα$_q$-deficient) as compared to control mice. The α$_1$-adrenergic control of lipolysis was nearly absent in adipocytes isolated from the transgenic mice in which Gα$_q$ was suppressed by the induction of antisense RNA (not shown).

demonstrates a virtual loss of the expression of $G_q\alpha$ in target, but not in nontarget tissues of only the transgenic mouse harboring the pPEPCKASGαq transgene, reflecting the activation of the PEPCK promoter at birth in the target tissues only. No changes in the expression of the other member of the $G_q\alpha$ family, $G_{11}\alpha$, were observed. Such nontarget tissues are ideal "controls" for measurements made of the lack of expression of the transgene (via RT-PCR), normal expression of the targeted protein (via immunoblotting), and as a measure of pleiotropic effects, if observed.

The mice harboring the pPEPCKASGαq transgene displayed an interesting phenotype characterized by increased weight, increased adipose tissue, and adiposity.[6] The target and nontarget cells of transgenic mice can be isolated and studied acutely for analysis of cell signaling. As an example, adipocytes were isolated from mice harboring the pPEPCKASGαq transgene ($G_q\alpha$-) as well as from same-sex littermates (CONTROL) and their activation of phospholipase Cβ examined (Fig. 3). The results demonstrate that in response to a 10-sec challenge with norepinephrine (NOR), vasopressin (VASO), phenylephrine (PHEN), or bradykinin (BRADY), adipocytes acutely prepared from control mice display an accumulation of inositol 3-phosphate (IP3). Adipocytes lacking $G_q\alpha$ isolated from the transgenic mice, in sharp contrast, display a poor response to these activators of phospholipase Cβ although harboring the wild-type level of $G_{11}\alpha$ expression. These data demonstrated unequivocally that $G_{11}\alpha$ and $G_q\alpha$ are not redundant with respect to mediating the activation of phospholipase Cβ.[6]

Expression of the pPEPCKASGi2α in transgenic mice, in contrast, led to the following alterations: (1) a robust transcriptional induction of the transgene at birth in liver, fat, and skeletal muscle, but not spleen, lung, and brain of the transgenic mice; (2) a suppression of the expression of $G_{i2}\alpha$ mRNA and protein in liver, fat, skeletal muscle, but no change in expression in the nontarget tissues; and (3) a markedly altered phenotype in which the mice were "runted" and displayed frank insulin resistance.[3,4] The analysis of loss-of-function mutants for other G protein subunits in these and other tissues requires the availability of suitable biochemical and cellular assays of G protein signaling in transgenic mice and/or cells derived from the tissues targeted by the transgene.

[24] Construction of Replication Defective Adenovirus That Expresses Mutant Gα$_s$ Q227L

By TARA ANN SANTORE and RAVI IYENGAR

Introduction

The adenovirus has been studied extensively as a model for molecular studies of mammalian DNA replication, transcription, and RNA processing.[1,2] Currently it is used as an expression vector for both gene delivery and protein expression in mammalian cells. There are several reasons that make the adenovirus a good choice as a tool for targeted protein expression. First, the DNA genome of 36,000 base pairs can be manipulated by recombinant DNA techniques. Second, the genome does not undergo rearrangement at a high rate. Third, the viral particle is fairly stable, and last, the virus replicates in permissive cells very efficiently and is able to produce 10^4 plaque-forming units (pfu) per infected cell, allowing the production of high-titer stocks.[3]

The first method developed to construct an adenovirus vector was developed by Graham and colleagues.[4] This method of construction utilizes two plasmids. The first plasmid contains the entire adenovirus genome with the deletion of the E1 gene. This deletion renders the virus replication defective. In order to replicate, the virus is propagated in a permissive cell line such as HEK-293 cells. This cell line is transformed by the adenovirus and expresses the E1 gene products. The E1a and E1b proteins are the proteins responsible for the packaging of the virus.[5] The second contains the gene to be inserted into the adenovirus and a region of homology to the adenovirus genome. This is so that *in vivo* homologous recombination can occur between the first and the second plasmid to yield an adenovirus with the inserted gene. To obtain an infectious adenovirus, these two plasmids are then transfected into low-passage HEK-293 cells. This approach is well established and its efficiency is low because of the inefficiency of the homologous recombination. This method has been improved with the use of Cre-mediated site-specific recombination between the two plasmids by the addition of a *loxP* site.[6] The plasmids are transfected into 293 cells expressing Cre recombinase to generate adenovirus vectors. This method yields a 30-fold increase in the recombination event when compared to the *in vivo* homologous recombination method.

[1] K. L. Berkner, *BioTechniques* **6**, 616 (1988).

[2] F. L. Graham and L. Prevec, *Methods Mol. Biol.* **7**, 109 (1991).

[3] F. L. Graham and A. J. Van der Eb, *Virology* **52**, 456 (1973).

[4] F. L. Graham, J. Rudy, and P. Brinkley, *EMBO J.* **7**, 2077 (1989).

[5] F. L. Graham, J. Smiley, W. C. Russell, and R. Nairn, *J. Gen. Virol.* **36**, 59 (1977).

[6] P. Ng, R. J. Parks, D. T. Cummings, C. M. Evelegh, U. Sankar, and F. L. Graham, *Human Gene Ther.* **10**, 2667 (1999).

The method that utilizes the Cre-mediated site-specific recombination has only recently been developed. At the time we constructed the adenovirus we chose a method that did not rely on a chance homologous recombination event *in vivo* of two plasmids because of the low efficiency and tedious preparation of the large plasmid that contains the adenovirus genome. The method we chose relied on DNA manipulation techniques. This method was developed in Robert Lefkowitz's laboratory.[7] The method utilizes one plasmid and the naked adenovirus genome instead of a plasmid containing the entire genome. Construction of the virus is achieved by the ligation of the gene of choice into a modified form of the plasmid pBluescript SK (+) (Stratagene, La Jolla, CA). This modified plasmid is a gift from Robert Lefkowitz's laboratory and is designated pSKAC. The modifications to this plasmid are the addition of the adenovirus terminal repeat (TR), the cytomegalovirus promoter (CMV), and the alfalfa mosaic virus translational enhancer (AMV). The Gα$_s^*$ is then ligated into this vector. The plasmid is then cut to yield a fragment that contains the adenovirus terminal repeat (TR), the cytomegalovirus promoter (CMV), and the alfalfa mosaic virus translational enhancer (AMV) followed by the inserted gene. This fragment will serve as the 5' end ("left end") of the adenovirus vector genome.

To obtain the 3' end of the genome ("right end"), extraction of the adenoviral DNA is carried out. The adenovirus used is the adenovirus type 5 (gift of Lefkowitz laboratory). In addition, the E1 gene is deleted and is replaced by the bovine growth hormone poly(A) site and a transcriptional terminator [bGH poly(A)]. A small viral preparation is made. Once the virus is obtained it is treated with SDS and proteinase K to remove the viral protein coat. The viral DNA is then digested to yield the "right end" of the virus genome. The next step is to ligate the "left end" with the "right end" to reconstitute a complete viral genome. The DNA is ligated overnight at a molar ratio of 3 : 1 ("left end" : "right end").

To produce infectious viral particles, the ligation reaction is then transfected into HEK-293 cells. It is important that low-passage cells be used because at higher passages expression of E1a and E1b is sometimes lost. The 293 cells are grown to 60% confluence are transfected with the ligation mixture using LipofectAMINE. Plaques begin to form 7 days later, and the entire plate is allowed to lyse to completion. The medium containing the lysed cells is then used to isolate individual viral clones.

Materials

Bacterial Cells

  Escherichia coli (One Shot, Invitrogen)

---

[7] M. H. Drazner, K. C. Peppe, S. Dyer, A. O. Grant, W. J. Koch, and R. J. Lefkowitz, *Clin Invest.* **99**, 288 (1997).

Growth Media

> LB: 10 g Tryptone, 5 g yeast extract, 10 g NaCl per liter
>
> SOC: 20 g Tryptone, 5 g yeast extract, 1.8 g glucose, 10 mM NaCl, 2.5 mM KCl, 10 mM MgCl$_2$, 10 mM MgSO$_4$ per liter
>
> YT plates: 16 g Bacto-tryptone, 10 g Bacto-yeast extract, 5 g NaCl, 15 g Bacto-agar

Mammalian Cells

> HEK-293 cells (Microbix, Canada)

Cell Culture Reagents

> Dulbecco's modified Eagle's medium, DMEM (Gibco-BRL, Gaithersburg, MD)
>
> Minimal essential medium, MEM (Gibco-BRL)
>
> Fetal bovine serum (Hyclone)
>
> LipofectAMINE (Gibco-BRL)
>
> Noble agar (Difco Labs)
>
> Falcon tissue culture plates, 3043, 12-well plates
>
> Falcon tissue culture plates, 3046, 6-well plates
>
> Falcon tissue culture plates, 3002, 60 mm plates
>
> Falcon tissue culture plates, 3025, 150 mm plate
>
> Falcon 50 ml tubes, 2098

Solutions

> dNTPs (Stratagene)
>
> Sequenase v 2.0
>
> STET: 500 mM EDTA, 1.0 M Tris pH 7.5, 8% sucrose, 5% Triton X-100
>
> Cesium chloride (cell culture grade) (Sigma, St. Louis, MO)
>
> CL-6B Sepharose Beads (Pharmacia, Piscataway, NJ)
>
> Wash Solution: 1× phosphate-buffered saline (PBS), 1% fetal calf serum (Gibco-BRL), 0.1% Tween 20, 0.1% dry milk, 0.1% ovalbumin
>
> T-PBS: 1× PBS, 0.05% Tween 20

Buffers

> PCR Buffer: 50 mM KCl, 10 mM Tris-HCl (pH 8.3), 1.0 mM MgCl$_2$
>
> 10× Ligation Buffer: 60 mM Tris-HCl, pH 7.5, 60 mM MgCl$_2$, 50 mM NaCl, 1 mg/ml bovine serum albumin, 70 mM 2-mercaptoethanol, 1 mM ATP, 20 mM DTT, 10 mM spermidine
>
> TAE: 40 mM Tris–acetate, 1 mM EDTA
>
> Resuspension Buffer: 10 mM Tris, pH 7.4, 5 mM MgCl$_2$

Virus Storage Buffer (VSB): 137 mM NaCl, 5 mM KCl, 10 mM Tris (pH 7.4), 1 mM MgCl$_2$

RIPA: 1× PBS, 1% Nonidet P-40 (NP-40), 0.5% sodium deoxycholate, 0.1% sodium dodecyl sulfate (SDS)

Sample Buffer: 1.0 ml glycerol, 0.5 ml 2-mercaptoethanol, 3.0 ml 10 SDS, 1.25 ml 1.0 M Tris, pH 6.7, 1–2 mg bromphenol blue

Transfer Buffer: 3.028 g/liter Trizma base, 14.415 g/liter glycine, 10% methanol

Blocking Buffer: 1.0 M glycine, 5% dry milk, 5% fetal calf serum, 1% ovalbumin, 1× PBS

Construction of Adenovirus Vector Expressing Gα_s^*

Tagging Gα_s^* with FLAG Epitope. The FLAG epitope is added to the N terminus of Gα_s^* (denotes the mutant form) by PCR amplification. Epitope tagging is carried out so that the protein expression can be monitored after virus infection. The template for the PCR reaction is pRc/CMV-α_s^{*}[8] and the primers used for the amplification are 5′-CCATCACACTGGCGGCCGCAAGCTTATGGACTACAAGG-ACGACGATGACAAGGGCTGCCTCGGGAACAGTAAGACCG-3′ (containing *Not*I and *Hin*dIII sites and the 8 amino acid FLAG epitope; Asp-Tyr-Lys-Asp-Asp-Asp-Asp-Lys) and 5′-CCTGCTCTAGAAGCTTTCAAATTTGGAACATCTA-AGC-3′ (containing a *Hin*dIII site). The reaction mixture (99 μl) contains PCR buffer, 200 μM of each dNTP, 1 μM of each primer, and 100 ng of template. The mixture is place in a Perkin-Elmer Cetus (Norwalk, CT) thermal cycler where the DNA is denatured for 2 min at 94°. After denaturing, 4 units/μl of *Taq* DNA polymerase (Promega, Madison, WI) is added to the reaction and the reaction mixture is topped with 50 μl of mineral oil to prevent evaporation. The reaction mixture is subjected to 25 cycles of denaturation (94°, 1 min), annealing (55°, 2 min), and extension (72°, 2 min). A final extension cycle of 7 min completes the reaction time. The reaction is electrophoresed using a 1% agarose gel containing TAE buffer. The desired band is then purified from the agarose gel by the Gene-Clean method (BIO 101, La Jolla, CA).

Insertion of Gα_s^* into Adenovirus Shuttle Vector

1. Ligate the fragment into the *Hin*dIII site of the vector pCR 2.1 (Invitrogen TA Cloning Kit). The reaction mixture contains 5 μl of sterile water, 1 μl of 10× ligation buffer, 2 μl of pCR 2.1 vector (25 ng/μl), 1 μl of PCR (polymerase chain reaction) product (50 ng), and 1 μl of T4 ligase. The reaction is then incubated at 12° for 4 hr and the ligated DNAs are transformed into competent *Escherichia coli*.

[8] J. H. Chen and R. Iyengar, *Science* **263,** 1278 (1994).

2. On ice, add 2 μl of 0.5M 2-mercaptoethanol to 50 μl of competent cells and incubate for 2 min on ice.
3. Add 1 μl of ligation reaction to the cells and incubate on ice for 30 min.
4. Heat shock the cells at 42° for 30 sec and then incubate on ice for 2 min.
5. Add 450 μl of SOC medium to the transformed bacteria and amplify at 37° for 1 hr.
6. Spread bacteria onto YT plates containing 100 μg/ml of ampicillin.
7. Prepare minipreparations of selected colonies.
8. Digest the DNA obtained from the minipreparations with *Hind*III.
9. Confirm the presence of the desired fragment by electrophoresis.
10. Prepare a culture of *E. coli* transformed by pCR2.1- $G\alpha_s^*$.
11. Digest the plasmid with *Kpn*I and *Xba*I to liberate the FLAG- $G\alpha_s^*$.
12. Run the reaction mixture on a 1% agarose gel in TAE purify the fragment using the Gene-Clean method of purification.
13. Digest the adenovirus shuttle vector, pSKAC, with *Kpn*I and *Xba*I and treat with calf intestinal phosphatase.
14. Ligate the FLAG- $G\alpha_s^*$ into the pSKAC vector using T4 ligase. This is illustrated in Fig. 1.
15. Transform the ligation mixture into bacteria and pick colonies.
16. Pick colonies and prepare minipreparation of the DNA. Carry out diagnostic digests to ensure that the insert is the correct size and in the correct orientation.
17. Colonies with the insert in the correct direction and with the correct size should be grown and frozen at −70° with glycerol.

Sequencing of Plasmid pSKAC-FLAG-$G\alpha_s^*$

1. For sequencing 3 ml cultures are grown. Microfuge 1.5 ml of the culture media at 13,000 rpm.
2. Resuspend the pellet in 100 μl of STET and lysozyme (2 μl of 50 mg/ml).
3. Mix well and boil for 2 min.
4. Spin tubes at 13,000 rpm for 15 min.
5. Using a sterile toothpick to remove the sticky pellet and add 80 μl of 2-propanol to the supernatant to precipitate the DNA.
6. Vortex the tubes and spin for 15 min.
7. Wash once with 200 μl of 70% (v/v) ethanol.
8. Air dry the pellet and resuspended in 60 μl TE (Tris-EDTA) plus 1 μl of 10 mg/ml RNase A and then incubate at 37° for 15 min.
9. To prepare the DNA for the sequencing reactions add 2 μl of 5N NaOH to 20 μl of each of the minipreparations. Bring the volume to 50 μl with TE and incubate at 37° for 1 hr to denature the DNA.
10. Following incubation, add 5 μl of 3.0 M sodium acetate and 110 μl of

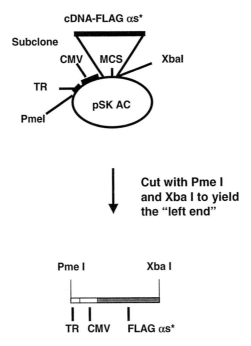

FIG. 1. Subcloning of the Gα_s into the adenoviral shuttle vector pSKAC and digestion to produce the required "left end." The FLAG epitope is added by PCR to the amino-terminal end of the Gα_s. The PCR product is first cloned into the TA cloning vector pCR 2.1 (Invitrogen). The fragment is then liberated by digestion with *Kpn*I and *Xba*I. The fragment is then ligated into the adenoviral shuttle vector pSKAC (generous gift from the Lefkowitz laboratory). The pSKAC-Gα_s is digested with *Pme*I and *Xba*I to yield the "right end" (5' end) of the adenoviral genome. This "right end" consists of the adenoviral terminal repeat, the CMV promoter, and the FLAG-tagged Gα_s.

ethanol to the tubes. Vortex and incubate on ice for 10 min. Spin the tubes for 10 min.

11. Wash the precipitated DNA once with 70% ethanol and air-dry the pellet.

12. Resuspend the pellet in 7 μl of water, 2 μl of 5× Reaction Buffer (component of Sequenase v 2.0), and 1 μl of primer (10 ng/μl) and incubate the DNA mixture at 65° for 2 min.

13. Cool the tubes slowly down to 35° on bench top and then place the tubes on ice.

14. While the DNA is cooling, aliquot the termination mixtures (ddGTP, ddATP, ddTTP, ddCTP) into tubes designated G, A, T, and C.

15. Label the DNA mixture by adding of 2 μl of diluted labeling mix, 1 μl of 0.1 M dithiothreitol (DTT), 0.5 μl [^{35}S]dATP, and 2 μl of diluted Sequenase v2.0. Mix the tubes and incubated at 37° for 10 min.

16. Following incubation, add 3.5 μl of labeled reaction to each of the termination tubes (G, A, T, and C) and mix and spin the tubes.
17. Incubate the reactions at 37° for 5 min and then add 4 μl of stop solution to each reaction tube.
18. Heat the samples at 75° for 2 min before loading them onto a 6% acrylamide gel. Once the samples are loaded, run the gel at 1500–1800 V at constant voltage until the dye front reaches the bottom of the gel.
19. Place the gel on Whatman (Clifton, NJ) paper and cover with plastic wrap.
20. Dry the gel on a gel dryer for 90–120 min at 80°.
21. Expose the gel to X-ray film in a cassette, keep at room temperature for 12–16 hr, and then develop.
22. Read the bands on the film to determine if the sequence is correct.

Rescue of Recombinant Adenovirus Vector Containing FLAG-Gα_s^*

1. Once the sequence is verified, digest the plasmid pSKAC-FLAG-Gα_s^* overnight with *Pme*I and *Xba*I to liberate the FLAG-Gα_s^*.
2. The fragment that is liberated is the "left end" of the adenovirus. To isolate this run the digest on an agarose gel and gel purified the fragment using Gene Clean.
3. Treat the purified fragment with calf intestinal phosphatase. This fragment contains the adenovirus map units 0.0–1.3, the CMV promoter, the AMV translational enhancer, and the FLAG-tagged Gα_s^*.
4. Ligate the "left end" fragment to the "right end" fragment using T4 ligase (the remaining portion of the adenovirus type 5 genome with a E3 deletion), overnight at 16° (see Fig. 2). This is done at a 3 : 1 molar ratio of fragment to vector and yields the adenovirus illustrated in Fig. 3.
5. Transfect the ligation mixture onto a single 60 mm dish (Falcon, 3002) of HEK 293 cells at a confluence of approximately 70%. It is important that Falcon plates be used when culturing 293 cells because the cells adhere best to these plates. LipofectAMINE should be used to transfect the DNA.
6. Wash the plate 2 times with DMEM medium without serum. Mix the ligation reaction and 12 μl of LipofectAMINE and incubate for 5 min at room temperature.
7. Add 1 ml of medium without serum to this mixture
8. Remove the medium from the 293 cells. Add the DNA/LipofectAMINE mixture to the plate of cells as shown in Fig. 4.
9. Transfer the plate to a 5% (v/v) CO_2/37° incubator and incubate for 4 hr. After this time check the health of the cells and add 2 ml of medium without serum to the cells incubated overnight.
10. The next day remove the supernatant and replace it with 5 ml of DMEM with 2% fetal bovine serum (FBS).

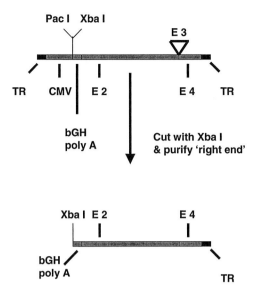

FIG. 2. Digestion of the type 5 adenovirus genome to yield the "right" end. The entire adenovirus genome is digested with *Xba*I to remove the 5′ portion of the genome. The 5′ terminal repeat and the CMV promoter are removed leaving an intact poly(A) tail followed by the remainder of the type 5 adenovirus genome with a E3 deletion. The "left end" (3′ end) is then gel purified.

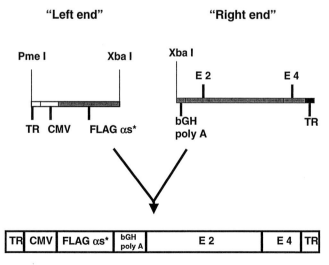

FIG. 3. Ligation of the "left end" to the "right end" of the adenovirus genome. The "left end" and the "right end" are ligated overnight with T_4 ligase. A molar ratio of 3 : 1 was used ("left end": "right end").

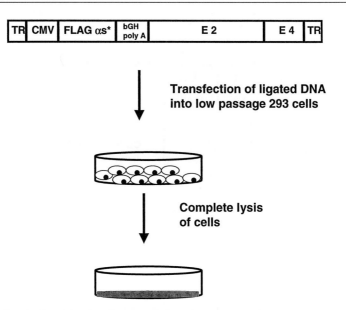

FIG. 4. Transfection of the adenovirus vector DNA and cell lysis. The ligation reaction is transfected into low-passage 293 cells. Lipofectamine is used to carry out the transfection. Plaques begin to form 7 days after transfection. The plates are allowed to completely lyse before removing the medium that contains the vector. This medium is then diluted and used to isolate individual vector clones by plaque formation.

11. Plaques will began to appear 7 days later.
12. The dishes should be incubated until the cytopathic effect (CPE, cell lysing) is completed.

Isolation of Individual Viral Clones

1. Individual viral clones are isolated by plaque purification. For this procedure, 293 cells are seeded into 6-well plates one day prior to infection (Falcon, 3046). The cells are ready for infection at 90% confluence.
2. Prepare the agar overlay before the actual infection of the cells. For this, 10 ml of sterile 1.3% Noble agar in water is melted and place in a 50° water bath. In addition to the agar, 10 ml of 2× MEM containing 4% FBS and 4% penicillin–streptomycin is warmed to room temperature.
3. Dilutions of the virus should be prepared in the range of 10^4 to 10^{-8}. Aspirate the medium off of the cells and add 200 μl of medium containing virus.
4. Gently pipette the medium onto the center of the well. For a control, add medium to only one of the wells. Return the plate back to the incubator for 5 min.
5. Following the incubation time, mix the melted agar with the warmed medium

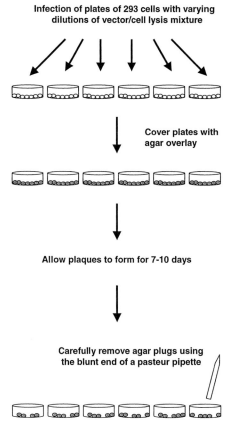

FIG. 5. Isolation of vector clones by plaque formation. For this step the medium from the lysed cells is diluted and used to transduce 293 cells. Following transduction, the cells are covered with an agar overlay (0.5% agar noble). The agar is allowed to solidify and the plates are incubated for 7–10 days, when plaques began to form. Well-isolated plaques are removed using the blunt end of a sterile Pasteur pipette. The agar plug is placed in 1× PBS to allow elution from the agar. The plugs are stored at −70°.

Immediately add 3 ml onto each well. Carefully drip the mixture down the side of the well so that the monolayer of cells is not disrupted. Return the plate back to the incubator.

6. Plaques should appear within 8 days. Mark the plaques on the underside of the plate by circling the clearings with a Sharpie marker.

7. To remove the plaques, stab the agar with the blunt end of a sterile Pasteur pipette as illustrated in Fig. 5. Then transfer the plugs to sterile cryovials containing phosphate-buffered saline (PBS).

8. Store the plugs at −80°.

9. To expand the clones, freeze–thaw the plugs three times.
10. Add the mixture is then to a 80% confluent 60 mm plate of 293 cells in DMEM with 2% FBS.
11. Scrape the plate and remove cells and medium when the CPE is completed.
12. Prepare large-scale preparations of the virus.

Purification and Titering of Adenovirus Vector

Purification

1. For the initial expansion plate 293 cells into 10 150-mm plates (Falcon, 3025).
2. Infect the 293 cells with the different clones isolated and then incubate the plates for 48 hr until the cells are completely lysed.
3. Collect both the medium and cells for each clone and stored at $-70°$ until needed.
4. Small preparations should be done of the vectors first to expand the virus. Divide the supernatant from the initial expansion over 5 150-mm plates.
5. For the large-scale preparations of subsequent virus vector, set up 120–150 150-mm plates and grow to 80% confluence. The 293 cells are infected at a multiplicity of infection (MOI) of 3 infectious virus particles per cell.
6. Feed the cells with DMEM containing 2% FBS.
7. After 36–48 hr of infection, harvest the cells by gentle scraping and then pipette the medium and cells into 250-ml centrifuge bottles.
8. Collect the cells by centrifuging the cells for 10 min at 2000 rpm in a GSA rotor (500g) at $4°$.
9. Resuspend the cell pellet in Resuspension Buffer and Freeze–thaw the cells three times.
10. Add RNase (100 μg/ml) to the tube and incubated at room temperature for 15 min.
11. Transfer the lysate to a new tube and spin in a SS 34 rotor for 5 min at 5000g at $4°$.
12. Transfer the supernatant to a new tube and avoid any removal of pellet debris.
13. Add 1.5 g of solid cesium chloride for every 10 ml of supernatant and dissolve gently at room temperature.
14. Prepare the cesium chloride gradient. For this two different mixtures of cesium chloride are prepared in sterile Virus Storage Buffer (VSB). Gradient solution I contains 39 g of solid CsCl brought to a total volume of 100 ml with VSB (1.3 g/ml). Filter the solution (to sterilize) into a 0.22 μm filter bottle. Gradient solution II contains 53 g of solid CsCl brought to a total volume of 100 ml with VSB (1.4 g/ml). Filter the solution (to sterilize) into a 0.22 μm filter bottle.

15. Prepare the gradient in an ultraclear disposable tube for the Sorvall TH 64 rotor. Divide the 10 ml of crude virus/CsCl mixture between two tubes.

16. Underlay this mixture with the 1.3 g/ml CsCl solution using a 9-inch Pasteur pipette. Fill the tube almost to the top (0.5 inch below the lip of the tube).

17. Then underlay the 1.4 g/ml CsCl solution under the 1.3 g/ml solution and mark the interface.

18. Centrifuge the tubes for 2.5 hr at 30,000 rpm in a Sorvall (Newton, CT) TH 64 rotor.

19. While this spins, prepare the spin columns for the next step as follows. The columns are prepared in 3 ml syringes. The Sepharose CL-6B beads are washed 7 times in the VSB to remove all residual ethanol and aliquotted into sterile 50 ml conical tubes. The beads are then stored at 4°. The tubes should contain 45 ml of settled beads in a total volume of 50ml VSB. The spin columns are prepared by placing 3 ml syringes with their plungers removed into sterile 50 ml tubes (Falcon, 2098). With sterile instruments a small amount of sterile glass wool is placed in the bottom of the syringe and packed down with the end of a sterile Pasteur pipette. The syringes are then filled with freshly vortexed Sepharose CL-6B beads and spun at 1000 rpm for 2 min in tabletop centrifuge (Beckman Instruments, Inc., Fullerton, CA). This yields a packed bed of about 2 ml of beads. One ml of VSB is added and the syringe is spun at 1000 rpm for 2 min. The recovery of the 1 ml of VSB in the bottom of the tubes is to check to ensure the columns are not blocked and that the beads do not leak through the glass wool.

20. The virus band should be recovered in a volume no greater than 1 ml. The band containing the infectious virus is formed at the 1.3–1.4 g/ml interphase. Remove the band with an 18-gauge needle connected to a 3 ml syringe.

21. Layer the virus preparation on top of a spin column that contains a 2 ml bed of Sepharose CL-6B beads and spin at 1000 rpm for 2 min. Collect the virus at the bottom of the tube and load the virus onto a second spin column in a fresh tube and spin at 1000 rpm for 1 min. This is done to ensure that all of the cesium chloride is removed. The volume of the final flow-through is then measured and is used to calculate the yield of the preparation later.

Titering Virus Stock

The titer of the stock is determined by absorbance at 260 nm (pfu/ml $= A_{260} \times$ dilution $\times 10^{10}$). A 1 : 50 dilution (20 μl) is the standard protocol used. The stock is adjusted to 1×10^{11} plaque-forming units (pfu)/ml in $1\times$ PBS/10% sucrose. The stock is aliquoted into sterile cryovials and stored at $-80°$. Adenovirus titers are also checked using plaque titration (limited dilutions) on HEK 293 cells to ensure accuracy. Each aliquot is thawed a maximum of two times.

Expression and Functional Characterization of ADV-Gα$_s^*$

Detection of FLAG Epitope by Western Blot Analysis. Western blot analysis is carried out to determine if the FLAG epitope is detectable by the anti-FLAG antibody (Kodak, Rochester, NY, and then sold to Sigma).

1. Grow plates of HEK 293 cells to 80% confluence. Then infect the plates at multiplicity of infections (MOIs) of 1, 5, and 10 with the ADV-G$_\alpha$s*.
2. Harvest the cells 36 hr after infection. Remove the medium and rinse the plates with 1× PBS at room temperature. Subsequent steps should be carried out at 4° using cold buffers.
3. Use RIPA buffer as the lysis buffer and add the following protease inhibitors immediately prior to use: 100 μg/ml phenylmethylsulfonyl fluoride (PMSF), 30 μl/ml aprotinin (Sigma), 1 mM sodium orthovanadate, 0.4 μg/ml leupeptin. Add 0.6 ml to each 100 mm plate.
4. Scrape the plates and place the lysate in a 1.5 ml microfuge tube. Add an additional 0.3 ml of lysis buffer to the plates, scrape, and place in the corresponding microfuge tube.
5. Incubate the tubes on ice for 30 min and then pass the cell lysate through a 21-gauge needle to shear the DNA.
6. Spin the samples at 13,000 rpm in a microcentrifuge for 20 min at 4°.
7. Remove the supernatant and place it in clean tubes.
8. Keep the samples on ice while protein concentrations are determined. Using the BCA protein assay reagent (Pierce), add 100 μl of sample to 2 ml of BCA working solution (50 parts Reagent A:1 part Reagent B). Use BSA standards (provided with the assay kit) of 0, 2, 5, 10, 20, and 50 μg alongside the samples for preparation of a standared curve of protein concentration. Bring the volume of the controls to 100 μl with lysis buffer and add 2 ml of the working solution to each of the standards.
9. Incubate the samples and standards at 37° for 30 min. Calculate the protein concentrations by measuring the absorbance at 562 nm using BSA as the standards.
10. Add fifty μg of each sample to 2× sample buffer as well as to 10 μl rainbow-colored protein molecular weight markers (RPN 756, Amersham, Life Science) and boil for 3 min before loading into a acrylamide gel. For this use a 10% separating gel (acrylamide:bisacrylamide, 30:0.8), 5% stacking gel with 0.1% SDS in either separating buffer (0.375 M Tris-Cl, pH 8.8) or stacking buffer (0.125 M Tris-Cl, pH 6.8).
11. Run the gels at approximately 100 V in the minigel apparatus (Hoefer, Pharmacia, Piscataway, NJ) for about 2 hr or until the dye front runs off of the gel.
12. Soaked a piece of nitrocellulose for each gel in transfer buffer for 30 min

prior to transfer. Soak the acrylamide gel for 15 min in transfer buffer prior to the transfer.

13. Transfer the proteins in a wet transfer apparatus (Hoeffer) for 1 hr at 70–80 V.
14. Wash the nitrocellulose membranes with fresh transfer buffer for 10 min.
15. Place the membranes in Blocking Buffer overnight at 4° on a platform shaker set on low.
16. Place the membranes in fresh blocking buffer with primary antibody. The primary antibody (anti-FLAG M2 monoclonal antibody, Sigma) is diluted at the concentration of 10 μg/ml. To minimize the amount of primary antibody used, place the membranes and wash solution containing antibody in a Seal-a-Meal bags on a platform shaker set on low for 4 hr at room temperature.
17. After incubation of the primary antibody, remove the membranes and wash the membranes (vigorous shaking) 3× for 5 min in Wash Solution at room temperature.
18. Wash the membranes for 15 min in T-PBS.
19. Place the membranes in secondary antibody (goat anti-mouse alkaline phosphatase, Roche) that is diluted in wash solution at a concentration of 1 : 7500. Incubate the blots at room temperature for 1 hr.
20. After incubation with secondary antibody, wash the membranes 3× for 15 min with T-PBS at room temperature.
21. Use enhanced chemiluminescence (ECL, Amersham) as described by the manufacturer's protocol for detection of the protein bands. The results are shown in Fig. 6.

Functional Characterization of Adenovirus Expressing Gα_s by Measurement of cAMP

The vector clones are then checked for their ability to raise intracellular cAMP levels. This is done in S49 cyc$^-$ cells. The S49 cyc$^-$ cell line is a lymphoma cell line that lacks G$_s$ but does contain adenylyl cyclase. These cells are transduced with the ADV-Gα_s^* and cAMP levels are measured.

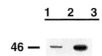

FIG. 6. Detection of the FLAG-tagged Gα_s^* protein by Western blotting. HEK-293 cells are infected with clones 1.2, 2.2, or no virus. Thirty-six hours after infection the cells are prepared for immunoblot analysis. Membranes are incubated with anti-FLAG antibodies. Detection is carried out by chemiluminescence (ECL).

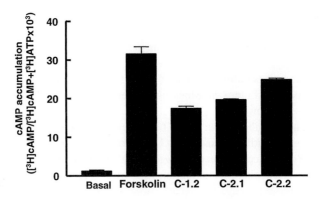

FIG. 7. Measurement of cAMP accumulation in S49 cyc⁻ cells after transduction by ADV-Gα_s. Cells were counted and cultured in 12-well dishes 10 hr prior to transduction by the vector. The cells were transduced by each of the clones at a MOI of 10, 12 hr prior to labeling. The cells were labeled with [³H]adenine(2 μCi/ml)(ICN)(2 μl) for 24 hr. The cells were incubated for 1 hr with the phosphodiesterase inhibitor Ro-20174 before measuring cAMP accumulation. The accumulation of cAMP was measured calculated by the ratio of [³H]cAMP/([³H]ATP + [³H]cAMP) ($\times 10^3$). Values are a mean of triplicate determination. Coefficient of variance was less than 10%.

1. Grow small-scale preparations of virus of the three clones (designated C-1.2, C-2.1, and C-2.2) and the preparations are titered by measuring the optical density and by plaque assays.
2. Plate cells in triplicate at a density of 5×10^4 cells per flask.
3. Infect the cells with the virus 12 hr prior to labeling. The S49 cyc⁻ cells are infected with each clone at MOI of 10. Check three of the clones isolated for their ability to raise cAMP levels.
4. Label the cells with [³H]adenine (2 μCi/ml)(ICN)(2 μl) for 24 hr. After incubation with label, wash the cells 2× with DMEM containing 20 mM HEPES.
5. Incubate the cells for 1 hr with the phosphodiesterase inhibitor Ro-20174 (0.3 mM; Calbiochem, La Jolla, CA) in DMEM containing 20 mM HEPES (3.5 ml) at 37°. Stimulate control cells with forskolin (10 μM) and incubate on ice for 30 min.
6. Remove the medium and add 1 ml of cold 5% TCA (trichloroacitic acid) containing 0.5 mM ATP and 0.5 mM cAMP to the cells.
7. Incubate the cells on ice for 30 min.
8. Vortex samples and add each sample to Dowex AG-50 columns.
9. Add 3 ml of water to the Dowex (placed on scintillation vials) and collect the elution (8 ml of scintillation fluid is added to the vials).
10. Place the Dowex columns on top of alumina columns and add 10 ml of water.

11. Remove the Dowex columns and place the alumina columns on a second set of scintillation vials.

12. Elute the cAMP off the alumina with the addition of 6 ml of 0.1 M imidazole, pH 7.5.

13. Add 12 ml of scintillation fluid to the second set of vials.

14. Accumulation of cAMP is measured by scintillation counting and is carried out in triplicates. Calculate cAMP accumulation by the ratio of $[^3H]cAMP/([^3H]ATP + [^3H]cAMP) (\times 10^3)$. The results of the assay are shown in Fig. 7. Of the three clones tested, all were able to increase intracellular cAMP levels demonstrating that the $Gα_s^*$ expression by the vector is functional. The negative control contains no vector and the positive control contains the addition of forskolin (an activator of adenylyl cyclase) to raise cAMP levels.

15. Columns are regenerated prior to use by the following method. The Dowex AG-50 columns are washed with 10 ml of distilled H_2O, 2 N NaOH, distilled H_2O, 2 N HCl, distilled H_2O, distilled H_2O, distilled H_2O. Alumina columns are washed with 10 ml of distilled H_2O, 0.1 M imidazole, pH 7.5, distilled H_2O right before use.

[25] Expression of Adenovirus-Directed Expression of Activated Gα_s in Rat Hippocampal Slices

By George P. Brown and Ravi Iyengar

Introduction

G-protein-coupled receptors (GPCRs), the largest family of cell surface receptors, transfer their signal to intracellular second messengers via heterotrimeric G proteins. Heteromeric G protein $α$ subunits (Gαs) determine the signaling specificity of the G protein heterotrimer. Gαs contain a distinct GPCR interaction domain, GTP binding domain, effector interaction domain, and an intrinsic GTPase activity. There are 20 genes that code for 22 distinct G proteins which are grouped into four families according to the second messengers that they activate. The cAMP and Ca^{2+} signaling pathways are among the downstream pathways that are regulated by Gα's. The G_s and G_i pathways regulate the cAMP pathway by directly activating or inhibiting adenylyl cyclase (AC), respectively. Gα proteins in the G_q family of G proteins increase intracellular calcium by activating PLC-$β$, producing IP_3.[1]

[1] A. G. Gilman, *Ann. Rev. Biochem.* **56,** 615 (1987).

The cAMP and Ca^{2+} signaling pathways are two of the most prominent signaling pathways in neurons, and thus, in modulating behavior. Activation of either the cAMP or Ca^{2+} pathways leads to the activation of a number of signaling molecules including PKA, calcium/calmodulin-dependent protein kinases (CAMKs), MAPK-1,2, and CREB.[2,3] Activation of MAPK-1,2 and CREB lead to the transcription of novel genes in response to a given stiumulus. Rapid transcriptional response to extracellular stimuli is thought to be one of the underlying factors in synaptic plasticity.[4,5] Synaptic plasticity describes the ability of the post-mitotic neuron to adapt to local changes in its environment and is thought to be the central underlying event in changes in animal behavior.

In order to understand the molecules important in modulating any animal behavior it is first necessary to understand the neuroanatomical locus responsible for the processing of the behavior. One of the most studied animal behaviors is long-term spatial memory. Surgical, behavioral, and electrophysiological experiments have identified the subcortical area called the hippocampus as being critically involved in the processing of long-term spatial memory in humans and rodents.[6] To date, efforts aimed at identifying the molecules underlying long-term spatial memory have mainly focused on gene deletion experiments by homologous recombination (knockout) or transgenic expression (under the control of a hippocampal-specific promoter) experiments engineered to create loss-of-function mutants in mice. These studies have revealed that a number of signaling components in the cAMP and Ca^{2+} signaling cascades in hippocampal neurons are necessary for long-term spatial memory formation in mice including PKA,[7,8] CaMKII,[9-11] and CREB.[12]

However, these types of genetic approach are not without drawbacks.[13,14] Namely, the resulting mutant mice constitutively express the transgene or lack the targeted gene their entire lives. This can give rise to compensatory changes

---

[2] T. R. Soderling, *Trends Biochem Sci.* **24**, 232 (1999).

[3] A. J. Shaywitz and M. E. Greenberg, *Ann. Rev. Biochem.* **68**, 821 (1999).

[4] S. Finkbeiner, S. F. Tavazoie, A. Maloratsky, K. M. Jacobs, K. M. Harris, and M. E. Greenberg, *Neuron* **19**, 1031 (1997).

[5] J. Curtis and S. Finkbeiner, *J. Neurosci. Res.* **58**, 88 (1999).

[6] H. Eichenbaum, *Ann. Rev. Psych.* **48**, 547 (1997).

[7] T. Abel, P. V. Nguyen, M. Barad, T. A. Deuel, E. R. Kandel, and R. Bourtchouladze, *Cell* **88**, 615 (1997).

[8] E. P. Brandon, M. Khuo, Y. Y. Huang, M. Qi, K. A. Gerhold, K. A. Burton, E. R. Kandel., G. S. McKnight, and R. L. Idzerda, *Proc. Natl. Acad. Sci. U.S.A.* **92**, 8851 (1995).

[9] M. Mayford, M. E. Bach, Y. Y. Huang, L. Wang, R. D. Hawkins, and E. R. Kandel, *Science* **274**, 1678 (1996).

[10] A. J. Silva, R. Paylor, J. M. Wehner, and S. Tonegawa, *Science* **257**, 206 (1992).

[11] K. P. Giese, N. B. Fedorov, R. K. Filipkowski, and A. J. Silva, *Science* **279**, 870 (1998).

[12] R. Bourtchuladze, B. Frenguelli, J. Blendy, D. Cioffi, G. Schutz, and A. J. Silva, *Cell* **79**, 59 (1994).

[13] M. R. Picciotto and K. Wickman, *Physiol. Rev.* **78**, 1131 (1998).

[14] E. M. Pich and M. P. Epping-Jordan, *Ann. Med.* **30**, 390 (1998).

that may directly or indirectly affect the observed behavioral changes. Additionally, in knockout experiments, the targeted gene may be critical for survival, thus creating a lethal phenotype upon targeted deletion. Lastly, these approaches require a significant investment of time to generate the desired mutant mice.

Advances in recombinant viral gene transfer technologies offer the potential of an alternative method for analyzing the roles of specific molecules in behavior. Unlike other genetic approaches, viral transfer generates region- and time-specific delivery of the desired gene, thus obviating much concern over secondary compensation. For viral gene transfer to serve as an adequate alternative to existing genetic approaches, the targeted gene must: (1) be functionally expressed in the target tissue, (2) modify the behavior in question, and (3) have a cellular expression pattern consistent with the neuroanatomical locus for the behavior tested.

In order to determine the adequacy of this approach for studying the molecular basis of various behaviors, we have tested the ability of the constitutively active, GTPase-deficient mutant of Gα_s (Gα_s^*) to modify long-term spatial memory in mice. Since many genetic experiments in mice have identified genes downstream of cAMP that are necessary for normal long-term spatial memory, we hypothesized that hippocampal expression of Gα_s^* should result in a gain of function.

Hippocampal Transduction of Adv-Gα_s^* Resulting in Increase in Basal PKA Activity

A type I recombinant, replication-deficient adenovirus capable of expressing either β-galactosidase (β-Gal) or N-terminal FLAG-tagged Gα_s^* is constructed and grown up to a titer of 1.11×10^{11} pfu (plague-forming unit)/ml in HEK-293 cells as described elsewhere in this volume (XX). Mice are anesthetized with pentobarbital (80 mg/kg i.p.) and placed in a small-animal stereotaxic surgical frame. Bilateral hippocampal 1 μl injections of Adv-Gα_s^* (2.2×10^{10} pfu/ml) are administered over 15 min targeting area CA1 using the following coordinates: A/P (anterior–posterior) +2.0 mm; M/L (medial–lateral) ±1.6 mm; and D/V (dorsal–ventral) −1.37 mm with respect to the Bregma suture, using a microprocessor-controlled infusion pump. Following the 15-min injection time an additional 15 min are allowed to elapse to minimize backflow of the virus via the injection site.

The animals are allowed to recover from surgery and 30 hr later were sacrificed. The hippocampi are dissected and homogenized in lysis buffer (50 mM Tris-HCl pH 7.5, 2 mM isobutylmethylxanthine, 40 mM dithiothreitol (DTT), 20 mM NaF, and 5 mM EDTA). PKA activity is determined by the ability of 10 μg lysate to phosphorylate the PKA-specific peptide Kemptide (50 μM in 100 μM ATP, 10 μM MgCl$_2$, 250 μg/ml BSA) in the presence and absence of 10 μM cAMP and 1 μM PKI (6–22) amide. Hippocampi from Gα_s^* treated animals display a significantly higher basal PKA activity ($p < 0.05$) with no change in total PKA activity as compared to controls as soon as 30 hr postinfection, indicating that the Gα_s^* is being efficiently and functionally expressed by the recombinant adenovirus (Fig. 1).

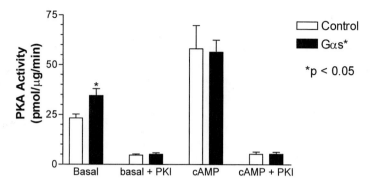

FIG. 1. Adenoviral expression of $G\alpha_s^*$ increases basal PKA activity. PKA activity was measured in extracts prepared from hippocampi of ADV-$G\alpha_s^*$-treated mice 30 hr after injection and control mice. Ten micrograms of total protein was used in each assay tube, and the phosphorylation of the Kemptide substrate was determined in the presence or absence of cAMP to measure total PKA activity the specific PKA inhibitor PKI for specificity. Hippocampi from $G\alpha_s^*$ treated mice have significantly greater basal PKA activity as compared to controls (Student's t-test; $p < 0.05$) while having no difference in the total (cAMP stimulated) activity.

Expression of $G\alpha_s^*$ Resulting in Increase in Long-Term Spatial Memory

To determine whether adenoviral gene transfer is capable of altering mouse behavior, groups of mice are injected with Adv-$G\alpha_s^*$ (2.2×10^7 pfu; $n = 9$) or Adv-β-Gal (2.2×10^7 pfu; $n = 9$) as described above, and their performance in the hidden platform version of the Morris water maze (MWM) is assessed.[15,16] The maze consists of a 1.3 m pool with a water temperature of 26°, in a room with fixed distal cues >1 m from the edge of the pool. Between 28 and 30 hours postinjection, mice were habituated to the task by being allowed to find the platform three separate times following random release from the middle of the pool. Two hours after habituation, mice began training. Training consisted of eight daily trials divided into two blocks of four, in which mice were required to swim to a circular platform 11 cm in diameter hidden below the water when randomly released from four drop points.

Figure 2A shows that both $G\alpha_s^*$ and β-Gal treated mice improve in their ability to locate the hidden platform through repetitive trials. Although $G\alpha_s^*$ treated mice perform better overall as compared to β-Gal treated mice ($p < 0.05$), there is not a significant difference between $G\alpha_s^*$ and β-Gal treated mice in any one trial block. Measuring the total distance traversed in search of the platform, however, is only one parameter for measuring spatial memory in the water maze. Therefore, we also

---

[15] R. Morris, *J. Neurosci. Methods* **11**, 47 (1984).
[16] H. P. Lipp and D. P. Wolfer, *Curr. Opin. Neurobiol.* **8**, 272 (1998).

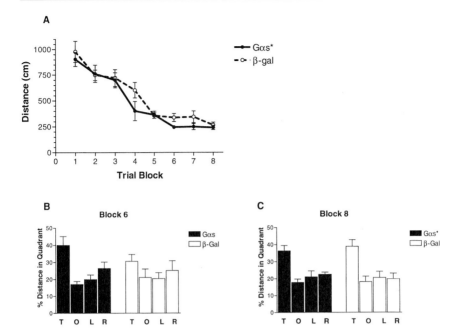

FIG. 2. Hippocampal $G\alpha_s^*$ expression improves spatial memory. Mice were treated with either ADV-$G\alpha_s^*$ or ADV-β-Gal and tested in the Morris water maze to assess spatial memory. (A) ADV- $G\alpha_s^*$ ($n = 9$) or ADV-β-Gal ($n = 9$) treated mice were trained with 8 trials a day in 2 blocks of 4 trials each. The average distance traveled to reach the submerged platform is plotted vs trial block. ANOVA with repeated measures showed that the overall performance of ADV-$G\alpha_s^*$ treated animals was significantly better than that of ADV-β-Gal treated control animals ($F[1, 7] = 3.93$, $p < 0.05$). However, paired comparisons between the 2 groups for each of the 8 trial blocks revealed no significant differences. (B) Results of a retention trial conducted after 6 trial blocks. An ANOVA showed that $G\alpha_s^*$-treated mice search selectively in the quadrant in which the platform had been placed during training following the 6th trial block ($F[3, 32] = 8.136, p < 0.001$), and β-Gal treated controls do not ($F[3, 32] = 1.003, p = 0.40$). Newman–Keuls multiple comparisons between the distance traveled by $G\alpha_s^*$-treated mice in the training quadrant as opposed to the other quadrant confirmed significant differences (T vs O, $p < 0.001$; T vs L, $p < 0.01$; T vs R, $p < 0.05$). (C) Following the 8th training block β-Gal treated control search selectively in the quadrant in which the target platform had been placed ($F[3, 24] = 7.603, p < 0.01$) as do $G\alpha_s^*$ treated mice ($F[3, 24] = 9.294, p < 0.001$). Newman–Keuls multiple comparison test between the distance traveled by the mice in the target qudrant as opposed to the target quadrant confirmed significant differences for both β-Gal treated mice (T vs O, $p < 0.01$; T vs L, $p < 0.01$; T vs R, $p < 0.01$) and $G\alpha_s^*$ treated mice (T vs O, $p < 0.001$; T vs L, $p < 0.001$; T vs R, $p < 0.001$).

examined the search strategy of treated mice in 60 sec retention trials in which the platform was removed from the pool.

$G\alpha_s^*$ treated mice demonstrated a statistically significantly better ability to retain spatial information learned in the trial blocks as determined by the retention trials (Fig. 2B). Retention trials following the sixth trial block reveal that $G\alpha_s^*$

treated animals search selectively in the target quadrant of the pool ($p < 0.001$) spending 40% of the total search distance in the target quadrant. β-Gal treated control animals, however, randomly search for the location of the hidden platform ($p = 0.40$). Following the eighth block of training, however, β-Gal treated animals did employ a selective search strategy ($p < 0.01$) and $G\alpha_s^*$ treated animals again displayed selective searching ($p < 0.001$) when probed in retention trials (Fig. 2C). Thus, elevations in hippocampal PKA activity appear to increase long-term spatial memory.

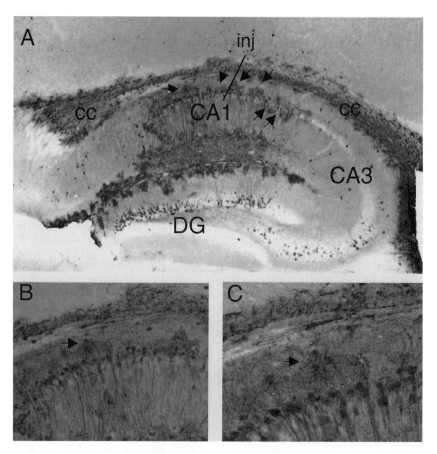

FIG. 3. Localization of adenoviral gene expression. (A) Magnification: ×40. Composite micrograph displaying the β-Gal immunoreactivity four days following injection. (B) Magnification (×100) of the boxed image in (A) detailing the expression of β-Gal in CA1 pyramidal neurons. (C) Magnification ×200 of the boxed area in B showing the transduction of nonneuronal cells. CA1, Area CA1; CA3, area CA3; DG, dentate gyrus; cc, corpus callosum; inj, plane of injection.

Immunocytochemical Localization

In order to determine the cellular localization of the adenovirally expressed genes we examined the expression pattern of β-Gal using immunocytochemistry with diaminobenzidine (DAB) as chromogen. Although we had put a FLAG tag on the N terminus of $G\alpha_s^*$, and we know that this protein is expressed and is functional, we were unable to detect the activated $G\alpha_s^*$ expression by immunohistochemistry using anti-FLAG antibodies. Hence, the cellular expression patterns were determined for β-Gal. Following testing in the MWM, mice were sacrificed, their brains were dissected and immersion-fixed in 4% paraformaldehyde for 16–18 hr, and 40 μm sagittal sections were taken. Four days after treatment, β-Gal is expressed in a mixture of neuronal and nonneuronal cells in the hippocampus. As expected from the injection coordinates, the primary area of viral transduction is in area CA1 of the hippocampus. Expression of β-Gal, however, is not limited to area CA1. Significant levels of β-Gal expression are also present in neurons in the dentate gyrus, area CA3, as well as glial cells surrounding the passing fibers of the corpus callosum and the hippocampal fissure. Within area CA1 we observe high-level expression of β-Gal in CA1 pyramidal cell bodies and dendrites (Figs. 3A and 3B). Additionally, under lower magnification large patches of immunoreactivity are observed in proximity to the CA1 pyamidal cell layer (arrows Fig. 3). Higher magnification reveals that these patches are due to the transduction of glial cells (Figs. 3B and 3C). Since glial cells may be capable of affecting synaptic plasticity by the secretion or uptake of unknown neuromodulators, it is unclear whether the observed changes in behavior are specifically due to neuron specific changes.

Conclusions

Viral transfer of activated G proteins into the rodent brain in theory has the potential of making targeted functional alterations in the living animal in a time- and region-specific manner. Replication-deficient adenovirus could be a good vector since it is possible to generate the high titers required to transduce a large number of cells. However, the lack of specificity with which the virus infects the neurons, with very substantial infection of the glial cells, makes it difficult to interpret the observed physiological and biological effects. Neuron-selective expression may be obtained by the use of neuron-specific promoters into the adenovirus. Even in this case, the physiological effects of the glial infection will have to be taken into account in interpreting the overall effects. Thus, replication-deficient adenovirus may have limted use as a gene transfer vector in the brain. Other viral vectors that may have a significantly better neuron/glia infection ratio could be better for such a purpose. Ultimately, if one is able to achieve neuron-selective transduction, viral vectors could prove to be an attractive alternative to traditional mouse genetic approaches because of their ability to acutely determine the effects of the gene of interest.

[26] Quench-Flow Kinetic Measurement of Individual Reactions of G-Protein-Catalyzed GTPase Cycle

By SUCHETANA MUKHOPADHYAY and ELLIOTT M. ROSS

Introduction

Fast and efficient G-protein-mediated signaling depends on the kinetic balance of the individual reactions of the GTPase cycle: GTP binding, hydrolysis of bound GTP, and dissociation of GDP. To understand the physiological regulation of G protein signaling, one must understand the regulation of each reaction and how they are kinetically synchronized. Each step of the GTPase cycle is determined both by the instrinsic activity of the Gα subunit and by the combined regulatory inputs of Gβγ, agonist-liganded receptors, and GTPase-activating proteins (GAPs). Conversely, the activation state of the Gα subunit determines its association with these and other regulatory proteins. Consequently, regulation of the individual reactions of the GTPase cycle is a complex function of many inputs and must be studied in this context.

The rates of partial reactions of the GTPase cycle were first estimated in the late 1970s from the kinetics of activation and deactivation of adenylyl cyclase in erythrocyte membranes[1]; these rates were measured directly with purified and reconstituted proteins a few years later.[2] These experiments were feasible because the G_s–adenylyl cyclase system is characterized by relatively slow activation and deactivation. In most G protein systems, the relevant reactions are usually too fast for manual measurement, with rates of activation and deactivation between 1 and $100\ s^{-1}$ ($t_{1/2} \sim 5$ ms–1 s).[3] Either optical triggers (caged reagents) or a rapid mixing device must be used to initiate the reaction, and the reaction must be followed either by continuous optical monitoring or by chemical assay following rapid mechanical quenching. While rapid mixing and fluorescence monitoring (stopped-flow) has contributed much to our understanding of the GTPase cycles of monomeric GTP-binding proteins, finding and applying suitable optical reporters for heterotrimeric G proteins has been problematic.[4] Hydrolysis of bound GTP has been measured for a soluble complex of Gα and GAP according to changes in the tryptophan fluorescence of Gα,[5] but receptor-promoted nucleotide exchange reactions and other receptor–heterotrimer systems have been unapproachable. We have found

---

[1] D. Cassel and Z. Selinger, *Proc. Natl. Acad. Sci. U.S.A.* **75,** 4155 (1978).
[2] D. R. Brandt and E. M. Ross, *J. Biol. Chem.* **261,** 1656 (1986).
[3] E. M. Ross and T. M. Wilkie, *Ann. Rev. Biochem.* **69,** 795 (2000).
[4] A. E. Remmers, *Anal. Biochem.* **257,** 89 (1998).
[5] K.-L. Lan, H. Zhong, M. Nanamori, and R. R. Neubig, *J. Biol. Chem.* **275,** 33497 (2000).

quench-flow, rapid mechanical initiation and quenching of the relevant reactions, to be a successful alternative.

Quench-flow kinetic assays use a mechanical mixing device to initiate fast reactions of interest, incubate them for set times, and then quench the reaction by addition of another reagent, all on the time scale of 1–10,000 ms. After final quenching, the reaction mixture is expelled to allow the assay of product or of remaining substrate. Quench-flow assays are generally considered less efficient than stopped-flow spectrophotometry because stopped flow provides a continuous reaction trace whereas quench flow requires a separate assay for each time point. However, quench flow is applicable when optical reporters are unavailable. Quench flow also provides absolute chemical quantitation of the reaction, which is often impossible with stopped flow. In the past, quench flow was often limited by primitive equipment that required large amounts of reagents and could only execute inflexible and simplistic mixing protocols. Modern computer-controlled, low-volume, multi-syringe mixers have overcome many of these problems and permit the sorts of assays necessary to monitor individually the binding, hydrolysis, and dissociation of guanine nucleotides on a physiologically reasonable time scale.

We describe here strategies and applications for using rapid quench-flow mixing to measure the key steps in the GTPase cycle under conditions where steady-state catalytic turnover is stimulated by both receptor and GAP (Scheme 1). Under these conditions the overall reaction cycle is fast ($k_{cat} \sim 2$ s^{-1}) and individual steps can occur with rates up to 25 s^{-1} ($t_{1/2} \sim 25$ ms). These procedures should be adequate for rates up to ~ 100 s^{-1}. The assays described here were developed using heterotrimeric G$_q$, m1 muscarinic cholinergic receptor, and either of two GAPs, PLC-β1 or RGS4.[6] The reactions of interest presumably reflect the activity of a complex of heterotrimeric G protein with both receptor and GAP.[7] Each is measured above the background of the slower basal reactions that reflect the intrinsic activities of the Gα subunit alone.

$$\text{G-GDP} \xrightarrow{k_{diss}} \text{G} \xrightarrow{k_{assoc}} \text{G-GTP} \xrightarrow{k_{hydrol}} \text{G-GDP} + \text{P}_i$$

SCHEME 1.

The assays include: (1) receptor-promoted dissociation of GDP from Gα (k_{diss}), which is rate-limiting for the cycle at physiological concentrations of GTP and which provides a simple introduction to the setup of quench-flow assays; (2) receptor-promoted GDP/GTP exchange (k_{exch}, a composite apparent rate

[6] S. Mukhopadhyay and E. M. Ross, *Proc. Natl. Acad. Sci. U.S.A.* **96,** 9539 (1999).

[7] G. H. Biddlecome, G. Berstein, and E. M. Ross, *J. Biol. Chem.* **271,** 7999 (1996).

constant), which both provides an overall rate for the activation phases of the cycle and allows direct determination of the rate of binding of GTP to unliganded Gα (k_{assoc}) at low GTP concentrations where GDP dissociation is not rate limiting; and (3) GAP-promoted hydrolysis of GTP bound to Gα(k_{hydrol}), the fastest rate we have measured, which demands a more complex mixing protocol and displays the versatility of the quench-flow approach.

Equipment and Reagents

Design of Quench-Flow Mixer

In addition to well-designed protocols that maximize recovery of products and minimize waste of reagents, quench-flow assays require an adequate rapid mixing apparatus. The design of these experiments require an instrument with four independently controllable syringes and small mixing and dead volumes. Protocols were developed on the Bio-Logic SFM4/Q quench-flow mixer (Fig. 1), which consists of four syringes (5 ml or 20 ml), three mixing chambers (~25 μl each), interchangeable delay lines (17–190 μl) to allow working with different reaction volumes, and an external ejection line which may also be used as a third delay line (50 μl for protocols described here). The syringes and associated loading valves are housed in a chamber connected to a circulating thermostatted water bath. Syringes deliver liquid upward into mixers and delay lines that are affixed to the top of the chamber and which are themselves thermostatted by water circulated from the bath. Reaction mixture exits from the top of the final mixer, where it can either be expelled directly or be directed into a flow-through cuvette for stopped-flow spectrophotometry. (The SFM4/Q also has a waste/collect valve,

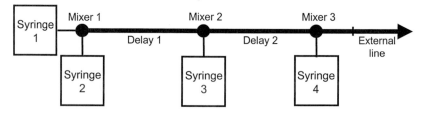

FIG. 1. Flow diagram of the SFM4Q quench flow mixer. Syringes 1 and 2 inject reagents into mixer 1. Solution leaves mixer 1, enters delay line 1, and continues to mixer 2, where it can be mixed with reagent from syringe 3. Solution from mixer 2 continues through mixer 3, where it can mix further with solution from syringe 4. After mixer 3, solution can be expelled for analysis (quench flow) or can be driven into a flow-through optical cell (not shown) for spectrophotometric analysis (stopped flow). Syringes may be either 5 ml or 20 ml; each is independently computer controlled by stepper motors. Extended delay lines (17–190 μl) may be inserted between mixers 1, 2, and 3. Mixer volumes are ~25 μl. The exit line is 50 μl in our configuration, but may be cut to size. The syringes, mixers, and delay lines are thermostatted.

but we did not use it to minimize loss of reagents in the valve's dead space.) The 5 ml syringes reliably deliver volumes as small as 20 μl ($\pm \sim 1$ μl) and the 20 ml syringe can deliver volumes as small as 40 μl. To conserve reagent volumes and to maximize speed, this apparatus does not use valves to separate the syringes from the flow lines. Consequently, the protocols below execute washing and line-clearing steps before initiating a reaction. Other quench-flow mixers may share all or some of these capacities and may be adapted for the assays described below.

Syringe movements are independently computer controlled by four stepper motors. Each sequential time period in the protocol, during which one or more syringes may inject liquid into the flow path or during which flow may stop to allow a reaction to proceed, is referred to as a "phase." Each phase is independently programmed. A protocol will typically consist of 5–10 phases, and all protocols described here are outlined phase by phase. The maximum reaction rate that can be measured in a quench-flow assay depends on the speed of mixing of reagents in the mixing chambers, on maximum flow rates, and on the various "dead volumes" of the apparatus. Dead volumes are the volumes through which a reagent must travel to reach the next mixing event; dead volume divided by flow rate equals dead time, the minimum time between the initiation and termination of a reaction. For the SFM4/Q mixer and the sorts of complex protocols described here, dead times limit measurable reaction rates to 100–200 s^{-1}.

General Guidelines for Designing Quench-Flow Protocols

A quench-flow assay protocol contains four basic stages: (1) The overall reaction pathway is allowed to come to overall steady state or the partial reaction of interest to equilibrium. For the GTPase cycle, the partial reactions are GDP dissociation and GTP association, which combine as GDP/GTP exchange, and hydrolysis of bound GTP. (2) The partial reaction of interest is isolated by quenching upstream reactions (with inhibitors or by radiochemical dilution). (3) The reaction mixture is incubated further to allow the partial reaction of interest to proceed. (4) The intermediate reaction is quenched and the mixture is expelled for analysis. To terminate the reaction, either the final quench is delivered by syringe 4 or the reaction mixture is ejected into an external tube that contains the final quench.

In two of the protocols described here, GDP dissociation and GDP/GTP exchange, only one mixing step is required to initiate the reaction because the starting intermediate, an equilibrium system of receptor–G protein vesicles, GTP (^{32}P-labeled or unlabeled) and agonist, is preformed in one syringe. The vesicles are mixed with nucleotide (unlabeled GTP to monitor [α-^{32}P]GDP dissociation; [α-^{32}P]GTP to monitor GDP/GTP exchange), incubated for a set time, and then quenched. In these protocols, the quench must be chilled and is external to the quench-flow mixer. In such two-step protocols with external quench, only three syringes are used. Both syringes 1 and 2 contain the reaction buffer, syringe 3 contains nucleotide (blanketing reagent), and syringe 4 contains vesicles (limiting reagent). It is also possible to perform these two-step protocols with an internal quench. In this case, syringe 1 contains buffer, syringe 2 contains excess nucleotide, syringe 3 contains the vesicles, and syringe 4 contains the quench.

For the last protocol, hydrolysis of $G\alpha$-bound $[\gamma\text{-}^{32}P]GTP$, the starting intermediate is formed during each determination, and the added step requires the use of all four syringes. Syringe 1 is again the reaction buffer and syringes 2 and 3 contain $[\gamma\text{-}^{32}P]GTP$ and vesicles, respectively. The reagents in syringes 2 and 3 are mixed for a fixed time to form the starting intermediate. Continued binding of $[\gamma\text{-}^{32}P]GTP$ is stopped by radiochemical dilution by unlabeled GTP from syringe 4, and hydrolysis is then allowed to occur in the final delay line. The mixture is then expelled into an external acid-quenching solution to terminate the reaction.

In designing the protocols, reagents are assigned to specific syringes and delay lines are chosen to minimize both the required volume of each valuable reagent and the distance each reagent needs to travel in the flow path, as well as to increase the efficiency of recovery of products in a reasonably small volume. These choices significantly determine data quality and the overall feasibility of an experiment. The specific choices depend on the performance of the mixing apparatus, the precision of delivery by the syringes, and the concentrations of reagents. However, some general rules apply to the basic design of all the protocols. First, syringe 1 is filled with assay buffer without protein or reactants; it is used to push the reaction mixture through the mixing chambers and out of the apparatus. This procedure saves other reactants, which would otherwise be wasted in pushing the reaction volume. The buffer in syringe 1 is also used to wash the flow path between reactions. Syringe 1 is therefore a 20 ml syringe that contains complete reaction buffer and must be refilled periodically during the course of multiple assays. Other syringes are usually 5 ml to optimize the precise injection of small volumes of reagents. Second, the final quench is placed in syringe 4, the last syringe, unless the final quench is external. Internal quenches potentially provide greater quenching speed than an external quench, but external quenches can be chilled, which makes them preferable for some assays. Third, reactants are loaded as close as possible to the exit line to decrease distance that each reagent must travel. This assignment minimizes the time (volume) needed to expel the the reaction mixture into the final quench. Last, the syringe that contains the reagent in stoichiometric excess (usually nucleotide) is placed before the syringe that contains the limiting reagent (receptor–G protein vesicles) because the excess reagent will be used to surround (blanket) the limiting reagent before, during, and after mixing (Fig. 2). Examples of blanketing are given in each protocol.

To vary incubation times, either when forming an intermediate or when incubating the reaction of interest, we completely stopped the flow of the reaction

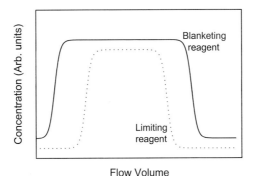

Flow Volume

FIG. 2. Blanketing the vesicles in the flow path to maintain constant concentration of added reagents. The graph shows the concentrations of a stoichiometrically limiting reagent (vesicles) and a second reactant (blanketing reagent; e. g., nucleotide) as a function of linear position (volume) in the flow path of the quench-flow mixer. When different solutions are injected into the flow path of a quench-flow mixer, turbulent mixing occurs at their edges such that concentrations change gradually. Higher flow rates both broaden these mixing zones and make them asymmetric because of frictional shear near the walls of the tubing. Mixing and asymmetry under experimental conditions can be estimated using stopped-flow optics to monitor the passage of dye solutions through a cuvette placed after mixer 3, and would show traces similar to those in the figure. In the protocols described here, the blanketing reagent is injected into the mixer before, during, and after injection of the limiting reagent, such that the concentration of the added reagent is constant. In this example, all of the limiting reagent (vesicles) may be collected with confidence that it reacted homogeneously with the other reagents. This principle is executed in each of the three protocols given in the text.

mixture for the desired period of time ("interrupted mode") rather than trying to match continuing flow rates to the volumes of the delay lines to achieve the same incubation intervals. Minimal reagent diffusion occurs even when incubating for periods as long as 1 min at 30°. The advantage of interrupted mode is that a single flow rate can be used, thus allowing the investigator to establish only one set of controls for the efficiency of mixing and washing, both of which depend on flow rate. This approach also avoids the need to change delay lines during an experiment and simplifies the design of protocols.

For reactions that are terminated by ejecting the reaction into an external quenching solution, the speed of quenching depends on the ejection flow rate, the volumes of the reaction mixture and the quench solution, and the shape of the collection vessel. We routinely check the external mixing rate by ejecting a pH-sensitive dye into a buffer that causes a color change, which is recorded by a video camera. Conventional video recorders have a raster scan rate of ~17 s^{-1}, which limits determination of the final mixing rate by this method. The dead time for external quenching can also be determined directly by externally quenching a control reaction with a known fast rate (dinitrophenyl acetate hydrolysis; see below) and extrapolating data to nominal zero time.

Optimizing and Controlling Reagent Flow, Mixing, and Recovery

Before initiating experiments, the investigator must optimize volumes and flow rates for washing the reaction line and for delivery, mixing and ejecting reagents. It is also necessary to measure the fractional recovery and dilutions associated with each of these procedures, and their reproducibility. First, determine the minimum and maximum volumes that can be delivered reproducibly and accurately from one syringe. The simplest way to do this is to fill each syringe with water, collect a range of programmed volumes and determine the actual expelled volumes. Next, determine the volume of buffer necessary to eject the reaction mixture. The ejection volume is determined by injecting experimentally appropriate volumes of a dye or radiochemical into the location in the flow path that will contain the reaction mixture, and then testing the amount of push buffer needed to expel it with reasonable recovery. The ejection volume is usually 2- to 3-fold greater than the volume of the reaction mixture plus the volume through which it must travel (Fig. 3). The flow rate of the push buffer (syringe 1) will alter the volume required to eject sample; higher flow rates are less efficient. Recovery of reagents from the flow path in a practical ejection volume will be less than 100%, and may be as little as 50–60% if it is necessary to minimize the ejection volume. Note that flow rate, reaction volume, and ejection volume

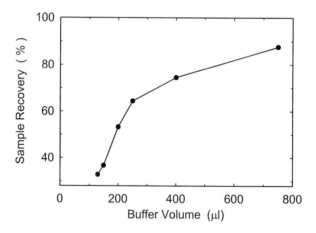

FIG. 3. Variation of recovery with the volume of push buffer used to expel the reaction mixture. A mock reaction mixture (40 μl) in the exit line after mixer 3 was expelled by injection of increasing amounts of buffer from syringe 3 at a flow rate of 1 μl/ms. Note that if there were no mixing and if expulsion were perfect, 100% of the reaction mixture should have been expelled by only 90 μl of buffer, the volume from mixer 3 to the exit port. Practically, 300 μl gives acceptable expulsion of the reaction. Residual reagents are purged from the flow lines by the wash step at the beginning of the next reaction cycle.

determine fractional recovery coordinately; these parameters must be optimized iteratively. Finally, record the fractional recovery of reaction mixture using these optimized ejection conditions. This value will be used to correct data obtained during experiments.

We recommend optimizing and then verifying mixing and quenching in actual protocols by measuring the base-catalyzed hydrolysis of O-acetyl-2,4-dinitrophenol ("dinitrophenyl acetate," DNPA) by NaOH. The reaction is quenched in dilute HCl and the hydrolysis product, 2,4-dinitrophenol, is measured according to its absorbance at 320 nm ($\epsilon = 5600\,M^{-1}\mathrm{cm}^{-1}$). The second-order rate constant for this reaction is 56 M^{-1} s^{-1} at 20°; the actual hydrolysis rate can be adjusted to the desired range by adjusting the NaOH concentration (20 mM NaOH at room temperature yields a pseudo-first-order rate of 1.12 s^{-1}). Use 2 mM HCl as the push buffer in syringe 1 (or 1 and 2), NaOH as the excess reagent, and DNPA as the limiting reagent. This mock reaction tests essentially every phase of the quench-flow protocol without wasting valuable reagents.

Materials

The reagents, preparations, and product assays necessary for the automated measurement of the partial individual reactions of the GTPase cycle are the same as those used for manual measurements of GTP hydrolysis and guanine nucleotide binding.[8] Purification of receptors, Gα_q and G$\beta\gamma$ and PLC-β1 and RGS4[6,7] used in the development of these protocols and their reconstitution into phospholipid vesicles[7,9,10] are described elsewhere, as are other aspects of the quantitation of GTPase reactions and GAP activity.[8]

Two aspects of the transient reaction measurements described here demand somewhat novel reagents. First, the purity of [γ-^{32}P]GTP must be greater than 99% for most hydrolysis measurements. Preparation and purification of [α-^{32}P]GDP and [γ-^{32}P]GTP are described elsewhere.[7,8] Second, both the concentrations of receptor and G protein and the receptor : G protein ratio in the reconstituted vesicles are crucial for forming and accumulating specific receptor–G protein–GAP intermediates in the GTPase cycle while maintaining low assay backgrounds. When determining k_{diss} and k_{exch}, the receptor : G protein ratio is usually 0.1–0.2. For determinations of k_{hydrol}, the receptor : G protein ratio must be greater than 0.3 to increase the fraction of G protein in a rapidly turning over complex of receptor, G protein, and GAP, as described elsewhere.[6]

---

[8] J. Wang, Y. Tu, S. Mukhopadhyay, P. Chidiac, G. H. Biddlecome, and E. M. Ross, in "G Proteins: Techniques of Analysis" (D. R. Manning, ed.), p. 123. CRC Press, Boca Raton, FL, 1999.

[9] G. Berstein, J. L. Blank, A. V. Smrcka, T. Higashijima, P. C. Sternweis, J. H. Exton, and E. M. Ross, J. Biol. Chem. 267, 8081 (1992).

[10] R. A. Cerione and E. M. Ross, Methods Enzymol. 195, 329 (1991).

Dissociation of GDP

The receptor-driven dissociation of GDP from the Gα subunit is the rate-limiting step in the GTPase cycle under physiological conditions and is also the most straightforward in terms of assay design and execution. The assay consists of only two mixing events: one to initiate dissociation of Gα–[α-^{32}P]GDP and a second to arrest dissociation at set times. We will therefore use this reaction as an example of how protocols for kinetic measurements are translated to execution in a rapid mixing apparatus.

Although the reaction of interest is the dissociation of newly formed GDP during steady-state GTP hydrolysis, its measurement is simplified by starting with [α-^{32}P]GDP already bound to the receptor–G protein vesicles. Free and Gα-bound [α-^{32}P]GDP are first allowed to reach receptor-catalyzed equilibrium in syringe 4 of the rapid mixing apparatus. Association of receptor and G protein in the vesicles depends on agonist and takes about 1 min at 30°.[7] Dissociation of [α-^{32}P]GDP is initiated by addition of excess nonradioactive GTP (or other guanine nucleotide) and is terminated by addition of antagonist and detergent followed by rapid chilling. Gα-bound [α-^{32}P]GDP in the quenched reaction is then determined by the standard nitrocellulose adsorption assay for G-protein-bound nucleotides.[2,7] The multiprotein complex from which rapid GDP dissociation is measured includes Gα–GDP, agonist-liganded receptor, and, presumably, G$\beta\gamma$. If GDP dissociation were measured during steady-state hydrolysis, the relevant catalytic complex would also include a GAP, but no effect of GAPs on GDP dissociation has been observed.[6] During this assay, some [α-^{32}P]GDP also dissociates from G$\alpha\beta\gamma$, but basal dissociation is slow and is insignificant in the time frame of these measurements.

Experimental Overview

For the measurement of receptor-catalyzed [α-^{32}P]GDP dissociation, syringe 1 contains assay buffer, which is used both for washing the system and for advancing the reagents (push buffer). Syringe 2 is not used, but it is also filled with push buffer because it shares a mixer with syringe 1. Syringe 3 contains unlabeled GTP (in assay buffer, including agonist) to initiate the dissociation reaction, and syringe 4 contains receptor-G protein vesicles. The vesicle suspension also contains agonist and [α-^{32}P]GDP, which are allowed to come to binding equilibrium in the syringe. The time needed for loading syringe 4 is more than adequate to allow agonist-driven association of receptor and G protein.[7] To measure dissociation of [α-^{32}P]GDP, vesicles in equilibrium with [α-^{32}P]GDP (syringe 4) and nonradioactive GTP (syringe 3) are mixed in mixer 3 (still with saturating agonist), and the reaction proceeds in the exit line (used as delay line 3). The unlabeled GTP blocks rebinding of [α-^{32}P]GDP and thereby allows observation of receptor-stimulated dissociation. After incubation in the ejection line for varying times, the

TABLE I
TIME VS VOLUME IN EXPERIMENTAL PHASES: DISSOCIATION OF GDP

Parameter	Phase 1	Phase 2	Phase 3	Phase 4	Phase 5	Phase 6	Phase 7
Time (ms)	20	630	2500	40	20	Vary	270
S1 (μl): Buffer	20	630					
S2 (μl): Buffer							
S3 (μl): GTP				40	20		270
S4 (μl): Vesicles	5				20		

reaction mixture is expelled into cold quenching buffer. Table I gives times of each phase and the volumes delivered by each syringe. In this protocol, all syringes move at a rate of 1 μl/ms (except for syringe 1 in phase 1). The diagrams below show the composition of the flow path during each of the sequential phases of the experiments.

Assay Solutions

(See Ref. 6.)

Reaction buffer: 20 mM Na–HEPES (pH 7.5), 0.1 M NaCl, 2 mM MgCl$_2$, 1 mM EGTA, 0.1 mg/ml bovine serum albumin (BSA), 1 mM dithiothreitol (DTT)

GTP: 200 μM GTP plus 1 mM carbachol in reaction buffer. The concentration is chosen to block any rebinding of [α-^{32}P]GDP. Saturating GDP may be used interchangeably.

Vesicles: Phospholipid vesicles that contain both G$_q$ and m1 muscarinic receptor[7] plus 1 mM carbachol and 300 μM [α-^{32}P]GDP (70–100 cpm/fmol), in reaction buffer. The concentration of vesicles in buffer and of proteins in the vesicles is adjusted to yield 1–3 nM receptor-coupled Gα_q and 0.2–0.8 nM receptor, with a receptor : Gα ratio of 0.1–0.2. GAPs may be included in this volume, but we have not detected an effect of a GAP in this assay.

Quench buffer: 20 mM Tris-Cl (pH 8.0), 0.1 M NaCl, 1 mM dithiothreitol, 1 mM GTP, 5 mM EDTA, 1 mM atropine, 1 mg/ml Lubrol PX (dodecylpolyethylene oxide ($n \sim 8.5$).

Filtration buffer: 20 mM Tris-Cl (pH 8.0), 0.1 M NaCl, 5 mM EDTA.

Because the walls of the Kel-F syringes provide significant thermal insulation, equilibrate all solutions to the assay temperature before loading the syringes. This caution applies to all quench-flow assays.

Individual Phases in the GDP Dissociation Assay

Phase 1 primes the lines between the syringes and the mixers to remove any buffer or other reagents that may have infiltrated since the previous injection sequence. This step, required by the valveless design of the mixer and the need to use small reaction volumes, ensures that subsequent low-volume injections contain only the intended solution. Syringe 1 is also moving to apply positive pressure because only a small volume is being delivered from syringe 4.

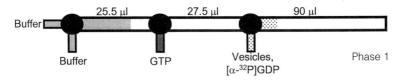

Phase 2 washes out the reaction line. The ejected volumes from phases 1 and 2 are collected separately and discarded as radioactive waste.

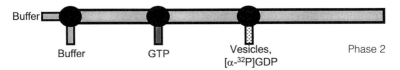

Phase 3 is a 2.5 s delay to allow the investigator to place the chilled collection tube, which contains the external quench, at the end of the exit line. The investigator should practice coordinating this step with the timed function of the mixing apparatus.

Phase 4 injects nonradioactive GTP (plus agonist) into the flow path before introduction of the vesicles. The concentration of GTP is arbitrary; it is used only to block rebinding of $[\alpha\text{-}^{32}P]GDP$. This is the beginning of the "blanket" where excess reagent surrounds ("covers") the limiting reagent to ensure thorough mixing. By injecting 40 μl of GTP, reagents from syringe 3 are pushed past mixer 3. (The center-to-center volume between mixer 2 and mixer 3 is 27.5 μl.) When vesicles are added from syringe 4 in the next phase, even the leading edge of the injected vesicles suspension will be exposed to the nonradioactive GTP already in the line from this step. Blanketing the stoichiometrically limiting reactant in this way is generally necessary because a bolus of liquid in the reaction line does not move as a perfect cylindrical slug with edges perpendicular to the direction of flow. Instead, friction causes liquid in the center of the channel to move faster, and thus distorts both the leading and trailing edges of the reaction volume (Fig. 2). Faster flow rates aggravate this phenomenon.

Buffer

Buffer   GTP   Vesicles,   Phase 4
                             [α-^{32}P]GDP

Phase 5 injects vesicles and mixes them with an equal volume of nonradioactive GTP. Syringe 3 and syringe 4 move simultaneously to thoroughly and rapidly mix vesicles in equilibrium with [α-^{32}P]GDP with nonradioactive GTP. Syringe 3 continues to inject GTP to fill delay line 2 with nonradioactive nucleotide to expose the trailing edge of the vesicles to GTP and completes the "blanketing" of the vesicles. Failure to blanket the end of the vesicle suspension can allow [α-^{32}P]GDP to reassociate with the vesicles and thus cause underestimation of both the extent and rate of GDP dissociation.

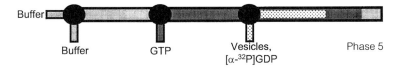

Buffer

Buffer   GTP   Vesicles,   Phase 5
                             [α-^{32}P]GDP

Phase 6 is the incubation time during which bound [α-^{32}P]GDP dissociates. There is no movement of the syringes.

Phase 7 expels the vesicles from the third delay line into the test tube that contains 100 μl of chilled quench buffer. Either syringe 1 (or 2) or syringe 3 can be used in this phase. For the configuration of the SFM4Q mixer that we used during these experiments, 270 μl was the minimum volume needed to expel most of the reaction mixture using syringe 3 at a flow rate of 1 ml/s. Liquid from phases 5–9 is usually collected together. For longer incubation times, it is possible to collect only phase 9. Immediately after phase 9 vortex and chill the quenched reaction mixture.

Buffer

Buffer   GTP   Vesicles,   Phase 7
                             [α-^{32}P]GDP

Analysis of Data

Data are analyzed as the time-dependent loss of G protein-bound [α-^{32}P]GDP. Incubation times (phase 6) are varied to cover the entire time course of dissociation, usually spanning the range of 0.2 to 5 times the half-time of the reaction. At least one pilot determination is thus required to establish the reaction time span to be sampled. We usually take about 25 time points for a process that displays a single exponential time course, enough to allow reliable determination of the amount of $[\alpha\text{-}^{32}\text{P}]\text{GDP}$ initially bound, the initial dissociation rate, and the final plateau. Multiple zero-time and long-time samples should be taken to constrain the limiting values. Zero-time samples are taken at the beginning and end of the experiment to control for any drift in background. It is advisable to take other time points in random order for the same reason. In addition, we recommend performing a manual determination of bound $[\alpha\text{-}^{32}\text{P}]\text{GDP}$ at zero and "infinite" time to confirm accuracy of the mechanical mixing.

To analyze the data, the amount of bound $[\alpha\text{-}^{32}\text{P}]\text{GDP}$ at time t (B_t) is fitted to a single exponential rate equation $B_t = B_c e^{-kt} + B_\infty$. The total $[\alpha\text{-}^{32}\text{P}]\text{GDP}$ bound at zero time ($B_c + B_\infty$), the first-order rate constant k_{diss}, and residual bound $[\alpha\text{-}^{32}\text{P}]\text{GDP}$ at the end of the reaction (B_∞) are all allowed to vary during the fit. The amount of $[\alpha\text{-}^{32}\text{P}]\text{GDP}$ that dissociates rapidly, B_c, reflects the amount of receptor-coupled $G\alpha$ in the injected vesicles, the initial relative saturation with $[\alpha\text{-}^{32}\text{P}]\text{GDP}$, and the efficiency of expulsion of the reaction mixture. Expulsion efficiency is reproducible among individual determinations and among experiments; totals can therefore be back-calculated to yield the amount of $[\alpha\text{-}^{32}\text{P}]\text{GDP}$ bound in the volume of vesicles originally injected. When establishing the assay, the reproducibility of all these parameters should be determined independently, and the flatness of the plateau should be checked with a few long-time determinations to confirm single-exponential behavior.

Receptor-Catalyzed Nucleotide Exchange

The reaction protocol used to determine the rate of GDP dissociation from $G\alpha$ can also be used to determine the overall rate of guanine nucleotide exchange by beginning with $G\alpha$ bound to unlabeled GDP and monitoring the binding of $[\alpha\text{-}^{32}\text{P}]\text{GTP}$, $[^{35}\text{S}]\text{GTP}\gamma\text{S}$, or other radiolabeled nucleotide. Bound $[\alpha\text{-}^{32}\text{P}]\text{GTP}$ is hydrolyzed, but the $[\alpha\text{-}^{32}\text{P}]\text{GDP}$ product remains bound. Nucleotide exchange, described by the rate constant k_{exch}, is a two-step process that includes dissociation of GDP (k_{diss}, Scheme 1) and association of another nucleotide (k_{assoc}). Both processes are regulated by receptor and $G\beta\gamma$; during steady-state GTP hydrolysis exchange is also facilitated indirectly by GAPs. In cells, the GDP/GTP exchange reaction is responsible for G protein activation; it both initiates and maintains signaling to effector proteins. At physiological concentrations of GTP ($\geq 10\mu M$), exchange is limited by the rate of dissociation of GDP from the receptor–$G\alpha$ complex. At lower GTP concentrations, exchange appears to be limited by the rate of association of GTP.

The k_{exch} assay, like the k_{diss} assay, consists of only two mixing events. The reaction begins when receptor–G protein vesicles in the presence of agonist are mixed with radiolabeled nucleotide. The reaction proceeds until it is quenched by chilling and addition of antagonist and detergent. Bound nucleotide is then measured by the standard nitrocellulose adsorption assay.

TABLE II
TIME VS VOLUME IN EXPERIMENTAL PHASES: RECEPTOR-CATALYZED NUCLEOTIDE EXCHANGE

	Phase 1	Phase 2	Phase 3	Phase 4	Phase 5	Phase 6	Phase 7
Time (ms)	20	630	2500	40	20	Vary	270
S1 (μl): Buffer	20	630					
S2 (μl): Buffer							
S3 (μl): [α-^{32}P]GTP				40	20		270
S4 (μl): Vesicles	5				20		

Experimental Overview

To measure nucleotide exchange, syringes 1 and 2 contain reaction buffer and may be used interchangeably for washing the lines and advancing reagents. Syringe 3 contains agonist and radiolabeled nucleotide. Syringe 4 contains the stoichiometric limiting reagent, receptor–G protein vesicles. Vesicles are equilibrated with agonist (and GAP, as desired). The protocol for the rapid mixing of the different reagents is essentially identical to the dissociation protocol described earlier. Each phase of the reaction is described below (see Table II).

Assay Solutions

Reaction buffer: As for GDP dissociation, above.

[α-^{32}P]GTP: Up to ~1 μM; 100–150 cpm/fmol; in reaction buffer plus 1 mM carbachol. Other radiolabeled nucleotides, such as [^{35}S]GTPγS, may be substituted to study their individual behaviors.

Vesicles: Phospholipid vesicles that contain both G$_q$ and m1 muscarinic receptor,[7] plus 1 mM carbachol, in reaction buffer. The concentration of vesicles in buffer and of proteins in the vesicles is adjusted to yield 1–3 nM receptor-coupled Gα_q and 0.2–0.8 nM receptor. The receptor : G protein ratio may be as low as 0.1 without sacrificing signal quality. Inclusion of GDP up to 100 nM had no effect on subsequent reactions in our hands.

Quench buffer: As for GDP dissociation, above.

Rinse buffer: As for GDP dissociation, above.

Individual Phases in GDP/GTP Exchange Assay

Phases 1 and 2 are the priming and washing steps described in the previous protocol. Phase 3 is a pause so the user can place the collection tube at the end of the exit line.

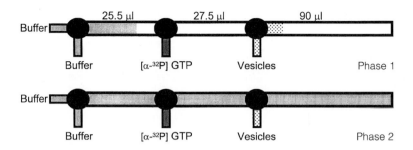

Phase 4 begins the blanket of $[\alpha-{}^{32}P]GTP$ that will surround the vesicles. Radiolabeled nucleotide is pushed from syringe 3 past mixer 3 to ensure complete mixing of reagents from syringe 3 and syringe 4 during the next phase. In general, we try to use the least possible amount of labeled nucleotide to minimize background. Because GDP/GTP exchange is nearly second-order at low concentrations of GTP (see Analysis, below), it is important to maintain a constant and known concentration of GTP to maintain the pseudo-first-order domain. The absolute concentration of the vesicles is less important, but must be reproducible.

Phase 5 mixes vesicles with an equal volume of radiolabeled nucleotide.

Phase 6 is the variable incubation time to allow nucleotide exchange. As with the dissociation reaction, several zero-time measurements should be made, both at the beginning and at the end of the overall experiment to confirm reproducible mixing, stability of reagents and proteins, etc.

Phase 7 expels the reaction mixture and into the chilled external quench solution.

Analysis of Data

The amounts of bound $[\alpha\text{-}^{32}P]GDP$ or $[^{35}S]GTP\gamma S$ accumulated at various incubation times are corrected for assay background and converted to molar amounts of bound nucleotide essentially as described for the dissociation of $[\alpha\text{-}^{32}P]GDP$, above. Bound nucleotide is then fit to a single exponential function to yield the apparent first-order rate constant, k_{exch}, and the amount of coupled $G\alpha$ subunit that undergoes rapid, receptor-catalyzed exchange. The amount of nucleotide bound at zero and infinite time should be confirmed by a manual assay.

The nucleotide exchange rate, k_{exch}, increases with increasing concentrations of GTP until the initial receptor-promoted dissociation of bound GDP becomes rate-limiting (Scheme 1). k_{exch} is therefore a complex function of both the receptor-catalyzed rate of GDP dissociation and the subsequent second-order binding of GTP to the receptor-bound, GDP-free $G\alpha$ subunit. At low GTP concentrations, k_{exch} is approximately equal to the pseudo-first-order association rate constant for GTP binding ($[GTP] \cdot k_{assoc}$), where k_{assoc} is the second-order association rate constant. At high GTP concentrations, where ($[GTP] \cdot k_{assoc}$) $\gg k_{diss(GDP)}$, k_{exch} approaches $k_{diss(GDP)}$.

Analysis of GDP/GTP exchange over a broad range of GTP concentrations should allow the determination of both the component constants $k_{diss(GDP)}$ and $k_{assoc(GTP)}$. In practice, we have not been able to extend these measurements to high concentrations of GTP where $k_{diss(GDP)}$ is rate-limiting because the assay background becomes intolerable.[6] Regardless, significant information can be obtained over a range of low GTP concentrations. First, plotting k_{exch} vs the concentration of GTP in this range should yield a straight line, indicating that the reaction is controlled by k_{assoc} and yielding the value of k_{assoc} as the slope of the line. Second, extrapolation of the line to $[GTP] = 0$ yields an estimate of the rate constant for dissociation of GTP ($k_{diss(GTP)}$), which has been difficult to obtain by other means because even receptor-promoted GTP dissociation from $G\alpha$ subunits is significantly slower than is GTP hydrolysis.

Hydrolysis of $G\alpha$-Bound GTP

The protocol below is designed to measure the rate of GAP-stimulated hydrolysis of $G\alpha$-bound GTP during the steady-state GTPase reaction (k_{hydrol}). Hydrolysis is potentially the fastest step in the GTPase cycle. We have measured rates up to 25 s^{-1} for hydrolysis of GTP bound to $G\alpha_q$ during steady-state hydrolysis.[6] Because the hydrolytic step may be faster than the GDP/GTP exchange steps that precede it, its measurement requires that the $G\alpha$–GTP reactant be created for each determination (time point). Therefore, each assay consists of three sequential steps: (1) The intermediate complex $G\alpha$–$[\gamma\text{-}^{32}P]GTP$ (presumably in a larger complex with $G\beta\gamma$, GAP, and receptor) is formed during steady-state hydrolysis. (2) Formation of the radiolabeled complex is stopped by radiochemical dilution of free $[\gamma\text{-}^{32}P]$GTP to establish a single kinetic pool of labeled reactant, which continues to hydrolyze. (3) Hydrolysis is terminated at fixed times to allow determination of the $[^{32}P]$ P_i product. The assay is most readily applicable to a system that contains purified receptor, G protein $(\alpha\beta\gamma)$, and GAP reconstituted in phospholipid vesicles, which minimizes background GTP hydrolysis and allows adjustment of the receptor : G protein ratio.

The key feature of this assay is preparation of the enzyme–substrate complex, $G\alpha$-GTP-GAP, within the quench-flow mixer. Because receptor and G protein do not associate significantly in reconstituted vesicles in the absence of agonist,[7] the vesicles and GAP are first equilibrated with agonist in the syringe before starting the assay. Binding of G protein to agonist-liganded receptor takes about 1 min at $30°$[7]; the complex is stable thereafter over the time needed for a series of assays ($\sim$30 min). In the first mixing step, $[\gamma\text{-}^{32}P]$GTP and vesicles (plus GAP) are mixed to allow the G protein–receptor complex to bind $[\gamma\text{-}^{32}P]$GTP. Determining the optimal time of incubation of $[\gamma\text{-}^{32}P]$GTP with vesicles is critical for obtaining usable data from the experiment. This step (phase 6) must be long enough to allow significant binding of $[\gamma\text{-}^{32}P]$GTP, which creates a pool of $G\alpha$-$[\gamma\text{-}^{32}P]$GTP, the amount of GTP hydrolysis that will be measured (the signal). This step must be short enough to minimize the accumulation of free $[^{32}P]P_i$ (background). The extent of binding is determined by k_{exch}, temperature, and the concentration of $[\gamma\text{-}^{32}P]$GTP. The time of 6 s suggested for phase 6 is a compromise and should be evaluated by the investigator. Determining the optimum mixing time between the $[\gamma\text{-}^{32}P]$GTP and the vesicles is tricky, especially in the presence of a GAP and at higher temperatures. Fortunately, once optimum conditions are decided, formation of $G\alpha$-$[\gamma\text{-}^{32}P]$GTP is highly reproducible if the concentration of GTP is well above the concentration of $G\alpha$.

After the enzyme–substrate complex, $G\alpha$-$[\gamma\text{-}^{32}P]$GTP, is formed in Phase 6, further binding of $[\gamma\text{-}^{32}P]$GTP is quenched and bound $[\gamma\text{-}^{32}P]$GTP is allowed to hydrolyze for set periods of time. The hydrolysis phase (phase 8) is begun by addition of nonradiolabeled nucleotide and antagonist. Excess nucleotide prevents binding of $[\gamma\text{-}^{32}P]$GTP directly and antagonist essentially terminates catalyzed nucleotide exchange. The hydrolysis reaction is quenched by ejecting the reaction mixture into chilled H_3PO_4.

Experimental Overview

The diagrams below show the composition of the flow path during each of the sequential phases of the experiments. Note the introduction between mixers 2 and 3 of a delay line in which the hydrolysis reaction takes place. Syringe 1 contains the buffer used to push the reaction mixture through the system. Syringe 2 contains $[\gamma\text{-}^{32}P]$GTP. Syringe 3 contains vesicles, GAP, and agonist. Syringe 4 contains unlabeled GTP and antagonist; the acidic charcoal quench is external. The mixing protocol is shown below (see Table III).

TABLE III
TIME VS VOLUME IN EXPERIMENTAL PHASES: HYDROLYSIS OF $G\alpha$-BOUND GTP

	Phase 1	Phase 2	Phase 3	Phase 4	Phase 5	Phase 6	Phase 7	Phase 8	Phase 9
Time (ms)	20	1000	2500	35	20	6000	50	Vary	500
S1 (μl): Buffer	20	1000			20		75		
S2 (μl): [γ-^{32}P]GTP				35					
S3 (μl): Vesicles					20				
S4 (μl): Antagonist, nonradioactive GTP	5	500		35	20		50		500

Assay Solutions

Reaction buffer: As for GDP dissociation, above.

[γ-^{32}P]GTP: 0.3–1 μM, 150–200 cpm/fmol, plus 1 mM carbachol, in reaction buffer. This solution should be kept at 0° until it is adjusted to the reaction temperature before loading into the syringe. Holding [γ-^{32}P]GTP at 30° will increase background because of nonenzymatic hydrolysis.

Vesicles: Phospholipid vesicles that contain both G_q and m1 muscarinic receptor,[7] appropriate GAP (generally at saturating concentration, e.g., 9 μM RGS4 or 20 nM PLC-β1 for G_q vesicles), and 1 mM carbachol, in reaction buffer. The concentration of vesicles in buffer and of proteins in the vesicles is adjusted to yield 3–10 nM receptor and 10–30 nM G_q. The molar ratio of receptor to coupled G protein in the vesicles must be greater than 0.3 because the amount of receptor limits the formation of the receptor–G protein complex that catalyzes rapid GTP hydrolysis. Excess uncoupled G protein simply adds to accumulation of background formation of [^{32}P] P$_i$.

Unlabeled GTP (or GDP): 100 μM, plus 100 μM atropine, in reaction buffer.

Quench solution: 10% slurry of charcoal (Norit A) in 50 mM H$_3$PO$_4$ at 0°. pH must be adjusted to $\leq$3 to rapidly quench hydrolysis.

Individual Phases in GTP Hydrolysis Assay

Phases 1 and 2 are the priming and washing steps described above. Phase 3 is a pause so the user can place the collection tube at the end of the exit line. The charcoal slurry should be suspended just before use.

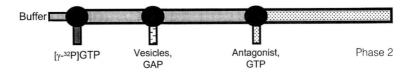

Phases 4–6 form the enzyme-substrate intermediate, $G\alpha$-$[\gamma$-$^{32}P]GTP$. In phase 4, $[\gamma$-$^{32}P]GTP$ is injected past mixer 2 to form the blanket that will surround the vesicles coming from syringe 3 in the next step. The unlabeled GTP from syringe 4 provides extra washing of the downstream flow path. Phase 5 mixes vesicles with $[\gamma$-$^{32}P]GTP$. As with the other reactions discussed above, more $[\gamma$-$^{32}P]GTP$ is also injected to form the end of the blanket that surrounds the vesicles. $[\gamma$-$^{32}P]GTP$ binds to $G\alpha$ during phase 6. The final concentration of $[\gamma$-$^{32}P]GTP$, 50–500 nM, influences both the amount of bound $G\alpha$-$[\gamma$-$^{32}P]GTP$ complex present at the beginning of the hydrolysis phase (phase 7) and the amount of background $[^{32}P]P_i$. It should be optimized according to the amount of coupled $G\alpha$, the efficiency of loading, and the $G\alpha$ hydrolysis rate.

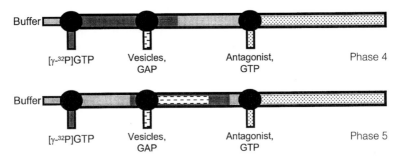

Phase 7 initiates the timed hydrolysis reaction (experimental zero time) by mixing the vesicles and $[\gamma$-$^{32}P]GTP$ with excess unlabeled GTP (40 μM final concentration, $\geq$100-fold isotopic dilution) and antagonist. Hydrolysis continues through Phase 8 for various times. For hydrolysis rate constants of 5–30 s^{-1}, incubation times may vary from 5 ms to 2 s to obtain data over the entire reaction time course.

Phase 9 expels the reaction mixture into cold acidic charcoal slurry (1.8 ml). Immediately after collection of the reaction mixture, the test tube is vortexed and kept on ice. The quenched reaction mixtures are centrifuged at 0–4° to precipitate the charcoal, and [^{32}P]P$_i$ in an aliquot of supernatant (1.0–1.4 ml as convenient) is measured by either liquid scintillation counting or Cerenkov counting.

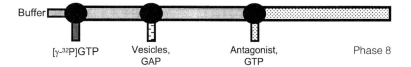

Analysis of Data

Data points should cover the time period of the hydrolysis event plus long times to establish the rate of background hydrolysis. Zero-time and long-time data should be collected several times during performance of an experiment. For each sample, the amount of [^{32}P]P$_i$ in the charcoal supernatant is corrected for the fraction of the total supernatant volume that was counted and radioactivity is converted to moles of GTP hydrolyzed according to the specific activity of the [γ-^{32}P]GTP in syringe 2. Also correct for the fraction of the total reaction mixture that is recovered in the volume expelled in phase 9 according to previous calibration of the quench-flow mixer.

GAP-stimulated [^{32}P]P$_i$ formation should follow a single exponential function with a rate equal to k_{hydrol} and a maximum equal to the amount of Gα-[γ-^{32}P]GTP formed during phase 6. At 30°, however, we usually detect a second, small, and much slower component of [^{32}P]P$_i$ formation that presumably reflects hydrolysis of [γ-^{32}P]GTP that has continued to bind after addition of unlabeled GTP. In this case, data can be fit either by two exponentials or by the relevant exponential function superimposed on a slower and apparently linear reaction.

[27] Analysis of Genomic Imprinting of G$_s\alpha$ Gene

By LEE S. WEINSTEIN, SHUHUA YU, and JIE LIU

Introduction

The α subunit of the heterotrimeric G protein G$_s$ (G$_s\alpha$) is critical for receptor-stimulated intracellular cAMP generation. The human and mouse G$_s\alpha$ genes (*GNAS1* and *Gnas,* respectively) are located within syntenic regions at 20q13 in human[1–3] and distal chromosome 2 in mouse.[4,5] Originally these genes were

[1] P. V. Gejman, L. S. Weinstein, M. Martinez, A. M. Spiegel, Q. Cao, W-T. Hsieh, M. R. Hoehe, and E. S. Gershon, *Genomics* **9**, 782 (1991).

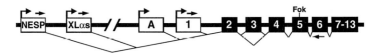

FIG. 1. Organization of the *GNAS1/Gnas* genes. Four alternative promoters are shown with their corresponding upstream exons as open boxes. The two most upstream promoters, which generate transcripts for NESP55 and XLαs, respectively, are separated from the rest of the gene by ~35 kb in humans. The exon A promoter generates a transcript of unknown function that probably does not encode a protein. Use of the exon 1 promoter generates transcripts encoding $G_s\alpha$. All four alternative upstream exons splice into exon 2. Exons 2 through 13, which are common to all of the major gene transcripts, are shown as black boxes (exons 7 through 13 as a single box). Long and short forms of $G_s\alpha$ are formed by alternative splicing of exon 3. The position and direction of RT-PCR primers (arrows) and position of the exon 5 *Fok*I polymorphism (Fok) used for assessing allele-specific expression of each human *GNAS1* transcript are shown. *Gnas* knockout mice were generating by targeting an insertion into exon 2.

shown to have 13 coding exons for $G_s\alpha$ (Fig. 1),[6] but they are now known to be much more complex, with three additional promoters and upstream exon regions which generate alternative transcripts by splicing onto a common exon (exon 2).[7–10]

Genomic imprinting is a phenomenon affecting a small number of autosomal genes in which expression of the gene product from one of the two parental alleles is markedly reduced or totally absent (for reviews, see Refs. 11–13). For some imprinted genes the paternal allele is poorly expressed (e.g., *H19, Igf2r*), whereas for other imprinted genes it is the maternal allele that is poorly expressed (e.g., *Snrpn, Igf2*). It is presumed that the modification that marks an allele as maternal or paternal must be erased in the primordial germ cell, reestablished in the male or female gametes, and maintained in somatic cells throughout development. At present, the most likely candidate for the modification that leads to imprinting

[2] V. V. N. Gopal Rao, S. Schnittger, and I. Hansmann, *Genomics* **10**, 257 (1991).

[3] M. A. Levine, W. S. Modi, and S. J. O'Brien, *Genomics* **11**, 478 (1991).

[4] C. Blatt, P. Eversole-Cire, V. H. Cohn, S. Zollman, R. E. K. Fournier, L. T. Mohandas, M. Nesbitt, T. Lugo, D. T. Jones, R. R. Reed, L. P. Weiner, R. S. Sparkes, and M. I. Simon, *Proc. Natl. Acad. Sci. U.S.A.* **85**, 7642 (1988).

[5] J. Peters, C. V. Beechey, S. T. Ball, and E. P. Evans, *Genet. Res.* **63**, 169 (1994).

[6] T. Kozasa, H. Itoh, T. Tsukamoto, and Y. Kaziro, *Proc. Natl. Acad. Sci. U.S.A.* **85**, 2081 (1988).

[7] A. Swaroop, N. Agarwal, J. R. Gruen, D. Bick, and S. M. Weissman, *Nucleic Acids Res.* **19**, 4725 (1991).

[8] Y. Ishikawa, C. Bianchi, B. Nadal-Ginard, and C. J. Homcy, *J. Biol. Chem.* **265**, 8458 (1990).

[9] R. H. Kehlenbach, J. Matthey, and W. B. Huttner, *Nature* **372**, 804 (1994).

[10] R. Ischia, P. Lovisetti-Scamihorn, R. Hogue-Angeletti, M. Wolkersdorfer, H. Winkler, and R. Fischer-Colbrie, *J. Biol. Chem.* **272**, 11657 (1997).

[11] M. Constancia, B. Pickard, G. Kelsey, and W. Reik, *Genome Res.* **8**, 881 (1998).

[12] M. S. Bartolomei and S. M. Tilghman, *Annu. Rev. Genet.* **31**, 493 (1997).

[13] S. M. Tilghman, *Cell* **96**, 185 (1999).

is DNA methylation.[11,14,15] This is based on the observation that most imprinted genes have differentially methylated regions (DMRs) that are methylated in only one parental allele. Allele-specific methylation within some of these DMRs has been shown to be established in the male or female gamete and maintained in the paternal or maternal allele throughout pre- and postimplantation development, making it likely that the methylation within these DMRs is critical for the establishment and maintenance of imprinting. These regions have been called core DMRs or imprint marks.

The first clue that *GNAS1* is an imprinted gene was provided by genetic studies showing that heterozygous inactivating *GNAS1* mutations lead to multihormone resistance when inherited from the mother but do not lead to hormone resistance when inherited from the father.[16,17] Imprinting of $G_s\alpha$ has been confirmed in mice[18] but still remains to be definitively proven in humans.[19,20] It is now established that two of the alternative upstream regions in *GNAS1* and *Gnas* are oppositely imprinted.[20–22] The most upstream promoter, which generates a transcript encoding the chromogranin-like protein NESP55, is only active in the maternal allele and is methylated in the paternal allele. In contrast, the $XL\alpha s$ promoter located 11 kb further downstream is only active in the paternal allele and methylated in the maternal allele.

Three types of observations establish that a gene is imprinted: (1) Distinct phenotypes result from either heterozygous inactivating mutations in the maternal or paternal allele of the gene or from maternal or paternal uniparental disomy (UPD) of the chromosomal region including the gene (UPD is the inheritance of a chromosome or subchromosomal region from a single parent). (2) Transcription is primarily (or totally) from one parental allele. (3) Regions within or in the vicinity of the gene are only methylated in one parental allele. In this chapter, we will summarize the experimental approaches used to establish all three criteria, with specific emphasis on the genes *GNAS1* and *Gnas*.

[14] E. Li, C. Beard, and R. Jaenisch, *Nature* **366,** 362 (1993).

[15] A. P. Bird, *Nature* **321,** 209 (1986).

[16] S. J. Davies and H. E. Hughes, *J. Med. Genet.* **30,** 101 (1993).

[17] L. S. Weinstein, *in* "G Proteins, Receptors, and Disease" (A. M. Spiegel, ed.), p. 23. Humana Press, Totowa, NJ, 1998.

[18] S. Yu, D. Yu, E. Lee, M. Eckhaus, R. Lee, Z. Corria, D. Accili, H. Westphal, and L. S. Weinstein, *Proc. Natl. Acad. Sci. U.S.A.* **95,** 8715 (1998).

[19] R. Campbell, C. M. Gosden, and D. T. Bonthron, *J. Med. Genet.* **31,** 607 (1994).

[20] B. E. Hayward, M. Kamiya, L. Strain, V. Moran, R. Campbell, Y. Hayashizaki, and D. T. Bonthron, *Proc. Natl. Acad. Sci. U.S.A.* **95,** 10038 (1998).

[21] B. E. Hayward, V. Moran, L. Strain, and D. T. Bonthron, *Proc. Natl. Acad. Sci. U.S.A.* **95,** 15475 (1998).

[22] J. Peters, S. F. Wroe, C. A. Wells, H. J. Miller, D. Bodle, C. V. Beechey, C. M. Williamson, and G. Kelsey, *Proc. Natl. Acad. Sci. U.S.A.* **96,** 3830 (1999).

Maternal vs Paternal *GNAS1/Gnas* Mutations Leading to Distinct Phenotypes

One hallmark of an imprinted gene is that heterozygous inactivating mutations of the gene in the maternal or paternal allele will lead to distinct phenotypes. This is because mutations in the active allele will lead to almost total absence of the gene product, whereas mutations in the inactive allele will have a minimal effect on its expression. Similarly UPDs will lead to abnormal phenotypes because of the presence of either two active or two inactive alleles. For example, maternal and paternal UPD of chromosome 15 leads to Prader-Willi and Angelman syndrome, respectively, due to the absence of paternally and maternally expressed imprinted genes.[23,24]

Heterozygous *GNAS1* mutations lead to Albright hereditary osteodystrophy (AHO), an autosomal dominant disorder characterized by short stature, obesity, and skeletal and mental defects.[17] In addition, maternal transmission of *GNAS1* mutations leads to resistance to several hormones that activate G_s-coupled pathways in their target tissues, such as parathyroid hormone (PTH) and thyrotropin (TSH), whereas paternal transmission does not lead to multihormone resistance [these two presentations are termed pseudohypoparathyroidism type 1a (PHP Ia) and pseudopseudohypoparathyroidism (PPHP), respectively]. If $G_s\alpha$ is poorly expressed (imprinted) from the paternal allele in hormone target tissues, then inactivating mutations in the normally active maternal allele would markedly reduce $G_s\alpha$ expression and hormone signaling while mutations in the normally inactive paternal allele would have few or no consequences.[17,18] Consistent with this model, the urinary cAMP response to administered PTH is markedly reduced in PHP Ia but is normal in PPHP.[25] $G_s\alpha$ expression in easily accessible tissues (e.g., erythrocytes, fibroblasts) is equally reduced by ~50% in both PHP Ia and PPHP,[25-29] indicating that $G_s\alpha$ is not imprinted in these tissues. Therefore this model would predict that $G_s\alpha$ is imprinted in a tissue-specific manner, as has been shown to be the case for other imprinted genes.[30-33]

[23] R. D. Nicholls, J. H. M. Knoll, M. G. Butler, S. Karam, and M. Lalande, *Nature* **342**, 281 (1989).

[24] S. Malcolm, J. Clayton-Smith, M. Nichols, S. Robb, T. Webb, J. A. Armour, A. J. Jeffreys, and M. E. Pembrey, *Lancet* **337**, 694 (1991).

[25] M. A. Levine, T. S. Jap, R. S. Mauseth, R. W. Downs, and A. M. Spiegel, *J. Clin. Endocrinol. Metab.* **62**, 497 (1986).

[26] Z. Farfel, A. S. Brickman, H. R. Kaslow, V. M. Brothers, and H. R. Bourne, *N. Engl. J. Med.* **303**, 237 (1980).

[27] M. A. Levine, R. W. Downs, Jr., M. Singer, S. J. Marx, G. D. Aurbach, and A. M. Spiegel, *Biochem. Biophys. Res. Commun.* **94**, 1319 (1980).

[28] M. A. Levine, C. Eil, R. W. Downs, Jr., and A. M. Spiegel, *J. Clin. Invest.* **72**, 316 (1983).

[29] Z. Farfel and H. R. Bourne, *J. Clin. Endocrinol. Metab.* **51**, 1202 (1980).

[30] Y. Jinno, K. Yun, K. Nishiwaki, T. Kubota, O. Ogawa, A. E. Reeve, and N. Niikawa, *Nature Genet.* **6**, 305 (1994).

[31] T. M. DeChiara, E. J. Robertson, and A. Efstradiadis, *Cell* **54**, 849 (1991).

[32] T. H. Vu and A. R. Hoffman, *Nature Genet.* **17**, 12 (1997).

We generated mice with an insertion within *Gnas* exon 2 that disrupts $G_s\alpha$ expression.[18] One advantage of studying knockout mice is that the phenotypic effect of maternal and paternal transmission of a single knockout can be easily examined by mating mutant females to wild-type males and wild-type females to mutant males to create maternal and paternal heterozygotes (denoted m−/+ and +/p− mice, respectively). In fact, both m−/+ and +/p− mice have distinct phenotypes,[18] strongly suggesting that *Gnas* is imprinted. The m−/+ and +/p− phenotypes might be due to lack of expression of maternally (e.g., $G_s\alpha$) and paternally (e.g., XLα s) expressed *Gnas* gene products. One phenotypic difference was the presence of PTH resistance in m−/+ but not in +/p− mice,[18] consistent with what is observed in PHP Ia and PPHP, respectively.

Abnormal phenotypes in mice genetically engineered with UPDs of specific chromosomal regions have identified regions that include imprinted genes.[34] Using this approach Cattanach and Kirk reported that maternal and paternal UPDs of a distal chromosome 2 region which includes *Gnas* produces distinct phenotypes, and they proposed that this is due to loss of expression of paternally and maternally expressed imprinted genes located within the region.[35] The maternal and paternal UPD phenotypes are very similar to the phenotypes observed in +/p− and m−/+ mice, respectively,[35,36] strongly suggesting that the UPDs may be caused by lack of expression of oppositely imprinted *Gnas* gene products.

Parental Allele-Specific Expression of *GNAS1*/*Gnas* Gene Products

The most common approach to determine the relative expression of a gene product from the maternal and paternal alleles of an imprinted gene requires a sequence polymorphism within the mRNA that allows one to distinguish the two alleles. In humans one is limited to naturally occurring polymorphisms within the mRNA. To be informative an individual must be heterozygous and at least one parent must be homozygous in order for parental origin to be assigned. For *GNAS1* a frequent *Fok*I polymorphism has been identified within exon 5 that is heterozygous in about half the population.[1] Genotyping is performed by amplifying an exon 5 fragment from genomic DNA by polymerase chain reaction (PCR) followed by *Fok*I digestion or direct sequencing of the PCR products. RNA from specific tissues is then amplified from informative subjects by reverse transcription-PCR (RT-PCR) and the relative expression of the *Fok*I+ and *Fok*I−alleles are determined. Since exon 5 is common to all *GNAS1* transcripts, the RT product needs to be amplified

[33] C. Rougeulle, H. Glatt, and M. Lalande, *Nature Genet.* **17**, 14 (1997).

[34] C. V. Beechey, B. M. Cattanach, and A. G. Searle, *Mouse Genome* **87**, 64 (1990).

[35] B. M. Cattanach and M. Kirk, *Nature* **315**, 496 (1985).

[36] C. M. Williamson, C. V. Beechey, D. Papworth, S. F. Wroe, C. A. Wells, L. Cobb, and J. Peters, *Genet. Res.* **72**, 255 (1998).

using a transcript-specific upstream primer complementary to the specific upstream exon and a common downstream primer complementary to exon 6 (Fig. 1).[20,21] For most tissues the products will appear as a doublet due to alternative splicing of exon 3.[6] If only the maternal allele is expressed, then *Fok*I will digest the RT-PCR products from *Fok*I$+^{mat}/-^{pat}$ subjects but will not digest the RT-PCR products from *Fok*I$-^{mat}/+^{pat}$ subjects. Paternal-specific expression will lead to the opposite result. Using this approach Hayward and colleagues established that the XLαs and NESP55 transcripts are expressed exclusively from the paternal and maternal alleles, respectively.[20,21] In these same experiments G$_s\alpha$ was shown to be expressed from both alleles. This is probably due to the fact that G$_s\alpha$ is imprinted in a tissue-specific and partial manner. In order to see evidence for imprinting of G$_s\alpha$ it is necessary to examine the specific tissues or cell types in which G$_s\alpha$ is imprinted. One limitation to examining allele-specific expression by *Fok*I digestion is that it is not very quantitative. Specifically, if both alleles are expressed to some degree, then a variable proportion of the RT-PCR products will be heteroduplexes containing one *Fok*I$^+$ and one *Fok*I$^-$ strand and these will be resistant to *Fok*I digestion. To more accurately determine the proportion of the two alleles within the RT-PCR product one can directly score the frequency of each allele by such methods as direct sequencing of subcloned fragments, allele-specific oligonucleotide hybridization, or RNase protection assays.

In mice the most common approach to studying the allele-specific expression of imprinted genes is to find sequence polymorphisms between *Mus musculus,* the most commonly used species of laboratory mouse, and other mouse species such as *M. castaneus, M. domesticus,* or *M. spretus* and to examine the relative expression of each allele after cross mating. However, to our knowledge no *Gnas* polymorphisms have as yet been identified between mouse species.

An alternative approach to studying allele-specific expression in imprinted genes is to examine the effect of maternal or paternal UPD on gene expression. For a maternally expressed gene its mRNA will be present in the setting of maternal UPD and absent in paternal UPD, whereas a paternally expressed gene will have the opposite expression pattern. If the gene is not imprinted, then expression should be unaffected by either UPD. Using both Northern analysis and RT-PCR Peters and colleagues showed that NESP55 mRNA is present in maternal UPD and absent in paternal UPD mice, while XLαs mRNA is present in paternal UPD and absent in maternal UPD mice.[22] This confirmed that in mice, as in humans, NESP55 and XLαs are expressed exclusively from the maternal and paternal alleles, respectively.

One can also take advantage of knockout mice to establish parental allele-specific expression by examining gene expression in m$-$/+ and +/p$-$ mice. For example, a maternally expressed gene will have no mRNA product (or possibly an abnormal fusion product) in m$-$/+ mice but will be expressed normally in +/p$-$ mice. In contrast, the expression of nonimprinted genes should be equally

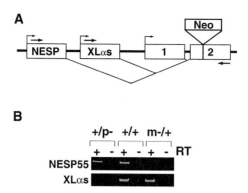

FIG. 2. RT-PCR using an either a NESP55- or XLαs-specific upstream primer and a common exon 2-specific downstream primer. (A) Diagram showing the position and direction of the NESP55- and XLαs-specific upstream primers (arrows above) and the common exon 2-specific downstream primer (arrow below). Alternative upstream exons and promoters for NESP55, XLαs, and G$_s\alpha$ (exon 1) are shown with splicing onto exon 2 indicated below. The position of the Neo cassette inserted into exon 2 is shown above. (B) Results of RT-PCR using either a NESP55-specific (above) or XLαs-specific (below) upstream primer. Mouse genotypes and the presence or absence of enzyme in the reverse transcription (RT) reaction are indicated above. RT-PCR was performed as previously described[41] using 1 μg of adrenal total RNA per sample and either a NESP55-specific (5′-GAGGAGAAGCAGCAGCACCGCTGCAAG-3′) or XLαs-specific (5′-CAGAAGCGCGC-AGATAAGAAACGCAGCAAGCTCATCGAC-3′) upstream primer and a common exon 2-specific downstream primer (5′-TCTCCGTTAAACCCATTAACATGCAGGATCCTCATCTGC-3′). The PCR profile consisted of an initial 3 min denaturation at 95°, followed by 35 cycles of annealing (56°, 30 sec), extension (72°, 90 sec), and denaturation (94°, 30 sec) and a final cycle with a 10 min extension. RT-PCR reactions were analyzed on 6% acrylamide gels (Novex, San Diego, CA) stained with ethidium bromide.

and partially reduced in m−/+ and +/p− mice. If Northern analysis is being used to quantitate gene expression it is important to be sure that the targeted mutation does not produce an aberrant transcript of similar size that is recognized by the probe. We have shown that G$_s\alpha$ expression in renal cortex and adipose tissue is significantly reduced in m−/+ and normal in +/p− mice, confirming that G$_s\alpha$ is preferentially expressed from the maternal allele in these tissues.[18] In contrast, G$_s\alpha$ expression was equally reduced by ~50% in many other tissues, indicating that the imprinting of G$_s\alpha$ is tissue-specific.

Using the *Gnas* knockout mouse model we have confirmed that NESP55 and XLαs gene products are oppositely imprinted (Fig. 2). RT-PCR was performed on adrenal gland RNA using upstream primers complementary to either the NESP55 or XLαs upstream exon and a common downstream primer complementary to exon 2. The targeted allele has a large insertion in exon 2 located between the upstream and downstream primers that in each case will prevent the amplification of the normal RT-PCR product. RT-PCR with the NESP55-specific upstream primer generated the expected product from the +/p− sample but no product from the m−/+ sample, indicating that NESP55 is only expressed from the maternal allele. In contrast, RT-PCR of the same samples using the XLαs-specific primer generated the expected product from the m−/+ sample but not from the +/p− sample, indicating that XLαs is only expressed from the paternal allele.

Parental Allele-Specific Methylation of *GNAS1/Gnas* Gene

DNA is methylated specifically on cytosines (C's) of CpG dinucleotides by various DNA methyltransferases (DNMTs), some of which are critical in establishing methylation in previously unmethylated CpG sites (*de novo* methylation) and others of which are critical in maintaining methylation in specific CpG sites in dividing somatic cells.[14,37,38] The role of methylation in the imprinting mechanism is established by the observations that imprinted genes have regions in which the two parental alleles are differentially methylated and that imprinting is disrupted in mice lacking DNMT1.[14]

Southern Analysis Using Methylation-Sensitive Restriction Enzymes

The first method developed to examine DNA methylation is Southern analysis of genomic DNA using restriction enzymes that are sensitive to methylation of a CpG within their recognition sites. For example, the enzyme *Hpa*II will only digest DNA at *Hpa*II sites that are unmethylated. In contrast the enzyme *Msp*I recognizes the same sequence as *Hpa*II, but will digest DNA at these sites regardless of whether or not they are methylated. Other methylation-sensitive enzymes include *Bss*HII, *Sma*I, *Sac*II, *Not*I, *Ngo*MI, *Fsp*I, *Hha*I, *Bst*UI, *Asc*I, and *Mlu*I. Unlike *Hpa*II, none of these enzymes have a methylation-insensitive isoschizomer.

Genomic DNA is first digested with an unrelated restriction enzyme that generates a restriction fragment of known length. If the enzyme produces a restriction fragment length polymorphism (RFLP), this is ideal because the enzyme will produce fragments of different size that allows one to determine the methylation status of each allele in a single experiment. A portion of the sample is then digested with a methylation-sensitive restriction enzyme known to have sites within the restriction fragment. If *Hpa*II is used, a separate digest with *Msp*I can be performed for comparision. These samples are separated on agarose gel, transfered to a membrane, and hybridized with a radiolabeled genomic DNA probe from the same region. Digestion with the initial enzyme alone will produce a fragment of expected length while addition of *Msp*I will completely digest this fragment to produce one or more

[37] M. Okano, D. W. Bell, D. A. Haber, and E. Li, *Cell* **99**, 247 (1999).
[38] F. Lyko, B. H. Ramsahoye, H. Kashevsky, M. Tudor, M. A. Mastrangelo, T. L. Orr-Weaver, and R. Jaenisch, *Nature Genet.* **23**, 363 (1999).

smaller bands. If the *Hpa*II sites are methylated in both alleles then the initial restriction fragment will be fully maintained. If both alleles are unmethylated the result will appear similar to that of the sample digested with *Msp*I. If only one allele is methylated then only half of the fragments will be digested with *Hpa*II and therefore one should see both the initial upper band and the lower band(s). If the first enzyme produces two bands due to an RFLP, then one can determine which allele is methylated since generally only one of the bands will be resistant to *Hpa*II digestion. Using this method on DNA samples from UPD mice and human parthenogenetic lymphocytes it has been established that the NESP55 and XLαs upstream regions are methylated only in the paternal and maternal alleles, respectively.[20–22] To illustrate this method we examined the methylation status of a portion of the mouse NESP55 upstream region in sperm.

Method. Genomic DNA was isolated from mouse renal cortex using the QIAamp Tissue kit (Qiagen, Valencia CA). Mouse spermatozoa aspirated from the ductus deferens were lysed in 20 mM Tris-HCl (pH 8.0), 20 mM EDTA, 220 mM NaCl, 80 mM dithiothreitol (DTT), 4% sodium dodecyl sulfate (SDS) with proteinase K (250 μg/ml) for 1 hr at 55°. Genomic DNA was isolated by phenol and chloroform extractions followed by ethanol precipitation. A fragment containing a portion of the NESP55 upstream region with the T7 promoter at one end was amplified from mouse genomic DNA by PCR using the primers 5′-GCAACTTTATAGGGCCCCATTG-3′ and 5′-TAATACGACTCACTATAGG-GAGGATCCATTCTCTTAGGTGCTCACC-3′ (Ref. 22) and a riboprobe was generated using the MAXIscript T7 *in vitro* transcription kit (Ambion, Austin TX). DNA samples (20 μg) were digested with *Hin*dIII alone, *Hin*dIII and *Msp*I, or *Hin*dIII and *Hpa*II (New England Biolabs, Beverly, MA). To ensure complete digestion the samples were digested with one enzyme at a time. For each enzyme a second aliquot was added after 2 hr of incubation, and the digestion was then allowed to continue overnight. Digested samples were separated by electrophoresis on a 1.5% agarose MS (Boehringer Mannheim, Indianapolis, IN) gel and transferred to Nytran filters (Schleicher & Schuell, Keene, NH). Filters were incubated with probes in QuikHyb hybridization solution (Stratagene, La Jolla, CA) at 68° for 1 hr and then washed twice with 2× SSC (1× SSC is 0.15 M NaCl plus 0.015 M sodium citrate), 0.1% (w/v) sodium dodecyl sulfate for 30 min at 25° and once with 0.1× SSC, 0.1% (w/v) sodium dodecyl sulfate for 1 hr at 68°. Filters were exposed to Kodak (Rochester, NY) Bio-Max MR films.

Results. Peters and colleagues showed using a similar probe that a 4.7 kb *Hin*dIII fragment was completely digested to a 470 bp fragment by *Hpa*II in DNA from maternal UPD mice but was completely resistant to *Hpa*II digestion in DNA from paternal UPD mice, confirming the presence of two closely spaced *Hpa*II sites in the NESP55 upstream region that are methylated exclusively in the paternal allele.[22] Similar digestion of genomic DNA from renal cortex of wild-type mice by *Hin*dIII and *Hpa*II produces both 4.7 kb and 470 bp bands, indicating the

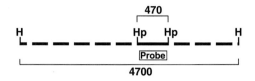

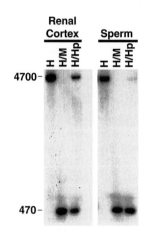

FIG. 3. Southern analysis using a methylation-sensitive restriction enzyme. A 4700 bp *Hin*dIII (H) restriction fragment from the NESP55 upstream region in mouse *Gnas* is depicted, showing the presence of two *Hpa*II (Hp) sites that are separated by 470 bp. The position of the sequence complementary to the riboprobe used in the Southern analysis is shown below. Genomic DNA samples from renal cortex (left lower panel) and sperm (right lower panel) were digested with *Hin*dII alone (H), *Hin*dIII and *Msp*I (H/M), or *Hin*dII and *Hpa*II (H/Hp) and after Southern blotting were probed with the riboprobe shown on the top panel. For both samples *Hin*dIII digestion produces the expected 4700 bp band and *Msp*I digestion produces only the 470 bp band, since it cuts at the *Hpa*II sites whether or not they are methylated. About half of the renal cortex sample is digested by *Hpa*II from 4700 to 470 bp, consistent with the presence of a methylated paternal and unmethylated maternal allele.[22] In contrast *Hpa*II almost completely digests the sperm DNA sample, indicating that methylation of these two *Hpa*II sites is not established during gametogenesis.

presence of both the methylated paternal and unmethylated maternal allele (Fig. 3). In contrast, the same enzymes almost completely digest DNA from sperm to the 470 bp band, indicating that the methylation is virtually absent in sperm DNA. *Msp*I completely digests both samples since it is not sensitive to methylation.

Bisulfite-Modified Genomic Sequencing

In this method (Fig. 4), genomic DNA is denatured and then treated with sodium bisulfite and hydroquinone at an acid pH, which converts all unmethylated C's to uracil sulfonate.[39] After removing excess salts and remaining bisulfite

---

[39] S. J. Clark, J. Harrison, C. L. Paul, and M. Frommer, *Nucleic Acids Res.* **22,** 2990 (1994).

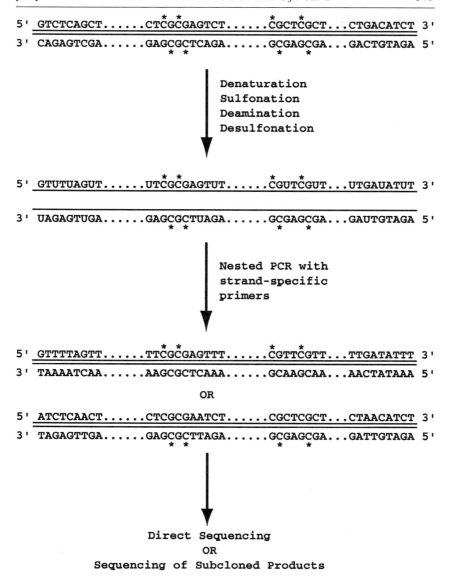

FIG. 4. Bisulfite-modified genomic sequencing. Genomic DNA is denatured and then treated with sodium bisulfite and hydroquinone, which converts unmethylated C's to uracil by sulfonation, deamination, and desulfonation. In contrast, methylated C's (denoted with asterisks) are protected from chemical conversion. The chemically modified samples are then amplified by two rounds of nested PCR using strand specific primers. In the final PCR products unmethylated C's (when sequencing in the $5' \rightarrow 3'$ direction) will be converted to T's while methylated C's will remain as C's in the sequence. The PCR products can be directly sequenced to assess the methylation status at each site averaged over the whole sample, or the products can be subcloned and sequenced to determine the methylation status of individual fragments.

anions, samples are treated with NaOH to desulfonate the samples. The net result is that unmethylated C's are converted to uracil while methylated C's are not. The samples are then amplified by PCR using nested primers and sequenced to determine which C's are converted to thymine (T) and are therefore unmethylated and which remained as C's and are therefore methylated. After bisulfite treatment the two strands are no longer complementary, and therefore primers must be designed for each individual strand (for an illustration of this point see Fig. 4). The products can be sequenced directly to determine the overall methylation status at each CpG site or can be subcloned to determine the methylation profile of individual PCR products.

The disadvantages of this method are that it is more technically difficult and time consuming, it only assesses the methylation status of a small region ($\sim$100–400 bp) in one experiment, and allele assignment is impossible if no polymorphism lies within the region sequenced. The advantages are the ability to examine the methylation of all CpG sites, not just those located within the recognition sites of methylation-sensitive restriction enzymes, and the ability to assess the methylation status in tissues or cells in which it is difficult to obtain large amounts of DNA, such as oocytes and blastocysts. Assessing the presence or absence of methylation in these latter tissues is critical for determining the developmental stage during which methylation is established and therefore helps to determine whether or not the methylation is a candidate imprint mark.

Method. The method used is similar to that described by Zeschnigk *et al.*[40] Samples in which large amounts of genomic DNA are available were first linearized by digesting 0.1 μg of the sample with a restriction enzyme that does not cut within the region being analyzed in a 100 μl reaction (in our example we used *Eco*RI). Small DNA samples such as those obtained from oocytes or blastocysts were used without prior digestion. DNA samples were denatured for 15 min at 37° by adding 11 μl of 3 M NaOH (final concentration, 0.3 M). To maximize denaturation, samples were then incubated at 95° for 3 min and immediately placed on ice. Sodium bisulfite (8.1 g) (Sigma, St. Louis, MO) was dissolved in 15 ml of water and then mixed with 1 ml of 40 mM hydroquinone (Sigma). The mixture was adjusted to pH 5 by adding 600 μl of 10 N NaOH. The denatured DNA sample (110 μl) was mixed with 1 ml of the bisulfite mixture and was incubated at 55° for 20 hr under mineral oil. Although others have found a 16 hr incubation to be sufficient,[39] we found that a longer incubation time produces better C to T conversion. Long incubation times ($>$36 hr) will lead to lower yields due to strand breakage.[39] The samples were then desalted using the Wizard DNA Clean-Up System (Promega, Madison, WI) and the eluted DNA (in 50 μl H_2O) was desulfonated

---

[40] M. Zeschnigk, B. Schmitz, K. Dittrich, B. Horsthemke, and W. Doerfler, *Hum. Mol. Genet.* **6,** 387 (1997).
[41] D. R. Warner, P. V. Gejman, R. M. Collins, and L. S. Weinstein, *Mol. Endocrinol.* **11,** 1718 (1997).

by adding 5.5 μl 3 M NaOH and incubating at 37° for 15 min. After the samples were neutralized by adding 55 μl of 6 M ammonium acetate, pH 7.0, the DNA was ethanol-precipitated, washed in 70% ethanol, dried, and redissolved in 20 μl water.

The bisulfite-modified DNA samples were next amplified by nested PCR. We find that performing nested PCR leads to a greater percentage of products with complete conversion of unmethylated C's to T's in the PCR product, possibly because the use of two sets of primers incorporating C to T changes leads to more selective amplification of DNA strands with complete chemical conversion. This can be assessed by examining the conversion of C's that are not within CpG dinucleotides, as these are unmethylated and should all be converted to T's. (When sequencing with the complementary primer one is looking for conversion of G's to A's.) In designing primers, one should try to select genomic sequences with a minimal number of CpG dinucleotides, and if these are within the primer sequence then a mixture of primers with C and T in the C position (G and A in the complementary downstream primer) should be used. Generally we amplify the sense strand using the following sets of upstream and downstream primers in the first PCR: 5′-GTAATTTTATAGGGTTTTATTG-3′ and 5′-ATCCATTCTCTTAAATACTCACC-3′ for the mouse NESP55 upstream region; 5′-GATTTAGATAGTTTGTTGTTGGTGT-3′ and 5′-AAACCCCACTC-CCCCCAATCAT-3′ for the mouse XLαs upstream region.[22] Nested upstream and downstream primers were 5′-GAGAGGATTAGTGGAGGTATTTTT-3′ and 5′-ACTCACCCTCTAACTCTACAAAAAAT-3′ for the NESP55 upstream region; 5′-GTGTTGGTGTTTATTTTTTGTGTT-3′ and 5′-ACCCAACAAATTACCCA-AAATACCA-3′ for the XLαs upstream region.[22]

PCRs were performed in 50 μl reaction mixtures containing 0.5 mM of each primer, 1.5 mM MgCl$_2$, dNTPs (200 μM each), and 2.5 U Taq DNA polymerase (Gibco/BRL, Rockville, MD). The PCR cycling profile consisted of an initial 5 min denaturation at 94°, followed by 35 cycles of denaturation (94°, 45 sec), annealing (65°, 45 sec), and extension (72°, 2 min) with a 10 min extension on the last cycle. In the initial PCR the template was 3–5 μl of bisulfite-treated genomic DNA and in the second PCR the template was 1 μl of a 50 : 1 dilution of the first PCR reaction.

Amplified fragments were gel purified and then either they were directly sequenced with the nested upstream primer using the THERMO Sequenase kit (Amersham Pharmacia Biotech, Piscataway, NJ), or the PCR products were subcloned into pCRII-TOPO by TA cloning (Invitrogen, Carlsbad CA) and individual clones were sequenced using the same primer (see Fig. 5). The advantage of direct sequencing is that one can get a general sense of the methylation status of each CpG. This approach is particularly useful when sites are completely methylated or unmethylated. Also, one is less likely to be confused by Taq sequencing errors or the occasional failure of unmethylated C's to convert to T since these are infrequent

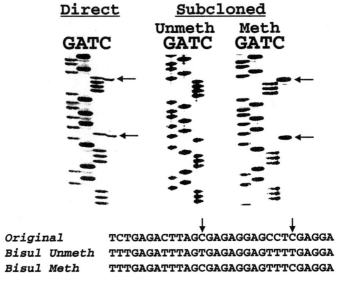

```
Direct              Subcloned
                 Unmeth    Meth
  GATC           GATC      GATC
```

Original	TCTGAGACTTAGCGAGAGGAGCCTCGAGGA
Bisul Unmeth	TTTGAGATTTAGTGAGAGGAGTTTTGAGGA
Bisul Meth	TTTGAGATTTAGCGAGAGGAGTTTCGAGGA

FIG. 5. Analysis of a portion of the mouse NESP55 upstream region by bisulfite-modified genomic sequencing. Mouse genomic DNA from renal cortex was chemically modified by bisulfite and a portion of the NESP55 upstream region[22] was amplified by nested PCR using primers specific for the sense strand as described in the text. The PCR products were either directly sequenced using the nested upstream primer (top left) or subcloned and then individual subclones were sequenced using the same primer (top right). The results of one clone derived from an unmethylated allele (Unmeth) and a second clone derived from a methylated allele (Meth) are shown. The original genomic DNA sequence as well as the sequence derived from the unmethylated and methylated alleles after bisulfite treatment are shown below. The methylated C's within CpG dinucleotides are indicated with arrows. All other C's were converted to T's in the final sequence.

and therefore will not be visible by direct sequencing. However, when sites are methylated on only a proportion of the alleles, it is impossible to quantitate the percent of methylated alleles by direct sequencing. The advantages of subcloning and sequencing are that one can determine the percent of alleles that are methylated at each site in a quantitative manner and determine whether or not methylation at different sites are present in the same alleles. Also, if a polymorphism is located within the sequence then one can directly determine the methylation status of each allele. By sequencing individual clones one will now see *Taq* sequencing errors and failure of individual unmethylated C's to convert to T. This will have a small effect on the final quantitation of methylated alleles at each site. Some clones will have many C's not within CpG dinucleotides that have not converted to T's, and these clones should be discounted from the analysis since the bisulfite conversion was incomplete (this also holds true for analysis by direct sequencing). It has been reported that for some sequences there is a subcloning bias such that one population of PCR fragments (representing either the methylated or unmethylated allele) will be preferentially subcloned, leading to quantitative results that do not reflect the true methylation status.[39] We have also found this to be the case in analyzing a portion of the XLαs upstream region. One can try to bypass this problem by attempting to subclone into a different vector.[39] In any case, we recommend performing direct sequencing in all experiments to confirm that the proportion of methylated alleles in the population of subcloned products reflects the true proportion of methylated alleles.

Conclusions

The *GNAS1/Gnas* genes are imprinted based on three criteria: mutations or UPDs that specifically disrupt the maternal or paternal allele lead to distinct phenotypes; several gene transcripts are expressed primarily or exclusively from only one parental allele; and regions of the gene are methylated in only one parental allele. Further studies will need to confirm whether or not $G_s\alpha$ is imprinted in specific tissues, such as the renal proximal tubules, in humans and, if so, what is the mechanism by which tissue-specific imprinting occurs. It will also be of interest to determine the mechanisms by which the gene has multiple oppositely imprinted transcripts, specifically whether the imprinting of each transcript is established independently or whether one specific region is critical for establishing imprinting throughout the *GNAS1* locus.

[28] Subcellular Localization of G Protein Subunits

By SUSANNE M. MUMBY

Introduction

G proteins are bound at the inner face of the plasma membrane where they are strategically positioned to interact functionally with membrane-spanning receptors and appropriate effectors. There is mounting evidence that G proteins and their signaling partners are organized in subdomains of the plasma membrane and that this compartmentalization may contribute to the speed and fidelity of signaling.[1,2] G proteins are not limited to the plasma membrane, however; activities and/or immunoreactive species have been found in other locations including

[1] R. R. Neubig, *FASEB J.* **8,** 939 (1994).

[2] C. Huang, J. R. Hepler, L. T. Chen, A. G. Gilman, R. G. W. Anderson, and S. M. Mumby, *Mol. Biol. Cell* **8,** 2365 (1997).

the Golgi apparatus, nuclei, mitochondria, endosomes, and the cytoplasm. More-over, G proteins may[3] or may not[4] change location in response to activation and/or delipidation. The techniques described here for subcellular fractionation and im-munofluorescence are useful for identifying the distribution and translocation of G proteins and other signaling proteins.

The first section below offers techniques to disrupt cells in order to begin fractionation. A simple method for a crude fractionation is then described for the separation of nuclei, membranes, and cytoplasm by differential centrifugation. A modification for obtaining nuclei freed of trapped membranes follows. Two methods are presented for isolation of membrane subdomains characterized by low buoyant density and enrichment in signaling proteins.

The second section describes a protocol for immunofluorescence that may be applied to permeabilized whole cells or plasma membranes isolated by sonication of cells adherent to coverslips. Preparation of coverslips, cells, and the sonicated plasma membranes is specified. Immunofluorescence offers a higher resolution picture of G protein distribution but suffers from an increased risk of artifactual results. Some risk is incurred because most methods that are strong enough to permeablize cells also can strip G proteins from their membrane locales.

Isoform specific antibodies are essential for specific detection of G proteins by Western immunoblotting or immunofluorescence. Many are commercially avail-able and methods for producing them have been published.[5,6] Commercial sources abound; they include (but are not limited to) Calbiochem (San Diego, CA), Chemi-con (Temecula, CA), NEN Life Science Products (Boston, MA), Gramsch Labora-tories (Schwahausen, Germany), Santa Cruz Biotechnology (Santa Cruz, CA), Sig-nal Transduction, Inc. (San Diego, CA), and Upstate Biotechnology (Lake Placid, NY). To search for a particular antibody, try http://www.antibodyresource.com.

Subcellular Fractionation

Disruption of Cells

In preparation for disruption, cells are rinsed with buffer to remove residual cul-ture medium that contains serum and secreted proteins. Suspended cells (scraped if they grow adherent to culture vessel) are concentrated by centrifugation and resuspended in a hypotonic buffer to make them more fragile and thus promote their disruption. Methods for disrupting cells are numerous. A preferable method is nitrogen cavitation because this approach can reduce the quantity of membranes

---

[3] P. B. Wedegaertner and H. R. Bourne, *Cell* **77,** 1063 (1994).

[4] C. Huang, J. A. Duncan, A. G. Gilman, and S. M. Mumby, *Proc. Natl. Acad. Sci. U.S.A.* **96,** 412 (1999).

[5] C. A. Chen and D. R. Manning, *in* "G Proteins: Techniques of Analysis" (D. R. Manning, ed.), p. 99. CRC Press, Boca Raton, FL, 1999.

[6] S. M. Mumby and A. G. Gilman, *Methods Enzymol.* **195,** 215 (1991).

trapped in the nuclear pellet. This method is recommended when the number of cells is not limiting since it is best to have cells at a density of at least 1×10^7/ml during cavitation. As little as 2 ml can be successfully subjected to cavitation in a model 4639 cell disruption bomb (Parr Instrument Company, Moline, Il). Alternatively cells may be homogenized (with a Dounce or Potter–Elvehjem homogenizer, 20–50 strokes) or sheared by multiple passages through a 25-gauge needle attached to a syringe. Keeping nuclei unbroken is key because if DNA is released it can trap more membranes than does a pellet of intact nuclei.

Crude Fractionation by Differential Centrifugation

This approach is useful to isolate a membrane fraction enriched in endogenous G proteins, to monitor release of G proteins from the membranes to cytosol, or to reveal the distribution of a mutant G protein expressed as a result of transfection. The recipes and definitions of abbreviations for the solutions are summarized in Table I. The volumes in the following protocol are for a single 150 mm diameter culture dish of adherent cells that are confluent (approximately 2.5×10^7 COS cells). These volumes are appropriate for use of a clinical type centrifuge, a variable-speed microfuge, and a tabletop ultracentrifuge (such as a TL 100, Beckman Instruments, Palo Alto, CA). A flow diagram of this procedure is shown in Fig. 1A. G proteins can be detected in the P1 and P2 fractions by Western immunoblotting (Fig. 2). Care should be taken to avoid formation of bubbles or foam.

1. Place culture dish on ice.
2. Remove culture medium by aspiration.
3. Rinse cells 2 times with 10 ml ice-cold PBS (taking care not to dislodge cells at this step).
4. Scrape cells into 5 ml of cold PBS + protease inhibitors.
5. Transfer the suspended cells to a centrifuge tube.
6. Rinse plate with 5 ml cold PBS + protease inhibitors and combine this material with cells in the tube (from step 5).
7. Pellet cells 500g for 5 min at 4°. (Transfer the supernatant fraction to a transparent tube or beaker. If this fraction is cloudy, then there are membranes present, indicating that a significant number of cells were prematurely disrupted and the yield of cytoplasmic and membrane proteins reduced. Membranes can be recovered by centrifugation of this fraction at 200,000g for 30 min at 4° and combined with others at step 18 below.)
8. Resuspend the cell pellet in 1 ml ice-cold HMEDS + protease inhibitors.
9. Allow cells to sit (and swell) on ice ≥10 min.
10. Disrupt cells by method desired (suggestions given above in previous section). If shearing or homogenization is employed, avoid introducing air bubbles.
11. Transfer homogenate to centrifuge tube.

TABLE I

SOLUTIONS FOR FRACTIONATIONS

Phosphate-Buffered Saline, pH 7.4 (PBS)
    140 mM Sodium chloride
    2.7 mM Potassium chloride
    1.5 mM Potassium phosphate, monobasic
    8 mM Sodium phosphate, dibasic
1000X protease Inhibitor Stocks
    10 mg/ml each in water
        Leupeptin
        Lima bean trypsin inhibitor
    10 mg/ml each in dimethylsulfoxide
        Phenylmethylsulfonyl fluoride (PMSF)
        N^{α}-*p*-Tosyl-L-lysine chloromethyl ketone (TLCK)
        N-Tosyl-L-phenylalanine chloromethyl ketone (TPCK)
Store at −20°
Make dilutions of stocks into buffer immediately before use.
All components available from Sigma (St. Louis, MO)
HEPES Buffer
    H   20 mM HEPES, pH 7.5–8.0
    M   2 mM Magnesium chloride
    E   2 mM EDTA
    D   2 mM Dithiothreitol (DTT)
    S   250 mM Sucrose
    K   10 mM Potassium chloride
First three components (HME) can be stored as 10× stock.
DTT and sucrose should be freshly added to HME.
MBS: MES-buffered saline
    25 mM 2-(*N*-Morpholino)ethanesulfonic acid (MES), pH 6.5
    150 mM NaCl
Tricine Buffer
    T   20 mM Tricine, pH 7.8
    E   1 mM EDTA
    S   250 mM Sucrose
Sucrose Stock [86% (w/v) = 65% (w/w) = 2.5 M]
Place 150 ml water in 600 ml beaker that is positioned in 60° water
    bath (on top of a combination hot plate/magnetic stirrer).
Add 428 g sucrose slowly to the water stirring in beaker.
Add more water to beaker (up to point between 475 and 500 ml lines).
Dissolution of sucrose takes approximately 90 min with stirring.
Transfer sucrose solution to graduated cylinder.
Use water (and/or an optional sodium azide solution [5 ml of 2% (w/v) as preservative)
    to rinse beaker and fill graduated cylinder to 500 ml mark.
Refractive index of final solution should be 1.4532.

A. Crude Fractionation

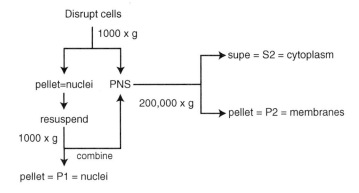

B. Membrane Free Nuclei

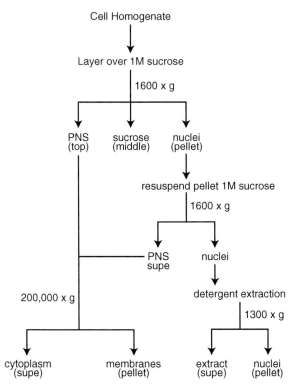

Fig. 1. Flow charts for crude fractionation by differential centrifugation (A) and an alternative to reduce contamination of the nuclear fraction by membranes (B).

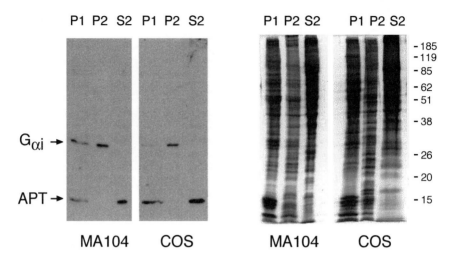

FIG. 2. Western blotting and Coomassie blue staining of crudely fractionated MA104 and COS cells. Cell type is indicated below each panel. Approximately 20 μg of protein per lane was resolved by SDS–PAGE. The cytoplasmic fractions were precipitated by 4 volumes of acetone and solubilized in sample buffer. Successful fractionation results in distinct patterns of protein staining between P1 (nuclear), P2 (membrane), and S2 (cytoplasmic) fractions (right two panels). Transfer of proteins to nitrocellulose and processing successively with antibodies specific for $G_{\alpha i}$ and acyl protein thioesterase (APT) are predominant in the P2 and S2 fractions, respectively (left two panels).

12. Rinse cell disruption vessel (bomb, homogenizer, or syringe) with minimal volume of HMEDS + protease inhibitors and combine this liquid with the homogenate in the centrifuge tube (from step 11).

13. Pellet nuclei and unbroken cells at 1000g, 4° for 5 min.

14. Transfer the supernatant fraction (postnuclear supernatant containing membrane and cytoplasmic proteins) to an ultracentrifuge tube.

15. Gently resuspend the pellet from step 13 with 0.2 ml HMEDS + protease inhibitors and repeat the 1000g centrifugation to recover more membranes and cytoplasm from the nuclear pellet.

16. Transfer the supernatant fraction and combine it with postnuclear supernatant from step 14. It is advisable to set aside a small portion of this combined fraction for analysis. Subject the remainder to ultracentrifugation at 200,000g for 30 min at 4°. This will separate membranes from cytoplasmic proteins.

17. During the ultracentrifugation of the postnuclear supernatant, gently resuspend the second 1000g pellet (from steps 15/16) in 0.1 ml HMEDS + protease inhibitors. This is designated the P1 or nuclear fraction.

18. Following the 200,000g centrifugation of the postnuclear supernatant, a pellet should be clearly discernible. Thoroughly and carefully transfer the

200,000g supernatant fraction to a storage tube. This tube is designated the S2 or cytoplasmic fraction.

19. The pellet may be rinsed with 0.2 ml of HMED + protease inhibitors to remove any residual soluble proteins if care is taken not to disturb it. Soak 200,000g pellet in 0.1 ml HMED + protease inhibitors for 10 min to aid in resuspension. Use a pipette tip (on a micropipettor set at 50 μl) to triturate the pellet into a uniform suspension of membranes (while meticulously avoiding the introduction of air bubbles).

20. Aliquot samples of each fraction for protein determination. Snap freeze the remainder of the samples in liquid nitrogen and store at $-80°$. Freeze/ thawing can cause nuclei to break and viscous DNA to be released. This DNA can be pelleted by centrifugation at 200,000g (Beckman TL100).

Notes: For Western immunoblotting of typically dilute cytoplasmic fractions (S2), it is usually necessary to concentrate proteins by precipitation with acetone or trichloroacetic acid. Acetone precipitation of nuclear fractions (P1) is recommended to avoid problems with viscous DNA released by SDS–PAGE sample buffer added directly to a nuclear fraction.

Modification of Crude Fractionation to Produce a Membrane-Free Nuclear Fraction

This approach can substantially reduce contamination of the nuclear fraction by membranes (which can occur in the procedure described above, Fig. 2). The presence of G proteins in the P1 fraction (Fig. 2) is often the result of such membrane contamination. This protocol is included here because some G proteins (and other proteins that act on them) are found associated with nuclei. The method (based on that by Chen et al.[7]) begins with a homogenate of 10^8 cells (four 150 mm dishes of confluent COS cells) disrupted by nitrogen cavitation in HMEDK buffer (Table I). The number of cells and volumes can be reduced if cells are disrupted by Dounce homogenization. Potassium in the buffer helps preserve the integrity of the nuclei. A flow diagram of this procedure is shown in Fig. 1B.

1. Layer 10 ml of cell homogenate onto 10 ml of cold 1 M sucrose in HMEDK buffer.
2. Centrifuge 1600g at $4°$ in swinging bucket rotor for 10 min to pellet nuclei.
3. Transfer 10 ml from top of tube to 50-ml tube and designate this the post-nuclear supernatant fraction. If you see a cloudy band near top of sucrose layer, transfer it to the same tube.
4. Remove and discard remaining sucrose layer from nuclear pellet.

---

[7] R.-H. Chen, C. Sarnecki, and J. Blenis, *Mol. Cell. Biol.* **12**, 915 (1992).

5. Resuspend nuclear pellet in 10 ml 1 M sucrose in HMEDK buffer (save an aliquot of these unwashed nuclei to analyze).
6. Centrifuge the resuspended nuclei in a swinging bucket rotor, 1600g, 5 min, 4°. Remove and combine the supernatant fraction with the postnuclear supernatant from step 3, mix, and set aside an aliquot to analyze.
7. Centrifuge the postnuclear supernatant at 200,000g, 4°, for 30 min. Remove the supernatant fraction carefully and thoroughly; designate this as the S2 or cytoplasmic fraction. Resuspend the pellet (P2 or membrane fraction) as described for the crude fractionation above (step 19).
8. During the centrifugation of the postnuclear supernatant fraction, resuspend nuclear pellet in 0.8 ml of HMEDK. Nuclei should be easily discerned under a light microscope.

A portion of the nuclei can be further freed of membrane and cytoplasmic material by brief extraction with a cocktail of ionic and nonionic detergents [1 part 10% deoxycholate, 2 parts 10% Nonidet P-40 (NP-40), w/w].[8] Add a 15% volume of the detergent cocktail to an aliquot of nuclear suspension, vortex vigorously 3 sec, and pellet the extracted nuclei at 1300g, 4°, 6 min. Resuspend nuclei in HMEDK if you desire that they remain intact. For analysis by Western immunoblotting, acetone precipitation of nuclear fractions may avoid the problem of viscous DNA released when SDS–PAGE sample buffer is added directly to nuclear fractions.

Membrane Subdomains

Many G proteins are found in subdomains of membranes that are characterized by low buoyant density and relative resistance to extraction by nonionic detergents. These subdomains are generally enriched in a number of signaling proteins, cholesterol, and sphingolipids and have generated keen interest because they may provide an organization of membrane components that are crucial for fidelity and efficiency of signal transduction. Methods for isolation of membrane subdomains vary, particularly with regard to the inclusion of detergent. Detergent is employed in the first protocol below, which is useful and convenient, but this use of detergent has been criticized for its potential to modify the structure and function of membrane subdomains during their isolation. Song *et al.* have published a detergent-free modification of this procedure, but it has not been as well characterized and the use of pH 11 buffer precludes measurement of most enzyme activities.[9] The second method is a neutral pH, detergent-free protocol that utilizes Percoll and OptiPrep gradients.[10] This procedure yields plasma membrane

---

[8] E. Holtzman and I. Smith, *J. Mol. Biol.* **17,** 131 (1966).
[9] K. S. Song, S. Li, T. Okamoto, L. A. Quilliam, M. Sargiacomo, and M. P. Lisanti, *J. Biol. Chem.* **271,** 9690 (1996).

domains that are highly enriched in signaling proteins and are useful for detecting enzyme activities.[2,11] Not included are methods for immunoisolation of caveolin-rich domains from cultured cells[12,13] or use of cationic silica particles to yield caveolae from perfused lung.[14,15]

Differential Detergent Extraction to Isolate Membranes of Low Buoyant Density. Membranes isolated by differential detergent extraction have been referred to by several terms including caveolae, rafts, and detergent-resistant membranes (DRMs). Although caveolae (which are morphologically identified as particular invaginations of the plasma membrane) are certainly present, there is not complete agreement among investigators that they are the exclusive components of low-density DRMs when whole cells are extracted with detergent. In addition, the yield of some caveolar proteins may be incomplete in DRMs. Two crucial aspects of this protocol are the temperature and the ratio of detergent to lipid (or protein). If this ratio or the temperature is too high, detergent resistant membrane proteins will be solubilized and will not rise in the gradient, as they should. A good marker for proper floating of membranes is caveolin (a protein marker for caveolae in most cell types), which can be detected by Western immunoblotting with commercially available antibodies (such as those from Transduction Laboratories, Lexington, KY). The bulk of total protein (soluble and higher density TX-100 resistant) remains at the bottom of the tube where the extract was loaded. The protocol below is adapted from two published procedures.[16,17] A flow diagram is shown in Fig. 3A and Western immunoblotting of fractions in Fig. 4. We have successfully scaled this protocol down 2.5- to 3-fold (by use of 1 plate of cells, 0.4 ml TX-100 extraction, a 3.6 ml sucrose gradient, and a Beckman SW60 rotor, not shown).

1. Rinse three 150 mm plates of confluent cells twice with ice-cold MES-buffered saline (MBS, Table I).
2. Scrape cells with 5 ml ice-cold MBS + protease inhibitors (per plate) and transfer to centrifuge tube (one tube for 3 plates of cells).
3. Rinse each plate with another 5 ml ice-cold MBS and transfer to the centrifuge tube.

[10] E. J. Smart, Y.-S. Ying, C. Mineo, and R. G. W. Anderson, *Proc. Natl. Acad. Sci. U.S.A.* **92,** 10104 (1995).
[11] P. W. Shaul, E. J. Smart, L. J. Robinson, Z. German, I. S. Yuhanna, Y. Ying, R. G. W. Anderson, and T. Michel, *J. Biol. Chem.* **271,** 6518 (1996).
[12] R. V. Stan, W. G. Roberts, D. Predescu, K. Ihida, L. Saucan, L. Ghitescu, and G. E. Palade, *Mol. Biol. Cell* **8,** 595 (1996).
[13] P. Oh and J. E. Schnitzer, *J. Biol. Chem.* **274,** 23144 (1999).
[14] B. S. Jacobson, J. E. Schnitzer, M. McCaffery, and G. E. Palade, *Eur. J. Cell Biol.* **58,** 296 (1992).
[15] J. E. Schnitzer, D. P. McIntosh, A. M. Dvorak, J. Liu, and P. Oh, *Science* **269,** 1435 (1995).
[16] M. P. Lisanti, Z. Tang, P. E. Scherer, and M. Sargiacomo, *Methods Enzymol.* **250,** 655 (1995).
[17] P. Liu and R. G. W. Anderson, *J. Biol. Chem.* **270,** 27179 (1995).

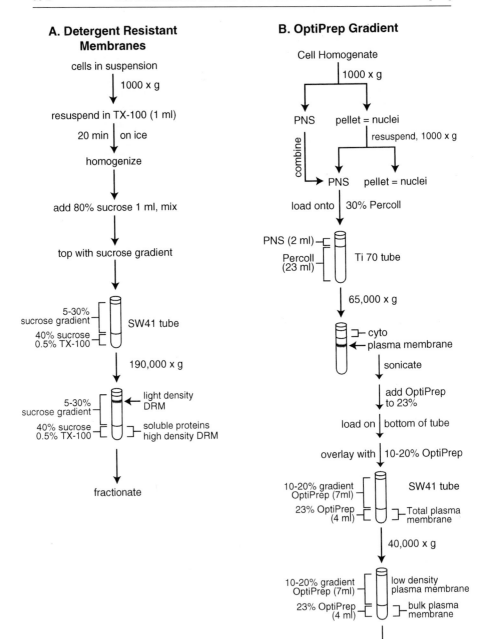

FIG. 3. Flow charts for isolation of membrane subdomains.

MDCK (α_O)

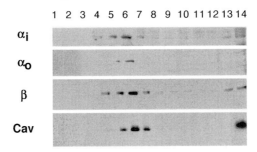

MDCK (vector)

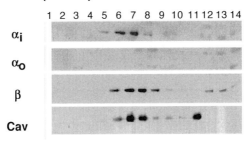

FIG. 4. Fractionation of endogenous and ectopically expressed G proteins and caveolin (cav) from TX-100 extracts of MDCK epithelial cells. Stably transfected MDCK cells that heterologously express α_o (upper set of panels) were compared with G418 resistant control cells that had been stably transfected with the empty vector (lower set of panels). Unlike the protocol in the text, in this case a 4-ml detergent extract was loaded at the bottom of the tube. Most of the cellular protein (visualized by Ponceau S staining of the blots, not shown) remained in the higher density fractions (numbered 8–14). The blots were probed with a monoclonal antibody to detect heterologously expressed α_o or with rabbit polyclonal antibodies to detect endogenous α_i, β, or caveolin (Cav). Reprinted from *Mol. Biol. Cell* **8,** 2365 (1997), with permission from the American Society for Cell Biology.

4. Pellet cells by centrifugation at 1000g, 10 min, 4°. If the supernatant is not clear, significant cell damage and loss is likely.
5. Disperse cell pellet gently by pipetting with 1 ml of 1% Triton X-100 in MBS. Avoid formation of bubbles.
6. Let cells incubate on ice 20 min, swirling the tube a couple of times during this period.
7. Homogenize cells with Dounce homogenizer, 10–12 strokes on ice. This homogenate should be ≥5 mg/ml protein to ensure flotation of detergent-resistant membranes in the sucrose gradient. Protein concentration can be determined without interference from detergent by Amido Black staining.[18]

8. Transfer homogenate to transparent 13-ml centrifuge tube.
9. Add an equal volume (1 ml) ice-cold 80% sucrose (w/v, made from 86% stock, Table I) in MBS + protease inhibitors to the homogenate, cover with Parafilm, and mix well by gentle inversion (for final concentration of 40% sucrose).
10. Form a detergent free 9-ml linear gradient of ice cold 5–30% sucrose in MBS on top of the detergent-containing 40% sucrose homogenate.
11. Centrifuge sample in SW41 rotor (Beckman Instruments, Palo Alto, CA): 39,000 rpm (190,000g, 16–24 h, 4°). Alternatively a Sorvall TH641 rotor may be used.
12. The light density DRMs may (or may not) be seen as a single light-scattering band in the upper half of the tube. Take equal volume fractions for analysis as shown in Fig. 4.

Detergent-Free OptiPrep Gradients. A flow diagram for this protocol developed by Smart *et al.*[10] is shown in Fig. 3B. The cells may be disrupted by nitrogen cavitation[2] or by homogenization as originally described. Plasma membranes are isolated in Percoll, sonicated, and then fractionated in an OptiPrep gradient. The bulk of the plasma membrane fragments remain at the bottom of this gradient where they were loaded, whereas the lighter density fragments, including caveolae, rise up into the lower density portion of the gradient. Protein concentration, Western blotting of G proteins and caveolin, and an activity assay for adenylyl cyclase are shown in Fig. 5.

1. Per gradient, prepare an ice-cold homogenate from approximately 10^8 cells (7–8 mg total protein, 4×150 mm dishes of cells) using 1 ml of TES + protease inhibitors (Table I; +1 mM dithiothreitol, optional).
2. Homogenize cells, 20 strokes in a 2-ml Potter–Elvehjem tissue grinder (Wheaton, from Fisher Scientific, Pittsburgh, PA).
3. Pellet nuclei and unbroken cells by centrifugation at 1000g, 10 min, 4°.
4. Transfer postnuclear supernatant fraction to a fresh 5-ml tube. Store on ice.
5. Resuspend the nuclear pellet with 1ml TES in the 2-ml size homogenizer.
6. Repeat centrifugation (step 3) of resuspended pellet in order to increase recovery of membranes.
7. Transfer the second postnuclear supernatant fraction and combine it with the first (from step 4). Mix and save small aliquots of combined postnuclear supernatant and nuclear pellet to assay.
8. Carefully layer combined postnuclear supernatant fractions (just under 2 ml in volume) over 23 ml of 30% (v/v) Percoll (Amersham Pharmacia Biotech, Piscataway, NJ) in TES in a Ti 70 ultracentrifuge tube.

---

[18] W. Schaffner and C. Weissmann, *Anal. Biochem.* **56**, 502 (1973).

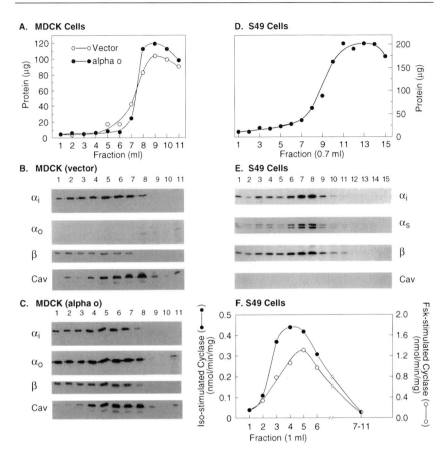

FIG. 5. OptiPrep gradient fractionation of detergent-free plasma membranes from MDCK epithelial or S49 lymphoma cells. Sonicated plasma membranes were brought to 23% OptiPrep in 4 ml and were placed at the bottom of the tube. A linear gradient of 20–10% OptiPrep was poured on top of the plasma membranes. Following centrifugation, either 1 ml (A, B, C, and F) or 0.7 ml fractions (D and E) were taken from the top of the tube. These fractions were analyzed for total protein content by Bradford assay (A and D), G protein or caveolin (Cav) content by Western immunoblotting (B, C, and E), or adenylyl cyclase activity (F). The specific activity of isoproterenol- (Iso-) and forskolin- (Fsk-) stimulated adenylyl cyclase in the individual fractions 1–6 and combined fractions 7–11 are shown in (F). Reprinted from *Mol. Biol. Cell* **8,** 2365 (1997), with permission from the American Society for Cell Biology.

9. Centrifuge Percoll-containing tube in a Beckman Ti 60 or Ti 70 rotor: 29,000 rpm (at average radius = 65,000g), 4°, 30 min.

10. Find a single light-scattering band of plasma membranes at 5.7 cm from bottom of tube. Remove 1.3 ml of material from top of tube and save as cytoplasmic fraction. With a Pasteur pipette, remove and discard remaining

material above the plasma membrane band. With a fresh Pasteur pipette transfer the plasma membrane band (approximately 2 ml) to a polyallomer Beckman SW41 or a Sorvall TH641 tube (ink-marked at level of 2-ml). If there are two light-scattering bands, take the upper one as plasma membrane.

11. If necessary bring plasma membrane volume up to the 2-ml mark with TES.

12. Fracture plasma membranes into smaller fragments by sonication with a 3 mm diameter probe placed midway within the plasma membrane suspension (Vibra Cell high-intensity ultrasonic liquid processor, model VC 60 from Sonics and Materials, Danbury, CT). Deliver 2 bursts of 50 joules/watt-sec, rest samples on ice for 2 min to cool, and repeat the sonication. Repeat this process again for a total of 6 bursts per tube. Set aside a small aliquot of sonicated plasma membranes.

13. Add 1.84 ml of 50% OptiPrep and 0.16 ml TES to 2 ml of sonicated plasma membrane suspension for final concentration of 23% OptiPrep (Accurate Chemical & Scientific, Westbury, NY). Cover tube with Parafilm and mix by inversion and brief vortexing.

14. Pour a 7 ml 20% to 10% (v/v) linear gradient of OptiPrep in TES on top of the 23% OptiPrep/plasma membranes.

15. Centrifuge the gradient in a SW41 rotor: 18,000 rpm (40,000g at average radius), 4°, 90 min.

16. Take 0.7- to 1.0-ml fractions for analysis (Fig. 5).

17. Protein concentrations may be determined by Bradford assay.[19] Mix 100 μl sample of each fraction with 5 ml of diluted (1 : 4) Bradford reagent (Bio-Rad, Hercules, CA) and measure the absorbance at 595 nm. OptiPrep interferes with other standard protein determination methods including the micro-Bradford assay.

18. Fractions usually require concentration by trichloroacetic acid or acetone precipitation for analysis by Western immunoblotting. Low yields from the precipitation step are often troublesome. Low concentration of protein in the upper fractions and the presence of Percoll in the lower fractions cause the problems. Addition of 5 μg bovine serum albumin (and/or deoxycholate at final concentration of 0.015%, w/v) to each precipitation sample can solve the problem of low protein concentration. Bovine serum albumin also serves well as a standard to monitor yield of the precipitation when samples are compared on a blot that is Ponceau S stained (before blocking the blot with protein-containing buffer and processing with antibodies). A method to manage the problem with Percoll is to add 200 μl of 2-fold concentrated SDS–PAGE sample buffer to each 1-ml fraction. Heat samples in boiling water bath for 5 min. The SDS in the sample buffer seems

---

[19] M. M. Bradford, *Anal. Biochem.* **72,** 248 (1976).

to promote formation of a Percoll pellet when spun in a microfuge. Precipitate the supernatant fraction from the microcentrifugation with 10% (w/v) trichloroacetic acid, rinse pellet with acetone, dry, and solubilize pellet with SDS–PAGE sample buffer.

Immunofluorescence

This higher resolution method for visualizing the subcellular location of G proteins can be performed on whole cells (Fig. 6) or plasma membranes (Fig. 7) prepared by sonication of cells attached to coverslips. The following protocols have been employed to support fractionation results[2] and to demonstrate membrane-delimited changes in G protein distribution.[4] Particular care must be taken to reduce the risk of artifactual results inherent in this approach, especially if the goal is to detect G proteins present at endogenous levels. Precautions include the method of fixation, avoidance of detergents and organic solvents, and quality control of antibodies.

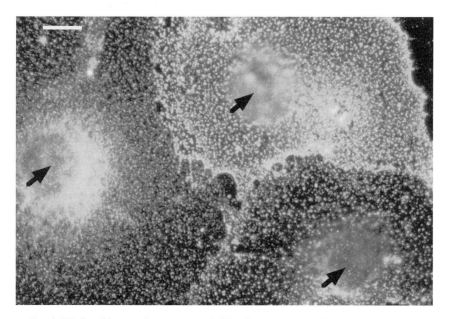

FIG. 6. Whole cell immunofluorescence. MA104 cells grown on coverslips were transiently transfected to express a constitutively activated form of α_{i1} (Q204L) and processed for immunofluorescence with affinity-purified B087 polyclonal antibodies specific for α_{i1}/α_{i2}.[4] The antibodies detect both the overexpressed mutant protein (upper right-hand cell) and endogenous α_i (the other two cells) in the same distribution. The arrows designate the position of nuclei. The punctate pattern of staining represents both plasma membrane subdomains[2] and intracellular vesicles (inferred from focusing at multiple focal planes). Bar: 2 μm.

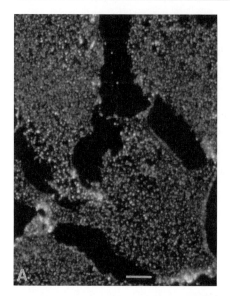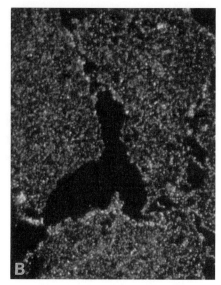

FIG. 7. Immunofluorescence of plasma membranes performed with antibodies specific for G protein subunits. *En face* views of the inner side of plasma membrane fragments were obtained by sonicating MA104 cells adherent to coverslips. Oregon Green-conjugated secondary antibodies were used to visualize primary antibodies to detect: (A) α_i subunits with B087 antibodies (10 μg/ml) or (B) β subunits with T20 antibodies (Santa Cruz Biotechnology, 1 μg/ml). The larger areas of the photographs that are devoid of fluorescent signal represent spaces where plasma membrane fragments are absent. Bar: 2 μm. Reprinted from *Mol. Biol. Cell* **8,** 2365 (1997), with permission from the American Society for Cell Biology.

Fixation should be performed with formaldehyde (Table II) rather than alcohols or acetone. Use of organic solvent solublizes membrane lipids and thus G proteins may be lost. Fixation with formaldehyde and permeabilization with saponin are suitable for retaining G proteins in whole cells but permeabilization is unnecessary for isolated plasma membranes. Use of Triton X-100 is not advised; it can cause solubilization (loss) of G proteins from membranes and/or nonspecific labeling of other compartments.

Suitability of primary antibodies is key (commercial sources are listed in the Introduction to this chapter). They should be characterized for reactivity among purified G proteins (for isoform specificity) and among total cell proteins by Western immunoblotting. A single reactive species in whole cell lysates is required for the highest level of confidence in immunofluorescence results (although this may not be possible if the antibody only reacts with naturally folded protein). Keep in mind that Western immunoblotting conditions are usually more stringent (presence of detergent throughout antibody incubations) and thus antibody reactivity can be "cleaner" than the conditions used for immunofluorescence. On the other hand,

TABLE II
SOLUTIONS FOR IMMUNOFLUORESCENCE

IF Buffer A
   10 mM Sodium phosphate buffer, pH 7.4
   144 mM NaCl
   2 mM MgCl$_2$
IF buffer A + BSA
   1% BSA (w/v) in IF buffer A
   (BSA, fraction V, Sigma, St. Louis, MO)
PBS/M
   PBS (Table I) containing 1 mM MgCl$_2$ pH 7.4
Sonication Buffer
   25 mM HEPES
   25 mM KCl
   2.5 mM Mg acetate
   pH 7.0
Formaldehyde
   Prepare 10% paraformaldehyde (w/v) stock in 50 mM phosphate buffer
     (once in solution it is termed formaldehyde).
   Heat water bath to 60° on hot plate/stirrer in fume hood.
  1. In fume hood, weigh out 10 g electron microscopy grade paraformaldehyde and add to
     50 ml of 0.1 M, pH 7.4 sodium phosphate buffer in beaker with magnetic stir bar.
  2. Place beaker in 60° water bath and begin stirring action.
  3. Add 30 ml of purified water to the beaker and cover loosely with aluminum foil.
  4. Heat contents of beaker to 60°.
  5. Add minimum volume of 1 N NaOH dropwise to clear solution.
  6. Check a drop of clarified solution with pH paper and compare result with
     phosphate buffer as control.
  7. If necessary add minimum volume 1 N HCl to pH solution to 7.4.
  8. Cool solution on ice
  9. Bring volume to 100 ml with purified water.
 10. Filter solution through 0.45-μm filter.
 11. Use solution fresh or store aliquotted at −20°. Dilute thawed formaldehyde
     solution to desired concentration and refilter through 0.45-μm filter before
     use. Discard thawed formaldehyde solution that is not used; do not refreeze.
     Dispose of formaldehyde properly as toxic waste.

---

denaturation of proteins (inherent in SDS–PAGE/Western immunoblotting) can expose potential immunoreactive sites that may not be exposed in the native protein fixed *in situ*. Usually affinity purification of polyclonal antisera is necessary for immunofluorescence because preimmune serum commonly gives substantial background staining. Fluorescently tagged secondary antibodies allow amplification and visualization of the primary antibody. We find Texas Red and Oregon Green (Molecular Probes, Eugene, OR), which can be visualized with filter sets for rhodamine and fluorescein, respectively, are superior to the latter fluorophores in resistance to photobleaching.

Preparation

Coverslips. Sterilize coverslips (18 mm diameter, number 1 thickness) by grasping each gently with fine forceps, dipping in 95% (v/v) ethanol, flaming, placing dozens of them in a glass petri dish, covering, and autoclaving on a dry cycle. Place individual sterilized coverslips in 12-well plate (using sterile forceps) and, if necessary, coat the coverslips to improve cell adherence and spreading. Coating can be especially important to improve yield of sonicated plasma membranes. Suggested coatings that can be tested and compared are poly(L-lysine) (Sigma, St. Louis, MO) 0.5 mg/ml in 0.1 M sodium borate, pH 8.2; fibronectin (Sigma St. Louis, MO); and laminin (Becton/Dickenson, Bedford, MA). From stocks of fibronectin and laminin at 1 mg/ml in Tris-buffered saline, dilute to 20 μg/ml in serum-free culture medium. Place coating solution on coverslip for 45 min at room temperature or overnight at 4°. Remove solution by aspiration and displace residual poly(L-lysine) by rinsing twice with water (unnecessary for protein coatings). Fibronectin and laminin are expensive; the working dilutions can be snap frozen and reused once. Introduce suspended cells over coverslips and allow cells to attach and spread for 1–3 days (time necessary to fully adhere and spread is cell type dependent). Be aware that coating of coverslips can influence cell morphology and/or rate of proliferation. Whole cell architecture is best viewed if cells are not confluent when immunofluorescence is performed.

Isolation of Plasma Membranes by Sonication. This protocol is derived from published procedures developed in R. G. W. Anderson's laboratory.[20,21] During this procedure the tissue culture plates should be kept on a glass plate placed on ice, except as noted. Keep cells and membranes wet at all times. To rinse or exchange buffers, remove liquid by aspiration from one well of cells and replace it immediately before removing liquid from the next well. Rinses are introduced into wells by a repeater pipette (such as model 4780, fitted with 12.5-ml Combitip from Eppendorf, Westbury, NY).

1. Rinse all wells of 12-well plate of cells (on coverslips) with 2 ml ice-cold PBS/M (Table II).
2. Rinse with 2 ml ice-cold sonication buffer (Table II) containing protease inhibitors (Table I) and 10 μM GDP, remove, and then add 5 ml of the same solution.
3. Transfer plate to a flat, height adjustable platform (such as a Lab Jack from Fisher Scientific, Pittsburgh, PA) and position a $\frac{1}{2}$-inch diameter sonicator probe into the liquid, 9 mm directly above the cover slip in one well. Sonicate

[20] M. S. Moore, D. T. Mahaffey, F. M. Brodsky, and R. G. W. Anderson, *Science* **236,** 558 (1987).
[21] K. H. Muntz, P. C. Sternweis, A. G. Gilman, and S. M. Mumby, *Mol. Biol. Cell* **3,** 49 (1992).

with a single burst of power* from an adjustable power sonicator (Vibra Cell, Sonics and Materials, Danbury, CT) and repeat for a second well of cells before returning the plate to the ice-chilled glass plate surface.

4. Rinse sonicated wells of cells three times with 2 ml ice cold sonication buffer containing protease inhibitors and 10 μM GDP, then again without protease inhibitors. Addition of each rinse should be done with the Combitip held at a tangent to the side of the well. This technique should produce a swirling action as the liquid is dispensed and thus will adequately remove cell debris.

5. Fix with 0.5 ml of cold 4% formaldehyde per well, 20 min, on ice.

6. Follow indirect immunofluorescence protocol below starting at step 2.

Antibody Dilutions. To optimize signal over background, useful concentrations of primary and secondary antibody must be determined empirically. Purified antibodies are generally effective in the range of 0.1–10 μg/ml. Dilute primary and secondary antibodies in IF buffer A + BSA (Table II) and remove any aggregates by centrifuging or filtering. Centrifuge at 4°, preferably in a tabletop ultracentrifuge at $\geq 100,000g$ for 30 min (otherwise in a microfuge). Carefully remove the supernatant fraction for use. Alternatively the dilution can be passed through a low protein-binding syringe filter such as Whatman 0.45-μm pore size PVDF (Clifton, NJ).

Indirect Immunofluorescence

Controls are essential; suggestions for them follow the protocol. During this procedure it is crucial to keep cells or membranes wet until directed to dry coverslips in step 12. When removing a solution, be prepared to replace it immediately with the next.

1. Fix cells or sonicated plasma membranes with 0.5 ml of filtered 4 or 8% formaldehyde (in phosphate buffer, Table II) for 15 min at 37° followed by 15 min at room temperature. The higher concentration of formaldehyde is desirable to retain cytoplasmic proteins.[21]

---

*The level of power necessary to retain the basal plasma membrane, while effectively removing other cellular material from the coverslip, is determined empirically. Test several power settings in duplicate and then following the rinse steps, stain the coverslips with Coomassie blue (0.25% Coomassie blue, w/v; 30% 2-propanol, v/v; 1% acetic acid, v/v, which is same solution used to stain proteins in an SDS–PAGE gel). Rinse the coverslips with PBS and view them by light microscopy at 125-fold magnification (ocular × objective magnification). If a sonication pulse is too weak the result will be mostly whole cells that are intensely stained. If the pulse is too great there will be no staining. If the pulse is just right, there will be faintly stained membranes visible and, at the edges of coverslips, perhaps a few intensely stained whole cells.

2. Remove fixative by aspiration and two room temperature incubations with PBS/M for 5 min each. Dispose of formaldehyde as toxic waste.

3. Permeabilize whole cells (but not sonicated plasma membranes) with 0.5 ml 0.1% saponin in IF buffer A for 5 min on ice. Alternatively, cells may be perforated with damp nitrocellulose.[22] If cells are fixed with 8% formaldehyde then this step is not necessary for localizing G proteins; the higher concentration of formaldehyde appears to both fix and permeabilize cells (unpublished data).

4. Quench any residual formaldehyde with two 10 min incubations with 1 ml serum-free culture medium (which contains amino acids that will react with formaldehyde) at room temperature.

5. Rinse with 1 ml IF buffer A + bovine serum albumin (BSA) at room temperature.

6. Block nonspecific immunoglobulin binding sites by incubating for 30 min with 0.5 ml 10% serum (preferably from animal species in which the secondary antibody was raised) in IF buffer A + BSA.

7. With sharply pointed forceps transfer coverslips (cell side down) onto a 55 μl drop of primary antibody (diluted in the same solution used to block nonspecific binding sites above). Multiple antibody drops may be arranged on a single piece of Parafilm that is placed on top of wet foam or paper towel kept inside of a plastic box (with lid). It is important to keep the antibody solution from drying during a room temperature incubation of 0.5–1.5 hr.

8. Transfer coverslips (cell side up) to a 12-well plate containing 1 ml of IF buffer A + BSA. Let soak for 5 min, replace with the same buffer, and soak again for $\geq$5 min. Plate may be shaken during washes if cells are well attached.

9. Transfer coverslips to a drop of diluted secondary antibody (such as Texas Red-labeled goat anti-rabbit, if the primary antibody was produced in rabbits) in the same manner as for the primary antibody (step 7). Incubate as above except cover plate to protect from light thus preserving the fluorescent antibody.

10. Repeat washes as in step number 8 plus two more with IF buffer A.

11. Using forceps, dip coverslip into a beaker of PBS and then a beaker of purified water.

12. Air dry coverslips cell side up and then place them cell side down onto approximately 15 μl of mounting medium such as Aqua Polymount (Polysciences, Inc., Warrington, PA).

13. View coverslips by fluorescence microscopy.

14. Store slides in the dark at 4°.

---

[22] N. T. Ktistakis, M. G. Roth, and G. S. Bloom, *J. Cell Biol.* **113**, 1009 (1991).

Controls. The number of controls performed for immunofluorescence of endogenous proteins should be maximized to reduce the risk of being misled by
artifacts. Suggestions include:

(a) If unpurified antiserum is used, compare it with preimmune serum from
the same animal.
(b) If purified antibody is used, compare it with immunoglobulin from the
nonimmunized animal of the same species.
(c) Compare action of primary antibody preincubated with either the correct
or incorrect immunogen peptide or protein. The correct peptide or protein
should prevent the fluorescent signal while the incorrect one should not.

Acknowledgments

I am indebted to many colleagues who have contributed in countless ways to my assembly of these
techniques and figures. In particular, I thank Chunfa Huang, Helen Aronovich, Erin Reid, Linda T.
Chen, Hsin Chieh (Calvin) Lin, Roger Sunahara, A. G. Gilman, Eric J. Smart, Richard G. W. Anderson,
and the late Kathryn H. Muntz. Special thanks go to Pingsheng Liu to whom I credit many helpful and
encouraging discussions in addition to a critical reading of the manuscript. This work was supported
by Grant GM50515 from the National Institute of General Medical Sciences.

[29] Fluorescence Approaches to Study G Protein Mechanisms

By DYKE P. MCEWEN, KYLE R. GEE, HEE C. KANG,
and RICHARD R. NEUBIG

Introduction

Guanine nucleotide binding proteins (G proteins) are heterotrimeric proteins
composed of an α subunit that binds and hydrolyzes GTP and a $\beta\gamma$ dimer.[1,2]
Heterotrimeric G proteins transduce signals from seven transmembrane spanning
receptors (GPCRs) to downstream effectors and are involved in regulating many
processes in cardiovascular, neural, and endocrine function.[3] Often, G protein function is studied either by using proteins purified from bacteria or insect expression
systems or by using membrane fragments isolated from mammalian cells.

[1] H. E. Hamm, *Cell Mol. Neurobiol.* **11**, 563 (1991).
[2] R. R. Neubig, M. C. Connolly, and A. E. Remmers, *FEBS Lett.* **355**, 251 (1994).
[3] A. E. Remmers, *Anal. Biochem.* **257**, 89 (1998).

Typically, nucleotide binding activity of the G protein is assessed using radio-labeled [^{35}S]GTPγS, a nonhydrolyzable GTP analog.[3,4] With the advent of high-throughput methodologies and the desire to reduce the cost and environmental concerns associated with radioactive waste, nonradioactive methods are desirable. Fluorescence methods usually do not require physical separation of bound and free ligand and they are also useful for studying fast processes in real time. In this chapter we discuss fluorescence approaches to study G protein mechanisms and describe the synthesis of a new class of fluorescent nucleotides based on the BODIPY moiety (Molecular Probes, Eugene, OR).

Intrinsic Fluorescence of G Proteins

Early fluorescence studies involving G proteins took advantage of the intrinsic tryptophan fluorescence of G proteins.[5–8] Addition of GTPγS and magnesium to Gα_o increases fluorescence emission intensity (λ_{ex} 280 nm and λ_{em} 340 nm) by about 60%.[5] Gα_o and Gα_t have two tryptophans and mutagenesis studies demonstrated that the activation-dependent change in intrinsic fluorescence was entirely dependent on a single tryptophan in the switch 2 region of the protein, namely Trp-212 and Trp-207 in Gα_o and Gα_t, respectively.[9,10] This approach has been used to study many aspects of G protein mechanisms.[6,7] Recently a stopped-flow fluorescence approach to measuring G protein deactivation and its acceleration by RGS proteins has been described[11] (see Protocol 1).

Advantages of the intrinsic fluorescence approach are its applicability to all G protein types. Also, the natural nucleotides (GTP or GDP) can be used since no chemical modification of the nucleotide is required. Disadvantages of intrinsic fluorescence are the low sensitivity due to modest fluorescence intensity of signals from tryptophan and small changes in fluorescence (10–60%). Excitation and emission are both in the UV range requiring special UV-transparent sample holders (e.g., quartz cuvettes). Finally, this approach is limited to highly purified systems since all proteins will contribute to intrinsic fluorescence signals.

Protocol 1: Measurement of G Protein α Subunit Deactivation Kinetics
  by Stopped-Flow Fluorescence

  Materials and Equipment

  Stock solution of purified G protein α subunit (minimum 1 μM, preferably
    5–50 μM)
  GTP stock solution: 1 mM GTP in H$_2$O stored frozen
  DTT stock solution: 100 mM Dithiothreitol (prepared fresh and stored on ice)

[4] D. J. Carty and R. Iyengar, *Methods Enzymol.* **237**, 38 (1994).
[5] T. Higashijima, K. M. Ferguson, P. C. Sternweis, M. D. Smigel, and A. G. Gilman, *J. Biol. Chem.* **262**, 762 (1987).
[6] T. Higashijima, K. M. Ferguson, M. D. Smigel, and A. G. Gilman, *J. Biol. Chem.* **262**, 757 (1987).

Solution 1: HED (50 mM Na–HEPES, 5 mM EDTA, 2 mM DTT, pH 8.0)

Solution 2: HED + 30 mM MgSO$_4$

Stopped-flow fluorimeter such as Applied Photophysics DX-17MV (Leatherhead, UK) with excitation 290 nm (2 nm slits for monochromator or 10 nm bandpass filter; emission: WG320 bandpass filter (Corion, Holliston, MA)

Procedure

1. Prepare Solutions 1 and 2 from the freshly prepared 100 mM DTT stock.
2. Dilute G protein α subunit into Solution 1 to prepare 1–2 ml of 400 nM Gα_o or 600 nM Gα_{i1} (16–50 μg).
3. Load the Gα subunits with GTP by incubating in Mg-free Solution 1 containing 2 μM GTP. The incubation time may need to be optimized for each type of Gα subunit (e.g., 20 min at 20° for Gα_o or 15 min at 30° for Gα_{i1}). *Note:* The high EDTA (5 mM) in this preloading solution reduces GTP hydrolysis during this incubation.
4. Return the GTP-loaded α subunits to ice for storage up to 2 hr.
5. Set the instrument at 20° (or the desired temperature). Configure optics and data collection software for optimal detection of intrinsic fluorescence from G protein at 200 or 300 nM (half of the starting concentration). Wavelengths should be 290 nm and 320 nm as indicated for instrumentation above. It is important to have the instrument ready before loading G protein samples due to denaturation and/or GTP hydrolysis which can occur at the higher temperatures present in the sample reservoir. *Note:* The measurement obtained is kinetics of the transition from the high fluorescence activated state of GTP-bound Gα to the lower fluorescence of GDP-bound Gα. Although this measurement correlates well with GTP hydrolysis, it is not directly measuring hydrolysis; thus we prefer to call it a "deactivation" rate.
6. Add any modulators of single-turnover GTPase, such as RGS proteins, to Solution 2 at twice the desired final concentration. Load this mixture into one sample syringe of the stopped-flow system and clear any bubbles that might accumulate. *Notes:* Solution 2 contains 30 mM MgSO$_4$, which will provide 10 mM free magnesium in the final reaction mixture. With the 200–300 nM final concentrations of Gα subunit, good results have been obtained up to 3–5 μM RGS4. Higher concentrations may be used but concomitantly higher Gα concentrations may be required since the RGS proteins also contribute to the intrinsic fluorescence.

[7] T. Higashijima, K. M. Ferguson, P. C. Sternweis, E. M. Ross, M. D. Smigel, and A. G. Gilman, *J. Biol. Chem.* **262**, 752 (1987).

[8] W. J. Phillips and R. A. Cerione, *J. Biol. Chem.* **263**, 15498 (1988).

[9] K.-L. Lan, A. E. Remmers, and R. R. Neubig, *Biochemistry* **37**, 837 (1997).

[10] E. Faurobert, A. Otto-Bruc, P. Chardin, and M. Chabre, *EMBO J.* **12**, 4191 (1993).

[11] K.-L. Lan, H. Zhong, M. Nanamori, and R. R. Neubig, *J. Biol. Chem.* **275**, 33497 (2000).

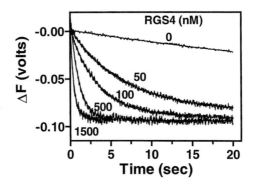

FIG. 1. Intrinsic fluorescence measurements of $G\alpha_o$ deactivation kinetics accelerated by RGS4. The rate of deactivation of GTP-preloaded hexahistidine (His_6) $G\alpha_o$ (200 nM final) was determined by stopped-flow fluorescence at 20° as described in Protocol 1. Data were collected for at least 100 sec in the absence of RGS and for 20–100 sec in the presence of RGS. The zero-time fluorescence subtracted from each trace ranged from 1.1 to 3.3 volts depending on the RGS concentration. Reprinted from K.-L. Lan, H. Zhong, M. Nanamori, and R. R. Neubig, *J. Biol. Chem.* **275,** 33497 (2000).

7. Load GTP-loaded $G\alpha$ into the other syringe, clear bubbles, and after a brief temperature equilibration period (3 min or less depending on the speed of temperature equilibration in the instrument) begin collecting stopped-flow "shots" to assess the kinetics of the single-turnover GTPase activity (see Fig. 1 for illustrative data). The composition of the magnesium-containing buffer can be changed after 4–5 good data traces have been obtained to study the effects of different concentrations or types of RGS protein on the $G\alpha$ deactivation kinetics.

8. Analyze averaged kinetic data by nonlinear least squares analysis with a one-phase exponential decay equation to determine rate constants (k_{deact}). Programs such as Graph Pad Prism (Graph Pad Software, San Diego, CA) and Origin (Northampton, MA) can be used.

9. Plot the rate of deactivation (k_{deact}) vs the concentration of modulator.

MANT Fluorophore

N-Methyl-3′-O-anthaniloyl(MANT)guanine nucleotide analogs in which the fluorescent group is introduced on the 2′- or 3′-hydroxyl group of the ribose ring were first described by Hiratsuka[12] in 1983. They were used extensively in the low molecular weight GTPase literature[13] before their utility for studies of

[12] T. Hiratsuka, *Biochim. Biophys. Acta* **742,** 496 (1983).
[13] J. F. Eccleston, K. J. M. Moore, G. G. Brownbridge, M. R. Webb, and P. N. Lowe, *Biochem. Soc. Trans.* **19,** 432 (1991).

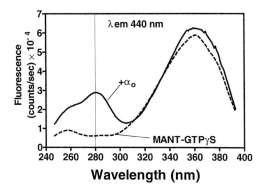

FIG. 2. Excitation spectrum of MANT-GTPγS on binding to Gα$_o$ MANT-GTPγS (500 nM) was incubated in the presence (solid line) or absence (dashed line) of 200 nM Gα$_o$ for 20 min at 20° in Solution 1. Excitation (λ_{em} 440 nm) spectra were measured in 200 μl of sample in a 5 mm round quartz cuvette in a PTI Alphascan fluorimeter. Slit widths were 2 × 2 nm. Data are expressed as counts per second.

heterotrimeric G proteins was recognized.[3] Several different MANT-labeled gua-nine nucleotide analogs have been prepared, with the most extensive studies using MANT–GTPγS.[3,14] The synthesis is relatively simple and has been described in detail by Remmers.[3]

MANT-labeled guanine nucleotides have two excitation maxima (360 and 260 nm) and an emission maximum at 440 nm. They show a significant increase in fluorescence on binding to low molecular weight GTPases.[13] In contrast, there is little increase in the fluorescence of MANT–GTPγS on binding to heterotrimeric G proteins (~6%) when the MANT moiety is excited directly (λ_{ex} 360 nm). However, a substantial increase (370%) is seen with 280 nm excitation (Fig. 2). The appearance of the new excitation peak at 280 nm for G protein bound MANT–GTPγS is consistent with resonance energy transfer from tryptophans in the G protein α_o subunit.[14]

Additional evidence for energy transfer from Gα$_o$ tryptophan to MANT–GTPγS is shown in Fig. 3.[3] Purified bovine brain G$_o$ (300 nM) is preequilibrated in HEDNML at 20° for 3 min prior to the addition of 500 nM MANT–GTPγS. Fluorescence is excited at 280 nm and emission measured via dual monochromators at 450 nm (Fig. 3A) and 340 nm (Fig. 3B). The MANT fluorescence (λ_{em} 450 nm) increases markedly as the nucleotide binds while tryptophan fluorescence (λ_{em} 340 nm) decreases with the same kinetics.[3] Therefore, MANT–GTPγS pro-vides a ready measure of guanine-nucleotide exchange with a large fluorescence enhancement on binding Gα$_o$.

[14] A. E. Remmers, R. Posner, and R. R. Neubig, $J.$ $Biol.$ $Chem.$ **269,** 13771 (1994).

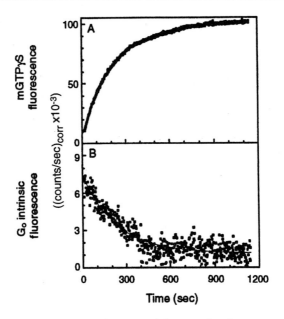

FIG. 3. MANT-GTPγS fluorescence increase and G_o tryptophan fluorescence quench. Heterotrimeric G_o purified from bovine brain (300 nM) was preequilibrated in HEDNML at 20° for 3 min followed by the addition of 500 nM MANT-GTPγS. Shown is the net fluorescence change at 450 nm (A) or 340 nm (B) (excitation 280 nm, dual emission). Background fluorescence was subtracted from A (60,000 counts/s) and B (87,000 counts/s). All data are corrected for variations in lamp output. Data were fit to a monoexponential association using Graph-Pad Prism. Half-times of MANT-GTPγS fluorescence increase and G_o tryptophan fluorescence decrease are 2.4 ± 0.2 and 2.5 ± 0.2 min, respectively. Reprinted from A. E. Remmers, *Anal. Biochem.* **257**, 89 (1998) with permission.

Finally, mutagenesis studies show that both Trp-189 and Trp-212 play a role in the MANT–GTPγS fluorescence increase[9] as opposed to the intrinsic fluorescence change which depends only on Trp-212. Thus using a λ_{ex} of 280 nm, the fluorescence of MANT–GTPγS in the absence of G protein is relatively low, allowing for easy detection of MANT–GTPγS binding. This has been useful a useful tool in studies of certain G proteins.[3,9]

Use of Fluorescent Nucleotides in G Protein Purification

In order to use a fluorescence method to quantitate amounts of a functional protein, it is necessary to demonstrate that the measurement provides a linear readout of the protein added. To date, the major method of measuring active G proteins has been the [^{35}S]GTPγS binding assay.[4] In Fig. 4 Remmers shows that the change in MANT–GTPγS fluorescence on addition of $G\alpha_o$ precisely parallels [^{35}S]GTPγS binding.[3]

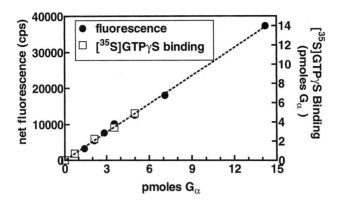

FIG. 4. MANT-GTPγS fluorescence is linearly dependent on G protein concentration. Myristoy-lated $G_{o\alpha}$ was initially quantitated by [^{35}S]GTPγS binding as described in text. [^{35}S]GTPγS-binding assay (square) was performed using 0–5 pmol myr$G_o\alpha$ and MANT-GTPγS-binding assay (circle) was performed using 0–14 pmol myr$G_o\alpha$ as described in text. Fluorescence from 500 nM MANT-GTPγS alone (8700 cps) in buffer has been subtracted from the total fluorescence. The line shown is a linear regression of the net MANT-GTPγS fluorescence. Reprinted from A. E. Remmers, *Anal. Biochem.* **257,** 89 (1998) with permission.

Protocol 2: Fluorescence Assay of G Protein by MANT-GTPγS

Materials

1. DTT Stock solution: 100 mM DTT (prepared fresh and stored on ice)
2. Solution 3: HENML buffer containing 50 mM HEPES, 1 mM EDTA, 0.1 M NaCl, 10 mM MgCl$_2$, and 0.1% Lubrol (or C12E10), pH 8.0
3. Solution 4: HEDNML buffer made from HENML buffer with 1 mM DTT
4. MANT–GTPγS stock solution (10–50 μM)

Procedure

1. Prepare fresh DTT stock solution and use it to make Solution 4.
2. Dilute MANT-GTPγS to a final concentration of 500 nM in Solution 4 and store on ice up to 2 hr before use. Prepare sufficient sample for 0.2 ml per measurement.
3. Add 20 μl of each sample to be measured to 200 μl of the MANT-GTPγS and incubate for the appropriate time (20 min at room temperature for $G\alpha_o$ or $G\alpha_s$ or 60 min at 30° for $G\alpha_i$).
4. Measure fluorescence with wavelengths 280 nm excitation and 440 nm emission. This can be done in a standard monochromator-based fluorime-ter in a 200 μl sample volume in 5 mm round (custom) or rectangular commercially available quartz cuvettes (Starna, Atascadero, CA).
5. Calculate net fluorescence by subtracting values from blank samples

containing MANT-GTPγS and 20 μl of the buffer(s) used to elute the column fractions.

6. If the fluorescence appears to have saturated (several fractions have the same high fluorescence) then dilute those samples 2- to 5-fold and recheck fluorescence.

Similarly, column elution profiles of myristoylated Gα_{i1} expressed in bacteria were determined by the increase in MANT–GTPγS fluorescence. As can be seen for both column profiles in Fig. 5, the MANT–GTPγS showed strongly enhanced fluorescence with fractions containing the Gα_{i1} protein, with little to no signal due to contaminating proteins.

There are several advantages to using the fluorescence-based binding assay to detect G-protein-containing fractions during purification. First, the relative rapidity of data collection is very useful during protein purification. Second, it is a nonradioactive method that is nearly as sensitive as the commonly used radioligand binding assay but does not require scintillation fluid or disposal of hazardous waste.

Limitations of the MANT–GTPγS method are: (1) the MANT moiety does reduce the affinity of the nucleotide for some types of Gα subunit (e.g., transducin) and it may not be suitable in those cases; (2) the measurements only report *relative* amounts of G protein; unlike radioligand methods, fluorescence does not provide exact nanogram or microgram amounts unless some fully active Gα subunits of the same type are available for use as a standard curve, and (3) the UV excitation of the MANT fluorophore requires special sample holders (e.g., quartz cuvettes), which are expensive so they must be washed and reused. Despite these limitations, the MANT–GTPγS binding assay is an easy, nonradioactive method for detection of G protein α subunits during purification.

BODIPY Fluorophore

As noted above, the UV excitation of the MANT–GTPγS causes technical difficulties for *in vitro* studies and virtually precludes *in vivo* or intact cell studies. Therefore, the recently described BODIPY FL GTPγS analog which was designed to measure hydrolysis via cleavage of the $\alpha-\beta$ phosphate bond[15] was of significant interest as a possible probe of guanine nucleotide binding to G proteins. BODIPY analogs have two major advantages over the MANT moiety. First, their excitation and emission wavelengths are in the visible range so there is no need for special optical components. Second, the fluorescence intensity of BODIPY probes is much greater than that of MANT-containing compounds. Third, the analogs with different emission wavelengths would permit detection in the presence of other

---

[15] A. Draganescu, S. C. Hodawadekar, K. R. Gee, and C. Brenner, *J. Biol. Chem.* **275,** 4555 (2000).

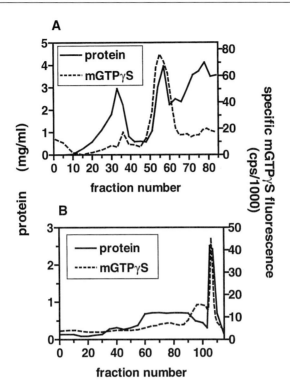

FIG. 5. Column profiles from $G\alpha_{i1}$ purification using the MANT-GTPγS-binding assay. Myristoylated $G\alpha_{i1}$ was expressed and bacterial extracts prepared from 4 liters of culture as described in text. (A) Extract (800 ml) was loaded onto a 120 ml HiLoad Q-Sepharose fast flow column, the column was washed with 400 ml TEPD buffer, and the proteins were eluted with a 0–300 mM NaCl gradient. Eleven-milliliter fractions were collected. Bradford protein assay (solid line) was performed on 2-μl aliquots and MANT-GTPγS fluorescence binding assay (dashed line) was performed on 20-μl aliquots of every third fraction. (B) G protein pooled from the HiLoad Q-Sepharose fast flow column was subjected to heptylamine-Sepharose chromatography as described in text. Fractions (9 ml) were collected and Bradford protein assay (solid line) was performed on 5-μl aliquots and MANT-GTPγS-binding assay (dashed line) was performed on 20-μl aliquots of every third fraction. Values are expressed as thousand counts per second (cps/1000) and a background fluorescence signal from MANT-GTPγS alone (7000 cps in A, 3000 cps in B) was subtracted from the total fluorescence to yield specific fluorescence. Reprinted from A. E. Remmers, *Anal. Biochem.* **257,** 89 (1998) with permission.

fluorophores (e.g., fluorescein, green fluorescent protein, or pairs of BODIPY-labeled compounds). It was not clear whether the bulky BODIPY moiety on the γ-thiol might sterically hinder binding, but as shown below the affinity for most G protein α subunit subtypes is sufficient for these studies. The synthesis of BODIPY FL GTPγS was described previously.[15] The synthesis of several additional BODIPY GTP analogs is illustrated in Fig. 6 and described in Protocol 3.

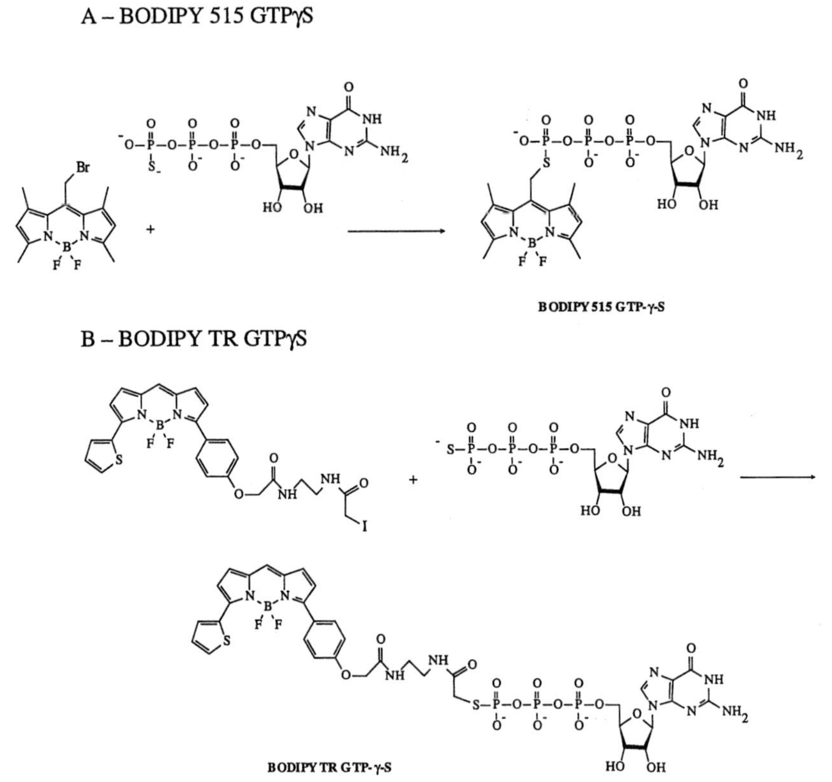

A – BODIPY 515 GTPγS

B – BODIPY TR GTPγS

BODIPY TR GTP-γ-S

FIG. 6. Synthesis of BODIPY GTP analogs.

Protocol 3: Synthesis of BODIPY GTP Analogs

BODIPY 515 GTPγS. To a solution of GTPγS, tetralithium salt (Sigma, 16.5 mg, 0.029 mmol) in 2 ml E-pure water is added 1 drop of saturated sodium bicarbonate solution, followed by a solution of 8-bromomethyl-4,4-difluoro-1,3,5,7-tetramethyl-4-bora-3a,4a-diaza-s-indacene (BODIPY 493/503 methyl bromide, 10 mg, 0.029 mmol) in 3 ml dioxane. The resulting red mixture is stirred overnight, whereupon silica gel TLC shows consumption of starting nucleotide [R_f 0.18, dioxane : 2-propanol : water : ammonium hydroxide (40 : 20 : 35 : 35, v/v)] and formation of the desired product (R_f 0.75). The volatiles are removed by rotary evaporation, and the residue is applied to a 2 × 20 cm column of Sephadex LH-20 as a mixture in 2 ml E-pure water. Elution with the same allows for separation of the product from residual starting dye (R_f 0.95). Fractions of 2–3 ml are collected, and the pure product fractions are combined and lyophilized to give the title compound as a sodium salt as 17 mg of a red-orange powder (68% yield): ^{1}H NMR (D$_2$O)

C – BODIPY FL GppNHp

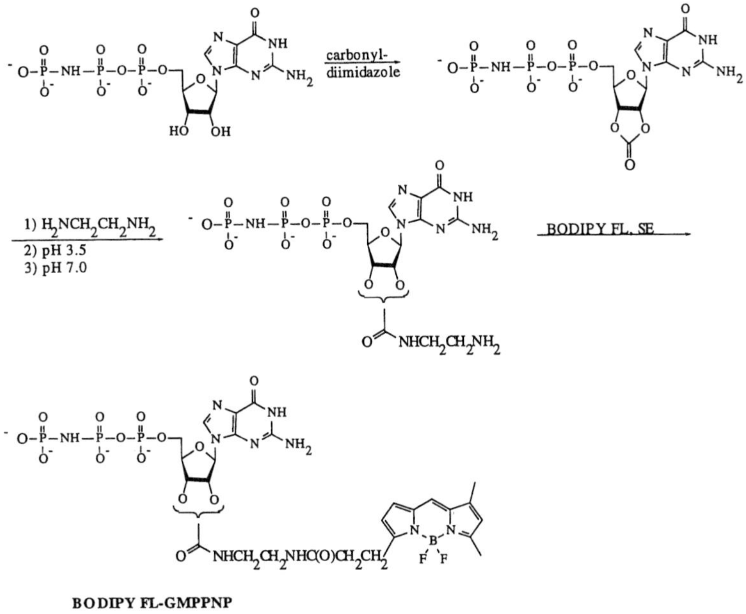

BODIPY FL-GMPPNP

FIG. 6. (*Continued*)

δ 7.95 (s, 1H, guanosine C8), 6.20 (s, 2H, dye CH), 5.95 (d, 1H, C1), 4.80 (s, 2H, CH$_2$-S), 4.53 (t, 1H, C3), 4.45 (t, 1H C2), 4.31 (m, 1H, C4), 4.24 (m, 2H, C5), 2.55 (s, 12H, dye CH$_3$); ^{31}P (D$_2$O) δ 8.5 (d, 1P), -9.5 (d, 1P), -21.8 (dd, 1P); absorption maximum 515 nm, emission maximum 528 nm, in pH 7.2 phosphate buffer solution.

BODIPY TR GTPγS. To a solution of GTPγS, tetralithium salt (Sigma, 5 mg, 0.01 mmol) in 1 ml E-pure water is added 1 drop of saturated sodium bicarbonate solution, followed by a solution of *N*-5-(((4-(4,4-difluoro-5-(2-thienyl)-4-bora-3*a*,4*a*-diaza-*s*-indacene-3-yl)phenoxy)acetyl)amino)pentyl)iodoacetamide (BODIPY TR cadaverine IA, 5 mg, 0.01 mmol) in 1 ml dioxane. The resulting red-purple solution is stirred overnight, whereupon silica gel TLC shows consumption of starting nucleotide [R_f 0.18, dioxane : 2-propanol : water : ammonium hydroxide (40 : 20 : 35 : 35, v/v)] and formation of the desired product (R_f 0.81). The volatiles are removed by rotary evaporation, and the blue residue is applied to a 2 × 20 cm column of Sephadex LH-20 as a solution in 1 ml E-pure water. Elution with the same allows for separation of the product from residual starting dye

(R_f 0.95). Fractions of 2 ml are collected, and the pure product fractions are combined and lyophilized to give the title compound as a sodium salt as 7 mg of a purple powder (70% yield): ^{1}H NMR (D$_2$O) δ 8.12 (d, 1H, dye thienyl CH), 8.08 (s, 1H, guanosine C8), 7.99 (d, 2H, dye phenyl CH), 7.48 (d, 1H, dye thienyl CH), 7.25 (s, 1H, dye C8 CH), 7.15 (t, 1H, dye thienyl CH), 7.09 (m, 2H, dye CH), 7.01 (d, 2H, dye phenyl CH), 6.81 (d, 1H, dye CH), 6.65 (d, 1H, dye CH), 5.99 (d, 1H, C1), 4.85 (s, 2H, CH$_2$-S), 4.60 (s, 2H, OCH$_2$C(O)N), 4.53 (t, 1H, C3), 4.45 (t, 1H C2), 4.31 (m, 1H, C4), 4.24 (m, 2H, C5), 3.35 (m, 2H, NHC$\underline{H}_2$), 3.20 (m, 2H, NHC$\underline{H}_2$), 1.6 (m, 4H, NHCH$_2$C$\underline{H}_2$CH$_2$C$\underline{H}_2$CH$_2$NH), 1.34 (m, 2H, NHCH$_2$CH$_2$C$\underline{H}_2$CH$_2$CH$_2$NH); ^{31}P (D$_2$O) δ 8.45 (d, 1P), -9.3 (d, 1P), -21.5 (m, 1P); absorption maximum 591 nm, emission maximum 615 nm, in pH 7.2 phosphate buffer solution.

BODIPY FL GTP. To a solution of 2′(3′)-*O*-(2-aminoethylcarbamoyl)GTP[16] (50 mg, 0.05 mM) in 1 ml of water is added a solution of 4,4-difluoro-5,7-dimethyl-4-bora-3a,4a-diaza-s-indacene-3-propionic acid, succinimidyl ester (25 mg, 0.06 mM) and triethylamine (20 μl, 0.14 mmol) in 1 ml of acetonitrile and the mixture is stirred at room temperature for 5 hr. The excess dye is removed by the extraction with ethyl acetate and the resulting aqueous residue is purified by Sephadex LH-20 chromatography eluting with water. The desired fractions are combined and lyophilized. The resulting lyophilized solid is redissolved in water (2 ml) and treated with Dowex 50WX8 (Na$^+$ form, 1.5 g). After stirring at room temperature for 1 hr the resin is removed by filtration. The resulting filtrate is lyophilized to give an orange solid as a mixture of 2′- and 3′-isomers[16]; one spot on TLC [R_f = 0.30, dioxane : 2-propanol : H$_2$O : NH$_4$OH(40 : 20 : 34 : 36)], ^{31}P NMR (D$_2$O) δ -8.83 (dd, 1P), -9.38 (dd,1P), -21.15 (t, 1P); absorption maximum 504 nm, emission maximum 512 nm in pH 7.2 phosphate-buffered saline. ^{1}H NMR indicates that the product is a roughly equal mixture of 2′- and 3′-isomers.

BODIPY FL-GMPPNP. 5′-Guanylylimidodiphosphate trisodium salt (200 mg, 0.34 mmol) is converted to the triethylammonium salt using a 2 × 10 cm column of Dowex 50WX8 (triethylammonium form). This is then stirred with tributylamine (800 μl, 3.40 mmol) for 1 hr. The mixture is exhaustively dried by repeated rotary evaporation from dry dimethylformamide (DMF, 4 × 10 ml) at 25°. A solution of 1,1′-carbonyldiimidazole (250 mg, 1.54 mmol) in 10 ml of DMF is added and the reaction mixture is stirred at an ice–water bath for 2 hr. Then the excess 1,1′-carbonyldiimidazole is quenched by the addition of 150 μl of absolute methanol. A solution of ethylenediamine (300 μl, 4.48 mmol) in 5 ml of DMF is added dropwise while the reaction mixture is stirred at an ice–water bath. After 30 min the resulting white precipitate is recovered by centrifugation and washed with DMF (3 × 10 mL). The precipitate is dissolved in 10 ml of water, adjusted

[16] T. L. Hazlett, K. J. Moore, P. N. Lowe, D. M. Jameson, and J. F. Eccleston, *Biochemistry* **32**, 13575 (1993).

to pH 3 with 1 M HCl solution. After stirring for 16 hr at 4° it is then adjusted to pH 7.0 with 1 M NaOH solution. It is then applied to a 2.5 × 20 cm DEAE-Sephadex A-25 column (HCO_3 form). It is eluted with a 2 liter linear gradient of 0.05 M to 0.5 M triethylammonium bicarbonate (pH 7.8). The desired fractions are then rotary evaporated to 2–3 ml. About 50 ml of methanol is added to the evaporated residue, which is then reevaporated to 2–3 ml. This procedure is repeated twice more and finally the evaporated residue is dissolved in water (5 ml). It is lyophilized to give 2′(3′)-O-(2-aminoethylcarbamoyl)GMPPNP as a white solid; one spot on TLC, positive to ninhydrin test [R_f = 0.15, dioxane : 2-propanol : water : ammonium hydroxide (40 : 20 : 34 : 36)]. To a solution of 2′(3′)-O-(2-aminoethylcarbamoyl)GMPPNP (17 mg, 0.017 mmol) in 1 ml of water is added a solution of 4,4-difluoro-5,7-dimethyl-4-bora-3a,4a-diaza-s-indacene-3-propionic acid, succinimidyl ester (12 mg, 0.013 mmol), and triethylamine (7 μl, 0.050 mmol) in 1 ml of acetonitrile and the mixture is stirred at room temperature for 5 hr. The excess dye is removed by extraction with ethyl acetate and the resulting aqueous layer is applied to a column (2.5 × 15 cm, Sephadex LH-20). It is eluted with water and the desired fractions are lyophilized. The lyophilized solid is dissolved in water (∼2 ml) and treated with Dowex 50WX8 (Na^+ form, 1.0 g). After stirring at room temperature for 1 hr the resin is removed by filtration. The resulting filtrate is lyophilized to give an orange solid as a mixture of 2′- and 3′-isomers; one spot on thin-layer chromatography (TLC; R_f = 0.28); absorption maximum 504 nm, emission maximum 512 nm, in pH 7.2 phosphate buffer solution.

Spectroscopic Analysis of BODIPY Nucleotide Binding to G Proteins

The excitation and emission spectra of BODIPY FL GTPγS in the presence and absence of Gα_o are shown in Fig. 7A. There was a dramatic enhancement in the fluorescence of BODIPY FL GTPγS upon binding Gα_o with no change in the maximum excitation or emission wavelengths (Fig. 7A). This enhancement of fluorescence was blocked by preincubation of Gα_o with 20 μM unlabeled GTPγS prior to addition of the BODIPY FL GTPγS. A comparison of four different nonhydrolyzable BODIPY GTP analogs is shown in Fig. 7B. BODIPY FL GTPγS and BODIPY 515 GTPγS gave the greatest increase in the presence of 200 nM Gα_o (5- to 6-fold). BODIPY FL GppNHp and BODIPY TR GTPγS showed smaller increases (2- to 3-fold). This is due in part to a significantly lower affinity of these analogs for Gα_o.[17] The fluorescence increase appears to be due to a displacement of the BODIPY moiety away from the guanine base as the base is sequestered in the G protein binding pocket. This relieves the intramolecular fluorescence quenching.[15,17]

---

[17] D. P. McEwen, K. R. Gee, H. C. Kang, and R. R. Neubig, *Anal. Biochem,* in press.

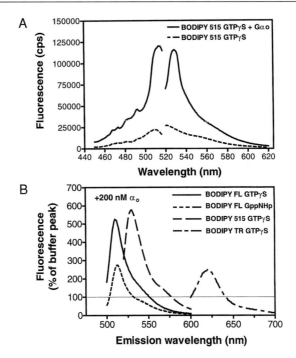

FIG. 7. Excitation and emission spectra of BODIPY GTP analogs bound to $G\alpha_o$. (A) BODIPY 515 GTPγS (50 nM) was incubated in the presence or absence of 200 nM $G\alpha_o$ for 20 min at 20° in Solution 1. Excitation (λ_{em} 530 nm) and emission (λ_{ex} 510 nm) spectra were measured in 500 μl of sample in a 2 × 10 mm cuvette in a PTI Alphascan fluorimeter. Slit widths were 2 × 2 nm. Data are expressed as counts per second (cps). (B) Emission spectra of four nonhydrolyzable BODIPY GTP derivatives are shown for excitation at: BODIPY FL analogs, 490 nm; BODIPY 515 analog, 510 nm; and BODIPY TR analog, 580 nm. Data are normalized to the peak fluorescence of the nucleotide in buffer without $G\alpha_o$. Spectra of HEM buffer alone have been subtracted from the data and contributed <3% of the total fluorescence.

The utility of the BODIPY FL GTPγS in the analysis of G proteins will depend on:(1) the range of G protein types/subtypes for which it shows a good binding affinity and (2) how adaptable the methodology is to high throughput techniques such as multiwell plates. We addressed both of these questions by use of Protocol 4.

Protocol 4: Fluorescent Nucleotide 96-Well Plate Binding Assay

Materials and Equipment

1. G protein subunits (α_o, α_{i1}, α_{i2}, α_s)
2. BODIPY FL GTPγS or BODIPY FL GppNHp stock solution (1 mM and 100 μM)

3. DTT Stock solution: 100 mM DTT (prepared fresh and stored on ice)
4. Solution 5: HEM buffer containing 50 mM HEPES, 1 mM EDTA, and 10 mM MgCl$_2$, pH 8.0
5. Solution 6: HEDM buffer made from HEM buffer with 1 mM DTT
6. Temperature controlled 96-well fluorescence plate reader (e.g., CytoFluor 4000, Biosystems, Foster City, CA)
7. Clear plastic Falcon 96-well plates

Procedure

1. Set the plate reader to the desired temperature (24° for G$_o$ and G$_s$ and 30° for G$_{i1}$ and G$_{i2}$) and configure wavelengths (excitation 485 nm, emission long pass filter >530 nm).
2. Dilute G protein samples to 10× the desired final concentration (final 5–100 nM) in Solution 6 preparing enough for 10 μl of the 10× sample per well. Place the G protein samples in an ice-cold plastic reservoir (if all samples will be the same) or a 96-well plate (if different concentrations or G proteins will be used) to permit rapid initiation of the fluorescence binding assay.
3. Distribute 90 μl of the BODIPY FL GTP analog (5–500 nM) into a separate 96-well plate and prewarm to the desired temperature for 5 min.
4. Make a blank reading of the plate with only the BODIPY FL GTP analog present.
5. Rapidly initiate the binding by addition of G protein using a multichannel pipette (e.g., Eppendorf Scientific, Inc./Brinkmann Instruments, Inc., Westbury, NY) and place the plate in the prewarmed fluorescence plate reader.
6. Measure fluorescence at 1 min intervals for time curves or incubate for a defined time (20 min for α_o or α_s or 60 min for α_{i1} or α_{i2}) and read the end-point values.
7. Analyze data by fitting time course data to an exponential or end-point data according to the experimental design.

Affinity and Specificity of BODIPY GTP Analogs for Different G Protein α Subunits

The time dependence of BODIPY FL GTPγS binding to Gα_o and Gα_{i1} is shown in Fig. 8. These experiments in a 96-well plate format deliberately used small amounts of Gα subunit (80 ng/well) to assess the feasibility of using the BODIPY GTP analogs in a high-throughput format. The signal/background was significantly better for Gα_o because of its higher affinity (see below) but clear time-dependent binding signals were readily observable with both Gα subunits. Improved signal/back ground is possible with higher concentrations of the Gα subunit and may also be possible with plates and/or plate readers with a lower background signal

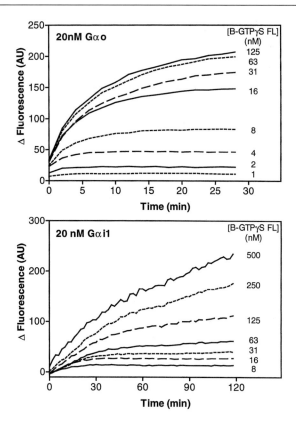

FIG. 8. Concentration dependence and time course of BODIPY FL GTPγS fluorescence changes in a 96-well plate format. The time course of fluorescence changes for the indicated concentrations of BODIPY FL GTPγS were measured in a 96-well plate reader as described in Protocol 4. Samples contained 20 nM $G\alpha_o$ at room temperature (top) or 20 nM $G\alpha_{i1}$ at 30° (bottom). Average blank fluorescence values (i.e., with no added G protein subunit) were 55, 99, 415, and 1455 units at 0, 16, 125, and 500 nM BODIPY FL GTPγS (i.e., blank fluorescence was essentially linear with nucleotide concentration).

(see legend to Fig. 8). Figure 9 shows a representative experiment measuring the K_D of BODIPY FL GTPγS for $G\alpha_o$, with the measured K_D values for four G protein alpha subunits summarized in Table I. BODIPY FL GTPγS looks to be more useful than BODIPY FL GppNHp because of the approximately 10-fold higher affinity of the G protein α subunits for BODIPY FL GTPγS.

Mastoparan-Induced Guanine Nucleotide Exchange

Because of the real-time nature of the fluorescence measurements, they are well suited to characterize proteins or small molecule compounds that act either as

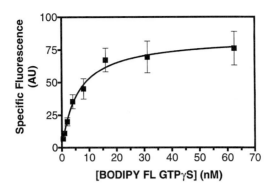

FIG. 9. Saturation binding analysis for BODIPY FL GTPγS binding to Gα_o in a 96-well plate format. Gα_o (20 nM) was incubated with the indicated concentrations of BODIPY FL GTPγS for 20 min at room temperature (24°). Excitation was at 485 nm, while emission was measured with a photomultiplier tube behind a 530 nm long-pass filter. Specific fluorescence was determined by subtracting values in the absence of Gα_o from the fluorescence with Gα_o present. At this low Gα concentration, nonspecific fluorescence (largely from the 96-well plate itself) represented 60–90% of the fluorescence, but the excellent reproducibility provided a clear specific binding signal. The curve shows a nonlinear least squares fit to a hyperbolic binding function.

guanine nucleotide exchange factors (GEF) or as guanine nucleotide dissociation inhibitors (GDI). Mastoparan, a toxin from wasp venom, has been shown by means of [^{35}S]GTPγS binding and MANT–GTPγS fluorescence[18] to enhance the rate of guanine nucleotide exchange on G_o heterotrimer in lipid vesicles by about 2.5-fold.[19] Using a reconstituted heterotrimer containing bacterially expressed myristoylated Gα_{i1} and bovine brain βγ in lipid micelles, McEwen et al. showed that 30 μM mastoparan accelerated the binding of BODIPY FL GTPγS to Gα_{i1}

TABLE I

BODIPY FL GTPγS INTERACTIONS WITH Gα_o, Gα_{i1}, Gα_{i2}, AND Gα_s

Gα subunit	GDP release, $t_{1/2}$[a] (sec)	Dissociation, $t_{1/2}$ (sec)	K_D[b]
G_o	75	1100	6
G_{i1}	1500	740	150
G_{i2}	1300	720	300
G_s	320	>1000[c]	70

[a] Half-times for the fluorescence increase are dependent on GDP release by the individual Gα subunits.

[b] K_D values were determined from fluorescence saturation analysis in 96-well plates (see Figs. 8 and 9).

[c] Not determined. BODIPY FL GTPγS dissociation was not determined for Gα_s, since dissociation was not seen due to high Mg^{2+}.

5-fold.[17] These data suggest that the BODIPY GTP analogs may be very useful in studying receptor-mediated guanine nucleotide exchange. The major advantage of using BODIPY rather than MANT nucleotides to study receptor-mediated guanine nucleotide exchange depends on the spectral properties. Since detection using the MANT moiety relies on tryptophan excitation at 280 nm with energy transfer emission at 450 nm, there will be substantial interference from other proteins in a receptor/G protein preparation while the visible wavelengths for BODIPY should exhibit much lower background fluorescence. Also, the fluorescence intensity of BODIPY FL GTP is much higher because of its much higher extinction coefficient (>60,000 vs 5600 for tryptophan). Both factors will be important in studies with lower concentrations and less highly purified G protein receptor systems.

Future Perspectives

Fluorescence techniques are becoming widely used and more useful with improvements in both instrumentation and fluorescent reagents. In this chapter we have outlined several fluorescence techniques that can be used to study mechanisms underlying G-protein-mediated signaling that may decrease the need for radioactive materials, scintillation fluids, and their accompanying hazards. The newly described BODIPY GTP fluorophores permit real-time measurements of guanine nucleotide binding and are readily adapted to multiwell methods with significant potential in high-throughput screening for G protein active compounds.

These methods also have significant potential that has yet to be shown in practice. Additional efforts will be required to develop optimized methods for use with receptor-activated G proteins because of the low concentrations of receptor available. Similarly, applications to G protein localization and/or movements in intact cell studies may be possible but will require additional development. Finally, none of the fluorescent nucleotides described here permit a fluorescence-based measurement GTPase activity. This is clearly an area that deserves attention.

In summary, we explore the strengths and weaknesses of a number of fluorescence methods and reagents which can be used to study G protein function. These tools are very useful; however, additional work should be done to extend their utility.

[18] S. M. Wade, M. K. Scribner, H. M. Dalman, J. M. Taylor, and R. R. Neubig, *Mol. Pharmacol.* **50**, 351 (1996).

[19] T. Higashijima, S. Uzu, T. Nakajima, and E. M. Ross, *J. Biol. Chem.* **263**, 6491 (1988).

[20] K. L. Lan, H. Zhong, M. Nanamori, and R. R. Neubig, *J. Biol. Chem.* **275**, 33497 (2000).

[30] Defining G Protein βγ Specificity for Effector Recognition

By E. J. DELL, TRILLIUM BLACKMER, NIKOLAI P. SKIBA, YEHIA DAAKA, LEE R. SHEKTER, RAMON ROSAL, EITAN REUVENY, and HEIDI E. HAMM

Introduction

Heterotrimeric G proteins play important roles in determining the specificity and temporal characteristics of intracellular responses to extracellular signals. On receptor activation, GTP binds the Gα subunit and the heterotrimer dissociates into free Gα and Gβγ subunits that can individually activate a number of effectors.[1,2] Free Gβγ is an activator of a large assortment of proteins, including phospholipases,[3] adenylyl cyclases,[4] ion channels,[5] G-protein-coupled receptor kinases,[6,7] and phosphoinositide 3-kinases.[8] Given this abundant and constantly expanding list of Gβγ effectors, it was important to determine the molecular basis of interaction between Gβγ and its effectors.

Our hypothesis was that Gβγ uses a common binding surface for its interaction with Gα and with its diverse effectors since the inactive form of the Gα subunit, Gα–GDP, can compete with Gβγ effectors and deactivate Gβγ-dependent signaling. Two regions on Gβγ, the switch interface (Gβ residues 57, 59, 98, 99, 101, 117, 119, 143, 186, 228, and 332) and the NH$_2$-terminal interface (Gβ residues 55, 78, 80 and 89), were shown in the crystal structure of the heterotrimer to be important for the interaction with Gα.[9,10] Our hypothesis was tested by mutational analysis of the Gβ residues that make contact with Gα–GDP (Fig. 1). The 15 Gα-binding residues of Gβ$_1$ were singly mutated to alanines and each Gβ$_1$ mutant was expressed with either H6Gγ$_1$ or H6Gγ$_2$ (Gγ with a NH$_2$-His$_6$, tag), two isoforms of the Gγ subunit, depending on which effector was to be tested.[11]

[1] H. E. Hamm, *J. Biol. Chem.* **273**, 669 (1998).

[2] E. J. Neer, *Cell* **80**, 249 (1995).

[3] S. G. Rhee and Y. S. Bae, *J. Biol. Chem.* **272**, 15045 (1997).

[4] R. K. Sunahara, C. W. Dessauer, and A. G. Gilman, *Ann. Rev. Pharmacol. Toxicol.* **36**, 461 (1996).

[5] T. Schneider, P. Igelmund, and J. Hescheler, *Trends Pharmacol. Sci.* **18**, 8 (1997).

[6] J. A. Pitcher, J. Inglese, J. B. Higgins, J. L. Arriza, P. J. Casey, C. Kim, J. L. Benovic, M. M. Kwatra, M. G. Caron, and R. J. Lefkowitz, *Science* **257**, 1264 (1992).

[7] N. J. Freedman and R. J. Lefkowitz, *Recent Prog. Hormone Res.* **51**, 319; discussion 352 (1996).

[8] B. Vanhaesebroeck, S. J. Leevers, G. Panayotou, and M. D. Waterfield, *Trends Biochem. Sci.* **22**, 267 (1997).

[9] D. G. Lambright, J. Sondek, A. Bohm, N. P. Skiba, H. E. Hamm, and P. B. Sigler, *Nature* **379**, 311 (1996).

[10] M. A. Wall, D. E. Coleman, E. Lee, J. A. Iniguez-Lluhi, B. A. Posner, A. G. Gilman, and S. R. Sprang, *Cell* **83**, 1047 (1995).

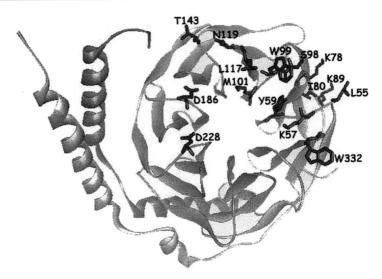

FIG. 1. Ribbon representation of Gβγ derived from the crystal structure with the Gα interacting amino acids shown that were mutated to alanines. Reproduced with permission from C. E. Ford *et al.,* *Science* **280,** 1271 (1998).

Several experiments were done to see if the mutants were expressed and functioned similarly to wild-type. (1) Expression of the $G\beta_1H6\gamma_1$ or $G\beta_1H6\gamma_2$ (herein referred to as $G\beta\gamma_1$ or $G\beta\gamma_2$) mutants was at similar amounts as in the wild-type as determined by Western blotting. (2) All mutated Gβγ dimers were folded properly as determined by a trypsin protection assay.[12] (3) The dimers were posttranslationally modified appropriately by isoprenylation and carboxymethylation as determined by mass spectrometry. (4) Gβγ mutants could form heterotrimers with transducin Gα-GDP ($G_t\alpha$-GDP) as determined by the ability of the Gβγ mutants to facilitate pertussis toxin-catalyzed adenosine diphosphate (ADP) ribosylation of $G_t\alpha$-GDP.[13] All mutants could support some level of ADP ribosylation, although Gβ mutants Ile80 Ala80 (I80A), K89A, L117A, and W332A showed reduced ability to form heterotrimers. Last, (5) all mutants could be activated by rhodopsin, but to a lesser extent than wild-type. This suggests that the Gβγ dimer may play a role in the receptor catalyze exchange of GDP for GTP in the Gα subunit.

Analysis of these mutants revealed the Gβ residues required for interaction with or activation of various effectors.[11] Those effectors tested were β-adrenergic

[11] C. E. Ford, N. P. Skiba, H. Bae, Y. Daaka, E. Reuveny, L. R. Shekter, R. Rosal, G. Weng, C.-S. Yang, R. Iyengar, R. J. Miller, L. Y. Jan, R. J. Lefkowitz, and H. E. Hamm, *Science* **280,** 1271 (1998).

[12] T. C. Thomas, T. Sladek, F. Yi, T. Smith, and E. J. Neer, *Biochemistry* **32,** 8628 (1993).

[13] T. Katada, K. Kontani, A. Inanobe, I. Kobayashi, Y. Ohoka, H. Nishina, and K. Takashashi, *Methods Enzymol.* **237,** 131 (1994).

receptor kinase 1 (βARK1, also known as GRK2), phospholipase C-β_2 (PLC-β_2), adenylyl cyclase 2 (AC2), calcium α1B subunit (CCα1B), and G-protein-coupled inward rectifying potassium channel (GIRK1/GIRK4). This chapter outlines the construction of the G$\beta\gamma$ mutants, the assays used to determine their ability to interact with or activate various effectors, and several possible strategic uses of these mutants.

Construction of Recombinant Baculovirus Vectors

Alanine substitutions of the amino acid residues in Gβ_1 that interact with the G$_t\alpha$ are generated using polymerase chain reaction (PCR)-based mutagenesis (QuikChange Kit, Stratagene, La Jolla, CA) with corresponding oligonucleotide primers containing the mutations and with bovine Gβ_1 cDNA inserted in pGEM4Z (Promega, Madison, WI) as template. The mutated Gβ_1 cDNAs are excised from pGEM4Z using the *Pst*I and *Xba*I restriction sites and cloned using standard methods into the baculovirus transfer vector, pVL1392, cut with the same restriction enzymes. All mutant cDNA sequences are confirmed by sequence analysis.

To attach a hexahistidine tag to the NH$_2$ terminus of Gγ (1 and 2), transducin Gα cDNA contained in the pHis$_6$ G$_t\alpha$ vector[14] is replaced with Gγ cDNA. A PCR fragment encoding intact Gγ amino acid sequence starting from the second Pro residue is generated. This PCR fragment contains a 5'-terminal *Pst*I site and a blunted 3' terminus. The PCR fragment is cut with *Pst*I and ligated with an *Nco*I-digested pHis$_6$ G$_t\alpha$ cDNA that has been filled with Klenow enzyme prior to a second digestion with *Pst*I. H6Gγ cDNA is subcloned into the baculovirus transfer vector pVL1393 using *Eco*RI and *Pst*I sites. The amino acid sequence encoded by the Gγ cDNA is preceded by an additional NH$_2$-terminal sequence, MAHHHHHHA.

Recombinant baculoviruses are generated by calcium phosphate transfection of *Spodoptera frugiperda* (Sf9) (Invitrogen) cells with recombinant transfer vectors according to manufacturer's protocol (BaculoGold DNA kit, Pharmingen).

Expression and Purification of Recombinant G$\beta\gamma$ and Mutants

The insect cell line Sf9 is used to express the recombinant proteins. Immunoblotting whole cell lysates from infected Sf9 cells with antibodies to Gβ (T-20, Santa Cruz Biotechnology) and antibodies to Gγ (P-19, Santa Cruz Biotechnology) is done to ensure correct protein expression of recombinant G$\beta\gamma$ proteins.

Expression and Purification

1. Sf9 cells (1 liter) are grown in suspension in Sf-900IISFM media (Gibco-BRL, Gaithersburg, MD) containing gentamicin (25 μg/ml) at 27° with

constant shaking (125 rpm). Approximately 2×10^6 cells/ml are infected with H6γ and either Gβ_1 or Gβ_1 mutant baculoviruses with a multiplicity of infection ranging from 2 to 10 depending on the virus used for transfection.

2. Cells are harvested 60 to 72 hr after infection by centrifugation at 1000g for 20 min at 4° and then washed twice with ice-cold phosphate-buffered saline buffer (PBS). Cell viability, as estimated by Trypan Blue staining, is used to determine when cells are harvested. Viability of Sf9 cells should be no less than 75–80%.

3. Cells are resuspended in 200 ml (10× volume of cell pellet) of ice-cold buffer A [20 mM Tris-HCl (pH 8.0), 150 mM NaCl, 1 mM MgCl$_2$, 10 mM 2-mercaptoethanol, 100 μg/ml Pefabloc SC (Boeringer Mannheim), 100 μM PMSF, 5 μg/ml aprotinin and leupeptin (each)] and cells are disrupted by homogenation (20–30 strokes) using a Potter Dounce homogenizer.

4. Genapol C-100 (Calbiochem, La Jolla, CA) is added to the homogenate to obtain a final concentration of 0.2% followed by gentle mixing for 1 hr at 4°.*

5. The homogenate is clarified by ultracentrifugation at 257,000g for 70 min at 4°.

6. The Genapol C-100 soluble supernatant (supernatant) is collected and adjusted to a final volume of 200 ml with Buffer A.

7. The supernatant is added to a 0.25 ml bed volume of nickel-NTA resin (Qiagen, Valencia, CA), which binds the His tag, preblocked with protease-free bovine serum albumin (2.5 mg/ml) in buffer A.

8. The samples are gently mixed at 4° for 1.5 hr.

9. The resin is pelleted at 500g for 2 min at 4° and washed with 5 ml of ice-cold buffer B [20 mM Tris-HCl (pH 8.0), 0.5 M NaCl, 20 mM imidazole, 10 mM 2-mercaptoethanol, 100 μg/ml of Pefabloc SC, 100 μM phenylmethylsulfonyl fluoride (PMSF), 5 μg/ml aprotinin and leupeptin (each)] to remove unbound proteins.

10. The recombinant proteins are eluted with 2 ml of ice-cold buffer C [0.1 M EDTA, 10 mM 2-mercaptoethanol, 100 μg/ml of Pefabloc SC (Boeringer Mannheim), 100 μM PMSF, 5 μg/ml aprotinin and leupeptin (each)].

11. The eluents are brought to a final volume of 4 ml with 20 mM Tris-HCl (pH 8.0).

---

[14] N. P. Skiba, H. Bae, and H. E. Hamm, *J. Biol. Chem.* **271**, 413 (1996).

*Genapol C-100 is a detergent used to dissociate the $\beta\gamma$ dimer from the membrane. When no detergent is used in purification, $\beta\gamma$ is distributed in the soluble, ~40%, and the particulate fraction, ~60%. The majority of $\beta\gamma$ in the combined fraction is posttranslationally modified, but not all (amount not determined). The slight contamination of unmodified $\beta\gamma$ has little effect on the subsequent assays. To obtain only modified $\beta\gamma$, clarify homogenate by ultracentrifugation at 257,000g for 70 min at 4° before adding Genapol C-100 (step 4) and then proceed using the pellet.

12. Using BioMax-30 microconcentrators (Millipore, Bedford, MA), the buffer is exchanged with 20 mM Tris-HCl (pH 8.0) and the purified protein is concentrated prior to storage at $-20°$.

The protein concentrations of recombinant proteins are determined with Bradford reagent (Pierce, Rockford, IL) and bovine serum albumin (Pierce) as a standard. The quality of purification is analyzed by comparing 20 μg of recombinant proteins to 5 μg of purified retinal G$\beta_1\gamma_1$ on SDS–polyacrylamide gel electrophoresis (SDS–PAGE) prior to staining gel with Coomassie blue to detect separated proteins. This purification scheme typically yields 200 to 500 μg of G$\beta\gamma$ dimers per 1×10^9 cells of estimated purity ranging from 30 to 50%.

Analysis of G$\beta\gamma$ Mutants with Various Effectors

PLC-β_2 Assay

G$\beta\gamma$ is an activator of PLC-β_2.[3,15] The effect of the mutants on PLC-β_2 activity is determined by quantitating the amount of inositol 1,4,5-triphosphate (IP3) produced in a reconstituted mixed detergent–phospholipid micelle containing purified H6PLC-β_2, G$\beta\gamma$ or mutant, and radiolabeled phosphatidylinositol 4,5-bisphosphate (PIP$_2$).

Purification of H$_6$PLC-β_2. With a multiplicity of infection equal to 10, Sf9 cells are infected with recombinant baculovirus containing H$_6$PLC-β_2 cDNA (kindly provided by Dr. Smrcka, University of Rochester, NY). The purification of H$_6$PLC-β_2 expressed in Sf9 cells is done essentially as described,[16] except harvested cells are lysed in ice-cold extraction buffer [20 mM Tris-HCl (pH 8.0), 0.1 mM EGTA, 0.1 mM EDTA, 37 mM sodium cholate] supplemented with 43 mM 2-mercaptoethanol, 1/500 dilution calpain inhibitor I (Roche Molecular Biochemicals, Indianapolis, IN), and 1/50 dilution of protease inhibitor cocktail (Pharmingen, San Diego, CA). Deoxyribonuclease I is not used during the metal-affinity chromatography nor is H$_6$PLC-β_2 further purified using a heparin-affinity column.

H$_6$PLC-β_2 Assay

1. Prepare mixed detergent–phospholipid micelles containing 58 μM PIP$_2$, 580 μM phosphatidylethanolamine, and [^{3}H]PIP$_2$ (2.5×10^{-3} mCi/assay) by sonication in 100 mM HEPES (pH 8.0), 6 mM EGTA, 2 mM deoxycholate, and 4 mM dithiothreitol (DTT).

2. After sonication, add acetylated bovine serum albumin (10 mg/ml) to the sonicated detergent–phospholipid micelles so that the final concentration of bovine serum albumin is 0.25 mg/ml. This is used to help stabilize the proteins in the micelle.

---

[15] M. Camps, A. Carozzi, P. Schnabel, A. Scheer, P. J. Parker, and P. Gierschik, *Nature* **360,** 684 (1992).

[16] V. Romoser, R. Ball, and A. V. Smrcka, *J. Biol. Chem.* **271,** 25071 (1996).

3. Add approximately 50 ng of $H_6PLC-\beta_2$ and 2 μM recombinant $G\beta\gamma$ dimers contained in 10 μl of 20 mM Tris-HCl (pH 8.0) to prepared detergent–phospholipid micelles (12.5 μl) on ice.

4. Add 60 mM $CaCl_2$ (2.5 μl) to start the reaction at room temperature and then immediately transfer the samples to a 30° bath for 15 min.

5. Terminate the reactions (25 μl) by the addition of chloroform/methanol/HCl (50 μl/50 μl/0.3 μl, 100 μl total), which separates [^{3}H]PIP$_2$ from the aqueous soluble [^{3}H]IP$_3$, and the subsequent addition of 1N HCl/5 mM EDTA (50 μl), which denatures and precipitates the proteins and chelates the excess calcium.

6. Centrifuge samples at 13,000 rpm for 5 min at room temperature.

7. Analyze a sample aliquot (125 μl) of the aqueous layer by liquid scintillation counting for soluble [^{3}H]IP$_3$ produced in the assay.

$G\beta\gamma$ Interaction with $\beta ARK1$

$G\beta\gamma$ mediates translocation of G-protein-coupled receptor kinases from the cytosol to the membrane. These G-protein-coupled receptor kinases can then in turn phosphorylate activated G-protein-coupled receptors and initiate receptor desensitization.[17] The ability of the $G\beta\gamma$ mutants to bind to $\beta ARK1$, a G-protein-coupled receptor kinase, is quantified using immunoprecipitation and immunoblotting.

Preparation of Cells

1. Seed COS7 cells into 10 cm tissue culture plates in Dulbecco's modified Eagle's medium (DMEM) containing 10% fetal calf serum (FCS) supplemented with 10 mM HEPES, pH 7.4, and antibiotics (10 U/ml penicillin and 10 μg/ml streptomycin).

2. Grow cells to a confluency of ~80% at 37° in a humidified 5% (v/v) CO_2 atmosphere.

Transfection of Cells

1. Mix cDNAs expressing $G\gamma 2$ (2 μg), $G\beta 1$ (wild-type or mutant; 2 μg) and $\beta ARK1$ (1 μg) with 35 μl LipofectAMINE (Gibco) in a 15 ml test tube containing 5 ml DMEM.

2. Add pcDNA3 empty vector to obtain a final DNA concentration of 10 μg DNA/transfection and incubate for 15 min at ambient temperature.

3. Wash cells 2× with warm DMEM (lacking serum or antibiotics).

4. Expose cells to transfection medium for 3–4 hr at 37° in a humidified 5% CO_2 atmosphere.

[17] R. H. Stoffel III, J. A. Pitcher, and R. J. Lefkowitz, *J. Membr. Biol.* **157**, 1 (1997).

5. Add 5 ml DMEM containing 20 mM HEPES, pH 7.4, 20% FCS and antibiotics and incubate cells for an additional 24 hr.

Immunoprecipitation and Immunoblotting

1. Transfer cells onto ice and wash 3× with ice-cold PBS.
2. Lyse cells for 10 min on ice in 1.0 ml CHAPS-HEDN buffer (CHAPS, 3-[(3-cholamidopropyl)dimethylammonio]-1-propanesulfonic acid; HEDN: 10 mM HEPES, pH 7.2, 1 mM EDTA, 1 mM dithiothreitol, and 100 mM NaCl).
3. Scrape cells and transfer lysates into 1.5 ml microcentrifuge test tubes.
4. Clear cellular debris by centrifugation at 14,000 rpm for 15 min at 4°.
5. Transfer cell lysates into new 1.5 ml microcentrifuge test tubes and add 50 μl of 50% slurry of protein A/G Sepharose (Calbiochem).
6. Add 15 μg of C5/1 anti-βARK1 monoclonal antibody[18] and rotate samples 2–3 hr at 4°.
7. Wash βARK1 immune complexes 2× with CHAPS buffer, resuspend the protein A/G-Sepharose bound immune complexes in 30 μl Laemmli sample buffer, and boil 5 min.
8. Fractionate samples on 4–20% SDS–PAGE and transfer onto nitrocellulose filter.
9. Expose filter to blocking buffer containing 4% (w/v) bovine serum albumin in wash buffer (150 mM NaCl, 10 mM Tris-HCl, pH 7.2, 0.05% Nonidet P-40, 0.05% Tween 20) for 1–2 hr at ambient temperature.
10. Blot filter with Gβ antibody (T-20, Santa Cruz Biotechnology) in blocking buffer (1 : 10,000 dilution) for 1 hr at ambient temperature then wash filter 3× with blocking buffer.
11. Add horseradish peroxidase-conjugated anti-mouse secondary antibody in blocking buffer (1 : 10,000) for 1 hr at ambient temperature, wash filters 3×, and visualize immune complexes on nitrocellulose by enzyme-linked chemiluminescence (Amersham, Piscataway, NJ).
12. Quantitate intensity of bands by scanning laser densitometry.

Calcium Channel Inhibition Assay

G$\beta\gamma$ inhibits the activity of many voltage-gated calcium channels[19–21] and has been shown to bind to and associate with the α subunit of the voltage-gated calcium

---

[18] M. Oppermann, M. Diversé-Pierluissi, M. H. Drazner, S. L. Dyer, N. J. Freedman, K. C. Peppel, and R. J. Lefkowitz, *Proc. Natl. Acad. Sci. U.S.A.* **93,** 7649 (1996).
[19] S. Herlitze, D. E. Garcia, K. Mackie, B. Hille, T. Scheuer, and W. A. Catterall, *Nature* **380,** 258 (1996).
[20] S. R. Ikeda, *Nature* **380,** 255 (1996).
[21] L. R. Shekter, R. Taussig, S. E. Gillard, and R. J. Miller, *Mol. Pharmacol.* **52,** 282 (1997).

channel.[22,23] The ability of $G\beta\gamma$ mutants to inhibit barium current carried through voltage-gated calcium channels is determined using HEK 293 cells expressing the $CC\alpha1B$ subunit and the $G\beta\gamma$ mutants. Inhibition of calcium channels by $G\beta\gamma$ mutants is quantified by determining the prepulse facilitation ratio.

Cloning of $G\beta_1$ into Mammalian Expression Vector. The mutant $G\beta$ subunits are subcloned into the mammalian expression vector pCMV5 (kindly provided by M. I. Simon, Caltech, Pasadena, CA). The $G\gamma_2$ subunit is expressed in pCDM8.1, (similarly provided by M. I. Simon). Each mutant was sequenced.

Preparation of Cells

1. A stable cell line G1A1 (HEK 293 cell line expressing human α_1, $\alpha_{2B}\delta$, and β_{1B} calcium channel subunit plasmids, which has been kindly provided by SIBIA Inc.,[24] is grown in plastic Falcon dishes in DMEM (Life Technologies, Gaithersburg, MD) containing 5% defined bovine serum (HyClone, Logan, UT) plus penicillin G (100 U/ml), streptomycin sulfate (100 μg/ml), and geneticin (500 μg/ml).
2. One day before recording, cells ($\sim$75% confluency) are dissociated by gentle trituration with a fire-polished Pasteur pipette and replated onto poly (L-lysine) coated glass coverslips (Sigma, St. Louis, MO).
3. Cells are cotransfected with plasmids containing the cDNAs for the G protein and CD8 using the standard calcium-phosphate precipitation technique[25] or transfection kit (Mammalian Transfection Kit; Stratagene, La Jolla, CA). Positive cells are those to which beads coated with the CD8 antibody adhere (Dynabeads M-450 CD8; Dynal, Lake Success, NY).

Calcium Current Recording

1. Use an extracellular buffer solution for whole-cell voltage-clamp experiments composed of (in mM): 160 tetraethylammonium chloride, 5 $CaCl_2$, 1 $MgCl_2$, 10 HEPES, 10 glucose; adjust pH to 7.4 with TEA $\cdot$ OH. The standard internal solution should consist of (in mM): 100 $CsCl_2$, 37 CsOH, 1 $MgCl_2$, 10 BAPTA, 10 HEPES, 3.6 MgATP, 1 GTP, and 14 phosphocreatine di-tris salt (N-(imino[phosphoamino]methyl)-N-methyglycine) (Tris$_2$CP), and 50 U/ml^{-1} creatine phosphokinase (CPK). Adjust the pH to 7.3 with CsOH. The osmolarity of the pipette solution should be

---

[22] M. De Waard, H. Liu, D. Walker, V. E. Scott, C. A. Gurnett, and K. P. Campbell, *Nature* **385**, 446 (1997).

[23] N. Qin, D. Platano, R. Olcese, E. Stefani, and L. Birnbaumer, *Proc. Natl. Acad. Sci. U.S.A.* **94**, 8866 (1997).

[24] M. E. Williams, P. F. Brust, D. H. Feldman, S. Patthi, S. Simerson, A. Maroufi, A. F. McCue, G. Velicelebi, S. B. Ellis, and M. M. Harpold, *Science* **257**, 389 (1992).

[25] F. Ausubel, R. Brent, R. Kingston, D. Moore, J. Seidman, J. Smith, and K. Struhl, *Current Protocols Mol. Biol.* **9**, 1 (1993).

300 mOsm/liter, and the osmolarity of the extracellular solution should be between 315 and 323 mOsm/liter.

2. Use a tight-seal whole-cell configuration of the patch-clamp technique[26] to record calcium currents. Make patch pipettes from soft, soda-lime capillary glass, coated with Sylgard (Dow Corning, Midland, MI), and have resistances of 1.8–3.5 MΩ when filled with internal solution. These patch pipettes are used to form GΩ seals on the surface of the HEK293 cells and the membrane is broken by applying gentle positive pressure through the patch pipette.

3. Make recordings at room temperature (21–24°). Currents are recorded using an Axopatch 1D (Axon Instruments, Foster City, CA) amplifier, filtered at 2 kHz by the built-in filter of the amplifier, and stored on the computer. Capacitative transients are canceled at 10 kHz, and their values are obtained directly, together with the series-resistance values from the settings of the Axopatch 1D amplifier. Apply series-resistance compensation between 40 and 80%. Correct leak by using a P/N protocol. The resting membrane potential of G1A1 cells should be around −55 mV. The input impedence should be greater than 1 GΩ.

4. Barium currents can be elicited by two 25-ms depolarizing voltage steps from a holding potential of −90 mV to +10 mV interspersed with a return to the resting potential for 65 ms. To relieve the G$\beta\gamma$ mediated inhibition, the two 25-ms steps are interspersed with a depolarizing pulse to +80 mV. The quotient of the amplitude of the barium current after the prepulse and before the prepulse is the prepulse facilitation ratio.

Adenylyl Cyclase Assay

G$\beta\gamma$ activates adenylate cyclase 2 (AC2)[27] and the effect of the various mutants is determined *in vitro* in the presence of constitutively activated G$_s\alpha$ [glutamine at residue 227 mutated to leucine (Q227L)]. G$\beta\gamma$ activates AC2 only after it is first activated by G$_s\alpha$, hence the use of Q227L in this assay. The activity of AC2 was assayed in a membrane preparation method first devised by Salomon *et al.*[28] This method quantitates cAMP production by using [α-^{32}P]ATP to generate [^{32}P]cAMP, followed by a two-column chromatography system that separates labeled cAMP from other labeled ATPs. The first column consists of negatively charged Dowex 50, which binds labeled cAMP, allowing the majority of other labeled components to be washed away with H$_2$O. The cAMP is then passed on to a second column of neutral alumina, which binds with the labeled cyclic nucleotide. A wash

[26] O. P. Hamill, A. Marty, E. Neher, B. Sakmann, and F. J. Sigworth, *Pflugers Arch.* **391,** 85 (1981).

[27] W. J. Tang and A. G. Gilman, *Science* **254,** 1500 (1991).

[28] Y. Salomon, C. Londos, and M. Rodbell, *Anal. Biochem.* **58,** 541 (1974).

with imidazole into the alumina column further eliminates unlabeled components. The labeled cAMP is then eluted with imidazole, collected, and scintillation counted. Eluting conditions can be found in Salomon's method. Our modification of Salomon's method consists of lowering assay volumes from 100 μl to 50 μl, enabling the decrease in use of radioisotopes, wash volumes, and column size.

Membrane Preparation

1. Membranes are prepared from *Trichoplusia ni* insect cells (Invitrogen, High Five) infected with a recombinant baculovirus containing AC2 cDNA (kindly provided by Ron Magnusun, Mt. Sinai, NY).
2. Constitutively active H_6-Q227L-$G_s\alpha$ (kindly provided by Tarun B. Patel, University of Tennessee) is expressed in *Escherichia coli* and purified to apparent homogeneity using nickel-NTA resin (Qiagen).[29]
3. *Trichoplusia ni* cells (Invitrogen, High Five) are infected with $H_6G\gamma_2$ recombinant baculovirus and either wild-type or mutant $G\beta_1$ baculoviruses.
4. Membranes are prepared from infected cells, and the protein concentration of the $G\beta\gamma$ expressed in these membranes is determined by protein immunoblotting with purified bovine brain $G\beta\gamma$ as a standard.
5. The $G\beta\gamma$-containing membranes are extracted in 50 mM HEPES, 1 mM EDTA, and 0.7% CHAPS.
6. Centricon microconcentrators (Amicon, Damers, MA) are used to reduce the detergent concentration to 0.14%.

cAMP Activity Assay

1. In a total assay volume of 50 μl, the following are added yielding these final concentrations: 20 mM NaHEPES (pH 8.0), 1 mM EDTA, 20 mM creatine phosphate, 0.2 mg/ml creatine phosphokinase, 0.02 mg/ml myokinase, 2 mM MgCl$_2$, $G\beta\gamma$ dimer membrane extract (0.5 to 0.6 μM), H_6-Q227L-$G_s\alpha$ (20 nM), AC2-containing membranes (5 μg), [α-^{32}P]ATP (0.1 mM, 2000–5000 cpm/pmol with 1 mM [^{3}H]cAMP, $\sim$10,000 cpm).
2. All assay mixtures are kept on ice to prevent reaction initiation.
3. The conversion reaction is carried out for 15 min at 33°.
4. The reaction is stopped with 2% SDS.
5. Column separation with Dowex 50 and neutral alumina is done.[28]
6. The quantitation of [^{32}P]cAMP is done with scintillation counting and performed in triplicate, where values are means of triplicate determinations.

Coefficient of variance is always less than 10% and all experiments are repeated three or more times with different batches of membranes, resulting in qualitatively similar results.

[29] M. P. Graziano and A. G. Gilman, *J. Biol. Chem.* **264**, 15475 (1989).

GIRK1/GIRK4 Assay

GIRK channels are mainly expressed as a heteromultimer of GIRK1/GIRK4 and GIRK1/GIRK2 in heart and brain, respectively.[30-32] The activation of these channels is mainly mediated by the G$\beta\gamma$ subunits[33-35] in concert with PIP$_2$ and sodium ions.[36,37] G$\beta\gamma$ is known to bind to the N and C termini of all GIRKs cloned to date[38-40] and causes them to gate, allowing potassium to flow down its chemical and electrical gradient. The capability of the Gβ mutants to activate GIRK conductances is measured in *Xenopus laevis* oocytes injected with RNAs for GIRK1/GIRK4 and G$\beta\gamma$ mutants.

Constructs. The channel cDNA is subcloned into a plasmid containing the 5′- and 3′-untranslated region of the *Xenopus laevis* β-globin gene. The untranslated regions enhance the translation and stability of the cRNA in the *Xenopus* oocyte, and therefore yield a robust gene expression.[41] The cDNA is then linearized using *Nhe*I and transcribed using T7 RNA polymerase using the following procedure.

cRNA Preparation. cRNA is transcribed using a home-assembled cRNA transcription kit using the following procedure:

cDNA template preparation

1. To 50 μl DNA (2–3 μg) add 5 μl of proteinase K (100μg/ml) and incubate at 37° for 2 hr.
2. Add 150 μl of distilled H$_2$O [diethyl pyrocarbonate (DEPC) treated] and 2 μl of glycogen.
3. Extract 2× with phenol/chloroform.
4. Add 20 μl of 3 M sodium acetate (pH 5.2).
5. Add 500 μl of clean 100% ethanol.

[30] G. Krapivinsky, E. A. Gordon, K. Wickman, B. Velimirovic, L. Krapivinsky, and D. E. Clapham, *Nature* **374,** 135 (1995).

[31] Y. J. Liao, Y. N. Jan, and L. Y. Jan, *J. Neurosci.* **16,** 7137 (1996).

[32] S. K. Silverman, H. A. Lester, and D. A. Dougherty, *J. Biol. Chem.* **271,** 30524 (1996).

[33] E. Reuveny, P. A. Slesinger, J. Inglese, J. M. Morales, J. A. Iniguez-Lluhi, R. J. Lefkowitz, H. R. Bourne, Y. N. Jan, and L. Y. Jan, *Nature* **370,** 143 (1994).

[34] K. D. Wickman, J. A. Iniguez-Lluhl, P. A. Davenport, R. Taussig, G. B. Krapivinsky, M. E. Linder, A. G. Gilman, and D. E. Clapham, *Nature* **368,** 255 (1994).

[35] D. E. Logothetis, D. Kim, J. K. Northrup, E. J. Neer, and D. E. Clapham, *Proc. Natl. Acad. Sci. U.S.A.* **85,** 5815 (1988).

[36] J. L. Sui, J. Petit-Jacques, and D. E. Logothetis, *Proc. Natl. Acad. Sci. U.S.A.* **95,** 1307 (1998).

[37] H. Zhang, C. He, X. Yan, T. Mirshahi, and D. E. Logothetis, *Nature Cell Biol.* **1,** 183 (1999).

[38] G. Krapivinsky, L. Krapivinsky, K. Wickman, and D. E. Clapham, *J. Biol. Chem.* **270,** 29059 (1995).

[39] A. Inanobe, K.-I. Morishige, N. Takahashi, H. Ito, M. Yamada, T. Takumi, H. Nishina, K. Takahashi, Y. Kanaho, T. Katada, and Y. Kurachi, *Biochem. Biophys. Res. Comm.* **212,** 1022 (1995).

[40] C. L. Huang, Y. N. Jan, and L. Y. Jan, *FEBS Lett.* **405,** 291 (1997).

[41] E. R. Liman, J. Tytgat, and P. Hess, *Neuron* **9,** 861 (1992).

6. Cool for 10 min at $-70°$ or 20 min at $-20°$.
7. Spin for 15 min at 14,000 rpm at $4°$.
8. Wash with 70% (v/v) clean ethanol (DEPC).
9. Resuspend in 15 μl of distilled H_2O (DEPC).

Transcription reaction

1. Reaction mixture: 15 μl DNA template, 10 μl methylated CAP (2.5 mM stock, Pharmacia), 10 μl rNTP (10 mM each nucleotide, Stratagene), 10 μl 5× Transcription buffer (Stratagene), 2 μl DTT (0.75 M), 1 μl RNasin (Promega), 1 μl T7 RNA polymerase (Stratagene).
2. Incubate for 3 hr at $37°$.
3. Add 180 μl of distilled H_2O (DEPC-treated).
4. Extract 1× with phenol, 1× with chloroform.
5. Add 150 μl of 5 M ammonium acetate.
6. Precipitate by adding 700 μl of 100% clean ethanol.
7. Chill $-70°$ 15 min, then spin for 15 min at 14,000 rpm at $4°$.
8. Wash with 70% (v/v) ethanol.
9. Spin for 15 min at 14,000 rpm at $4°$.
10. Resuspend pellet in 30 μl of distilled H_2O (DEPC).

The cRNA integrity and concentration can be determined by running an aliquot on a formaldehyde gel and absorbance measurement, respectively.

Oocyte Preparation. Oocytes are surgically removed from *Xenopus laevis* frogs anesthetized with 0.15% (w/v) Tricaine (Sigma, St. Louis, MO). Oocytes are defolliculated by shaking in nominally Ca^{2+}-free solution containing (in mM): NaCl, 96; KCl, 2; $MgCl_2$, 1; HEPES, 5; and 2 mg/ml Type 1 collagenase (Worthington, Lakewood, NJ) pH 7.4 until >50% of the oocytes are defolliculated, ~1 hr at room temperature.

The oocytes are then washed in collagenase-free solution (above) that also contains 1 mM $CaCl_2$. Fully defolliculated stage 5–6 oocytes are selected and microinjected with 50 nl cRNA mixture containing 1–3 ng of rat GIRK1[42,43] and GIRK4,[30] and 100–500 pg of bovine $G\gamma_2$ and $G\beta_1$ mutant using Drummond Nanoject microinjector.

Two-Electrode Voltage Clamp (TEVC)

1. Record currents through the expressed channels by the two-electrode voltage clamp (TEVC) technique using a CA1-B amplifier (Dagan Corp.)[33]

[42] N. Dascal, W. Schreibmayer, N. F. Lim, W. Wang, C. Chavkin, L. DiMagno, C. Labarca, B. L. Kieffer, C. Gaveriaux-Ruff, D. Trollinger, H. Lester, and N. Davidson, *Proc. Natl. Acad. Sci. U.S.A.* **90,** 10235 (1993).
[43] Y. Kubo, E. Reuveny, P. A. Slesinger, Y. N. Jan, and L. Y. Jan, *Nature* **364,** 802 (1993).

TABLE I
EFFECTOR RESPONSE TO VARIOUS $\beta\gamma$ MUTANTS

Mutation	βARK	PLC-β_2	AC2	GIRK	Calcium channel
L55	—[a]	↑[b]	↓[c]	↓	↑
K57	—	—	↓	—	—
Y59	↑	—	N/A	—	—
K78	—	N/A	↓	↓	↓
I80	—	↓	—	↓	↑
K89	↑	↓	↓	↓	—
S98	N/A	↑	N/A	—	—
W99	—	↓	↓	↓	—
M101	—	↓	↓	—	↓
L117	↓	↓	↓	—	—
N119	—	↓	↓	—	↓
T143	↓	↓	—	—	↓
D186	↑	↓	↓	—	↓
D228	—	↓	↓	↓	—
W332	↑	↓	↓	—	↓

[a] Same as wild-type.
[b] Greater than wild-type.
[c] Less than wild-type.

Data acquisition and analysis is done using the pCLAMP 6.04 software package (Axon Instruments).

2. Fill electrodes (thin borosilicate with filament, Warner GC120TF-10), which have a resistance of 0.1–0.6 MΩ, with 3 M KCl. A high K$^+$ external solution is used containing (in mM): KCl, 90; MgCl$_2$, 2; HEPES, 10, pH 7.4 (KOH).

3. Hold oocytes at -80 mV to inactivate endogenous nonselective cationic channels of the oocyte.

4. Current-to-voltage relationship can be extracted by either giving a ramp voltage from -100 mV to $+50$ mV at 0.5 mV/ms or a set of step voltage pulses from -100 mV to $+50$ mV at 10 mV increments for 250 ms.

In general, oocytes expressing GIRK1/GIRK4 channel and G$\beta_1\gamma_2$ display a large inwardly rectifying currents. In this external solution, the reversal potential of the current is ~-10 mV. In some cases, the leak can be subtracted by subtracting the currents recorded in the presence of 3 mM BaCl$_2$ or by scaling the outward current (mainly leak) and subtracting it from the inward current at -80 mV that mainly resembles the current flowing through the GIRK channels.

Discussion

This chapter, describing the construction and analysis of the alanine mutations of the $G\alpha$-interacting $G\beta$ residues, provides the initial framework for determining how $G\beta\gamma$ subunits interact with and regulate a diverse group of effectors. Our studies suggest that the effector interaction regions are clustered on $G\beta$ such that they partially overlap one another. This overlap allows for $G\alpha$ to be the key regulator of $G\beta\gamma$ signaling and suggests that subunit dissociation is crucial for $G\beta\gamma$ signaling. Further analysis of these mutants with other known and new effectors could lead to an even better understanding of the molecular basis of interaction of $G\beta\gamma$ with various proteins.

More importantly though, this chapter and the research herein has provided us with several new tools with which to study signaling through heterotrimeric $G\beta\gamma$ proteins. Because the different $G\beta\gamma$ mutants have various effects on the different effectors (Table I), they can be strategically used to define $G\beta\gamma$/effector interaction and activation. Several examples of possible uses for these mutants are followed. (1) They could be used in a cell to discern the internal response to a G-protein coupled receptor (GPCR) that is signaling through the $G\beta\gamma$ subunit. For example, the different mutants could be used to determine if the increase in intracellular calcium is due to activation through PLC or calcium channels. (2) In a similar manner, they could be used to map further downstream responses to a GPCR by determining which effector and therefore second messenger is needed to propagate the signal for the response to occur. (3) Since several different mutants activate different effectors better than wild-type (Table I), the downstream responses of an effector "over"-activated by these mutants can be studied. In the case of mutants that do not activate their effectors, it will be interesting in future studies to determine if they still bind their effectors with high affinity. If they do, they may be useful dominant negative tools to study the importance of that effector pathway in downstream physiological responses to signals.

Acknowledgment

This work was supported by Grants EY06062 (H.E.H.) and EY10291 (H.E.H.) from the National Institutes of Health.

[31] Ribozyme-Mediated Suppression of G Protein γ Subunits

By JANET D. ROBISHAW, QIN WANG, and WILLIAM F. SCHWINDINGER

Introduction

With the identification of at least 16 α, 5 β, and 12 γ subunit genes,[1] it has become increasingly important to decipher which G protein $\alpha\beta\gamma$ subunit complexes actually exist in cells and to identify their roles in particular signaling pathways. *In vitro* strategies have revealed that many combinations are physically able to form G protein $\alpha\beta\gamma$ trimers. However, while providing important structure–function information, such reconstitution approaches fall critically short of establishing roles of specific G protein $\alpha\beta\gamma$ subunit complexes in particular signaling pathways in the intact cell setting. Increasingly, reverse genetic approaches are being employed to fill this gap, including the use of ribozymes to suppress gene expression, via inactivation of their target mRNAs. Compared to antisense RNA, ribozymes offer significant advantages.[2–4] Ribozymes operate as site-specific ribonucleases, resulting in the degradation of the target mRNAs. Moreover, binding of ribozymes to their target mRNAs is often more stringent than to antisense RNA. Finally, controls for ribozyme activity can be made by substituting nucleotides in the catalytic core, thereby producing inactive ribozyme. The purpose of this article is to provide some guidelines and protocols for the use of ribozymes in identifying the functions of closely related members of G protein γ subunit family. Some of the caveats involved in designing ribozymes will be discussed. In addition, introducing ribozymes into cells, constructing appropriate controls, and discussing issues encountered with the use of ribozymes in cell culture will be considered.

Ribozymes

First discovered by Cech[5] and Altman,[6] ribozymes are RNA molecules that have the ability to both recognize and cleave other RNA molecules in a sequence-specific fashion. These properties make ribozymes powerful tools for studying

[1] E. H. Hurowitz, J. M. Melnyk, Y. J. Chen, H. Kouros-Mehr, M. I. Simon, and H. Shizuya, *DNA Res.* **7,** 111 (2000).

[2] A. Renato Muotri, L. da Veiga Pereira, L. dos Reis Vasques, and C. F. Martins Menck, *Gene* **237,** 303 (1999).

[3] D. J. Gaughan and A. S. Whitehead, *Biochim. Biophy. Acta* **1445,** 1 (1999).

[4] M. Kruger, C. Beger, and F. Wong-Staal, *Methods Enzymol.* **306,** 207 (1997).

[5] T. R. Cech, A. J. Zaug, and P. J. Grabowski, *Cell* **27,** 487 (1981).

[6] C. Takada-Guerrier and S. Altman, *Science* **223,** 285 (1984).

the functional consequences of suppressed gene expression. Among the several types of ribozymes, the hammerhead ribozyme has been the most extensively characterized.

Structure

Figure 1A shows the consensus sequence for a hammerhead ribozyme. Nearly all naturally occurring hammerhead ribozymes cleave 3' to a GUC sequence motif in the target mRNA.[7] Mutagenesis studies have revealed that alterations in the first, and particularly third, position are tolerated, leading to the generally accepted GUH sequence motif (H = A, C, U).[8] However, within the dictates of the target mRNA sequence, most researchers try to design ribozymes to target the GUC sequence motif since it is the most efficiently cleaved.[9] With regard to the ribozyme, there are two types of domains: (1) the flanking domains that confer specificity of binding of the ribozyme to the target mRNA; and (2) the catalytic domain that mediates the specific cleavage of the consensus GUH sequence motif.

The flanking domains determine the specificity of the ribozyme for the target mRNA. Since these regions are not required for catalysis, they can be designed to improve stability, facilitate hybridization, and position the catalytic domain of the ribozyme in reference to the GUH sequence motif within the target mRNA. Competing factors determine the ideal length of the flanking domains. On the one hand, the longer the flanking domains, the more specific the ribozyme is likely to be. On the other hand, the longer the flanking sequences, the less efficient the ribozyme is likely to be. For catalytic cleavage to occur, the ribozyme must be able to rapidly associate and dissociate from the target mRNA. Under model conditions, this occurs with flanking arms of 5 to 8 nucleotides.[10] However, this number of nucleotides is not considered sufficient to ensure specificity of the ribozyme in a cellular or animal setting. Assuming the haploid human genome contains approximately 3×10^9 bases, the prediction is that any sequence of 17 nucleotides or longer would have a high probability of being unique.[11] In our experience, a good balance of these competing factors is provided by designing the ribozymes for the γ_2, γ_5, γ_7, and γ_{12} subunits with a longer arm of 14 nucleotides and a shorter arm of 8 nucleotides. Another strategy to circumvent this problem is the use of asymmetric hammerhead ribozymes that are characterized by one long arm with the catalytic domain on the 5' end.[10]

The catalytic domain contains two stretches of conserved sequences: 5'-CUGANGA and 5'-GAAA, with mutagenesis studies identifying invariant residues

[7] G. Bruening, *Methods Enzymol.* **180**, 546 (1989).

[8] R. Perriman, A. Delves, and W. L. Gerlach, *Gene* **113**, 157 (1992).

[9] T. Shimayama, S. Nishikawa, and K. Taira, *Biochemistry* **34**, 3649 (1995).

[10] C. Hammann and M. Tabler, *Methods: Companion Methods Enzymol.* **18**, 273 (1999).

[11] A. D. Branch, *TIBS* **23**, 45 (1998).

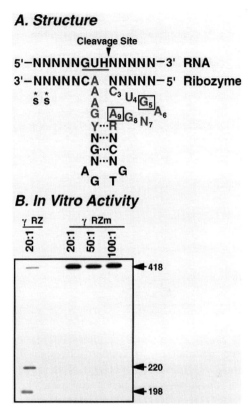

FIG. 1. Structure and function of hammerhead ribozyme. (A) Consensus hammerhead ribozyme structure, where N represent any nucleotide; H represents nucleotides C, U, or A. The ribonucleotides are shown in gray, whereas positions that can be substituted with deoxyribonucleotides are shown in black. Residues essential for cleavage activity are boxed. The nucleotide numbering scheme is based on that of Hertel *et al.*[13] (B) Cleavage activity of a hammerhead ribozyme *in vitro*, comparing the wild-type ribozyme (γRZ) to a mutant hammerhead ribozyme harboring boxed substitutions in the catalytic core (γRZm).

required for cleavage. In this regard, the G5 and A9 residues have been found to be important for full catalytic activity.[12] In addition to the role of the 2′-OH group as a nucleophile in the cleavage reaction, the amino group of the G5 residue is important. Thus, substitutions of the G5 and A9 residues have been found to abolish the cleavage activity (shown boxed in Fig. 1A). As shown in Fig. 1B, the

[12] N. Usman, L. Beigelman, and J. A. McSwiggen, *Curr. Opin. Struct. Biol.* **6,** 527 (1996).

[13] K. J. Hertel, A. Pardi, O. C. Uhlenbeck, M. Koizumi, E. Ohtsuka, S. Uesugi, R. Cedergren, F. Eckstein, W. L. Gerlach, R. Hodgson, and R. H. Symons, *Nucl. Acids Res.* **20,** 3252 (1992).

cleavage activity of a ribozyme can be measured in the test tube. In this example, the γ_7 ribozyme is targeted to a GUC sequence found at position +3 to +5 in relation to the translational start site of the γ_7 mRNA. Upon addition of the γ_7 ribozyme, the *in vitro* synthesized γ_7 RNA transcript is cleaved into two fragments of the expected sizes for cleavage at the targeted site. This result indicates that the ribozyme cleaves the γ_7 RNA transcript in a site-specific fashion. By contrast, on addition of a mutant ribozyme harboring substitutions of the G5 and A9 residues in the catalytic core, no cleavage of the γ_7 RNA transcript is observed even at very high ribozyme-to-template ratios. This result demonstrates that mutant ribozyme lacks the ability to cleave the γ_7 RNA transcript. This is a particularly useful control to test the specificity and provide information on the ribozyme's mechanism of action.

G Protein γ Subunits as Targets of Ribozymes

In vitro studies have shown that G protein $\beta\gamma$ subunits of varying composition exhibit differences in their abilities to interact with receptors and effectors.[14–19] *In vivo* studies have confirmed and extended these results,[20–22] demonstrating the G protein γ subunit composition has important ramifications for the fidelity of receptor–effector coupling. Consistent with this functional diversity, the γ subunits are members of a large, multigene family. To date, 12 different γ subunits have been identified.[1,23] Since their individual functions are not yet known, the γ subunits have been designated by number in order of cloning. Comparison of their predicted amino acid sequences reveals a relatively high degree of diversity concentrated in the amino-terminal region. Based on this diversity, we have divided members of the γ subunit family into three classes, with each class showing less than approximately 50% identity to members of the other classes.[23] Another level of diversity is imparted to the γ subunits by the nature of their post-translational modifications. Three sequential modifications have been identified for

[14] O. Kisselev and N. Gautam, *J. Biol. Chem.* **268,** 24519 (1993).

[15] P. Butkerait, Y. Zheng, H. Hallak, T. E. Graham, H. A. Miller, K. D. Burris, P. B. Molinoff, and D. R. Manning, *J. Biol. Chem.* **270,** 18691 (1995).

[16] R. A. Figler, M. A. Lindorfer, S. G. Graber, J. C. Garrison, and J. Linden, *Biochemistry* **36,** 16288 (1997).

[17] M. Richardson and J. D. Robishaw, *J. Biol. Chem.* **274,** 13525 (1999).

[18] N. Ueda, J. A. Iñiguez-Lluhi, E. Lee, A. V. Smrcka, J. D. Robishaw, and A. G. Gilman, *J. Biol. Chem.* **269,** 4388 (1994).

[19] C. S. Myung, H. Yasuda, W. W. Liu, T. K. Harden, and J. C. Garrison, *J. Biol. Chem.* **274,** 16595 (1999).

[20] C. Kleuss, H. Scherubl, J. Hescheler, G. Schultz, and B. Wittig, *Science* **259,** 832 (1993).

[21] Q. Wang, B. Mullah, C. A. Hansen, J. Asundi, and J. D. Robishaw, *J. Biol. Chem.* **272,** 26040 (1997).

[22] Q. Wang, B. K. Mullah, and J. D. Robishaw, *J. Biol. Chem.* **274,** 17365 (1999).

[23] E. A. Balcueva, Q. Wang, H. Hughes, C. Kunsch, Z. Yu, and J. D. Robishaw, *Exp. Cell. Res.* **257,** 310 (2000).

these proteins ending with a carboxyl-terminal CAAX motif (where C, cysteine; A, aliphatic amino acid; and X, any amino acid): (1) the addition of a prenyl group to the cysteine residue; (2) the proteolytic cleavage of the final three residues; and (3) the methylation of the newly exposed cysteine residue at the carboxyl terminus. Depending largely on the amino acid in the -X position, either a C_{15}-farnesyl or a C_{20}-geranylgeranyl group has been shown to be added to the γ subunit. Thus, the γ subunit family is beginning to rival the α subunit family in terms of both diversity and multiplicity. As a result, there is a great need to develop and apply new approaches to assess the importance of particular G protein γ subtypes in signaling systems in the intact cell setting. One such approach is the use of ribozymes to specifically inhibit the expression of individual γ subtypes and then to assess the functional consequences on receptor signaling pathways.[21,22]

Selection of Target Site

The use of ribozymes requires the identification of cleavage sites within the target mRNA molecules that are able to form hybrid complexes between the two molecules. Prediction of accessible sequences is often difficult because mRNA molecules form higher order structures. The challenge is even greater is using ribozymes to target members of multigene families that show a high degree of sequence conservation, such as the G protein γ subunit family. Currently, the process is still largely empirical.

As a first step, one or more cleavage site(s) within the γ mRNA has to be selected. In general, each γ mRNA has multiple cleavage sites from which to choose (Table I). In order of decreasing preference, these are GUC > GUA > GUU > AUH > CUH ~ UUH.[9] Several considerations have to be taken into account in the selection of the region of the γ mRNA to be targeted. Is the cleavage site unique or is it present in other γ mRNAs? Are the flanking sequences around the cleavage site unique or they present in other γ mRNAs? Since ribozymes may act in an antisense fashion, the obvious preference is to utilize those sequences that show least conservation with other γ mRNAs. Finally, are the flanking sequences around the cleavage site present in unrelated mRNAs in the database?

Another issue relates to the accessibility of the cleavage site. Since not all areas of the γ mRNA are equally accessible to hybridization with the ribozyme, it is necessary to confirm that the selected motif is efficiently cleaved. Therefore, if multiple cleavage sites are present, one strategy is to construct multiple ribozymes that target different regions scattered throughout the γ mRNA and test the ability of the ribozymes to reduce γ mRNA expression in cell culture. Alternatively, another approach is to use computer programs to predict the secondary structure of the γ mRNA. In a retrospective analysis, two RNA folding programs[24,25] were used

[24] M. Zuker, *Methods Mol. Biol.* **25**, 267 (1994).
[25] M. Amarzguioui and H. Prydz, *Cell. Mol. Life Sci.* **54**, 1175 (1998).

TABLE I
ALIGNMENT OF NUCLEOTIDE SEQUENCES FOR HUMAN G PROTEIN γ SUBUNITS[a]

```
GNGT1   -------------------AUGCCAGUAAUCAAUAUUGAGGACCUGACAGAAAAGGACAAAUU  44
GNG11   -------------------AUGCCUGCCCUUCACAUCGAAGAUUUGCCAGAGAAGGAAAAACU  44
GNGT2   ----------------------AUGGCCCAGGAUCUCAGCGAGAAGGACCUGUU  32
GNG12   --------------AUGUCCAGCAAAACAGCAAGCACCAACAAUAUAGCCCAGGCA------AG  44
GNG7    ----------------------AUGUCAGCCACUAACAACAUAUGCCCAGGCC------CG  32
GNG3    --------AUGAAAGGUGAGACCCCGGUGAACAGCACUAUGAGUAUUGGGCAAGCA------CG  50
GNG4    --------AUGAAAGAGGGCAUGUCUAAUAACAGCACCACUAGCAUCUCCCAAGCC------AG  50
GNG2    -------------------AUGGCCAGCAACAACACCGCCAGCAUAGCACAAGCC------AG  38
GNG8    ----------------AUGUCCAACAACAUGGCCAGGACAUGCCGAGGCC------CG  35
GNG5    --------------AUGUCUGGCUCUCCAGCGGUCGCCGCUAUG------AA  32
GNG10   -----------------------AUGUCCUCCGGGGCUAGCGCGAGCGCCCUG------CA  32
GNG13   ------------------------AUGGAGGAGUGGGACGUGCCACAGAUG------AA  29

GNGT1   GAAGAUGGAAGUUGACCAGCUCAAGAAAGAAGUGACACUGGAAAGAAUGCUAGUUUCCAAAUGUU  109
GNG11   GAAAAAUGGAAGUUGAGCAGCUUCGCCAAAGAAGAUUGCAGAGACAACAAGUGUCUAAAAUGUU  109
GNGT2   GAAGAUGGAGGUGGAGCAGCUGAAGAAAGAAGUGAAAAACACAAGAAUUCCGAUUUCCAAAGCGG  97
GNG12   GAGAACU---GUGCAGCAGUUAAGAUUAGAAGCCUCCAUUGAAAGAAUAAAAGGUUUCGAAGGCAU  105
GNG7    GAAGCUG---GUGGAACAGCUACGCAUAGAAGCCGGGAUUGAGCGCAUCAAGGUCUCCAAAGCGG  94
GNG3    CAAGAUG---GUGGAACAGCUUAAGAUUGAAGCCAGCUUGUGUCGGAUAAAGGUGUCCAAGGCAG  112
GNG4    GAAAGCU---GUGGAGCAGCUAAAGAUGGAAGCCUGUAUGGACAGGGUCAAGGUCUCCCAGGCAG  112
GNG2    GAAGCUG---GUAGAGCAGCUUAAGAUGGAAGCCAAUAUCGACAGGAUAAAGGUGUCCAAGGCAG  99
GNG8    CAAGACG---GUGGAACAGCUGAAGCUGGAGGUGAACAUCCACCGCAUGAAGGUGUCGGCGCAG  99
GNG5    GAAAGUG---GUUCAACAGCUCCGGCUGGAGGCGCGGACUCAACAUGACCCUCUCAAAGGCCUCCAGGCAG  94
GNG10   GCGCUUG---GUAGAGCAGCUCAAGUUGGAGGCUGGCGUGGAGAGGAUCAAGGUCUCUCAGGCAG  94
GNG13   GAAA---GAGGUGGAGAGCCUCAAGUACCAGCUGGCCUUCCAGCGGGAGAUGGCGUCCAAGACCA  89
              **   *      *          *                *          **  *

GNGT1   GUGAAGAAGUAAGAGAUUACGUUGAAGAACGAUCUGGCGAGGAUCCACUGGUAAAGGGCAUCCCA  174
GNG11   CUGAAGAAAUAAAGAACCUAUAUUGAAGAACGUUCUGGAGAGGAUCCUCUAGUAAAGGGAAUUCCA  174
GNGT2   GAAAGGAAAUCAAGGAGUACGUGGAGCCCAAGCAGGAAACGAUCCUUUGUCUGAUAGGAAUUACCA  162
GNG12   CAGCGGACCUCAUGUCCUACUGUGAGGAACAUGCCAGGAGUGACCCUUUGCUGAUAGGAAUUACCA  170
GNG7    CGGUCUGACCUCAUGAGCUACUGUGAGCAACAUGCUCGGAACGACCCCUGCUGGGUCGGAGUCCCU  159
GNG3    CAGCAGACCUGAUGGACUUACUGUGAUGCCCACGCCUGUGAGGAACCCUCACUACCCCGUGUGCC  179
GNG4    CCGCGGACCUCCUGGCCUACUGUGAGGCUCACGUGCGGGAAGAUCCUCUCAUCAUUCCAGUGCCU  177
GNG2    CUGCAGAUUUGAUGGCCUACUGUGAAGCACAUGCCAAGGAAGAACCCCUCCUGACCCCUGUUCCG  164
GNG8    CAGCGGAACUCCUGGCUUUCUGCGAGACGCAUGCCAAAGAUGACCCGCUGGUGACAGCCAGUACCCC  164
GNG5    CUGCAGACUUGAAACAGUUCGUCUGCAGAGAUCUCAACACAGCCCUCUGCUGAAAGGCAUCGUCU  169
GNG10   CUGCAGAGCUUCAACAGUACUGUGUAUGCAGAAGGCCUGCAAGGAUGCCCUGCUGGUGGGUGUUCCA  169
GNG13   UCCCGCGAGCUGCUGAAGUGGAUCGAGGACGGGAUCCCCAAGGACCCCUCCUGAACCCCGACCUG  154
              **   *       *                      **   *   *   *

GNGT1   GAGGACAAAAAUCCCUUCAAGGAGCUCAAAGGAGGC---UGUGUGAUUUCAUAA  225
GNG11   GAAGACAAGAACCCCUUUAAAGAA---AAAGGCAGC---UGUGUUAUUUCAUAA  222
GNGT2   GAGGACAAGAACCCCUUCAAGGAGCUC---AAGGUCUGAUAAGCUGA  210
GNG12   ACUUCAGAAAAACCCUUUCAAGGAU---AAAAAAACU---UGCAUCAUCUUAUAG  219
GNG7    GCCUCGGAGAACCCCUUUAAGGAC---AAGAAACCU---UGUAUUAUUUUAUAUAA  207
GNG3    ACUUCGGAGAACCCCUUCCGGGAG---AAGAAGUUCUUCUGUGCUCUCCUCUGA  228
GNG4    GCAUCAGAAAACCCUUCGCGAG---AAGAAGUUGUACCAUUCUCUAA  228
GNG2    GCUUCAGAAAACCCGUUUAGGGAG---AAGAAGUUUUUCUGUGCCAUCCUUUAA  216
GNG8    GCCGCGGAGAACCCCUUCCGCGAC---AAGCGCCUCUUUUGUGUUCUGCUCUGA  213
GNG5    UCAAGUACAAAUCCCUUCGACCCC---CAGAAAGUC---UGUUCCUUUUUGUAG  207
GNG10   GCUGGAAGUGAACCCUCCCGGGAG---CCUAGAUCC---UGUGCUUUACUCUGA  207
GNG13   AUGAAGAACAACCCAUGGGUGGAA---AAGGGCAAA---UGCACCAUCCUGUGA  204
              **   ** **  *                     **       *       *
```

[a] The shading indicates residues shared by more than half of the γ subunits. Possible GUC cleavage sites are shown underlined and targeted GUC cleavage sites used to make the γ_5, γ_7, and γ_{12} ribozymes are shown black boxed. The targeted GUC cleavage site used to construct the γ_2 ribozyme is located in the 3′ untranslated region (not shown).

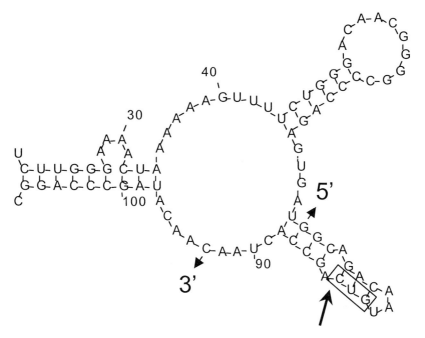

FIG. 2. Plot of the human G protein γ_7 mRNA. Secondary structure of the 5' end of the γ_7 mRNA was predicted using the MFOLD program.[24] The GUC site on the stem–loop structure is shown boxed.

to examine the secondary structure of the γ_7 mRNA. Both programs predicted a stem–loop structure in the vicinity of the γ_7 ribozyme cleavage site (Fig. 2), which was stable as judged by its presence in both optimal and suboptimal secondary structure predictions. Targeting of a stem–loop structure, where the ribozyme overlaps both single-stranded and double-stranded regions, has been suggested as a method to optimize ribozyme efficiency.[25] While progress is being made in the development of computer programs, it is still difficult to reliably predict those regions of mRNA that are most accessible. Also, the small size and sequence conservation of many γ mRNAs dictate those regions that can be targeted. In summary, it is our experience that the first method involves significant up-front effort and cost but is the most certain method. Moreover, our experience indicates that if the GUH sequence motif happens to be in region of translation initiation, the ribozyme has an excellent chance of working. Fortuitously, this region of the γ mRNAs shows the greatest sequence divergence (Table I). Also, cleavage within this region is very unlikely to produce truncated γ mRNAs that retain residual activity.

Chemical Synthesis of Ribozymes

   Once a suitable target site has been identified, the ribozyme is engineered so that it hybridizes to the nucleotide sequence flanking the GUH triplet within the

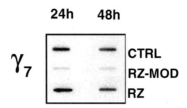

FIG. 3. Suppression of γ_7 mRNA is dependent on ribozyme modifications. Total RNA was prepared from control and ribozyme-treated cells at 24 or 48 hr after transfection and levels of γ_7 mRNAs were quantitated by Northern slot-blot analysis. RZ-MOD represents the chimeric DNA–RNA ribozyme containing two phosphorothioate linkages at the 3' end, whereas RZ represents the chimeric DNA–RNA ribozyme alone.

target mRNA. Although expressed ribozymes can be produced by introduction of the ribozyme sequence into an expression vector with a strong promoter of RNA synthesis, the focus of this chapter is chemically synthesized ribozymes. The use of chemically synthesized ribozymes has several advantages. First, the small size of the hammerhead ribozyme makes them relatively easy to make and to use. Moreover, the ability to incorporate various modifications results in a ribozyme with enhanced stability.[2–4] An inherent problem with RNA ribozymes is that they are rapidly metabolized by RNases in both serum and cells. Since RNases require the 2'-hydroxyl group found in RNA but not in DNA for their mode of action, one strategy for increasing stability is to replace all of the noncatalytic positions in the RNA ribozymes with DNA that lacks the 2'-hydroxyl group (ribonucleotides shown in gray and deoxyribonucleotides shown in black in Fig. 1). The stability of the chimeric DNA–RNA ribozyme can be further increased by additional modifications that protect from exonuclease activity. As shown in Fig. 3, the incorporation of two phosphorothioate linkages at the 3' end of a chimeric DNA–RNA ribozyme directed against the γ_7 mRNA greatly increases its stability compared to the chimeric DNA–RNA ribozyme alone. This agrees with other studies showing much of the degradation of ribozymes is due to a 3'-exonuclease activity.[26] The greater stability of the phosphorothioate-modified chimeric DNA–RNA ribozyme correlates well with its ability to suppress the level of the γ_7 mRNA in cells (Fig. 3). Finally, the small size of the ribozyme allows direct delivery into cells at high levels by either microinjection or transfection.

Chimeric DNA–RNA ribozymes are chemically synthesized on Applied Biosystems Model 394 DNA/RNA synthesizer at 1-μmol scale[27] (Applied Biosystems Division, PerkinElmer, Foster City, CA). Reagents, including ribonucleoside and deoxyribonucleoside phosphoramidites, 1 μmol RNA synthesis column, are obtained from Applied Biosystems Division. Phosphorothioate linkages are

[26] O. Heidenreich, F. Benseler, A. Fahrenholz, and F. Eckstein, *J. Biol. Chem.* **269**, 2131 (1994).
[27] B. Mullah and A. Andrus, *Nucleosides Nucleotides* **15**, 419 (1996).

introduced by sulfurization with [^{3}H]-1,2-benzodithiol-3-one 1,1-dioxide. The ribozyme is autocleaved on the synthesizer using ammonium hydroxide/ethanol (3 : 1) for 2 h and exocyclic amine-protecting groups are removed by heating at 65° for 3 h. The ribozyme is desilylated with neat triethylamine trihydrofluoride (10 μl per OD unit) and desalted by addition of water (2 μl per OD unit) and *n*-butanol (100 μl per OD unit). After cooling for 10 min in dry ice, the solution is centrifuged for 10 min and the precipitated ribozyme is collected. Following synthesis, a complete characterization of the ribozyme is critical, including *in vitro* demonstration of ribozyme cleavage and *in vivo* assessment of ribozyme efficacy on the levels of target RNA/protein in cultured cells or animals.

In Vitro Analysis of Ribozyme Activity

It is simple to test ribozyme activity in a cell-free system. Typically, the substrate RNA transcript is produced by *in vitro* transcription. As nearly as possible, the length of the substrate RNA transcript should approximate the full-length product since the rate of cleavage is dependent on its size (i.e., long RNA transcripts are cleaved much more slowly than short transcripts because of secondary structure considerations). For cleavage, the substrate RNA transcript and ribozyme are mixed in different ratios in low-salt buffer in the presence of magnesium at varying temperatures. The resulting cleavage products are analyzed by electrophoresis in a denaturing gel. As nearly as possible, physiological conditions should be employed in the *in vitro* cleavage reaction in order to predict the efficacy of the ribozyme in the *in vivo* setting. In the event that the ribozyme has poor activity, it is preferable to optimize the ribozyme itself by varying the length of the flanking arms or choosing a different target site rather than increasing the magnesium concentration or temperature of the cleavage reaction.

In Vitro Transcription of Substrate RNA

In vitro substrate RNA transcripts are generated from RNA polymerase promoter containing vectors, such as pBSK plasmids, containing the individual human γ cDNAs.[28] The substrate DNA templates are linearized with restriction enzymes located immediately downstream of the substrate sequence to minimize the amount of vector sequence in the RNA transcripts. The γ RNA transcripts of appropriate length are then transcribed with T3 RNA polymerase in accordance with the manufacturer's instructions (Promega, Madison, WI) in the presence of [α-^{32}P]CTP (50 μCi of a 10 mCi/ml, 3000 Ci/mmol stock; DuPont NEN, Boston, MA). Typically, the *in vitro* transcription reaction contains the following amounts of reagents in an autoclaved 1.5 ml microcentrifuge tube at room temperature: 1 μg DNA template; 4 μl 5× transcription buffer; 1 μl each rATP, rGTP, rUTP, rCTP at 2.5 mM;

---

[28] K. Ray, C. Kunsch, L. M. Bonner, and J. D. Robishaw, *J. Biol. Chem.* **270,** 21765 (1995).

5 μl [α-^{32}P]CTP at 10 mCi/ml (DuPont NEN); 1 μl dithiothreitol (DTT) at 100 mM; 20 units RNasin; 1 μl T3 RNA polymerase at ~20 units/μl; and RNase-free water to a total volume of 20 μl. After mixing, the components are collected at the bottom of the tube by brief centrifugation. The reaction is incubated at 37° for 1 to 2 h. The DNA template is degraded by addition of 1 μl of RNase-free DNase I (2 units/μl) and incubation at 37° for 15 min. The RNA transcript is obtained by extraction with an equal volume of buffer-saturated phenol : chloroform at a 1 : 1 ratio, precipitated by addition of 0.5 vol of ammonium acetate (7.5 M) and 2.5 vol of 100% ethanol, and chilled at −70° for 30 min. The precipitate is collected by centrifugation for 20 min and washed with 70% (v/v) ethanol, and the air-dried pellet is resuspended in 20 μl of RNase-free water. After purification, the quantity is determined by measuring the optical density of the RNA. Unless indicated otherwise, all reagents are obtained from Promega (Madison, WI).

Cleavage Reaction

Typically, the cleavage reaction is performed as follows. The ribozyme and radiolabeled substrate RNA transcript are denatured in separate, autoclaved 1.5 ml microcentrifuge tubes by heating in 100 mM Tris-HCl (pH 7.0–8.5) for 1 min at 95° followed by chilling on ice. The contents of the two tubes are then mixed together to achieve the desired ratio of ribozyme and substrate RNA transcript. Using 20–50 nM of the radioloabeled substrate RNA transcript, the ratio of ribozyme to target RNA is generally varied over a 1000-fold range. The reaction is started by the addition of 1 μl of MgCl$_2$ at 200 mM, and incubated for 1 h at 37–50°C. The reactions are stopped by addition of 1 volume of stop solution (95% formamide, 60 mM Na-EDTA, pH 8.0, 0.1% bromphenol blue, and 0.1% xylene cyanol), denatured by heating for 2 min at 95°, and then chilled on ice. The resulting cleavage products are analyzed by electrophoresis on a 5–8% polyacrylamide gel containing 7 M urea and visualized by autoradiography on Biomax MS film (Eastman Kodak Corp., Rochester, NY).

The cleavage products of reactions for four synthetic ribozymes targeted against the G protein γ_{12}, γ_5, γ_2, and γ_7 RNA transcripts are shown in Fig. 4. Because the γ subunits are members of a large, multigene family, it is necessary to validate that each ribozyme cleaves its corresponding RNA transcript in a gene-specific manner. Accordingly, the specificity of each ribozyme is assessed against the other γ RNA transcripts. As shown in panel A, the sizes of the γ RNA transcripts vary from 418 to 1632 bases. No cleavage of the γ RNA transcripts occurs in the absence of the appropriate ribozyme (Fig. 4A, Blank). However, on addition of the appropriate ribozyme, cleavage of the corresponding γ RNA transcript into fragments of the expected sizes is observed (Fig. 4B–E, as indicated by arrows). Of particular importance, no cleavage of the noncorresponding γ RNA transcripts occurs. Taken together, these results demonstrate the ability of these four synthetic ribozymes to specifically cleave their own templates in both a site- and gene-specific manner.

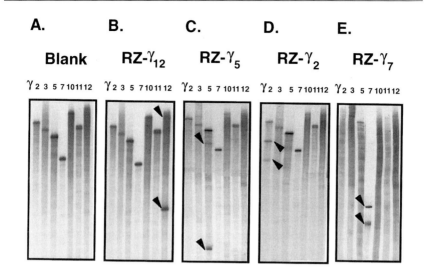

A.   **B.**   **C.**   **D.**   **E.**

Blank   **RZ-γ_{12}**   **RZ-γ_5**   **RZ-γ_2**   **RZ-γ_7**

γ 2 3 5 7 10 11 12 γ 2 3 5 7 10 11 12 γ 2 3 5 7 10 11 12 γ 2 3 5 7 10 11 12 γ 2 3 5 7 10 11 12

FIG. 4. Gene specificity of ribozymes targeted against several γ RNA transcripts *in vitro*. The specificities of the ribozymes were confirmed by demonstrating the ability of each ribozyme to cleave the targeted γ RNA transcript but not the other γ RNA transcripts in the test tube. *In vitro* synthesized γ RNA transcripts were incubated in the presence (RZ-γ) or absence (Blank) of the indicated γ ribozyme for 1 h at a ribozyme : template ratio of 20 : 1, and the production of cleavage products (indicated by arrows) was analyzed on polyacrylamide/urea gels.

Ribozyme Delivery into Cells

Although important, cell-free assays cannot predict the ability of the ribozyme to function in cells. This is because there are many factors in the cell that may influence the ribozyme's association or cleavage of the target mRNA. Some factors may affect ribozymes equally, such as stability and compartmentalization issues. Others may affect ribozymes differently, such as accessibility of the GUH site in the target mRNA. For these reasons, it is necessary to introduce the ribozyme into cells, where its impact on target mRNA and protein levels can be determined.

Since their anionic nature prevents efficient diffusion through the cell membrane, chemically synthesized ribozymes must be introduced into cells by microinjection or transient transfection. Microinjection directs the ribozyme to a specific intracellular compartment, thereby enhancing the association of the ribozyme and target mRNA. However, this technique presents difficulties in terms of verifying suppression of the target mRNA and protein, and restricting the types of functional assays that can be performed to those amenable to single cells. On the other hand, transfection permits the ribozyme to be introduced into a large fraction of the cell population. The most common form of transfection involves the use of cationic liposomes.[2-4] In particular, LipofectAMINE (Life Technologies, Inc., Grand Island, NY) has been successfully employed to introduce ribozymes into a variety of cultured cells. Moreover, its use appears to circumvent the problem of degradation of ribozymes in endocytic vesicles.[2]

Selection of Cell Line

In order to be useful, the cell line must express the target mRNA and protein of interest, and must be readily transfected with ribozyme. For these purposes, human embryonic kidney (HEK) 293 cells have proven to be ideal.[21,22] These cells express multiple receptors that regulate either adenylylcyclase or phospholipase C activity, thereby allowing the use of ribozymes to test the hypothesis that receptors converging on a common effector may utilize G proteins composed of the same α subunit but distinct $\beta\gamma$ subunits. Consistent with their varied complement of signaling pathways, these cells also express a number of G protein γ subunits at both the mRNA and protein levels, thereby allowing the use of ribozymes to specifically test the hypothesis that the γ subunits contribute to the specificity of signaling pathways in the intact cell setting. HEK 293 cells (American Tissue Culture Collection, Manassus, VA) are grown in Dulbecco's modified Eagle's medium (DMEM, Life Technologies, Inc.) supplemented with 10% fetal calf serum (FCS, Life Technologies, Inc.) in a humidified incubator equilibrated with 10% (v/v) CO_2. Cells are plated at 8.5×10^6 cells/100 mm dish (for analysis of protein suppression) or at 1.4×10^4 cells/well of a 6-well plate (for analysis of RNA suppression, cAMP determination, and inositol phosphate accumulation), and are transfected when cells reach 60 to 80% confluency.

Transfection

HEK 293 cells are efficiently transfected with LipofectAMINE, with typically 70% of cells showing fluorescence following transfection with green fluorescent protein (GFP). Prior to transfection, HEK 293 cells are preincubated with serum-free DMEM medium for 1 h. This is necessary because serum is replete with RNases that degrade ribozymes. Cells are then transfected for 5 h at 37° with fresh serum-free medium containing premixed ribozyme (2 μM) and LipofectAMINE (15 μg/ml). At 5 h posttransfection, heat-inactivated FCS is added to a final concentration of 6%. At varying times after transfection, the dishes are supplemented with additional ribozyme (0.5 μM) such that the total concentration of added ribozyme is 4 μM. Depending on the stability of the ribozyme and the half-life of the target mRNA/protein, repeated addition of ribozyme and/or longer time courses may be required.

In Vivo Analyis of Ribozyme Activity

Ribozymes are typically taken to be specific when: (1) cell viability is maintained; and (2) the levels of the target mRNA and its associated protein fall much more than those of the control mRNAs and proteins. Since high concentrations of cationic liposomes and/or liposome–ribozyme complexes may adversely affect viability, it is important to include the proper controls to ensure that any observed effect is not due to nonspecific toxicity. In this regard, nonspecific toxicity can be assessed by comparison of growth rates of untreated cells, cells treated with liposomes, cells treated with a mixture of liposomes and active ribozyme, and cells treated with a mixture of liposomes and inactive ribozyme containing a point substitution in the catalytic domain. Also, although it may seem obvious, it is essential to confirm the ability of the active ribozyme to inhibit the expression of the target mRNA and protein. In this regard, several issues warrant consideration. The active ribozyme should reduce the expression of the γ mRNA and protein in a dose-dependent fashion. In this regard, the failure of the inactive ribozyme containing a point substitution in the catalytic domain to cause these changes is particularly convincing. Moreover, the active ribozyme should not inhibit the expression of other gene products, including closely related members of the γ subunit family as well as unrelated gene products involved in "housekeeping" functions. Since there is a limit on the number of gene products that can be examined, a homology search of the nucleic acids databases should be carried out to rule out other possible mRNAs that may cross-react. Finally, the amounts of target and control mRNAs and proteins should be measured in untreated cells, cells treated with liposomes alone, cells treated with a mixture of liposomes and active ribozyme, and cells treated with a mixture of liposomes and inactive ribozyme. For these purposes, easy methods of measuring mRNA and protein levels are needed.

Northern Slot Blotting Analysis

To determine the effect of ribozyme exposure on γ mRNA levels, a 3 μg aliquot of total RNA from control and ribozyme treated cells is slot-blotted and hybridized with γ subunit gene-specific probes. Total RNA is prepared from the transiently transfected cells using the RNeasy kit, as recommended by the manufacturer (Qiagen Inc., Valencia, CA). Before use, the specificity of γ subunit probes is confirmed by Southern slot-blot analysis against different human γ cDNAs. In addition, a probe against elongation factor (EF1a) is used as control: 5'-CGTTGAAGC CTACATTGTCC-3'. All probes are 3' end-labeled with terminal deoxynucleotidyl transferase (Promega) and $[\alpha\text{-}^{32}\text{P}]$dATP [10 mCi/ml, 3000Ci/mmol, DuPont NEN]. The membrane is prehybridized with QuikHyb hybridization solution (Stratagene, La Jolla, CA) for 15 min at 30° and then hybridized with labeled probe for another 2 h at 30 to 38°. After hybridization, the membrane is washed twice with 2× SSC [1× SSC : 0.15 M sodium chloride, 0.015 M sodium citrate, and 0.1% sodium dodecyl sulfate (SDS)] at room temperature and then washed once with 0.5× SSC and 0.1% SDS for 30 min at 38°. The hybridization signals are analyzed from at least three separate experiments and quantitated by PhosphorImager SI analysis

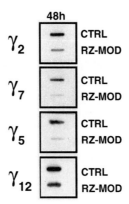

FIG. 5. Efficacy of ribozymes *in vivo*. Total RNA was prepared from control and ribozyme treated cells at 48 h after transfection and levels of indicated γ mRNAs were quantitated by Northern slot-blot analysis. RZ-MOD represents the chimeric DNA–RNA ribozyme containing two phosphorothioate linkages at the 3' end.

(Molecular Dynamics, Sunnyvale, CA), with the relative amounts of control and target mRNAs in ribozyme treated cells expressed as a percentage of their levels in control cells. The specific action of the ribozyme is demonstrated if the target γ mRNA is significantly reduced relative to control proteins including closely related γ mRNAs.

The Northern blot analysis for synthetic ribozymes designed to specifically target the γ_2, γ_5, γ_7, and γ_{12} subunits is shown in Fig. 5. When introduced into HEK 293 cells, three of the four ribozymes markedly suppress the levels of their corresponding γ mRNAs based on Northern slot-blotting analysis. This reinforces the important point that ribozymes are likely to be of broad use in selectively reducing the levels of γ mRNAs, and thus in establishing the functional roles of γ subunits in particular receptor signaling pathways. The γ_{12} ribozyme has proven to be only marginally effective in suppressing the level of the γ_{12} mRNA (Fig. 5). Such marginal suppression might result from several causes that are explored in more detail below.

Western Blotting Analysis

It is imperative to confirm that the ribozyme-induced loss of the target γ mRNA is paralleled by suppression of the protein. For this purpose, membranes from control and ribozyme-treated cells are evaluated by immunoblot analysis at varying times after transfection. To prepare membranes, cells washed with phosphate-buffered saline are lysed in a low-salt buffer (2 mM magnesium chloride, 1 mM EDTA, 20 mM HEPES, pH 7.0, 10 mM DTT, 1 mM aminoethylbenzene-sulfonyl fluoride, 1 μg/ml pepstatin A, and 1 mM benzamidine) on ice by several passages through a 25 gauge needle. The cell lysate is centrifuged at 250,000g for

30 min at $4°$ and the particulate fraction is extracted with 1% sodium cholate on a rocker plate in the cold room overnight. The cholate soluble fraction is collected and the protein concentration is determined. To determine the effect of ribozyme exposure on the target γ protein level, equal amounts (30 to 50 μg) of the cholate soluble fractions from control and treated cells are resolved on a 15% polyacrylamide–SDS and transferred to Nitro-plus nitrocellulose (0.45-μm pore size, Micron Separations Inc., Westborough, MA), using a high-temperature transfer procedure.[29] Following transfer, the nitrocellulose is cut along the 30 kDa marker: the higher molecular weight blot is probed with one of the β subtype specific antibodies[22] and the lower molecular weight blot is probed with one of the γ subtype-specific antibodies,[21,29] using ^{125}I-labeled goat anti-rabbit antibody for detection. The immunodetectable bands are analyzed from at least three separate experiments and quantitated by PhosphorImager SI analysis, with the relative amounts of control and target proteins in ribozyme treated cells expressed as a percentage of their levels in control cells. The specific action of the ribozyme is demonstrated if the target γ protein is significantly reduced relative to control proteins including closely related γ proteins. Importantly, the ribozyme-induced loss of the γ protein can result in a coordinated loss of the associated β protein(s).[21,22] In this regard, the $\beta\gamma$ dimer has been shown to undergo synthesis and assembly in the cytosol. If sufficient γ protein is not available to associate with the β protein, then the β protein accumulates in the cytosol where it is targeted for degradation. This mechanism appears selective and accounts for the specific loss of the β_1 protein that is observed in cells treated with ribozyme to suppress the expression of the γ_7 subunit.[22]

Detailed methods of the high-temperature transfer method and the use of these antibodies are provided elsewhere.[29] Briefly, the nitrocellulose blot is blocked for 1 h in high-detergent Blotto (50 mM Tris-HCl, pH 8.0, 80 mM sodium chloride, 2 mM calcium chloride, 5% nonfat powdered milk, 2% Nonidet P-40, and 0.2% SDS) and then incubated for 1 h with the primary antibody in high-detergent Blotto. After three successive 10-min washes with high-detergent Blotto, the nitrocellulose blot is incubated for 1 h with ^{125}I-labeled goat anti-rabbit F(ab')$_2$ fragment in high-detergent Blotto (1×10^5 dpm/ml, New England Nuclear Corp.). After three successive 10-min washes with high-detergent Blotto, the nitrocellulose blot is washed twice for 10 min each with 50 mM Tris-HC1, pH 8.0, 80 mM sodium chloride, and 2 mM calcium chloride. The nitrocellulose blot is then subjected to autoradiography by exposure to Biomax MS film and the intensities of the immunodetectable bands are quantified by PhosphorImager SI analysis.

Functional Assays

The final step in the *in vivo* analysis is to determine how the ribozyme-mediated loss of the γ subunit affects the integrity of receptor signaling pathways. HEK

[29] J. D. Robishaw and E. A. Balcueva, *Methods Enzymol.* **237**, 498 (1994).

293 cells respond to numerous receptor agonists to activate G proteins to regulate effectors including adenylylcyclase, phospholipase C, and kinases,[21] thereby providing a system in which to identify the roles of the γ subunits in particular signaling pathways. Of particular interest, isoproterenol and prostaglandin E_1 increase cAMP production; thrombin, endothelin, carbachol, and ATP stimulate inositol phosphate accumulation; and lysophosphatidic acid and bradykinin activate mitogen-activated protein kinases. The specific role of a γ subunit is demonstrated if ribozyme-induced loss of the γ subunit results in significant attenuation of one of these signaling pathways. In this regard, several issues warrant consideration. First, ribozyme usage cannot distinguish between G protein α $\beta\gamma$-mediated effects on these signaling pathways. Since the receptor requires both the G protein α and $\beta\gamma$ subunits for productive interaction, the ribozyme-mediated loss of the γ subunit would lead to reduced assembly of the required G protein heterotrimer. This would compromise receptor–G protein interaction, thereby disrupting both α-and $\beta\gamma$-mediated effects. Second, the receptor–G protein content often exceeds that required for maximal functional response.[30] Therefore, ribozyme-induced loss of the γ subunit would not necessarily result in a similar degree of attenuation of the signaling pathway in which it is involved.[21] Detailed methods of the measurement of these receptor signaling pathways are beyond the scope of this chapter.

Potential Problems

Although ribozymes have been employed successfully, the efficiency of these ribozymes appears to be variable[2–4] (for example, Fig. 5). The parameters determining ribozyme activity *in vivo* are just starting to be understood.[2–4] One of the most critical factors seems to be the association step between the ribozyme and its target mRNA, which is greatly influenced by the accessibility of the GUH sequence motif in the target mRNA. Since this is often difficult to predict, one strategy to increase the likelihood of success is to design a series of ribozymes directed against different GUH sequence motifs in the target mRNA. Particularly effective, these ribozymes can be used together to simultaneously recognize and cleave several sites in the target mRNA, thereby mitigating the problem of accessibility as well as increasing specificity.

Another potential problem is that the ribozyme approach may suppress the level of the target mRNA, but may not be nearly as efficacious in suppressing the level of target protein. Again, this could result from several causes. First, if a particular γ protein has a very long half-life, the approach of transiently transfecting the ribozyme into cells might not suppress mRNA expression for long enough to allow a reduction in protein content to be seen. In this case, it is possible to stably introduce a gene encoding the ribozyme into cells to produce a sustained

---

[30] B. Zhu, *J. Pharmacol. Toxicol. Methods* **29,** 85 (1993).

expression of the ribozyme. Currently, retroviral vectors are most commonly used in both cell culture and animals.[2-4] They have several advantages, including wide host range, stable integration into dividing host genome, and the fact that absence of viral gene expression reduces the chance of immune response in animals.

Summary

Efforts to determine the sequence of the human genome have resulted in sequence information on thousand of genes. Now, the challenge is to determine the functions of this myriad of genes, including those encoding the G protein subunit families.[1] In this chapter, we describe the successful use of ribozymes to inactivate mRNAs expressed from the G protein γ subunit genes. Ribozymes are unique in that they can inactivate specific gene expression, and thereby can be used to help identify the function of a protein or the role of a gene in a functional cascade. Compared to other means of identifying the role of a gene (i.e., transgenic or knockout animals), ribozymes are specific and relatively easy to use. Moreover, ribozymes are able to discriminate closely related, or even mutated, sequences within gene families.[21,22,31,32] Thus, in addition to elucidating functions, ribozymes have the potential to be used in treating genetic disorders associated with mutations of G protein subunits.

Acknowledgment

This work was supported by NIH Grant GM58191 awarded to J.D.R.

[31] K. Moelling, B. Strack, and G. Radziwill, *Recent Results Cancer Res.* **142,** 63 (1996).
[32] A. Persidis, *Nature Biotechnol.* **15,** 921 (1997).

Section IV

G Protein Structure and Identification of Functional Domains

[32] Use of Scanning Mutagenesis to Delineate Structure–Function Relationships in G Protein α Subunits

By CATHERINE H. BERLOT

Introduction

In order to transmit signals from cell surface receptors to intracellular effector proteins, G protein α subunits must perform a series of functions. These include binding to the correct receptors, responding to activated receptors by replacing GDP with GTP, binding to and modulating the functions of specific effector proteins, and hydrolyzing GTP to GDP to terminate signaling. Some functions, such as guanine nucleotide binding and hydrolysis, share a conserved mechanism and involve conserved residues. Other functions, such as interactions with specific receptors and effectors, are regulated by both conserved and divergent residues. G protein α subunit function depends on a number of interrelated components. For instance, GTP binding is required for activating conformational changes that are necessary for effector modulation. However, since the various functions can be measured separately, it is possible to apply mutagenesis approaches to test the importance of specific residues for each step in the GTPase cycle.

There is a high degree of homology among α subunits: at least 40% identical amino acids overall, with 60–90% identity within subfamilies.[1] This high degree of conservation is reflected in the very similar X-ray crystal structures of α_t,[2] α_{i1},[3] and α_s.[4] Comparisons of the structures of active and inactive α subunits show three conserved regions (switches I–III) that change conformation during the GTPase cycle.[5,6] The similarities of the structures of α subunits, coupled with differences in the specificities of their interactions with other proteins, make it possible to apply mutagenesis approaches to identify the molecular basis of the specificities, as well as to elucidate the mechanisms by which these interactions regulate function.

Scanning mutagenesis can be used to investigate how α subunits contact other proteins such as receptors or effectors, how the specificity of these interactions is

[1] M. I. Simon, M. P. Strathmann, and N. Gautam, *Science* **252,** 802 (1991).

[2] J. P. Noel, H. E. Hamm, and P. B. Sigler, *Nature* **366,** 654 (1993).

[3] D. E. Coleman, A. M. Berghuis, E. Lee, M. E. Linder, A. G. Gilman, and S. R. Sprang, *Science* **265,** 1405 (1994).

[4] R. K. Sunahara, J. J. G. Tesmer, A. G. Gilman, and S. R. Sprang, *Science* **278,** 1943 (1997).

[5] D. G. Lambright, J. P. Noel, H. E. Hamm, and P. B. Sigler, *Nature* **369,** 621 (1994).

[6] M. B. Mixon, E. Lee, D. E. Coleman, A. M. Berghuis, A. G. Gilman, and S. R. Sprang, *Science* **270,** 954 (1995).

determined, and how these protein–protein interactions regulate function. It is easiest to start by asking how specificity is determined, which can be addressed using chimeric α subunits containing portions of α subunits with different specificities and simple screening assays in transiently transfected tissue culture cells. Once the important regions have been identified, the mechanism by which they regulate function can be addressed by analyzing the effects of point mutations using more sophisticated functional assays in stable cell lines or biochemical assays using recombinant α subunits expressed in and purified from *Escherichia coli* or insect cells.

In this chapter, I will describe how specific types of mutagenesis can be used to answer particular questions about G protein α subunit function and how mutant α subunits can be characterized in transiently transfected tissue culture cells to determine the effects of the mutations on interactions with effectors and receptors. Controls for the specificity of the effects of mutations will also be discussed. Then, I will describe how mapping mutations onto the α subunit crystal structures can help to determine the mechanism by which they cause changes in function. I will show how this analysis can lead to further experiments that will more clearly define the roles of the mutated residues. Finally, a few general principles that have emerged from studying the structure–function relationships of some of the best-characterized α subunits will be summarized. These principles can be applied to characterizing additional α subunits, including the many novel gene products that are emerging from the genome sequencing projects.

Mutagenesis Approaches

The mutagenesis strategy to be used depends on the question being asked. To identify the α subunit residues that specify interaction with another protein, homolog-scanning mutagenesis is the method of choice.[7] This type of mutagenesis involves replacing residues in an α subunit with those of a homologous α subunit that has a clear functional difference. Because the α subunit sequences and structures are so highly conserved, clusters of homolog substitutions and even large swaps of homologous sequence in chimeras generally preserve structural integrity. Thus, making chimeric proteins containing portions of two α subunits with distinguishable functions is the best way to initially localize important functional regions. For instance, a single α_{i2}/α_s chimera localized the region of α_s that specifies activation of adenylyl cyclase in response to β-adrenergic receptor stimulation to the carboxyl terminal 40% of the molecule.[8] The same chimeric proteins can be used to localize functional regions in both of the parent α subunits. For instance,

---

[7] B. C. Cunningham, P. Jhurani, P. Ng, and J. A. Wells, *Science* **243**, 1330 (1989).
[8] S. B. Masters, K. A. Sullivan, R. T. Miller, B. Beiderman, N. G. Lopez, J. Ramachandran, and H. R. Bourne, *Science* **241**, 448 (1988).

α_{i2}/α_q chimeras have localized the effector-specifying regions of both α_{i2} and α_q.[9] Homolog scanning, however, will not identify functionally important regions that are conserved among α subunits. To identify α subunit residues involved in conserved contacts and functions, alanine-scanning mutagenesis[10] is the method of choice. Substitutions with alanines test the importance of a residue's side chain, but do not tend to disrupt the overall structure of a protein. Alanine substitutions can be introduced in small clusters, but generally the clusters need to be smaller than when homologs are substituted.

Chimeras

Homolog-scanning mutagenesis can be initiated by substituting large segments of α subunit sequence with that of a homologous α subunit to produce a chimeric α subunit. Chimeric α subunits can be generated by subcloning using available or engineered restriction endonuclease sites or by overlap extension using the polymerase chain reaction.[11] To increase the likelihood that chimeric α subunits will be properly folded, it is best to make chimera junctions within regions of sequence that are conserved between the two α subunits.

One method for producing chimeras that ensures that the junctions will be within conserved regions of sequence utilizes homologous recombination in *E. coli* (Fig. 1).[9,12] The method involves subcloning two α subunit cDNAs head to tail in an expression vector, with at least one (preferably two) unique restriction endonuclease sites in between. The plasmid is then linearized at the restriction endonuclease sites and transformed into bacteria. Transformants are derived either from uncut plasmids or from plasmids that recircularized by undergoing homologous recombination *in vivo*. Colonies are screened by restriction mapping to distinguish between tandem insert constructs and recombinant α subunits and to roughly localize chimera junctions. The precise crossover points are then identified by DNA sequencing. Applying this technique to the α_{i2} and α_q cDNAs resulted in six unique chimera junctions within regions containing 8–20 bases of sequence identity.[9]

Homolog-Scanning Mutagenesis

After functionally important segments of α subunit sequence have been identified using chimeras, substituting homologs for clusters of nonconserved residues

---

[9] R. Medina, G. Grishina, E. G. Meloni, T. R. Muth, and C. H. Berlot, *J. Biol. Chem.* **271**, 24720 (1996).

[10] B. C. Cunningham and J. A. Wells, *Science* **244**, 1081 (1989).

[11] R. M. Horton, H. D. Hunt, S. N. Ho, J. K. Pullen, and L. R. Pease, *Gene* **77**, 61 (1989).

[12] C. H. Berlot, *in* "G Proteins: Techniques of Analysis" (D. R. Manning, ed.), p. 39. CRC Press LLC, Boca Raton, FL, 1999.

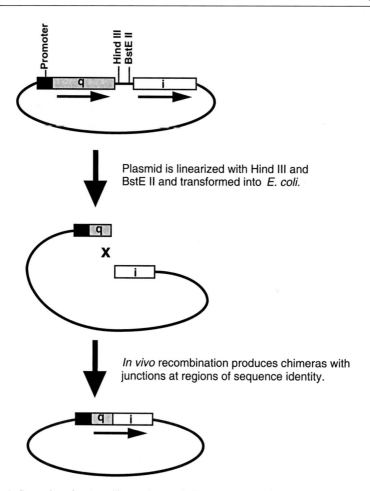

FIG. 1. Generation of α_q/α_{i2} chimeras in *E. coli*. The cDNAs encoding α_q and α_{i2} are subcloned in tandem in the 5' to 3' orientation downstream from the cytomegalovirus (CMV) promoter in the expression vector pcDNAI/Amp (Invitrogen) with two unique restriction enzyme sites, *Hind*III and *Bst*EII, in between. The plasmid is then linearized by digestion with these two restriction enzymes, and transformed into *E. coli*. To produce α_{i2}/α_q chimeras, the α_{i2} cDNA is cloned in front of the α_q cDNA. Reproduced with permission from C. H. Berlot, *in* "G Proteins: Techniques of Analysis" (D. R. Manning, ed.), p. 39. CRC Press LLC, Boca Raton, FL, 1999.

within these segments can further localize the critical residues. For instance, within a 121-residue α_s segment shown from a chimera to specify activation of adenylyl cyclase, 78 residues are not conserved between α_s and α_{i2} and therefore were potentially involved in specifying effector activation. Of these residues, 63 were changed in small clusters using 19 constructs.[13] Alanines were substituted when

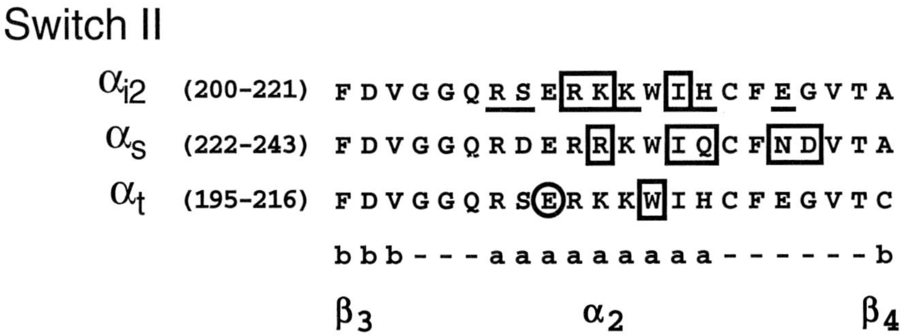

Switch II

α_{i2} (200–221) F D V G G Q R S E R K K W I H C F E G V T A

α_s (222–243) F D V G G Q R D E R R K W I Q C F N D V T A

α_t (195–216) F D V G G Q R S E R K K W I H C F E G V T C

 b b b - - - a a a a a a a a a - - - - - - b

 β_3 α_2 β_4

FIG. 2. Comparison of effector-interacting residues of α_{i2}, α_s, and α_t in switch II. Residue numbers of α_{i2}, α_s, and α_t in the switch II region are indicated in parentheses. Mutations of boxed residues impair effector interaction. Mutations of underlined residues do not impair effector interaction. Mutation of the circled glutamate residue in switch II of α_t causes constitutive activation of PDE. Data for α_{i2} and α_s are from G. Grishina and C. H. Berlot, *J. Biol. Chem.* **272**, 20619 (1997) and C. H. Berlot and H. R. Bourne, *Cell* **68**, 911 (1992). Data for α_t are from E. Faurobert, A. Otto-Bruc, P. Chardin, and M. Chabre, *EMBO J.* **12**, 4191 (1993) and R. Mittal, J. W. Erickson, and R. A. Cerione, *Science* **271**, 1413 (1996). Reproduced with permission from G. Grishina and C. H. Berlot, *J. Biol. Chem.* **272**, 20619 (1997).

homologs were not available (see next section). Six of the constructs exhibited significantly reduced abilities to activate adenylyl cyclase. Substitutions in three of these produced a specific loss of function in that they did not affect α subunit expression or a GTP-dependent conformational change that can be measured using a trypsin-resistance assay (see below).

Alanine-Scanning Mutagenesis

Alanine substitutions are useful for testing residues in highly conserved regions such as the conformational switch regions. If a particular α subunit function is dependent on the activation state of the α subunit, it is very likely that one or more of these conformational switch regions is involved. These regions are so highly conserved that their importance may be missed if only homolog-scanning mutagenesis is employed. For example, conserved residues in switch II are important for the effector interactions of α_{i2}, α_s, and α_t (Fig. 2).

Alanine substitutions are also useful for testing residues for which homologs are not available, such as when the sequence of a particular α subunit represents an insertion of sequence relative to other α subunits. For instance, in the scanning mutagenesis study of α_s referred to in the previous section, we substituted α_{i2} homologs for α_s residues when possible and alanine residues for α_s residues

---

[13] C. H. Berlot and H. R. Bourne, *Cell* **68**, 911 (1992).

within a 13-residue segment (residues 324–336) that has no counterpart in the α_{i2} sequence.

In order to avoid causing nonspecific structural problems, alanine substitutions should be introduced in clusters of no more than three residues at a time. Because of the unique constraints on the backbone conformation of glycines and prolines, these residues are generally not mutated to alanines. Also, alanine substitutions of surface-exposed residues (see below) are most likely to be tolerated. If a cluster of alanine substitutions interferes with protein folding, a smaller cluster, or individual substitutions, should be tested.

Advantages of Employing Both Homolog- and Alanine-Scanning Mutagenesis

It is helpful to test the effects of both homolog and alanine substitutions, because the results can differ and this can lead to a greater understanding of the role of the wild-type residue. For instance, if a homolog substitution causes a defect not seen with an alanine substitution, this may indicate that the defect caused by the homolog substitution is not due to the loss of a particular side chain, but rather to the introduction of a disruptive one. For example, substitution of Asn-167 in α_s with the homologous α_{i2} residue, arginine, causes a decrease in receptor-mediated activation.[14] However, substitution of the same residue with alanine does not have an effect. Because the defect caused by substituting arginine for Asn-167 is corrected by simultaneously substituting asparagine for Asn-254 (see below), the defect in this case appears to be due to an incompatibility between an arginine at position 167 and an asparagine at position 254.

Functional Analysis of Mutant α Subunits

Epitope Tags

With some exceptions, such as cyc^- S49 lymphoma cells, which lack α_s,[15] it is difficult to find a cell line that lacks the α subunit of interest. Therefore, since it is important to determine the stability and expression level of mutant constructs (see below), it is advisable to work with epitope-tagged α subunits that can be distinguished from the endogenous α subunits. Since the amino and carboxyl termini of α subunits are important for interactions with $\beta\gamma$ and receptors, and for membrane attachment, it is advisable to use an internal epitope. Placing an epitope from an internal region of polyoma virus medium T antigen, referred to as the EE epitope,[16] in an exposed loop connecting α helices αE and αF in the helical domain

[14] G. Grishina and C. H. Berlot, *J. Biol. Chem.* **273**, 15053 (1998).

[15] B. A. Harris, J. D. Robishaw, S. M. Mumby, and A. G. Gilman, *Science* **229**, 1274 (1985).

[16] T. Grussenmeyer, K. H. Scheidtmann, M. A. Hutchinson, W. Eckhart, and G. Walter, *Proc. Natl. Acad. Sci. U.S.A.* **82**, 7952 (1985).

leaves the effector- and receptor-interacting abilities of α_s,[17] α_{i2},[18] and α_q[17] intact. The site is generated by mutating α_s residues DYVPSD (189–194) to EYMPTE, α_{i2} residues SDYIPTQ (166–172) to EEYMPTE, and α_q residues SYLPTQ (171–176) to EYMPTE (single-letter amino acid code). Monoclonal antibodies to the EE epitope can be obtained from BAbCO (Richmond, CA, MMS-115).

Screening Assays in Transiently Transfected Cells

Since localization of functionally important α subunit residues entails screening a large number of chimeric and mutated α subunits, it is essential to utilize simple and rapid functional assays for the initial characterizations. Transiently transfected HEK-293 cells (American Type Culture Collection, Manassus, VA) are useful for these screening assays because the signaling pathways of G_s, G_i, and G_q can be easily distinguished. In these cells α_{i2} but not α_q inhibits adenylyl cyclase, and α_q but not α_{i2} activates phospholipase C.[9] Also, the β and γ subunit[19] and adenylyl cyclase[20] isoforms expressed in HEK-293 cells have been identified.

HEK-293 cells can be easily and inexpensively transfected using DEAE dextran as described previously.[12] With this method of transfection, the levels of expression are proportional to the amount of transfected plasmid over a range of plasmid doses. Therefore, constructs that express at different levels can be analyzed at the same level of protein expression by adjusting the amount of plasmid used in the transfection. Stimulation of adenylyl cyclase by a transfected α subunit can be measured in cells labeled with [^{3}H]adenine.[12] Inhibition of adenylyl cyclase by a transfected α subunit can be measured in cells that are co-transfected with a constitutively activated α_s mutant.[21] Stimulation of phospholipase C can be measured in cells labeled with [^{3}H]inositol.[9]

When investigating the determinants of effector-modulating ability, it is helpful to introduce a mutation that causes constitutive activation, so that receptor-independent function can be easily detected above the activities of α subunits endogenous to the transfected cells. Constitutive activation due to inhibition of GTP hydrolysis can be produced by substituting cysteine for a conserved arginine at positions 201 in α_s,[22] 179 in α_{i2},[23] and 183 in α_q,[24] or leucine for a conserved glutamine at positions 227 in α_s,[25] 205 in α_{i2},[23] and 209 in α_q.[26]

[17] P. T. Wilson and H. R. Bourne, *J. Biol. Chem.* **270**, 9667 (1995).

[18] A. M. Pace, M. Faure, and H. R. Bourne, *Mol. Biol. Cell* **6**, 1685 (1995).

[19] Q. Wang, B. K. Mullah, and J. D. Robishaw, *J. Biol. Chem.* **274**, 17365 (1999).

[20] P. A. Watson, J. Krupinski, A. M. Kempinski, and C. D. Frankenfield, *J. Biol. Chem.* **269**, 28893 (1994).

[21] G. Grishina and C. H. Berlot, *J. Biol. Chem.* **272**, 20619 (1997).

[22] C. A. Landis, S. B. Masters, A. Spada, A. M. Pace, H. R. Bourne, and L. Vallar, *Nature* **340**, 692 (1989).

[23] Y. H. Wong, A. Federman, A. M. Pace, I. Zachary, T. Evans, J. Pouysségur, and H. R. Bourne, *Nature* **351**, 63 (1991).

When investigating α subunit–receptor interactions, an effector readout assay may be used, but it is necessary to establish that the mutations do not alter effector interaction. Introducing a constitutively activating mutation makes it possible to test for receptor-independent effector modulation. To investigate receptor-mediated activation in HEK-293 cells, the cells should be co-transfected with plasmids expressing the α subunit and a receptor that is not endogenously expressed. Cells that are transfected with two plasmids tend to take up both or neither of them.[27] Therefore, untransfected cells will be invisible in the assay even though they express the transfected α subunit endogenously. Alternatively, to study interactions between α_s and receptors, cyc^- S49 lymphoma cells, which lack α_s,[15] can be used. Methods for transient and stable expression of α subunit constructs in S49 lymphoma cells and assays for characterizing these constructs are provided in [19] in this volume.[27a]

Controls for Specificity

Since mutations that disrupt function may do so by impairing protein folding, it is important to establish that the effects of mutations are specific and not global. Proper folding of transiently expressed α subunits can be tested for by determining whether they can assume a GTP-dependent activated conformation that is resistant to trypsin cleavage. This assay has been described in detail previously.[12] In brief, α subunits are transiently expressed in HEK-293 cells using DEAE-dextran, and membranes are prepared. Trypsin digestion is performed in the presence and absence of GTPγS. The no-trypsin control demonstrates whether the mutations alter expression levels. Under the conditions of the assay, in the presence of GTPγS, trypsin removes a short segment from the amino terminus of the α subunit, but leaves most of the protein intact (Fig. 3). However, in the absence of GTPγS, trypsin degrades the α subunit to small fragments not seen on SDS polyacrylamide gels. Including an epitope tag in the α subunit constructs, as described above, ensures that only transfected constructs, rather than endogenous α subunits, are visualized in this assay.

Since trypsin resistance in the presence of GTPγS requires both nucleotide exchange and GTP-dependent conformational changes, GTPγS-dependent protection in the trypsin assay generally indicates that an α subunit is properly folded. For mutants with decreases in GTPγS-dependent trypsin resistance, a direct test

---

[24] B. R. Conklin, O. Chabre, Y. H. Wong, A. D. Federman, and H. R. Bourne, *J. Biol. Chem.* **267,** 31 (1992).

[25] S. B. Masters, R. T. Miller, M. H. Chi, F.-H. Chang, B. Beiderman, N. G. Lopez, and H. R. Bourne, *J. Biol. Chem.* **264,** 15467 (1989).

[26] D. Wu, C. H. Lee, S. G. Rhee, and M. I. Simon, *J. Biol. Chem.* **267,** 1811 (1992).

[27] J. Marshall, R. Molloy, G. W. Moss, J. R. Howe, and T. E. Hughes, *Neuron* **14,** 211 (1995).

[27a] C. H. Berlot, *Methods Enzymol.* **344,** [19] 2002 (this volume).

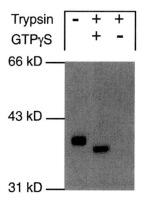

FIG. 3. Example of trypsin resistance assay. HEK-293 cells (12.5×10^6) were transfected with 2 μg/10^6 cells of plasmid expressing α_{i2}RCEE, which contains the R179C activating mutation and the EE epitope. Membranes were prepared, treated with trypsin, and immunoblotted using an anti-EE monoclonal antibody, as described [G. Grishina and C. H. Berlot, *J. Biol. Chem.* **272**, 20619 (1997)]. The first lane is the control (no trypsin). The second and third lanes show the result of trypsin digestion in the presence or absence, respectively, of GTPγS.

of nucleotide binding on protein expressed in and purified from *E. coli* or insect cells is required to determine if the activation defect is secondary to a nucleotide binding defect.

Determining Role(s) of Functionally Important Residues

Analysis in Structural Context

Once a region of sequence is shown by mutagenesis to be important for a specific function, determining where the critical residues map onto the three-dimensional α subunit structures, and whether they are solvent-exposed or buried, can help to elucidate the role of the region. X-ray coordinates of the α subunit structures are available from the Protein Data Bank (http://www.rcsb.org/pdb/). These coordinates can be displayed using graphics programs such as RasMol (http://mc2.cchem.berkeley.edu/Rasmol/), which can be run on a Macintosh or a PC. By displaying the coordinates as a space-filling model and highlighting a residue with a specific color, it is easy to determine the degree to which the residue is surface-exposed. In addition, calculation of the fractional accessibility of the residue can be performed.[28] When structures of complexes between α subunits and other proteins such as $\beta\gamma$, effectors, or RGS proteins are available, comparing

[28] B. Lee and F. M. Richards, *J. Mol. Biol.* **55**, 379 (1971).

the accessibilities of α subunit residues in the complex to their accessibilities in the absence of their binding partners can help to identify contact residues.

Residues that are solvent-exposed are potentially important for protein–protein interactions. Mapping noncontiguous regions of sequence onto the α subunit structures may reveal that they actually form a surface in three dimensions. For instance, three regions of α_s sequence that are not adjacent in the primary sequence, but which map onto adjacent loops in the structure of α_s, are involved in activation of adenylyl cyclase.[13,29]

Residues that are buried may be important for nucleotide binding and/or for transmitting intramolecular signals. For example, substitution of a cluster of residues in the $\beta6/\alpha5$ loop of α_s disrupts receptor-mediated activation.[30] Separately mutating the buried and the surface-exposed residues showed that substitutions of the buried, but not the surface-exposed, residues cause defects. Because the buried residues are near the guanine ring of the bound nucleotide, this region most likely plays a role in regulating the nucleotide affinity of α_s rather than serving as a receptor contact site.

Testing for Functionally Important Interactions between Residues

Analysis of the α subunit X-ray crystal structures can reveal interactions between residues. The role of a functionally important residue that is in direct contact with other residues may be to communicate with them. To determine if an interaction between a particular pair of residues plays a role in a specific function, the effect of individually mutating each of the pair should be compared with that of simultaneously mutating both of them. If two residues participate in a functionally important interaction, simultaneous substitutions of both of them will produce a nonadditive effect. For instance, combining two mutations may produce the same defect as either mutation alone does if both of the mutations disrupt the same interaction. In contrast, if two residues contribute independently to a specific function, simultaneous mutations will generally produce additive effects.[31]

The most convincing evidence for an intramolecular interaction is a special case of nonadditivity in which the defect caused by one mutation is actually suppressed by simultaneously mutating a second residue with which it interacts. This can happen if the paired substitutions reintroduce a functionally important interaction that was disrupted by the individual substitutions. Such intramolecular complementation can be achieved either by simultaneously substituting homologs for each of the residues or by swapping the two residues.

---

[29] J. J. G. Tesmer, R. K. Sunahara, A. G. Gilman, and S. R. Sprang, *Science* **278,** 1907 (1997).

[30] S. R. Marsh, G. Grishina, P. T. Wilson, and C. H. Berlot, *Mol. Pharmacol.* **53,** 981 (1998).

[31] J. A. Wells, *Biochemistry* **29,** 8509 (1990).

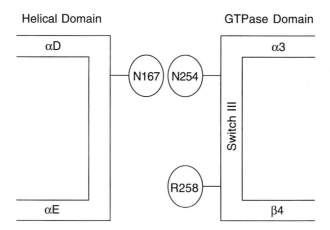

FIG. 4. Testing for functionally important interactions at the domain interface. This figure schematically represents two adjacent loops in the α subunit structure, the αD/αE loop in the helical domain and switch III in the GTPase domain. Individually replacing Asn-167 or Arg-258 in $α_s$ with the homologous $α_{i2}$ residue decreases receptor-mediated activation of G_s. Simultaneously substituting $α_{i2}$ homologs for Asn-167 and Asn-254 restores receptor-mediated activation, indicating a functionally important interaction between the residues at these two positions. In contrast, simultaneous homolog substitutions of Asn-167 and Arg-258 produce an additive decrease in receptor-mediated activation, indicating that the two residues do not interact and that the substitutions disrupt two independent interactions.

Interactions between α subunit residues across the interface of the GTPase and helical domains have been demonstrated by both of the above types of intramolecular complementation. In one case, substituting Asn-167 in the helical domain of $α_s$ with the homologous $α_{i2}$ residue produced a defect in receptor-mediated activation that was suppressed when Asn-254 in switch III of the GTPase domain, with which it interacts, was simultaneously replaced with its $α_{i2}$ homolog (Fig. 4).[14] In the second case, swapping two conserved charged residues that interact across the domain interface of $α_s$, Asp-173 in the helical domain and Lys-293 in the GTPase domain, corrected a defect in receptor-mediated activation caused by individually mutating each of them.[32] These results suggest that appropriate matching of residue pairs at each of these positions plays a role in receptor-mediated activation.

In contrast, simultaneously substituting $α_{i2}$ homologs for Asn-167 and the switch III residue, Arg-258, which are near each other, but not in direct contact, caused an additive defect in receptor-mediated activation (Fig. 4).[14] Thus, the defects caused by mutating each of them are due to disruptions of distinct interactions. Since the defect caused by the Arg-258 substitution is suppressed when the entire helical domain of α_s is simultaneously replaced with that of α_{i2}, residue(s) other than Asn-167 are involved in interacting with this switch III residue.

The approach of simultaneously mutating pairs of interacting residues can also be applied to testing the functional importance of interactions between α subunit residues and those of $\beta\gamma$, effectors, or RGS proteins, when the structures of the complexes are available. In this case, it is most feasible to test whether pairs of alanine substitutions have additive effects or whether swapping residues suppresses the defects caused by individual substitutions. Even in the absence of a structure of such a complex, it may be possible to identify regions of contact by testing the effects of pairs of mutations within functionally important regions of the α subunit and the other protein. This approach was applied to studying interactions between mutants of the catalytic subunit of cAMP-dependent protein kinase and substituted peptide substrates.[33] The results were later confirmed by an X-ray crystal structure of a peptide inhibitor bound to the catalytic subunit of this kinase.[34]

General Principles of α Subunit Function Derived from Scanning Mutagenesis

Regulation of Effector Modulation

In order for G protein signaling to be specific and to be dependent on hormonal stimulation, interactions between α subunits and their effectors must meet two criteria. First, each α subunit needs to specifically recognize the correct effector and regulate it in the appropriate manner. Second, interaction with the effector should only take place when the α subunit is in the GTP-bound active conformation. For the α subunits characterized so far, these two functional requirements are met by distinctive types of residues. Divergent residues outside of the conformational switch regions determine the specificity of effector interaction and regulation, while conserved residues in the switch regions signal GTP-dependent effector binding. For example, activation of adenylyl cyclase by α_s is specified by nonconserved residues in switch II and in the adjacent $\alpha3/\beta5$ loop.[13,35] Inhibition of adenylyl cyclase by α_{i2}, on the other hand, is specified by nonconserved residues in the $\alpha3$ helix and the $\alpha4/\beta6$ loop.[21] However, as mentioned earlier, highly conserved residues in switch II regulate the GTP-dependence of the interactions of both α_s and α_{i2} with adenylyl cyclase (Fig. 2). Therefore, when investigating effector interaction, chimeras and homolog substitutions will be most useful for identifying the residues that specify which effector is regulated and the nature of

---

[33] C. S. Gibbs and M. J. Zoller, *Biochemistry* **30,** 5329 (1991).

[34] D. R. Knighton, J. Zheng, L. F. Ten Eyck, N.-H. Xuong, S. S. Taylor, and J. M. Sowadski, *Science* **253,** 414 (1991).

[35] G. Grishina and C. H. Berlot, *Mol. Pharmacol.* **57,** 1081 (2000).

	S54N	D173K	N167R	R231H	R258A	E259D	α3/β5	K293D	R389P
Receptor Affinity			↑	N	↑		↑		↓
βγ Affinity		N		N	N	N		↓	
Activation by AlF₄⁻	↓	↓	↓	↓	↓	↓	N	↓	N
GDP Dissociation	↑		↑	N	↑	N	N		
GTP stability	↓			↓		↓	N		
GTP Hydrolysis					↑		N		

FIG. 5. Effects of α_s mutations that decrease receptor-mediated activation. Arrows pointing up indicate increases. Arrows pointing down indicate decreases. N indicates no difference from wild-type α_s. Data for S54N are from J. D. Hildebrandt, R. Day, C. L. Farnsworth, and L. A. Feig, *Mol. Cell. Biol.* **11**, 4830 (1991). Data for D173K and K293D are from J. Codina and L. Birnbaumer, *J. Biol. Chem.* **269**, 29339 (1994). Data for N167R are from G. Grishina and C. H. Berlot, *J. Biol. Chem* **273**, 15053 (1998) and C. H. Berlot, unpublished results (1999). Data for R231H are from T. Iiri, Z. Farfel, and H. R. Bourne, *Proc. Natl. Acad. Sci. U.S.A.* **94**, 5656 (1997). Data for R258A are from G. Grishina and C. H. Berlot, *J. Biol. Chem.* **273**, 15053 (1998); D. R. Warner, G. Weng, S. Yu, R. Matalon and L. S. Weinstein, *J. Biol. Chem.* **273**, 23976 (1998); and D. R. Warner and L. S. Weinstein, *Proc. Natl. Acad. Sci. U.S.A.* **96**, 4268 (1999). Receptor affinity was measured using an α_s construct containing four substitutions (N254D, M255L, I257L, R258A), but only the R258A mutation caused a defect in receptor-mediated activation. Data for E259D are from D. R. Warner, R. Romanowski, S. Yu, and L. S. Weinstein, *J. Biol. Chem.* **274**, 4977 (1999). Data for the $\alpha 3/\beta 5$ construct, which contains five substitutions of α_{i2} homologs for α_s residues (N271K/K274D/R280K/T284D/I285T), are from G. Grishina and C. H. Berlot, *Mol. Pharmacol.* **57**, 1081 (2000). Data for R389P are from K. A. Sullivan, R. T. Miller, S. B. Masters, B. Beiderman, W. Heideman, and H. R. Bourne, *Nature* **330**, 758 (1987).

the regulation, whereas alanine substitutions within the switch regions will reveal the residues that regulate the GTP dependence of effector interaction.

Regulation of Receptor-Mediated Activation

Receptor-mediated activation of G protein α subunits is a complex process that requires receptor binding, receptor-stimulated GDP release and GTP binding, and dissociation of $\alpha \cdot$ GTP from the receptor and $\beta\gamma$. Defects in receptor-mediated activation can be due to alterations in any of these steps, as demonstrated in Fig. 5, which shows the effects of a panel of α_s mutants that cause decreases in receptor-mediated activation. Decreased receptor-mediated activation can be caused by either increased or decreased receptor affinity, or may be associated with normal receptor affinity. Since α and $\beta\gamma$ must be associated in order to interact with receptors, decreased receptor-mediated activation may be due to decreased affinity for $\beta\gamma$. Receptor-mediated activation requires that the nucleotide binding site be in an appropriate conformation, and alterations in this site are often reflected by a decreased ability to be activated by AlF_4^- that is associated with other defects in nucleotide handling. If the duration of the activated state is decreased by an increased rate of GTP hydrolysis, the degree of activation attained by receptor-stimulated nucleotide exchange will be diminished. Therefore, it is desirable to test each of these functions to precisely determine the role of a residue that is important for receptor-mediated activation.

Studies of α_s mutants with defects in receptor-mediated activation suggest that interaction between activated receptors and G proteins leads to nucleotide exchange by initiating a series of conformational changes within the α subunit. The two potential receptor contact sites in α_s, the extreme carboxyl terminus, which contains Arg-389,[36] and the $\alpha3/\beta5$ loop[35] (Fig. 5), are distant from the nucleotide binding site. However, binding of activated receptors to α_s appears to lead to an interaction between switch II, which binds to $\beta\gamma$, and switch III. The importance of this interaction for receptor-mediated activation is demonstrated by the effects of disrupting an interaction between Arg-231 in switch II[37] and Glu-259 in switch III[38] (Fig. 5). The helical domain presents a barrier to nucleotide release, but activated receptors appear to overcome this barrier by initiating conformational changes that are transmitted across the domain interface. These conformational changes require matching of the residue in position 167 in the $\alpha D/\alpha E$ loop of the helical domain with that in position 254 in switch III of the GTPase domain[14] and of the residue in position 173 in the $\alpha D/\alpha E$ loop with that in position 293 in the $\beta5/\alpha G$ loop of the GTPase domain[32] (Fig. 5). It will be interesting to determine the degree to which these conclusions generalize to receptor-mediated activation of other α subunits.

Acknowledgments

I thank Thomas Hynes for helpful discussions and critical reading of the text. Work from the author's laboratory was supported by the National Institutes of Health Grant GM50369. The author is an Established Investigator of the American Heart Association.

[36] K. A. Sullivan, R. T. Miller, S. B. Masters, B. Beiderman, W. Heideman, and H. R. Bourne, *Nature* **330,** 758 (1987).
[37] T. Iiri, Z. Farfel, and H. R. Bourne, *Proc. Natl. Acad. Sci. U.S.A.* **94,** 5656 (1997).
[38] D. R. Warner, R. Romanowski, S. Yu, and L. S. Weinstein, *J. Biol. Chem.* **274,** 4977 (1999).

[33] Development of Gs-Selective Inhibitory Compounds

By Christian Nanoff, Oliver Kudlacek, and Michael Freissmuth

Introduction

For many years, the bacterial exotoxins of *Bordetella pertussis* and *Vibrio cholerae* have been invaluable tools in the study of G-protein-dependent signaling pathways. Pertussis toxin, in particular, continues to be widely employed because blockage of a biological response by pertussis toxin is considered formal proof of the involvement of a member of the family comprising $G_i/G_o/G_t$ heterotrimers. Cholera toxin is somewhat less useful: ADP-ribosylation of an internal ARG (R187/188 or R201/202 depending on the splice variant of $G\alpha_s$) impairs the GTPase activity of $G\alpha_s$ and constitutively activates the protein. Active $G\alpha_s$ is subject to down-regulation by an ill-defined mechanism. Thus the consequence of cholera toxin treatment is biphasic; initially, $G\alpha_s$-dependent signaling is stimulated, but long-term exposure causes the levels of membrane-bound $G\alpha_s$ to decline and hence the cellular response to G_s-coupled receptors is lost. This biphasic action is seen in many—but not all—cells and, in addition, the time course is variable. In view of these limitations and because there are no toxins available to inactivate members of the G_q family (G_q, G_{11}, G_{14-16}) as well as the G protein heterotrimers G_{12}, G_{13}, and G_z, compounds that selectively inhibit individual G proteins are clearly of interest as experimental tools. In addition, these inhibitors may be ultimately relevant to pharmacotherapy. Although this concept is, of course, received with skepticism, there are arguments why G proteins can per se be considered as drug targets, and these have been summarized elsewhere.[1,2]

General Considerations

In the cycle of activation and deactivation the G protein α subunit interacts sequentially with a series of reaction partners[3]: (i) In the basal state, GDP-liganded $G\alpha$ combines with the $\beta\gamma$-dimer which results in mutual inactivation. (ii) The active (agonist-liganded) receptor binds to the heterotrimer and catalyzes the nucleotide exchange reaction (GTP for GDP on the $G\alpha$) which leads to subunit dissociation. (iii) $G\alpha$ or $G\beta\gamma$ or both transfer the signal to an effector. (iv) In many

[1] C. Höller, M. Freissmuth, and C. Nanoff, *Cell. Mol. Life Sci.* **55**, 257 (1999).

[2] M. Freissmuth, M. Waldhoer, E. Bofill-Cardona, and C. Nanoff, *Trends Pharmacol. Sci.* **20**, 237 (1999).

[3] J. R. Hepler and A. G. Gilman, *Trends Biochem. Sci.* **17**, 383 (1992).

cases, Gα also interacts with an RGS-protein (regulator of G protein signaling, GTPase-activating protein), which accelerates the turn-off reaction.[4] Thus, binding sites are sequentially exposed on the α subunit and these may be targeted by inhibitors. The G protein inhibitor ought to be selective for a specific G protein α subunit and to efficiently block the signal transfer from a receptor to an effector. Based on these considerations, some of the binding sites on the α subunit are unlikely to be useful targets; the guanine nucleotide binding pocket, for instance, may be targeted, but this will result in nonspecific inhibition.[5,6] Similarly, an inhibitor that prevents binding of Gα to G$\beta\gamma$ is unlikely to be selective and, in addition, its usefulness is questionable for the following reason: blockage of subunit association prevents receptor-dependent signaling (the receptor requires the heterotrimer) but it actually increases basal signaling, because the mutual inactivation of Gα and G$\beta\gamma$ resulting from heterotrimer formation is eliminated. Based on these considerations, the effector or the receptor binding site on Gα can be postulated to be candidate sites that are most likely to be targeted by useful inhibitors. This prediction has been verified with a series of suramin analogs that disrupt receptor–G protein coupling and effector regulation (see below). Finally, to be useful in experiments in intact cells and animals, the inhibitor ought to be membrane-permeable. This obstacle has not yet been overcome with synthetic inhibitors; however, it does not appear to represent an insurmountable problem, because low molecular weight G protein activators exist that can overcome the membrane barrier. These include the wasp venom mastoparan and related peptides[7] as well as alkyldiamines and -triamines.[8]

Assays for Inhibitors of Gα_s

In the search for selective G protein inhibitors, Gα_s is a good starting point because its biochemical properties are well understood and can be readily measured: Gα_s exchanges guanine nucleotides spontaneously with a measurable rate; it couples to an array of receptors, several of which, in particular the β-adrenergic receptors, are extremely well characterized; its effector molecule is adenylyl cyclase; and several methods are available to measure cAMP formation, both in membrane preparations[9] and in intact cells.[10] Moreover, membranes prepared from the

[4] J. R. Hepler, *Trends Pharmacol. Sci.* **20**, 376 (1999).

[5] M. Hohenegger, C. Nanoff, H. Ahorn, and M. Freissmuth, *J. Biol. Chem.* **269**, 32008 (1994).

[6] C. Nanoff, S. Boehm, M. Hohenegger, W. Schütz, and M. Freissmuth, *J. Biol. Chem.* **269**, 31999 (1994).

[7] E. M. Ross and T. Higashijima, *Methods Enzymol.* **237**, 26 (1994).

[8] B. Nürnberg, W. Togel, G. Krause, R. Storm, E. Breitweg-Lehmann, and W. Schunack, *Eur. J. Med. Chem.* **34**, 2 (1999).

[9] R. A. Johnson and Y. R. Salomon, *Methods Enzymol.* **195**, 3 (1991).

[10] Y. R. Salomon, *Methods Enzymol.* **195**, 22 (1991).

cyc^- variant of S49 mouse lymphoma cells allow for a simple biochemical complementation assay that is specific for Gα_s.[11] S49 cyc^- cells are deficient in Gα_s but express adenylyl cyclase and β_2-adrenergic receptors; hence, guanine nucleotide- and receptor-dependent regulation of cAMP formation is absent but can be reconstituted upon addition of exogenous Gα_s. Thus, three approaches can be employed to investigate the properties of Gα_s-selective inhibitors, namely the effect of the compounds on (i) the basal rate of nucleotide exchange, (ii) on receptor–G protein coupling, and (iii) the block of effector regulation. (iv) Finally, direct binding assays can be employed provided that α subunits are available in highly purified form and in reasonably large quantities. Because Gα_s[12] (and other α subunits including myristoylated versions[13]) can be expressed in *Escherichia coli* and purified from bacterial lysates, the amount of protein is not limiting, if the affinity of the inhibitor is in the low micromolar range.

Assays of Guanine Nucleotide Exchange

The rate-limiting step in G protein activation is the release of GDP prebound to Gα. Thus any compound that inhibits the dissociation of GDP is, by definition, a G protein inhibitor. The rate of GDP release can be determined either directly, by prelabeling the G protein with radioactive GDP using [^{3}H]GDP, [α-^{32}P]GDP, or [α-^{32}P]GTP, or indirectly, by determining the rate of association of [^{35}S]GTPγS (which is limited by GDP release).

Measuring the rate of [^{35}S]GTPγS binding is straightforward. Purified recombinant Gα_s[12] is diluted in HEPES–NaOH (pH 7.6), 1 mM EDTA, 1 mM DTT (dithiothreitol), 0.01% Lubrol to yield a final concentration of $\sim$40 to 140 ng/10 μl (corresponding to $\sim$1–3 pmol/assay) and held on ice. The reaction is started by adding 4 volumes (i.e., 40 μl/assay) of prewarmed buffer containing HEPES–NaOH (pH 7.6), 1 mM EDTA, 1 mM DTT, 12.5 mM MgSO$_4$, 1.25 μM [^{35}S]GTPγS ($\pm$ the test compound of interest), 0.01% Lubrol. The time points and the incubation temperature are chosen based on the known exchange rates of the individual Gα. In the case of Gα_s, the basal rate of GDP release is reasonably fast, i.e., in the range of 0.13 to 0.3 min^{-1} at 20°, for the short and the long splice variants, respectively[14] (Gα_o also releases GDP rapidly; in the case of the three forms of Gα_i exchange is considerably slower, and 30° is a more convenient temperature). The reaction is stopped by withdrawing 50 μl aliquots and diluting these immediately with 2 ml of ice-cold stop buffer of suitable ionic strength (100–125 mM NaCl), pH (7–8), Mg^{2+} concentration (10–25 mM), and unlabeled guanine nucleotide (e.g., 10–100 μM GTP). In the presence of $\geq$1 mM free Mg^{2+}, binding

[11] E. M. Ross, A. C. Howlett, K. M. Ferguson, and A. G. Gilman, *J. Biol. Chem.* **253**, 6401 (1978).
[12] M. P. Graziano, M. Freissmuth, and A. G. Gilman, *Methods Enzymol.* **195**, 192 (1991).
[13] M. E. Linder and S. M. Mumby, *Methods Enzymol.* **237**, 254 (1994).
[14] M. P. Graziano, M. Freissmuth, and A. G. Gilman, *J. Biol. Chem.* **264**, 409 (1989).

of GTPγS is quasi-irreversible (i.e., the off-rate is essentially zero); the dilution into cold buffer and the excess of guanine nucleotide prevents any further binding. Thus, the data points can be collected and all samples can be conveniently filtered at the end of the incubation. Bound nucleotide is trapped on nitrocellulose filter (Schleicher & Schuell, Keane, NH BA85 or comparable) using a suitable manifold for manual filtration. The filters are rinsed with 15 to 20 ml of buffer (stop buffer lacking GTP). The blank is determined in the absence of any Gα (if 3–5 × 10^5 cpm/assay are employed, about 0.1% of the total activity is typically retained as the filter blank). The following additional considerations are worth mentioning: (i) The specific activity can be adjusted appropriately (5–20 cpm/fmol). It is generally not wise to add more than 1 × 10^6 cpm/assay because the blank will increase substantially; if the sensitivity is to be increased, it is preferable to reduce the concentration of GTPγS (to 0.1 μM) and to keep the amount of radioactivity constant. (ii) The presence of Lubrol during the incubation is per se not required because isolated α subunits are soluble; there are nevertheless several arguments why it is preferable to use detergents. First, the presence of detergents prevents adsorptive losses of the proteins; these are particularly pronounced at low protein concentrations and occur regardless of whether silanized glass or polypropylene tubes are employed. Second, there are many instances in which it is desirable to compare the properties of the isolated α subunit with those of the heterotrimer. With few exceptions (e.g., retinal $\beta\gamma$-dimers) wild-type forms of $\beta\gamma$ dimers are not soluble in the absence of detergent. It is therefore wise to carry out experiments under comparable conditions because both the presence of detergents and their final concentrations affect the rate of nucleotide exchange.[15] Lubrol may be substituted with 8 mM CHAPS (or 12 mM cholate). (iii) Nitrocellulose filters have the major drawback that they cannot be used with automated filtration apparatuses. However, there is no substitute to nitrocellulose; untreated glass fiber filters do not efficiently retain G proteins in solution. Soaking the filters in polyethyleneimine (PEI) greatly increases the retention of soluble proteins. However this approach cannot be employed because PEI-coated glass fiber filters bind [^{35}S]GTPγS (and other nucleotides) in an essentially quantitative manner.

For GDP-release experiments, [α-^{32}P]GDP and [α-^{32}P]GTP are interchangeable because wild-type versions of G protein α subunits hydrolyze GTP more rapidly than they release GDP (k_{cat} being $\geq 10\,k_{off,GDP}$). Hence, at any given time point, essentially all protein bound nucleotide is [α-^{32}P]GDP, even if the protein is incubated in the presence of [α-^{32}P]GTP. Assay conditions are in principle similar to those outlined above. Gα_s is prelabeled with either [α-^{32}P]GDP/GTP or [^{3}H]GDP for 30 min at 20° in small volume (e.g., at 0.1–1 μM Gα_s and 1–10 μM radiolabeled GDP) to obtain the desired specific activity (e.g., 10–30 cpm/fmol Gα_s). In the case of Gα_s, loading is enhanced if the incubation with [α-^{32}P]GTP is done in the presence of high Mg^{2+} (e.g., 10 mM MgSO$_4$) because millimolar

---

[15] D. J. Carty and R. Iyengar, *Methods Enzymol.* **237**, 38 (1994).

Mg^{2+} enhances GDP release in $G\alpha_s$; $MgCl_2$ is less effective because Cl^- slows GDP release.[16] Obviously, one has to account for the carryover of Mg^{2+} into the final reaction mixture. Thereafter, the protein is held on ice (to prevent denaturation). Dissociation is initiated by diluting the protein (1–2 pmol prelabeled $G\alpha_s$/assay) into prewarmed buffer (for composition see above) containing excess unlabeled GTP (10–100 μM) ± the desired concentration of the test compound. At suitable time points (see preceding paragraph), aliquots are withdrawn and diluted into 2 ml ice-cold stop buffer (pH 7–8; 100–130 mM NaCl, 10 mM $MgCl_2$, 10 mM NaF, 20 μM $AlCl_3$). The presence of $NaF/AlCl_3$ is of paramount importance. AlF_4 substitutes for the γ-PO_4 of GTP, the $Mg \cdot GDP \cdot AlF_4$-liganded α subunit corresponds to the transition state,[17] and the release of GDP from this complex is extremely slow. Thus, similar to the binding assay for [^{35}S]GTPγS, filtering can be done after all data points have been collected.

It is obvious that the association rate of GTPγS and the dissociation rate of GDP ought to be similar; this is exemplified by the data summarized in Fig. 1. It is also evident from Fig. 1A and Fig. 1C that the addition of 1 μM suramin slows both the rate of [^{35}S]GTPγS and the rate of GDP release. The effect of suramin on the rate of [^{35}S]GTPγS binding to $G\alpha_o$ is less pronounced (Fig. 1B). A comparison of several suramin analogs shows that suramin has a modest selectivity for $G\alpha_s$ (when compared to $G\alpha_o$ and $G\alpha_{i-1}$).[18] The determination of [^{35}S]GTPγS binding is simpler than the assay for [α-^{32}P]GDP or [^{3}H]GDP release, for [^{35}S]GTPγS binding requires only one incubation step. In addition, because binding is quasi-linear if the incubation time is kept below the first half-life of association, a single time point suffices to evaluate the inhibitory effect of a test compound. Finally, the signal is very robust in [^{35}S]GTPγS binding assays. Thus, employing inhibition of GTPγS-binding to $G\alpha_s$ and to $G\alpha_{i-1}$, we have identified two suramin analogs that are selective for $G\alpha_s$, namely NF503 and NF449.[19,20]

Receptor–G Protein Coupling

For the functional analysis of the interaction between receptors and G proteins two simple types of binding assays are suggested that allow for examining the sequential microscopic steps which lead from ligand binding to the catalytic G protein turnover promoted by the active receptor. The G protein activation cycle is initiated by the binding of an agonist ligand to the receptor; formation and stability of the high-affinity ternary complex that consists of agonist, receptor, and G protein

[16] T. Higashijima, K. M. Ferguson, and P. C. Sternweis, *J. Biol. Chem.* **262,** 3597 (1987).

[17] S. R. Sprang, *Ann. Rev. Biochem.* **66,** 639 (1997).

[18] M. Freissmuth, S. Boehm, W. Beindl, P. Nickel, A. P. Ijzerman, M. Hohenegger, and C. Nanoff, *Mol. Pharmacol.* **49,** 602 (1996).

[19] M. Hohenegger, M. Waldhoer, W. Beindl, B. Böing, P. Nickel, C. Nanoff, and M. Freissmuth, *Proc. Natl. Acad. Sci. U.S.A.* **95,** 346 (1998).

[20] NF449 is commercially available through Calbiochem (La Jolla, CA).

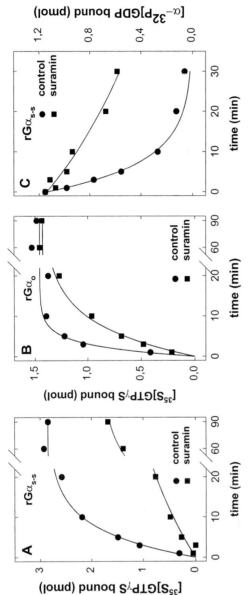

Fig. 1. Time course of [^{35}S]GTPγS binding to Gα_s (A) and to Gα_o (B) and of [α-^{32}P]GDP-release from Gα_s (C) in the absence and presence of suramin. (A and B) Recombinant Gα_{s-s}[12] (the short splice variant of Gα_s) and r(recombinant) Gα_o[13] were purified from bacterial lysates. The reaction mixture (50 μl/assay) contained rGα_{s-s} (150 ng) or rGα_o (70 ng), buffer [HEPES · NaOH (pH 7.6), 1 mM EDTA, 1 mM DTT (dithiothreitol), 0.01% Lubrol], 10 mM MgSO$_4$, and 1 μM [^{35}S]GTPγS (specific activity = 20 cpm/fmol). The incubation was carried at 20° for the indicated time intervals in the absence (●) and presence (■) of 1 μM suramin. The calculated rate constants were for rGα_s 0.15 and 0.026 min^{-1} and for rGα_o 0.38 and 0.12 min^{-1}, in the absence and presence of suramin, respectively. (C) Recombinant Gα_{s-s} was prelabeled with [α-^{32}P]GTP (1 μM, specific activity = 30 cpm/fmol) for 30 min at 30° in the same buffer as used for panels A and B. Subsequently, [α-^{32}P]GDP-liganded rGα_{s-s} (1.2 pmol/assay) was diluted into prewarmed buffer (50 μl final volume) containing 100 μM unlabeled GTP and incubated at 20° for the indicated time intervals in the absence (●) and presence (■) of 1 μM suramin. It is evident that the off-rates for GDP (0.14 and 0.023 min^{-1} in the absence and presence of suramin, respectively) are comparable to the on-rates for [^{35}S]GTPγS determined in panel A. For details on the conditions for filtration see text. Reproduced with permission from Ref. 18.

is a direct measure of the specificity that governs the coupling reaction.[21] Once the complex is formed, it is sufficiently stable such that a binding assay with an agonist radioligand can be carried out until equilibrium. In order to disrupt receptor/G protein coupling, the test compound has (i) to render inaccessible the G protein domains that contribute to the receptor/G protein interface (e.g., pertussis toxin), (ii) to keep the α subunit in the guanine nucleotide-bound form, or (iii) to prevent the association of the heterotrimer.

The second assay is used to determine the ability of the activated receptor to promote G protein turnover; in addition to the above, the variables determining the outcome include the G protein activation switch that is prompted by the active receptor and for which the formation of the high-affinity conformation is the prerequisite. For a detailed description of assays determining the receptor-dependent G protein activation see a previous volume of this series.[22–24]

Radioligand binding experiments are straightforward given that the experimental setup (radioligand, binding substrate, filtration manifold, radiation counters) is generally available. For testing G protein antagonists, transfected cell lines expressing a G$_s$ coupled receptor offer certain advantages over membranes prepared from native tissue, in particular because they avoid systematic errors due to cellular inhomogeneity and because several receptors can be compared in the identical cellular background. When using these cells, however, receptor/G protein coupling may be influenced by the receptor expression level that is achieved upon transfecting cells with foreign cDNA.[25] Therefore, efforts should be directed at selecting cell clones that give a receptor expression level similar to that found in native tissue. Drastic overexpression will lead to aberrant G protein coupling and may give rise to misleading interpretation of the data.[25]

It is important to verify out that the test compound does not interfere with binding of the receptor–ligand to the ligand pocket. This is also true if receptor-promoted GTPase or GTPγS-binding is used as a test bed for G protein inhibitors. If a radiolabeled antagonist is available, this is assessed by equilibrium binding carried out in the presence of increasing concentrations of the test compound. Alternatively, a radiolabeled agonist can be used in membranes treated with GTPγS. This approach requires that the low-affinity state of the (uncoupled) receptor still binds the agonist with an acceptable affinity (<10 nM) that withstands separation of free and bound radioligand by filtration.

If a radiolabeled full-agonist ligand is available, formation of the high-affinity ternary complex can be directly monitored. A G protein inhibitor ought (i) to block

[21] M. Waldhoer, A. Wise, G. Milligan, M. Freissmuth, and C. Nanoff, *J. Biol. Chem.* **274,** 30571 (1999).

[22] T. Wieland and K. H. Jakobs, *Methods Enzymol.* **237,** 3 (1994).

[23] P. Gierschik, T. Bouillon, and K. H. Jakobs, *Methods Enzymol.* **237,** 13 (1994).

[24] K.-L. Laugwitz, K. Spicher, G. Schultz, and S. Offermanns, *Methods Enzymol.* **237,** 283 (1994).

[25] T. Kenakin, *Pharmacol. Rev.* **48,** 413 (1996).

formation of the high-affinity complex and (ii) to destabilize the preformed complex. It is however, important to remember, that detergents also uncouple receptors from G proteins by lowering their mutual affinity.[26] A detergent-like effect is best ruled out using a series of closely related compounds with similar amphiphilic properties. Dissociation of preformed high-affinity complex is examined as follows: Binding of the agonist ligand is allowed to proceed to equilibrium. Then, the test compound (or vehicle) is added and the decay in binding is followed. The dissociation rate in the presence of the test compound should be compared to the effect induced by adding an excess of unlabeled ligand (100 K_D) or of a nonhydrolyzable guanine nucleotide (e.g., 1 μM GTPγS). Dissociation induced by the G protein antagonist is not expected to occur if the compound acts allosterically (i.e., at a site distant from the receptor interface) to inhibit the release of GDP. This is because in the high-affinity ternary complex the guanine nucleotide binding site is empty; however, a test compound that directly binds to the receptor interface with sufficient affinity may disrupt the complex.[27]

If a suitable agonist radioligand is not available, the essential information can also be obtained by use of antagonist radioligands.[19] In competition experiments using an unlabeled agonist that displaces a labeled antagonist the competition curve has a biphasic shape; at low concentrations (phase 1) the agonist displaces the antagonist radioligand from the G-protein-coupled high-affinity conformation of the receptor. Displacement by high agonist concentrations reflects binding to the uncoupled low-affinity state of the receptor. The addition of guanine nucleotides (GTP analogs) abolishes the high-affinity state and renders the displacement curve monophasic. Similarly, in the presence of a G protein inhibitor, formation of the high-affinity state ought to be reduced and the curve shifted to the right. Using the competition binding approach, however, only a rough estimate of the potency of the G protein inhibitor may be obtained; this can be assessed by performing antagonist binding in the presence of a fixed concentration of agonist (one that competes predominantly for the high-affinity state of the receptor). Increasing the concentration of the test compound should gradually reverse binding inhibition due to the agonist. For the G_s-selective inhibitors NF449 and NF503, we have carried out these experiments in parallel on prototypical receptors (coupled to either G_i/G_o, G_s, or G_q) to confirm their selectivity.[19]

Effector Regulation

As mentioned above, the $G\alpha_s$-deficient S49 cyc^- membranes provide an ideal assay system to assess effector regulation by $G\alpha_s$. The assay relies on the robust stimulation of adenylyl cyclase that can be observed after addition of

---

[26] F. Roka, L. Brydon, M. Waldhoer, A. D. Strosberg, M. Freissmuth, R. Jockers, and C. Nanoff, *Mol. Pharmacol.* **56,** 1014 (1999).

[27] W. Beindl, T. Mitterauer, M. Hohenegger, A. P. Ijzerman, C. Nanoff, and M. Freissmuth, *Mol. Pharmacol.* **50,** 415 (1996).

exogenous Gα_s. Because the membrane lipids act as effective detergent buffers, the choice of detergents is not too important (although the enzymatic activity of adenylyl cyclase is higher in Lubrol than in CHAPS or cholate); guanine nucleotide- and AlF$_4$-dependent activation of cAMP formation can readily be observed even if Gα_s is added at high detergent concentrations that partially solubilize the components. For receptor-dependent stimulation, the concentration of detergent is more important. It is best kept below the critical micellar concentration or—in the case of Lubrol—at a final level where it does not exceed the amount of lipid. It is also important to note that reaction mixtures employed in adenylyl cyclase assays often contain bovine serum albumin. This must be omitted: it may confound the results, because acidic compounds bind tightly to albumin. Suramin, for instance, binds to albumin with a stoichiometry of >1 (there are several high- and low-affinity sites). Finally, one has to rule out that the test compounds have a direct effect on adenylyl cyclase by blocking catalysis. It is a further advantage of the S49 cyc^- membranes that any potential site of action can be separately assessed because the catalytic activity of adenylyl cyclase can be stimulated directly with forskolin; this control is shown for the G$_s$-selective inhibitor NF449 in Fig. 2A. Because G$_i$ is present in S49 cyc^- membranes, forskolin-stimulated cAMP formation can be inhibited by guanine nucleotides. Figure 2A also shows that NF449 does not impair the ability of GTPγS to inhibit cAMP formation, a finding consistent with the observation that NF449 does not affect the guanine nucleotide-exchange reaction of Gα_i.[19] In contrast, receptor- and GTP-dependent regulation of adenylyl cyclase by Gα_s is blocked by NF449 (Fig. 2B). The inhibition of GTP-dependent activation of adenylyl cyclase by Gα_s is accounted for by the suppression of GDP release[18,19] (see also Fig. 1); the block of receptor-dependent stimulation is due to inhibition of receptor–G protein coupling (see above). Finally, suramin analogs also directly impair the ability of Gα_s to interact with adenylyl cyclase.[18,19]

Direct Binding of [^{3}H]Suramin to Gα_s

The observations summarized above show that suramin analogs exert three types of effects on G protein α subunits (i.e., inhibition of GDP release, inhibition of effector interaction, and uncoupling of the receptor). These may arise from binding of the compound to multiple sites on the α-subunit or may reflect the action at a single binding site. In order to address this issue, we have determined the binding of [^{3}H]suramin[28] to purified Gα_s by equilibrium dialysis; half-maximum inhibition of the on-rate for [^{35}S]GTPγS or of the off-rate for [α-^{32}P]GDP was seen at $\sim$0.2 μM suramin.[18] A filter binding assay is therefore unlikely to be useful because this affinity estimate predicts a rapid off-rate; this consideration is also true for separation of bound and free ligand by gel filtration. We have therefore

---

[28] [^{3}H]Suramin is commercially available through Moravek Biochemicals (Brea, CA) or Hartmann Analytics (Braunschweig, FRG).

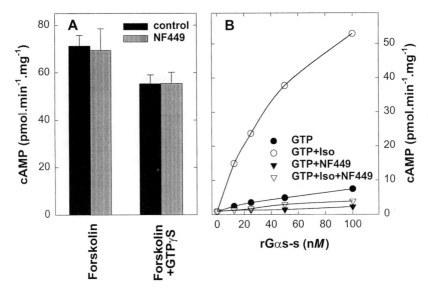

FIG. 2. Effect of the suramin analog NF449 on adenylyl cyclase activity in S49 cyc^- membranes following stimulation with forskolin (A) and with exogenously added recombinant $G\alpha_{s\text{-}s}$ (B). (A) S49 cyc^- membranes (25 μg/assay) were stimulated with 100 μM forskolin or with 100 μM forskolin and 1 μM GTPγS (to activate G_i endogenous to the membranes) in the absence and presence of 10 μM NF449. The reaction was carried for 30 min at 20° in a substrate solution (final volume 100 μl) containing 20 mM HEPES–NaOH (pH 7.6), 2 mM MgCl$_2$, 0.05 mM [α-^{32}P]ATP (specific activity 300–400 cpm/pmol), 0.5 mM creatine phosphate, 1 mg/ml creatine kinase, 1 mM rolipram; [^{32}P]cAMP was resolved from the substrate by sequential chromatography on Dowex AG50-X8 and neutral alumina.[9] Data are means ± SEM ($n = 4$). (B) The indicated amounts of recombinant $G\alpha_{s\text{-}s}$ [diluted in 20 μl HEPES · NaOH (pH 7.6), 1 mM EDTA, 1 mM DTT, 0.01% Lubrol] were added to cyc^- membranes [25 μg/20 μl 20 mM HEPES · NaOH (pH 7.6), 1 mM EDTA, 2 mM MgCl$_2$] and held on ice for 30 min. Thereafter, the reaction was carried out as in (A) with 10 μM GTP ($\bullet$,$\blacktriangledown$) or with 10 μM GTP and 10 μM isoproterenol ($\bigcirc$,$\triangledown$) in the absence ($\bullet$,$\bigcirc$) and presence of 10 μM NF 449 ($\triangledown$,$\blacktriangledown$). Data are means from duplicate determinations and were reproduced twice with similar results.

employed a modified equilibrium dialysis method. To obtain a robust signal-to-noise ratio, a large difference between ligand in the ultrafiltrate (free ligand) and in the retentate (sum of free + bound ligand) is desirable; this is best achieved if the macromolecular binding partner is present at a concentration that is in excess of the expected K_D. We have therefore incubated $G\alpha_s$ at ~0.6 μM (i.e., 2.7 μg/0.1 ml final volume) in buffer (50 mM HEPES–NaOH, pH 7.6, 1 mM EDTA, 1 mM DTT) containing [^{3}H]suramin at total concentrations covering the range from 0.25 to 10 μM (specific activity 250–10,000 cpm/pmol). The incubation was carried out for 30 min at 25°; an ultrafiltrate (~20 μl) was generated by briefly (~10–15 s at 4000g) centrifuging the reaction volume over Millipore Ultrafree-MC micro-concentrators (nominal cutoff: $M_r \approx 30,000$). We have found that the reaction can

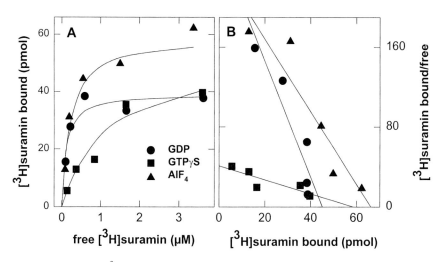

FIG. 3. Binding of [³H]suramin to purified recombinant Gα$_{s-s}$. The binding reaction was carried out in a final volume of 0.1 ml containing buffer (mM : 50 HEPES–NaOH, pH 7.6, 1 EDTA, 1 DTT), [³H]suramin (total concentration ranging from 0.25 to 10 μM; specific activity 250–10,000 cpm/pmol), 2.7 μg rGα$_{s-s}$, 10 μM GDP (●), the combination of 10 μM GDP, 20 μM AlCl$_3$, 10 mM NaF, and 10 mM MgSO$_4$ (▲), and 10 μM GTPγS + 10 mM MgSO$_4$ (■); in the last case rGα$_{s-s}$ was preincubated in the presence of 50 μM GTPγS and 20 mM MgSO$_4$ for 30 min at 30° to fully exchange prebound GDP for GTPγS. The incubation with [³H]suramin lasted for 30 min at 25°; an ultrafiltrate (~20 μl) was generated by briefly (~15 s at 4000g) centrifuging the reaction volume over Millipore Ultrafree-MC microconcentrators (nominal cutoff: M_r 30,000); the amount of radioactivity in the retentate (sum of free + bound ligand) and in the ultrafiltrate (free) was determined. Shown are the means of duplicate determinations in a representative experiment where the three curves were generated in parallel. The data were fitted to a saturation isotherm yielding the following estimates (mean ± SD; $n = 3$): K_D, 0.17 ± 0.06, 0.22 ± 0.10, and 1.24 ± 0.06 μM; stoichiometry, 0.7 ± 0.05, 1.0 ± 0.03, 0.9 ± 0.04 mol/mol; for GDP-bound, AlF$_4$-liganded, and GTPγS-bound rGα$_{s-s}$, respectively.

be carried out directly in the microconcentrator. The amount of radioactivity in the retentate (sum of free and bound ligand) and in the ultrafiltrate (free ligand) is determined to calculate the amount of bound [³H]suramin. The data summarized in Fig. 3 show that the stoichiometry of binding was close to 1 : 1. It was lower, namely ~0.7, in the GDP-bound conformation, which is due to the notorious instability of Gα subunits. In the active GTPγS-bound and GDP · A1F$_4$-bound conformation, the protein is stabilized and the stoichiometry approached unity. In addition, the affinity was higher in the GDP-bound form than if the protein had been preactivated in the presence of GTPγS. This is to be expected; since suramin increases the affinity of the α subunit for GDP (by reducing its off-rate; see Fig. 1), thermodynamic considerations dictate that the presence of GDP increases the affinity of the protein for [³H]suramin. It is also worth pointing out that the affinity measured directly by equilibrium dialysis ($K_D = 0.17$ μM) is similar to that determined by inhibition of the on-rate for GTPγS or the off-rate for GDP (IC$_{50}$ = 0.24 μM).[18] In contrast, it was somewhat surprising that suramin bound more tightly to GDP · A1F$_4$-liganded Gα_s (which, as mentioned earlier, mimics the transition state) than to GTPγS-bound Gα_s; the crystal structures reveal few—if any—differences between these two conformations in G protein α subunits.[17] However, RGS proteins are also capable of discriminating between the GTPγS-bound conformation (to which they bind, in several instances, with modest affinities) and the transition state; i.e., they bind tightly to the AlF$_4$ · GDP conformation.[4]

Conclusions

A detailed discussion of the properties of various suramin analogs as G protein inhibitors is beyond the scope of this chapter. Suffice it to say that the active compounds exert three types of effects, namely inhibition of GDP release,[18] inhibition of receptor–G protein coupling,[26,27,29,30] and inhibition of effector regulation.[18] Presumably, these actions are interrelated and can be accounted for by the interaction of the compounds with a single binding site. Two properties make suramin potentially useful as inhibitors of Gα subunits: there are analogs that are selective for individual α subunits, e.g., Gα_s,[19] and some compounds can be shown to differentiate between individual receptor–G protein complexes that contain the identical α subunit.[30] The major drawback is currently the fact that the compounds are not membrane permeable. Clearly, the biochemical assays are available and it should be possible to overcome this obstacle in the future. Finally, the structure of a complex has been solved in which Gα_s is bound to a dimer formed by the catalytic domains of adenylyl cyclase.[31] Thus, a structural model is available that may guide the search for improved inhibitors of Gα_s.

Acknowledgments

Work from the authors' laboratory was supported by grants from the Austrian Science Foundation (FWF-12125 and FWF-12750).

[29] R.-R. C. Huang, R. N. Dehaven, A. H. Cheung, R. E. Diehl, R. A. F. Dixon, and C. D. Strader, *Mol. Pharmacol.* **37**, 304 (1990).

[30] M. Waldhoer, E. Bofill-Cardona, G. Milligan, M. Freissmuth, and C. Nanoff, *Mol. Pharmacol.* **53**, 808 (1998).

[31] J. J. Tesmer, R. K. Sunahara, A. G. Gilman, and S. R-Sprang, *Science* **278**, 1907 (1997).

[34] Characterization of Deamidated G Protein Subunits

By William E. McIntire, Kevin L. Schey, Daniel R. Knapp,
Jane Dingus, and John D. Hildebrandt

Introduction

Heterotrimeric G proteins, composed of an α subunit and a $\beta\gamma$ dimer, are responsible for initiating intracellular processes based on specific extracellular stimuli.[1] Molecular diversity in the α, β, and γ subunits[2] contributes to the dynamic nature of the signaling pathways with which these proteins are involved. Such diversity can result from different genes within a family, alternative splicing of a particular gene, or posttranslational processing of the protein product. Advances in molecular techniques have allowed the identification of numerous G protein subunit isoforms at the level of DNA and RNA. However, the techniques for characterization of covalent modifications of the protein subunits themselves have not progressed with the same speed. Identification of a covalent modification often involves purification of substantial quantities of protein, and if the modification has not yet been characterized, several avenues of investigation may need to be pursued.

This chapter describes the characterization of two $G_o\alpha$ isoforms that differ at one of two Asn residues at the C terminus,[3,4] a modification resulting from deamidation of Asn to Asp. This modification was not previously known to exist in G protein α subunits. Description and discussion of the separation and purification of G protein α subunits by chromatographic techniques can be found elsewhere in this volume.[4a] Here, the use of sodium dodecyl sulfate (SDS)–polyacrylamide gels supplemented with urea will be described as a method for detection of heterogeneity in a population of α isoforms. Next, methods for localization of structural difference between G protein α isoforms will be described, involving limited tryptic digestion, chemical derivatization, and mass spectrometry techniques. Finally, sequencing techniques using both tandem mass spectrometry and Edman degradation will be described and compared for use in the identification of potential deamidation site(s).

[1] A. G. Gilman, *Biosci. Rep.* **15,** 65 (1995).

[2] J. D. Hildebrandt, *Biochem. Pharmacol.* **54,** 325 (1997).

[3] W. E. McIntire, K. L. Schey, D. R. Knapp, and J. D. Hildebrandt, *Biochemistry* **37,** 14651 (1998).

[4] T. Exner, O. N. Jensen, M. Mann, C. Kleuss, and B. Nurnberg, *Proc. Natl. Acad. Sci. U.S.A.* **96,** 1327 (1999).

[4a] J. Dingus, W. E. McIntire, M. D. Wilcox, and J. D. Hildebrandt, *Methods Enzymol.* **344,** [13] 2002 (this volume).

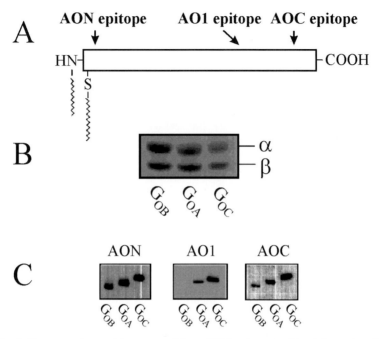

FIG. 1. Linear map of α_o, and immunoreactivity and SDS–polyacrylamide gel electrophoretic mobility of α_o isoforms. (A) Map of the α_o protein sequence, its known sites of modification by myristoyl (at the N terminus of the protein after removal of Met) and palmitoyl (via a thioester linkage at a Cys residue near the N terminus) groups, and the approximate immunoreactive sites of antibodies used in these studies. (B) Coomassie blue-stained gel of purified G protein heterotrimers (3 μg) separated on an 11% SDS–polyacrylamide gel. (C) Immunoblotting of G_o proteins with subunit antibodies AON (0.5 μg of each G_o protein), AO1 (1.5 μg of each G_o protein), and AOC (3 μg of each G_o protein) after transfer from an 11% SDS–polyacrylamide gel in the presence of 6 M urea. Adapted from W. E. McIntire, J. Dingus, K. L. Schey, and J. D. Hildebrandt, *J. Biol. Chem.* **273**, 33135 (1998).

Resolution of $G_o\alpha$ Subunits by Urea/SDS–PAGE

Theory

Many of the G protein α subunits can be resolved by traditional SDS–PAGE, depending on the percent acrylamide and cross-linker present. Methods described here utilize three site-specific antibodies, AON, AO1, and AOC (Fig. 1A), as well as Coomassie blue staining, to visualize the electrophoretic mobility of the α_o isoforms. Generation of antibodies is described elsewhere.[5] Closely related subunits, such as the α_o isoforms, migrate as one band at approximately 39 kDa on standard SDS–PAGE gels (Fig. 1B). The addition of urea to SDS–PAGE (urea/SDS–PAGE)

---

[5] W. E. McIntire, J. Dingus, K. L. Schey, and J. D. Hildebrandt, *J. Biol. Chem.* **273**, 33135 (1998).

was demonstrated by Scherer *et al.*[6] to separate multiple α_o isoforms, depending on the tissue investigated. Deamidated proteins could be expected to have a faster rate of migration when analyzed with urea/PAGE in the absence of SDS because an extra negative charge at the deamidation site would provide increased attraction to the anode. However, in the case of α_o isoforms, SDS has typically been present with urea, and the deamidated species migrates at a slower rate (Fig. 1C). This may indicate differences in detergent binding or protein unfolding.[7]

SDS–Polyacrylamide Gels with 6 M Urea

The procedure for analysis of G protein subunits by SDS–PAGE in the presence of 6 M urea is described here in some detail because such gels need to be made in the laboratory and are not available commercially. The methods described, however, are based, with modifications, on the traditional method of Laemmli.[8] These recipes are specific for a homemade gel apparatus using 14 cm $\times$ 17 cm glass plates with 0.75 mm spacers, but can of course be adjusted as needed depending on the apparatus. These gels require approximately 10 ml separating buffer and 3 ml stacking buffer.

Reagents. Acrylamide, bisacrylamide, sodium dodecyl sulfate (SDS), ammonium persulfate (APS), $N,N,N'N'$-tetramethylenediamine (TEMED), 2-mercaptoethanol, glycine, and pyronin Y are from Bio-Rad. (Hercules, CA). Tris-base, glycerol, and urea are from Sigma-Aldrich (St. Louis, MO). Isobutanol is from Fisher (Pittsburgh, PA).

Solutions

1. 30% Acrylamide stock solution (30% acrylamide 0.8% bisacrylamide) (100 ml):

   | Acrylamide | 30.0 g |
   | Bisacrylamide | 0.8 g |

   Dissolve in water to a final volume of 100 ml, and store at 4°. (*Note:* Acrylamide is a neurotoxin and is readily absorbed through the skin. Gloves should be worn when handling all acrylamide solutions, and caution should be taken not to inhale the power when preparing acrylamide solutions.)
2. 4X Separating Gel Buffer (100 ml): 1.5 M Tris-HCl, pH 8.8, and 0.4% SDS. (Final concentrations in the gel are 375 mM Tris-HCl, pH 8.8, and 0.1% SDS.)

[6] N. M. Scherer, M. J. Toro, M. L. Entman, and L. Birnbaumer, *Arch. Biochem. Biophys.* **259**, 431 (1987).

[7] T. E. Creighton, *J. Mol. Biol.* **137**, 61 (1980).

[8] U. K. Laemmli, *Nature* **227**, 680 (1970).

Tris–base	18.2 g
SDS	0.4 g

Dissolve in 80 ml water and adjust pH to 8.8 with concentrated HCl. Bring to a final volume of 100 ml, and store at room temperature.

3. 2X Stacking Gel Buffer (100 ml): 125 mM Tris-HCl, pH 6.8, 0.2% SDS. (Final concentrations in the gel are 62.5 mM Tris-HCl, pH 6.8, and 0.1% SDS.)

Tris–base	3.03 g
SDS	0.2 g

Dissolve in 80 ml water and adjust pH to 6.8 with concentrated HCl. Bring to a final volume of 100 ml and store at room temperature.

4. 12.5% Ammonium Persulfate (APS). Weigh out 125 mg APS and bring to 1 ml with water. Store at 4° no more than 1 week.

5. 2X Sample Buffer: 125 mM Tris-HCl, pH 6.8, 2% SDS, 20% glycerol, 0.002% pyronin Y. (Final concentration of the components in the sample will be 62.5 mM Tris-HCl, pH 6.8, 1% SDS, 10% glycerol, 0.001% pyronin Y, 2.5% 2-mercaptoethanol.)

Tris–base	1.51 g
SDS	2 g

Dissolve in 50 ml water, then adjust pH to 6.8 with concentrated HCl. Add 20 g glycerol, and bring to 95 ml with water. Add 2 mg pyronin Y and dissolve completely. Store Sample Buffer without 2-mercaptoethanol at room temperature. Just prior to use, add 50 μl 2-mercaptoethanol to 950 μl sample buffer.

6. Water-Saturated Isobutanol: Mix equal volumes (40ml each) of water and isobutanol in a 100 ml bottle. Shake, and then allow to separate. The upper layer contains water-saturated isobutanol.

Gel Preparation. Below are protocols for the preparation of urea/SDS–PAGE gels, using as an example a 10% separating gel. This percent gel is usually excellent for separating G protein α_o subunits, but a slightly higher (11%) or lower (9%) percentage may be better in certain instances, which must be determined empirically. Generally, even a 1% change in the amount of acrylamide can have a substantial effect on positions of the proteins on the gel.

1. Preparation of 10% Separating Gel (10 ml): Set up gel with two 14 cm × 17 cm glass plates and 0.75 mm spacers. Weigh out 3.6 g of urea, and add 2.5 ml of 4× Separating Gel Buffer, and 3.33 ml of 30% acrylamide stock solution. Mix gently until the urea in completely dissolved, and bring to 10 ml with water. (If the gel apparatus is not self-sealing, the spacers and

plates should be sealed with polymerized acrylamide. Remove 1 ml of 10% gel solution and add 4 μl TEMED and mix. Add 8 μl 12.5% ammonium persulfate and mix again. Quickly add the solution down each side of the gel apparatus, allowing it to diffuse between the plates and the spacer until about 5 mm of the acrylamide is in the bottom of the assembled apparatus between the plates. It will polymerize rapidly.) Initiate gel polymerization by adding 5 μl TEMED and mix gently but well. Add 120 μl 12.5% ammonium persulfate and mix again. Pour 8.5 to 9.0 ml solution immediately between the plates, and layer 1 ml of water-saturated isobutanol across the top of the gel. Allow to polymerize for at least 45 minutes. Gel is polymerized when a clear double interface forms between the isobutanol and the gel. Pour off the water-saturated isobutanol and wash the surface of the gel extensively until the smell of the isobutanol is completely gone.

2. Preparation of 5% Stacking Gel (10 ml): Insert sample comb for forming sample wells between the gel plates. Weigh out 2.16 g of urea; add 3 ml of 2× Stacking Gel Buffer and 1 ml of 30% acrylamide solution. Mix gently until the urea in completely dissolved, and bring to 6 ml with water. Initiate gel polymerization by adding 4 μl TEMED and mixing gently. Add 100 μl 12.5% ammonium persulfate and mix again. Pour solution carefully between the plates, making sure that any bubbles are removed.

Sample Preparation

Protein samples are mixed with sample buffer to attain the following concentrations of components: 62.5 mM Tris-HCl, pH 6.8, 1% SDS, 10% glycerol, 0.001% pyronin Y, and 2.5% 2-mercaptoethanol. The gel will look better after visualization if, before loading, all samples run are adjusted to the same buffer, salt, and detergent concentrations. Before loading into wells, the samples are heated at 95° for 5 min, then centrifuged briefly to remove condensation formed during heating.

Electrophoresis and Visualization of Protein

The upper and lower chambers of the apparatus used to run the gel are filled with electrode buffer containing 25 mM Tris, 192 mM glycine, and 0.1% SDS; this solution is not adjusted for pH, but the pH should be between 8.2 and 8.6. Gels are run at 150 V (constant voltage) until the indicator dye reaches the bottom of the gel. Several methods are available for visualization of protein, depending on the purity and quantity of the protein. These include staining of the gel itself with silver or Coomassie blue, and transfer of the gel to nitrocellulose followed by immunoblotting with G protein α specific antibodies[9] and staining with silver or Coomassie blue.

[9] W. E. McIntire, J. Dingus, M. D. Wilcox, and J. D. Hildebrandt, *J. Neurochem.* **73,** 633 (1999).

Discussion and Comments

The use of urea/SDS–PAGE was of great advantage in separating the modified forms of α_o. The degree of separation of α_{oA} and α_{oC}, given that they differ by only 1 Δ_a in mass, is surprising. We hypothesize that this is due to differences in detergent binding of the two proteins due to the Asp in α_{oC} (with an added negative charge), vs the Asn in α_{oA}. The fact that this difference is only apparent in the presence of urea suggests that neither protein is completely denatured in SDS alone. Alternatively, the urea may affect some aspect of the tertiary structure of the proteins that is different when Asp is in positions 346 or 347, compared to Asn. This difference in electrophoretic behavior is not dependent on the protein being intact, because it is also observed in proteolytic fragments containing the deamidation sites (see below). The degree of separation of α_{oA} and α_{oC} can be increased by running gels for a longer than normal time (i.e., by running the tracking dye off the end of the gel).

Limited Tryptic Digestion of G Protein α Subunits: Functional Conditions

Theory

G protein α subunits share a conformational "switch" that is nucleotide dependent. The active conformation occurs when the α subunit binds GTP, and this is maintained when the nonhydrolyzable GTP analog, GTPγS, is used. If the nucleotide is replaced with GDP, the α subunit reverts to an inactive conformation. These conformations can be distinguished biochemically by their susceptibility to tryptic cleavage.[10] In the active conformation with GTPγS bound, only the N terminus of the protein is vulnerable to proteolysis by trypsin (Fig. 2A). However, when GDP is bound, an additional cleavage site is exposed, resulting in the production of a 17 kDa C-terminal fragment, and an internal 25 kDa fragment (Fig. 2B). Antibodies recognizing the N and C termini of the proteins will specifically recognize these fragments (see Fig. 2) and can be used to generally localize the sites of modification of the protein. The conditions for the limited proteolysis of G proteins under activating and nonactivating conditions are described, as are procedures for evaluating the proteolytic products on SDS–PAGE/urea gels, and by mass spectrometry.

Trypsin Digestion

Reagents. Tris–base, EDTA, dithiothreitol (DTT), NaCl, guanosine 5'-diphosphate (GDP), and TPCK-treated trypsin are obtained from Sigma-Aldrich.

[10] J. W. Winslow, J. R. Van Amsterdam, and E. J. Neer, *J. Biol. Chem.* **261,** 7571 (1986).

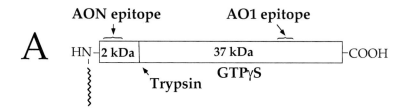

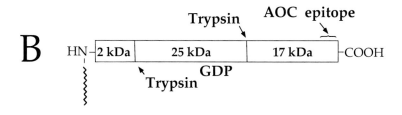

C

```
  1  MGCTLSAEER AALERSKAIE KNLKEDGISA AKDVKLLLLG AGESGKSTIV
 51  KQMKIIHEDG FSGEDVKQYK PVVYSNTIQS LAAIVRAMDT LGIEYGDKER
101  KADAKMVCDV VSRMEDTEPF SPELLSAMMR LWGDSGIQEC FNRSREYQLN
151  DSAKYYLDSL DRIGAADYQP TEQDILRTRV KTTGIVEIHF TFKNLHFRLF
201  DVGGQRSERK KWIHCFEDVT AIIFCVALSG YDQVLHEDET TNRMHESLML
251  FDSICNNKFF IDISIILFLN KKDLFGEKIK KSPLTICFPE YTGSNTYEDA
301  AAYIQAQFES KNRSPNKEIY CHMTCATDTN NIQVVFDAVT DIIIANNLRG
351  CGLY
```

FIG. 2. Location of trypsin cleavage sites of α_o in the presence of Mg^{2+} and GTPγS or GDP. Maps of the trypsin digestion patterns of α_o in the presence of Mg^{2+} and GTPγS (A) or GDP (B) showing the preferential sites of trypsin cleavage and the sites of antisera reactivity is illustrated in Fig. 1A. (C) Boxed residues are the sites of trypsin cleavage giving rise to the fragments in the presence of GDP or GTPγS. N346 and N347, the sites of deamidation of α_{oA} to produce α_{oC}, are underlined in the sequence. Adapted from W. E. McIntire, K. L. Schey, D. R. Knapp, and J. D. Hildebrandt, *Biochemistry* **37,** 14651 (1998).

Acetonitrile and *n*-propanol, 2-propanol, and TFA are obtained from Burdick & Jackson. Guanosine 5''-*o*-(3-thiotriphosphate) (GTPγS) and Thesit (dodecyl-polyethylene glycol ether) are from Roche Molecular Biochemicals (Mannheim, Germany).

Digestion Conditions. Purified heterotrimeric G_o protein or α_o subunit is incubated in buffer (20 mM Tris-HCl, pH 8.0, 0.1% Thesit, 1 mM EDTA, 1 mM DTT, 150 mM NaCl, 25 mM MgCl$_2$) containing either 1 μM GTPγS or 100 μM

GDP, and TPCK-treated trypsin (1 : 25 w/w for GTPγS or 1 : 50 w/w for GDP). Incubation times for samples with GTPγS bound α_o and GDP bound α_o are 30 and 45 min at 32°, respectively.

Analysis by Urea/SDS–PAGE

Digestion of α_o subunits is stopped by heating in sample buffer for 5 min. Separation of digests by urea/SDS–PAGE is performed as described above. Analysis of the tryptic fragments is based on the differential electrophoretic mobilities of the intact α_o isoforms. As illustrated in Fig. 3A, the difference in mobilities is retained

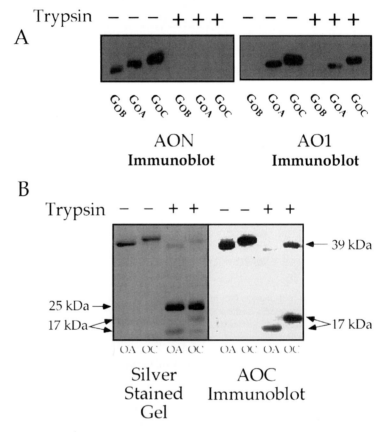

FIG. 3. Limited tryptic digest of G_o isoforms in the presence of GTPγS or GDP. (A) Immunoblot of a urea/SDS–polyacrylamide gel using the AON and AO1 antibodies, before and after trypsin digestion in the presence of Mg^{2+} and GTPγS as described. (B) Silver-stained urea gel and immunoblot of urea gel with AOC antiserum before and after trypsin digestion in the presence of Mg^{2+} and GDP as described. Adapted from W. E. McIntire, J. Dingus, K. L. Schey, and J. D. Hildebrandt, *J. Biol. Chem.* **273,** 33135 (1998).

in the 37 kDa fragment in the GTPγS digest; this fragment retains immunoreactivity with the AO1 antibody, but not the AON N-terminal antibody. After further tryptic digestion with GDP (Fig. 3B), the 17 kDa C-terminal fragment retains a difference in mobility, shown in the immunoblot with the AOC C-terminal antibody. Silver staining (Fig. 3B) demonstrates that the 25 kDa fragments have essentially the same mobility.

Antibodies specific to different regions of the α subunit are quite useful in these types of experiments. Immunoblotting simplifies the interpretation of stained gels, which can be complicated by the presence of other proteins, such as $\beta\gamma$ dimers, and fragments thereof.

Analysis by ESI-MS

Mass analysis by electrospray ionization-mass spectrometry (ESI-MS) of α_o isoforms and their limiting tryptic fragments is similar in theory to their analysis by urea/SDS–PAGE, i.e., comparison of the molecular weight of two proteins. However, the mass accuracy and resolution of ESI-MS allows direct comparison of observed mass with that predicted from known cDNA sequence. This allows direct evaluation of the presence of possible posttranslational modifications. The instrument used in these studies was a Finnigan LCQ ion trap mass spectrometer where the sample can be injected into the instrument as it elutes from an HPLC column. Highly purified α_o isoforms are separated on a 4.6×30 mm Aquapore phenyl column [PerkinElmer Instruments (Shelton, CT)] with a linear gradient of 90% (v/v) aqueous, 10% (v/v) acetonitrile with 0.1% (v/v) trifluoroacetic acid (TFA) (buffer A) going to 90% n-propanol, 10% (v/v) acetonitrile with 0.1% (v/v) TFA (buffer B) over 20 min at 0.5 ml/min. Absorbance is measured at 214 nm, with fractions collected at 1 min intervals. HPLC fractions are stored at $-20°$ for further analysis. Simultaneously, 5% of the HPLC effluent is analyzed in real time by the Finnigan LCQ ion trap mass spectrometer with an electrospray ionization (ESI) source. Figure 4 illustrates the ESI mass spectra and deconvoluted mass spectra (obtained with the manufacturer's analysis software) for α_{oA} and its deamidated analog, α_{oC}. Masses of both α_o isoforms are not significantly different from each other, or the predicted mass of the bovine α_o protein, assuming myristoylation of the N terminus following cleavage of the N-terminal Met.

In the case of limited tryptic proteolysis, digestions were performed immediately before separation by HPLC, with termination of the reaction achieved by dilution with HPLC buffer A. Tryptic fragments from GDP bound α_o are separated as described for the intact α_o isoforms, except the gradient is over 40 min and buffer B is 75% acetonitrile, 25% 2-propanol with 0.095% TFA. Predicted masses for the fragments are described in Fig. 5A, and the ESI and deconvoluted mass spectra for the 25 kDa and 17 kDa tryptic fragments of α_{oA} and α_{oC} are shown in Figs. 5A and B, respectively.

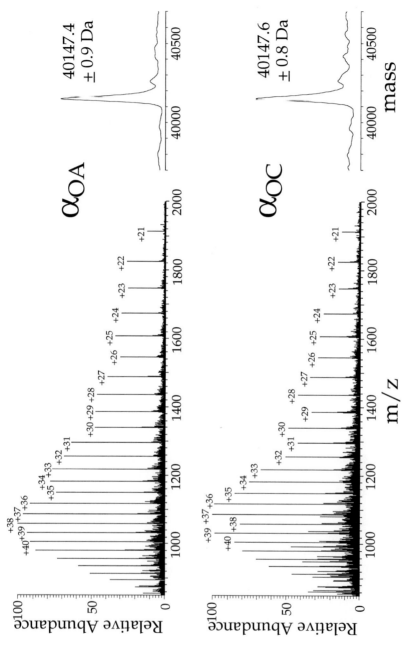

FIG. 4. Analysis of intact α_{oA} and α_{oC}. Intact α_o subunits are purified by HPLC and passed in-line to the LCQ mass spectrometer. ESI mass spectra obtained from the LCQ during the separation are shown for the α_{oA} and α_{oC} isoforms. At right are deconvoluted mass spectra with predicted masses of the observed ions. Calculated mass and S.E. are based on the labeled charge states in the spectra on the left. Adapted from W. E. McIntire, J. Dingus, K. L. Schey, and J. D. Hildebrandt, *J. Biol. Chem.* **273**, 33135 (1998).

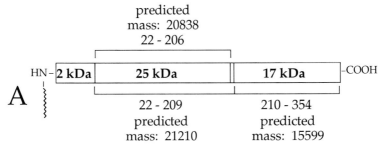

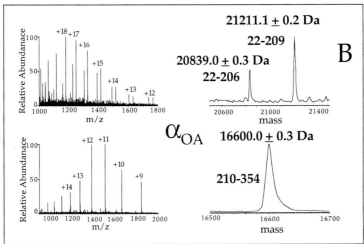

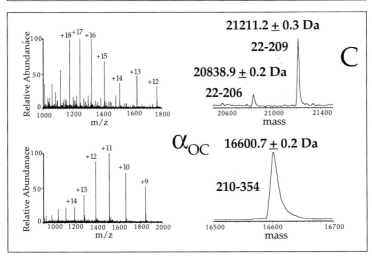

Discussion

Although no significant mass differences are observed between α_{oA} and α_{oC}, or their respective tryptic fragments, the data are critical, especially when taken in the context of the urea/SDS–PAGE experiments, which do demonstrate structural differences between the isoforms. The expected mass accuracy of the LCQ is ±0.01% or 2 Da for a 20 kDa protein. Since the mass difference resulting from deamidation is only 1 Da, the mass accuracy of the mass spectrometer used in this work (about 0.01%) is not sufficient to distinguish such modifications on proteins or large protein fragments. In effect, these results place upper limits on the mass difference of the observed structural differences in the two proteins. Therefore, the mass spectral data, when considered in light of the electrophoretic mobility differences, are consistent with deamidation as a structural difference between the 17 kDa tryptic fragments of α_{oA} and α_{oC}. These techniques demonstrate the utility of large predictable proteolytic fragments for eliminating regions of the α subunit that do not appear to have structural differences. The benefit is that the procedures outlined below can focus on a smaller, more manageable region of the α_o protein.

Proteolytic Digestion of G Protein α Subunits: Denaturing Conditions

Theory

After structural variation between G protein α subunits is localized to a specific region of the protein, identification of potential deamidation site(s) is facilitated by complete proteolytic digestion of the α_o isoforms. The smaller peptides produced from such digestions are more amenable to mass spectrometric analysis and sequencing. The complete digestion of the protein requires it to be denatured prior to proteolysis.

Method

Reagents. Tris, EDTA, DTT, NaCl, CaCl$_2$, sequencing grade chymotrypsin, and TPCK-treated trypsin are obtained from Sigma-Aldrich. Acetone, acetonitrile, *n*-propanol, 2-propanol, and TFA are obtained from Burdick & Jackson. Thesit is from Roche Molecular Biochemicals.

Fig. 5. ESI mass spectra from the LCQ analyses of digests of α_{oA} and α_{oC}. (A) Map of the α_o protein showing the predicted fragments generated by trypsin digestion in the presence of GDP and the corresponding predicted masses for the 25- and 17-kDa fragments (seen by SDS–PAGE). (B) Analysis of the two peaks from the α_{oA} separation. On the left are the ESI mass spectra showing all charge states of the fragments; on the right are deconvolution mass spectra with predicted masses based upon trypsin cleavage. (C) Analysis of the two peaks from the α_{oC} separation. On the left are the ESI spectra showing all charge states of the fragments; on the right, the deconvoluted mass spectra are shown. Average and S.E. of estimates are calculated from the multiple charge states in the ESI mass spectra. Adapted from W. E. McIntire, J. Dingus, K. L. Schey, and J. D. Hildebrandt, *J. Biol. Chem.* **273,** 33135 (1998).

Procedure. Tryptic digests were performed on 3 nmol of purified α_{oA} and α_{oC} isoforms, denatured by boiling for 5 min followed by acetone precipitation.[11] Pellets are resuspended with 150 μl of digestion buffer (20 mM Tris-HCl, pH 8.0, 1 mM EDTA, 1 mM DTT, 100 mM NaCl, 0.1% Thesit), containing TPCK-treated trypsin (1 : 50, w/w); the sample is incubated overnight at 32°.

For chymotryptic digests, 1.5 nmol of purified G$_o$ or α_o isoforms is precipitated with acetone and resuspended with 40 μl of digestion buffer (20 mM Tris-HCl, pH 8.0, 1 mM EDTA, 1 mM DTT, 10 mM CaCl$_2$, 0.1% Thesit) containing sequencing-grade chymotrypsin (1 : 50, w/w). Incubation time is 30 min at 25°.

Analysis of Peptides by ESI-MS

Separation of tryptic peptides was accomplished with a 4.6 × 30 mm Aquapore C$_8$ RP300 column (PerkinElmer Instruments), using a linear gradient of 90% water, 10% acetonitrile with 0.1% TFA (buffer A) going to 75% acetonitrile, 25% 2-propanol with 0.095% TFA (buffer B) over 75 min at 0.5 ml/min. The same protocol is used for the chymotryptic digests, with the gradient being run over 50 min.

Analysis is performed as described above (Limited Tryptic Digestion of G Protein α Subunits: *Analysis by ESI-MS*), except in some instances, HPLC fractions were collected at 30 sec intervals in order to analyze closely eluting peptides. In addition, mass spectra collected during HPLC separations are scanned for predicted masses of the tryptic peptides generated from the 17 kDa region of the α_o isoforms (Fig. 6A). Selected ion chromatograms with a window of 1 m/z unit are generated from these scans using 7-point boxcar smoothing (Figs. 6A and B). Since ESI-MS generates ions with varying numbers of charges, $[M + xH]^{x+}$, the selected ion chromatogram for any given peptide is typically displayed for the most abundant charge state. The molecular ion nomenclature is defined as $[M + xH]^{x+}$, where M is the mass of the peptide, and x is the number of protons, and therefore the charge, on the peptide. One important point is that mass alone cannot unambiguously identify a proteolytic peptide derived from a protein. For example, two or more peptide cleavage products from a protein may be similar in mass, or a contaminating protein may contain a similar peptide cleavage product. The identity of the peptides from the 17 kDa region of the α_o isoforms is confirmed by sequence analysis using ESI-MS/MS or Edman degradation, as discussed below.

From the data in Fig. 6, it is clear that over 90% of the predicted tryptic peptides from the 17 kDa region are present in both α_{oA} and α_{oC} digests, but there do not appear to be significant differences in the elution times of the peptides. Therefore, additional methods must be employed to probe for structural differences between the tryptic peptides from α_{oA} and those from α_{oC}.

[11] M. D. Wilcox, J. Dingus, E. A. Balcueva, W. E. McIntire, N. D. Mehta, K. L. Schey, J. D. Robishaw, and J. D. Hildebrandt, *J. Biol. Chem.* **270**, 4189 (1995).

A

P1: 210 - 243   KKWIHCFEDVTAIIFCVALSGYDQVLHEDETTNR*

P2: 244 - 278   MHESLMLFDSICNNKFFIDTSIILFLNKKDLFGEK*

   279 - 281   IKK

P3: 282 - 311   SPLTICFPEYTGSNTYEDAAAYIQAQFESK*

P4: 312 - (349) NR (+ P5)

P5: 314 - 349   SPNKEIYCHMTCATDTNNIQVVFDAVTDIIIANNLR*

   350 - 354   GCGLY

B

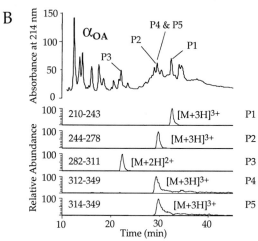

C

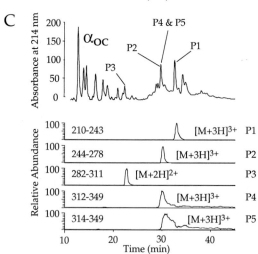

Chemical Derivatization of Peptides

Theory

Since deamidation of tryptic peptides of the size described above (Fig. 6) requires better than unit mass resolution, it can be challenging to identify such modifications with certainty even using ESI-MS. One approach to circumvent this difficulty is to use differential chemical derivatization to compare tryptic peptides from α_{oA} and α_{oC} . Methyl esterification is a chemical reaction in which a methyl group is covalently attached to acidic groups on peptides. Although the deamidation of Asn to Asp only results in a difference of 1 Da, the presence of the carboxyl group on Asp represents an additional site for methyl esterification, a 14 Da modification easier to resolve by mass spectrometry. Application of this approach presumes one knows the putative sequence of the proteins/peptides being analyzed, allowing identification of potential sites of methyl esterification. Analysis of the methyl esterification of the tryptic peptides of α_{oA} and α_{oC} can be performed by matrix-assisted laser desorption ionization (MALDI) mass spectrometry because of its high sensitivity and the predominance of singly charged ions in the mass spectrum.

Method

Reagents. Tris(2-carboxyethyl)phosphine (TCEP) is obtained from Pierce (Rockford, IL); α-cyano-4-hydroxycinnamic acid is from Sigma-Aldrich; and HCl is from Fisher. Acetonitrile and TFA are obtained from Burdick & Jackson.

Procedure. The procedure used for methyl esterification is based on the method described by Knapp.[12] In the case described here, HPLC fractions from the separation of the tryptic digests of α_{oA} and α_{oC} detailed above were analyzed (Fig. 6). Three separate fractions for each protein were collected: one containing peptide

---

[12] D. R. Knapp, *Methods Enzymol.* **193,** 314 (1990).

FIG. 6. Separation of tryptic peptides generated from α_{oA} and α_{oC} and identification of fractions containing fragments from the 17 kDa C-terminal domain. (A) Sequence of α_o broken into major peptides generated by tryptic cleavage and labeled P1, P2, P3, P4, and P5. Two short sequences (279–281 and 350–354) are not analyzed here. P4 and P5 differ only in the first two residues in P4 that are missing in P5. Asterisks indicate the sites of carboxyl groups that would be available for methyl esterification. (B) Data for the separation and characterization of α_{oA}. (C) Data for the separation and characterization of α_{oC}. Digests were generated and separated by HPLC as described. Separations were monitored by absorbance at 214 nm and by in ESI-MS on a Finnigan LCQ ion trap mass spectrometer. For each separation the 214 nm absorbance profile is shown along with the selected ion chromatograms for five major peptides generated from the 17 kDa C-terminal fragment, showing the abundance of the respective m/z ions during the elution. Predicted m/z values used for selected ion chromatograms ($\pm$ 0.5 m/z) are the following: 282–311 $[M+2H]^{2+} = 1673.9$, 244–278 $[M+3H]^{3+} = 1385.3$, 312–349 $[M+3H]^{3+} = 1437.3$, 314–349 $[M+3H]^{3+} = 1347.2$, and 210–243 $[M+3H]^{3+} = 1328.5$. Adapted from W. E. McIntire, K. L. Schey, D. R. Knapp, and J. D. Hildebrandt, *Biochemistry* **37,** 14651 (1998).

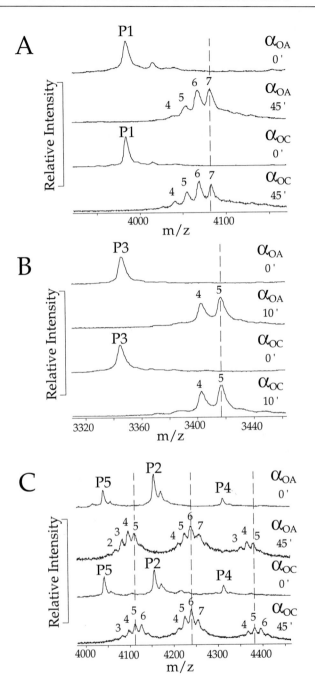

P1, one containing peptide P3, and one containing a mixture of peptides, P2, P4, and P5. Samples are lyophilized, reduced with 1.5 μl 100 mM aqueous TCEP for 20 min at room temperature, and lyophilized again. Peptides are esterified with 2 N methanolic HCl, prepared by the dropwise addition of 150 μl of acetyl chloride into 1 ml of ice-cold methanol while stirring. Precautions, such as the use of a fume hood, eye protection, and protective clothing, should be taken in the preparation of methanolic HCl, as the reaction can be violent. Fresh solutions should be prepared daily. Approximately 1.5 μl of 2 N methanolic HCl is added to the dried samples, which are incubated at room temperature for various times, and then lyophilized. Typical reaction times are on the order of 30–60 min, particularly if the reaction is not allowed to go to completion.

Analysis by MALDI-MS

Mass analysis is performed using a Voyager-DE MALDI mass spectrometer (PerSeptive Biosystems). Methyl esterification samples are solubilized in 1 μl of n-propanol : water (1 : 1), containing 0.1% TFA and mixed with 1 μl of 50 mM α-cyano-4-hydroxycinnamic acid (Aldrich), in 70% acetonitrile, 0.1% TFA. Samples are spotted on a sample plate and allowed to crystallize at room temperature. Masses are internally calibrated by including peptides or proteins of known mass (renin substrate, 1760.1 Da; cytochrome c, 12,361 Da) with the samples. Typically 256 laser shots are averaged to produce a mass spectrum.

In Fig 7. are shown the MALDI mass spectra of the derivatized peptides from the digest analyzed in Fig. 6. That experiment analyzed digests of the 17 kDa region of α_{oA} and α_{oC}. These digests contained five identifiable peptides. Their sequences are given in Fig 6A, and potential sites of methyl esterification are marked with an asterisk. These peptides have either five (peptides P3, P4 and P5) or six (peptides P1 and P2) potential sites of esterification. The conditions described above do not completely methyl esterify the peptides, but instead lead to a population of peptides with different numbers of methyl groups added (Fig. 7). It is apparent in Fig. 7 that the peptides derived from α_{oA} and α_{oC} are the same for P1 (Fig. 7A), P2 (Fig. 7C), and P3 (Fig. 7B), but differ for P4 and P5 (Fig. 7C). Specifically, in Fig. 7C there are additional peaks in the α_{oC} spectrum compared to the α_{oA} spectrum resulting from the derivatization of peptides P4 and P5, but not P2. These extra peaks are consistent with a deamidation site on the peptide

FIG. 7. Methyl esterification of tryptic peptides generated from the 17 kDa fragments of α_{oA} and α_{oC}. (A) MALDI mass spectra of P1 peptides from α_{oA} and α_{oC} before and after (45 min) methyl esterification. Numbers above peaks indicate the number of methyl groups that were incorporated into each peptide. (B) Same as in (A), but for P3. (C) Same as in (A), but for P2, P4, and P5, which were all recovered in the same fraction. Dotted lines align the methylated peptides with the largest predicted number of derivatization sites. Adapted from W. E. McIntire, K. L. Schey, D. R. Knapp, and J. D. Hildebrandt, *Biochemistry* **37,** 14651 (1998).

comprising residues 314–349 of α_{oC}. Although this technique identifies a peptide of α_{oC} with a potential deamidation site, it does not differentiate which of the five Asn or one Gln residues has the modification.

Notes and Discussion

Direct analysis of deamidated peptides by MALDI mass spectrometry can be done with mass spectrometers capable of very high accuracy and resolution. For example, a deamidated peptide from an α_o isoform was demonstrated to have a 1 Da increase in mass using a reflector type MALDI time-of-flight mass spectrometer.[4] Although most modern mass spectrometers can resolve a 1 Da difference in smaller peptides under optimal conditions, chemical derivatization can be a useful tool for larger peptides suspected to be deamidated.

Several potential problems exist with this technique. Although absolutely pure samples are not required, the number of molecular species is increased after methyl esterification, which may complicate spectra that contain several masses in the same m/z range. Also, residues such as Met are prone to oxidation, which increases the mass of a peptide by 16 Da; this mass difference can be confused with the addition of a methyl group (14 Da). In the analysis in Fig. 7, the P2 peptide has six potential sites of methyl esterification, but seven sites are observed for peptides from either α_{oA} or α_{oC}. The most likely explanation for this is that the Met at the N terminus of the P2 peptide was oxidized in either peptide. Oxidation can be limited with the use of a reducing agent such as TCEP. Finally, the esterification reagent can convert amides to methyl esters (although at a much slower rate than for carboxylic acids). Therefore the time course of the experiment should be monitored and comparisons made at the same time point. In the case described here, this concern was minimized by using reaction times and conditions at which the reaction was incomplete.

Sequencing of Deamidated Peptides

Theory

Tandem mass spectrometry (MS/MS) and Edman degradation are the two most widely used techniques for obtaining sequence information from proteins. This information is most easily obtained from small peptides (such as proteolytic cleavage fragments) from the region of interest in the protein. Peptides in the range of 500–2000 Da are optimal. Separation of these peptides is also desirable, for example, by HPLC. In MS/MS, a peptide is ionized and selected by its m/z ratio, and the selected mass is fragmented along the peptide backbone by collision with an inert gas. This produces fragment ions with charges at the N terminus, which are called b ions, and with charges at the C terminus, called y ions.[13] The pattern of b and y ions in a tandem mass spectrum can be interpreted to obtain the amino acid sequence of a peptide. In Edman degradation, the N-terminal residue of a peptide is derivatized and hydrolyzed, then analyzed by HPLC. This process is repeated as each N-terminal residue is disclosed by the hydrolysis.

One of the advantages of MS/MS is the ability to characterize covalent modifications of peptides in terms of mass, whereas Edman degradation in many instances is blocked by covalent modifications. An excellent example of this is isoaspartic acid. Depending on the mechanism of deamidation of Asn, isoaspartic acid may be a significant product along with Asp.[14] Because of the structural differences between these two residues, sequencing by Edman degradation is blocked by isoaspartic acid, but not by Asp. However, Edman degradation has an advantage in its ability to more easily differentiate amino acids of the same nominal mass such as Leu/Ile (113 Da) and Lys/Gln (128 Da). The similar masses of these residues may lead to ambiguity with the interpretation of MS/MS data. Ideally, both of these techniques can be used to complement one another in the analysis of deamidated peptides.

Analysis of C-Terminal Chymotryptic Peptides by HPLC and MS

Following the procedures described above, chymotryptic fragments of α_{oA} and α_{oC} were separated by HPLC; both absorbance at 214 nm and current for selected ions are recorded (Figs. 8A and B). Although there are many peaks in the HPLC chromatograms in Fig. 8A, selected individual ion currents can be extracted from mass spectra termed selected ion chromatograms (SIC). The selected ion chromatograms in Fig. 8B are specific for the C-terminal peptide 337–354. This specific screening in complex mixtures of peptides is a major advantages of mass spectrometry. Two peaks for the C-terminal peptide from α_{oC} in Fig. 8B are evident; however, this heterogeneity is clearly not evident in the UV absorbance spectrum. Mass spectra in Fig. 8C of the C-terminal peptides from α_{oA} and α_{oC} also illustrate that both peaks from α_{oC} in Fig. 8B are approximately 0.5 m/z units larger than the peak from α_{oA} in Fig. 8B. Since the m/z values are for ions with two charges, denoted by the nomenclature $[M + 2H]^{2+}$ in the legend of Fig. 8, the resultant mass difference between the peptides from α_{oC} and the peptide from α_{oA} is 1 Da. This supports deamidation as the difference between α_{oA} and α_{oC}.

ESI-MS/MS

Peptides from α_{oA} and α_{oC} were analyzed on a Finnigan LCQ ion trap mass spectrometer with a standard ESI source either directly from the HPLC separation described above, or by concentration of HPLC fractions and performing nanospray. The nanospray source was constructed in-house and utilized a pulled

---

[13] K. Biemann, *Annu. Rev. Biochem.* **61**, 977 (1992).
[14] H. T. Wright, *Protein Eng.* **4**, 283 (1991).

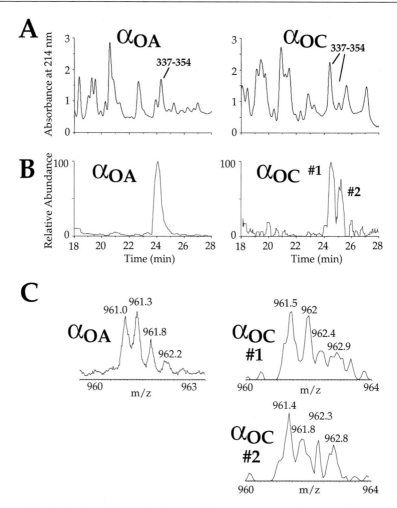

FIG. 8. Analysis of the C-terminal chymotryptic peptides of α_{oA} and α_{oC} by HPLC and LCQ mass spectrometry. (A) Separation of chymotryptic digests of α_{oA} and α_{oC} showing the absorbance at 214 nm; indicated are peaks corresponding to the elution of the C-terminal peptide 337–354. (B) Selected ion chromatograms for the predicted monoisotopic $[M + 2H]^{2+}$ ion for 337–354 for α_{oA} ($m/z = 961.0$) and the $[M + 2H]^{2+}$ ion for 337–354 for α_{oC} ($m/z = 961.5$), considering the one deamidation site predicted from Fig. 2. Note the resolution of the 337–354 peptide into two peaks, #1 and #2, for α_{oC}. (C) Isotopic cluster of the $[M + 2H]^{2+}$ ions for peaks in B: α_{oA} and α_{oC} #1 and α_{oC} #2. Note that the isotopic cluster is shifted 0.5 m/z units higher in α_{oC} for both peak #1 and peak #2, consistent with one deamidation site. Adapted from W. E. McIntire, K. L. Schey, D. R. Knapp, and J. D. Hildebrandt, *Biochemistry* **37**, 14651 (1998).

glass capillary, situated approximately 1 mm from the heated metal capillary, with an applied voltage of 1500 V (see elsewhere in this volume).[15] The capillary was loaded with 1 to 2 μl of sample solubilized in 47% water/47% methanol/ 6% acetic acid, and flow (typically) was initiated by application of high voltage, at a flow rate of 10–100 nl/min. Tandem mass spectrometry was achieved through the selection and fragmentation of the precursor ion of interest. Tandem mass spectra were then collected for the fragment ions. The precursor ion was selected with a window of 2 m/z units. Predicted m/z values were generated with MacBioSpec software v. 1.0.1 (PE SCIEX Instruments, 1992, Thornhill Ontario, Canada).

Sequence data from α_{oA} are consistent with the predicted sequence including two Asn residues located at 346 and 347 in the α_{oA} sequence (Fig. 9A). In contrast, sequence data from peaks #1 and #2 of the α_{oC} digest in Fig. 8B identify Asp residues at positions 346 and 347, respectively, instead of Asn residues, with no other difference from the analogous peptide from α_{oA}. This sequence difference is reflected in the 1 Da larger observed mass beginning with ions b_{10} and y_9 in the MS/MS data for the α_{oC} peptide #1 (Fig. 9B), compared to the α_{oA} peptide (Fig. 9A). These ions implicate Asn-346 as the deamidation site in α_{oC} peptide #1. Alternatively, the MS/MS data for α_{oC} peptide #2 (Fig. 9C) describe a b_{10} ion one Da lower than the b_{10} ion in α_{oC} peptide #1 (Fig. 9B), suggesting residue 346 in α_{oC} peptide #2 is the predicted Asn. In addition, ions beginning with b_{12} and y_9 in α_{oC} peptide #2 are 1 Da higher than the α_{oA} peptide, implicating residue 347 as the deamidation site in α_{oC} peptide #2. These data characterize the identity of α_{oC} as resulting from conversions of Asn-346 or Asn-347 in α_{oA} into Asp residues in α_{oC}.[3]

Edman Degradation

Aliquots of HPLC fractions from the separation of chymotryptic digests of α_o isoforms (above) were also sequenced by Edman degradation on an ABI 494 Protein Sequencer. In preparation for peptide sample, a glass fiber filter is coated with BioBrene (Applied Biosystems), and two cycles are run to remove any free amino acids from the filter. Approximately 100 μl of HPLC sample is applied to the filter in 50 μl increments, and dried under nitrogen. The filter is then introduced into the instrument for sequence analysis. After cleavage of each individual residue, the absorbance of the 3-phenyl-2-thiohydantoin (PTH) amino acid derivative is measured during separation by HPLC. Chromatograms for each cycle are corrected for background absorbance. Integration of peak area for each derivatized residue is performed after baseline is adjusted to the base of all peaks in the chromatogram.

Sequence data summarized in Fig. 10 show Asp at residues 346 and 347 of α_{oC} peptides #1 and #2, respectively, in agreement with the MS/MS data. One characteristic of Edman data is the tailing off of signal from the previous cycle.

[15] K. L. Schey, M. Busman, L. A. Cook, H. E. Hamm, and J. D. Hildebrandt, *Methods Enzymol.* **344,** [41] 2002 (this volume).

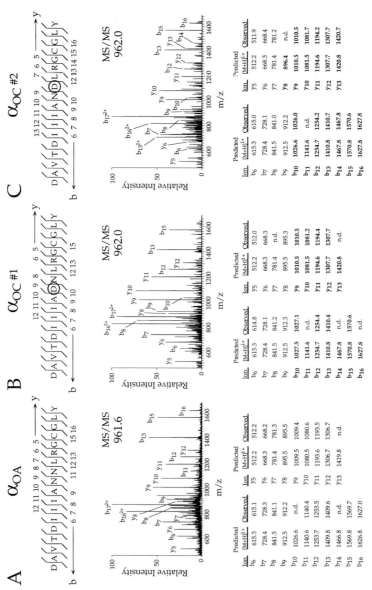

A αoA — DAVTDIIIANNLRGCGLY

MS/MS 961.6

Ion	Predicted [M+H]+	Observed
b6	615.3	615.1
b7	728.4	728.3
b8	841.5	841.1
b9	912.5	912.2
b10	1026.6	n.d.
b11	1140.6	1140.4
b12	1253.7	1253.5
b13	1409.8	1409.6
b14	1466.8	n.d.
b15	1569.8	1569.7
b16	1626.8	1627.0

Ion	Predicted [M+H]+	Observed
y5	512.2	512.1
y6	668.3	668.2
y7	781.4	781.3
y8	895.5	895.5
y9	1009.5	1009.4
y10	1080.5	1080.6
y11	1193.6	1193.5
y12	1306.7	1306.7
y13	1419.8	n.d.

B αoC #1 — DAVTDIIIADNLRGCGLY

MS/MS 962.0

Ion	Predicted [M+H]+	Observed
b6	615.3	614.8
b7	728.4	728.1
b8	841.5	841.2
b9	912.5	912.3
b10	1027.5	1027.1
b11	1141.6	n.d.
b12	1254.7	1254.4
b13	1410.8	1410.4
b14	1467.8	n.d.
b15	1570.8	1570.6
b16	1627.8	

Ion	Predicted [M+H]+	Observed
y5	512.2	512.0
y6	668.3	668.3
y7	781.4	n.d.
y8	895.3	895.3
y9	1010.5	1010.5
y10	1081.2	1081.2
y11	1194.4	1194.4
y12	1307.7	1307.7
y13	1420.8	n.d.

C αoC #2 — DAVTDIIIANDLRGCGLY

MS/MS 962.0

Ion	Predicted [M+H]+	Observed
b6	615.3	615.0
b7	728.4	728.1
b8	841.5	841.0
b9	912.5	912.2
b10	1026.6	1026.0
b11	1141.6	n.d.
b12	1254.7	1254.2
b13	1410.8	1410.7
b14	1467.8	1467.8
b15	1570.8	1570.6
b16	1627.8	1627.8

Ion	Predicted [M+H]+	Observed
y5	512.2	511.9
y6	668.3	668.4
y7	781.4	781.2
y8	896.4	n.d.
y9	1010.5	1010.5
y10	1081.5	1081.7
y11	1194.6	1194.2
y12	1307.7	1307.7
y13	1420.8	1420.7

FIG. 9. Sequencing by tandem mass spectrometry of the C-terminal peptides obtained from chymotrypsin digests of αoA and αoC. MS/MS analysis of the 337–354 peptide for αoA and αoC. (A) MS/MS spectra of the 337–354 peptide from αoA described in Fig. 8. (B) Same as (A), but for peak #2 from αoC. (C) Same as (A), but for peak #1 from αoC. Masses for the predicted and observed b and y ions are illustrated for αoA, αoC peak #1, and αoC peak #2. The observed b and y ions in αoC peak #1 and αoC peak #2 are compatible with deamidation at Asn-346 and Asn-347, circled in the sequences above. n.d. indicates not determined. Adapted from W. E. McIntire, K. L. Schey, D. R. Knapp, and J. D. Hildebrandt, Biochemistry **37**, 14651 (1998).

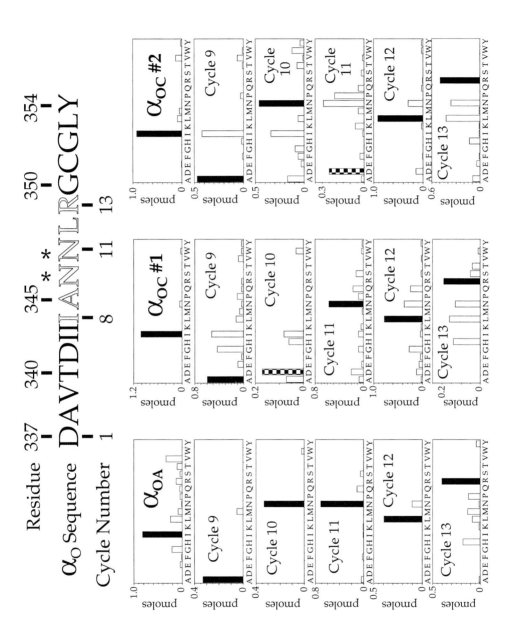

For example, in cycle 11 of α_{oC} peptide #2, there is a relatively high level of Asn. It is, however, lower than the level from cycle #10. In addition, levels of Asp are almost absent in cycle #10 of α_{oC} peptide #2, so the interpretation would be that the Asp signal in cycle #11 is significant.

Notes and Discussion

From the results presented here, there are several potential pieces of data to support or establish sites of deamidation as posttranslational modifications of proteins. The first is altered electrophoretic mobility in the presence of urea, although this altered behavior may be protein specific and dependent on conformational effects on the protein affecting detergent binding or other properties. The second is altered mass detected by ESI of intact proteins or isolated fragments. The associated 1 Da change may likely be difficult to document, however, unless fragments of optimal size are characterized. The third is altered susceptibility to chemical derivatization of acidic groups produced by deamidation, such as susceptibility to methyl esterification. This leads to mass change associated with modification, easily detected by many mass spectrometry techniques. This approach is potentially complicated by other modifications with similar mass increases, such as oxidation, or potential derivatization of other residues, such as Asn residues themselves.

Finally, and ultimately, deamidation can best be verified by direct sequencing of peptide fragments by either tandem mass spectrometry, or by Edman degradation. Ideally there is complementary information in obtaining data by both approaches. The derivatized amino acids generated by Edman sequencing for Asn and Asp are easily differentiated. However, carryover of amino acids sequenced from residues before or after the site of deamidation may complicate interpretation. In addition, the presence of an isoaspartate residue may prevent sequencing through the site of deamidation. These results can be compared to those obtained by sequencing by tandem mass spectrometry. In this case, precise data are required to differentiate Asn from Asp, which differ by only 1 Da. On the other hand, both aspartate and isoaspartate will be observed by this method, giving data complementary to Edman sequencing.

FIG. 10. Sequencing by Edman degradation of the C-terminal peptides obtained from chymotrypsin digests of α_{oA} and α_{oC}. Across the top is the predicted sequence of α_{oA} for residues 337–354, corresponding to the chymotryptic peptide isolated in the experiment in Fig. 8. Fractions containing this peptide from α_{oA} or two peptides with the mass of deamidated analogs recovered from the digest of α_{oC} were sequenced by Edman degradation as described. Shown in the bar graphs are the amino acid signals at each cycle corrected for background and individual signal intensity of the derivatized amino acids. Data for cycles 8–13 are shown the corresponding residues in α_{oA} are denoted in the sequence at the top by outlined letters. Black bars show the corrected signal intensity for the expected residue from the predicted sequence in each cycle. Hatched bars show the signals for Asp where an Asn was expected in the two α_{oC} peptides. Asterisks in the sequence above mark the sites of Asn variably identified in the sequencing records. Adapted from W. E. McIntire, K. L. Schey, D. R. Knapp, and J. D. Hildebrandt, *Biochemistry* **37,** 14651 (1998).

Acknowledgments

This work was supported in part by NIH Grant NS38534 and by institutional support of the Mass Spectrometry and Protein Sequencing and Peptide Synthesis Facilities of the Medical University of South Carolina. The authors thank Ms. Rebecca Ettling for Edman sequencing work on this project.

[35] Determining G Protein Heterotrimer Formation

By YONGMIN HOU, VANESSA CHANG, and N. GAUTAM

Heterotrimeric G proteins ($\alpha\beta\gamma$) relay signals from cell surface receptors to intracellular targets. The formation of a heterotrimer between the α and $\beta\gamma$ subunits is essential for efficient G-protein interaction with a receptor.[1] The affinity of an α subunit type for a $\beta\gamma$ complex can therefore have a direct impact on the efficiency with which a receptor will activate a heterotrimer made up of those subunit types. Although results from *in vitro* binding experiments indicate that β_5 selectively binds to α_q but not other types of α subunits,[2] such differences have not in general been demonstrated in direct binding assays.[3] The affinity differences among different combinations of α and $\beta\gamma$ subunits may, however, be relatively small but with dramatic consequences for receptor activation and modulation of particular signaling pathways.

Three different methods have been previously utilized for examining G protein heterotrimer formation. (i) Direct protein binding assays as mentioned above: it is possible to examine direct binding only in cases where one of the proteins is tagged or an antibody is available for coimmunoprecipitation. A disadvantage of this method is that relatively high concentrations of the proteins need to be used to facilitate detection. Subtle differences in affinity may therefore remain unrevealed. (ii) The most widely used assay is based on the enhancement by the $\beta\gamma$ complex of pertussis toxin-catalyzed ADP-ribosylation of the α subunit.[4] Among the disadvantages of this method are that (a) the precise mechanism for this reaction remains unknown; (b) the concentrations of subunits required for this assay are high (micromolar range); and (c) this method cannot directly assess the relative affinities of the $\beta\gamma$ complex for the α subunit since $\beta\gamma$ complex function is catalytic.[4]

[1] N. Gautam, G. B. Downes, K. Yan, and O. Kisselev, *Cell Signal.* **10,** 447 (1998); B. K. Fung, *J. Biol. Chem.* **258,** 10495 (1983).

[2] J. E. Fletcher, M. A. Lindorfer, J. M. DeFilippo, H. Yasuda, M. Guilmard, and J. C. Garrison, *J. Biol. Chem.* **273,** 636 (1998).

[3] O. Kisselev and N. Gautam, *J. Biol. Chem.* **268,** 24519 (1993).

[4] N. Ueda, J. A. Iniguez-Lluhi, E. Lee, A. V. Smrcka, J. D. Robishaw, and A. G. Gilman, *J. Biol. Chem.* **269,** 4388 (1994).

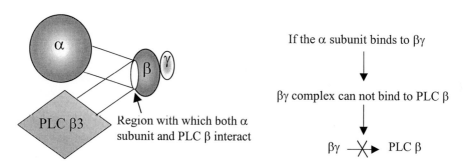

FIG. 1. Diagrammatic representation of the conceptual basis for measuring G protein heterotrimer formation using PLC-β. The β subunit has overlapping sites for interaction with the α subunit and PLC-β. Thus heterotrimer formation between the α subunit and the $\beta\gamma$ complex prevents $\beta\gamma$ complex interaction with PLC-β, leading to the inhibition of $\beta\gamma$-stimulated PLC enzyme activity.

Consistent with the known pitfalls of this assay, experiments with this procedure have shown that α_o does not discriminate between $\beta_1\gamma_1$ and $\beta_1\gamma_2$ in an ADP ribosylation assay but does so in a more sensitive assay based on α subunit GTPase inhibition (below).[4] (iii) A third approach for measuring G protein heterotrimer formation is the GTPase inhibition assay, which relies on the ability of the $\beta\gamma$ complex to slow down the steady-state rate of GTP hydrolysis by an α subunit. This procedure seems to work efficiently with αs but not with α_i or α_o unpublished results from our laboratory [Y. Hou and N. Gautam (2000)] and Ref. 4.

A sensitive and reliable method for determining G protein heterotrimer formation can therefore be a great aid in analyzing G protein function. The approach described below was based on the inhibition of $\beta\gamma$ stimulation of phospholipase C-β isozymes (PLC-β) by the α subunit. This inhibition occurs because sites on the $\beta\gamma$ complex which bind the α subunit overlap with those which bind PLC-β[5] (Fig. 1). The method is simple, highly sensitive (requiring nanomolar subunit concentration), and quantitative. It can be directly used for $\beta\gamma$ subunit interaction with α types-$\alpha_{i/o}$, α_s, and α_{13} families that cannot activate PLC-β enzyme. It could also be used for measuring $\beta\gamma$ interaction with G_q if a mutant form of PLC-β is used, as will be discussed later.

[^{35}S]GTPγS Binding Assay of Purified Recombinant α Subunits

The expression and purification of recombinant G protein α_{i2} and α_o subunit from bacteria are described in Ref. 6. The purified α_i or α_o subunit is over

[5] C. E. Ford, N. P. Skiba, H. Bae, Y. Daaka, E. Reuveny, L. R. Shekter, R. Rosal, G. Weng, C. S. Yang, R. Iyengar, R. J. Miller, L. Y. Jan, R. J. Lefkowitz, and H. E. Hamm, *Science* **280**, 1271 (1998).

[6] M. E. Linder, C. Kleuss, and S. M. Mumby, *Methods Enzymol.* **250**, 314 (1995).

90% pure as assessed by laser densitometric scanning of Coomassie blue stained proteins on sodium dodecyl sulfate (SDS) gels. [^{35}S]GTPγS incorporation by G protein α subunit is used to determine the functional proportion of α subunit and its concentration. A typical reaction mixture in 20 μl of GTPγS binding assay contains 20 mM Na–HEPES (pH 8.0), 100 mM NaCl, 1 mM EDTA, 1 mM dithiothreitol (DTT), 100 nM [^{35}S]GTPγS ($\sim$200,000 cpm/min/assay), 0.5 mg/ml bovine serum albumin (BSA), and the required amount of MgCl$_2$. The αi subunit binds GTPγS much slower than αo and requires a higher concentration of Mg^{2+} ($\geq$25 mM).[7] We generally use 5 mM MgCl$_2$ for αo and 30–50 mM MgCl$_2$ for αi. The mixture is incubated at 30° for 60 min. The reaction is terminated by adding 100 μl of ice-cold stopping solution containing 100 μM GTP, 20 mM Na–HEPES (pH 8.0), 100 mM NaCl, 1 mM EDTA, and 1 mM DTT. The samples are then filtered through nitrocellulose membranes (HAWP 025, Millipore, Bedford, MA) under vacuum. The membranes are immediately washed with 10 ml ice-cold stopping solution and bound radioactivity determined in a liquid scintillation counter.

Expression and Purification of Recombinant G Protein $\beta\gamma$ Complex from Baculovirus–Insect Cell System

The method described here is used for purification of G protein $\beta\gamma$ complex from 2 liter Sf9 (spodoptera frugipisda ovary) cells. It is modified from Ref. 8. The procedure is performed at 4° unless otherwise specified. To purify recombinant $\beta\gamma$ proteins from Sf9 cells, we mix hexahistidine (His$_6$)-αi, β, and γ recombinant baculoviruses and add to 2 liter Sf9 cells (1–1.5 × 10^6/ml). Harvest cells 48 hr later by centrifugation at 2000g for 5 min and resuspend the cells in 120 ml ice-cold lysis buffer (Table I) and fresh protease inhibitor cocktail [21 μg/ml phenylmethylsulfonyl fluoride (PMSF), 21 μg/ml N-tosyl-L-phenylalanine chloromethyl ketone, and 21 μg/ml sodium-p-tosyl-L-lysine chloromethyl ketone]. Lyse cells by nitrogen cavitation (Parr bomb) at 800 psi for 15 min. Centrifuge cell lysates at 2000g for 10 min to remove cell debris and nuclei. The supernatants are centrifuged at 100,000g for 30 min and the resultant pellets are resuspended using a homogenizer in 75 ml of wash buffer (Table I) containing fresh protease inhibitor cocktail as above. Repeat the centrifugation step at 100,000g as above. Freeze membrane pellets in liquid nitrogen for storage at −80°. Next day thaw the membranes and resuspend them in 72 ml of wash buffer containing fresh protease inhibitor cocktail. To lyse the membrane protein, add 8 ml of 10% sodium cholate to it and rotate the mixture for 1 hr at 4°. Centrifuge at 1000,000g for 30 min. Dilute the supernatants (membrane extract) with 400 ml buffer A (Table I) and add 2 ml Ni-NTA beads. Rotate the mixture for 1 hr at 4°. Collect the beads by centrifugation for 2 min at 500g and remove most of supernatant. Pack the beads into a column

[7] D. J. Carty and R. Iyengar, *Methods Enzymol.* **237,** 38 (1994).
[8] T. Kozasa and A. G. Gilman, *J. Biol. Chem.* **270,** 1734 (1995).

TABLE I
BUFFERS FOR PURIFICATION OF G PROTEIN $\beta\gamma$ SUBUNITS

Buffer	Lysis	Wash	A	A2	A3	C	Q
HEPES–NaOH (pH 8.0) (mM)	50	50	20	20	20	20	20
NaCl (mM)	100	50	100	300	50	50	100
MgCl$_2$ (mM)	3	3	1	1	1	1	3
2-Mercaptoethanol (mM)	10	10	10	10	10	10	10
EDTA (mM)	0.1						1
C12E10 (%)				0.5	0.5		
Cholate (%)						0.2	1
Imidazole (mM)						10	250
CHAPS (%)							0.7
GDP (μM)	10	50	50	50	50	50	

(2 ml) and wash the column with 40 ml of buffer A2 (Table I). Wash the column with 10 ml of buffer A3 (Table I) before it is warmed to room temperature for 15 min. Wash the column with additional 10 ml buffer A3 prewarmed at 30°. Elute the $\beta\gamma$ protein with 10 ml buffer E (20 mM HEPES–NaOH at pH 8.0, 0.1 mM EDTA, 50 mM MgCl$_2$, 10 mM 2-mercaptoethanol, 50 mM NaCl, 1% cholate, 30 μM AlCl$_3$, 10 mM NaF, 50 μM GDP) which is prewarmed at 30° and collect 1 ml fractions. If desired, His$_6$-α_i can be also eluted with buffer C (Table I) in 1 ml fractions. Analyze the fractions through SDS–PAGE followed by Coomassie blue staining. In some cases where certain nonspecific proteins were observed, additional Ni-NTA beads were added to remove them. Pool the peak fractions of $\beta\gamma$ protein and dialyze them overnight in buffer Q (Table I). Concentrate the dialyzed protein using Centricon columns (YM30, Amicon, Danvers, MA) and quantify the protein using densitometry scanning. Purified $\beta_1\gamma_2$ protein with this procedure is usually over 95% pure.

$\beta\gamma$ Complex Stimulated PLC-β Activity

To examine if the purified $\beta\gamma$ complex is functional, we determined its activity on one common effector, PLC-β_3, basically using a previously published procedure.[9] PLC-β_3 catalyzes hydrolysis of the lipid phosphatidylinositol-4,5-bisphosphate (PIP2) to inositol-1,4,5-triphosphate (IP3) and diacylglycerol. Radiolabeled lipid substrate [^{3}H]PIP$_2$, is mixed with PLC-β3 and $\beta\gamma$ complex. Reaction is initiated by addition of calcium and incubation is at 30°. Reaction is stopped by addition of trichloroacetic acid. Addition of BSA to this mixture precipitates proteins and lipids. The product [^{3}H]IP$_3$ is soluble and remains in the supernatant. [^{3}H]IP$_3$ production is measured by counting radioactivity in the supernatant.

[9] V. Romoser, R. Ball, and A. V. Smrcka, *J. Biol. Chem.* **271**, 25071 (1996).

Preparation of Stocks for PLC Assay

Preparation of Stock Lipid Mixture. We obtain phosphatidylinositol 4,5-bisphosphate (PIP$_2$) and phosphatidylethanolamine (PE) from Avanti Polar Lipids, Inc. (Alabaster, AL). [^{3}H]PIP$_2$ is purchased from New England Nuclear (5 μCi per 500 μl). Lipids are stored at $-20°$ in lipid storage vials from Avanti Polar Lipids.

The concentrations of each lipid in the final stock mixture is 150 μM PIP$_2$, 600 μM PE, and [^{3}H]PIP$_2$ to give a total of ~400 cpm/μl stock lipid mix. The amount of [^{3}H]PIP$_2$ which gives a certain number of cpm should be empirically determined; however, we find that 0.045 μCi [^{3}H]PIP$_2$ per 100 μl stock lipid mix is usually sufficient. The lipids are packaged in chloroform or other organic solvents. We remove and mix lipids together in a glass Pyrex tube with a Hamilton syringe. The syringe is washed well with chloroform between different lipids. We evaporate organic solvents off the lipids by placing the lipid mixture beneath a thin stream of nitrogen gas, and then resuspend dried lipids in 1× assay buffer to give the appropriate lipid concentrations. The mixture is then sonicated in an ultrasonic water bath cleanser (Branson model 1510, Danbury, CT) for 10 min to suspend lipids in buffer and form lipid vesicles. During this time, suspension of dried lipids into solution will be visible, and the final solution will be translucent and slightly milky. We have found that adding the detergent Alconox to our ultrasonic water bath increases strength of sonication.

Preparation of Stock PLC. PLC-β_3 is diluted into 1× PLC buffer at a stock concentration of 0.25 ng/μl (see Table II).

TABLE II
BUFFERS USED FOR ASSAY PLC-β3 ACTIVITY

1× assay buffer		CaCl$_2$ solution (1 ml)	
50 mM Na–HEPES, pH 7.2		2× assay buffer	500 μl
3 mM EGTA		0.1 M CaCl$_2$	168 μl
80 mM KCl		H$_2$O	322 μl
1 mM DTT		0.1 M DTT	10 μl
1× $\beta\gamma$ buffer		2× assay buffer	
50 mM Na-Hepes, pH 7.2		100 mM Na–HEPES, pH 7.2	
3 mM EGTA		6 mM EGTA	
1 mM EDTA		160 mM KCl	
5 mM MgCl$_2$		2× $\beta\gamma$ buffer	
100 mM NaCl		100 mM Na–HEPES, pH 7.2	
1 mM DTT		6 mM EGTA	
1× PLC buffer (1 ml)		2 mM EDTA	
2× assay buffer	500 μl	10 mM MgCl$_2$	
0.1 M DTT	10 μl	200 mM NaCl	
10 mg/ml BSA	300 μl		
H$_2$O	190 μl		

Preparation of Stock βγ. βγ is diluted in 1× βγ buffer to stock concentrations desired. Stock concentrations(s) will be diluted 6-fold in final reaction volume. We usually make a series of stocks which will give us a range of final concentrations from 1 to 20 nM βγ.

PLC Assay Procedure

Assay conditions are essentially as described in Romoser *et al.*[9] Reactions are done in 60 μl volumes in reaction buffer containing 50 mM Na–HEPES, pH 7.2, 67 mM KCl, 17 mM NaCl, 0.83 mM MgCl$_2$, 3mM EGTA, 0.17 mM EDTA, 1 mM DTT, 1 mg/ml BSA, and 2.8 mM CaCl$_2$. Aliquot 20 μl stock lipid mix and 20 μl stock PLC-β3 solution into each assay tube and then add 10 μl stock βγ solution to each sample. Add 10 μl CaCl$_2$ solution to each sample to initiate reactions. Pipette up and down to mix calcium and other reagents well, then place assay tube on ice until all samples are finished.

Transfer all samples at once to a 30° water bath and incubate for 15 min. Place samples back on ice and stop reactions by addition of 200 μl 10% trichloroacetic acid to each sample. Add 100 μl 10mg/ml BSA solution to each sample and vortex to precipitate lipids and proteins. Spin samples in a tabletop microcentrifuge for 10 min at 7000 rpm. Remove 300 μl of supernatant and measure radioactivity in a scintillation counter. It is important to make sure that PLC-β enzyme does not hydrolyze more than 30% of the total substrate in the assay. Otherwise, one needs to reduce time of reaction, amount of PLC, or amount of calcium.

As indicated in Fig 2, PLC-β_3 activity is linearly increased with increasing βγ concentrations, up to 15 nM βγ, and is apparently saturated at 20 nM βγ concentration. This result is similar to previous reports and confirms that our purified recombinant βγ protein preparation is functional.

Measurement of G Protein Heterotrimer Formation

To form G protein heterotrimer, a fixed concentration of $\beta_1\gamma_5$ subunit was mixed with various concentration of α subunit in a wide range of α : βγ ratios (1 : 32, 1 : 16, 1 : 8, 1 : 4, 1 : 2, 1 : 1). Initially, 360 nM $\beta_1\gamma_5$ protein was incubated with increasing concentrations of α subunit (11.25, 22.5, 45, 90, 180, 360 nM) in ice for 30 min in a 10 μl buffer carrying 20 mM HEPES (pH 8.0), 100 mM NaCl, 2 mM MgCl$_2$, 0.1 mM EDTA, 1 mM DTT, and 0.5 mg/ml BSA. Five μl of this mixture was then diluted 10 times to a total 50 μl buffer containing 50 mM Na–HEPES (pH 7.2), 3 mM EGTA, 1 mM EDTA, 5 mM MgCl$_2$, 100 mM NaCl, and 1 mM DTT. Ten μl of this diluted samples containing βγ (6 nM final concentration) and various concentrations of α subunit were then added to a total 60 μl PLC reaction buffer containing [^{3}H]PIP$_2$ substrate and enzyme for determining PLC β3 activity as described above.

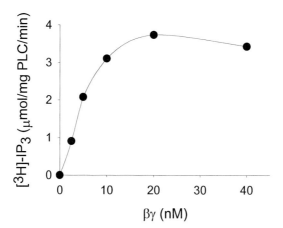

FIG. 2. PLC-$\beta 3$ stimulation by purified $\beta_1\gamma_5$. $\beta_1\gamma_5$ was expressed and purified from baculovirus–Sf9 cell system as described in text. Indicated concentration of $\beta_1\gamma_5$ was incubated with a buffer containing [^{3}H]PIP$_2$, PLC-β_3 for 15 min at 30°. [^{3}H]IP3 production was measured by scintillation counting. Representative result from three independent experiments.

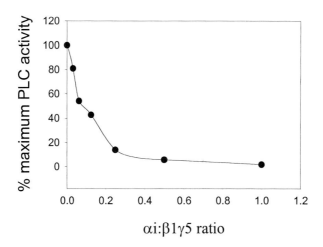

FIG. 3. The heterotrimer formation between α_{i2} and $\beta_1\gamma_5$ complex inhibits $\beta_1\gamma_5$ stimulated PLC-β_3 activity. The heterotrimer formation between $\beta_1\gamma_5$ and increasing concentration of α_{i2} subunit is described in text. The ratio of $\alpha_{i2} : \beta_1\gamma_2$ is indicated (X axis). PLC-β_3 activity stimulated by the $\beta\gamma$ subunits was measured to monitor heterotrimer formation as stated earlier and also in Fig. 2. The activity stimulated by free $\beta_1\gamma_5$ serves as a positive control representing maximum PLC activity (100%) whereas the enzyme activity stimulated by αo subunit alone was the negative control. Experiments have been repeated at least three times and representative results are shown.

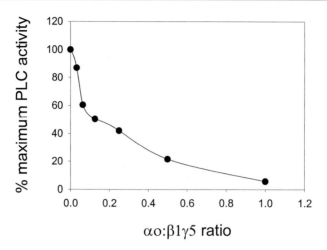

FIG. 4. The inhibitory effect of α_o on $\beta_1\gamma_5$ stimulation of PLC-β_3 activity. Increasing concentrations of α_o subunit were incubated with $\beta_1\gamma_5$ subunit (6 nM final concentration) to form the heterotrimer. All other procedures were the same as described in Fig. 3.

As shown in Fig. 3 and 4, increasing concentrations of αi or αo protein lead to a dramatic decrease in $\beta_1\gamma_5$-promoted PLC-β_3 activity. At the 1 : 1 ratio of $\alpha : \beta\gamma$ which would be the ideal condition for forming heterotrimer, the PLC β activity is almost completely wiped out, approaching the level of the negative control in which $\beta\gamma$ subunit is absent. This result indicates that the different α and $\beta\gamma$ subunits tested form heterotrimers at equal concentrations of each subunit.

As with $\beta\gamma$ subunit, α_q can also activate PLC-β activity. Thus, this approach cannot be directly used to measure the efficiency of heterotrimer formation between α_q and $\beta\gamma$ complex. However, it has been shown that a PLC$\beta2\Delta$, a deletion mutation lacking a C-terminal region necessary for stimulation by α_q, retains the ability to be stimulated by $\beta\gamma$ subunit.[10] This mutant form of PLC-$\beta2$ could be used to measure αq inhibition of $\beta\gamma$ stimulated PLC-$\beta2$ activity. The assay for stimulation of PLC-$\beta2$ activity follows essentially the same protocol as that for PLC-$\beta3$ except 1 ng of PLC-$\beta2$ is used per assay compared to 5 ng PLC-$\beta3$ per assay.[9] Although the method has not been tested with all α subunit types, it should be possible to determine the efficiency of heterotrimer formation between all known α subunit types and $\beta\gamma$ types using this assay.

Acknowledgment

We thank Dr. A. Smrcka, University of Rochester, for purified PLC-$\beta3$ protein and valuable discussions.

[10] D. Illenberger, F. Schwald, D. Pimmer, W. Binder, G. Maier, A. Dietrich, and P. Gierschik, *EMBO J.* **17,** 6241 (1998).

[36] Use of Peptide Probes to Determine Function of Interaction Sites in G Protein Interactions with Effectors

By Elizabeth Buck and Ravi Iyengar

Introduction

Protein regions involved in protein–protein recognition can function in a variety of ways to enable communication between two interacting proteins. Regions may be general binding domains. In this manner, regions could serve in adhesion, acting as the glue holding two partners together, while not playing any role in the transfer of signal information. General binding domains may also be involved in antiadhesion, preventing a transient protein–protein interaction from becoming too tight. Other protein regions may act as signal transfer regions because they have the ability to regulate an effector protein through direct interactions that initiate allosteric changes in the effector. Finally, some regions of a protein might be involved in mediating an interaction because they are involved in intramolecular interactions necessary for the maintenance of structure while making no direct contact with another protein.

Understanding the roles that specific regions of a protein perform to enable regulation of an effector protein is a major step toward being able to formulate concrete principles underlying how proteins communicate with each other. This information is also the cornerstone to our design of therapeutics that can successfully inhibit or stimulate specific protein interactions in disease-relevant signal transduction cascades. Experimental methods used to locate important sites of protein–protein contact include NMR,[1] crystallographic experiments,[2] and site-directed mutagenesis.[3] However, while being able to identify protein sites important for interactions with an effector, each of these experimental approaches fails to generate information necessary to functionally resolve each of these individual regions. For example, cocrystallization experiments can pinpoint protein regions involved in intermolecular interactions but cannot reveal the strength of the interaction, nor can they discern the type of function that the interaction plays. Site directed mutagenesis has proved to be useful in identifying specific amino acids which are relevant to a particular protein interaction, but cannot be used to determine if the

[1] J. Song and F. Ni, *Biochem. Cell Biol.* **76,** 177 (1998).
[2] S. Jones and J. Thornton, *Proc. Natl. Acad. Sci. U.S.A.* **93,** 13 (1996).
[3] C. E. Ford, N. P. Skiba, H. Bae, Y. Daaka, E. Reuveny, L. R. Shekter, R. Rosal, G. Weng, C. S. Young, R. Iyengar, R. J. Miller, L. Y. Jan, R. J. Lefkowitz, and H. E. Hamm, *Science* **280,** 1271 (1998).

amino acid functions in a intra- or intermolecular fashion. Site-directed mutagenesis experiments also cannot address functional roles, nor can they address the issue of sufficiency. Thus, these experimental methods fail to provide critical information necessary to determine how protein regions function in protein–protein communication. Peptides encoding regions of importance for protein–protein interactions have been used to isolate sites of direct protein–protein contact.[4] Herein, we describe how the use of peptides can be extended to functionally resolve these regions of direct contact.

Using Peptides to Address Function

The use of peptides to assign functions to direct protein contact domains can be used in concert with other methodology such as crystallography and site-directed mutagenesis to study how specific regions of a protein mediate protein–protein recognition and regulation. By testing the effect that a peptide has on the regulation of a given protein, whether that protein be a receptor or an effector enzyme, functions may be assigned. If a peptide region can inhibit a protein–protein interaction but not have any effect on the activity of the effector by itself, then that specific region which the peptide encodes is likely to be a general binding domain which functions in making a sufficient interaction with a partner protein but is not involved in the transfer of signal information to the effector protein. On the other hand, if a peptide region can stimulate the activity of an effector protein, then that specific region which the peptide encodes is likely to function as a signal transfer region by making an interaction with the effector protein that is capable of transferring signal information to that effector. If a peptide encoding a region of a protein shown to be important from site-directed mutagenesis data has no effect on either basal or stimulated activity, then it is possible that the region makes no direct contact with the effector. This would support the idea that the protein region functions in a structural manner to maintain either global or local structure necessary for effective interaction and regulation of the effector enzyme.

Crystallographic experiments[5] as well as site-directed mutagenesis experiments[3,6–9] have been used to study how various regions of the $G\beta$ subunit of the heterotrimeric G protein coordinate regulation of effectors. These studies have

[4] M. M. Rasenick, M. Watanabe, M. B. Lazarevic, S. Hatta, and H. E. Hamm, *J. Biol. Chem.* **269,** 21519 (1994).

[5] J. Sondek, A. Bohm, D. G. Lambright, H. E. Hamm, and P. B. Sigler, *Nature* **379,** 369 (1996).

[6] M. P. Panchenko, K. Saxena, Y. Li, S. Charnecki, P. M. Sternweis, T. F. Smith, A. G. Gilman, T. Kozasa, and E. J. Neer, *J. Biol. Chem.* **273,** 28298 (1999).

[7] Y. Li, P. M. Sternweis, S. Charnecki, T. F. Smith, A. G. Gilman, E. J. Neer, and T. Kozasa, *J. Biol. Chem.* **273,** 16265 (1998).

[8] I. Garcia-Hinguere, C. Gaitatzes, T. F. Smith, and E. J. Neer, *J. Biol. Chem.* **273,** 9041 (1998).

[9] K. Yan and N. Gautam, *J. Biol. Chem.* **272,** 2056 (1997).

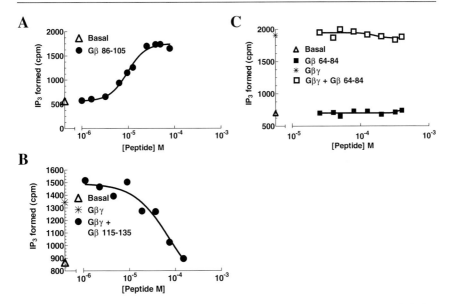

FIG. 1. Effects of peptides from various functionally important regions of $G\beta$. (A) General binding domain. Effect of varying concentrations of $G\beta$ 115–135 peptide on $G\beta\gamma$ stimulation of PLC-β2 [reproduced with permission from E. Buck *et al., Science* **283**, 1332 (1999). Copyright 1999 American Association for the Advancement of Science]. (B) Signal transfer region. Effect of $G\beta$ 86–105 peptide on PLC-β2 basal activity. (C) Structurally important region. Effect of $G\beta$ 64–84 peptide on basal and $G\beta\gamma$ stimulation of PLC-β2 [reproduced with permission from E. Buck and R. Iyengar, *J. Biol. Chem.* **276**, 36014 (2001)]. All results are typical of at least three independent experiments.

been useful in isolating regions of the $G\beta$ subunit which are involved in effector regulation but fail to provide direct evidence relating to how these specific regions function in $G\beta$ effector regulation. Using peptides encoding various functionally important regions of the $G\beta$ subunit we have shown that the functions of regions can be resolved and separately identified.[10,11] For $G\beta$ stimulation of the effector phospholipase C-β_2, PLC-β2, one region of $G\beta$, $G\beta$ 115–135, acts as a general binding domain because the peptide encoding this region can inhibit $G\beta\gamma$ stimulation of PLC but displays little activity on its own (Fig. 1A). On the other hand, the region of $G\beta$, $G\beta$ 86–105, acts as a signal transfer region because a peptide encoding this region can stimulate enzyme activity on its own, indicating that this region of $G\beta$ makes sufficient contact with PLC to initiate allosteric changes in PLC necessary for stimulation of activity (Fig. 1B). Another region of $G\beta$, $G\beta$ 64–84, was shown, by mutagenesis experiments, to be involved in regulation of PLC.[3,7,9] However, residues in this domain make direct intra-molecular contact

[10] E. Buck, J. Li, Y. Chen, G. Weng, S. Scarlata, and R. Iyengar, *Science* **283**, 1332 (1999).
[11] E. Buck and R. Iyengar, *J. Biol. Chem.* **276**, 36014 (2001).

with another region of $G\beta$, $G\beta$ 86–105, which we have found to be a signal transfer region on $G\beta$ for stimulation of PLC-$\beta2$.[5] We tested the $G\beta$ 64–84 peptide region on basal and $G\beta\gamma$ stimulation of PLC and found it to have no effect on either activity[11] (Fig. 1C). These data support the notion that the $G\beta$ 64–84 domain functions by making an intramolecular contact with the $G\beta$ 86–105 signal transfer region rather than by making a direct contact with PLC.

Selecting Peptide Regions

Simply dividing a protein randomly into short peptide regions is an ill-advised strategy for the design of peptides as this scenario might not present critical amino acids in the proper framework for observing a potential interaction. A good strategy is to design peptides from regions which encode a structural unit of the protein under study. When designing peptides from the $G\beta$ subunit we often select each to encode a structural feature of the $G\beta$ landscape. For example, the $G\beta$ 321–340 peptide codes for a region on $G\beta$ comprising two β strands and a hairpin turn.[11] This increases the likelihood that the resulting peptide will be structurally similar to the protein domain. Hot spot regions identified from structural experiments or site-directed mutagenesis experiments can also be used as the basis for the design of peptides. For example, if site-directed mutagenesis experiments identify a single residue or groups of residues as important for regulation of an effector, those residues can be chosen as the core region for the peptide. Flanking the core region of the peptide by several residues on the amino- and carboxy-terminal ends can decrease the flexibility of the core region, thus increasing the opportunity for a potential interaction. We often will work with peptides of around 20 amino acids in length. Designing peptides from protein regions with no basis as to which residues might play part in direct interactions is more difficult and would probably require testing multiple overlapping peptides from a single region. Here it would probably be best to start with a longer peptide region, at least 30–40 amino acids in length, and then truncate this region if an effect is seen.

Truncating Peptide Regions

Although selecting peptides of at least 15–20 amino acids in length is the best strategy to identify functionally important regions, once a region is found with a particular function, whether it be a general binding domain or a signal transfer region, the peptide region can be truncated in size to isolate the functional core of the domain. For the $G\beta$ 86–105 signal transfer region, the 20 amino acid peptide could be reduced to a critical core of six amino acids, $G\beta$ 96–101, which was the minimum peptide sequence that could still elicit a response from PLC (10) (Fig. 2). The identification of critical core regions at a protein–protein interface that play distinct roles in signaling is the basis for the development of a therapeutic

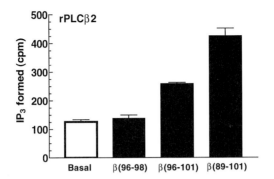

FIG. 2. Effects of truncated peptides from $G\beta$ 86–105 region. Effects of 600 μM $G\beta$ 96–98, $G\beta$ 96–101, and $G\beta$ 89–101 peptides on PLC-β_2 activity. Values are given as mean ± SEM of three experiments. [Reproduced with permission from E. Buck *et al., Science* **283,** 1332 (1999). Copyright 1999 American Association for the Advancement of Science.]

peptidomimetic to agonize or antagonize a specific protein–protein interaction.[12] From this point, a combinatorial peptide library can be constructed using the protein region as the parent peptide.[13] This library can be screened to determine the best consensus motif for interacting with an effector. This approach can aid in further identifying the critical determinants within the peptide region and select for substitutions that result in increased potency and affinity.

Designing Control Peptides: Scrambling of Peptide Regions vs Substitutions within Peptide Regions

Testing the sequence specificity of a peptide region toward a certain target enzyme can be accomplished by using either scrambled peptides or substituted peptides. For a scrambled peptide the amino acid composition is identical to the wild-type peptide, but the sequence has been scrambled. In this approach the sequence should be scrambled in such a fashion that the amino acid characteristics are altered from those of the wild-type peptide. For example, if a certain peptide has a sequence motif of two acidic residues separated by a stretch of four hydrophobic residues, then in designing a scrambled peptide care should be taken that this motif is disrupted. This approach is in contrast to scrambling the peptide in a Monte Carlo fashion where each amino acid is treated as distinct. Ensuring that sequence motifs are disrupted in the scrambled peptide is the best way of preventing the scrambled peptide from displaying the same basic structure as the wild-type peptide and thus preventing the scrambled peptide from interacting with the target enzyme in the

[12] A. S. Ripka and D. H. Rich, *Curr. Opin. Chem. Biol.* **2,** 441 (1998).
[13] M. A. Gallop, R. W. Barrett, W. J. Dower, S. P. A. Fodor, and E. M. Gordon, *J. Med. Chem.* **37,** 1233 (1994).

same fashion. For example, in one attempt to design a scrambled $G\beta$ peptide, $G\beta$ 86–105, the peptide was scrambled in a Monte Carlo fashion. However, the effect on the effector enzyme PLC-β_2 of the $G\beta$ 86–105 scrambled peptide was similar to that of the wild-type peptide. On looking at a modeled structure of the scrambled peptide compared to the wild-type peptide we found that both had a similar structure, mostly β sheet. Therefore, in this case Monte Carlo scrambling did not sufficiently disrupt the structure of this $G\beta$ region.

For a substituted peptide control, the sequence is roughly similar to that of the wild-type peptide, but the composition has been altered at one or more positions. This approach has proved very successful for our studies with $G\beta$ peptides and their effect on PLC-β_2. Often, substitution of one critical amino acid can disrupt the behavior of the peptide. This approach also has the benefit that it identifies critically important amino acids.

Substitutions within Peptide Regions: Identifying Critical Amino Acids

Substitution of amino acid positions within the peptide can aid in determining specific residues of importance within a particular domain. For example, in the $G\beta$ 86–105 signal transfer region peptide, substitution of specific resides has identified which amino acids function in regulating the affinity of this signal transfer region.[10] Substitution of R96 to Ala decreased the apparent affinity of this region for PLC-β_2 but left the maximal efficacy of PLC-β_2 activity unchanged (Fig. 3). This demonstrates the role of R96 to be important for the potency of this signal transfer region.

Substitution of amino acids in a signal transfer region peptide can be used to determine whether an amino acid position within the peptide region is relevant

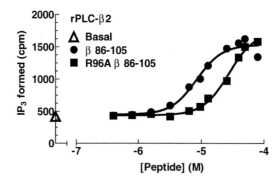

FIG. 3. Effects of substituted peptide from $G\beta$ 86–105 region. Effects of varying concentrations of $G\beta$ 86–105 and $G\beta$ 86–105 R96A-substituted peptide on PLC-β_2 activity. Results are typical of at least three experiments. [Reproduced with permission from E. Buck *et al., Science* **283,** 1332 (1999). Copyright 1999 American Association for the Advancement of Science.]

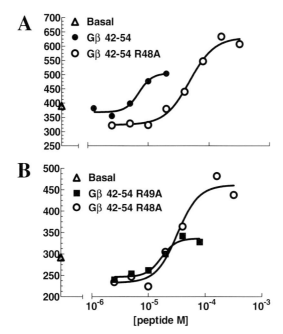

FIG. 4. Resolving functions of individual amino acids. Effects of varying concentrations of $G\beta$ 42–54 and $G\beta$ 42–54 R48A-substituted peptide (A) or $G\beta$ 42–54 R49A-substituted peptide and $G\beta$ 42–54 R48A-substituted peptide (B) on PLC-β_2 activity. All results are typical of at least three experiments. [Reproduced with permission from E. Buck and R. Iyengar, *J. Biol. Chem.* **276**, 36014 (2001).]

for regulating the affinity of the region toward the interacting protein or if the position is relevant for controlling the efficacy of enzyme stimulation. For example, substitution of $G\beta$ R48 to Ala within the $G\beta$ 42–54 region resulted in a decrease in the apparent affinity that was accompanied by an increase in the maximal effect of PLC stimulation[11] (Fig. 4A). R48, therefore, seems to be important for regulating both affinity and the signal transfer capabilities. On the other hand, substitution of $G\beta$ R49 to Ala within the $G\beta$ 42–54 signal transfer region resulted in a decrease in the apparent affinity for PLC-β_2 but no change in the efficacy of PLC stimulation (Fig. 4B). Thus, R49 appears to largely regulate affinity but has little role in direct signal transfer. This peptide approach thus demonstrates that functions can be resolved even at the level of individual amino acids.

Substitutions within Peptide Regions: Identifying Critical Features of Amino Acids

Each amino acid displays a variety of characteristics such as size, charge, hydrophobicity, and hydrogen bonding potential which give it its own unique

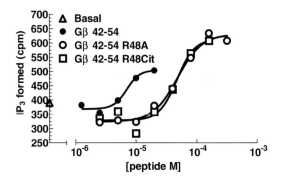

FIG. 5. Resolving functional characteristics of individual amino acids. Effects of varying concentrations of $G\beta$ 42–54, $G\beta$ 42–54 R48A-substituted peptide, and $G\beta$ 42–54 R48Cit-substituted peptide on PLC-β_2 activity. All results are typical of at least three experiments. [Reproduced with permission from E. Buck and R. Iyengar, *J. Biol. Chem.* **276**, 36014 (2001).]

identity. However, not all characteristics of an amino acid might be important for its interaction with an effector. For example, if an Ile can be substituted with an Ala with no change in effect, then hydrophobicity is most likely the predominant characteristic of this residue. We have tested peptides with different types of substitutions at single amino acids to resolve the specific characteristic of the amino acid that is critical to convey its effect. For residue R48 of $G\beta$ we sought to determine whether the charge of this amino acid or the size of its side chain was the major determinant for binding and signal transfer. So, we tested a $G\beta$ 42–54 peptide which was substituted at position 48 with citrulline, an amino acid which bears overall structural similarity with Arg but is uncharged.[11] We found that the $G\beta$ 42–54 R48Cit peptide had a similar effect to the $G\beta$ 42–54 R48A-substituted peptide, indicating that the charge of the side chain at this position is likely the critical determinant of its binding and signal transfer (Fig. 5).

Conclusions

Using peptides to discern how protein contact sites perform in protein–protein communication has widespread uses. These include protein interactions in intracellular signaling cascades and ligand–receptor interactions at the cell surface. Being able to assign functional roles to specific residues of a protein as well as understanding how various amino acid properties at these crucial points convey changes in activity are critical elements that lie at the heart of designing peptidomimetic therapeutics. This level of understanding for protein–protein contact also makes possible the idea of reengineering proteins with altered abilities to communicate with other proteins within the cell.

Acknowledgments

This research is supported by NIH Grant DK-38761. E.B. was a predoctoral trainee of the Molecular Endocrinology Training Grant DK-07135.

[37] Protein Interaction Assays with G Proteins

By Guangyu Wu,[*] Michael L. Bernard, and Stephen M. Lanier

Introduction

As part of a multifaceted approach to define molecular events at proximal stages of signal processing by G-protein-coupled receptor (GPCR) systems, we focused on two experimental strategies. These strategies grew out of the hypothesis that signal processing by GPCRs involves a ill-defined signal transduction complex that includes multiple accessory proteins distinct from receptor, G protein, and effector. One strategy was based on early observations concerning the transfer of signal from receptor to G protein and focused on a functional readout involving G-protein activation.[1–4] This approach resulted in the partial purification and characterization of a G-protein activator and the AGS group of proteins (activator of G-protein signaling) that activate heterotrimeric G-protein signaling pathways in the absence of a typical receptor. A second experimental approach utilized subdomains of receptor or G-protein regulators as bait to identify interacting proteins and define sites of interaction and the selectivity of the protein interaction.[5–8]

[*] Dr. Wu's current address is: Department of Pharmacology and Experimental Therapeutics, Louisiana State University, Health Sciences Center, 1901 Perdido St., New Orleans, LA 70112.

[1] M. Sato, C. Ribas, J. D. Hildebrandt, and S. M. Lanier, *J. Biol. Chem.* **271,** 30052 (1996).

[2] M. Cismowski, A. Takesono, C. Ma, J. S. Lizano, S. Xie, H. Fuernkranz, S. M. Lanier, and E. Duzic, *Nature Biotech.* **17,** 878 (1999).

[3] A. Takesono, M. J. Cismowski, C. Ribas, M. Bernard, P. Chung, S. Hazard III, E. Duzic, and S. M. Lanier, *J. Biol. Chem.* **274,** 33202 (1999).

[4] M. Cismowski, C. Ma, C. Ribas, X. Xie, M. Spruyt, J. S. Lizano, S. M. Lanier, and E. Duzic, *J. Biol. Chem.* **275,** 23421 (2000).

[5] G. Wu, J. G. Krupnick, J. L. Benovic, and S. M. Lanier, *J. Biol. Chem.* **272,** 17836 (1997).

TABLE I
PROTEINS INTERACTING WITH G PROTEINS OR G-PROTEIN-COUPLED RECEPTORS

Protein	Characteristic
14-3-3ζ	Binds to third intracellular loop of α_2-adrenergic receptors
AGS 1	Increases GTPγS binding to G_i and G_o; binds $G_i\alpha$
AGS2	Receptor-independent activator of heterotrimeric G-proteins; binds to $G\beta\gamma$
AGS3	Binds to GDP-bound $G_i\alpha$ and $G_t\alpha$; inhibits GTPγS binding to $G\alpha$
β-APP, Presenilin I	Bind to and activate $G_o\alpha$
Arrestins	Binds to phosphorylated/activated receptor; signal termination; signal processing
Calcyon	Binds to C-terminal tail of D1-dopamine receptors; calcium signaling
Caveolins	Influence G-protein activity
EIF2bα	Binds to C-terminal tail of α_{2A}-, α_{2B}-, α_{2C}-, and β_2-adrenergic receptors
GAP-43	Binds to and activates $G_o\alpha$
$G\beta\gamma$	Binds to third intracellular loop of α_2-adrenergic and M_2, M_3 muscarinic receptors
Grb2/Nck	Binds to third intracellular loop of D4-dopamine receptor
GRIN1	Binds to activated $G_i\alpha$, $G_o\alpha$, and $G_z\alpha$
Homer proteins	Bind to IP3 receptor/C-terminal tail of metabotropic glutamate receptors
Jak2	Binds to C-terminal tail of the AT1 receptor
LGN	Binds to $G_i\alpha$
NHERF	Binds to C-terminal tail of the β_2-adrenergic receptor
Nucleobindin	Binds to $G_i\alpha_3$
Pcp2	Binds to and activates $G_o\alpha$
Phosducin	Binds $G\beta\gamma$ and impedes formation of heterotrimer
PINS[a]	Binds to G-proteins and is involved in asymmetric cell division in *D. melanogaster*
RAMPs	Bind to receptor; role in receptor trafficking/ligand recognition
RaplGap	Binds to $G_i\alpha$, $G_o\alpha$, $G\alpha_z$
RhoA, ARF	Small G-proteins that coimmunoprecipitate with M_3-muscarinic receptors
RGS proteins	GTPase activating proteins for G_{i-}/G_{o-} or G_{q-}
Spinophilin	Binds to third intracellular loop of D2-dopamine receptor
Src	Complexed with β_2-adrenergic receptor via arrestin

[a] PINS is the AGS3/LGN homolog in *D. melanogaster*. LGN and AGS3 are two distinct proteins with conserved structure that exhibit 66% sequence homology.

Protein interaction technology (i.e., yeast two-hybrid screens, biochemical approaches such as GST fusion proteins, coimmunoprecipitation) has certainly proven its utility in the field of signal transduction and has resulted in the identification of new regulatory molecules with clear biological significance (Table I). An overview of such technology is provided in Ref. 9. The use of protein interaction technology has also allowed detailed analysis of molecular interactions between various signaling molecules. This chapter presents a fairly straightforward approach to evaluate the interaction of proteins with purified heterotrimeric G proteins, purified G-protein subunits, and G proteins in detergent extracts of cells or tissues.

[6] G. Wu, J. L. Benovic, J. D. Hildebrandt, and S. M. Lanier, *J. Biol. Chem.* **273**, 7197 (1998).

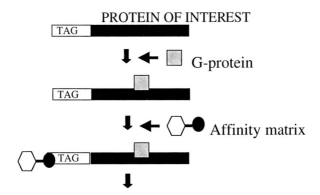

PROTEIN OF INTEREST

Wash affinity matrix to eliminate nonspecific interactions

Solubilization of protein complex and SDS-PAGE

Transfer to PVDF membrane and immunoblot

Fig. 1. Flow chart for protein interaction assay using GST fusion proteins.

General Considerations

Proteins or subdomains of proteins of interest (i.e., intracellular domains of G-protein-coupled receptors) are generated as "tagged" proteins by various techniques [i.e., glutathione S-transferase (GST) fusion proteins, pGEX vectors from Promega (Madison, WI); polyhistidine tag (His), pQE vectors from Qiagen (Valencia, CA)] (Fig. 1). The basic idea is to provide a handle by which one can use an affinity matrix (glutathione matrix for GST fusion proteins; nickel-charged matrix for His-tagged proteins, i.e., ProBond from Invitrogen) to "pull out" the protein of interest along with bound proteins. The proteins bound to the tagged protein on the resin are then identified by microsequencing, mass spectrometry, or immunoblotting. The interaction assays can be done with purified G-protein subunits or heterotrimer as well as with cell/tissue lysates. Since there are generally good, specific antisera to each G-protein subunit, it is fairly straightforward to identify bound G-protein subunits by immunoblotting. There are several factors to consider including protein size, solubility, and placement of tag. Generally, GST fusion proteins are generated by inserting the cDNA for the protein of interest

[7] G. Wu, G. S. Bogatkevich, Y. V. Mukhin, J. D. Hildebrandt, J. L. Benovic, and S. M. Lanier, *J. Biol. Chem.* **275**, 9026 (2000).

[8] M. Bernard, Y. K. Peterson, P. Chung, J. Jourdan, and S. M. Lanier, *J. Biol. Chem.* **276**, 1585 (2001).

[9] E. M. Phizicky and S. Fields, *Microbiol. Rev.* **59**, 94 (1995).

into a vector containing GST such that the protein coding region is in frame with the carboxyl terminus of GST. There is a short linker between GST and the fused protein that contains specific recognition sites for a protease allowing cleavage of the GST and protein of interest. One may also insert phosphorylation sites within the linker region allowing the generation of radiolabeled protein.[10] The His tags can be attached at the amino or carboxyl terminus of the protein of interest by either using commercial vectors containing polyhistidines in the multiple cloning sites or generating a polyhistidine tag with oligonucleotides.

The GST fusion proteins thus have an additional mass of 26,000 (size of GST), whereas the His-tagged proteins do not carry this extra mass. For larger proteins (i.e., >20,000) His tags are likely preferred just in terms of handling the protein and expressing the protein in reasonable amounts in bacteria. One may also select one "tag" over another based on the solubility properties of the protein of interest such that one tag may allow better recovery of protein. Finally, the resins used to isolated "tagged" proteins along with their binding partners are quite different and the degree of nonspecific binding to a nickel-charged resin or glutathione resin clearly varies from one protein to another. Ideally, it would be best to evaluate the protein interactions with the same protein using different tags, but this can become labor intensive and laboratories generally become comfortable with one system of protein tagging. There are several other motifs used to put a handle on proteins, including epitope tags, maltose binding protein fusions (New England Biolabs, Beverly, MA), and biotinylation.[9,11,12] However, the GST and His tags are easy to use and are generally the method of choice for routine analysis of protein interactions.

Materials

Source of Template for Fusion Protein

The full-length rat M_3-MR was kindly provided by Dr. Tom Bonner (Laboratory of Cell Biology, National Institutes of Mental Health, Bethesda, MD). Rat AGS3 was cloned from a rat brain cDNA library.[3]

G Protein

Bovine brain G-protein heterotrimer is purified as described.[13] The brain G-protein preparation contains approximately 70–80% G_{oA}, ~10% G_{oC}, ~5% G_{oB}, 5–10% G_{i1}, and <5% G_{i2}. Greater than 95% of the heterotrimer is functional based

[10] D. Ron and H. Dressler, *Biotechniques* **13**, 866 (1993).
[11] R. E. Kohnken and J. D. Hildebrandt, *J. Biol. Chem.* **264**, 20688 (1989).
[12] J. E. Cronan, Jr., *J. Biol. Chem.* **265**, 10327 (1990).
[13] J. Dingus, M. Wilcox, and J. D. Hildebrandt, *Methods Enzymol.* **237**, 457 (1994).

on the amount of $[^{35}S]GTP\gamma S$ binding. $G\alpha$ and $G\beta\gamma$ subunits in the bovine brain G-protein preparation are subsequently separated as described.[14] G_{oA} is readily separated from other $G\alpha$ subunits in the bovine brain preparation in sufficient quantities for routine protein interaction studies. The $G\beta\gamma$ subunit preparation contains $G\beta_{1-5}$ isoforms. Retinal G_t is also another source of purified G-protein subunits.[15] Thus $G\beta\gamma$, $G_o\alpha$, and $G_t\alpha$ are readily obtained from brain and/or retinal preparations. Additional $G_i\alpha$ isoforms and specific combinations of $G\beta\gamma$ are generally obtained following expression in Sf9 (*Spodoptera frugiperda* ovary) cells.[16–18] Purified $G\alpha$ subunits can also be obtained following expression in bacteria,[19,20] although there are issues with proper posttranslational modifications.

Chemicals/Compounds

Guanosine diphosphate (GDP) and guanosine 5'-O-(3-thiotriphosphate (GTPγS) (Boehringer-Mannheim, Indianapolis, IN); Lubrol, Sigma, St. Louis, MO). Thesit (polyoxyethylene-9-lauryl ether) (Boehringer-Mannheim) can also be used in place of Lubrol as the latter is no longer available.

Supplies

Polyvinylidene difluoride (PVDF) membranes (Gelman Sciences, Ann Arbor, MI); pGEX-4T-1 vector, glutathione Sepharose CL-4B and *Escherichia coli* BL21-DE3 (Pharmacia Biotech, Piscataway, NJ); isopropyl-β-D-thiogalactoside (IPTG) (Sigma).

Antisera

Antiserum to the carboxyl-terminal 10 amino acids of $G\beta_1$, which recognizes $G\beta_{1-4}$, is generated as described.[21] Polyclonal $G_i\alpha_3$ antiserum generated against the carboxyl-terminal 10 amino acids is kindly provided by Dr. Thomas W. Gettys (Pennington Biomedical Research Institute).[22] Antiserum to the amino-terminal

[14] M. D. Wilcox, J. Dingus, E. A. Balcueva, W. E. McIntire, N. D. Mehta, K. L. Schey, J. D. Robishaw, and J. D. Hildebrandt, *J. Biol. Chem.* **270**, 4189 (1995).

[15] D. G. Lambright, J. Sondek, A. Bohm, N. P. Skiba, H. E. Hamm, and P. B. Sigler, *Nature* **379**, 311 (1996).

[16] S. G. Graber, R. A. Figler, and J. C. Garrison, *Methods Enzymol.* **237**, 212 (1994).

[17] S. G. Graber, R. A. Figler, V. K. Kalman-Maltese, J. D. Robishaw, and J. C. Garrison, *J. Biol. Chem.* **267**, 13123 (1992).

[18] E. M. Parker, K. Kameyama, T. Higashijima, and E. M. Ross, *J. Biol. Chem.* **266**, 519 (1991).

[19] E. Lee, M. E. Linder, and A. G. Gilman, *J. Biol. Chem.* **237**, 146 (1994).

[20] S. M. Mumby and M. E. Linder, *Methods Enzymol.* **237**, 254 (1994).

[21] M. Makhlouf, S. H. Ashton, J. D. Hildebrandt, N. Mehta, T. W. Gettys, P. V. Halushka, and J. A. Cook, *Biochem. Biophys. Acta* **1312**, 163 (1996).

[22] T. W. Gettys, T. A. Fields, and J. R. Raymond, *Biochemistry* **33**, 4283 (1994).

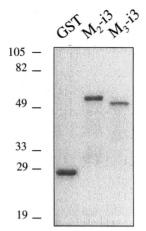

Coomassie blue stain of GST
and GST fusion proteins

FIG. 2. Expression of receptor subdomains as GST-fusion proteins. The third intracellular loop of the M_2- and M_3-muscarinic receptors was generated as a GST fusion protein as described in the text. Figures adapted from Refs. 5 and 6.

16 amino acids of $Go\alpha$ is kindly provided by Dr. Graeme Milligan (Department of Biochemistry and Pharmacology, University of Glasgow, Scotland, UK). Purified GA antibody, which selectively recognizes $G_i/G_o\alpha$, is kindly provided by Drs. Paul Goldsmith, Andrew Shenkar, and Allen Spiegel.[23]

GST Fusion Proteins

Procedures for purification of GST fusion proteins are well described [Pharmacia Biotech's Web site (www.apbiotech.com); see subheadings-Technical Support Literature, Recombinant Protein Expression & Purification, Gene Fusion Expression Systems. A similar source of information for His-tagged proteins can be found at www.invitrogen.com/maunals.html; see file name www.invitrogen.com/pdf_manuals/proresin_man.pdf. The *E. coli* strain BL21-DE3 is recommended for expression of GST fusion proteins, as it is deficient in proteases. Before performing a large-scale purification of fusion protein it is suggested to establish optimal conditions for the expression of specific GST fusion proteins. The protocol described below is used in our laboratory for preparation of variety of GST fusion proteins (Fig. 2).

[23] P. Goldsmith, K. Rossiter, A. Carter, W. Simonds, C. G. Unson, R. Vinitsky, and A. Spiegel, *J. Biol. Chem.* **263**, 6476 (1988).

The BL21-DE3 strain is transformed with the GST fusion protein vector or vector alone (to generate GST for control experiments in protein interactions studies) by standard procedures and a single colony propagated in 2 ml of $2\times$ YTA (16 g pepticase tryptone, 10 g yeast extract, 5 g NaCl, pH 7.0, dissolved in 1 liter) at $37°$ for 12 hr. For protein expression, 2 ml of the overnight culture is added to 200 ml of $2\times$ YTA medium in a 1 liter flask and grown at $30°$ with shaking to an OD_{600} of 1–1.5. Add 100 mM IPTG to a final concentration of 100 μM and continue incubation for an additional 3–4 hr. Pellet the bacteria by centrifugation at $6000g$ and resuspend the pellet in 10 ml of buffer I [140 mM NaCl, 2.7 mM KCl, 10 mM Na$_2$HPO$_4$, 1.8 mM KH$_2$PO$_4$, pH 7.3, 1 mM dithiothreitol (DTT), 5 mM EDTA, 5 mM EGTA, 5 mM benzamidine, 0.5 mM phenylmethylsulfonyl fluoride (PMSF)] with a 10 ml pipette followed by sonication (Branson Sonifier 250, setting 2, Danbury, CT) on ice for 30 sec.

Pellet the bacteria debris by centrifugation ($12,000g$, 10 min, $4°$) and recover the supernatant. The pellet is washed and resuspended in 10 ml of buffer I as before and saved for gel analysis. An aliquot of the supernatant (200 μl) is removed for later evaluation and 0.2 ml of glutathione Sepharose 4B (50% slurry in buffer I) is added to the supernatant. Incubate with gentle rotation at $4°$ for 1 hr. Pellet the slurry by centrifugation ($500g$, 5 min, $4°$) and take a 200 μl aliquot of the supernatant (200 μl) for later evaluation. Wash the resin with 2×5 ml each of buffer I, buffer I containing 500 mM NaCl, and again buffer I at $4°$. Extensive washing is required to remove bacterial proteins nonspecifically bound to the resin. For hydrophobic proteins, one may consider including detergent in buffer I during lysis, resin washing, and protein elution from the resin.

After the final wash the resin is resupended in an equal volume of buffer I and an aliquot (25 μl) taken for later analysis. For separation of fusion protein and resin, the GST fusion protein is eluted from the glutathione affinity matrix by incubation with 10 mM glutathione in 50 mM Tris pH 7.5 at $24°$ with gentle rotation for at least 2 hr. Glutathione must be removed prior to use in the interaction assays, and this is achieved by ultrafiltration using a Centricon-3 concentrator (Amicon, Beverly, MA) (molecular weight cutoff 3000). The concentration of purified GST fusion protein is determined by the BioRad protein assay system. Protein interaction studies can be done with the GST fusion protein bound to the affinity matrix. However, each GST fusion protein is made at variable amounts and thus there will most likely be different amounts GST and the GST fusion protein on same amounts of the resin limiting comparative analysis and controls. GST itself is usually used as a control for the specificity of any observed protein interactions and GST is almost always more efficiently made by the bacteria than are the fusion proteins. Thus, elution of the bound GST and GST fusion protein from the matrix with subsequent desalting and concentration by ultrafiltration allows for a tighter control on the relative amounts of the fusion proteins and controls used in each experiment.

For each preparation of GST fusion protein, aliquots of the lysed bacterial pellet, the lysate (pre and post resin), the resin (pre and post elution), and the desalted elution are evaluated by SDS–PAGE and Coomassie blue staining. The above-described procedure yields from 0.5 to 1 mg of GST fusion protein. The efficiency of protein expression for different GST fusion proteins can be variable. Several manipulations can be used to optimize expression including the temperature ($30°$ vs $37°$) and/or time for bacterial growth during induction. For difficult proteins, one may also evaluate expression in different bacterial strains.

Preparation of Cell/Tissue Lysates

Rat brain is homogenized in 3 ml buffer/gram tissue of lysis buffer [50 mM Tris-Cl pH 8.0, 150 mM NaCl, 5 mM EDTA, 1% Nonidet P-40 (NP-40)]. Confluent 100 mm dishes of cells are washed with cell washing solution (137 mM NaCl, 2.6 mM KCl, 1.8 mM KH_2PO_4, 10 mM Na_2HPO_4) and resuspended in 1 ml/dish of lysis buffer by homogenization. Following a 1 hr incubation on ice at $4°$, the cell or tissue homogenate is centrifuged at $27,000g$ for 30 min. Supernatants are collected and spun at $100,000g$ for 1 hr at $4°$ to generate a detergent-soluble fraction. The supernatant is immediately processed for protein interaction assays. Protein concentrations are determined by a Bio-Rad (Hercules, CA) protein assay.

Method of Assay

1. Preparation of Affinity Matrix

Wash the required amount of glutathione Sepharose with 2×5 volumes of binding buffer II (20 mM Tris-HCl, pH 7.5, 0.6 mM EDTA, 1 mM dithiothreitol, 70 mM NaCl, 0.01% Lubrol) at $4°$ with pelleting in a microfuge. Resuspend resin in buffer II to final concentration of 50% (packed resin : volume of buffer II, 1 : 1) and store at $4°$ until use.

2. Incubation of G Protein or Cell Lysates with GST Fusion Proteins

As discussed in the materials section above, heterotrimeric G proteins and resolved G-protein subunits can be obtained by various procedures. Generally, both heterotrimeric G proteins and free Gα subunits (i.e., following expression in bacteria or Sf9 cells) are purified in the presence of GDP. The presence of GDP would stabilize the interaction of Gα and G$\beta\gamma$. The presence of GTPγS (plus 5 mM $MgCl_2$) would promote subunit dissociation (with the exception of transducin where nucleotide exchange generally requires receptor activation). In addition, the conformation of Gα changes upon binding different nucleotides. Thus, for protein interaction studies with heterotrimeric G-proteins, one would conduct experiments in the presence of GDP or GTPγS to determine if the protein of interest interacts with G-protein subunits in the context of heterotrimer or only when Gα and G$\beta\gamma$ are free from each other. Along this line, one can also conduct protein interaction experiments with free Gα subunits in the presence of GDP, GDP plus AlF$_4$ or GTPγS to determine conformation-dependent binding of Gα to the protein of interest. Similar manipulations may be done in protein interaction experiments using cell/tissue lysates. To promote a particular nucleotide bound state of G-proteins, the preparation is generally preincubated with nucleotide for 30 min at 24° prior to initiating the protein interaction protocol described below. The optimal conditions (i.e., temperature and time of incubation) for such a preincubation depend on the basal rate of nucleotide exchange for the Gα of interest.

(a) GST and GST fusion proteins (300 nM) are incubated with 30–100 nM G-protein heterotrimer or G-protein subunits in a total volume of 250 μl of buffer II at 4° for 40–90 min with gentle rotation. GST is processed in parallel for each interaction assay providing internal controls for specificity of observed interactions.

(b) For interaction assays with cell/tissue lysates, GST and the GST fusion protein (100–300 nM) are incubated with cell/tissue lysate (1 mg protein) for 1 hr at 24° in a total volume of 250 μl of 1% NP40 lysis buffer.

(c) Twenty five μl of glutathione Sepharose slurry (50% in buffer II) is added to the incubation mixture in (a) or (b) and incubation continued for an additional 20 min at 4°.

(d) Allow the resin to pellet by gravity for 2 min at 4° and then spin the sample for 5 sec in a PicoFuge centrifuge (Stratagene, La Jolla, CA). The gentle handling of the affinity matrix at this point and in the washes is a key to reduce nonspecific interaction of G-protein subunits with the resin.

(e) Add 500 μl of buffer I to the resin and mix gently by inverting the reaction tube 3–5 times. Sediment the affinity matrix as above and repeat buffer I wash twice for a total of three washes.

3. *Evaluation of G-Protein Interaction by Immunoblotting*

The washed protein complex retained on the affinity matrix is solubilized in 50 μl of Laemmli gel sample buffer, separated by 10% SDS–PAGE, and transferred to PVDF membrane in a semidry electrophoretic transfer cell. The membrane transfer is then incubated in blocking solution [i.e., milk or bovine serum albumin (BSA)] and processed for immunoblotting. Immunoblots are processed by standard procedures using a chemiluminescence-based detection system. Signal detection on film requires careful attention to exposure time and the linearity of the signal. For such experiments it is critical to evaluate multiple times of film exposure for each experiment, particularly when concentration-dependent studies are undertaken. In the latter case, the signals can be semiquantitated by densitometric analysis.

However, the best way to evaluate the interactions is with an optical detection system (i.e., Bio-Rad Fluoro-S-Max Imager) that allows accurate quantitation of the generated intensities.

For reblotting with different antisera, membranes are stripped by incubation in stripping buffer (62.5 mM Tris-HCl, pH 7.5, 2% SDS, and 100 mM 2-mercaptoethanol) for 30 min at 50° with subsequent washing (4 × 15 min) in normal immunoblot buffer. With gentle handling and good antibodies, the same blot can be reprobed at least five times.

The membrane transfers used for immunoblotting are evaluated for the relative amounts and quality of GST fusion proteins by either immunoblotting with anti-GST antibody or amido black staining of proteins to control for sample processing in individual experiments. Immunoblotting with GST antibody can be done at any time, but the amido black stain is best done with a blot that has not undergone exposure to stripping buffer and reprobing. For amido black staining membranes are placed in stain solution [0.2% weight/volume naphthol blue-black (Sigma), 45% methanol, 10% glacial acetic acid, 45% H$_2$O, v/v] and gently rocked at room temperature for 30 minutes. The stained membranes are then placed in destain solution (45% methanol, 10% glacial acetic acid, 45% H$_2$O, v/v) 4 × 15 min or until protein bands are visible. If the membrane will be probed again with antibody, the membrane should be stripped after removal from the destain solution.

Applications

Interaction of Receptor Subdomains with G Proteins

For most members of the superfamily of G-protein-coupled receptors, the third intracellular (i3) loop and the carboxyl-terminal tail of the receptor are key sites for signal propagation and termination and these receptor domains also exhibit the greatest variability in size among different subfamilies of these receptors. The largest i3 domains (100–240 amino acids) are found in a subgroup of receptors coupled to the G$_i$, G$_o$, and/or G$_q$ family of G proteins (i.e., muscarinic, α-adrenergic) contrasting with the shorter i3 loop (~20 amino acids) in the photoreceptor rhodopsin or β_2-adrenergic receptor (AR) (~50 amino acids). The juxtamembrane regions of the i3 loop are critical for G-protein coupling and there are discrete motifs in the i3 loop and/or carboxyl-terminus region involved in signaling specificity, receptor trafficking, and receptor regulation.

We used the large i3 loop of muscarinic receptor (MR) and α_2-AR as probes to screen bovine brain cytosol for interacting proteins[5] and evaluate their interaction with purified G proteins.[6,7] This work led to the surprising observation that G$\beta\gamma$ was capable of binding to the i3 loop. The binding of G$\beta\gamma$ to the i3 loop was inhibited by Gα (Fig. 3). We could also use the protein interaction assay to determine the relative affinity of the receptor subdomain for G$\beta\gamma$ in comparison to

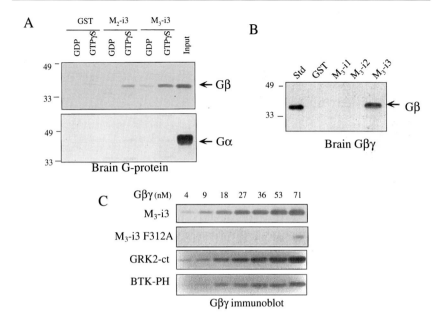

FIG. 3. Interaction of the third intracellular loop of the M₃-muscarinic receptor with Gβγ. (A) Heterotrimeric brain G protein (30 nM) was incubated (4° for 90 min) with the control GST, M₂-MR i3, or M₃-MR i3 affinity matrix (~5 μg of protein) in a total volume of 250 μl of buffer containing either 10 μM GDP or 10 μM GTPγS plus 5 mM Mg²⁺. The resins were washed and bound proteins visualized by immunoblotting of membrane transfers following SDS–PAGE. The membrane transfers were first incubated with Gβ antisera (upper panel) and then stripped for reprobing with G₀α antisera (lower panel). The numbers to the left of the gel indicate the migration of standards of known molecular weight × 10⁻³. GST, Glutathione S-transferase, Input, 200 ng of brain G-protein without resin incubation. (B) GST fusion proteins (~5 μg of protein) were incubated with Gβγ (30 nM) and samples processed as described above using Gβ antisera. (C) Binding of Gβγ to M₃-MR i3, M₃-MR i3 F312A, GRK2-ct, or Btk-PH fusion proteins. Increasing concentrations of purified brain Gβγ (4–71 nM) were incubated with the M₃-MR i3, M₃-MR i3 F312A, GRK2-ct, or Btk-PH affinity matrix (~5 μg of protein). The F312A mutation disrupts Gβγ binding to the M₃-MR i3 loop.[7] Similar results were obtained in three experiments for A–C using different preparations of fusion protein and brain G-protein. Figures adapted from Refs. 5, 6, and 7.

other Gβγ binding proteins (Fig. 3C). However, it is difficult to determine accurate affinities by these assays or to compare substantially different proteins for maximum binding as it is likely that only a certain portion of the GST fusion protein is in the right conformation for protein binding. In addition, nonspecific interaction of G-protein subunits with the affinity matrix can occur at higher concentrations.

To further localize the Gβγ binding domain in the M₃-MR, the M₃-MR i3 loop containing the Gβγ binding motif was further truncated at the amino and carboxyl

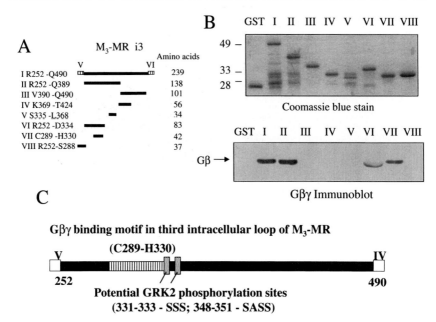

FIG. 4. Localization of a Gβγ binding domain in the M_3-MR i3 loop. The M_3-MR R^{252}-Q^{490} i3 loop peptide was progressively truncated at the amino and carboxyl terminus to generate the GST fusion protein constructs illustrated in (A) and the top panel of (B). Each of the constructs were evaluated in Gβγ binding assays as described[7] (lower panel in B). These experiments were repeated 5–6 times with identical results. The aberrant migration of immunoreactive Gβ in the construct VI lane is due to comigration with the fusion protein itself. (C) Location of Gβγ binding sites and GRK2 phosphorylation sites on the M_3-MR R^{252}-Q^{490} i3 loop peptide. Figures adapted from Ref. 7.

terminus to generate constructs encoding different regions of the M_3-MR i3 loop. Each construct was evaluated for Gβγ binding. This strategy localized the Gβγ-binding motif to C^{289}-H^{330} within the M_3-MR i3 loop (Fig. 4). We then generated full-length receptor constructs lacking Gβγ binding motifs and evaluated their function/regulation in the context of the intact cell. Receptor constructs lacking the Gβγ binding motifs effectively coupled to G_q to increase intracellular calcium in response to agonist but they were deficient in agonist-induced internalization of the receptor.[7]

One of the advantages of this protein interaction technique, in contrast to yeast two-hybrid screens or expression library probing, is the ability to evaluate multicomponent interaction. This is of particular interest for Gβγ as it interacts with multiple signaling proteins including calcium channels, G-protein receptor kinase 2 (GRK2), selected proteins containing plekstrin homology domains,

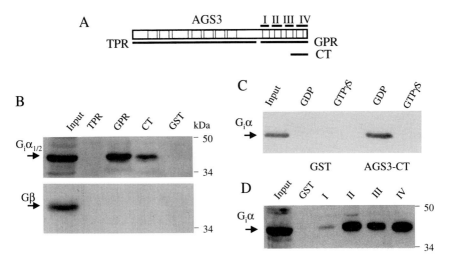

FIG. 5. AGS3 domains interacting with G proteins. Subdomains of AGS3 (A) were generated as GST fusion proteins and purified following expression in bacteria for protein interaction studies. Lysates were prepared from DDT$_1$-MF2 cells and 1 mg of lysate protein was incubated with 300 nM GST–AGS3 fusion proteins as previously described.[3,8] Membrane transfers of bound proteins were probed with G-protein subunit antisera. (B) TPR M1-I462, GPR P463-S650, CT M577-S650 (C) CT M577-S650 and (D) GPR-I P463-E501, GPR-II S516-L555, GPR-III G563-T602, GPR-IV T602-S650. Similar results were obtained in 3–5 individual experiments using different batches of lysate. The input lane contains 1/10 of the lysate volume used in each individual interaction assay. Protein interaction assays in (B) and (D) contained 30 μM GDP. Protein interaction assays in (C) contained 30 μM GDP or 30 μM GTPγS plus 25 mM MgCl$_2$. Figures adapted from Refs. 3 and 8.

GIRKs, Bruton's tyrosine kinase, adenylyl cyclase type II, and AGS2 (activator of G-protein signaling). Indeed, protein interaction studies with G$\beta\gamma$, G-protein-coupled receptor kinase 2, and the M$_3$-i3 fusion protein suggest the formation of a ternary complex in which G$\beta\gamma$ serves as an adaptor.[6]

Interaction of G Proteins with AGS3

AGS3 (Activator of G-protein signaling 3) was isolated in a yeast-based functional screen for receptor-independent activators of heterotrimeric G-proteins.[3] AGS3 (AF107723, calculated molecular weight 72,049) consists of two functional domains separated by a linker in the middle of the protein (Fig. 5). The amino-terminal half of the molecule exists as a series of tetratricopeptide repeats (TPRs). The carboxyl-terminal half of AGS3 contains the second functional domain consisting of four ~20 amino acid repeats. These repeat sequences were termed the G-protein regulatory motif (GPR) based upon their functionality and their presence in other proteins that interact with and/or regulate heterotrimeric

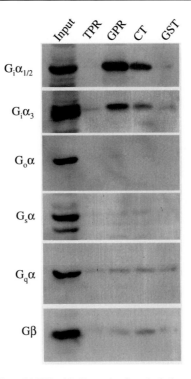

FIG. 6. Selective interaction of AGS3 with G proteins from brain lysates. One milligram of lysate protein from rat brain was incubated with 300 nM GST–AGS3 fusion protein (TPR M1-I462, GPR P463-S650, CT M577-S650). All interactions were done in the presence of 30 μM GDP. The input lanes represent 1/10 of the lysate used in each interaction assay. Membrane transfers of bound proteins were probed with the indicated antisera with intervening stripping of the blot. The $G_i\alpha_3$ antibody exhibits some cross-reactivity with $G_o\alpha$ that likely accounts for the broad immunoreactive band observed in the input lane. Similar results were obtained in 2–3 separate experiments. Figures adapted from Ref. 8.

G-proteins.[3] The domains of AGS3 interacting with $G_i\alpha$ proteins was determined in protein interaction assays following the expression of AGS3 domains as GST fusion proteins. We generated the amino-terminal half of AGS3 (AGS3-TPR, M1-I462) and the carboxyl-terminal half of AGS3 (AGS3-GPR, P463–S650) as GST fusion proteins (Fig. 5). The AGS3-TPR, AGS3-GPR and the 74 amino acid carboxyl terminus (AGS3-CT, M577-S650) isolated in the original yeast functional screen were incubated with DDT$_1$-MF2 cell lysates and proteins bound to the AGS3 subdomains identified by immunoblotting of gel transfers (Fig. 5). The Gα binding domains of AGS3 were found in the carboxyl terminal half of the protein (Figs. 5A, 5B). The TPR domains of AGS3 did not interact with Gα or

$G\beta\gamma$ (Fig. 5B). The GST-AGS3GPR fusion protein preferentially interacted with the GDP-bound conformation of $G_i\alpha$ (Fig. 5C).

Further analysis indicates that each of the GPR domains in the carboxyl half of AGS3 appears capable of binding $G\alpha$. Each GPR motif was generated as a GST-fusion protein and evaluated in protein interaction assays using DDT_1-MF2 lysates (Figs. 5A, 5D). Each GPR motif bound $G\alpha$, although GPR I, at least in this context, bound less $G\alpha$ than did GPR II-IV (Fig. 5D). These data also suggest that AGS3 is capable of binding multiple $G\alpha$ subunits consistent with a putative role of AGS3 as a scaffolding protein within a larger signal transduction complex.

The interaction of AGS3 with $G\alpha$ was selective for different G-protein families. For these studies we used crude tissue/cell lysates and purified $G\alpha$ subunits.[8] The AGS3-GPR GST fusion protein was incubated with rat brain lysate and bound proteins identified by immunoblotting with $G\alpha$-specific antisera. AGS3-GPR effectively bound $G_i\alpha_{1-3}$, but not $G_s\alpha$, $G_o\alpha$, $G_q\alpha$, or $G\beta\gamma$ (Fig. 6). Based on the comparison of the signal intensity in the input versus sample lane, it is estimated that AGS3-GPR binds $\sim$20–30% of the total $G_i\alpha$ protein in the lysate sample. Similar results were obtained in DDT_1-MF2 cell lysates. Each of the protein interaction experiments in the tissue/cell lysates was done in the presence of GDP, which would stabilize heterotrimeric $G\alpha\beta\gamma$; however, immunoblotting with G-protein β subunit antisera indicated that AGS3 was complexed with $G_i\alpha$ in the absence of $G\beta$ (Fig. 6). Thus, either AGS3 effectively promoted subunit dissociation or there is a population of $G_i\alpha$ that exists free of $G\beta\gamma$. The selectivity of AGS3 for different G-proteins was also observed using purified $G\alpha$ subunits.[8]

Acknowledgments

This work was supported by Grants (NS24821, MH5993-S.M.L.) from the National Institutes of Health. The authors thank John D. Hildebrandt (Medical University of South Carolina) for valuable discussions and for providing G-protein antisera and purified bovine brain G protein, Graeme Milligan for $G_o\alpha$ antisera, Thomas W. Gettys (Pennington Biomedical Research Institute) for $G_i\alpha_3$ antisera, and Drs. Goldsmith, Shenkar, and Spiegel for GA antibody.

[38] Evolutionary Traces of Functional Surfaces along G Protein Signaling Pathway

By Olivier Lichtarge, Mathew E. Sowa, and Anne Philippi

Introduction

The characterization of active sites in proteins is an important problem in biology. Proteins carry out nearly all their functions through these specialized surfaces where the conformational, dynamic, and electrochemical properties of specific amino acids define which ligands they may bind and which transformations may be exerted on them. This preeminent role translates into widespread interest in active site identification and in elucidating how their constituent residues contribute to function. This is useful, for example, to design drugs or functional mimics,[1–3] to impart novel properties to engineered molecular scaffolds by transplanting these sites,[4] and to develop generic biosensors.[5] Moreover, a better understanding of active sites can help manipulate protein interactions and dissect cellular pathways.[6–8]

Despite these broad applications, the characterization of active sites lags far behind the exponential growth of sequence and structure databases. One reason is that active sites can remain cryptic in protein structures that are determined without all their ligands. A deeper difficulty is that not all the interfacial residues contribute to an interaction.[9–11] Thus in order to determine the origin of affinity and specificity in molecular detail, it is necessary to perform mutational analysis.[12,13]

[1] I. D. Kuntz, J. M. Blaney, S. J. Oatley, R. Langridge, and T. E. Ferrin, *J. Mol. Biol.* **161**, 269 (1982).

[2] P. J. Whittle and T. L. Blundell, *Annu. Rev. Biophys. Biomol. Struct.* **23**, 349 (1994).

[3] A. K. Patick, H. Mo, M. Markowitz, K. Appelt, B. Wu, L. Musick, V. Kalish, S. Kaldor, S. Reich, D. Ho, and S. Webber, *Antimicrob. Agents Chemother.* **40**, 292 (1996); erratum: *Antimicrob. Agents Chemother.* **40**, 1575 (1996).

[4] C. Vita, J. Vizzavona, E. Drakopoulou, S. Zinn-Justin, B. Gilquin, and A. Menez, *Biopolymers* **47**, 93 (1998).

[5] H. W. Hellinga and J. S. Marvin, *Trends Biotechnol.* **16**, 183 (1998).

[6] T. Clackson, M. H. Ultsch, J. A. Wells, and A. M. de Vos, *J. Mol. Biol.* **277**, 1111 (1998).

[7] T. Pawson, *Nature* **373**, 573 (1995).

[8] D. J. Peet, D. F. Doyle, D. R. Corey, and D. J. Mangelsdorf, *Chem. Biol.* **5**, 13 (1998).

[9] B. C. Cunningham and J. A. Wells, *J. Mol. Biol.* **234**, 554 (1993); erratum: *J. Mol. Biol.* **237**, 513 (1993).

[10] K. H. Pearce, Jr., M. H. Ultsch, R. F. Kelley, A. M. de Vos, and J. A. Wells, *Biochemistry* **35**, 10300 (1996).

[11] G. Schreiber and A. R. Fersht, *J. Mol. Biol.* **248**, 478 (1995).

[12] L. L. Sengchanthalangsy, S. Datta, D. B. Huang, E. Anderson, E. H. Braswell, and G. Ghosh, *J. Mol. Biol.* **289**, 1029 (1999).

[13] S. J. Davis, E. A. Davies, M. G. Tucknott, E. Y. Jones, and P. A. van der Merwe, *Proc. Natl. Acad. Sci. U.S.A.* **95**, 5490 (1998).

Unfortunately, constructing mutants and assaying activity are laborious tasks that are also protein specific and resource intensive, restricting their use to a fraction of all available protein structures.

By contrast, computational analysis is more amenable to large-scale application. Some algorithms already usefully identify pockets or rough patches on the protein surface where small ligands preferentially bind,[14,15] and electrostatic charge computations can yield clues to RNA or DNA binding sites.[16] Predictions remain difficult, however, in the absence of recognizable geometric or chemical motifs. This is the case for the large interfaces typical in proteins from macromolecular cellular networks, which are indistinguishable from their surroundings based on hydrophobic character, solvation potential, planarity, protrusion, or accessible surface area.[17,18]

The evolutionary trace method aims to facilitate active site characterization by combining an algorithmic approach with the experimental strategy of mutational analysis. It does so by categorizing natural sequence variations in terms of the evolutionary divergences of related proteins, thereby establishing an association between residue variation and functional changes.

Principles of Evolutionary Trace Method

The basic hypothesis of the method is that protein active sites evolve through variations on a conserved architecture. If so, active sites from divergent proteins may be expected to have two evolutionary components: one that is invariant, and one that is specific to each functional class. Correspondingly, there should be two types of functionally important residues. The first type should be mostly invariant and thereby define the fundamental stereochemical architecture underlying activity. The second type should be invariant *within* functional classes but variable *among* them. These so-called *class-specific* residues could thus impart unique functions to the proteins in the family.

This model leads to a simple procedure to identify class specific residues, illustrated in Fig. 1. First, homologs of a protein of interest are gathered, aligned, and separated into functional subgroups so that the invariant residues of each group are identified in a consensus sequence. Second, these consensus sequences are compared to reveal positions that are invariant within each class but variable among them. These are the class-specific positions, or residues. By construction, their variations are always associated with a change in function, which

[14] J. Liang, H. Edelsbrunner, and C. Woodward, *Protein Sci.* **7**, 1884 (1998).

[15] F. K. Pettit and J. U. Bowie, *J. Mol. Biol.* **285**, 1377 (1999).

[16] B. Honig and A. Nicholls, *Science* **268**, 1144 (1995).

[17] L. L. Conte, C. Chothia, and J. Janin, *J. Mol. Biol.* **285**, 2177 (1999).

[18] S. Jones and J. M. Thornton, *J. Mol. Biol.* **272**, 121 (1997).

Tree-based Genomic Analysis

The Evolutionary Trace Method

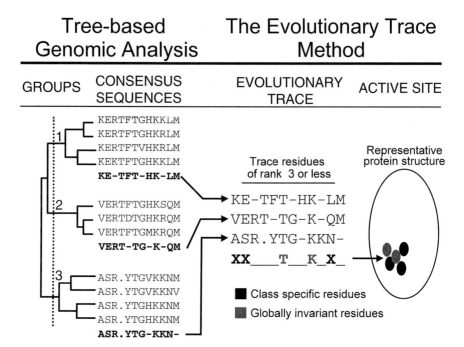

FIG. 1. The Evolutionary trace method. The first step in the evolutionary trace (ET) is to align all relevant amino acid sequences from which a sequence identity tree, or dendrogram, can be attained. In step 2, the tree is divided into groups, and the invariant residues in each group define its consensus sequence. In step 3, consensus sequences are compared, producing a trace sequence. Residue positions that have conserved residues within each group, but among the groups have different identities are called *class specific* (for example, positions 1, 2, and 11). Positions that have completely conserved amino acids across all family members are called *invariant* (positions 6 and 9). The *rank* of a residue is the minimum number of groups into which the tree has to be partitioned in order for that residue position to become class specific (for example, position 1 has rank 3). These trace residues, both class specific and invariant, are finally mapped onto a representative three-dimensional structure in step 4. If these residue cluster on the structure, then this site is considered to be of evolutionary importance and is likely an active site on the protein.

is the *sine qua non* of functionally important residues. In the last step class-specific residues are mapped onto a representative structure of the protein family. If they cluster, this indicates an evolutionary privileged site where variations are strictly linked to functional differentiation, as would be expected from an active site.

This procedure is not well defined, however, until we describe how to divide a family of proteins into functional classes. This is not trivial given that database searches can retrieve tens or even hundreds of related proteins whose functions have never been tested in the laboratory. This problem is resolved through a second hypothesis, namely, that a sequence identity tree is a good approximation of a

functional classification. This is plausible because proteins with very similar sequences will have diverged relatively recently and should therefore have more
closely related functions than proteins with weaker sequence similarity. In practice, sequence identity relationships provide a sensible estimate of functional relationships as can be seen from the sequence identity tree (dendrogram) of Gα
subunits (Fig. 2) which correctly separates the functionally distinct Gα subclasses
G$_i\alpha$, G$_o\alpha$, G$_s\alpha$, G$_t\alpha$ and G$_q\alpha$ into different branches.

Consequences of Tree Usage

The use of a sequence identity tree has several advantages. First, it completes
the remaining step in Fig. 1, so that the evolutionary trace is a fully defined algorithmic procedure. In practice, standard software such as PILEUP (distributed
through GCG), or CLUSTALW[19] and PHYLIP[20] provide sequence alignments
and identity trees. Although these programs may not generate perfectly identical
trees, most differences will be confined to nodes that are near the leaves rather than
the root of the tree. Since these terminal nodes contribute little to a trace, these
variations have little impact. If sequence identity drops below 25–30%, however,
nodes that are closer to the root and even the alignments may not be robust. The
simplest solution to this difficulty is to narrow the analysis to subfamilies within
which sequence similarity is higher. If a trace over the full family is still desired, it
may then be possible to align subfamilies following the method we discuss below
for GPCRs.

Second, the tree establishes a natural hierarchy among class-specific residues
that reflects the relative impact of their variations during evolution. The hierarchy
is derived by computing successive traces as the protein family is progressively
divided into more classes, defined by the branches of the tree. Thus, the first trace
is computed with the entire family in one group. The second trace is done with
the family divided into two classes defined by the first two branches in the tree.
The third trace is done with the family divided into the three groups defined by the
first three branches in the tree. This is repeated up until the family is divided into
N classes, where N is the total number of sequences. In this process, every residue
eventually becomes class specific, but some do so when the family is divided into
fewer branches than others. By definition, a residue's *evolutionary rank* (rank for
short) is the minimum number of branches into which it is necessary to divide the
family for this residue to become class specific. Thus a residue of rank k will be
variable within one of the first $k - 1$ branches of the tree, but it will be invariant
in each of the first k branches. Since nodes nearer to the tree root reflect the most
profound evolutionary splits, residues ranked low are correlated with the most
fundamental features of the protein's function. As the evolutionary rank number

[19] J. Thompson, D. Higgins, and T. Gibson, *Nucleic Acids Res.* **22,** 4673 (1994).
[20] J. Felsenstein, *Cladistics* **5,** 164 (1993).

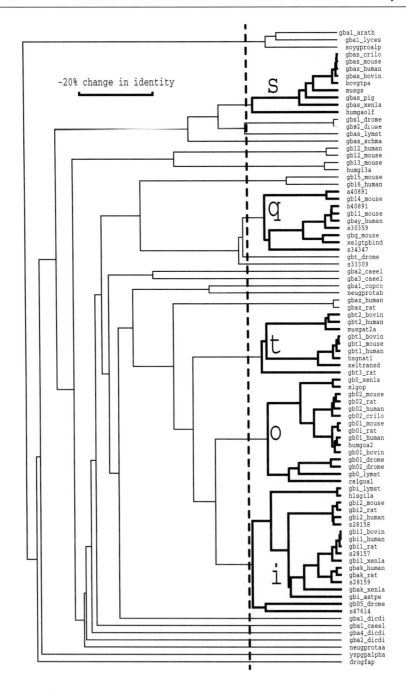

~20% change in identity

grows, class specificity is associated with evolutionary divergences of less and less significant, until at some rank threshold they become so trivial that class specificity loses significance. That threshold is identifiable because at that point class specific residues start to map randomly onto the surface.

Third, ET's use of the evolutionary tree follows a strategy that is closer to experimental mutational analyses than to typical computational methods. The latter are based on reasoning by analogy, that is, the analysis of protein X depends on recognizing that it bears sequence motifs also found in, say, protein A, and therefore X has some of the properties of A. Mutational analysis uses a different paradigm. It constructs variants X', X'', X''', and assays whether they are functionally different from X. This creates a causal link between residues and function. ET also links sequence variations with functional differences, using evolutionary divergence, or lack thereof, as its "virtual" functional assay.

If we accept that tree branch points are virtual assays and that ET performs retrospective mutational analysis based on evolutionary experiments, then the fourth consequence of tree usage is that ET will in fact often benefit from more mutations and more assays than are typically available in the laboratory. First, this is because the comparison of sequence yields a large number of pairwise variations (ET's equivalent of mutations). Second, and this is a crucial, because a tree with N proteins has $N-1$ branch points. Since each branch point is equivalent to a virtual assay, even if only a third to a half of these are above the noise threshold, $(N-1)/3$ is far more than the handful of assays typically available in the laboratory.

Initial Control Studies

Before we discuss the uses of ET in the context of G protein signaling, it is useful to briefly review control studies that show that class-specific residues exist and in fact cluster at active sites. In SH2 and SH3 modular signaling domains, and in the zinc finger domain of intracellular hormone receptors, the optimally ranked class-specific residues form a tight cluster on the protein structure while the remaining protein surface is free of ET signal.[21,22] These clusters match the interfaces known from structures of the protein–ligand complex. In the SH3 analysis, a discrepancy between the predicted and the observed structural interface reproduces that found by alanine scanning mutagenesis. Thus the evolutionary mutational experiments embodied in ET analysis match those from the laboratory and add functional insights to the protein structure.

FIG. 2. Sequence identity tree of G_α subunits. The sequences of 88 G-protein α-subunits were retrieved from SwissProt and aligned using the GCG program PILEUP. The resulting dendrogram is shown here with the dashed line indicating the tree partition where the five principal G_α subgroups are separated. Since functional subgroups can be attained from appropriate division of the dendrogram, identity trees are used to approximate a functional classification of the members of a protein family.

Comparison with mutational studies further suggests that evolutionary rank is linked to functional specificity. One example is SH2 domain mutations. Those at low-rank residues eradicate activity; mutations of higher rank residues modulate activity without destroying it outright; and mutations of the highest ranked residues have no effect on function. Another example is in the zinc finger domain of nuclear hormone receptors where low-ranked class-specific residues fall into two qualitatively different groups. The first group appears to define the essential characteristic of DNA binding. It contains residues with ranks between 1 and 7 that either are invariant or undergo at most one substitution and contact the most conserved bases of the DNA response element. In contrast, the second group appears to define the specific response element that will be recognized. It consists of positions with ranks between 8 and 21 that undergo many and highly nonconservative substitutions, and which contact DNA bases that are themselves so variable as to fall outside the consensus sequence.

These examples show that some active sites evolve through variations on a conserved structural framework and that sequence identity trees yield reasonable approximations of functional classifications. We now discuss how evolutionary data can help locate binding sites, identify functional determinants, and propose models of quaternary structure in proteins from the G-protein-signaling pathway.

Evolutionary Traces along G Protein Signaling Pathway

Active Sites in Gα Subunits

In the Gα family, the evolutionary trace identifies a likely receptor interface whose role is confirmed by mutational analysis. These and other studies severely constrain the receptor–G protein interaction and lead to a model of the quaternary structure of the GPCR–G protein complex.

ET analysis, performed on the tree in Fig. 2,[23] yields three clusters of class-specific residues. One is at the nucleotide binding cleft, and the other two form distinct functional surfaces on opposite sides of the Gα Ras-like domain. The first of these two surfaces, A1, comprises 17 residues from the distal two-thirds of helix α_5, the sixth β-strand (β_6), the α_4/β_6 loop, the N-terminal ends of β_4 and β_5 and the C-terminal tail, following the nomenclature of Noel et al.[24] The second surface, A2, is larger with 32 residues distributed on either side of helix α_2 and strands β_1, β_2, helix α_3, and loops β_3/α_2, β_4/α_3, and β_1/α_2. On the strength of data linking the C-terminal tail to receptor specificity,[25] site A1 was initially thought to

[21] O. Lichtarge, H. R. Bourne, and F. E. Cohen, *J. Mol. Biol.* **257,** 342 (1996).

[22] O. Lichtarge, K. R. Yamamoto, and F. E. Cohen, *J. Mol. Biol.* **274,** 325 (1997).

[23] O. Lichtarge, H. R. Bourne, and F. E. Cohen, *Proc. Natl. Acad. Sci. U.S.A.* **93,** 7507 (1996).

[24] J. P. Noel, H. E. Hamm, and P. B. Sigler, *Nature* **366,** 654 (1993).

[25] B. R. Conklin and H. R. Bourne, *Cell* **73,** 631 (1993).

be the receptor interface leaving A2 as the logical interface with $G\beta$.[26] The $G\alpha\beta\gamma$ trimer structure[27] shows that indeed the $G\alpha$–$G\beta$ interface spans nearly two-thirds of A2.

The role of site A1 in receptor coupling was further confirmed by Onrust *et al.*,[28] who constructed 100 $G_t\alpha$ mutants with either single (92) or double (8) replacement of residues with alanine. Each mutant was classified either as wild-type or as having impaired coupling to the receptor based on two assays: (1) a decrease in the rate of $G_t\alpha$ hydrolysis by trypsin, or (2) a decrease in $G_t\alpha$ binding to photoactivated rhodopsin (the receptor responsible for the physiological activation of transducin). Figure 3 displays, on both sides of the $G_t\alpha$ structure, the comparison between the predicted importance of residues and the effect that alanine substitutions has on GPCR coupling. The overall agreement, shown in black, is 68% (sensitivity, 75%; specificity 65%; $p = 0.002$) indicating that nearly 7 out of 10 times ET analysis correctly anticipated whether an alanine would or would not alter function. This outcome may underestimate ET's predictive accuracy, however, because many of the residues that were predicted to be important but yet produced no functional change upon alanine substitution, shown in light gray, were located at the nucleotide binding cleft or at the $G\beta\gamma$ binding site. Therefore, it is likely that some of these residues are important for signaling, but not as measured by assays aiming to detect GPCR interaction. At site A1, where the experimental assay and ET analysis are most likely to measure the same effect, the positive predictive value of ET rises to 79%. This discrepancy highlights an important difference between mutational analysis based on evolution and that based on the laboratory experiments: the former is sensitive to all the functions of a protein, whereas the latter will focus on the specific properties tested for by the assays.

The agreement between evolutionary and laboratory-based mutational analysis strongly suggests that A1 is an interface to the G-protein-coupled receptor (GPCR), and this knowledge can be incorporated into a model of the GPCR-G protein complex shown in Fig. 4. This model is also based on the lipid modifications of the N terminus of $G\alpha$ and of the C terminus of $G\gamma$, which suggest that both interact with the membrane; on contacts between the third intracellular loop of the receptor with the C terminus[29] and the N terminus of $G\alpha$,[30] and also with $G\beta\gamma$[31]; and between the

---

[26] H. R. Bourne, *Philo. Trans. R. Soc. Lond. B Biol. Sci.* **349**, 283 (1995).

[27] M. A. Wall, D. E. Coleman, E. Lee, J. A. Iniguez-Lluhi, B. A. Posner, A. G. Gilman, and S. R. Sprang, *Cell* **83**, 1047 (1995).

[28] R. Onrust, P. Herzmark, P. Chi, P. D. Garcia, O. Lichtarge, C. Kingsley, and H. R. Bourne, *Science* **275**, 381 (1997).

[29] E. Kostenis, B. Conklin, and J. Wess, *Biochemistry* **36**, 1487 (1998).

[30] J. M. Taylor, G. G. Jacob-Mosier, R. G. Lawton, A. E. Remmers, and R. R. Neubig, *J. Biol. Chem.* **269**, 27618 (1994).

[31] G. Wu, J. L. Benovic, J. D. Hildebrandt, and S. M. Lanier, *J. Biol. Chem.* **273**, 7197 (1998).

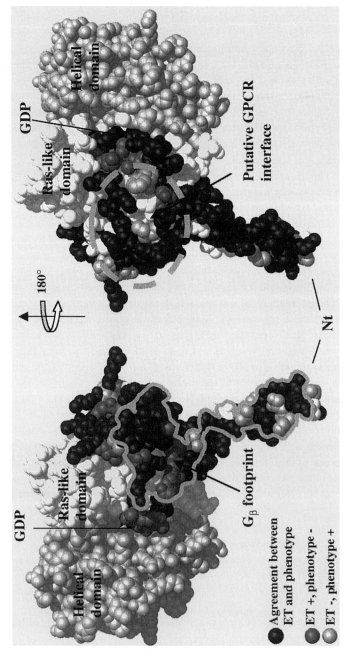

FIG. 3. ET prediction and mutational analysis agree on GPRC binding sites on Gα. Comparison of important residues determined by trace analysis on the G-protein family shown in Fig. 2 and those experimentally determined to be functionally important residues,[28] displayed in the structure of Gα·GDP. There is agreement at VV residues, shown in black. XX residues were false negatives, at this functional resolution (in medium gray), and YY were false positives (in light gray). As discussed in the text many of the latter are likely to be important, but for other functions than coupling to the receptor. Nt, Amino-terminus of Gα.

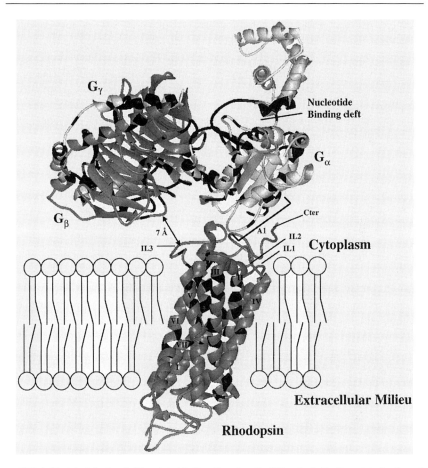

FIG. 4. Model of rhodopsin/G-protein binding. Based on the ET analysis the G-protein family and the available structural and experimental data, a quaternary model for the interaction between the G-protein and the receptor (in this case rhodopsin) can be generated. The receptor is shown in gray and is oriented in a cartoon of the lipid bilayer with the intracellular space oriented at the top of the figure. The G protein is colored as follows: Gα, white; Gβ, gray; Gγ white. Black denotes trace residues on all four proteins.

fourth intracellular loop and $G\beta\gamma$.[32] This configuration creates a favorable charge interaction between the bilayer and the $G\alpha\beta\gamma$ complex,[33] and it shows that the short intracytoplasmic loops of the receptor cannot reach the nucleotide binding cleft. Thus exchange of GDP for GTP must be triggered "at a distance" through an

[32] W. J. Phillips and R. A. Cerione, *J. Biol. Chem.* **267,** 17032 (1992).
[33] D. G. Lambright, J. Sondek, A. Bohm, P. S. Nikolai, H. E. Hamm, and P. B. Sigler, *Nature* **379,** 311 (1996).

allosteric pathway. Pending a definitive structure of a GPCR–G protein complex, this model also provides a useful context in which to discuss experimental studies of their interactions. In particular, it is interesting that intracellular loop 2 is the most likely contact between the receptor and site A1, while intracellular loop 3 is wedged between $G\alpha$ and $G\beta\gamma$, where it is accessible to the N- and C-terminals of $G\alpha$ and to $G\beta\gamma$. The fourth intracellular loop of the receptor can interact with the $G\beta\gamma$ and the first intracellular loop is farthest from the G protein, consistent with the paucity of any data on its role in G protein coupling.

Active Sites in RGS Proteins

Downstream from the receptor, the $G\alpha$·GTP complex activates effectors, enzymes, and ion channels, until it reverts to its inactive $G\alpha$·GDP state. The intrinsic rate of GTP hydrolysis by $G\alpha$ is too slow, however, to account for the rate at which signaling is turned off. The regulators of G protein signaling (RGS) proteins[34–38] reconcile this difference by binding to and stabilizing the $G\alpha$ catalytic switch regions,[39] increasing the rate of $G\alpha$·GTP hydrolysis. The existence of RGS proteins in general and the diversity of family members indicates that regulation of RGS proteins may add yet another level of control in G protein signaling. For example, the inactivation of $G_t\alpha$ (the G protein of vision) via RGS9 (the physiological GAP for $G_t\alpha$) is enhanced in the presence of the γ subunit of the cGMP phosphodiesterase (the effector which $G_t\alpha$ activates). However, PDEγ inhibits RGS4, RGS16, GAIP, and the RGS9 subfamily members RGS6 and RGS7. In order to understand how the RGS domain may be regulated by effectors[40–42] or other factors,[43–45] we conducted an ET analysis of 42 members of the RGS family and mapped the results to the only RGS structure available at the time, RGS4.[39]

A trace of the RGS family identifies a large cluster of both invariant and class-specific residues on the surface of the representative RGS4 domain (Fig. 5). These residues have ranks below 20 while the rest of the protein's surface remains

---

[34] N. Watson, M. E. Linder, K. M. Druey, J. H. Kehrl, and K. J. Blumer, *Nature* **383,** 172 (1996).

[35] D. M. Berman, T. Kozasa, and A. G. Gilman, *J. Biol. Chem.* **271,** 27209 (1996).

[36] D. M. Berman and A. G. Gilman, *J. Biol. Chem.* **273,** 1269 (1998).

[37] K. M. Druey, K. J. Blumer, V. H. Kang, and J. H. Kehrl, *Nature* **379,** 742 (1996).

[38] C. W. Cowan, W. He, and T. G. Wensel, *Prog. Nucl. Acid Res. Mol. Biol.* **65,** 341 (2000).

[39] J. J. Tesmer, D. M. Berman, A. G. Gilman, and S. R. Sprang, *Cell* **89,** 251 (1997).

[40] W. He, C. W. Cowan, and T. G. Wensel, *Neuron* **20,** 95 (1998).

[41] V. Y. Arshavsky and M. D. Bownds, *Nature* **357,** 416 (1992).

[42] J. K. Angleson and T. G. Wensel, *J. Biol. Chem.* **269,** 16290 (1994).

[43] T. Benzing, M. Yaffe, T. Arnould, L. Sellin, B. Schermer, B. Schilling, R. Schreiber, K. Kunzelmann, G. Leparc, E. Kim, and G. Walz, *J. Biol. Chem.* **275,** 28167 (2000).

[44] B. Zheng, D. Chen, and M. Farquhar, *Proc. Natl. Acad. Sci. U.S.A.* **97,** 4040 (2000).

[45] A. Kovoor, C. K. Chen, W. He, T. G. Wensel, M. I. Simon, and H. A. Lester, *J. Biol. Chem.* **275,** 3397 (2000).

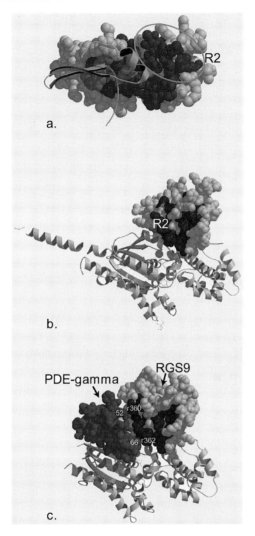

FIG. 5. ET analysis of the RGS family reveals two distinct active sites. (a) The Gα interaction surface on the RGS domain is correctly identified by ET (region on Gα that principally contacts the RGS domain is shown as secondary structure) and is composed of 4 invariant and 6 class-specific residues (only 1 contact residue was not identified by ET at this resolution, rank = 20). A second evolutionarily privileged site, termed *R2*, is located in close proximity to the RGS/Gα catalytic interface but does not directly contact Gα. (b) The complex of RGS4-$G_{i1}\alpha \cdot GDP \cdot AlF_4^-$ shows that R2 is exposed above the RGS/Gα interface and could function as a binding site for other factors to bind and modulate RGS activity. (c) The ternary complex of the RGS9-1 core domain, $G_{t/i1}\alpha \cdot GDP \cdot AlF_4^-$, and the C-terminal 38 amino acids from the effector subunit PDEγ, reveals that the effector binds to Gα along site R2 with which it contacts at residues 360 and 362.

free of signal until rank 23, suggesting this site is functionally important. This is consistent with the known RGS4-$G_{i1}\alpha \cdot$GDP$\cdot$AlF$_4^-$ structure[39] since 10 of the 11 RGS residues at the RGS4-Gα interface fall within this cluster.[46] The remaining 7 residues, if taken by themselves, form a second, smaller cluster, *R2,* that extends beyond the Gα-binding site and whose function is unknown *a priori.*

Two observations suggest that site R2 is an interface whereby the effector could influence RGS domain activity. First, amino acids in this region vary in a manner that is consistent with the unique activity of distinct RGS proteins in the presence of the PDEγ. Specifically, in proteins inhibited by PDEγ, the residues at RGS4 position 117 are acidic, and at position 124 they are either polar or hydrophobic, but these residues are hydrophobic (L) and basic (K), respectively, in RGS9 which is enhanced by PDEγ. Second, in the Gα–RGS complex, site R2 lies in near contiguity with to a part of cluster A2 in Gα that (a) does not interact with Gβ and (b) contains residues linked to PDEγ interaction.[47] Thus in order to influence GTPase activity, the effector is likely to bind the RGS–Gα complex by spanning part of A2 and R2 (Fig. 5).[46]

Structural data now supports a role for R2 in mediating these interactions. The structure of the catalytic core domain of RGS9 in complex with both $G_{t/i1}\alpha \cdot$GDP$\cdot$AlF$_4^-$ and the C-terminal 38 amino acids of PDEγ reveals that PDEγ V66 contacts R2 at class-specific residue RGS9-W362 (RGS4-126) (Fig. 5c). A second R2 residue RGS9-R360 (RGS4-124) is in close proximity to PDEγ D52.[47] Moreover, other R2 residues in the $\alpha5/\alpha6$ connecting loop lie parallel to the effector binding site on Gα, suggesting that they could play in role in positioning the RGS domain for interactions with both the effector and the effector bound Gα.

Allostery and Specificity in RGS

Mutational analysis reveals that residues at site R2 effect the activity of the RGS domain, even though they do not directly contact G$_t\alpha$. A series of mutations were made in RGS7, the protein with the most closely related RGS domain to RGS9, yet potently inhibited by PDEγ.[48] Class-specific residues in RGS7 corresponding to RGS4 positions 77 (RGS7-348), 117 (RGS7-387), and 124 (RGS7-394) were mutated singly, doubly, and all together to their corresponding amino acids from RGS9. Both E387L/P394R and the triple mutants were remarkable for reduced basal activity, down to a level slightly less than the PDEγ inhibited form of the wild-type RGS7. Addition of PDEγ caused no further inhibition suggesting the mutants were constitutively inhibited as if in the presence of PDEγ, as shown in Fig. 6. The mutation at residue 348 has little effect by itself, and the triple mutant

[46] M. E. Sowa, W. He, T. G. Wensel, and O. Lichtarge, *Proc. Natl. Acad. Sci. U.S.A.* **97,** 1483 (2000).

[47] K. C. Slep, M. A. Kercher, W. He, C. W. Cowan, T. G. Wensel, and P. B. Sigler, *Nature* **409,** 1071 (2001).

[48] M. Sowa, W. He, K. Slep, M. Kercher, O. Lichtarge, and T. Wensel, *Nat. Struct. Biol.* **8,** 234 (2001).

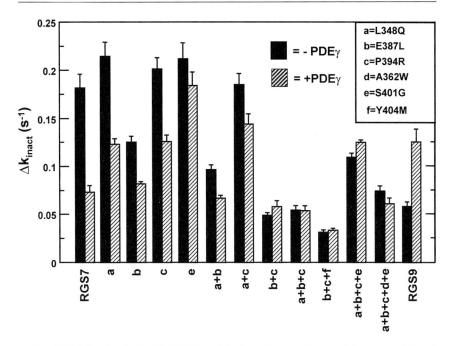

FIG. 6. Mutational analysis of the RGS domain indicates Trace residues participate in regulation of RGS domain activity. Mutations were made in the RGS7 catalytic core domain at residues in site R2 (Figs. 4a and 4b), replacing these residues with their corresponding amino acids from RGS9. Mutant proteins were expressed with a GST tag and purified via GSH affinity chromatography. The proteins were then assayed for their ability to increase the rate of GTP hydrolysis by $G_t\alpha$ in either the absence (black bars) or presence (hashed bars) of PDEγ using the method of Cowan et al.[54] Δk_{inact} was calculated as: $\Delta k_{inact} = [k_{inact} (RGS + G_t\alpha) - k_{inact} (G_t\alpha)]$, with k_{inact} calculated by fitting the time course of GTP hydrolysis to: $\%GTP$ hydrolyzed $= 100(1-\exp[-k_{inact}\cdot time])$. Mutations are labeled with letters for simplicity.

behaves nearly the same as the E387L/P394R mutant. Thus, class-specific residues 387 and 394 are critical for regulating RGS7 domain activity, and changes at these positions are sufficiently drastic as to alter RGS conformation or dynamics in the manner that appears to mimic the PDEγ inhibited form of the wild-type protein.

In order to determine which residues controlled the RGS-specific enhancement or inhibition effect of PDEγ, additional mutations were made at the RGS/Gα interface. Class-specific residues RGS7-A396, S401, and Y404 were targeted because they differed markedly between RGS7 and RGS9. The triple mutant E387L/P394R/Y404M remains insensitive to PDEγ, but has a reduced activity as compared to E387L/P394R (Fig. 6). Mutant L348Q/E387L/P394R/S401G has basal activity that is nearly that of wild-type RGS9 with PDEγ, and on PDEγ addition is slightly enhanced to match exactly the activity of RGS9 with PDEγ. Adding the mutation A396W to produce the mutant L348Q/E387L/P394R/A396W/S401G creates a protein that now is slightly inhibited in the presence of PDEγ, possibly due to structural restrictions imposed by surrounding residues (Fig. 6). Thus, residue S401 in RGS7 appears to be a critical determinant of the direction of the PDEγ effect on Gα. However, S401 requires the assistance of 387 and 394 since when S401 is mutated alone, the protein remains inhibited by PDEγ, indicating an allosteric relationship between 387/394 and 401.[48]

A network of residues connects 387/394 to Gα contact residue 401 through the α_5/α_6-connecting loop. Residues 387 is located N-terminal to the α_5/α_6-connecting loop in which lies P394. This loop is critical for the GTPase accelerating activity of the RGS domain[47,49] and is composed almost entirely of class-specific residues, consistent with an important role for the entire protein family. Residues at positions corresponding to 387 and 394 may exert their influence by communicating through this loop to the catalytic interface, with specificity determined by the amino acids that comprise both the α_5/α_6-connecting loop and the RGS/Gα interaction surface.

These results illustrate that the knowledge of class-specific residues may help direct mutagenesis to identify and unravel the interplay of functional elements in a multiprotein complex, including an intraprotein allosteric pathway.

Signal Transduction in G-Protein-Coupled Receptors

The last example focuses on the membrane domain of G-protein-coupled receptors (GPCRs), where ET helps identify which residues mediate general signal transduction properties and which are responsible for ligand-specific functions. This distinction is possible because ligands are extremely diverse in size and character, whereas G proteins are much more conserved and couple to receptors in both a one-to-many and many-to-one fashion. It follows that ligand binding should be highly specific while signal transduction and G protein coupling is likely to be more generic. To distinguish the functional determinants responsible for these distinct aspects of GPCR function, the approach is to identify positions that are important to all receptors and compare them to those that are important in a given subfamily.

In order to find the global functional determinants, GPCRs were selected broadly, including 58 opsins, 58 adrenergic receptors, 63 chemokine-related receptors, and 30 olfactory receptors, all in Class A, as well as 33 secretin-related receptors, from Class B. Before an evolutionary trace can be computed on all these receptors, it is necessary to align them, which is difficult because members of Class B have no sequence homology and traditionally cannot be aligned to members of Class A. We now describe a novel procedure to identify and align remote homologs based on the correlation of evolutionary ranks at cognate residues.

These correlations are obtained as follows. The five receptor families are traced separately, to assign an evolutionary rank at every position of their seven

[49] M. Natochin, R. L. McEntaffer, and N. O. Artemyev, *J. Biol. Chem.* **273,** 6731 (1998).

TABLE I
SPEARMAN RANK-ORDER CORRELATION COEFFICIENTS FOR GPCRs[a]

Correlation	OP-AD	OP-TH	AD-TH	OP-OL	AD-OL	TH-OL	Average global
TM1	0.09	0.28	0.3	0.33	0.34	0.12	0.24
TM2	0.07	0.32	0.32	0.12	0.45*	0.36	0.27
TM3	0.58***	0.40*	0.40*	0.45**	0.05	0.49**	0.40*
TM4	0.21	0.21	0.41*	0.56**	0.36	0.37	0.35
TM5	0.56**	0.64***	0.64***	0.41*	0.28	0.21	0.46*
TM6	0.52**	0.38*	0.37*	0.04	0.22	−0.38	0.19
TM7	0.41*	0.54**	0.33	0.19	0.65***	0.34	0.41*
Overall	0.43***	0.43***	0.49***	0.35***	0.39***	0.33***	0.40***

[a] The Spearman rank-order correlation coefficients are shown for the indicated comparisons of GPCR classes (OP, opsins; AD, adrenergic; TH, chemokine-related; OL, olfactory; TM, transmembrane helix; *, $p < 0.05$; **, $p < 0.01$; *** $p < 0.001$). The average global correlation is the average of all comparisons for a given helix. The overall correlation was determined by first concatenating the ET results for all 7 helices within each group and then calculating the correlation between the indicated groups.

transmembrane helices, TM1 to TM7. Residues are then assigned an *order* based on their ET rank. For example, if in opsins residue position 1 has an ET rank of 15, which is the fifth overall lowest rank in the sequence, then position 1 is assigned the evolutionary rank order of 5. When this is repeated in each receptor family, it becomes possible to compute the Spearman rank-order correlation between any two. A perfect correlation would mean that the relative rank of cognate residues was perfectly matched between two families. A correlation of 0, however, would indicate that cognate residues have evolutionary ranks that are completely independent. As might be expected, the evolutionary ranks among Class A receptors are positively correlated, presumably reflecting a common origin, but the correlations are modest and not equal in all helices, consistent with significant functional divergences (Table I).

Importantly, rank-order correlation is a sensitive indicator that two groups are correctly aligned. This is shown in Fig. 7, where the misalignment of visual and adrenergic receptors by up to ±4 positions is associated with a decrease in their correlation. The minimum is at ±2, because low-ranked residues are mostly internal. Hence when the internal residues of one helix (low-ranked) are compared to lipid-facing (high-ranked) residues in the other, at ±2, the correlation is least. Interestingly, even when the helices are back in phase at ±4, the correlation does not fully recover. Thus, the amphipathic nature of low versus high ranked residues is not sufficient alone to yield the maximum correlation of evolutionary ranks. The difference, although small, should reflect that cognate residues involved in similar functions have an additional degree of rank correlation.

a.

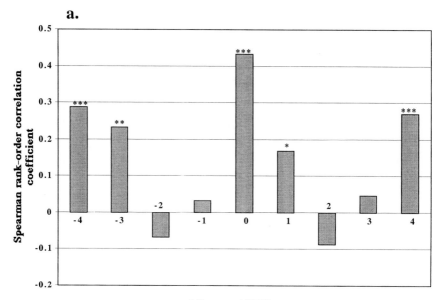

Alignment Shift

b.

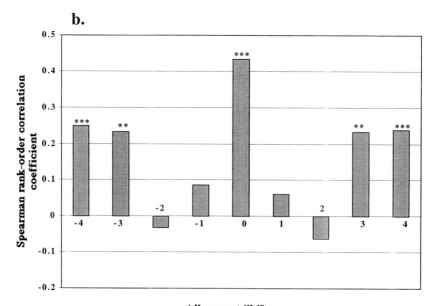

Alignment Shift

FIG. 7. GPCR correlation. (a) Spearman rank-order correlation coefficients are shown for the comparison between the opsin and adrenergic receptors as alignments are shifted by n residues (x axis). The

TABLE II
ALIGNMENT BETWEEN BOVINE VISUAL RHODOPSIN AND HUMAN
PARATHYROID HORMONE RECEPTOR

Correlation	Source	Sequence
TM1	OPS:	QFSMLAAYMFLLIMLGFPINFLTLYVTVQ
	PTH:	VFDRLGMIYTVGYSVSLASLTVAVLILAY
TM2	OPS:	LNYILLNLAVADLFMVFGGFTTTLYT
	PTH:	RNYIHMHLFLSFMLRAVSIFVKDAVL
TM3	OPS:	TGCNLEGFFATLGGEIALWSLVVLAIERYVVVCK
	PTH:	AGCRVAVTFFLYFLATNYYWILVEGLYLHSLIFM
TM4	OPS:	AIMGVAFTWVMALACAAPPLVGW
	PTH:	LWGFTVFGWGLPAVFVAVWVSVR
TM5	OPS:	SFVIYMFVVHFIIPLIVIFFCYGQLVFTV
	PTH:	GNKKIIWIIQVPILASIVLNFILFINIVRVL
TM6	OPS:	VTRMVIIMVIAFLICWLPYAGVAFYIFT
	PTH:	LLKSTLVLMPLFGVHYIVFMATPYTEVS
TM7	OPS:	IFMTIPAFFAKTSAVYNPVIYIMMNK
	PTH:	VQMHYEMLFNSFQGFFVAIIYCFCNG

It is now possible to align receptors from Class A and B so as to maximize its evolutionary rank order correlation with GPCRs from Class A. The result is shown in Table II, which describes the proposed alignment between bovine visual rhodopsin and the human parathyroid hormone receptor (PTH). As shown in Fig. 7b, these alignments yield a correlation between Class B and Class A receptors over all seven helices that is comparable to that among Class A receptors. Moreover, alternative alignments shifted by up to ± 4 residue positions yield significantly smaller correlations.

Three lines of evidence support this alignment. First, it maximizes sequence identity between Class A and Class B GPCRs, despite their profound lack of sequence similarity. This is significant because rank order correlation and sequence identity are independent of each other, as can be established from the lack of correlation between the degree of identity of cognate residues and the extent to which their rank orders are correlated.[50] Second, experiments based on the TM3 and TM6 alignments successfully reproduced double histidine mutations in the

---

correlation is best when there is no shift ($x = 0$), and it is least when internal residues are compared to lipid-facing residues ($x = \pm 2$). (b) Correlation coefficients are shown for Class A versus Class B comparisons. The results are similar to those shown in (a) and illustrate that correlation can be used as a means to gauge accurate alignments of proteins based on the abstract notion of evolutionary importance rather than sequence similarity (*, $p < 0.05$; **, $p < 0.01$; ***, $p < 0.001$).

PTH receptor that had created a Zn-dependent switch in opsin.[51,52] Third, the generic functional determinants of GPCR signaling predicted by a trace of Class A and B receptors, aligned as proposed, is consistent with the literature. Specifically, mutations at residues with rank-order below the 15th percentile disrupt signaling in a multitude of GPCRs, whereas mutations of residues ranked above the 85th percentile have few consequences and these are ligand specific.

Lastly, it is possible to compare generic versus specific determinants of signaling. This is shown in Fig. 8, where the 15% best-ranked residues from opsins, adrenergic, chemokine, olfactory, and secretin-related receptors taken together and from visual rhodopsin alone are mapped onto the $C\alpha$ trace of rhodopsin[53] in black and white, respectively, with overlaps in gray. Most trace residues are internal, and concentrate toward the cytoplasmic, G-protein-coupling side of the transmembrane domain. Remarkably, class-specific residues that are unique to visual rhodopsins (white) cluster around the retinal and then form a pathway extending toward the G protein through a series of van der Waals interactions between residues from TM3 and TM6. At the cytoplasmic third of the TM domain, this pathway spreads out into an intricate network of interactions involving residues from TM1 (N55), TM2 (N73, I75, L76), TM3 (E134, R135), TM6 (V254, P267), and TM7 (N302, P303, Y306, N310). Nearly all these residues are functionally important globally as well, and as expected, this gray cluster is immediately adjacent to the extracellular loops of rhodopsin, consistent with an intimate role in G protein activation.

Limitations and Future Direction

These examples show how evolutionary tree information may be used to identify functional sites in proteins; to tease out functional specificity determinants that can be targeted for experiments; to model protein–protein interactions; and to identity remote homologies. The variety of these applications reflects the richness of the "virtual functional assays" embedded in the tree, which links changes in sequence to changes in function during evolution. Since this approach exploits the experiments carried out by evolution, it follows that the utility of ET in its most direct applications will be limited by how much evolutionary information exists in a protein family.

If a protein has too few homologs, or if its evolutionary history is recent, then invariance of residues within branches will not be statistically significant. Scattered

[50] A. Philippi, S. Madabushi, E. Meng, A. Falicov, R. L. Dunbrack, S. R. Coughlin, F. E. Cohen, H. R. Bourne, and O. Lichtarge, submitted (2001).

[51] S. P. Sheikh, T. A. Zvyaga, O. Lichtarge, T. P. Sakmar, and H. R. Bourne, *Nature* **383,** 347 (1996).

[52] S. P. Sheikh, J. P. Villardarga, T. J. Baranski, O. Lichtarge, T. Iiri, E. C. Meng, R. A. Nissenson, and H. R. Bourne, *J. Biol. Chem.* **274,** 17033 (1999).

[53] K. Palczewski, T. Kumasaka, T. Hori, C. A. Behnke, H. Motoshima, B. A. Fox, I. Le Trong, D. C. Teller, T. Okada, R. E. Stenkamp, M. Yamamoto, and M. Miyano, *Science* **289,** 739 (2000).

[54] C. W. Cowan, T. G Wensel, and V. Y Arshavsky, *Methods Enzymol.* **315,** 524 (2000).

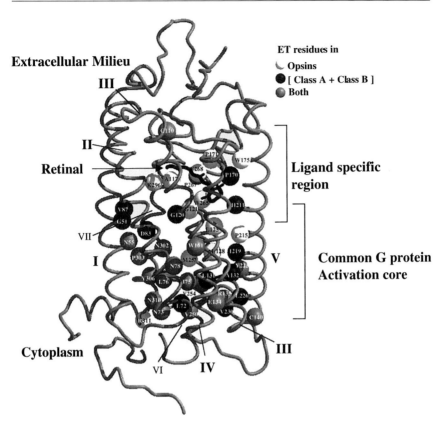

FIG. 8. Global and ligand specificities in GPCRs. Comparison between residues in the bottom 15th rank-order percentile from the visual opsin family and from selected receptors from [Class A + Class B], shown in the rhodopsin structure.[53] Residues that are unique to opsins, in white, form a cluster around the retinal moiety with a narrow extension toward the G protein. This extension then mushrooms into a network of interaction that involves residues from all TMs and that extends to the intracellular loops. These residues, in gray, are important to both the opsins and the other members of Class A and B included in this analysis (see text). A few residues, in black, are in the bottom 15th rank percentile for [Class A + Class B] receptors, but not for opsins.

trace signal on the protein surface is the hallmark of this situation and it occurs, in our experience, when the number of sequences falls below 12 to 15. Fortunately, nearly 75% of proteins in the structural database have at least that many homologs, and as ever more organisms undergo whole-genome shotgun sequencing, most protein families should grow large enough to support an ET analysis. Proteins that are uniquely mammalian may still remain too invariant for analysis, especially those that do not include modular domains that themselves have a longer evolutionary histories. As the number of predicted mammalian genes is revised downward from 100,000 to 40,000 or fewer, ET may still prove applicable to a majority of human proteins. Another limitation may also arise from the occasional active site that does not evolve through random variation and natural selection such as the antigen binding sites of antibodies. These are exceptions, however, and other active sites in the same proteins should evolve through the usual germline mutations and selection and therefore should be detectable through ET analysis.

ET also requires careful oversight by the user, who must fine-tune the input and interpret the output in its biological context. In particular, the sequences that are taken into account must truly share functions of interest. For example, a naive search for Gα sequences will also retrieve small GTP-binding proteins and even members of the EF-Tu family. An alignment of these proteins will be laden with gaps, and it will not identify the Gα binding sites to the GPCR, G$\beta\gamma$, RGS proteins, and effectors. This null result is not a failure. In fact, it is accurate since Gα, Ras, and EF-TU do not share these common binding sites, and it indicates that ET analysis should be refocused on an evolutionary subbranch that includes only Gα homologs. Database searches may also return protein fragments, mutant protein constructs, and functionally aberrant homologs, such as viral oncogenes. These should be at least identified if not eliminated from the ET analysis.

These issues illustrate that careful input selection is mandatory if the computation and the result are to be biologically meaningful. This is true of all computational procedures and it is no different from the careful preparation of reagents and adherence to assay protocols required of laboratory experiments. Toward the large-scale application of ET, it will be important in the future to help select and refine the input efficiently by adding statistics, heuristics, and automated iterative refinement steps.

Acknowledgments

This work was supported by a grant from the American Heart Association (O.L.) and by the W. M. Keck Center for Computational Biology and National Library of Medicine (LM07093, M.E.S.).

[39] Discovery of Ligands for $\beta\gamma$ Subunits from Phage-Displayed Peptide Libraries

By ALAN V. SMRCKA and JAMIE K. SCOTT

Heterotrimeric G protein $\beta\gamma$ subunits bind to many different effector proteins.[1] The detailed protein–protein interactions involved in $\beta\gamma$ binding to these targets are not well understood. One proposed model for interaction between $\beta\gamma$ subunits and effectors suggests that each effector may have shared and unique binding sites on the $\beta\gamma$ surface. Evidence in favor of this hypothesis is that activation of all effectors is blocked by GDP bound α subunits, yet specific site-directed mutants can be found that block activation of some effectors while having little effect on others.[2,3] That unique binding interactions may exist for various effectors suggests that specific ligands could be identified that bind to some of these unique interaction sites and thus specifically inhibit the activation of some effectors by $\beta\gamma$ subunits.

One method for defining sequences in effectors that bind to $\beta\gamma$ subunits has been to use peptides derived from the effectors in competition-based assays of effector activation.[4,5] Since $\beta\gamma$ subunits bind to peptides derived from various effectors, we reasoned that combinatorial phage-displayed peptide library screening could identify novel ligands for binding to G protein $\beta\gamma$ subunits at both common and unique effector interaction sites. One such screen in our hands yielded a family of peptides that further defines the protein requirements for interaction at specific sites on $\beta\gamma$ subunits.[5a] The peptides have a relatively high affinity for $\beta\gamma$ subunits and selectively interfere with certain $\beta\gamma$-mediated pathways. The phage bearing the peptides provide a convenient tool for comparing the $\beta\gamma$ binding sites of various effectors to determine if they are shared or unique. Ultimately such ligands could lead to the development of specific small-molecule inhibitors of $\beta\gamma$ subunit mediated pathways. Many phage display methods have been published elsewhere.[6,7] Here we describe many of these methods as they specifically relate

[1] D. E. Clapham and E. J. Neer, *Ann. Rev. Pharmacol. Toxicol.* **37**, 167 (1997).

[2] H. E. Hamm, *J. Biol. Chem.* **273**, 669 (1998).

[3] Y. Li, P. M. Sternweis, S. Charnecki, T. F. Smith, A. G. Gilman, E. J. Neer, and T. Kozasa, *J. Biol. Chem.* **273**, 16265 (1998).

[4] J. Chen, M. DeVivo, J. Dingus, A. Harry, J. Li, J. Sui, D. J. Carty, J. L. Blank, J. H. Exton, R. H. Stoffel, J. Inglese, R. J. Lefkowitz, D. E. Logothetis, J. Hildebrandt, and R. Iyengar, *Science* **268**, 1166 (1995).

[5] G. Krapivinsky, M. E. Kennedy, J. Nemec, I. Medina, L. Krapivinsky, and D. E. Clapham, *J. Biol. Chem.* **273**, 16946 (1998).

[5a] J. K. Scott, S. F. Huang, B. P. Gangadhar, G. M. Samoriski, P. Clapp, R. A. Gross, R. Taussig, and A. V. Smrcka, *EMBO J.* **20**, 767 (2001).

[6] G. P. Smith and J. K. Scott, *Methods Enzymol.* **217**, 228 (1990).

to analysis of G protein $\beta\gamma$ subunits. These methods also have the potential to be directed toward defining ligands for other signal transduction molecules.

General Phage and Bacterial Methods

f88-4 Phage Display System

The phage display system for these studies uses the phage f88 4, which is a derivative of fd-tet which itself is in the M13 family of filamentous phage.[8] In this system individual peptides or peptide libraries are fused with the N terminus of the pVIII coat protein. Each phage particle contains 2000–2500 copies of the pVIII coat protein. The DNA sequences for the peptides to be displayed are inserted at the 5' end of a second copy of gene VIII in the phage genome under the control of a separate promoter (*tac*) such that expression of the peptide/pVIII fusion proteins is restricted to 1–10% of the total coat proteins. This multivalent display of roughly 20–200 copies of the peptide per phage particle was crucial to successfully finding ligands that bound to $\beta\gamma$ subunits. Because any single peptide may have relatively low affinity for $\beta\gamma$, the combined effect yields a particle with a high avidity for the immobilized target. As will be described, if binding of the phage to $\beta\gamma$ is assessed under conditions that do not allow for multivalent interactions, no binding to $\beta\gamma$ subunits is detected (see section on phage ELISA). Compositions of solutions are given in the appendix at the end of this article.

F88-4 phage contain an inducible tetracycline-resistant element inserted in the origin of replication for $(-)$strand synthesis which renders the f88-4 genome replication deficient. As such, cells infected with f88-4 virions are tetracycline resistant and since f88-4 is replication deficient they do not form large plaques after infection of *Escherichia coli*. For propagation of phage clones, instead of selecting plaques, cells infected with virus are plated on tetracycline and isolated colonies of *E. coli* infected with phage are selected. Tetracycline resistance under the control of the *tet* repressor is induced with low levels of tetracycline followed by selection of infected *E. coli* at a higher tetracycline concentration.

Methods

Preparation of Starved K91 E. coli Cells for Infection with f88-4. Filamentous phage infect *E. coli* strains displaying the sex pilus encoded by the F episome. The following procedure maximizes display of the F pilus and concentrates the cells for optimal infection.

Streak K91 strain of *E. coli* cells (CGSC#4616; http://cgsc.biology.yale.edu) from a glycerol stock onto NZY agar and incubate overnight at 37°. Pick an isolated

---

[7] C. F. Barbas, J. E. Buss, J. K. Scott, and G. J. Silverman, "Phage Display: A Laboratory Manual." Cold Spring Harbor Laboratory Press, Cold Spring Harbor, NY, 2000.

[8] L. L. C. Bonnycastle, J. S. Mehroke, M. Rashed, X. Gong, and J. K. Scott, *J. Mol. Biol.* **258,** 747 (1996).

colony from the plate and transfer to a 14 ml snap-cap culture tube containing 2 ml of NZY medium. Incubate the culture overnight at 37° in a shaker at 250 rpm. In the morning inoculate 20 ml of NZY medium with 200 μl of the overnight culture in a 125 ml flask. Shake at 250 rpm at 37° until the cells reach OD_{600} of 0.45 (start to measure the OD after about 1 h and 20 min). Reduce the speed of the shaker to 100 rpm for 10–20 min. Measure the OD_{600} again, which should be between 0.5 and 0.6. Transfer the culture to a 50 ml sterile conical tube and centrifuge at 600g for 10 min at 4° or room temperature. Suspend the cells in 20 ml of sterile 80 mM NaCl and shake at 100 rpm in a 125 ml culture flask at 37° for 45 min to starve the cells. Transfer the mixture to a 50 ml sterile conical tube and centrifuge at 850g for 10 min at 4°. Gently suspend the cells in 1 ml of 4° NAP buffer. Keep the cells on ice until ready for use. The cells can be stored on ice for up to 1 week but are best if used when freshly prepared.

Propagation and Small-Scale Preparation of Phage Particles. Dilute the desired phage (clones or pools) to 1×10^2–1×10^4 phage particles/μl in TBS–gelatin and mix with 10 μl of starved cells in an 1.5 ml microfuge tube and incubate at room temperature for 15 min (always use aerosol resistant tips when pipetting phage). Add 500 μl NZY medium containing 0.2μg/ml tetracycline and incubate at 37° for 30 min in a shaker with the tube inverted in a beaker. Add 250 μl of NZY medium containing 15 μg/ml tetracycline and spread 100 μl on NZY agar containing 40 μg/ml tetracycline and incubate overnight at 37°.

To amplify phage for a binding experiment or sequencing, isolated colonies are selected from the NZY-tetracycline plates and inoculated into 1.5 ml of NZY medium containing 15 μg/ml tetracycline in a 14 ml culture tube and grown overnight at 37° at 250 rpm. After removal of *E. coli* cells by centrifugation, 1.2 ml of the supernatant is transferred to a fresh microfuge tube and the phage particles are partially purified by addition of 180 μl of polyethylene glycol (PEG)/NaCl followed by vortexing and incubation for 2 h on ice. Samples are centrifuged for 40 min at 12,000 rpm in microfuge at 4°. The supernatant is aspirated and the pellet is suspended in 100 μl of TBS. The tubes are heated to 70° for 30 min to kill residual *E. coli* cells and centrifuged for 10 min at 12,000 rpm at 4°, and the supernatant is removed to a new tube. The final phage concentration from this procedure is generally around 1×10^{10} phage particles/μl. Phage particles should be stored at 4° for several months or can be stored for a longer term at -20° if glycerol is added to 50%. For confirmation of the sequence of the phage at the same time as the phage particles are prepared, 3 ml cultures are grown and 1.5 ml of the culture supernatant is used to prepare ssDNA for sequencing using Qiagen M13 single-stranded (ss) DNA purification kit, and 1.2 ml is used for phage particle preparation.

Phage concentrations can be estimated by agarose gel electrophoresis of the purified phage and ethidium bromide staining of the phage DNA. A standard curve is generated with f88-4 phage that had been grown in large scale, purified twice by PEG precipitation and quantitated spectrophotometrically[6] (100 ng

DNA $\sim 1 \times 10^{10}$ phage particles). Two μl of the phage preparation is diluted to 8 μl with TBS followed by addition of 2 μl of 5× phage lysis mix. The sample is heated to 70° for 20 min followed by electrophoresis of the entire sample on a 1% agarose gel containing 4× GBB and 0.5 μg/ml ethidium bromide with f88-4 phage standard corresponding to a range of 5×10^9–4×10^{10} phage particles loaded on the gel. The gel is run for 4 h at 20 V.

Phage concentrations can also be estimated by titering the phage and given that the ratio of total phage particles to infectious phage particles is about 20 : 1. Starved cells are infected with phage serially diluted in TBS gelatin as described above. Rather than spreading 100 μl of the mixture on the plate, spot 15 μl of 4–5 dilutions on separate areas of a 100 mm dish with NZY-40 μg/ml tetracycline in agar. If the plates are put at 37° overnight the colonies often become too large and merge with each other, making it impossible to count the colonies. To overcome this, the plates are incubated overnight at room temperature in the dark. In the morning the plates are placed in a 37° incubator and the growth is monitored throughout the day until the colonies reach a size that is easily discernable.

Preparation of $\beta\gamma$ Subunits for Screening or ELISA

Prior to screening or phage enzyme–linked immunosorbent assay (ELISA) the $\beta\gamma$ subunits are immobilized in the wells of a 96-well microtiter dish. For screening we have used $\beta\gamma$ subunits immobilized via covalently attached biotin. For the ELISA we have successfully used both immobilization via biotin or immobilization of the $\beta\gamma$ directly to the plastic well of the dish.

$\beta\gamma$ Biotinylation

Sf9 Culture and Membrane Preparation. Biotinylated $\beta_1\gamma_2$ subunits are prepared using a modification of the method developed by Dingus *et al.*[9] Sf9 (*Spodoptera frugiperda* ovary) insect cells grown in suspension are infected with His$_6$-α_{i1}, β_1, and γ_2 with modifications to what has been previously described by Kozasa and Gilman.[10] The sf9 cells are grown in 800 ml Sf 900 II medium (Gibco-BRL, Gaithersburg, MD) in a 2 liter culture flask with shaking at 125 rpm at 27–28°. The cells are infected at a density of 2–3 × 10^6 cells/ml with 7.5 ml of His$_6$, α_{i1}, 5 ml β_1, and 2.5 ml γ_2 viruses at approximately 1 × 10^8 pfu (plague-forming units)/ml each. The insect cells from the 800 ml culture are harvested by centrifugation in 500 ml culture bottles in a Beckman JA 10 rotor at 2400 rpm. The cells are suspended in PBS (10 mM KPO$_4$, pH 7.4, 150 mM NaCl) and transferred to a 50 ml conical culture tube and centrifuged at 3000g for 30 min at 4°. The

---

[9] J. Dingus, M. D. Wilcox, R. Kohnken, and J. D. Hildebrandt, *Methods Enzymol.* **237,** 457 (1994).
[10] T. Kozasa and A. G. Gilman, *J. Biol. Chem.* **270,** 1734 (1995).

supernatant is discarded and the cell pellets either frozen in liquid N_2 and stored at $-70°$ for later use, or suspended for immediate processing in 15 ml of 4° lysis buffer [50 mM HEPES, pH 8.0, 0.1 mM EDTA, 100 mM NaCl, 10 mM 2-mercaptoethanol, 10 μM GDP, and 100 μg/ml phenylmethylsulfonyl fluoride (PMSF)]. If the cells are removed from storage at $-70°$ they should be suspended in room temperature lysis buffer to facilitate thawing of the cell pellet. The suspension is frozen and thawed 4 × by alternately plunging the 50 ml tube containing the suspension into liquid N_2 and thawing in a 37° water bath. The lysate is then diluted to 100 ml with lysis buffer and the particulate fraction containing membranes is harvested by centrifugation at 100,000g for 45 min at 4°. The supernatant is discarded and the membranes are suspended in 60 ml of extraction buffer, 50 mM HEPES, pH 8.0, 3 mM MgCl$_2$, 50 mM NaCl, 10 mM 2-mercaptoethanol, 10 μM GDP and 100 μg/ml PMSF, using a Dounce homogenizer.

Extraction and Partial Purification of G Protein Heterotrimer. Proteins are extracted from the membrane fraction by addition of cholate to 1% to the suspended membrane fraction with slow stirring at 4° for 1 h. Detergent-insoluble particulate matter is removed by centrifugation at 100,000g for 45 min at 4°. The supernatant is diluted 5-fold with 20 mM HEPES, pH 8.0, 100 mM NaCl, 0.5% polyoxyethylene 10 lauryl ether (C$_{12}$E$_{10}$), 1 mM MgCl$_2$, 10 mM 2-mercaptoethanol, 10 μM GDP, 100 μg/ml PMSF, and loaded onto a 2 ml Ni-NTA agarose column at 0.5 ml/min overnight. The column is washed with 80 ml of 20 mM HEPES, pH 8.0, 1 mM MgCl$_2$, 10 mM 2-mercaptoethanol, 10 μM GDP, 300 mM NaCl, 5 mM imidazole, 0.5% C$_{12}$E$_{10}$, and 100 μg/ml PMSF. Heterotrimeric His$_6$-$\alpha_1\beta_1\gamma_2$ is eluted from the column with six successive 2 ml aliquots of 20 mM HEPES, pH 8.0, 100 mM NaCl, 0.1% C$_{12}$E$_{10}$, 10 μM GDP, and 150 mM imidazole. The eluted fractions are assayed for protein using an amido black protein assay[11] and the fractions are separated on a 12% sodium dodecyl sulfate (SDS)–polyacrylamide gel and stained with Coomassie blue. Fractions containing the largest amount of $\beta\gamma$ are used for the biotinylation reaction.

Biotinylation of $\alpha_i\beta_1\gamma_2$ Heterotrimer and Purification of Biotinylated $\beta\gamma$ Subunits. The His$_6$-$\alpha_i\beta_1\gamma_2$ eluted from the Ni-NTA-agarose is diluted to 1 mg total protein/ml with 20 mM HEPES, pH 8.0, 1 mM EDTA, 1 mM DTT, 100 mM NaCl, 10 μM GDP, and the final detergent concentration is adjusted to 0.05% C$_{12}$E$_{10}$. NHS-LC-biotin (Pierce, Rockford, Il) is added from a 20 mM stock in dimethyl sulfoxide (DMSO) to give a final concentration of 1 mM. The reaction is allowed to proceed for 30 min at room temperature followed by addition of 10 mM ethanolamine pH 8.0 from a 200 mM stock and incubation on ice for 10 min. The sample is diluted to 90 ml with dilution buffer: 20 mM HEPES, pH 8.0, 100 mM NaCl, 10 μM GDP, and 0.5% C$_{12}$E$_{10}$. Two ml of washed Ni-NTA agarose is added and incubated with mixing overnight at 4°. The mixture is poured through a column

[11] W. Schaffner and C. Weissmann, *Anal. Biochem.* **56,** 502 (1973).

and washed with 20 ml of dilution buffer. The column is warmed to room temperature for 15 min and washed with 5 ml of 30° wash buffer, 20 mM HEPES, pH 8.0, 100 mM NaCl, 10 μM GDP, and 1% cholate. $\beta\gamma$ subunits are eluted with 5 successive 2ml aliquots of 30° wash buffer plus 10 mM MgCl$_2$, 10 mM NaF, and 30 μM AlCl$_3$. The eluted $\beta\gamma$ is detected by electrophoresis of the fractions on a 12% SDS–polyacrylamide gel followed by staining with Coomassie blue. Biotinylation is confirmed by electrophoresis and blotting with streptavidin–horseradial peroxidase (HRP) and by showing that all of the biotinylated $\beta\gamma$ could be bound to streptavidin agarose. For screening of libraries and ELISA assays the b-$\beta\gamma$ is stored in the elution buffer. To test the viability of the $\beta\gamma$ in other assays the biotinylated $\beta\gamma$ may have to be exchanged into another buffer.

Libraries

Procedures for construction of phage displayed peptide libraries in f88-4 have been described. For detailed procedures for preparation of the libraries see (Ref. 7). Some libraries are available from Dr. George Smith's laboratory including an f88-4 linear 15-mer library and several types of cysteine constrained libraries as well as wild-type f88-4 virus (see Smith laboratory Web site for details (http://www.biosci.missouri.edu/smithgp/)]. Other phage display libraries are available commercially but use pIII-based peptide display (New England Biolabs, Beverly, MA).

Amplification of Libraries

Inoculate a 125 ml culture flask containing 15 ml NZY with a single fresh colony of K91 cells and shake overnight (250 rpm) at 37°. In a 2 liter culture flask, inoculate 300 ml NZY with 6 ml of the overnight K91 culture. Shake at 250 rpm until the cell concentration reaches OD$_{600}$ of 0.45. Slow the shaker to 100 rpm for 10 min to allow the cells to regenerate their pili. Measure the OD$_{600}$; it should not be over 0.65 (aim for 0.55–0.65). At OD$_{600}$ of 0.6, the cell concentration equals ~4 × 10^8 cells/ml, yielding at total of ~1.2 × 10^{11} cells. Transfer the cells to two 250 ml centrifuge bottles, and centrifuge them at 1000g for 10 min. Pour off the supernatant, briefly centrifuge the bottles again, and remove the remaining supernatant. Gently suspend each pellet with 150 ml 80 mM NaCl and transfer to a sterile 500 ml culture flask. Incubate the mixture for 60 min at 37° with shaking at low speed (50 rpm). Transfer the cells to two 250 ml centrifuge bottles, and spin them at 1100g for 10 min at 4°. Resuspend each pellet in 5 ml ice-cold NAP buffer, then transfer the cells to a 14 ml snap-cap tube. Wash each bottle out in sequence with a single aliquot of 1 ml of NAP buffer and transfer this to the snap-cap tube. This should give a final concentration of 10^{10} cells/ml in 11 ml. Store on ice at 4°.

Remove 30 μl of the starved cells to a 1.5 ml microfuge tube containing 10^6 f88-4 control phage particles to serve as a positive control infection. Add 2 × 10^{10}

phage particles from the library to be amplified to the remaining cells and mix by gentle inversion [this should give a multiplicity of infection (MOI) of ~5 cells per phage particle]. Let the infections stand for 10 min at room temperature with occasional gentle swirling. Briefly centrifuge at 1000g for 1 min at 4° on a tabletop centrifuge to bring down the droplets on the walls of the tube. Mix the cells with a 1 ml pipette. Add 0.5 ml of infected cells to each of 22 125-ml flasks with each containing 45 ml NZY and 0.2 μg/ml tetracycline. Shake the cultures at 250 rpm at 37° for 30–45 min to induce tetracycline resistance. Add 333 μl of 2 mg/ml tetracycline to each culture to bring the tetracycline concentration to 15 μg/ml. Mix and remove 20 μl samples from a few cultures for titering. This will quantitate the number of infected cells that will produce the library and hence the number of phage clones comprising the amplified library.

Shake the cultures at 250 rpm overnight (20 h) at 37°. Combine the library cultures in four 250 ml centrifuge bottles. Centrifuge at 2400g for 10 min at 4°. Pour off the supernatant into clean bottles, being careful not to disturb the cell pellet, and recentrifuge at 6000g for 10 min at 4°. Carefully pour the supernatant into tared, 250-ml centrifuge bottles and note the culture volume in each bottle (1 g = 1 ml). To each bottle, add 0.15 volume polyethylene glycol (PEG)/NaCl. Screw on caps tightly, and mix throughly by inverting the bottles gently ~100 times. Incubate the mixtures on ice for ≥4 h on ice or overnight at 4°. Pellet the phage by centrifuging at 6000g for 40 min at 4°. Pour off and discard the supernatant, being careful not to disturb the pellet. Remove the residual supernatant by briefly recentrifuging the bottles, tilting each bottle so that the pellet is opposite the remaining supernatant, and aspirating the residual supernatant with a P1000 Pipetman. Add 7.5 ml TBS to each bottle and shake at 150 rpm in a 37° incubator for ~30 min to dissolve the pellets. Centrifuge the bottles briefly to drive the solution to the bottom of each bottle. Transfer the solution from two bottles to a single tube, yielding two tubes. Rinse the bottles with another 7.5 ml TBS and add to each tube, as before. Balance the tubes with TBS and mix the phage thoroughly by inversion. Centrifuge the tubes at 10,000–20,000g for 10 min at 4° to clear the supernatants. Transfer the supernatants to fresh, tared tubes; note the volumes. Add 0.15 volume PEG/NaCl to each tube, and invert gently ~100 times. Allow the phage to precipitate by incubating the tubes on ice for ≥1 h (a heavy precipitate should be evident at the end of that time). Collect the precipitated phage by centrifuging at 10,000g for 40 min at 4°. Carefully and completely remove the supernatant. Add 10 ml TBS to each tube, and dissolve the phage pellet by gently vortexing, then allowing the pellet to soften at room temperature for ~1 h. Vortex again, and briefly centrifuge to drive the solution down. If the phage are to be further purified on a CsCl density gradient, add only 5 ml TBS to each tube and resuspend the phage as above. Combine the two supernatants into a single round-bottom polypropylene centrifuge tube. At this point, the phage can be heat treated to kill any remaining cells. Incubate the tubes for 30 min in a 70° water bath. (*Note:* The heat-treatment step is optional and may denature complex proteins displayed by some phage.) This step is not necessary if the phage are to be CsCl purified. Clear the supernatants by centrifuging the tubes at 10,000–20,000g for 10 min at room temperature or 4°. Pour the cleared supernatant from each tube into a 15-ml polypropylene snap-cap tube and store at 4° in the dark.

To determine the concentration and the yield of phage particles, treat an aliquot of phage with 5× Lysis mix, then dilute the mixture in 1× Lysis mix and run the samples on a 1.2% agarose gel in 4× GBB, using a known amount of control f88-4 phage treated in the same way as a standard. The concentration of phage particles can be more accurately assessed by spectrophotometric analysis; however, this is better done with CsCl-purified phage. The final concentration of phage should not exceed $\sim$3 × 10^{13}/ml, so once the phage concentration is known, it should be adjusted accordingly with TBS. To impede cell growth, the solution can be adjusted to a final concentration of 0.02% (w/v) NaN$_3$ (sodium azide, a highly toxic poison), or 20 mM Na$_2$EDTA. The phage can be stored long term in 50% (v/v) sterile glycerol at $-18°$.

Screening of Phage-Displayed Peptide Libraries against G Protein $\beta\gamma$ Subunits

Biotinylated $\beta\gamma$ subunits are immobilized in the wells of a microtiter plate coated with streptavidin. Prior to screening, the libraries are preadsorbed to wells of a plate containing 1 μg of streptavidin to reduce the possibility of obtaining streptavidin binders from the screen. To determine the degree to which specific $\beta\gamma$ binding clones are enriched with each stage of panning, the libraries are also screened in wells that do not have immobilized $\beta\gamma$ and equal numbers of f88-4 wild-type phage are screened in $\beta\gamma$ subunit-containing wells in parallel with the actual library screen against immobilized $\beta\gamma$.

Preabsorption of Libraries with Streptavidin

For each screening condition one well of a 96-well microtiter plate (Corning/Costar, Corning, NY) is coated with streptavidin (Sigma, St. Louis, MO) by incubating the plate overnight at 4° with 35 μl of TBS containing 1 μg of streptavidin. The streptavidin solution is removed and 200 μl of 2% dialyzed bovine serum albumin (BSA) in TBS is added to each well and incubated for 30 min at 4°. The BSA solution is then removed and the plate is washed 3 times with TBS. For each wash throughout this procedure a wash bottle is used to completely fill the wells being careful not to cross contaminate the wells, and the wells are emptied by shaking into a sink followed by slapping the plate face down on a paper towel to remove any remaining wash solution. Immediately after slapping the last wash from the plate add 10^3–10^4 equivalents of each phage library and 2% BSA in TBS/1.5% Tween 20 to bring the final volume/well to greater than 35 μl/well and the final concentration of Tween 20 to 0.5%. For example, if the library has a diversity of 1×10^8 clones, 10^{11}–10^{12} phage particles should be added to each well. Add the BSA/TBS/Tween to each well, then add the library using filtered pipette tips to transfer the library samples from their stock solutions. Incubate the plate 3–4 h at room temperature with rocking. Transfer the preadsorbed phage solutions to microfuge tubes.

Screening Libraries

A second plate that has been coated with streptavidin as described above is used for screening the library. The streptavidin solution is removed by aspiration and the wells are washed once with TBS. To each well is added 200 μl of 2% BSA in TBS followed by incubation at 37° for 1 h. The wells are washed three times with TBS/0.5% Tween 20 and 35 μl of biotinylated 50 nM $\beta\gamma$ subunit is added in TBS/0.5% Tween 20 for 40 min at 4°. After the incubation with b-$\beta\gamma$, 10 μl of 1 mM biotin in TBS/Tween 20 containing 2% BSA is added to each well and incubated for 10 min at 4° to block unbound biotin binding sites on streptavidin.

Remove the $\beta\gamma$ solution and wash the plate three times with TBS/Tween 20. Transfer the preadsorbed libraries to the wells containing $\beta\gamma$ subunits and incubate 2 h at room temperature. After the 2 h incubation aspirate the phage from the wells with a pipettor and wash the wells six times with PBS/Tween 20 with a 2 min incubation for each wash. After the last wash add 35 μl phage elution buffer (0.1 M HCl adjusted to pH 2.2 with glycine, 1 mg/ml BSA) to each well and mix with a pipette. Incubate 10 min at room temperature. Remove the eluate, transfer to another 96-well plate, and neutralize with 5–7 μl of 1 M Tris, pH 9.1. Verify that the pH has been neutralized to pH 7–8.5.

To amplify the eluted phage and to assess the yield of phage from the screen, K91 *E. coli* cells are infected with the eluted phage and known amounts of the control phage. Add 10 μl of starved K91 cells (made that same day) to the eluted phage from the control samples, the screening samples, and 10^6 particles of f88-4 phage vector and allow the infection to go for 15 min. The control samples are used solely for titering purposes and calculation of phage yields and they are not amplified for further rounds of screening. To induce tetracycline resistance add 140 μl of LB/0.26 μg/μl tetracycline to each infection reaction. Cover the plate with a lid and place in a humidified plastic box and shake for 30 min at 37°. After the tetracycline resistance genes are induced add 20 μl of LB containing 148 μg/ml tetracycline.

At this point the yield of phage from the selection is determined by titering 5 μl from each infection. After the first round of panning the phage yield should be about 10^{-4}–10^{-5}% (10^4 or 10^5 phage particles if 10^{11} phage particles are used for the screening). The titering is done in a range of 1 : 10 to 1 : 10,000 dilutions of phage in LB according to the procedure described in the section on propagation and preparation of phage particles. The remainder of the cells are incubated at 37° with shaking for 42 h to amplify the eluted phage. To calculate the percent phage yield, the efficiency of the infection is determined from the control infection with 10^6 particles of f88-4 (infective units derived from the titer/10^6 input phage particles). The infection efficiency is then used to determine the actual input tranducing units (TU) for each library (input TU equals the phage particles added to each well for the screening times the infection efficiency). The percent yield of the panning step is calculated by dividing the output TU determined from titering the eluted phage by the input TU (calculated above). The degree of enrichment is calculated by comparing the phage yields from the control panning (no $\beta\gamma$ in the well or $\beta\gamma$ screened with f88-4) with the phage yield from the actual screening well. At early stages of panning there may be very little enrichment of phage relative to control samples without biotinylated $\beta\gamma$, but with subsequent panning steps the yield of phage should be enriched relative to a control panning. This gives an idea of whether the panning steps have been successful.

After the 42-h amplification period the samples of amplified phage samples are harvested, the *E. coli* cells removed by centrifugation, and the supernatant is treated at 70° for 20 min to kill residual bacteria. The sample is centrifuged again to remove dead cells. Twenty μl of each culture supernatant is mixed with 90 μl of 60% glycerol and stored at$-20°$. Twenty μl of the supernatant is used for gel analysis with known amounts of control phage as described in the section on propagation and small-scale preparation of phage particles. This will help determine how much of the phage to use for the next round of panning.

Rounds 2, 3, and 4. Follow the same procedures as for round 1 with following exceptions:

1. Since the number of clones in the pools of enriched phage is relatively small compared to the number going into the first round of panning, fewer phage can be used in the subsequent rounds of screening. Use 10- to 100-fold fewer input phage than was used in the round 1 screening while keeping the volume at 35 μl and the concentration of Tween 20 in the preabsorptions as close to 0.5% as possible.

2. For each screen use the same amount of F-88 control phage as is used for the panning of the amplified phage pools. The phage from the control pannings in round 1 are not supposed to be used for anything but titering and determining the phage enrichment. Do not use these for subsequent rounds of panning.

3. Consider performing ELISA on eluted amplified phage pools to determine if there are increases in binding compared to vector-control phage. The highest sequence diversity and, possibly, the best binders can be present in rounds that precede rounds having yields that are above background.

After the third round of screening the amplified pools of selected phage should bind to $\beta\gamma$ subunits. The pools should be checked by performing an ELISA in the presence and absence of immobilized $\beta\gamma$. The binding should be compared with the binding of wild-type f88-4. If significant binding is observed above the f88-4 and -$\beta\gamma$ controls, the pools are diluted and used to infect K91 cells and spread on NZY tetracycline plates to give well-separated colonies as described in the section on propagation and small-scale preparation of phage particles. Individual colonies representing individual phage clones with unique peptide sequences are picked and grown overnight in 3 ml of media. The culture (1.5 ml) is used for phage preparation and purification by PEG precipitation as described. These clones of phage are then tested for binding for $\beta\gamma$ in the ELISA assay with the appropriate controls. If significant binding is observed, single-stranded DNA is prepared from the remaining 1.5 ml of culture supernatant using a Qiagen M13 single-stranded DNA preparation kit (or other method) for DNA sequencing.

Results from Screening. The phage selected from the libraries that bound to $\beta\gamma$ subunits had multiple sequences with various levels of homology to one another. For our analysis we screened several different types of libraries of various lengths and that had differing internal disulfide constraints.[8] Some of these sequences grouped into families are shown in Fig. 1. We have shown that synthetic peptides constructed on the basis some of these sequences are able to inhibit $\beta\gamma$ subunit-mediated regulation of phospholipase C-β (PLC-β) and phosphatidyl inositol 3-kinase γ but not adenylate cyclase I or a Ca^{2+} channel.[5a] Our model is that these peptides are binding at a unique site utilized by some effectors but not others.

Construction of Phage Displaying Selected Peptides

Phage bearing specific peptide sequences can be constructed by inserting complementary oligonucleotides encoding the desired sequence in frame with the gene encoding the second copy of the pVIII coat protein. Also encoded in the oligonucleotide is a portion of the leader sequence and *Pst*I and *Hin*dIII cloning sites.

Leader cleavage site
↓
M   L   S   F   A   N   V   P   A   E   G   D   D
.......ATG CT<u>A AGC TT</u>T GCC AAC GTC C<u>CT GCA G</u>AA GGT GAT GAC
           HindIII   PstI

5′-AGCTTTGCC XXXXXXXXXGCTGCA-3′
       AACGG XXXXXXXXXCG

The export signal sequence LSFA has to remain intact, NV are lost, and the phage sequence resumes with AAE.

For cloning of the insert, the double-stranded replicative form of the f88-4 viral DNA must be purified by standard methods[6] and cleaved with *Hin*dIII and *Pst*I.

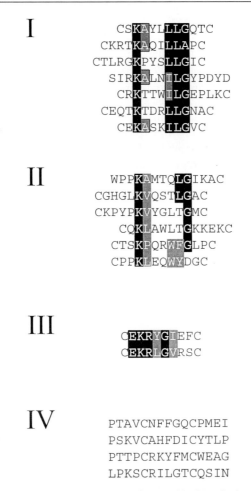

FIG. 1. Alignments of sequences obtained by random peptide phage display screening. Sequences were placed into four groups based on sequence similarity. Black boxes indicate identities and gray boxes indicate conservative substitutions. Reproduced from J. K. Scott *et al., EMBO J.* **20,** 767 (2001) with permission of Oxford University Press.

The oligonucleotides are annealed and then ligated into the cleaved f88-4 vector DNA. The ligated DNA is transformed into MC1061 *Escherichia coli* by electroporation or other methods. After electroporation, tetracycline resistance is induced by addition of 1 ml of NZY with low tetracycline (0.2 mg/ml) and shaking for 1 h at 37°. Two hundred μl of cells is then plated on NZY agar containing 40 μg/ml tetracycline and incubated overnight at 37°. Isolated colonies are picked and grown overnight in 3 ml NZY medium with 15 μg/ml tetracycline. Of the overnight culture, 1.5 ml is used for preparation of single-stranded phage

DNA (Qiagen M13 DNA preparation kit) followed by sequencing to confirm in frame fusion of the insert. The sequencing primer we have used has the sequence 5'-CTGAGTTCATTAAGACG-3'. The remainder of the culture is saved and used to prepare phage particles as described in the section on propagation and small-scale preparation of phage particles.

Based on our estimates of the affinity of the selected peptides for $\beta\gamma$ subunits we would predict that most peptides with an affinity of 100 μM or greater would bind in this assay. On the other hand we have fused peptide sequences from K^+ channels described by Krapivinsky et al.[5] but have been unable to detect binding. We suspect that there are steric constraints placed on the peptide by fusion to the pVIII coat protein that prevents interaction with $\beta\gamma$. To overcome this problem we are inserting spacer sequences between the coat protein and the displayed peptide so that the peptides will be more available on the surface of the phage for interaction with the $\beta\gamma$ subunits.

Analysis of Binding to $\beta\gamma$ Subunits using Phage ELISA

To examine phage binding to $\beta\gamma$, the $\beta\gamma$ subunits are immobilized in the wells of a microtiter plate. Selected phage particles representing single clones or pools of clones are then allowed to bind to the immobilized $\beta\gamma$ followed by washing and detection of the bound phage with an antiphage antibody. The $\beta\gamma$ subunits either can be immobilized directly or can be bound to immobilized streptavidin via covalently attached biotin. The biotinylation approach ensures that the $\beta\gamma$ is in an active conformation when it is immobilized, increasing the chances that sites required for binding are available to the peptide. Binding of the $\beta\gamma$ subunits directly to the plate could lead to partial denaturation of the protein and exposure of nonspecific sites, but the procedure is simpler because the reactions involved in biotinylation of $\beta\gamma$ do not have to be performed.

Another ELISA method that has been used to detect binding for phage to target molecules is to immobilize the phage in the bottom of the well and detect binding of soluble target with an antibody directed against the target molecule. This approach measures the ability of the target molecule to bind monovalently to the immobilized phage since the analyte in solution can only bind to one immobilized phage particle at a time. When the target is immobilized, since the f88-4 phage has multiple displayed peptides it can bind to multiple immobilized target molecules simultaneously, resulting in increased avidity. We have been unable to detect binding of $\beta\gamma$ to the phage that we have selected when the phage was immobilized, indicating that the use of a polyvalent phage display system is essential for detecting phage binding to $\beta\gamma$ subunits with any of the peptides we have identified.

ELISA Assay

One μg of streptavidin is added to each well to be tested in a 96-well microtiter dish in 40 μl of TBS. Control wells are set up that either do not contain $\beta\gamma$ subunits or will be screened with wild-type f88-4 phage with no peptide displayed. These wells should also be coated with streptavidin. Each condition is tested in duplicate. After overnight incubation with streptavidin at 4° the streptavidin solution is aspirated and replaced with 35 μl of a solution of TBS containing 0.1–0.5% Tween 20 and 2% BSA for 1 h at 4°. After 1 h the BSA solution is aspirated and the wells are washed 3 times at with TBS 0.1–0.5% Tween 20. For each wash a wash bottle is used to fill the wells and the wells are emptied by shaking into the sink followed by slapping the plate face down on a paper towel to remove any remaining wash solution. After washing, 50–100 ng of b-$\beta\gamma$ in 35 μl of TBS-Tween 20 is added to the streptavidin-coated wells and incubated for 1 h at 4°. After incubation, the $\beta\gamma$ solution is aspirated, the wells washed three times with TBS/0.1–0.5% Tween 20 and 1×10^{10} phage are incubated with the immobilized $\beta\gamma$ in TBS/0.1–0.5% Tween 20 for 1–4 h at 4°. Overall, the ELISA assays for phage binding are relatively insensitive to phage number but it is still a good idea to use a standard known amount of phage for each binding assay (see Fig. 2). For each incubation 35 μl of TBS–Tween 20 is added to the well followed by direct addition of phage in 1–2 μl. For competition ELISAs, peptides or other competitors are added to the wells just

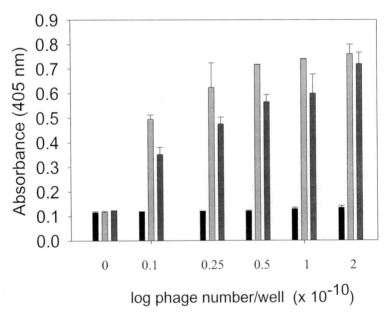

FIG. 2. Binding of phage displaying peptides to $\beta\gamma$ subunits at various concentrations of phage particles. Fifty nM biotinylated $\beta\gamma$ was bound to streptavidin-coated 96-well plate and various amounts of phage clones displaying two different peptide sequences were analyzed in the phage ELISA assay as is described in the text. Black bars are wild-type f88-4 phage controls; gray bars are two different clones displaying $\beta\gamma$ binding peptides.

prior to addition of the phage and the incubations only go for 1 h. After washing three times to remove unbound phage, anti-M13 antibody (Amersham Pharmacia Biotech, Piscataway, NJ) linked to horseradish peroxidase is added to each well in TBS/0.1–0.5% Tween 20 with 2% BSA (1 : 5000 dilution) and incubated at 4° for 1 h. The wells are washed three times with TBS–Tween 20, 40 μl of ABTS is added, and the color reaction allowed to proceed for 5–30 min. The extent of the color reaction is monitored in a microplate reader at 405 nm.

With nonbiotinylated $\beta\gamma$ subunits 0.05–1 μg of $\beta\gamma$ is immobilized in the well directly in TBS instead of streptavidin. The $\beta\gamma$ must diluted at least 10-fold into TBS prior to addition to the well since detergent in the $\beta\gamma$ solution will inhibit binding to the well. All other steps in the procedure are the same. A comparison of the results of these two methods is shown in Fig. 3.

Competition ELISA. One of the powerful uses of the displayed $\beta\gamma$ binding peptides on phage is the ability to test whether other molecules to compete for the binding of the phage to $\beta\gamma$ subunits. For example, peptides derived from other effectors could be tested for their ability to block the binding of a particular phage clone. If the peptide does compete for binding of the phage then it must be binding to a site that overlaps with the binding site for the peptide that is displayed on the phage. If the peptide does not block the interaction but is known to bind to $\beta\gamma$ subunits then it must be binding at a different site on the $\beta\gamma$ surface. If, for example, the phage clone used for the analysis displays a sequence known to bind to a PLC-β interaction site, any peptides that compete for binding of this phage must be binding to a site that is shared by PLC-β. If the peptide does not compete for phage binding but does compete for $\beta\gamma$-mediated PLC activation, then it binds at site shared by PLC-β but distinct from the epitope shared by the phage-borne sequence. In theory any peptide from any G protein $\beta\gamma$ subunit coupled effector could be fused to this phage and competition analysis could be used to determine which effector peptides share this particular interaction site.

The procedure for the competition ELISA is identical to what has been described for the phage ELISA above, except peptides or other competitors are added just prior to addition of the phage and the incubations with phage are only 1 h.

Analysis of Isolated Peptides in Functional Assays

The phage ELISA assay is a convenient assay for monitoring the ability of a phage-displayed peptide to bind to $\beta\gamma$ subunits in a semi high-throughput format, but the method has some disadvantages. One major disadvantage is that the assay does not give any quantitative information about the affinity of the peptide for the target molecule. Additionally, it is important to test the binding of the displayed peptide outside the context of the phage coat protein to be assured that the displayed sequence is all that is required for binding. There are a number of biophysical methods for measuring synthetic peptide interactions with proteins, including

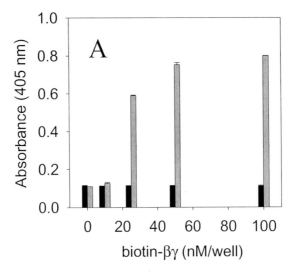

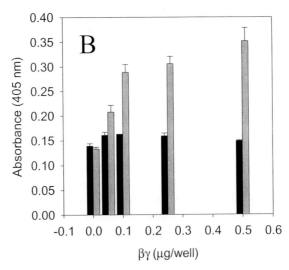

FIG. 3. Analysis of binding of phage displayed peptides to $\beta\gamma$ subunits immobilized either directly or via a biotin tether. (A) Biotin $\beta\gamma$ or (B) directly bound $\beta\gamma$. Black bars are f88-4 and gray bars are f88-4 displaying a $\beta\gamma$ binding peptide.

fluorescence polarization and surface plasmon resonance. Another approach that we have taken is to test the isolated synthetic peptides in competition-based effector assays where the peptide may compete with the activation of an effector by $\beta\gamma$ subunits. One system that we routinely use for this analysis is based on the ability of $\beta\gamma$ subunits to activate PLC-β isoforms *in vitro*. Other assays where

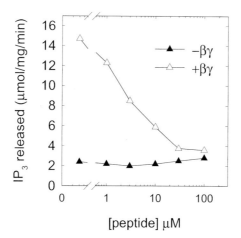

FIG. 4. Phage displayed peptides inhibit activation of PLC-$\beta_2\beta\gamma$ subunits. A synthetic peptide with the sequence SIRKALNILGYPDYD was included in phospholipase C assays at the indicated concentrations. Reactions included either no $\beta_1\gamma_2$ (▲) or 100 nM $\beta_1\gamma_2$ (△) and 1 ng of PLC-β_2; 100 nM free Ca^{2+} and were for 5 min. Each data point is the mean of duplicate determinations. Reproduced from J. K. Scott *et al., EMBO J.* **20**, 767 (2001) with permission of Oxford University Press.

competition with $\beta\gamma$ can be tested include assays of PI 3-kinase γ activation and adenylate cyclase regulation. Here we describe the assay to test if peptides derived from phage display can inhibit the ability of $\beta\gamma$ subunits to activate PLC-β_2. We test the peptides over a range of concentrations in the presence and absence of a fixed concentration of $\beta\gamma$. The assay in the absence of $\beta\gamma$ is to test whether the peptide has any direct effects on the PLC activity or substrate vesicles. An example of a concentration curve for inhibition of PLC activation by $\beta\gamma$ subunits by a phage display derived peptide is shown in Fig. 4.

Phospholipase C Peptide Competition Assay

The total reaction volume for each assay is 60 μl. Each of the following components is added separately to the final reaction in this order:

1. 20 μl Lipid vesicles containing both unlabeled phosphatidylinositol 4,5-bisphosphate (PIP$_2$) and [^{3}H]PIP$_2$ as well as phosphatidylethanolamine (PE). This assay is based on the ability of PLC to release [^{3}H]IP$_3$ into the aqueous phase.

2. 10 μl of solution containing PLC β as well as other necessary reaction components.

3. 10 μl of peptide solution at the desired concentration.

4. 10 μl $\beta\gamma$ subunit containing solution.

5. 10 μl Ca^{2+} solution (calcium is required for the enzymatic activity of the PLC).

Lipid Vesicles. PIP$_2$ (Avanti) at 50 μM and PE (Avanti, Alabaster, AL) at 200 μM are the final target concentrations for these lipids in each reaction. Vesicles are made at 150 μM PIP$_2$ and 600 μM PE because 20 μl will be diluted into a final volume of 60 μl. We also need to add [*inositol*-2-^{3}H(N)]PIP$_2$ ([^{3}H]PIP$_2$) (Dupont NEN, Boston, MA) such that each assay has 4000–8000 cpm. A general calculation to give the volume of stock [^{3}H]PIP$_2$ to be added is to estimate the total volume of lipid vesicle solution needed for the experiment and multiply by 0.03 to give the volume of stock [^{3}H]PIP$_2$ to be added. For example, if we make 450 μl of lipids we would multiply 450 × 0.03 to get 13.5 μl of ^{3}H-labeled PIP$_2$ to be added. *Note:* Unlabeled stock lipids are stored in glass vials (Kimble, Vineland, NJ) with chloroform-resistant screw caps (Kimble). We have noticed loss of [^{3}H]PIP$_2$ cpm if it is stored in these vials so we aliquot the [^{3}H]PIP$_2$ into 0.5 ml Nensure vials obtained from New England Nuclear. All the lipids are stored in desiccators at −20°. The lipids in chloroform are measured and transferred from the stock vials to a glass Pyrex tube with a Hamilton syringe. The lipids are dried under a stream of N$_2$ gas and sonication buffer is added at the final volume that has been calculated. Sonicate 5 min in a sonicating water bath. We find that if 20% Alconox liquid detergent is added to the water bath the sonication efficiency is greatly increased. The final lipid vesicles should be a translucent homogenous suspension without any visible particulate matter and all the lipid suspended from the bottom of the tube. Ten μl of the vesicle suspension is analyzed by liquid scintillation counting to verify that each assay will receive the proper number of counts per minute of [^{3}H]PIP$_2$.

PLC Mixture. The phospholipase C in this assay will be used at a concentration of 1–5 ng (depends on the assay) per sample. We usually use PLC-β_2 but PLC-β_3 can also be used. Each sample will get 10 μl of PLC solution. Make up enough PLC solution for all the reactions by multiplying the total number of reactions by 10 and adding some extra. Mix one-half volume of 2× assay buffer, 2 mM dithiothreitol (DTT), 6 mg/ml BSA, and 0.6% octyl-β-D-glucopyranoside (OG), the required amount of PLC, and sufficient water to achieve the final desired volume.

$\beta\gamma$ Solution. The stock $\beta\gamma$ is in stored in a buffer that contains 1% OG. We usually exchange the $\beta\gamma$ from the buffer used for purification from Sf9 cells into the $\beta\gamma$ storage buffer by binding the $\beta\gamma$ to a small (500 μl) column of hydroxyapatite (BioRad), washing with 5 ml of 20 mM HEPES, pH 8.0, 1 mM DTT, 100 mM NaCl, and 1% OG and eluting with the same buffer containing 200 mM KP$_1$, pH 8.0. Ten μl of incubation solution containing $\beta\gamma$ is added to each reaction tube and therefore we need to make the concentrations of $\beta\gamma$ 6× the final concentration that is desired. The $\beta\gamma$ to be added to the assay is mixed first in 1× incubation buffer with 1 mM DTT. For peptide competition experiments we usually test the peptide at various concentration in the presence and absence of 50–100 nM $\beta\gamma$. So we make up two solutions, and one that contains 6× $\beta\gamma$ (600 nM for a final of 100 nM) and one containing an equivalent volume of buffer that the $\beta\gamma$ was stored in. For example, if we have 33 μM $\beta\gamma$ stock and we want 150 μl of 600 nM $\beta\gamma$

we would need to add 75 μl 2× Inc. Buffer, 8 μl $\beta\gamma$, 1.5 μl 0.1 M DTT, 65.5 μl water. The solution without $\beta\gamma$ is the same except 8 μl of $\beta\gamma$ storage buffer would be added instead of $\beta\gamma$. If the $\beta\gamma$ needs to be diluted before it is added to the incubation buffer, this should be done in $\beta\gamma$ storage solution.

Peptide Solutions. Peptides are diluted in water to 6 times the desired final concentrations. The stock peptide solutions are made at 1–5 mM and the concentration verified using the peptide bond absorbance at 215 nm. We usually buy the peptides at 80–95% purity. If the peptide needs to be oxidized for activity the peptide is dissolved at 1 mg/ml, the pH of the peptide solution is adjusted to pH 7–8, and the solution is incubated overnight at room temperature with exposure to air.

Ca^{2+} *Solutions.* Make up 1 ml of fresh calcium solution every time 500 μl 2× assay buffer, 168 μl 0.1 M CaCl$_2$, 322 μl doubly distilled H$_2$O, 10 μl 0.1 M DTT. The calcium can be varied depending on the pH of the reaction and the amount of final calcium concentration desired.

Assay Procedure. Assays should include at least two blank reactions that contain everything except the calcium solution. The counts per minute from the blank are a measure of the [^{3}H]PIP$_2$ that is hydrolyzed prior to the reaction and are subtracted from all of the values obtained in the presence of Ca^{2+}.

Aliquot 20 μl of lipid vesicle solution into 4 ml plastic reaction tube (with tubes sitting in an ice bath). Add the other reagents in the order described above including the calcium solution. To start the reaction transfer the tubes to a 30° water bath [leave the blanks on ice and immediately add trichloroacetic acid (TCA)]. After 5–30 min transfer the tubes back to the ice bath and add 200 μl ice-cold 10% TCA to each tube to stop the reactions. Add 100 μl 10 mg/ml BSA, vortex, and centrifuge 5 min at 2000g at 4°. Pipette out 300 μl of supernatant and transfer to a scintillation vial, add 4 ml of scintillation fluid, and analyze by liquid scintillation counting. The amount of [^{3}H]IP$_3$ released should not exceed 40% of the total [^{3}H]PIP$_2$ cpm in the assay because the assay is saturated at this point. If the counts per minute released are either too high or too low the reaction can be adjusted by changing the amount of PLC in the assay, changing the reaction time, or changing the final calcium concentration in the assay. Results of a typical peptide competition experiment are shown in Fig. 4.

Summary

We have described a method using polyvalent peptide display on filamentous phage that can be used to identify ligands that bind to G protein $\beta\gamma$ subunits. Also described is how to construct phage that have known $\beta\gamma$ binding sequences fused to the coat protein to allow a competition analysis to be performed. Once selected or constructed, these phage-bearing $\beta\gamma$-binding peptides are powerful tools for mapping interaction sites for $\beta\gamma$ binding proteins and can be used to begin to dissect the unique modes of binding for individual $\beta\gamma$ subunit-regulated effectors.

Appendix: Solutions for Bacterial and Phage Manipulation

Many of these solutions are take directly from Ref. 6.

ABTS solution: 22 mg 2,2′-Azinobis(3-ethylbenthiazoline-6-sulfonic acid) (Sigma), 38.6 ml 0.2 M Na$_2$HPO$_4$, 61.4 ml 0.1 M citric acid, add 1/1000 volume of 30% (w/v) H$_2$O$_2$ before use

2% BSA: 2 g BSA in 100 ml 1× TBS

NZY medium: 21 g NZY broth powder (Gibco-BRL) 1000 ml distilled H$_2$O, autoclave for 15 min.

NZY agar medium: 21 g NZY broth powder, 1000 ml distilled H$_2$O, 15 g Bacto-agar; autoclave for 15 min

NZY/Tc agar medium: Autoclaved NZY agar medium, add Tc (tetracycline) at 40 μg/ml

NZY/Tc medium: NZY medium, add Tc at 20 μg/ml

PEG/NaCl: 100 g Polyethylene glycol (PEG) 8000, 116.9 g NaCl, 475 ml distilled H$_2$O, dissolve with heating if necessary

TBS (10× stock): 1.5 M NaCl, 0.5 M Tris-HCl (pH 7.5)

TBS/0.1% Tween: 100 ml 1× TBS, 0.1 ml Tween

TBS/gelatin: 0.1% (w/v) gelatin in TBS, autoclave

Tetracycline (Tc) (20-mg/ml stock): 40 mg/ml in water and filter sterilize into an equal volume of glycerol (autoclaved), store at $-20°$

Virus Lysis mix (5×): 2 g SDS, 18 ml H$_2$O, 2 ml 40× GBB, 40 mg bromphenol blue, 20 ml glycerol 100%

GBB (40× stock): 142.4 g Tris, 45.94 g sodium acetate anhydrous (or 76.16 g trihydrated), 18.83 g Na$_2$-EDTA-2H$_2$O. Adjust pH to 8.3 with acetic acid and take to final volume of 700 ml. Store at room temperature

Solutions for PLC Assay

2× Assay Buffer: 100 mM HEPES pH 7.2, 6 mM EGTA, 160 mM KCl, pH to 7.2

2× Incubation Buffer: 100 mM HEPES pH 7.2, 2 mM EDTA, 6 mM EGTA, 200 mM NaCl, 10 mM MgCl$_2$, pH to 7.2

Lipid Sonication Buffer: 750 μl 2× assay buffer pH 7.2, 235 μl doubly distilled H$_2$O, 15 μl 0.1 M DTT

[40] Exploring Protein–Protein Interactions by Peptide Docking Protocols

By Gezhi Weng

Introduction

The forces that govern protein folding and enable proteins to interact with one another are the same. These include noncovalent interactions such as hydrophobic interactions, ionic interactions, and hydrogen bondings. All of these interactions originate from electromagnetic force. The descriptions of these variations of electromagnetic interactions can be very complicated in situations where the quantum effects may not be negligible and when large number of atoms are involved. Solvent screening effects reduce the intensity and shorten the range of electrostatic interactions produced by the charged residues on the interacting proteins. Short-range van der Waals interactions are major contributors to the binding affinity.[1] Therefore protein–protein interactions occur through surface contact. Experimental evidence indicates that such interactions are often local. Interacting regions can be defined in terms of short stretches of primary sequences or restricted regions within the three-dimensional structure. Often interactions between two proteins can be mimicked by protein domains or even by small peptides encoding the local interacting regions. Small peptides of 15 to 30 amino acids are easy to synthesize and have been useful tools in studying protein–protein interactions. We have used peptides to study interactions between G-protein $\beta\gamma$ subunits and their effectors. In this article I present one of these studies to show how synthetic peptides and molecular modeling approaches can be used to locate regions involved in protein–protein interactions.

G proteins control many signal transduction pathways. $G\beta\gamma$ can regulate several intracellular effectors such as adenylyl cyclases, the PLC-β subfamily, Ca^{2+} and K^+ channels, and receptor kinases such as β-adrenergic receptor kinase. Biochemical experiments in our laboratory showed that a peptide encoding amino acids 956–982 of adenylyl cyclase 2 (AC2 peptide) blocked the interactions between $G\beta\gamma$ and several of its effectors.[2] Further studies were performed in order to map the peptide contact region on $G\beta\gamma$. Identification of this contact region could also define the AC2 interacting region on $G\beta\gamma$.

We designed a cross-linking experiment to covalently attach the AC2 peptide to $G\beta\gamma$.[3] The results showed that the peptide could be cross-linked to $G\beta$ only

[1] S. Jones and J. M. Thornton, *J. Mol. Biol.* **272**, 121 (1997).

[2] J.-Q. Chen, M. DeVivo, J. Dingus, A. Harry, J. Li, D. J. Carty, J. L. Blank, J. H. Exton, R. H. Stoffel, J. Inglese, R. J. Lefkowitz, D. E. Logothetis, J. D. Hildebrandt, and R. Iyengar, *Science* **268**, 1166 (1995).

when Gβ was not interacting with the Gα subunit. We concluded that the peptide interacting region on the Gβ subunit was occluded by the Gα subunit in the heterotrimer. At the time that this experiment was completed, the crystal structures of the Gβγ and Gαβγ heterotrimer were published.[4–6] Instead of using partial digestion and mass spectroscopy to locate the peptide interacting domain on Gβ surface, we chose to use a molecular modeling approach. First we identified the reactive residue on Gβ that was cross-linked to the peptide. A three-dimensional model of the peptide was constructed. By examining the surface properties of the Gβ and the peptide structure, we developed a docking model which revealed the putative effector interacting domains on Gβγ. The conclusions drawn from this model were confirmed by further experiments and by the crystal structure of phosducin bound to Gβ.[7]

Calculation of Residue Solvent Accessibility

The cross-linker we used in these experiments was SIAB [N-succinimidyl (4-iodoacetyl)aminobenzoate]. One end of this molecule reacts with the NH_2 group and the other end with the SH group. The length of the linker is about 24 Å. The AC2 peptide contains several NH_2 groups but has no SH groups. Under the neutral pH conditions at which the cross-linking was performed, the most reactive -NH_2 group was the N-terminal primary amine. Therefore it is most likely that SIAB cross-linked the N terminus of the peptide to an SH group of a cysteine residue on the Gβ surface.

The solvent accessibility of protein residues defines surface residues. Using a test sphere with the dimension of a water molecule, the residue solvent accessibility is calculated by touchable surface area of the "rolling" test sphere over the protein surface. Once the three-dimensional structure of a protein is known, the residue solvent accessibility can be determined. There are several programs on the Web that can calculate the solvent accessibility of protein residues. These programs all yield similar results. GETAREA 1.1 can be found at www.scsb.utmb.edu/cgi-bin/get_a_form.tcl. This program is simple to use and the output is easy to understand. Another program, ASC, can be found at www.bork.embl-heidelberg.de/ASC. In our studies, we used a program called GEPOL[8] to calculate the solvent

---

[3] G. Weng, J. Li, J. Dingus, J. D. Hildebrandt, H. Weinstein, and R. Iyengar, *J. Biol. Chem.* **271,** 26445 (1996).

[4] D. G. Lambright, J. Sondek, A. Bohm, N. P. Skiba, H. E. Hamm, and P. B. Sigler, *Nature* **379,** 311 (1996).

[5] J. Sondek, A. Bohm, D. G. Lambright, H. E. Hamm, and P. B. Sigler, *Nature* **379,** 369 (1996).

[6] M. A. Wall, D. E. Coleman, E. Lee, J. A. Iniguez-Lluhi, B. A. Posner, A. G. Gilman, and S. R. Sprang, *Cell* **83,** 1047 (1995).

[7] R. Gaudet, A. Bohm, and P. B. Sigler, *Cell* **87,** 577 (1996).

[8] C.-E. Silla, I. Tunon, and J. L. Pascual-Ahuir, *J. Comput. Chem.* **12,** 1077 (1991).

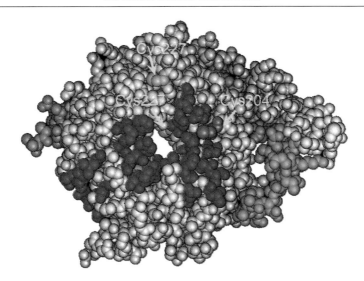

FIG. 1. The Gβγ surface facing Gα. The Gα contacting residues are colored in magenta. The Gγ subunit is in cyan. Cystines are colored in yellow. The three half-exposed cystines are indicated by arrows. Cys-204 in orange is the only Gα contacting cysteine.

accessibility of cysteine residues on Gβ in order to determine the cross-linker attachment point.

To locate the cross-linked SH-containing cysteine residue on the Gα occluded surface of Gβ, we utilized the X-ray structure of Gβγ.[4] There are a total of 14 cysteines in the Gβ subunit. The solvent accessible surface for all the cysteine residues was determined by using the GEPOL program.[8] Most of the 14 cysteines of Gβ are buried. Only three of them, Cys-204, Cys-271 and Cys-233, are half exposed. The location of these residues was examined by visualizing the Gβγ structure (Fig. 1). Cys-233 is located in the central channel of the Gβ propeller structure; both Cys-204 and Cys-271 are on the Gα facing surface. However, only Cys-204 makes contact with Gα in the Gαβγ heterotrimeric structure.[4] The side-chain SH group of Cys-204 is more exposed than that of Cys-271 and is completely shielded by Gα in the heterotrimer. We therefore inferred that Cys-204 must be the cross-linked residue on Gβ.

The determination of the cross-linking point provided an important constraint in locating the peptide interacting region on Gβ. To further define the AC2 peptide binding region, we modeled the structure of the peptide.

Secondary Structure Prediction

Primary, secondary, and tertiary structures provide different levels of information about the architecture of a protein. The first step to model a protein from a known sequence is to obtain its secondary structure. The secondary structure information is encoded primarily by the protein sequence. Various methods have been used to extract this information from known structures and use them to predict the secondary structure of proteins of unknown structure. There is a good collection of programs that predict secondary structure and predict structure in general at the Web site http://www.bmm.icnet.uk/people/rob/CCP11BBS/. One of the popular programs for predicting secondary structure is PHDsec, which was developed by Rost and Sander.[9] This program uses a neural network algorithm and protein structures in the Protein Data Bank as trainings set to calculate secondary structures. Evolutionary information was incorporated to improve predictions.[10] This program is available as a Web service at the site http://dodo.cpmc.columbia.edu/predictprotein/. Using this program, we predicted that the residues 964–982 of AC2 peptide form an α helix, while residues 956–963 are an unordered loop. This prediction turned out to be accurate when the predicted structure was compared to the X-ray crystal structure that was subsequently published.[11]

Protein Surface Visualization

Although proteins normally consist of several hundred amino acids, only part of their primary sequence encode regions are important in protein–protein interactions. These regions are usually exposed protein surfaces. Examination of the three-dimensional structures of protein complexes in the Protein Data Bank (PDB) showed that the interacting surfaces have geometric complementarity which, together with electrostatic interactions, determines the structure and the specificity of the complex.[12] For a small flexible peptide to interact with a relatively large and rigid protein, the electrostatic interactions can be particularly important in steering the collision and the subsequent contact. There are several programs with included modules that allow the visualization of the electrostatic features on a protein surface. InsightII (Molecular Simulation Inc., San Diego, CA) has a Delphi module that calculates surface potential by solving electrostatic Poisson–Boltzmann equations. The surface electrostatic potentials can be displayed as a color-coded surface map. MolMol, a free program developed by Wüthrich's group in ETH Zurich, also can draw surface potential maps. The details of this program can be found at http://www.mol.biol.ethz.ch/wuthrich/software/molmol. All these surface electrostatic potential rendering algorithms are based on the work of Barry Honig. One of the most popular surface-rendering programs, GRASP, was developed in Barry Honig's laboratory.[13] Besides displaying surface electrostatic

[9] B. Rost, C. Sander, and R. Schneider, *Comput. Appl. Biosci.* **10,** 53 (1994).

[10] B. Rost and C. Sander, *Proteins* **19,** 55 (1994).

[11] J. J. G. Tesmer, R. K. Sunahara, A. G. Gilman, and S. R. Sprang, *Science* **278,** 1907 (1997).

[12] S. Jones and J. M. Thornton, *Proc. Natl. Acad. Sci. U.S.A.* **93,** 13 (1996).

[13] A. Nicholls, K. A. Sharp, and B. Honig, *Proteins* **11,** 281 (1991).

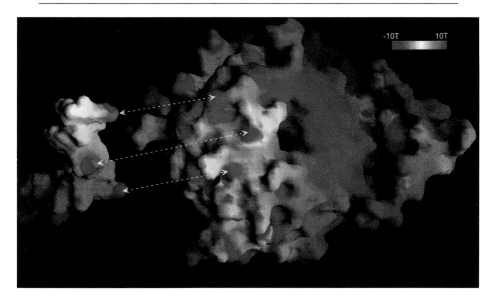

FIG. 2. The surface electrostatic potentials for Gβγ and for AC2 peptide. The complementary spots are indicated by the three arrows. These complementary properties are used in guiding the peptide docking.

potential, GRASP also calculates surface curvature and displays it as a color-coded surface map. The detailed information can be obtained at the Web site http://trantor.bioc.columbia.edu/.

Figure 2 shows the surface electrostatic potential of Gβ and the AC2 peptide as calculated by GRASP. The red color represents negative electrostatic potential, while the blue color represents positive electrostatic potential. There are three charged residues on the helical part of the peptide. Arg-964 and Lys-982 show up as two blue spot on the surface, while Glu-974 shows up as a red spot. We assumed that the helical part of the peptide is relatively stable. The electrostatic steering effect during the molecule collision is mainly determined by the electrical charge on this part of the molecule.

The Gα contacting surface of Gβ is facing the reader. Most of this surface has a negative potential. It is unlikely that the peptide could dock onto this part of the Gβ surface. However, just to the left of the Gα facing surface is a patch of electrostatic potential that is complementary to that of the peptide (Fig. 2). We docked the peptide on to this part of Gβ.

Protein Docking

Assuming the electrostatic long-range interactions to be the guiding force in the initial docking of the peptide to its binding site, we used a manual docking

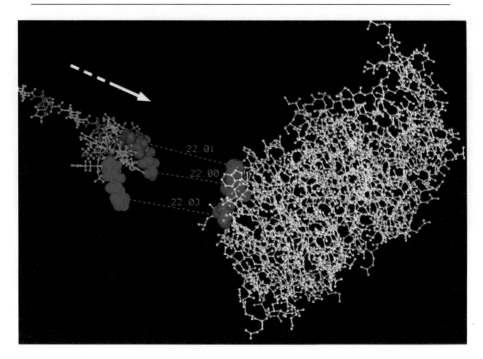

FIG. 3. Peptide docking using the guidance of distance constraints. The surfaces of the three pairing residues are represented by van der Waals surfaces. Blue colors indicate positively charged residues while red colors indicate negatively charged residues.

procedure to dock the AC2 peptide on the $G\beta$ surface. The residues that contribute to the complementary surface electrostatic potentials were identified. The distances of the three complementary residue pairs on two molecules were monitored when docking was performed manually using INSIGHTII (Molecular Simulations Inc., San Diego, CA) (Fig. 3). The protein surface was represented by a space-filled model during the docking to avoid obvious clashes. Once the helical part of the peptide was docked on the $G\beta$ surface, the N terminus of the peptide was turned toward Cys-204 on $G\beta\gamma$ surface by changing the torsional angles of the residues on the peptide N-terminal loop. After this initial docking, energy minimization (1000 steps using conjugate gradient) followed by short-run molecular dynamics simulation was performed with the DISCOVER package inside INSIGHTII on the peptide structure with all the $G\beta\gamma$ residue fixed. The distance between the AC2 peptide and the Cys-204 on $G\beta$ was set to 24, the length of the cross-linker. All the calculations were carried out under this distance constraint. The most favorable structure for the docked peptide was selected and further relaxed with a second round of energy minimization. The docked structure is shown in Fig. 4).

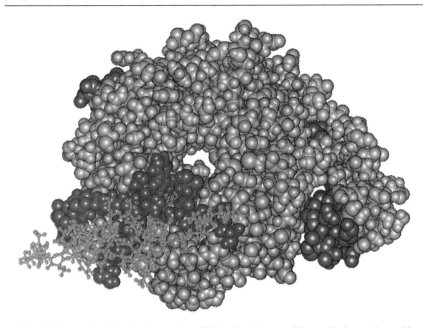

FIG. 4. The peptide docked on the surface of G$\beta\gamma$. Gγ is in cyan. The peptide interacting residues are colored in magenta. The AC2 peptide is in green.

The complementary electrostatic potential used in guiding the docking is presented on the relatively stable helical structure of the peptide. Under most circumstances the complementary patches of positive and negative charges, especially for those presented in loop regions, may not readily be identified. Therefore programs are available to automate the docking process. These include Affinity, INSIGHT's molecular docking module; SYBYL, another molecular modeling program (Tripos Inc. St. Louis) has the FlexiDock module. There are other docking procedure available through the World Wide Web: Dock, at http://www.uni-paderborn.de/~1st/HotDock/index.html; GOLD, at http://panizzi.shef.ac.uk/cisrg/gareth/gold/gold.html; and AutoDock, which applies simulated annealing to the docking process, and allows several torsional degrees of freedom in a flexible ligand to be searched. Detailed information about this program can be found at http://www.scripps.edu/pub/olson-web/doc/autodock/. Most of these automated procedures, however, still require expert knowledge and are often more suitable to

TABLE I
INTERACTING RESIDUES ON Gβ

Predicted AC2 interacting residues on Gβ	Phosducin interacting residues on Gβ[a]
His-91, Pro^{94}- Trp^{99}, Leu^{117}-Asn^{119}, Tyr-124, Arg^{129}-Asn^{132}, Arg^{134}-Val^{135}, Thr^{143}-Gly^{144}	Arg-42, Gln-44, Arg^{46}-Thr-^{47}, Lys-57, Tyr-59, Gln-75, Ser^{98}–Trp^{99}, Met-101, Leu-117, Tyr-145, Gly-162, Asp-1867, Met-188, Cys-204, Asp-228, Asn-230, Asp-246, Thr-274, Asp-290, Phe-292, Arg-304, Gly^{310}-His^{311}, Arg-314, Trp-332

[a] From Ref. 7.

dock small molecules into known active sites or surface cavities. To dock a peptide with more than 10 amino acids on a relatively flat protein surface still remains a challenge and needs to be implemented on a case-by-case basis.

Predictions by the Docking Model and Experimental Testing

We used the docking model (Fig. 4) to obtain the contact points between the Gβ and the AC2 peptide. Figure 4 shows the "footprint" of the AC2 peptide on Gβ. The regions of Gβ predicted to interact with the AC2 peptide are shown in magenta. Some of these residues (in boldface type) also interact with phosducin (Table I) which is a Gβ interacting protein and may function as a Gβ effector.

We reasoned that the predicted contact residues on Gβ could be involved in communicating signals to effectors. To test this idea, we synthesized peptides encoding the footprint sequences on Gβ to determine whether these peptides can interfere with the Gβγ regulation of adenylyl cyclases. Two peptides were designed based on the predicted contact region between Gβ and the AC2 peptide. The first peptide Gβ86-105 includes the stretch of residues 91–99 predicted by the docking model to be important for effector interactions. The second peptide Gβ115-135 included the footprint residues of 117–119 and 129–135. Further experiments showed that both the peptides blocked Gβγ stimulation of AC2 and the Gβγ inhibition of AC1. These two peptides, however, showed different characteristics toward modulating signals. The Gβ86-105 peptide can stimulate AC2 in the absence of Gβγ while the Gβ115-135 peptide blocked Gβγ stimulation of AC2 but had no effect when presented alone. Besides the modulation of Gβγ effects on adenylyl cyclases, these peptides also had dramatic effects on Gβγ stimulation of PLC-β, another Gβγ effector. Gβ86-105 peptide stimulated the activity PLC-β2 to the full level of Gβγ stimulation while Gβ115-135 blocked the Gβγ stimulation of PLC-β$_2$ but had no effect on PLC-β$_2$ when present alone. Detailed studies on the differences of these two peptides led to the classification of a signal transfer region (STR) and a general binding region (GBR).[14] Further detailed mapping of the PLC-β interacting surface of G$\beta\gamma$ revealed multiple STRs and GBRs (Buck and Iyengar, unpublished result).

Mutagenesis studies are also consistent with these findings using synthetic peptides.[15] In one of these studies, all the Gα contacting residues were mutated to alanine individually. Mutations S98A, W99A, M101A, L117A, N119A fall within the AC2 peptide footprint region. These mutations blocked G$\beta\gamma$ stimulation of AC2 activity and altered G$\beta\gamma$ modulation of several other downstream effectors. These experimental results indicate that the effector interaction regions on G$\beta\gamma$ are partially overlapping. By blocking parts of these overlapping regions from interacting with effectors, the binding and dissociation of Gα subunits can control the signal flow to several downstream effectors.

Utility of Molecular Modeling in Analyzing Protein–Protein Interactions

Molecular modeling is an efficient way of studying protein–protein interactions. It provides detailed information about how protein residues interact with each other at the atomic level. However, the accuracy of computer modeling is limited by the approximation and assumptions made in the process. The details of the molecular models are sometimes difficult to verify. Table I shows that the effector interacting regions predicted by the model contain more continuous residues than those seen from X-ray structure. This might be an inaccuracy produced by the docking procedure. Also, the model did not show contact of AC2 peptide to M101 of the Gβ. Later experiments using Gβ peptides, however, showed that substituting M101 by Asn eliminated the ability of the peptide to stimulate AC2.[16] Under normal temperature, interaction between proteins is a dynamic process. The contact points may not be well presented by "frozen" coordinates. The interactions of M101 and Y145 with AC2 peptide did show up in dynamic simulations.

Despite the precautions that one needs to take to deal with the interpretation of detailed residue contacts, modeling provides a good starting point for further experiments. Expert knowledge is still very important in guiding molecular modeling. The rapid development of high-speed computers and improvement of algorithms should soon incorporate such expert knowledge into programs and the modeling process may be automated so as to permit usage by nonspecialists. The rapid improvement in the speed and computing capabilities of desktop computers

---

[14] J. Buck, M. L. Sinclair, L. Schapal, M. J. Cann, and L. R. Levin, *Proc. Natl. Acad. Sci. U.S.A.* **96,** 79 (1999).

[15] C. E. Ford, N. P. Skiba, H. Bae, Y. Daaka, E. Reuveny, L. R. Shekter, R. Rossal, G. Weng, C.-S. Yang, R. Iyengar, R. J. Miller, L. Y. Jan, R. J. Lefkowitz, and H. E. Hamm, *Science,* in press (1998).

[16] Y. Chen, G. Weng, J. Li, A. Harry, J. Pieroni, J. Dingus, J. D. Hildebrandt, H. Weinstein, and R. Iyengar, *Proc. Natl. Acad. Sci. U.S.A.* **94,** 2711 (1997).

and the widespread use of the Internet provide an easy platform for the scientist with limited programming and computer knowledge to use molecular modeling tools. Many programs used for molecular modeling are available as Web services, allowing scientists to submit data and obtain results quickly and easily. With these developments, I expect in the future that molecular modeling will be used more frequently and will be widely accepted as one of the standard approaches to study protein–protein interactions.

[41] Structural Characterization of Intact G Protein γ Subunits by Mass Spectrometry

By KEVIN L. SCHEY, MARK BUSMAN, LANA A. COOK, HEIDI E. HAMM, and JOHN D. HILDEBRANDT

Introduction

The development of two ionization methods, matrix-assisted laser desorption ionization (MALDI)[1] and electrospray ionization (ESI),[2] has revolutionized the field of mass spectrometry for studies of biomolecular structure. Initially, accurate molecular weight measurements by MALDI-MS or ESI-MS provided a "high-resolution sodium dodecyl sulfate–polyacrylamide gel electrophoresis (SDS-PAGE)" technique; however, the combination of these ionization techniques, ESI in particular, with tandem mass spectrometry (MS/MS) has provided a powerful tool for peptide structure determination.[3] Indeed, the unprecedented speed, accuracy, and sensitivity of MS/MS for peptide sequencing has made mass spectrometry the premier tool in proteomics research.[4] The advantages of peptide sequencing by MS/MS compared to conventional Edman sequencing include the ability to sequence N-terminally blocked peptides, the ability to identify sites and structures of posttranslationally modified peptides, and the attomole sensitivity.[5] One limitation of this approach is the size of peptides that fragment in a mass spectrometer to produce sequence information: normally fewer than 20–25 amino acids or approximately 2500 Da. Therefore, a typical sequencing experiment requires

---

[1] M. Karas and F. Hillenkamp, *Anal. Chem.* **60,** 2299 (1988).

[2] J. Fenn, M. Mann, C. K. Meng, S. F. Wong, and C. M. Whitehouse, *Science* **246,** 64 (1985).

[3] M. Wilm, A. Shevchenko, T. Houthaeve, S. Breit, L. Schweigerer, T. Fotsis, and M. Mann, *Nature* **379,** 466 (1996).

[4] J. Yates III, *J. Mass Spectrom.* **33,** 1 (1998).

[5] C. J. Barinaga, C. G. Edmonds, H. R. Udseth, and R. D. Smith, *Rapid Commun. Mass Spectrom.* **3,** 160 (1989).

large proteins to be cleaved to produce peptides below this mass limit and the peptides are separated by high-performance liquid chromatography (HPLC) prior to, or on-line with, tandem mass spectrometry.

More recently, the mass limit for obtaining peptide sequence information has been expanded in MS/MS experiments by fragmenting multiply charged ions generated by ESI.[6] Remarkably, it was shown that low-energy collision-induced dissociation (CID) experiments provided structural information for intact proteins as large as immunoglobulin G (IgG, 150,000 Da).[7] Although fragmentation was observed, the accessibility of sequence information, i.e., data interpretation, was complicated by the possible presence of multiple charge states for fragments arising from cleavage at any amide bond. However, a few concepts have been developed from the interpretation of MS/MS spectra of smaller peptides to assist in the challenges of interpreting fragmentation of large proteins. First, it has been noted that in larger proteins, one often observes a stepwise fragmentation at the N terminus, resulting in a series of low charge state (often singly charged) N-terminal ions.[8] Second, multiply charged C-terminal ions are often seen for cleavages at the N-terminal side of proline residues.[9,10] The utility of Fourier transform ion cyclotron resonance (FT-ICR) instruments equipped with ESI sources has been demonstrated where ultrahigh resolution measurements of CID product ion masses allowed charge state determination, thereby greatly simplifying spectral interpretation.[10,11] This approach has yielded stretches of sequence information for proteins as large as albumin.[8] An ion trap tandem mass spectrometer has been employed for large protein dissociation where ion–ion chemistry was used to simplify the fragment ion charge state distributions and thus simplify interpretation.[12]

Previous MALDI-MS analysis of G protein γ subunits (described in elsewhere in this volume[12a]) resulted in accurate molecular weight determinations which, based on known posttranslational modifications, provided the identity of four proteins isolated from bovine brain.[13] Two additional studies on the bovine $G\gamma_2$ and the $G\gamma_5$ subunits provided detailed analysis of proteolytic fragments by MS and MS/MS which allowed unambiguous identification of C- and N-terminal

[6] P. E. Andren, M. R. Emmett, and R. M. Caprioli, *J. Am. Soc. Mass Spectrom.* **5**, 867 (1994).

[7] R. Feng and Y. Konishi, *Anal. Chem.* **65**, 645 (1993).

[8] J. A. Loo, C. G. Edmonds, and R. D. Smith, *Anal. Chem.* **63**, 2488 (1991).

[9] J. A. Loo, C. G. Edmonds, and R. D. Smith, *Anal. Chem.* **65**, 425 (1993).

[10] M. W. Senko, S. C. Beu, and R. W. McLafferty, *Anal. Chem.* **66**, 415 (1994).

[11] R. Chen, Q. Wu, D. W. Mitchell, S. A. Hofstadler, A. L. Rockwood, and R. D. Smith, *Anal. Chem.* **66**, 3964 (1994).

[12] T. G. Schaaff, B. J. Cargile, J. L. Stephenson, Jr., and S. A. McLuckey, *Anal. Chem.* **72**, 899 (2000).

[12a] L. A. Cook, M. D. Wilcox, J. Dingus, K. L. Schey, and J. D. Hildebrandt, *Methods Enzymol.* **344**, [16] 2002 (this volume).

[13] M. D. Wilcox, K. L. Schey, J. Dingus, N. D. Mehta, B. S. Tatum, M. Halushka, J. W. Finch, and J. D. Hildebrandt, *J. Biol. Chem.* **269**, 12508 (1994).

modifications.[14,15] In this report we present a rapid method for the determination of posttranslational modifications and sequence information based on ESI-MS/MS analysis of intact γ subunits. Sample preparation, instrumental considerations and operation, and data interpretation are described.

Methods

Sample Preparation

Two sample preparation protocols have been developed for ESI-MS/MS analysis of intact γ subunits using either acetone precipitation of G-protein heterotrimers or HPLC-separated γ subunits. Bovine brain heterotrimeric G proteins are isolated by the method of Sternweis and Robishaw[16] as modified by Kohnken and Hildebrandt.[17] Transducin is purified from bovine retina according to the method of Mazzeni et al.[18] Protein samples consist of purified G-protein heterotrimers in various buffer/detergent solutions at concentrations of approximately 2–4 mg/ml. The buffer/detergent systems examined to date are 0.7% (w/v) CHAPS, 0.1% (w/v) Thesit, and 0.7% (w/v) CHAPS/0.1% (w/v) Thesit.

1. Acetone Precipitation. The detergent-solubilized protein is acetone precipitated with 9 volumes of cold acetone for 20 min on ice. The solution is centrifuged at 14,000 rpm for 10 min at 4°. The supernatant is removed and the residual acetone is blown off with a stream of argon. Pellets are resuspended in water/methanol/acetic acid (48/48/4).

2. HPLC Purified γ Subunits. The method of Cook et al.,[15] described elsewhere in this volume,[12a] is used to fractionate γ subunits after activation of purified heterotrimeric G proteins. On-line ESI-MS/MS is accomplished by directing the HPLC effluent [water/acetonitrile/2-propanol/trifluoroacetic acid (TFA)] into the mass spectrometer. Alternatively, fractions are collected and speed-vacuum dried. Dried fractions are suspended in 5–10 μl of methanol/water/1- or 2-propanol/acetic acid (50/26/20/4, v/v/v/v).

Tandem Mass Spectrometry

A diagram depicting the MS/MS experiment and peptide fragmentation scheme is shown in Fig. 1. The first event in the experiment is the ionization of the peptides present in the sample (Fig. 1a). In the case of G-protein γ subunits, the proteins are ionized by electrospray ionization, which produces a series of multiply

[14] M. D. Wilcox, K. L. Schey, M. Busman, and J. D. Hildebrandt, *Biochem. Biophys. Res. Commun.* **212**, 367 (1995).

[15] L. A. Cook, K. L. Schey, M. D. Wilcox, J. Dingus, and J. D. Hildebrandt, *Biochemisty* **37**, 12280 (1998).

[16] P. C. Sternweis and J. D. Robishaw, *J. Biol. Chem.* **259**, 13806 (1984).

[17] R. E. Kohnken and J. D. Hildebrandt, *J. Biol. Chem.* **264**, 20688 (1989).

[18] M. R. Mazzeni, J. A. Malinski, and H. E. Hamm, *J. Biol. Chem.* **266**, 14072 (1991).

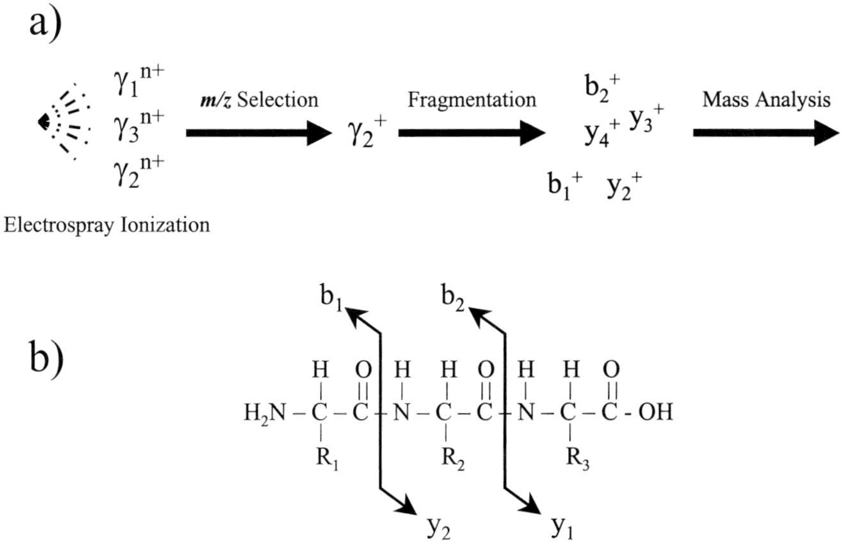

a)

Electrospray Ionization

b)

FIG. 1. Schemes showing (a) the events in a tandem mass spectrometry experiment and (b) the peptide fragmentation patterns and nomenclature.

charged ions ($[M + nH]^{n+}$ where $n = 4$–14) due to the attachment of a number of protons to the proteins. The peptide of interest is then selected, according to its m/z (where m is the mass of the ion and z is the charge on the ion), by the first stage of mass spectrometry. The selected ions, termed precursor ions, then collide with inert collision gas argon or helium and, as a result, fragment. The fragment ions, termed product ions, are then mass analyzed in the second stage of mass spectrometry. Peptides fragment predominantly and predictably at amide bonds under low-energy collision-induced dissociation (CID) conditions as depicted in Fig. 1b. Peptide MS/MS nomenclature assigns N terminus-containing fragment ions as b ions and C terminus-containing fragment ions as y ions.[19] Subscripts in ion labels define the number of amino acids from the respective termini present in the fragment. Because the amino acid residue masses are unique, with the exceptions of isoleucine/leucine and lysine/glutamine, the mass differences between fragment ions are easily interpreted to provide sequence information. This predictability has led to a number of interpretation algorithms including those which interpret raw data (Sequest)[20] and those which employ user-generated sequence tags (e.g., PeptideSearch) defined as a short section of sequence derived from manual interpretation of the data.[21]

[19] K. Biemann and H. A. Scoble, *Science* **237**, 992 (1987).
[20] J. Eng, A. L. McCormack, and J. R. Yates III, *J. Am. Soc. Mass Spectrom.* **5**, 976 (1994).
[21] http://www.mann.embl-heidelberg.de/Services/PeptideSearch/PeptideSearchIntro.html

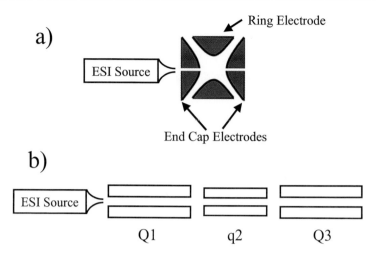

FIG. 2. Diagram of (a) the Finnigan LCQ ion trap mass spectrometer and (b) the Sciex triple quadrupole mass spectrometer indicating the first quadrupole mass analyzer (Q1), the second quadrupole collision cell (q2), and the third quadrupole mass analyzer (Q3).

Two instruments were used for the ESI-MS/MS experiments: a Finnigan LCQ ion trap mass spectrometer and a SCIEX API-III triple quadrupole mass spectrometer (now Applied Biosystems). Figure 2 shows the geometries of these two instruments. The triple quadrupole instrument achieves the two mass analysis steps of the tandem mass spectrometry experiment in two separate quadrupole mass analyzers (Q1 and Q3); therefore, the experiment is referred to as "tandem in space." In contrast, the mass analysis stages in the ion trap instrument occur in the same mass analyzer and this experiment is referred to as "tandem in time." Standard source conditions were employed for both electrospray ionization sources. In addition, a custom-built nanospray ionization source is employed on the LCQ instrument.

1. Ion Trap MS Analysis. In on-line HPLC-ESI/MS/MS experiments, the effluent from the LC is split and a portion is sent directly into the LCQ instrument at a flow rate of 50–100 μl/min (water, 2-propanol, acetonitrile, 0.1% trifluoroacetic acid). The instrument is operated in automated mode where the most abundant ions in the ESI mass spectra are isolated and fragmented. An LCQ collision energy of 45% is used and an isolation window of 2.5 Da is employed.

For nanospray analysis, collected HPLC fractions are resuspended as described above and loaded into borosilicate glass capillaries pulled to tips with internal diameters of approximately 1–10 μm. After sample loading (1 μl), the sample is placed onto the nanospray source where a wire is inserted into the back end of the capillary to supply the electrospray voltage. Typically 1–2 kV is applied to the solution. The MS/MS conditions are applied as described above. The nanospray source allows signal averaging for up to 30 min for 1 μl of sample.

Multiple stages of fragmentation (MS^n) are afforded by the ion trap technology as ions are efficiently trapped and manipulated in the "tandem in time" device. Electrical manipulation of the ions can proceed as long as ions remain in the trap, i.e., are not ejected, and this time is typically milliseconds but can be as long as seconds to minutes. Fragment ions observed in the MS/MS data of Gγ subunits are selected and subsequently fragmented in an MS^3 experiment using the same collisional activation conditions as for MS/MS. Further stages of analysis can be carried out if a sufficient number of ions remain in the trap, with a practical limit for the work described here being MS^4.

   2. Triple Quadrupole MS Analysis. Acetone-precipitated or HPLC-purified γ subunits are suspended as described above and infused utilizing a Harvard Apparatus Model 11 syringe pump (Holliston, MA) into the nebulization assisted ESI source of the API-III instrument at a rate of 5 μl/min. The first quadrupole mass analyzer (Q1) is operated with 1–2 Da resolution to select the precursor ion. The collision quadrupole (q2) is filled with a 90/10 mixture of argon/nitrogen to a collision gas thickness of 480 ($\times 10^{12}$ molecules/cm^2) and operated at a collision energy of 70 eV. The third quadrupole mass analyzer (Q3) is operated with 1–2 Da resolution. Typically 5–30 scans are acquired to produce an MS/MS spectrum.

Results

   Figure 3 displays and ESI mass spectrum of acetone-precipitated bovine transducin. Multiply charged molecular ions are clearly present for the transducin γ subunit, γ_1, from $[M + 13H]^{13+}$ to $[M + 7H]^{7+}$ providing a measured molecular mass of 8330.6 $\pm$ 2.1 Da. This value is determined from the ESI mass spectrum by multiplying the measured m/z values by the calculated charge and substracting the number of protons and averaging over the observed charge states. The measured molecular mass for Gγ_1, 8330.6 $\pm$ 2.1 Da, is consistent with the known posttranslational modifications of γ_1: C-terminal truncation, farnesylation, methylation; and N-terminal cleavage of Met and subsequent acetylation (8331.7 Da predicted). Both sample preparation methods produce abundant ESI signals; however, HPLC-purified signals typically provide stronger signals, most likely due to the lack of contaminating salts, detergents, and other G-protein subunits. Molecular ions having charges from 6 to 9 gave the most abundant and interpretable tandem mass spectra, as described below.

   The $[M + 8H]^{8+}$ ion at m/z 1042.4 from the γ_1 ESI mass spectrum was selected and fragmented in an MS/MS experiment to produce the MS/MS spectrum shown in Fig. 4 (triple quadrupole instrument). The most abundant fragment ion appears at m/z 1161 and corresponds to the commonly observed loss of the farnesyl group (f = $C_{15}H_{24}$, molecular weight 204)[22] and loss of one charge, i.e., $[M + 7H-f]^{7+}$. Singly charged b ions are apparent in the low mass range while the cluster of signals around the molecular ion represent large multiply charged b and y ions,

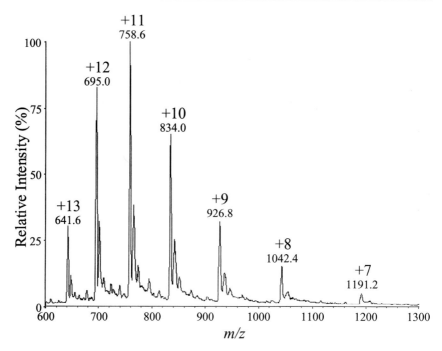

FIG. 3. Electrospray ionization mass spectrum of bovine γ_1 indicating the assigned charged states above each signal.

the most prominent of which are due to cleavage on the N-terminal side of proline residues. In addition, the C-terminal containing y ions also lose the isoprenyl group, e.g., $[y_9 - f]^+$ and $[y_{14} - f]^{2+}$. The sequence of transducin is shown at the top of Fig. 8 with the observed fragmentations indicated. Although assignment of these signals is complicated by the possibility of ion having multiple charge states, the known sequence assisted in the assignment and allowed specific patterns of fragmentation to be established. These fragmentation patterns were used in the assignment of MS/MS signals for other $G\gamma$ subunits.

The $[M + 7H]^{7+}$ ion at m/z 1108 from bovine γ_2 subunit was selected and fragmented in the triple quadrupole instrument. As was observed for transducin, the most abundant fragments in the MS/MS spectrum (Fig. 5) are due to loss of the isoprenyl group, in this case a geranylgeranyl group (gg = $C_{20}H_{32}$, molecular weight 272), and fragmentation N-terminal to proline residues (y_{16}^{2+}). Again note

[22] L. M. G. Heilmeyer, Jr., M. Serwe, C. Weber, J. Metzger, E. Hoffman-Posorske, and H. E. Meyer, *Proc. Natl. Acad. Sci. U.S.A.* **89,** 96554 (1992).

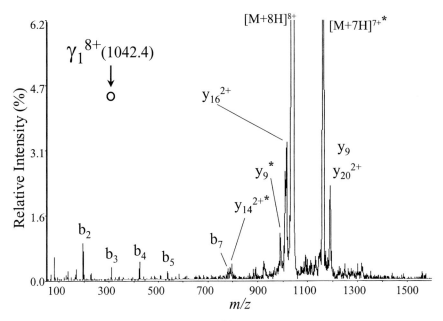

FIG. 4. Tandem mass spectrum of the $[M + 8H]^{8+}$ ion, m/z 1042.4, from bovine γ_1 acquired on the triple quadrupole instrument. The asterisk indicates loss of the farnesyl group. The arrow and the open circle under the selected m/z indicate a product ion scan to analyze all product (fragment) ions of the selected ion.

the cluster of large multiply charged fragment ions around the molecular ion signal and the loss of the isoprenyl group from abundant y ions.

The same $[M + 7H]^{7+}$ ion at m/z 1108 from bovine γ_2 subunit was selected and fragmented in the LCQ instrument and a very similar spectrum was obtained (Fig. 6). The same pattern of large multiply charged ions appears around the molecular ion and the loss of the isoprenyl group with loss of a charge is an abundant ion in the spectrum. However, the low-mass b ions are not abundant and due to the electronic requirements for ion storage in this instrument, data below approximately 27% of the precursor ion mass are not recorded. A significant advantage of the ion trapping instruments, however, is the ability to use multiple stages of fragmentation. In an MS^3 experiment on the bovine brain γ_2 subunit, the $[M + 7H]^{7+}$ ion at m/z 1108 was selected and fragmented followed by selection/fragmentation of the y_{14}^{2+} fragment ion (m/z 994.0). The second generation fragment ion spectrum was subsequently recorded and is displayed in Fig. 7. Clearly richer fragmentation patterns, and therefore sequence data, can be obtained in this experiment at the expense of higher sample consumption. The sequence of the γ_2 subunit and observed cleavage sites are depicted in Fig. 8.

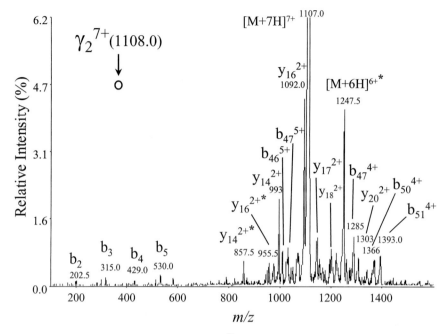

FIG. 5. Tandem mass spectrum of the $[M + 7H]^{7+}$ ion, m/z 1108, from bovine γ_2 subunit acquired on the triple quadrupole instrument. The asterisk indicates loss of the geranylgeranyl group.

Discussion

A relatively simple method has been developed to generate electrospray ionization signals from heterotrimeric G-protein γ subunits and hence measure accurate molecular weights. Furthermore, a tandem mass spectrometry strategy has been employed to obtain information on sequence and posttranslational modification. Samples can be provided as purified γ subunits or as acetone precipitates of G-protein heterotrimers. Generally, the purified γ subunit samples provided better signals and therefore more detailed information was obtained from these samples.

Two different tandem mass spectrometers were used for these experiments, a triple quadrupole instrument and an ion trap instrument. Both produced quality data and provide information on the structures of G-protein γ subunits as evidenced by the rich fragmentation patterns observed. The triple quadrupole has the advantage of scanning to low masses while the ion trap has the advantage of carrying out MS^n experiments. The MS^n experiment can be used to identify the structure of fragment ions in the MS/MS spectra, which can be challenging given the multiple possibilities for fragmentation sites and the multiple possibilities for the number of charges on an ion. The triple quadrupole instrument also has the ability to perform scans where the two mass analyzers are scanned with a given mass offset

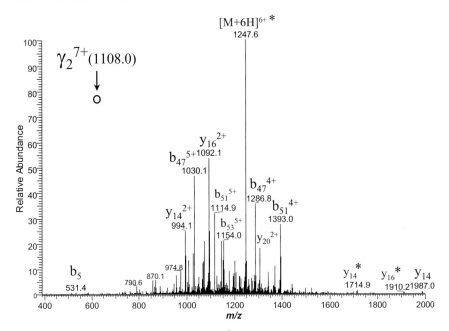

FIG. 6. Tandem mass spectrum of the $[M + 7H]^{7+}$ ion, m/z 1108, from bovine γ_2 subunit acquired on the ion trap instrument. The asterisk indicates loss of the geranylgeranyl group.

(neutral loss scan) or a given mathematical relationship (functional relationship scan) between the two analyzers.[23] This latter scan mode can potentially be used to identify all geranylgeranylated or farnesylated proteins in a single experiment by scanning for all ions that lose the isoprenyl group and a charge. The nanospray ionization source played a significant role in obtaining the highest quality data on small amounts of protein samples by allowing extensive signal averaging on small sample sizes.

There are three main classes of fragment ions common to all intact $G\gamma$ MS/MS spectra: (1) loss of the isoprenyl group with accompanying loss of a single charge, (2) predominant cleavage at the N-terminal side of proline bonds, and (3) large, multiply charged b and y ions surrounding the molecular ion region. This common pattern of fragmentation establishes rules for interpretation and outlines a strategy for the use of this experiment in $G\gamma$ structural characterization. Clearly, this approach can be used to identify the isoprenyl group present on the protein and obtain small stretches of sequence data. For known $G\gamma$ sequences, fragment

[23] M. Vincenti, J. C. Schwartz, R. G. Cooks, A. P. Wade, and C. G. Enke, *Org. Mass Spectrom.* **23,** 579 (1988).

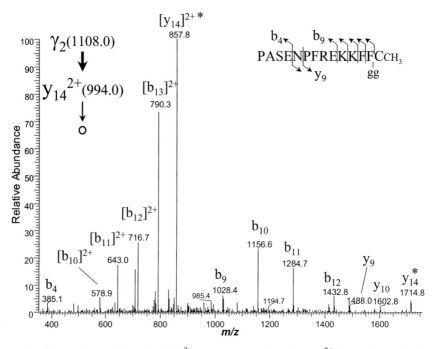

FIG. 7. Second-generation tandem (MS^3) mass spectrum of the selected y_{14}^{2+} fragment ion originating from the $[M + 7H]^{7+}$ ion, m/z 1108, from bovine γ_2 subunit acquired on the ion trap instrument. The asterisk indicates loss of the geranylgeranyl group. The thick arrow indicates a fixed m/z loss (1108.0 → 994.0). The thin arrow and open circle indicate a second-generation product ion scan to analyze for all fragments of m/z 994.0.

ions due to cleavage at proline residues can be used as a "mass spectrometric enzyme" to map the protein and identify posttranslational modifications predicted by molecular weight measurements. More specifically, since each $G\gamma$ subunit has a unique series of proline cleavage sites, modified $G\gamma$ subunits can easily be identified either by matching the mass of the proline fragment ions to a known $G\gamma$ sequence (identifies an N-terminal modification) or by looking for a systematic mass shift of the proline fragment ions (identifies a C-terminal modification). Further confirmation of a predicted sequence can be accomplished by interpretation of the large multiply charged b and y ions.

Conclusions

A method for sample preparation and mass spectrometric analysis has been developed for the characterization of intact G protein γ subunits. Rules for data interpretation and the expected structural information have been established based

γ1 SEQUENCE

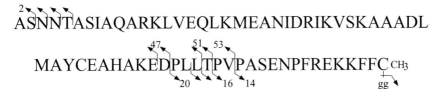

γ2 SEQUENCE

FIG. 8. Sequences of bovine γ_1 and γ_2 subunits indicating the fragments observed by tandem mass spectrometry.

on this initial work. This approach should prove valuable as more Gγ subunits are identified with considerable heterogeneity of structure.

Acknowledgments

We thank the National Ocean Service Center for Coastal Environmental Health and Biomolecular Research at Charleston, SC for use of the SCIEX API III. This work was supported in part by NIH Grant HL 07260 (postdoctoral fellowship for M.B.), a Young Investigator Award from the American Society for Mass Spectrometry to K.L.S., and NIH RO1 DK37219 to J.D.H.

Section V

RGS Proteins and Signal Termination

[42] Quantitative Assays for GTPase-Activating Proteins

By ELLIOTT M. ROSS

G protein signaling is fundamentally kinetic. The rate of receptor-stimulated GTP binding determines the rate of signal initiation, and the rate of hydrolysis determines the rate of termination. The balance of these rates determines signal amplitude. GTPase-activating proteins (GAPs) accelerate GTP hydrolysis up to 2000-fold, and thereby confer the speed needed for reliable cellular signaling. A G protein GAP can perform one or more diverse regulatory functions depending on its kinetic behavior, regulation, and subcellular localization.[1,2] First, a GAP will increase the speed with which signaling terminates on removal of agonist, and GAP activity is clearly necessary to reconcile the slow rate of hydrolysis of Gα-bound GTP *in vitro* with the rapid signal termination displayed in cells. GAPs can also inhibit signaling by decreasing the steady-state level of active Gα-GTP in the face of constant stimulation of GDP/GTP exchange by receptor. Last, complex effects of a GAP on the dynamics of the overall GTPase cycle can lead to two initially unsuspected effects: the damping of basal (background) G protein activation with little effect on the agonist-stimulated signal and/or the enhancement of selectivity of a G protein among multiple receptors. GAP activity may be detected in cells or tissues as increases in the rate of deactivation of a G protein-mediated signal on removal of agonist or as inhibition of the steady-state amplitude of an agonist-initiated signal. Diverse biochemical regulators can also produce these effects, of course, and *in vitro* determination of GAP activity is therefore desirable. The mechanism of G protein GAP activity has been reviewed.[2]

Assays for G protein GAP activity are all based on the reaction pathway for GAP-accelerated GTP hydrolysis shown in Scheme I.

$$G\alpha\text{--}GTP \quad \xrightarrow{k_h} \quad G\alpha\text{--}GDP + P_i$$
$$k_1 \downarrow\uparrow k_{-1}$$
$$GAP\text{--}G\alpha\text{--}GTP \quad \xrightarrow{k_{gap}} \quad GAP + G\alpha\text{--}GDP + P_i$$

SCHEME I

A GAP binds reversibly to the GTP-liganded Gα subunit and promotes the rapid hydrolysis of GTP in the GAP–Gα–GTP complex (i.e., $k_{gap} \gg k_h$). P_i dissociates essentially immediately, and most GAP assays monitor the production of the P_i product with time. It is possible, but usually less convenient, to measure the conversion of Gα–GTP to Gα–GDP by chromatography of bound nucleotide,

---

[1] E. M. Ross, *Rec. Prog. Hormone Res.* **50**, 207 (1995).
[2] E. M. Ross and T. M. Wilkie, *Annu. Rev. Biochem.* **69**, 795 (2000).

as is routine in assays of GAPs for small monomeric G proteins. Dissociation of the GAP from the Gα–GDP product is fast and is not shown separately, although the binding of phospholipase (PLC)-β^1 to Gα_q during steady-state GTP hydrolysis may be stable enough to persist over tens of seconds on the surface of a membrane bilayer.[3]

Two general approaches are used to measure GAP activity. The simplest and most quantitative assay is to measure the rate of hydrolysis of Gα-bound GTP in a single enzymatic turnover. This assay is usually performed by binding [γ-^{32}P]GTP to a Gα subunit under conditions where hydrolysis is minimal and then monitoring hydrolysis as release of [^{32}P]P$_i$ under optimal assay conditions.[4,5] "Single turnover" refers to the hydrolysis of a single molecule of GTP by each molecule of Gα; the GAP itself turns over multiple Gα–GTP complexes during the assay and thus acts catalytically. Single-turnover assays are preferred for assaying GAP activity in crude preparations (during purification, for example), for routinely standardizing the concentration of active GAP under defined conditions, or for studying the mechanism of GTP hydrolysis and its acceleration. The second general sort of GAP assay is to measure the increase in a G protein's steady-state GTPase activity.[6] Steady-state GAP assays depend on receptors to catalyze GDP/GTP exchange, which would otherwise be rate limiting for the overall GTPase reaction cycle. If exchange is relatively fast, acceleration of the hydrolytic step by a GAP will appear as an increase in the steady-state GTPase rate. The steady-state assay provides information on the role of the GAP in modulation of the overall GTPase cycle, reflects the importance of the membrane as an organizing structure, and measures activities under conditions closer physically to those encountered in cells. In some cases, the steady-state assay can detect GAP activity that is below the limits of sensitivity of a single-turnover assay. Rates of the individual reactions of the GTPase cycle can also be measured during steady-state hydrolysis (elsewhere in this volume[6a]). As is true for any biochemical measurement, the choice of GAP assay and the analysis of the data depend on the investigator's goals. GAP assays may be used to identify novel GAPs, to monitor GAP purification or standardize concentration, to study the regulation of GAP activity by allosteric ligands or covalent modification, or to evaluate the affinities and selectivities with which GAPs regulate different G proteins. The answer you get depends on how you ask the question.

[3] G. H. Biddlecome, G. Berstein, and E. M. Ross, *J. Biol. Chem.* **271**, 7999 (1996).

[4] J. Wang, Y. Tu, J. Woodson, X. Song, and E. M. Ross, *J. Biol. Chem.* **272**, 5732 (1997).

[5] It is also possible to measure hydrolysis of GTP bound to a Gα subunit fluorometrically because the GTP-activated state displays a higher intrinsic tryptophan fluorescence than does the GDP-bound state. [T. Higashijima and K. M. Ferguson, *Methods Enzymol* **195**, 321 (1991); T. Higashijima, K. M. Ferguson, P. C. Sternweis, M. D. Smigel, and A. G. Gilman, *J. Biol. Chem.* **262**, 762 (1987)].

[6] G. Berstein, J. L. Blank, D.-Y. Jhon, J. H. Exton, S. G. Rhee, and E. M. Ross, *Cell* **70**, 411 (1992).

[6a] S. Mukhopadhyay and E. M. Ross, *Methods Enzymol.* **344**, [26] 2002 (this volume).

Single-Turnover GAP Assays

To perform a single-turnover GAP assay, $[\gamma\text{-}^{32}P]GTP$ is first bound to the Gα target, residual nucleotide is usually removed, and hydrolysis of Gα-bound GTP is monitored either over time or at a single time point. GAP activity is thus measured as acceleration of the hydrolytic rate. This is the preferred assay for most applications.

Hydrolysis of Gα-bound GTP should follow a monoexponential time course (Scheme I, Fig. 1). The observed first-order rate constant, k_{app}, reflects a combination of k_h, the intrinsic hydrolytic rate of the Gα subunit, and k_{gap}, the GAP-stimulated rate, weighted according to the concentration of the GAP and its affinity for the Gα–GTP complex (k_{-1}/k_1).[7] GAPs act catalytically. Each GAP molecule sequentially binds multiple Gα–GTP molecules to promote their hydrolysis, and k_{app} is proportional to the concentration of GAP for GAP concentrations well below that of Gα–GTP (Fig. 1B).

Reliable quantitation of GAP activity in single-turnover assays is based on the analogy of Scheme I to the Briggs–Haldane formalism for an "enzyme" that catalyzes conversion of a Gα–GTP "substrate" to Gα–GDP and P_i products. GAP-independent hydrolysis during the assay is subtracted as background. This analogy allows application of the Michaelis–Menten equation, defining $K_m = (k_{-1} + k_{gap})/k_1$ and $V_{max} = [GAP] \cdot k_{gap}$ (Scheme I).[8] V_{max}, in units of moles of P_i produced per unit time, is a measure of total GAP activity. When k_{gap} is less than k_{-1}, the value of K_m approximates the affinity of the GAP for the particular Gα–GTP substrate because $K_m \sim K_s = k_{-1}/k_1$, and provides a measure of GAP–Gα selectivity that is independent of absolute hydrolysis rates. The value k_{gap}/K_m is an overall measurement of GAP activity,[8] but reflects both relative activity of the GAP and the hydrolytic capacity of the Gα substrate. Overall, this Michaelian analysis is the preferred method of data presentation for describing GAP activity.

A unit of GAP activity, like any enzyme unit, is defined as the amount of GAP that will elevate k_{app} (V_{max}) by a fixed amount, usually 1 min^{-1}.[4] The number of GAP units displayed by a particular concentration of a GAP depends on the assay conditions, but such a unit is a reliable measure of activity in that it scales linearly

---

[7] For Scheme I in rapid binding equilibrium that defines $K_1 = k_1/k_{-1}$, the differential rate equations can be solved to yield a simple first-order equation for the concentration of phosphate released $[P]$ as a function of time.

$$\frac{[P]_t}{[P]_\infty} = 1 - e^{-k_{app}t}$$

For $[GAP] \ll 1/K_1$, the apparent first-order rate constant k_{app} is defined as

$$k_{app} = \frac{k_h + k_{gap}K_1[GAP]}{1 + K_1[GAP]}$$

and k_{app} will increase linearly with $[GAP]$.

[8] A. Fersht, "Enzyme Structure and Mechanism." W. H. Freeman and Company, New York, 1977.

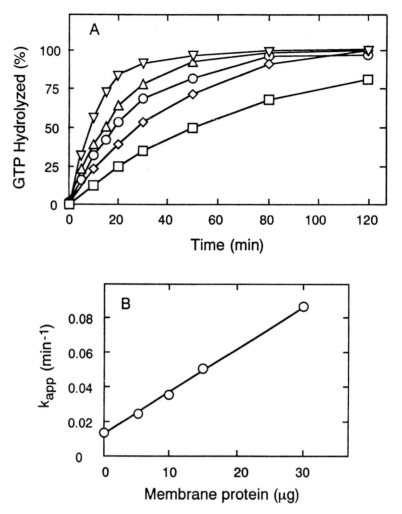

FIG. 1. Quantitative detection of G_z GAP activity in a particulate fraction from steer cerebral cortex. (A) $G\alpha_z$-[γ-^{32}P]GTP was incubated at 15° with a suspension of brain membranes, and release of [^{32}P]P$_i$ was monitored as described in the text. Assays contained 125 fmol of $G\alpha_z$-[γ-^{32}P]GTP and either 5 μg ($\diamond$), 10 μg ($\bigcirc$), 15 μg ($\triangle$), or 30 μg (∇) of untreated membranes or 15 μg of boiled membranes ($\square$). Hydrolysis approached 100% and approximated a family of monophasic exponential curves. Parallel binding experiments showed that loss of bound [γ-^{32}P]GTP equaled production of [^{32}P]P$_i$, indicating that dissociation of [γ-^{32}P]GTP was negligible. The boiled membranes had no effect compared with a buffer control. (B) The data shown in panel A were well fit by a single exponential rate equation to determine k_{app}, which is shown plotted vs the amount of membrane added. Activity in 30 μg of membranes is about equal to that of 10 fmol of purified RGSZ1. Hence, each molecule of GAP turned over about 10 molecules of $G\alpha_z$-[γ-^{32}P]GTP. Reproduced with permission from J. Wang, Y. Tu, J. Woodson, X. Song, and E. M. Ross, *J. Biol. Chem.* **272**, 5732 (1997).

with GAP concentration over a wide range of conditions. It is therefore useful for quantitating a GAP, particularly during purification. Note that a GAP unit is defined with reference to a single Gα–GTP substrate and set of assay conditions. Most importantly, a GAP unit reflects both the intrinsic GTPase activity of the particular Gα subunit and the relative stimulation of hydrolysis ("-fold increase"). For example, 1.5 units of a G_z GAP would increase k_{app} for $Gα_z$ from 0.015 min^{-1} to 1.515 min^{-1}, about 100-fold. An equal increment in the k_{app} of $Gα_i$, from 3 min^{-1} to 4.5 min^{-1}, would be only a 50% increase.[4]

Single-turnover assays depend absolutely on obtaining either reliable initial rates—while the reaction is linear with time—or a complete enough description of the entire reaction time course to allow calculation of the rate constant k_{app}. GAP activity must also be linear with increasing amounts of the GAP protein. Many published assays have grossly underestimated G protein GAP activity because the investigators ignored these points. This section presents methods for single-turnover GAP assays. The method for G_z GAP activity is presented first and in greatest detail as an example. The following section on GAPs for G_i, G_o, and G_s is a variation, and sections on G_q and G_t describe their unique problems. Procedures for purifying G protein subunits[9] and other ancillary techniques are described in this and previous volumes (see also Wang et al.[10]).

Prototypical Single-Turnover Assay for $Gα_z$ GAP Activity

$Gα_z$ hydrolyzes bound GTP more slowly than does any other Gα subunit, with $k_{hyd} \sim 0.014$ min^{-1}. This facilitates the preparation of $Gα_z$–[γ-^{32}P]GTP and allows its relatively stable storage during extended experiments.

Preparation of $Gα_z$-Bound [γ-^{32}P]GTP. Before beginning the binding reaction, prepare a 0.5× 10 cm (2 ml) gravity-feed column of Sephadex G-25 in buffer B [25 mM Na–HEPES (pH 7.5), 1 mM EDTA, 0.1% Triton X-100, 3 mM dithiothreitol, 2.5 μM GTP]. To bind [γ-^{32}P]GTP, incubate purified $Gα_z$, 1–60 pmol $Gα_z$ per planned assay, in a total of 200 μl of buffer B plus $\sim 3 \times 10^7$ cpm of [γ-^{32}P]GTP[11] for 20 min at 30°. Chill the reaction and determine total radioactivity

[9] E. Lee, M. E. Linder, and A. G. Gilman, *Methods Enzymol.* **237**, 146 (1994).

[10] J. Wang, Y. Tu, S. Mukhopadhyay, P. Chidiac, G. H. Biddlecome, and E. M. Ross, *in* "G Proteins: Techniques of Analysis" (D. R. Manning, ed.), p. 123. CRC Press, Boca Raton, FL, 1999.

[11] [γ-^{32}P]GTP can be purchased or easily synthesized by the method of R. A. Johnson and T. F. Walseth, *Adv. Cyclic Nucleotide Res.* **10**, 135 (1979). Purity is 98–99% in either case. This is adequate for single-turnover assays in which Gα-bound [γ-^{32}P]GTP is separated from free [γ^{32}P]GTP and [^{32}P]P$_i$. For steady-state assays, however, contamination with [^{32}P]P$_i$ must be below 0.5%, and is below 0.1% with freshly purified [γ-^{32}P]GTP. We purify [γ-^{32}P]GTP by anion-exchange HPLC with a 10–700 mM gradient of KP$_i$, pH 7.0. KP$_i$ concentrations depend on the anion exchanger.[3] Adjust elution rates and gradient using unlabeled nucleotide and a UV absorbance monitor so that the [γ-^{32}P]GTP (up to 5 mCi per preparation) is eluted in less than 0.5 ml with greater than 99.9% purity. Storage at $-80°$ with inclusion of 10 mM Tricine maintains purity for 3 weeks.

in a 2 μl volume to calculate the specific activity of the $[\gamma\text{-}^{32}\text{P}]\text{GTP}$. The amounts specified here should give about 60 cpm/fmol. Before continuing, check $G\alpha_z$-bound nucleotide in a 2 μl sample of incubation mix by the standard nitrocellulose filter assay.[12] About 25% of the $G\alpha_z$ should have $[\gamma\text{-}^{32}\text{P}]\text{GTP}$ bound. The remainder is bound to GDP.[13] $G\alpha_z\text{-}[\gamma\text{-}^{32}\text{P}]\text{GTP}$ may be stored at $-80°$ at this point.

Load the binding mixture onto the gel filtration column at 0–4° and wash the column with 350 μl of cold buffer B. Elute $G\alpha_z\text{-}[\gamma\text{-}^{32}\text{P}]\text{GTP}$ with an additional 750 μl of buffer B. (Adjust elution volumes for the volume and geometry of the column.) Dilute the $G\alpha_z\text{-}[\gamma\text{-}^{32}\text{P}]\text{GTP}$ to a predicted concentration of $\sim$0.2 pmol $G\alpha_z\text{-}[\gamma\text{-}^{32}\text{P}]\text{GTP}/70$ μl. This concentration is designed to give about 3 nM $G\alpha_z\text{-}[\gamma\text{-}^{32}\text{P}]\text{GTP}$ in the assay, just above the K_m for brain G_z GAP,[4] and may be altered according to the experiment. Add 0.05 volume of 0.1 M GTP (5 mM final), 0.05 volume of 0.2 mg/ml albumin (10 μg/ml final), and 0.01 volume of 0.68 M MgCl_2 (calculated free Mg^{2+} concentration of 1 mM; see below).[14] These volumes can be modified to allow for addition of other agents, to vary the concentration of $G\alpha\text{-}[\gamma\text{-}^{32}\text{P}]\text{GTP}$, or to allow for addition of GAP in larger volumes.

Before starting the assay, check bound $[\gamma\text{-}^{32}\text{P}]\text{GTP}$ again by nitrocellulose binding assay of a 70 μl aliquot, and check residual $[^{32}\text{P}]\text{P}_i$ by the charcoal precipitation assay described below. Background $[^{32}\text{P}]\text{P}_i$ should be less than 5% of total $G\alpha_z$-bound radioactivity. $G\alpha_z\text{-}[\gamma\text{-}^{32}\text{P}]\text{GTP}$ is relatively stable at 0° and in the absence of Mg^{2+} ($t_{1/2} \sim 4.5$ hr). Regardless, prepare and isolate $G\alpha_z\text{-}[\gamma\text{-}^{32}\text{P}]\text{GTP}$ just before use to maximize yield and minimize free $[^{32}\text{P}]\text{P}_i$. In our experience, dissociation of $[\gamma\text{-}^{32}\text{P}]\text{GTP}$ is much slower than its hydrolysis under all common conditions.

GAP-Stimulated Hydrolysis of $G\alpha_z\text{-}[\gamma\text{-}^{32}\text{P}]\text{GTP}$. G_z GAP activity is measured at 15°, where the unstimulated rate of hydrolysis of bound GTP $k_h = 0.014$ min^{-1} ($t_{1/2} = 50$ min) (Fig. 1). The assay can potentially be carried out over any time span, but 10 min is usually convenient and produces a low and reproducible level of unstimulated hydrolysis ($\sim$10% of total). Assays are initiated by adding the $G\alpha_z\text{-}[\gamma\text{-}^{32}\text{P}]\text{GTP}$ in 70 μl to prewarmed polypropylene tubes that contain the GAP (or buffer as a blank), Mg^{2+}, and any other additives in a total volume of 10 μl (80 μl total assay volume). Individual volumes of the components can be

[12] D. R. Brandt and E. M. Ross, *J. Biol. Chem.* **261**, 1656 (1986).

[13] In this and other single-turnover GAP assays, the $G\alpha\text{-}[\gamma\text{-}^{32}\text{P}]\text{GTP}$ substrate is substantially contaminated with $G\alpha$-GDP that was either not bound to $[\gamma\text{-}^{32}\text{P}]\text{GTP}$ originally or was formed by hydrolysis after binding. Although competition for GAP binding by $G\alpha$-GDP is not a problem because of its low affinity, the presence of the $G\alpha$–GDP should not be forgotten. For example, residual $G\alpha$ can chelate stoichiometric amounts of the $G\beta\gamma$.

[14] Many algorithms and computer programs are available to calculate the concentrations of free divalent cations in the presence of one or more chelators.[4,10] Under these assay conditions, values of K_d for Mg^{2+} of EDTA and GTP of 1 μM and 100 μM have yielded calculated values of $[\text{Mg}^{2+}]_{\text{free}}$ within 10% of the concentrations determined experimentally.[4]

varied as needed to accommodate other additions. Controls in each experiment should include a sample with no added GAP to provide a parallel measure of basal hydrolysis (k_h) (although this rate is quite reproducible) and a reagent blank ($G\alpha_z$-[γ-^{32}P]GTP plus charcoal, no incubation, no GAP) to correct for contaminating [^{32}P]P_i, counter background, inefficient precipitation of GTP, etc.

Each assay is terminated by addition of 920 μl of a 5% slurry of activated charcoal in H_3PO_4, pH 3 (Norit A or equivalent) that is continually stirred at 4°. Assay tubes are vortexed and kept on ice until the experiment is completed, and all are centrifuged at 1500g for 10 min. [^{32}P]P_i in a convenient volume of supernatant, usually 600 μl, is then determined by either Cerenkov or liquid scintillation counting.

The assay medium described here, buffer B plus 1 mM Mg^{2+}, is optimal for members of the RGSZ subfamily of RGS proteins and for RGS4. The GTP in the buffer is used to protect any free [γ-^{32}P]GTP (carried over from the binding reaction or dissociated from $G\alpha_z$) from hydrolysis by contaminating nucleoside triphosphatases. This caution is particularly important for crude preparations of GAPs. In our experience, most GAPs are not sensitive to the concentration of Mg^{2+} over the range of 1–1000 μM, although activity declines above 1 mM. RGSZ1 (Ret-RGS1) displays a distinct activity optimum at 1 mM free Mg^{2+}, however, and we recommend maintaining this concentration routinely. Because the assay contains 1 mM EDTA and 5 mM GTP, it is important to calculate the final concentration of free Mg^{2+} carefully, taking into consideration any Mg^{2+} or chelator that may be added as a component of the GAP dilution buffer. BSA is added both to increase recovery of $G\alpha_z$-[γ-^{32}P]GTP during gel filtration and to stabilize protein during the assay. A detergent is required to maintain the solubility of $G\alpha$ subunits and many GAPs; Triton X-100 was found to be the best detergent for G_z GAP. Components of the assay cocktail may be changed as necessary for other GAPs, as suggested by variations described in other assay protocols.

Calculation of Data. GAP activity is most simply calculated as the initial (linear) hydrolytic rate (moles [γ-^{32}P]GTP hydrolyzed per minute) after subtraction of the "no GAP" control value.[15] This format is simple and direct and should be used when GAP activity is analyzed according to the Michaelis–Menten equation (V_{max}, K_m). Data can also be presented as k_{app} or in GAP

---

[15] To calculate the specific activity of $G\alpha_z$–[γ-^{32}P]GTP, divide the radioactivity in the sample of the [γ-^{32}P]GTP binding cocktail by the amount of GTP in the same volume. This value is valid for the $G\alpha_z$–[γ-^{32}P]GTP complex because $G\alpha_z$ bound to GDP will not interfere in the assay and is therefore irrelevant. After subtracting radioactivity in the reagent blank (no GAP, no incubation) from that in the experimental samples, use the specific activity to convert radioactivity in each sample to moles of $G\alpha_z$–[γ-^{32}P]GTP hydrolyzed. Remember to correct for the fraction of the charcoal supernatant sampled, usually 600 μl of a total of 1000 μl. $G\alpha_z$–[γ-^{32}P]GTP is measured in the binding assay performed at the beginning of the experiment.

units (see above). To calculate the net hydrolysis rate constant k_{app} from the amount of $[\gamma\text{-}^{32}P]GTP$ hydrolyzed, the equation in footnote 7 is rearranged to give $k_{app} = (-1/t) \ln(1-[P]_t/[P]_\infty)$, where $[P]_t/[P]_\infty$ is the fraction of $[\gamma\text{-}^{32}P]GTP$ hydrolyzed during the assay time t. Alternatively, a complete hydrolysis time course can be fit to yield k_{app} (Fig. 1). k_{app} increases linearly with the concentration of GAP (for low concentrations) and is therefore a good measure of GAP activity (Fig. 1B).

Note that data are usable only when total hydrolysis is both (1) significantly above the "no GAP" background level, such that incremental GAP-stimulated hydrolysis can be determined accurately, and (2) significantly below complete hydrolysis of all the Gα–GTP substrate. Do not use data where $[P]_t/[P]_\infty$ exceeds 80%, because small errors in determining this fraction cause large errors in the calculated GAP activity. We have observed that efficient G_z GAPs increase the hydrolytic rate for $G\alpha_z$–GTP about 600-fold, with $K_m \sim 2$ nM.[16]

Modification of Single-Turnover Assays for $G\alpha_i$, $G\alpha_o$, and $G\alpha_s$

GAP assays for $G\alpha_i$, $G\alpha_o$, and $G\alpha_s$ are designed and executed essentially as described for $G\alpha_z$, but the intrinsically faster hydrolytic rates of these Gα subunits require that $G\alpha\text{-}[\gamma\text{-}^{32}P]GTP$ be purified and used promptly after preparation and that the assay itself be adjusted to account for the higher k_{gap} values that are achieved. Modifications suggested below have been applied routinely to $G\alpha_i$ and $G\alpha_o$, and should also be applicable to $G\alpha_s$.

To prepare $G\alpha\text{-}[\gamma\text{-}^{32}P]GTP$, incubate Gα subunit (usually 50 pmol) for 15–20 min at 30° with $[\gamma\text{-}^{32}P]GTP$ (1 μM or a 0.5 μM stoichiometric excess over Gα, whichever is greater, usually 20 cpm/fmol) in 100 μl of Buffer C [50 mM Na–HEPES (pH 7.5), 0.05% dodecyl polyoxyethylene ($n\sim9$; Lubrol PX, Genapol 24-L-75, etc.), 1 mM dithiothreitol (DTT), 5 μg/ml albumin] plus 10 mM EDTA. Because G_i is usually stored in the presence of $MgCl_2$, 10 mM EDTA is needed to reduce free Mg^{2+} to below 10 nM and thereby inhibit hydrolysis during loading. Chill the reaction mixture on ice and *immediately* purify $G\alpha\text{-}[\gamma\text{-}^{32}P]GTP$ by centrifugal gel filtration on a 3 ml column of Sephadex G-25 in buffer C plus 10 mM EDTA at 0–4°. Incubation time for binding $[\gamma\text{-}^{32}P]GTP$ is not critical, but the $G\alpha\text{-}[\gamma\text{-}^{32}P]GTP$ complex must be chilled, purified, and used as quickly as possible! About 5–10% of bound $[\gamma\text{-}^{32}P]GTP$ is hydrolyzed per 10 min at 0°, which decreases the substrate concentration and increases the assay background. Design experiments so that a reasonably small number of assays is performed with one batch of $G\alpha\text{-}[\gamma\text{-}^{32}P]GTP$ to keep its concentration acceptably constant for all the assays. Measure the specific activity of the $[\gamma\text{-}^{32}P]GTP$ and the concentration of Gα-bound $[\gamma\text{-}^{32}P]GTP$ as described for $G\alpha_z$.

[16] J. Wang, A. Ducret, Y. Tu, T. Kozasa, R. Aebersold, and E. M. Ross, *J. Biol. Chem.* **273**, 26014 (1998).

Hydrolysis of Gα-bound GTP is measured as described for Gα_z–[γ-^{32}P]GTP. Because other Gα subunits hydrolyze bound GTP 50- to 100-fold faster than does Gα_z, the assay is executed over a shorter period of time and, usually, at a lower temperature, 0–10°. Note that hydrolysis rates of Gα subunits may vary idiosyncratically with temperature; extrapolating rates from one temperature to another is dangerous. Regardless, it is important to choose assay time and temperature and the concentration of GAP so that an acceptable amount of Gα–[γ-^{32}P]GTP is hydrolyzed (more than basal, less than 75%). Our routine assay medium is Buffer C plus 1 mM GTP and 1 mM free Mg^{2+}. The concentration of free Mg^{2+} is achieved by controlling for chelation by EDTA and GTP (see above).[14] Optimal assay conditions must be established by the investigator. Assays are terminated, usually after 30 sec, by addition of 950 μl of a 5% (w/v) charcoal slurry in 50 mM H$_3$PO$_4$, pH 3, and radioactivity is determined in a 600 μl sample of the supernatant. GAP activity for Gα_o, Gα_i, or Gα_s is calculated exactly as described above for Gα_z.

Modification of Single-Turnover Assays for Gα_q

Because members of the G$_q$ family bind GTP slowly, it is not possible to load wild-type Gα_q with [γ-^{32}P]GTP fast enough to avoid complete hydrolysis, but G$_q$ GAP activity can be measured in a single-turnover assay using the hydrolysis-defective R183C mutant.[17] Gα_qR183C hydrolyzes bound GTP about 0.7% as fast as wild-type, which allows loading with [γ-^{32}P]GTP, and retains sensitivity to the GAP activity of both RGS proteins and PLC-β. The Gα_qR183C GAP assay provides a convenient way to quantitate G$_q$ GAP activity during purification or to standardize the activity of a Gα_q GAP; it has been described in detail elsewhere.[17] This GAP assay is not nearly so straightforward as those for other Gα subunits. First, meaningful conclusions clearly depend on the assumption that the GAP responses of Gα_qR183C are unaltered from those of wild-type Gα_q. Further, both spontaneous hydrolysis and dissociation of [γ-^{32}P]GTP are significant relative to GAP-stimulated hydrolysis, and appropriate corrections must be applied. In addition, hydrolysis time courses are not monophasic and frequently do not go to completion. The magnitude of the initial GAP-stimulated phase is crucially dependent on the identity and concentration of detergent. Investigators should evaluate the complete time course of GAP-stimulated hydrolysis and restrict assays to the period over which hydrolysis can be fit to a single exponential, where the first-order rate constant k_{app} increases linearly with the concentration of GAP, and where total hydrolysis is less than 50% complete.

---

[17] P. Chidiac and E. M. Ross, *J. Biol. Chem.* **274,** 19639 (1999).

Single-Turnover Assay for G_t

A variation of the single-turnover assay has been used by the groups of Arshavsky and Wensel[18,19] to study the GAP activity of RGS proteins on G_t, which is essentially impossible to load with $[\gamma\text{-}^{32}P]GTP$ in the absence of rhodopsin or a related receptor. In this assay, photoreceptor membranes (disks) are washed to remove G_t and as many other proteins as possible. Rhodopsin is then activated by light and substoichiometric G_t is added. The G_t releases bound GDP when it binds the bleached rhodopsin. The reaction is initiated by addition of an amount of $[\gamma\text{-}^{32}P]GTP$ substoichiometric to G_t, and the time course of its hydrolysis is monitored using charcoal precipitation of substrate as described above. This method is useful to demonstrate the presence of endogenous GAP activity in the disk membranes (RGS9/Gβ5) or in preparations of exogenous proteins, and to measure the potentiation of GAP activity by the γ subunit of cyclic GMP phosphodiesterase. It probably underestimates GAP activity substantially, however. The assay strategy assumes that initial binding of $[\gamma\text{-}^{32}P]GTP$ is fast and that chemical hydrolysis is uniquely rate-limiting, both of which are uncertain. Skiba *et al.* have measured GAP-stimulated hydrolysis rates for G_t GTP that are much higher than those measured with the older assay and which are commensurate with those measured for G_i and G_q.[20] Interested readers are referred to this work.

GAPs with High K_m

Values of K_m for different GAPs and their Gα-GTP substrates vary from $\sim$2 nM to $\sim$1 μM.[20–22] High values of K_m reflect in part the high rates of GAP-promoted GTP hydrolysis ($K_m = (k_{-1} + k_{gap})/k_1$, Scheme I), but low affinity for the Gα–GTP substrate (high K_s) may also contribute. It is therefore often impossible to perform single-turnover GAP assays at concentrations of Gα–$[\gamma\text{-}^{32}P]GTP$ well above the K_m, either because background hydrolysis is high or because substrate with appropriate specific activity cannot practically be prepared at adequate concentrations. Similarly, low values of V_{max} may necessitate using high concentrations of the GAP. In such situations, K_m and V_{max} can still be determined by measuring GAP activity as a function of GAP concentration at several low concentrations of the Gα–GTP substrate. Assay protocols are otherwise identical to those described above, and data are analyzed according to an alternative derivation of the Briggs–Haldane equation for steady-state enzyme activity.[8] When the concentration of Gα–GTP is

[18] J. K. Angleson and T. G. Wensel, *Neuron* **11**, 939 (1993).

[19] V. Y. Arshavsky and M. D. Bownds, *Nature* **357**, 416 (1992).

[20] N. P. Skiba, J. A. Hopp, and V. Y. Arshavsky, *J. Biol. Chem.* (2000).

[21] D. M. Berman, T. M. Wilkie, and A. G. Gilman, *Cell* **86**, 445 (1996).

[22] B. A. Posner, S. Mukhopadhyay, J. J. Tesmer, A. G. Gilman, and E. M. Ross, *Biochemistry* **38**, 7773 (1999).

low, the initial rate of GTP hydrolysis (after subtraction of background) is defined as:

$$v = [G\alpha - GTP]k_{gap}\left(\frac{[GAP]}{K_m + [GAP]}\right)$$

Thus, initial rate is proportional to the concentration of Gα–GTP and "saturates" at $V_{max} = k_{gap}[G\alpha-GTP]$ as the GAP concentration increases above K_m. Practically, it is also possible to assay GAPs reproducibly even when K_m is unmeasurably high. If activity is reproducible over a reasonable range of Gα–GTP concentrations, plots of rate vs concentration will have the slope k_{gap}/K_m, which is the virtual second-order reaction rate constant for the productive interaction between GAP and substrate.[8]

Monitoring Dissociation of Gα–[γ-^{32}P]GTP

When first establishing a single-turnover GAP assay, it is important to verify both that the Gα–[γ-^{32}P]GTP complex is stable throughout the assay period and that no [^{32}P]P$_i$ is formed from hydrolysis of free [γ-^{32}P]GTP (dissociated from Gα or carried over from the loading reaction). The most convenient control is to confirm that loss of Gα-bound [γ-^{32}P]GTP measured in a nitrocellulose filter binding assay[23] exactly equals the formation of [^{32}P]P$_i$ during the assay interval (Fig. 1A). This control should be performed periodically when using impure GAP preparations or when adding any reagent that might alter nucleotide dissociation. Definitive demonstration that [^{32}P]P$_i$ is derived quantitatively from Gα-bound [γ-^{32}P]GTP is to measure production of Gα-bound [α-^{32}P]GTP from Gα-bound [α-^{32}P]GTP directly. For this assay, Gα–[α-^{32}P]GTP is prepared exactly as described above for [γ-^{32}P]GTP.[24] The GAP assay is also performed similarly, but the reaction is quenched with cold buffer and Gα is trapped by binding to a nitrocellulose filter.[4,23] The filter is dissolved in 1 ml of cold acetone, and bound [α-^{32}P]GTP and [α-^{32}P]GTP are resolved by thin-layer chromatography on polyethyleneimine cellulose in a solvent composed of 0.75 M Tris base plus 0.45 M HCl.[23,25] Radioactivity in each spot is quantitated by Cerenkov counting.

Competitive Inhibition by Gα Subunits

Depending on its affinity for the GAP, a second Gα subunit bound to GTP or a GTP analog will competitively inhibit binding of the Gα-[γ-^{32}P]GTP substrate in a single-turnover GAP assay. Such competition allows the interaction of a GAP with diverse Gα subunits to be analyzed according to standard enzymologic

[23] D. R. Brandt and E. M. Ross, *J. Biol. Chem.* **260,** 266 (1985).
[24] Commercial [α-^{32}P]GTP (>98% pure) is acceptable.
[25] B. R. Bochner and B. N. Ames, *J. Biol. Chem.* **257,** 9759 (1982).

strategies for measuring competitive inhibition).[10] If the concentration of the $G\alpha-[\gamma^{-32}P]GTP$ substrate and its K_m are known, then the concentration dependence of inhibition will directly yield the affinity of binding of the competing $G\alpha$ ($K_d = K_i = IC_{50}/[0.1 + [S]/K_m)]$. This approach is particularly useful when comparing $G\alpha$ subunits with different intrinsic values of k_h or k_{gap}, for studying the effects of mutation or covalent modification of a $G\alpha$ on GAP binding, where the K_m for a $G\alpha$–GTP complex is too high to allow convenient preparation of high concentrations of $G\alpha-[\gamma^{-32}P]GTP$ substrate, or where the $G\alpha$ is bound to a nucleotide other than GTP.

Once assay conditions have been established for a particular GAP and $G\alpha-[\gamma^{-32}P]GTP$ substrate, K_i is determined by evaluating the apparent K_m at increasing concentrations of competitive inhibitor.[8] Plotting the apparent K_m against the concentration of inhibitory $G\alpha$ should yield a straight line whose slope is equal to K_m/K_i. More simply, GAP assays are performed at a fixed concentration of $G\alpha-[\gamma^{-32}P]GTP$ substrate (near or just above K_m) in the presence of a range of inhibitor concentrations adequate to yield fractional inhibition of 10–90%. Fractional inhibition should follow a simple saturation function, with half-maximal inhibition occurring at IC_{50}. This method is also valuable in that relative values of IC_{50} for different inhibitors can be compared even without knowledge of K_m.

For inhibition assays, the concentration of native inhibitor should be determined directly according to nitrocellulose filter binding assays.[23] Note that the binding of AlF_4^- to $G\alpha$-GDP is relatively unstable. If the $G\alpha$–GDP–AlF_4 is isolated before the assay, it must be used quickly to avoid dissociation.[26,27] Alternatively, fixed concentrations of $G\alpha$–GDP plus 20 μM $AlCl_3$ and 10 mM NaF can be added directly to the assay medium.[4] However, because single-turnover GAP assays always contain some $G\alpha$–GDP that is formed by hydrolysis of the $G\alpha-[\gamma^{-32}P]GTP$ substrate during loading, the excess $Al^{3+}F^-$ will bind this $G\alpha$–GDP and thus form another potent inhibitor. Thus, if Al/F is not removed before assay, its additional inhibitory activity must be subtracted before inhibition by the alternative $G\alpha$–GDP-AlF_4 is evaluated.[4]

Steady-State GAP Assays

Because steady-state GTPase activity depends on both GTP hydrolysis and GDP/GTP exchange, GAP activity can be observed and measured as stimulation of a G protein's steady-state GTPase activity. Stimulation is significant only if the nucleotide exchange rate is sufficiently fast, however. In most cases, GTPase

[26] B. Antonny and M. Chabre, *J. Biol. Chem.* **267**, 6710 (1992).
[27] D. M. Berman, T. Kozasa, and A. G. Gilman, *J. Biol. Chem.* **271**, 27209 (1996).

TABLE I
STEADY-STATE GTPASE ACTIVITY: POTENTIATIVE
STIMULATION BY RECEPTOR AND GAP

Compound	$-PLC$	$+PLC$
Atropine	0.14	0.25
Carbachol	0.41	14.4

[a] GTPase depends on both GDP/GTP exchange and GTP hydrolysis steady-state GTPase activity was measured in phospholipid vesicles that contain m1 muscarinic cholinergic receptor and G_q, with or without 5 nM phospholipase C-β (PLC) and in the presence of either atropine (antagonist) or carbachol (agonist). Stimulating nucleotide exchange alone (agonist) increases activity only about 3-fold to where hydrolysis becomes rate-limiting. Stimulating hydrolysis alone (with PLC) has little effect because exchange is already rate-limiting. Both agents combine to accelerate steady-state hydrolysis about 100-fold. Data, expressed as moles of P_i released per mole of G_q per min, are recalculated from Biddlecome.[34]

activity is limited by the rate of GDP release and GTP binding, such that increasing the rate of hydrolysis itself has little effect on steady-state rates (Table I). To observe GAP activity in a steady-state assay, GDP/GTP exchange must therefore be accelerated by a receptor or receptor-mimetic peptide.[3]

Steady-state GAP assays are most useful for studying the coordinated activities of receptor, G protein, and effector during signal transduction, their mechanisms and regulation. In cells, GAPs stimulate the steady-state GTPase activity of a coupled receptor–G protein system during concurrent stimulation of GDP/GTP exchange by agonist-bound receptor and multiple modulating inputs. These interactions are complex and depend both on a network of protein–protein interactions and on the balance of the rates of many partial reactions. Acceleration of hydrolysis can alter the nucleotide exchange activity of receptors, increased availability of Gα-GTP can alter GAP activity, and location of all the proteins at a membrane surface will have an impact on steady-state GTPase activity. GAPs interact with G$\beta\gamma$, receptors, and other regulatory proteins in addition to Gα, and steady-state assays can be used to evaluate these interactions. Measurement of GAP activity at steady state thus gives the most applicable information about how well a GAP can interact with other signaling components and what its effects on signal output will be: enhanced turn-off rate, attenuated signal, enhanced receptor selectivity, etc. Steady-state GAP assays are also useful when it is difficult to prepare the Gα-GTP substrate for a single turnover assay but where a coupled receptor-G protein preparation is available. Last, steady-state assays are much more sensitive for detection of GAP activities than are single turnover assays.[22,28]

Steady-state assays are less useful for determining a GAP's absolute activity or affinity for substrate because GDP/GTP exchange can become rate-limiting as the hydrolysis reaction is accelerated, and maximal stimulation of hydrolysis cannot be observed. Relative stimulation of steady-state GTPase by a GAP cannot be interpreted as proportional acceleration of the hydrolysis step. In addition, accelerated hydrolysis can itself increase the rate of receptor-catalyzed GDP/GTP exchange and alter its mechanism,[3] further complicating kinetic analysis of GAP effects. The overwhelming reason to study GAP activity during steady-state GTP hydrolysis is that this is the situation in which GAPs act in cells: on the surface of a membrane, as part of a rapidly cycling multiprotein complex that includes receptor and (probably trimeric) G protein, and in kinetic "competition" with the receptor to determine the fractional activation of the $G\alpha$ subunit.

Receptor–G Protein Vesicles

Steady-state GTPase measurements must be made when both receptor and G protein are incorporated in either natural membranes or reconstituted phospholipid vesicles. Because natural membranes contain enormous nucleoside triphosphatase activity, membrane-based GTPase assays produce low signals over high background; purified and reconstituted systems are the practical alternative. The methods described below are applicable to the G_q, G_i, and G_s families and to appropriate adrenergic and muscarinic cholinergic receptors. These methods should be generally applicable, but each investigator must validate conditions to the receptor and G protein of interest.

Methods for purifying receptors, working with phospholipids, and reconstitution of heterotrimeric G proteins and receptors into unilamellar phospholipid vesicles are outside the scope of this chapter. We have reviewed methods for reconstituting components of G protein signaling pathways[29,30] and described in detail methods for preparing large unilamellar phospholipid vesicles containing several different receptors and heterotrimeric G proteins.[3,13,31,32] For most signaling studies, vesicles are prepared with receptor and heterotrimeric G protein in an approximately 1 : 10 molar ratio, similar to that in cell membranes. Changing

[28] T. Ingi, A. M. Krumins, P. Chidiac, G. M. Brothers, S. Chung, B. E. Snow, C. A. Barnes, A. A. Lanahan, D. P. Siderovski, E. M. Ross, A. G. Gilman, and P. F. Worley, *J. Neurosci.* **18,** 7178 (1998).
[29] R. A. Cerione and E. M. Ross, *Methods Enzymol.* **195,** 329 (1991).
[30] E. M. Ross, *in* "The β-Adrenergic Receptor" (J. P. Perkins, ed.), p. 125. The Humana Press, Inc., Clifton, NJ, 1991.
[31] E. M. Parker, K. Kameyama, T. Higashijima, and E. M. Ross, *J. Biol. Chem.* **266,** 519 (1991).
[32] S. K.-F. Wong and E. M. Ross, *J. Biol. Chem.* **269,** 18968 (1994).

this ratio will have predictable effects on signaling reactions. Equimolar receptor and G protein are more appropraite in a GAP assay, where the intent is to
accelerate GDP/GTP exchange and thus make hydrolysis as close as possible to
rate-limiting. We also routinely use a $G\alpha : G\beta\gamma$ ratio of less than 1, usually 0.5,
to increase the fractional recovery of $G\alpha$ during reconstitution and to enhance
receptor–G protein coupling. Note that all protein components must be quantitated after reconstitution because recoveries are not uniform. It is advisable to
prepare large batches of vesicles so that a single batch can be assayed for its content of receptor and G protein, stored under argon in small aliquots at $-80°$, and
used for multiple experiments. We thaw vesicles only once for use in signaling
assays.

Steady-State GTPase Assays

GTPase assays are carried out at $30°$ in a total volume of 30–50 μl in polypropylene tubes for 15 min or less, according to the total GTPase activity and the
needs of the experiment. It is generally convenient to prepare a cocktail of assay reagents and to add to it the vesicles, GAP, and any other effectors. GTPase
reactions frequently display a pronounced lag before reaching steady state, probably a consequence of the time needed for receptor and G protein to associate in
the membrane.[3] Therefore, preincubate all components, including nonradioactive
GTP and agonist/antagonist, for 2–4 min at $30°$ before initiating the assay by
addition of $[\gamma$-^{32}P]GTP. Preincubation of the GAP with the vesicles should also be
evaluated. Soluble GAPs may be added directly to the assay medium, but GAPs
that require detergents for solubility must be coreconstituted with receptor and
G protein during formation of the vesicles.[16]

The assay buffer contains 20 mM Na HEPES (pH 8.0), 0.1 M NaCl, 1 mM
EDTA, 2 mM MgCl$_2$ ($\sim$ 1 mM free Mg^{2+}), 1 mM dithiothreitol, 0.1 mg/ml
albumin, 1–10 μM $[\gamma$-^{32}P]GTP (concentration depends on observed K_m; specific
activity $\sim$10 cpm/fmol), 0.1 mM adenyl-5′-yl imidodiphosphate [App(NH)p], and
receptor agonist or antagonist. Under these conditions, it is usually convenient to
add vesicles that contain 50–200 fmol of G protein. The concentration of GAP
to be assayed is determined by its EC$_{50}$, the minimum detectable signal, and the
total amount of GTP hydrolyzed during the assay period. App(NH)p is used to
inhibit any contaminating nucleoside triphosphatases and may be omitted if background activity is absent. EGTA may replace EDTA to buffer free Ca^{2+} as desired,
with appropriate adjustment to maintain the concentration of free Mg^{2+}. We have
not seen effects of micromolar Ca^{2+} on GTPase or GAP activities. HEPES buffer
is preferred because it decreases background.

GTPase assays are terminated by addition of 950 μl of a cold 5% slurry of
charcoal in H$_3$PO$_4$, pH 3, and $[^{32}$P]P$_i$ is determined in the supernatant. Background
can be high in steady-state GTPase assays, particularly if $[\gamma$-^{32}P]GTP is not fresh.

Both zero-time and zero-protein controls should be run routinely. They should give identical results, and the background value should be subtracted from experimental values before data are analyzed.

Evaluation of Steady-State GTPase Data

Calculate $[^{32}P]P_i$ formed in the total reaction volume, after subtraction of background $[^{32}P]P_i$, according to the specific activity of the $[\gamma-^{32}P]GTP$ substrate. GTPase rates are expressed in terms of moles of GTP hydrolyzed per unit time. Further normalization to the amount of $G\alpha$ (mol GTP/min/mol $G\alpha$) is useful if other parameters are constant—concentration of receptor, lateral density of the proteins in the bilayer, etc.[33] In general, a complete data set consists of four values: GTPase activities measured the presence of agonist, GAP, both, and neither. Data should ideally be expressed as a molar turnover number: moles of GTP hydrolyzed per minute per mole of $G\alpha$ (determined in a parallel $[^{35}S]GTP\gamma S$ binding assay). Because even unliganded receptors can often promote GDP/GTP exchange at a low level, a better estimate of activity at the basal GDP/GTP exchange rate can sometimes be obtained by adding receptor ligands that inhibit exchange catalyst activity ("inverse agonists").

GAP activity during steady-state GTP hydrolysis can be evaluated at various mechanistic or descriptive levels. Routinely, relative (-fold) stimulation of agonist-stimulated GTPase activity at saturating GAP concentration is a reliable measure of a GAP's net activity.[22] However, as the GAP accelerates steady-state GTPase activity by increasing the hydrolytic rate, GTPase activity will reach a maximum when GDP/GTP exchange becomes rate-limiting. A low exchange rate will therefore put a somewhat artifactual upper limit on the observable GAP activity.

This change in the predominantly rate-limiting step can be inferred from both from the small effect of GAPs observed in the absence of agonists and from the change in agonist potency in the presence and absence of a GAP. Without agonist, a GAP has little effect; the same is true for an agonist without and with a GAP (Table I). In the absence of a GAP, the EC_{50} for agonist will frequently be well below the K_d (apparent "spare receptors") because stimulation of GDP/GTP exchange by only a small number of agonist-bound receptors increases GTPase activity to the point where hydrolysis of $G\alpha$-bound GTP becomes rate-limiting. In the presence of a GAP, the EC_{50} for agonist approximates the agonist–receptor K_d because more agonist-bound receptors will be required to keep up with hydrolysis.[6] Similarly, the potency of a GAP in a steady-state assay, its EC_{50} at saturating agonist, may

---

[33] Not all of the G protein in reconstituted vesicles is accessible to receptors, presumably for geometric reasons.[29,30] The pool of "coupled" G protein, usually 30–50% of the total, is defined as that which undergoes rapid GDP/GTP exchange in response to agonist. For most mechanistic studies, the concentration of coupled G protein should be used to calculate reliable turnover numbers and specific activities.

or may not approximate its K_d for binding to the Gα–GTP complex. GAPs and receptors can influence each other's activities in complex ways.[3] However, when rate constants for the individual reactions of the GTPase cycle were used to predict steady-state K_m and V_{max} according to the Briggs–Haldane equation,[8] the predicted values were consistent with those measured directly at steady state.[34]

The interactions of these kinetic parameters, determined by careful parallel monitoring of hydrolytic and GDP/GTP exchange rates, can give considerable mechanistic information about the coordinated action of receptor and GAP on the activation of the G protein. Such studies form the basis for analyzing secondary regulatory effects: allosteric modulators, covalent modification of the proteins, and modulation of protein concentration.

Acknowledgments

I am indebted to Gloria Biddlecome, Suchetana Mukhopadhyay, Yaping Tu, and Jun Wang for their many contributions to the development of these methods.

[34] G. H. Biddlecome, "Regulation of Phospholipase C-β_1 by G$_q$ and m1-Muscarinic Cholinergic Receptor," Ph.D. dissertation. The University of Texas Southwestern Medical Center at Dallas (1997).

[43] Analysis of RGS Proteins in *Saccharomyces cerevisiae*

By GINGER A. HOFFMAN, TIFFANY RUNYAN GARRISON, and HENRIK G. DOHLMAN

Introduction

Saccharomyces cerevisiae (baker's yeast) can be employed by regulators of G-protein signaling (RGS) researchers for a variety of purposes. Yeast can be used to screen for novel RGS proteins.[1] Yeast provide a simple readout of RGS function, and thus are ideal for assessing function of candidate RGS proteins from other organisms. In addition, yeast can be used to screen for mutations or novel regulators of RGS proteins. Yeast possess several attributes that greatly aid in these applications: (1) They can be grown quickly and in vast quantities. This is helpful for large-scale genetic screens, protein purification, and biochemical analysis.[2]

[1] K. M. Druey, K. J. Blumer, V. H. Kang, and J. H. Kehrl, *Nature* **379,** 742 (1996).
[2] T. R. Garrison, D. M. Apanovitch, and H. G. Dohlman, *Methods Enzymol.* **344,** [44] 2002 (this volume).

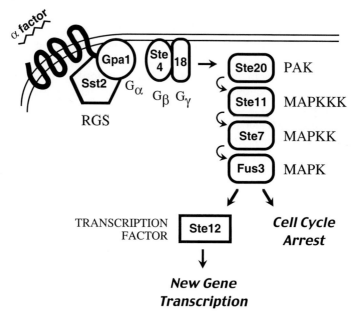

FIG. 1. Diagram of the yeast mating pathway. The α-factor pheromone binds to a cell surface receptor (Ste2) that in turn promotes GTP binding to Gα (Gpa1) and dissociation of Gα from the G$\beta\gamma$ subunits (Ste4, Ste18). Free $\beta\gamma$ activates a downstream cascade of protein kinases (Ste20, Ste11, Ste7, Fus3) leading to mating and growth arrest. Signal inactivation is accelerated by the RGS protein (Sst2), a GTPase activating protein for Gpa1.

(2) They can exist as haploids, greatly simplifying identification and characterization of recessive mutations. (3) They are readily manipulated genetically; overexpression or disruption of genes is easily obtained.

Yeast exist either in the haploid or diploid state. The transition from the former to the later is accomplished by cell fusion, or mating. Haploid cells signal readiness to mate by secreting either **a**-factor or α-factor pheromone, depending on the "sex" of the haploid cell. This mating pheromone binds to a G-protein-coupled receptor expressed by a putative mating partner. Pheromone binding induces dissociation of G$\beta\gamma$ (Ste4/Ste18) from Gα (Gpa1), enabling G$\beta\gamma$ to activate a mitogen-activated protein kinase (MAPK) cascade. This in turn elicits a number of cellular responses that prepare the cell for mating, including cell cycle arrest and the induction of genes important for fusion (Fig. 1). One of the induced genes is the yeast RGS protein, Sst2, which accelerates the hydrolysis of GTP to GDP by Gpa1. By doing so, Sst2 promotes reassociation of Gα and G$\beta\gamma$ and termination of $\beta\gamma$-dependent activation of the MAPK cascade. In this way, Sst2 diminishes levels of new gene transcription and growth arrest and thereby completes a negative feedback loop.

Sst2 is the founding member of the RGS family and possesses significant functional homology to mammalian RGS proteins. For example, RGS function can be restored to Sst2 knockout cells by expression of one of several mammalian RGS proteins. RGS1, RGS4, and RGS16 can substantially restore function, and RGS2 and RGS3 can partially restore function.[1,3] It seems that the larger RGS proteins such as RGS5 and RGS9 do not successfully replace Sst2.[3,4]

This chapter explores how to utilize yeast for analysis of either Sst2 or mammalian RGS proteins *in vivo* and is geared toward investigators who are new to working with yeast. The first section addresses expression of RGS proteins in yeast: how to chose an expression vector, how to transform the vector into yeast, and how to check for expression. The second part looks at two principal bioassays of RGS function: monitoring growth arrest and measuring new gene transcription. Finally, various considerations for setting up a functional screen for RGS regulators are presented.

I. Expression of RGS Proteins in Yeast

Choice of Expression Vectors

For successful expression in yeast, an RGS protein should be cloned into a vector containing the elements discussed below.

1. *Nutritional marker.* This allows for the selection of the vector in yeast. Most yeast strains commonly used for research possess loss-of-function mutations in several nutritional genes. In order to survive, they must either have the nutrients supplied in their media or contain a vector with a wild-type nutritional gene (the nutritional marker). Therefore, before transformation yeast should be grown in YPD [1% yeast extract (Difco, Sparks, MD), 2% Bacto-peptone (Difco), 2% dextrose],[5] which contains all necessary nutrients. Once transformed with the plasmid, the cells should be grown in synthetic complete (SC) medium containing all nutrients except the one required for plasmid selection. A carbon source, usually dextrose, is also added (SCD medium). SCD consists of 0.17% yeast nitrogen base without amino acids and ammonium sulfate (Difco), 0.5% ammonium sulfate, 2% (w/v) dextrose, plus any required nutrients (amino acids, uracil).[6] The medium is referred to by the nutrient it is lacking; for example, SCD-Ura lacks uracil. Either YPD or SCD media can be supplemented with 2% (w/v) agar to make solid media.

   Common nutritional markers include *LEU2* (required for synthesis of leucine), *URA3* (uracil), *HIS3* (histidine), and *TRP1* (tryptophan). Yeast have a doubling time of approximately 90 min in YPD media and 2–4 hr in

[3] C. Chen, B. Zheng, J. Han, and S. C. Lin, *J. Biol. Chem.* **272**, 8679 (1997).

[4] G. Hoffman, unpublished observations.

[5] F. Sherman, *Methods Enzymol.* **194**, 3 (1991).

[6] F. M. Ausubel, R. Brent, R. E. Kingston, D. D. Moore, J. G. Seidman, J. A. Smith, and K. Struhl, "Current Protocols in Molecular Biology." John Wiley & Sons, Inc., New York, 1999.

SCD media, depending on the nutrient that is "dropped out." We have found that our yeast strains double the fastest when containing a plasmid marked with *LEU2*. Thus this is generally our first choice of nutritional marker.

2. *amp^R gene*. The *amp^R* gene confers resistance to ampicillin, allowing the vector to be amplified and maintained in *Escherichia coli*.[6]

3. *Sequence element that allows for autonomous replication in yeast*. The most common are the *CEN* and the 2-μm elements.[7] The *CEN* element is a portion of a yeast centromere and allows the vector to segregate much like a chromosome, maintaining 1–2 copies per cell. In contrast, the 2-μm element allows the cell to produce a large number of vectors, up to hundreds per cell. Choice of the replication element has a bearing on the level of protein expression, as discussed below.

4. *Promoter*. Different promoters can be used to adjust levels of expression (see below). It is important to ensure that the promoter does not contain a cryptic ATG since yeast (unlike mammals) tend to readily mistake it for the start of the open reading frame.[7]

5. *Transcription initiation sequence*. This is included in virtually all promoter elements and consists of a TATA box and other upstream activating sequences. In general, the transcription initiation sequence does not differ between yeast and other organisms.[7]

6. *Transcription termination sequence*. Although these do vary between yeast and mammals, many mammalian termination sequences function properly in yeast.[7] Additionally, many commercially available yeast expression vectors contain a yeast termination sequence.

7. *Translation initiation sequence*. This is not necessary in every circumstance, but may prove helpful. The optimal sequence is: (A/Y)A(A/T)AATG, where Y is a pyrimidine and ATG is the start codon of the open reading frame.[8]

Levels of Expression

There are two main ways to adjust the level of protein expression in yeast: by modifying the promoter, and by adjusting plasmid copy number. Although expression level is not directly proportional to copy number, it is possible to substantially increase expression using a vector with a high copy number, such as a 2-μm vector. There is also a vector that contains a defective version of *LEU2* (Clontech, Palo Alto, CA). When grown in SCD-Leu, yeast containing this vector can only make enough leucine to survive if the vector is maintained at an extremely high copy number. In our experience, this defective *LEU2* vector in combination with a strong promoter provides the highest possible expression.

[7] J. C. Schneider and L. Guarente, *Methods Enzymol.* **194**, 373 (1991).
[8] A. M. Cigan and T. F. Donahue, *Gene* **59**, 1 (1987).

Different promoters provide different options for regulating protein expression. The *SST2* promoter can be used for pheromone-inducible expression of RGS proteins. With this promoter, little expression will occur unless α-factor is added to the media (2.5 μM α-factor during 1–2 hr of log phase growth works well). If overexpression of an RGS protein is desired, the constitutively active promoter from the alcohol dehydrogenase gene (*ADH1*) is a good all-purpose choice. High levels of expression can also be achieved using inducible promoters such as *GAL1/10* and *CUP1*.[7,9] The *GAL1/10* promoter is a bidirectional promoter that is induced by galactose and repressed by dextrose. This ability to repress the *GAL1/10* promoter allows for more precise control over expression. One limitation of this promoter is that yeast grow more slowly when they are using galactose as a carbon source. Often it is necessary to first grow the cells in dextrose, centrifuge, and resuspend in medium containing galactose. Another option is to grow the cells in raffinose and then add the galactose to the medium to induce expression. However, raffinose does not repress the *GAL1/10* promoter, so this method does not prevent basal expression. The *CUP1* promoter is another commonly used inducible promoter and is activated by copper (or silver). Unlike the *GAL1/10* promoter, this promoter allows for a range of expression levels determined by the amount of copper added. Copper concentrations from 0.2 mM to 0.5 mM are commonly used.[9] An excess of copper should be avoided because it can be toxic (the toxic dose depends on the strain). In contrast to the *GAL1/10* promoter, the *CUP1* promoter cannot be repressed. In fact, there is often substantial basal activation in the absence of any added copper.

Introduction of DNA into Yeast

There are two methods commonly used to introduce DNA into yeast, PLATE and TRAFO. The PLATE method yields anywhere from 50 to 300 colonies per microgram of plasmid DNA.[4] The TRAFO method is much more efficient, usually producing around 10^6 transformants per μg of plasmid DNA.[10] However, the PLATE method requires much less time and is almost always sufficient to introduce a plasmid (or two) into a yeast strain. The efficiency gained from the TRAFO method is generally only necessary for specialized applications such as transformation of a DNA library (e.g., for genetic screens—see below).

Keep in mind that when introducing heterologous RGS proteins into yeast, it is preferable to use a yeast strain deleted for *SST2*, so that the measurement of RGS function is not complicated by the presence of an endogenous RGS.

---

[9] T. Etcheverry, *Methods Enzymol.* **185**, 319 (1990).

[10] R. Agatep, R. D. Kirkpatrick, D. L. Parchaliuk, R. A. Woods, and R. D. Gietz, *Elsevier Trends Journals Technical Tips Online (http://tto.trends.com)* P01525 (1998).

PLATE Transformation. (See Ref. 11.) PLATE is an acronym of the names of ingredients in the transformation solution: polyethylene glycol (PEG), lithium acetate, Tris, and EDTA.

Use sterile technique and sterile solutions throughout this method.

1. In a 15 ml culture tube, inoculate 2–3 ml of the appropriate medium with a single colony of the yeast strain to be transformed. This is the starter culture. Various wild-type and mutant yeast strains can be obtained from Research Genetics (Huntsville, AL) or ATCC (Manassas, VA).
2. Grow the starter culture at 30° with shaking (250 rpm) until it reaches saturation. This takes anywhere from 1 to 6 days, depending on the strain and the medium.
3. Place 0.5 ml of the saturated culture in a sterile microfuge tube.
4. Collect the cells by centrifuging at 16,000g at 22° for 30 sec.
5. Aspirate the supernatant.
6. Add 10 μl of sonicated salmon sperm DNA (10 mg/ml stock) (Stratagene, La Jolla, CA). The DNA must be single-stranded, which can be achieved by boiling for 5 min (a 100° heat block works well) and immediately chilling on ice. The DNA only needs to be boiled every 3 to 4 times it is used (as long as it remains on ice when thawed).
7. Add 1–2 μg of the plasmid DNA to be transformed and vortex.
8. Add 500 μl of PLATE solution [40% PEG 3350 (w/v), 100 mM Lithium acetate, 10 mM Tris, pH 7.5, 0.4 mM EDTA]. Mix by inverting or by pipetting gently. Do not vortex.
9. Leave at room temperature or at 30° for 24–48 hr.
10. Collect the cells by centrifuging at 16,000g at 22° for 30 sec.
11. Aspirate the supernatant.
12. Resuspend the cells in 200 μl of sterile water by pipetting gently and thoroughly.
13. Spread on solid media that will select for the plasmid.
14. Incubate at 30° until colonies appear. This takes around 2 to 6 days, depending on the yeast strain and the plasmid's nutritional marker. In our experience, it usually takes $\sim$2 days for *LEU2* plasmids and $\sim$6 days for *URA3* plasmids.
15. Restreak 2–3 transformed colonies onto a new plate.

TRAFO Transformation. (See Ref. 10.)
Use sterile technique and sterile solutions throughout this method.

1. Grow a starter culture at 30° with shaking (250 rpm) until it reaches saturation.

---

[11] R. Elble, *Biotechniques* **13**, 18 (1992).

2. Determine the cell density using a hemocytometer. The density can also be determined by measuring the $OD_{600 \, nm}$. Keep in mind, however, that the relationship between cell density and absorbance is strain specific.

3. Use the starter culture to inoculate 50 ml of media to a cell density of 5×10^6 cells/ml.

4. Grow the culture at 30° with shaking for 3–5 hr until it reaches a density of 2×10^7 cells/ml. It is usually sufficient to grow the cells for 5 hr without determining the actual density. The transformation efficiency remains constant for 3 to 4 cell divisions.

5. Transfer the culture to a 50 ml conical tube and centrifuge at 2000g at 22° for 10 min.

6. Pour off the supernatant and resuspend the cells in 25 ml of water.

7. Centrifuge at 2000g at 22° for 10 min.

8. Pour off the supernatant and resuspend the cells in 1 ml of 100 mM lithium acetate.

9. Transfer to a 1.5 ml microfuge tube and centrifuge at 16,000g at 22° for 15 sec.

10. Remove the supernatant using a micropipette.

11. Resuspend the cells in 100 mM lithium acetate to a final volume of 500 μl. The cells usually occupy a volume of 100 μl.

12. For each transformation, pipette 50 μl of the cell suspension into a microfuge tube.

13. Centrifuge at 16,000g at 22° for 15 sec.

14. Remove the supernatant using a micropipette.

15. Add the following ingredients in the order listed:
    240 μl 50% PEG 3350 (w/v)
    36 μl 1 M Lithium acetate
    10 μl Single-stranded sonicated salmon sperm DNA (10 mg/ml stock)
       (see step 6 of the PLATE protocol above)
    0.1–10 μg Plasmid DNA to be transformed
    Sterile water, to a final volume of 360 μl

16. Vortex each tube vigorously for approximately 1 min. Make sure the cells have been completely mixed.

17. Incubate at 30° for 30 min.

18. Incubate at 42° for 30 min. This incubation time varies among different strains and can be optimized.

19. Centrifuge at 3000–5000g at 22° for 15 sec.

20. Remove the supernatant with a micropipette.

21. Gently resuspend the cells in 1 ml of water by pipetting.

22. Spread 2 to 200 μl of the resuspended cells on solid medium that will select for the plasmid. For volumes of cells less than 200 μl, use sterile water to adjust to a final volume of 200 μl.

23. Incubate at 30° until colonies appear (2–6 days).

24. Restreak 2–3 transformed colonies onto a new plate.

Storing Yeast Strains

Yeast strains, transformed or untransformed, can be maintained as colonies on solid media at 4° with restreaking every 2 to 4 weeks. Alternatively, strains may be stored at −80° indefinitely.[5] This is preferable in that it reduces the likelihood of accumulating spontaneous mutations.

Use sterile technique and sterile solutions throughout this method.

1. Grow a starter culture at 30° with shaking (250 rpm) until it reaches saturation.
2. In a 1.8 ml cryotube, mix 0.5 ml of the saturated culture with 0.5 ml of YPD containing 20% (v/v) glycerol.
3. Flash freeze the tube in liquid nitrogen and store at −80°.
4. To use, chip out a few pieces of the frozen stock using a sterile pipette tip or sterile toothpick and streak onto a plate containing the appropriate solid medium.
5. Allow the yeast to grow at 30° until colonies appear (2–6 days).

Checking for Expression

Whole cell extracts can be produced from yeast cultures that have reached $OD_{600\ nm} \sim 1$. The extracts can then be subjected to SDS–PAGE and the protein of interest can be visualized by immunoblotting.

Use sterile technique and sterile solutions in steps 1 to 3.

1. Using a saturated starter culture, inoculate 25 to 30 ml of appropriate medium in a 125 ml flask. Since it is often difficult to estimate the growth rate of yeast, it is helpful to start several 25 ml cultures, each with a different dilution of the starter culture (e.g., 1 : 100, 1 : 300, 1 : 900).
2. Grow at 30° shaking (250 rpm) until the $OD_{600\ nm} \sim 1.0$ (this is usually done overnight). When growing several strains at once, it is likely that they will all reach $OD_{600\ nm} \sim 1.0$ at different times. If desired, sodium azide (1 M stock in water, diluted to a final concentration of 10 mM) can be added to a culture once it reaches an $OD_{600\ nm} \sim 1.0$. The culture can then be placed on ice until the others are ready.
3. Transfer to a 50 ml conical tube and centrifuge for 10 min at 2000g at 4°.
4. Resuspend each sample in 1 ml of 10 mM sodium azide and place on ice.
5. Calculate the volume of resuspended cells that would translate to an $OD_{600\ nm}$ reading of 10. For example, this would equal 1 ml if 10 ml of culture at $OD_{600\ nm} = 1.0$ had been centrifuged and resuspended. This step is necessary to equalize the amount of cells (and protein) in a given volume of whole cell extract.

6. Transfer the calculated volume of resuspended cells to a microfuge tube and centrifuge at 16,000g at 22° for 1 min.
7. Aspirate the supernatant.
8. Resuspend the pellet in 200 μl of 1× SDS–PAGE sample buffer.[6]
9. Immediately place in a 100° heat block for 10 min.
10. Allow the tube to cool and add 200 μl of glass beads (Sigma, St. Louis, MO).
11. Vortex at high speed at 22° for 2 min. Invert after the first minute. Several tubes can be vortexed at the same time by using a foam tube floater to hold them together.
12. Using a 21-gauge needle, poke a hole in the bottom of each tube and place it into a new microfuge tube.
13. Centrifuge at 2000g at 22° for 10 sec to expel the liquid into the bottom tube, leaving the glass beads in the top tube.
14. Discard the glass beads and centrifuge the bottom tube at 16,000g at 22° for 2 min. This sediments any insoluble material.
15. Transfer the supernatant to a new microfuge tube. Store at −20°.
16. When ready to use, heat at 37° for 10 min, vortex, and centrifuge at 16,000g at 22° for 1 min. Keep in mind that repeated freezing and thawing can degrade the protein sample.
17. Immunoblots can be performed using standard methods.[6]

II. Assay of RGS Function *in Vivo*

Two assays of RGS function in yeast are described below. Both measure RGS activity by looking at downstream consequences of exposure to α-factor: the β-galactosidase assay measures new gene transcription, and the halo assay measures growth arrest (Fig. 1). The halo assay is often the only one chosen by researchers because it is extremely quick and easy to perform. It is important to realize, however, that the β-galactosidase assay can provide additional information by measuring an early event affected by RGS proteins; new gene transcription is typically monitored after 90 min of α-factor treatment. The halo assay, on the other hand, measures growth arrest after 24–48 hr of α-factor exposure. Additionally, the β-galactosidase assay is more quantitative than the halo assay. Although halo assay response is usually correlated with β-galactosidase activity, situations have been observed where the two measures are uncoupled, i.e., decreased RGS function as detected by β-galactosidase does not necessarily translate to decreased RGS function by halo assay.

Halo Assay

(See Ref. 12.) This assay ascertains the extent of growth arrest caused by different concentrations of α-factor. A small volume of the strain to be tested is

---

[12] G. F. Sprague, Jr., *Methods Enzymol.* **194,** 77 (1991).

sst2Δ

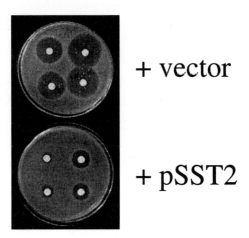

+ vector

+ pSST2

FIG. 2. Halo assay results. The top plate shows an SST2 knockout strain (*sst2Δ*) transformed with an empty plasmid ("vector"). The bottom plate depicts the same strain transformed with a plasmid expressing Sst2 under the control of its own promoter ("pSST2"). The α-factor amounts are (counterclockwise from the upper left): 5 μg, 15 μg, 45 μg, and 60 μg.

mixed with molten agar, poured onto a plate of solid medium, and allowed to develop into a lawn of cells. Immediately after the agar/cell mixture is poured, paper disks containing different amounts of α-factor are placed on the plate. The α-factor diffuses away from the paper disk producing a zone of growth inhibition (the "halo"). The size of the halo is proportional to the sensitivity of the cells to pheromone (Fig. 2).

Impaired RGS function leads to increased activity of the mating pathway, making it more likely that cells will respond to a low concentration of α-factor. Thus, cells with impaired RGS function tend to yield bigger halos. Inversely, enhanced RGS function makes the cells less sensitive to α-factor and results in smaller halos. In this way, the size of the halo is a measure of RGS function.

Use sterile technique and sterile solutions throughout this method.

1. Grow a starter culture at 30° with shaking (250 rpm) until it reaches saturation. It is important to make sure all strains to be tested are completely saturated, since cell density can affect the size of the halo.
2. Microwave a sterilized solution of 1% agar until melted and place in a 55° to 60° water bath. Don't leave unattended when microwaving since the agar tends to boil over suddenly. Make sure the agar equilibrates to 55° to 60° before starting the assay; otherwise it will kill the cells.

3. Distribute several paper disks (Difco) into a sterile petri dish, and spot either 5 μl or 15 μl of synthetic α-factor (1–5 mg/ml) onto each disk. In some instances, it may be beneficial to use more disks per plate to obtain a wider range of α-factor concentrations. This step should be done no more than 1–2 hr before use.

4. Aliquot 2 ml of sterile water into a 5 or 14 ml plastic tube with a pop-off cap (e.g., Falcon, Bedford, MA).

5. Immediately before performing the assay, transfer a small volume of the saturated starter culture into the tube containing water. For strains grown in SCD, 100 μl of starter culture should be used; for strains grown in YPD, 10 μl should be used.

6. Add 2 ml of 1% agar, cap, invert a few times, and pour onto a prewarmed plate containing the appropriate solid media. Swirl to cover the plate evenly. This step should be done quickly to prevent the 1% agar from solidifying. It is convenient to do two tubes at a time, one in each hand. Bubbles can usually be eliminated by poking with recently flamed sterile forceps.

7. On the plate, place one paper disk containing 5 μl α-factor, and one disk containing 15 μl α-factor. When doing several assays, it is helpful to use a paper template for consistent placement of the disks on each plate.

8. Place the plate at 30° until a lawn of cells appears. Halos can usually be seen clearly after 24 hr for most strains. It is often desirable to let the cells grow longer (up to several days) to get a denser lawn and to observe any changes that might occur over an extended period of time.

9. Differences in halo size are normally big enough to be detected by eye. It is also possible to measure halo size and plot this value vs log[α-factor].

β-Galactosidase Reporter Gene Assay (Liquid Form)

(See Ref. 13.) The β-galactosidase reporter gene assay measures RGS activity in terms of new gene transcription. The mating cascade directly activates a transcription factor that induces genes containing a PRE (pheromone responsive element) in their promoter. For this assay, cells transformed with a vector containing a PRE (typically from *FUS1*) fused to the *lacZ* gene are exposed to a range of α-factor concentrations and assayed for β-galactosidase produced. This is accomplished by detection of the fluorescence produced by cleavage of FDG, a β-galactosidase substrate (Fig. 3). We find this method to be more reproducible and far less cumbersome than one previously described using ONPG as the substrate.[12]

Use sterile technique and sterile solutions in steps 1 to 3.

1. Grow a starter culture at 30° shaking (250 rpm) until it reaches saturation. The cells should be transformed with a yeast expression vector containing

[13] V. A. Rakhmanova and R. C. MacDonald, *Anal. Biochem.* **257**, 234 (1998).

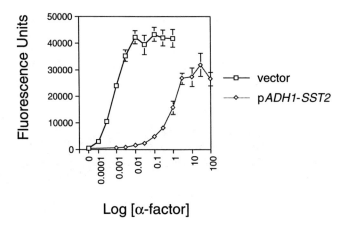

FIG. 3. β-Galactosidase reporter gene assay (liquid form) results. Two strains were assayed for β-galactosidase activity over a range of α-factor concentrations. An *SST2* knockout strain (*sst2*Δ) was transformed with an empty plasmid ("vector"), or a plasmid expressing *SST2* under the control of the *ADH1* promoter ("p*ADH1-SST2*").

the *lacZ* gene under the control of the *FUS1* promoter. Alternately, a strain can be used with the *FUS1-lacZ* reporter integrated into the genome. These materials are available from a number of yeast laboratories, including our own.

2. Using the saturated starter culture, inoculate 5 to 25 ml of the appropriate medium.

3. Grow at 30° shaking (250 rpm) until the $OD_{600\ nm} \sim 0.8$ (this is usually done overnight). This is the most difficult part of the assay: getting different strains to reach $OD_{600\ nm} \sim 0.8$ at the same time. The best way to handle this is to start a 2 ml starter culture about 3–4 days before the assay. The night before the assay, start a 10 to 25 ml intermediate culture. The morning of the assay, measure the absorbance of all of the intermediate cultures and dilute them down to an $OD_{600\ nm}$ of 0.2 in prewarmed medium. The strains will now only have to go through two doublings and the amount of variance between them should be reduced. If the strains still reach $OD_{600\ nm} \sim 0.8$ at different times, it is acceptable to put strains that have reached $OD_{600\ nm} \sim 0.8$ on ice while waiting for the others. Recheck the absorbance of all cultures before proceeding.

4. Aliquot 10 μl of α-factor at the appropriate concentrations into 96-well plates. Each strain should be tested with 10 to 12 different concentrations of α-factor in triplicate. A good set of final α-factor concentrations for testing an *sst2*Δ mutant strain is 0 μM, 0.00003 μM, 0.0001 μM, ... , 0.3 μM. For strains expressing a mammalian or yeast RGS protein, the

final concentrations should be 100- to 300-fold higher. The concentration of α-factor added to each well must be 10-fold higher than the desired final concentration. Keep a frozen stock of α-factor and make a new set of dilutions each time the assay is performed, since α-factor is subject to degradation on repeated freeze–thaw cycles or exposure to room temperature. A multichannel pipettor and a plastic pipettor basin (Fisher Scientific, Pittsburgh, PA) are helpful for aliquoting solutions.

5. Add 90 μl of cells to each well.
6. Incubate 90 min at 30°, shaking gently.
7. During the incubation, prepare the FDG solution.

   Solution #1: 1 mM FDG stock diluted in 25 mM PIPES (pH 7.2).
   Solution #2: 5% Triton X-100 diluted in 250 mM PIPES (pH 7.2). Mix Solution #1 and Solution #2 in equal amounts just prior to use, and pour into a clean pipettor basin.
   FDG Stock: 10 mM FDG (Fluorescein di-β-D-galactopyranoside, Molecular Probes, Eugene, OR) in dimethyl sulfoxide (DMSO) (FDG should be stored in DMSO at $-20°$; it is more stable in solution).

8. After the 90 min incubation, add 20 μl FDG solution per well. Shake plates gently and briefly.
9. Cover in aluminum foil and incubate at 37° until a bright yellow color appears in some of the wells. This can take from 10 to 90 min. Do not incubate longer than 90 min.
10. Stop the reaction by adding 20 μl of 1 M Na$_2$CO$_3$ per well. Shake the plates gently and briefly.
11. Read the plates with a fluorescence multiwell plate reader using an excitation of 485 nm and an emission of 530 nm.
12. Read the absorbance and normalize for cell density. Alternatively, use the final absorbance values [obtained immediately before aliquoting cells into the 96-well plate (step 3)] to normalize for cell density.

III. Screening for RGS Activators/Inhibitors in Yeast

As mentioned in the Introduction, one benefit of yeast is their suitability for large-scale genetic screens, such as for regulators and mutants that affect RGS activity. Screens can identify mutations in RGS proteins that negatively or positively affect their ability to function. They can also reveal novel proteins that regulate RGS proteins.

The basic premise of any such screen is that colonies can be manipulated in some way (mutagenized, transformed with a library, etc.) and then quickly tested for alterations in RGS activity by measuring changes in new gene transcription and/or growth arrest. For example, a screen designed to discover negative regulators of RGS function would look for increased G protein signaling and therefore enhanced gene transcription and growth arrest. The opposite would apply in a screen for positive regulators of RGS function.

Alterations in growth arrest can be assessed for several colonies at a time by looking for growth on an α-factor-impregnated plate. This is made by spreading a volume of α-factor [e.g., 300 μl of a known concentration of α factor onto 30 ml of solid medium (100 × 15 mm plate)] and incubating at 30° overnight. Changes in gene transcription, on the other hand, are easily measured using a β-galactosidase filter assay. The filter assay differs from the liquid β-galactosidase assay in a few ways. It assays several hundreds of colonies at a time and uses a different substrate (X-Gal) which allows for the visual detection of a blue-colored product. Although this assay is not quantitative, it is easy and quick. For example, DiBello et al. screened for mutagenized strains that were supersensitive to α-factor by selecting for colonies that turned blue in the presence of a low dose of α-factor.[14] This led to the discovery of a Gpal allele that is completely resistant to RGS action.

β-Galactosidase Filter Assay

1. Start with a plate of yeast colonies to be tested (often the result of the transformation of a library or a round of mutagenesis). These colonies should contain the lacZ gene under the control of the FUS1 promoter.
2. For each plate to be tested, place a circle of 3MM Whatman (Clifton, NJ) paper into each of two empty petri dishes. The Whatman paper should cover the entire bottom of the petri dishes.
3. Add 2.5 ml of α-factor to the Whatman paper at the bottom of one of the petri dishes. Use a concentration of α-factor that produces a different response in wild-type vs mutant strains. This must be determined empirically. If there are any bubbles, smooth them out with forceps.
4. Label a circular piece of nitrocellulose (NC) filter the size of the petri dish (GelmanSciences, BioTrace NT 82mm, Ann Arbor, MI) with a ballpoint pen.
5. Pick up the NC filter with tweezers and place it carefully across the plate of yeast colonies so that there are no bubbles or wrinkles in the filter. Let sit for 1–2 min.
6. While the NC filter is sitting on the yeast colonies, dip a needle in india ink and poke holes through the filter and into the solid medium. This enables positive colonies on the filter to be matched with the colonies on the plate from which they were lifted.

---

[14] P. R. DiBello, T. R. Garrison, D. M. Apanovitch, G. Hoffman, D. J. Shuey, K. Mason, M. I. Cockett, and H. G. Dohlman, J. Biol. Chem. **273**, 5780 (1998).

7. Transfer the NC filter to the petri dish containing α-factor, colony side up.
8. Cover the dish and seal with Parafilm. Put at 30° for 2.5 hr.
9. During the 2.5 hr incubation, aliquot 2.5 ml of Z buffer + X-Gal into the remaining petri dish containing Whatman paper.

> Z buffer + X-Gal:
> 60 mM Na$_2$HPO$_4$
> 40 mM NaH$_2$PO$_4$·H$_2$O
> 10 mM KCl
> 1 mM MgSO$_4$·7H$_2$O
> 39 mM 2-mercaptoethanol
> 1 mg/ml X-Gal
> Adjust pH to 7.0.

Z buffer without X-Gal and without 2-mercaptoethanol can be stored at room temperature. X-Gal should be stored as a 100 mg/ml stock in dimethylformamide (DMF) at −20°.

10. Remove the plates from 30°, lift the NC filter off the Whatman paper, and place on an aluminum foil "boat." Then, slowly lower into an ice bucket containing liquid nitrogen. Freezing the cells in this way serves to permeabilize the membrane and allow the subsequent entry of X-Gal, the substrate for β-galactosidase. The aluminum foil boat avoids having to maneuver the NC filter with tweezers. (The NC gets very brittle when frozen.)
11. After about 10 sec, carefully raise the boat to remove the NC filter from the liquid nitrogen.
12. Allow the NC filter to warm to room temperature (about 1 min).
13. Place the NC filter, colony side up, in the petri dish containing Z Buffer + X-Gal.
14. Cover the dish and seal with Parafilm. Put at 30° overnight. Blue color may start to appear after a few hours.

Conclusions

The methods above offer a basic set of guidelines for the researcher wishing to use yeast for RGS protein analysis. They include the main protocols needed for expressing heterologous (or endogenous) RGS proteins in yeast and detailed descriptions of the two most commonly used assays for measuring RGS function. In addition, they encourage the use of yeast for functional screens of RGS activity and provide some suggestions for starting such screens.

[44] Purification of RGS Protein, Sst2, from *Saccharomyces cerevisiae* and *Escherichia coli*

By TIFFANY RUNYAN GARRISON, DONALD M. APANOVITCH,
and HENRIK G. DOHLMAN

Introduction

The mating process of *Saccharomyces cerevisiae* (hereafter referred to as yeast) is initiated by the secretion of peptide pheromones that bind to seven transmembrane receptors (Ste2, Ste3) on cells of the opposite type. In yeast, as in other organisms, the intracellular signal is transmitted through the associated heterotrimeric G protein. The ligand-bound receptor promotes binding of GTP to the Gα subunit (Gpa1), lowering its affinity for the Gβ/γ dimer (Ste4, Ste18), and freeing Gβ/γ to interact with the effector molecules including a MAP kinase cascade. The Gα subunit hydrolyzes the GTP to GDP, returning the G protein to its inactive, heterotrimeric form. G proteins are a target of desensitization through the action of RGS proteins (regulators of G protein signaling).[1-3] The founding member of the RGS family is the yeast protein Sst2.[4,5] Desensitization is thought to occur as a result of Sst2's acceleration of the GTPase activity of the Gα subunit.[6,7]

The characterization of Sst2 has been enhanced by the ability to integrate biochemical and genetic techniques available using yeast. Genetic screens can be employed to discover mutations that abolish or enhance the *in vivo* function of Sst2. Biochemical analysis of Sst2 purified from its native system can identify stimulus-dependent changes in the protein, such as *in vivo* posttranslational modifications. The modification can then be blocked by mutating that particular residue. The mutants of Sst2 generated through such genetic and biochemical means can be expressed in a null background at near endogenous levels, allowing an accurate assessment of the mutant's activity *in vivo*. Finally, alterations in the enzymological activity of the mutants can be detected using assays of RGS function *in vitro*. In contrast to many other biological systems, Gpa1 and Sst2 are known to interact *in vivo*, so a cognate Gα–RGS pair can be studied.

[1] L. De Vries, M. Mousli, A. Wurmser, and M. G. Farquhar, *Proc. Natl. Acad. Sci. U.S.A.* **92,** 11916 (1995).
[2] M. R. Koelle and H. R. Horvitz, *Cell* **84,** 115 (1996).
[3] H. G. Dohlman and J. Thorner, *J. Biol. Chem.* **272,** 3871 (1997).
[4] R. K. Chan and C. A. Otte, *Mol. Cell. Biol.* **2,** 21 (1982).
[5] R. K. Chan and C. A. Otte, *Mol. Cell. Biol.* **2,** 11 (1982).
[6] P. R. DiBello, T. R. Garrison, D. M. Apanovitch, G. Hoffman, D. J. Shuey, K. Mason, M. I. Cockett, and H. G. Dohlman, *J. Biol. Chem.* **273,** 5780 (1998).
[7] D. M. Apanovitch, K. C. Slep, P. B. Sigler, and H. G. Dohlman, *Biochemistry* **37,** 4815 (1998).

We have developed two techniques for the purification of Sst2 (Fig. 1A). The first method was designed to help identify posttranslational modifications of Sst2 *in vivo*. Since the isolated protein did not need to have biochemical activity, we employed denaturing conditions for the purification, to preserve all posttranslational modifications, and to prevent the addition of artifactual modifications during the purification process. We have used this approach, in conjunction with electrospray ionization mass spectrometry, to identify a site of pheromone-dependent

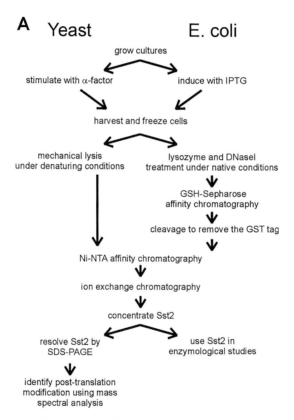

FIG. 1. Purification of Sst2. (A) The purification of Sst2 from yeast and *Escherichia coli* consists of many similar steps (center column). The major differences between the two procedures are also highlighted, with the steps unique to the yeast purification listed on the left and to the *E. coli* purification on the right. (B) His-tagged Sst2 was purified from yeast cells by Ni-NTA affinity chromatography followed by ion-exchange chromatography. Sst2 (arrows) purified from α-factor stimulated cells migrates as a doublet of molecular mass 82 and 84 kDa (top, silver stain). An identical gel was immunoblotted using anti-Sst2 antibodies, to confirm that the two bands represent Sst2 (bottom, Sst2 Ab). (B) Reproduced from T. R. Garrison, Y. Zhang, M. Pausch, D. Apanovitch, R. Aebersold, and H. G. Dohlman, *J. Biol. Chem.* **274,** 36387 (1999).

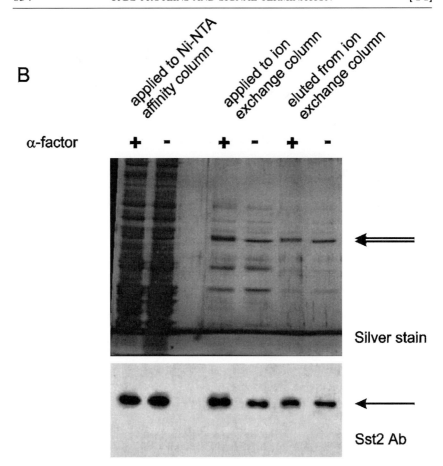

FIG. 1. (*Continued*)

phosphorylation.[8] The second method was designed to produce sufficient functional protein for enzymological studies. We used this approach to demonstrate that Sst2 accelerates the GTPase activity of its cognate $G\alpha$ subunit, Gpa1.[7] Although we designed the procedures for the purification of Sst2, they can easily be adapted to analyze other signaling proteins from yeast or higher eukaryotes.

Construction of Expression Vectors

Both purification procedures use a hexahistidine (His$_6$) tag added to *SST2*. Neither a N- nor C-terminal His$_6$-tag affects the activity of Sst2, given that either

[8] T. R. Garrison, Y. Zhang, M. Pausch, D. Apanovitch, R. Aebersold, and H. G. Dohlman, *J. Biol. Chem.* **274,** 36387 (1999).

can complement a *SST2* gene deletion. Two procedures have been used for the purification of Sst2 from *Escherichia coli* (*E. coli*) using a version of the protein tagged at the N or C terminus. To produce an *E. coli* plasmid containing *SST2* with a C-terminal his-tag, a *Nco*I restriction site is introduced during PCR amplification of the coding sequence. A *Nco*I/*Bst*BI fragment of the PCR product is combined with a *Bst*BI/*Bam*HI fragment from pBS-SST2-GST (provided by Ken Blumer, Washington University St. Louis, MO) in the corresponding *Nco*I/*Bam*HI sites of the *E. coli* expression vector, pQE60 (Qiagen, Valencia, CA). The *Bst*BI cuts once in the coding sequence of SST2 and the SST2-GST construct contains a mutant *Bam*HI site that replaces the stop codon of *SST2*. The *SST2*-His$_6$ sequence from pQE60 is excised using *Nco*I/*Hin*dII and ligated into the corresponding sites in pET21d (Stratagene, La Jolla, CA). To produce an *E. coli* expression plasmid containing *SST2* with an N-terminal His$_6$-tag and glutathione *S*-transferase (GST) tag, PCR amplification is used to introduce *Bam*HI and *Eco*RI sites into an N-terminally His$_6$-tagged *SST2*. The product is introduced into pGEX2T (Amersham Pharmacia Biotech, Piscataway, NJ) containing a GST tag followed by a thrombin cleavage site producing pGEX2T–GST–thrombin cleavage site–His$_6$–SST2.

To produce a yeast expression plasmid, pQE60-SST2-his is digested with *Bss*HII and *Hin*dIII. This fragment is cloned into the corresponding sites of pAD4M-SST2-GST (pAD4M provided by Peter McCabe, Onyx Pharmaceuticals, Richmond, CA; SST2-GST is excised from pAB-SST2-GST and introduced into the *Sal*I/*Sac*I sites of pAD4M), replacing the GST-tag with a His$_6$-tag. The *Bss*HII cuts once in the coding sequence of SST2 and *Hin*dIII cuts after both affinity tags. The pAD4M-SST2-his expression plasmid is a 2μ vector (high copy) and uses the constitutive ADH1 promoter.[9]

Comments

Overexpression of Sst2 curtails the pheromone response.[10] Since one of our objectives is to identify pheromone-dependent modifications, we need to achieve an expression level that will provide ample protein for purification, but will not appreciably diminish signaling. All of the plasmids we have tested are 2μ vectors. The *ADH1* promoter produces significantly higher expression than the *GAL1/10* promoter. The expression from the *ADH1* promoter on a vector containing the nutritional marker, *URA3*, and a defective *leu2* gene produces the highest level of expression. However, the expression from this plasmid totally blocks the pheromone response.

We have shown that either a His$_6$-tagged or a GST-tagged Sst2 (N- or C-terminal) can complement an *SST2* gene deletion. Furthermore, the presence of a N- or C-terminal His$_6$-tag does not affect GAP activity.[7,11] Initially, we had difficulty

---

[9] G. Hoffman, T. R. Garrison, and H. G. Dohlman, *Methods Enzymol.* **344**, [43] 2002 (this volume).

[10] H. G. Dohlman, J. Song, D. Ma, W. E. Courchesne, and J. Thorner, *Mol. Cell. Biol.* **16**, 5194 (1996).

[11] L. Kallal and R. Fishel, *Yeast* **16**, 387 (2000).

with the expression of full-length, affinity-tagged, Sst2 in *E. coli*. Expression of a C-terminal Sst2–GST fusion protein in *E. coli* has not been successful. Numerous attempts have been made to express either N- or C-terminal His$_6$-tagged Sst2, at either 37° or 24° or induced with isopropylthiogalactoside (IPTG) concentrations that range from 10 μM to 1.0 mM. Results from immunoblot analysis of the various expression conditions suggest that at high temperatures or high IPTG concentrations the protein is insoluble or degraded. Even at lower temperatures and IPTG concentrations the protein is barely detectable on a Coomassie stained SDS–PAGE gel. In an attempt to increase the solubility of Sst2, we have constructed an expression plasmid with an N-terminal GST tag located prior to the His$_6$-tag of Sst2. The expression increases dramatically from 0.1 mg/liter to 1 mg/liter with an increase in solubility. The GST-his-SST2 plasmid provides superior expression; although we have not tested this fusion protein, another group has demonstrated GAP activity using a N-terminally His$_6$-tagged Sst2. [11] In this chapter we provide procedures for the purification of GST-his-Sst2 from *E. coli* as well as Sst2-his from *E. coli* and yeast.

Denaturing Purification of Sst2 from Yeast

We have purified His$_6$-tagged Sst2 from stimulated and nonstimulated yeast using denaturing conditions. This procedure produces greater than the 1 μg of protein needed for solid phase extraction capillary electrophoresis ionization tandem mass spectrometry. This procedure can easily be adapted for the purification of other signaling proteins.

Yeast Strain

The yeast strain BJ2168 (MAT**a** *ura3-52 leu2-1 trp1-63 prb1-1122 prc1-407 pep4-3*) has several vacuolar proteases mutated and is often used to reduce degradation during the purification process. However, in initial experiments we have found that the nondenaturing purification of Sst2-GST (pAD4M-SST2-GST) does not produce sufficient protein for mass spectral analysis. We have found that Sst2 expressed in our *SST2* deletion strain, YDM400, undergoes an increased level of pheromone-independent phosphorylation detected as two bands on an SDS–PAGE gel (82 kDa and 84 kDa). Purification from nonstimulated BJ2168 provides a single, 82 kDa band of Sst2.

Solutions

Sodium azide is used to arrest the growth of the yeast. It can be stored at room temperature as a 1 M solution. All the purification solutions contain the urea buffer (6 M urea, 100 mM Na$_2$H$_2$PO$_4$, pH 8, 10 mM Tris, pH 8, 10 mM 2-mercaptoethanol). The stock 8 M urea solution is made the day before the purification and deionized by stirring overnight at 4° with AG501-X8(D) resin (Bio-Rad, Hercules, CA). The other stock solutions are stored at room temperature: 1 M Tris, pH 8, 1 *M* Na$_2$H$_2$PO$_4$, pH 8 (pH is adjusted with NaOH), 5 *M* NaCl, 2 *M* imidazole, pH 8. The working solutions are prepared on the day of the purification and are degassed followed by filtration. The α-factor is dissolved in sterile water at a concentration of 1 mg/ml and stored in 1 ml aliquots at $-20°$. We obtain α-factor from a peptide synthesis facility (Keck Facility, Yale University).

Growth of Yeast

1. Prepare a starter culture by inoculating 10 ml of selective medium with a single colony of strain BJ2168 transformed with plasmid pAD4M-SST2-his.
2. Grow the starter culture with shaking (250 rpm) at 30° to saturation (approximately 36 hr).
3. Add 1 ml of the starter culture to each 6 × 1 liter of fresh medium in 4 liter Erlenmeyer flasks. For convenience, the dilution can be adjusted to alter the length of time it take for the cultures to reach the correct density.
4. Grow the cultures with shaking (250 rpm) at 30° for 10–12 hr, until an OD$_{600}\approx0.8$ is reached.
5. To produce Sst2 that contains stimulus-dependent modifications, treat the cultures with 2.5 μM α-factor (4.125 ml of a 1 mg/ml stock) and grow with shaking (250 rpm) at 30° for 60 min (final OD$_{600}\approx1$). Add sterile water to the control, nonstimulated cultures.

Harvesting Yeast

All procedures are carried out at 0–4°.

1. Place the cultures at 4° while harvesting the cells. Centrifuge for 10 min at 3840*g*. Pour off the supernatant and add more culture on top of the cell pellet. Repeat until all the cells are harvested. If the cells must be left at 4° for an extended period of time, add sodium azide to the culture to a final concentration of 10 m*M* and store at 4° until harvesting.
2. Resuspend each pellet in 5 ml of 10 m*M* sodium azide. Move the cells to a 50 ml conical tube. Rinse the centrifuge bottles with approximately 10 ml of 10 m*M* sodium azide and add it to the 50 ml tube.
3. Centrifuge for 10 min at 1000*g*. This second centrifugation step helps reduce the volume of liquid added to the cells before freezing them.
4. Pour off the supernatant and resuspend the cells in approximately 10 ml of 10 m*M* sodium azide.
5. Flash freeze the cells by dribbling the slurry of cells directly into liquid nitrogen. The cells can be recovered by pouring off all the liquid nitrogen and transferring the cells to a piece of paper. The paper can be used to funnel the cells into a 50 ml tube for storage. Be sure to work quickly or the cells will begin to thaw.
6. Store the cells at $-80°$.

Lysis of Yeast Expressing Sst2-His

1. In a stainless steel bead beater chamber (Biospec, Bartlesville, OK) thaw cells from 6 liters of culture in approximately 250 ml of urea buffer containing 15 mM imidazole and 250 mM NaCl. As a less efficient alternative, the cells can be lysed by vortex with glass beads in 50 ml conical tubes.
2. Stir until thoroughly mixed.
3. Add glass beads (Sigma, St. Louis, MO) until the bead beater chamber is full. Close the chamber using the reservoir and top.
4. Place the chamber on the motor and cool the chamber by placing ice and salt in the reservoir. The cooling is necessary since the friction produced by the bead beater can increase the temperature of the lysate by up to 10° during each minute of operation.
5. Lyse by running the motor in 10 × 30-sec pulses, once every 90 sec.
6. Remove the supernatant and transfer to a 250 ml centrifuge bottle.
7. To recover the rest of the supernatant, punch approximately 6 holes in the bottom of 50 ml conical tubes using a 21-gauge needle. Place each tube containing the holes into another, intact 50 ml tube. Transfer the glass beads from the bead beater to the top tubes with the holes and recover the lysate by centrifuging at 110g for 2 min at room temperature. The glass beads will remain in the top 50 ml tube and can be washed according to Biospec's directions and reused in future purifications.
8. Thoroughly resuspend any pellet created by the low-speed spin, and transfer the lysate from the 50 ml tubes to the 250 ml centrifuge bottle.
9. Solublize the protein by incubating for 60–90 min by slowly rocking the centrifuge bottle at room temperature.
10. Clarify the lysate by centrifuging at 3840g for 20 min, followed by paper filtration using Whatman (Clifton, NJ) No. 1. Transfer the supernatant to a new 250 ml centrifuge bottle.

Ni-NTA Affinity Chromatography

All steps are carried out at room temperature.

1. To the lysate add 3 ml of Ni-NTA superflow resin (Qiagen) that has been equilibrated, according to the manufacturer's directions, with urea buffer containing 10 mM imidazole and 250 mM NaCl. To equilibrate the resin, wash it 4 × 15 ml of the same buffer.
2. Allow the proteins to batch bind for 60–90 min by gently rocking the bottle. Rotate the bottle periodically to prevent settling.
3. Centrifuge at 110g for 2 min to recover the resin.
4. Remove most of the supernatant, being careful not to disturb the resin.

5. Resuspend the resin in the remaining supernatant (30–50 ml) and pack an HR 10/10 column (Amersham Pharmacia Biotech) following the manufacturer's directions. Instead of using gravity to settle the resin, the process is accelerated by pulling the liquid through the column using a 60 ml syringe connected to the outlet tubing. Make sure not to dry out the resin. All of the purification steps can be performed using a BioLogic system (Bio-Rad, Hercules, CA); however, a peristaltic pump would be adequate.

6. Wash with 10 column volume (CV) of urea buffer containing 10 mM imidazole and 250 mM NaCl at 1.5 ml/min.

7. Wash with 10 CV of urea buffer at 1 ml/min. It is important to remove the salt before eluting the protein since the next step is ion-exchange chromatography.

8. Elute Sst2 using 10 CV of urea buffer containing 75 mM imidazole at 1 ml/min. Eluting using imidazole instead of changing the pH conditions, which is often used to elute denatured proteins, preserves any pH-labile modification. There should be a trailing peak of a low-intensity increase in absorbance (OD$_{280}$) that corresponds to Sst2. The fractions are analyzed by resolving by SDS–PAGE and visualizing the proteins by silver staining (BioRad) and immunoblotting using an antibody for Sst2. Alternatively, an antibody to the His$_6$-tag can be used (Qiagen). Because Sst2 elutes in a broad peak, we pool all the elution fractions for the next step.

Comments

An advantage of using a His$_6$-tagged protein is that urea does not interfere with the interaction of histidine and nickel, so the purification can be carried out under denaturing conditions. The denaturing purification preserves any posttranslational modifications of Sst2. These conditions also increase the yield by solubilizing more of the peripherally membrane-associated Sst2, which is refractory to most nonionic detergents.[10] The addition of an incubation step for 60–90 min after lysing the cells also increases the amount of soluble Sst2. The addition of 2-mercaptoethanol to the buffers reduces the aggregation of Sst2. Although the batch binding step can be continued past the specified 60 min, we found that when the binding step was extended to overnight the Sst2 aggregated, even with 2-mercaptoethanol present. The addition of imidazole and NaCl to the buffers helps reduce nonspecific binding but does not result in a substantial loss in the yield of Sst2. The concentrations of imidazole and NaCl were determined empirically by examining the fraction produced from a step gradient of imidazole and a linear gradient of NaCl. The final protein is approximately 35% pure after the affinity purification (Fig. 1B). The two most prominent contaminants are significantly smaller than Sst2 and are largely reduced by the second purification step. However, since the protein mixture is resolved by SDS–PAGE before submission for mass spectral analysis, the ion-exchange chromatography may not be necessary.

Ion-Exchange Chromatography

All steps are carried out at room temperature.

1. Pool all the fractions from the elution step in a 50 ml conical tube and add 2 ml of Mono Q resin (Amersham Pharmacia Biotech) that has been equilibrated, according to the manufacturer's directions, in urea buffer. To equilibrate the resin wash 4 × 10 ml of urea buffer.
2. Allow the protein to batch bind for 60 min by gently rocking the tube. Rotate the tube periodically to prevent settling.
3. Centrifuge at 110g for 2 min to recover the resin.
4. Remove most of the supernatant, being careful not to disturb the resin.
5. Resuspend the resin in the remaining supernatant and pack an HR 10/10 column (Amersham Pharmacia Biotech) using the same method as described above for the Ni-NTA resin. A BioLogic system (Bio-Rad) can be used; however, any system capable of producing linear gradients would be adequate.
6. Wash with 5 CV of urea buffer at 1 ml/min.
7. Wash with 7.5 CV of urea buffer at 0.5 ml/min.
8. Wash with 7.5 CV of a linear gradient of 0 to 60 mM NaCl in urea buffer at 0.5 ml/min.
9. Wash with 7.5 CV of urea buffer containing 60 mM NaCl at 0.5 ml/min.
10. Elute with 15 CV of urea buffer containing 1 M NaCl at 0.5 ml/min. There should be a sharp peak of a low-intensity increase in absorbance (OD$_{280}$) that corresponds to Sst2.
11. The peak fractions are identified using SDS–PAGE to resolve the proteins, visualized by silver staining (BioRad), and confirmed by immunoblotting using an antibody to Sst2 or the His$_6$-tag (Qiagen). Most of the purified Sst2 elutes in the fractions corresponding to the increase in absorbance. These peak fractions are pooled and concentrated to a final volume of 500 μl using an Ultrafree-30 filter (Millipore, Bedford, MA) according to the manufacturer's directions. Sst2 needs to be concentrated before submission for mass spectral analysis, but there is no need to exchange the buffer.

Comments

Essentially all of the Sst2 will bind to the Mono Q resin, in contrast to the binding to the Ni-NTA resin. The concentration of NaCl used to wash the column and to elute the protein is determined empirically by analyzing the fractions after a linear gradient from 0 to 1 M NaCl. Washing with 60 mM NaCl does not remove a significant portion of Sst2. In an effort to minimize the amount of Sst2 that is lost, the 60 m*M* NaCl is reached gradually using a linear gradient from 0 to 60 m*M* NaCl. The Ultrafree-30 filters can be used to exchange Sst2 into another buffer by several cycles of concentrating and diluting it into the new buffer, according to the manufacturer's directions. However, the Ultrafree filters will concentrate some detergents such as SDS. Millipore can provide a list of compounds that do not pass freely through the filter. Moreover, there is a loss in yield when the buffer is exchanged. The final Sst2 is approximately 90% pure as assessed by SDS–PAGE and silver staining (Fig. 1B).

Determination of Protein Concentration

The purification of Sst2 from yeast will yield 1–5 μg of protein per liter of culture. The yield is not sufficient for typical protein assays such as the Lowry or Bradford methods. To estimate the concentration of this small amount of protein we resolve the sample on an 8% SDS–PAGE gel (Mini-PROTEAN, Bio-Rad) and visualize by silver staining (Silver Stain Plus, Bio-Rad). Known amounts of a protein standard such as bovine serum albumin (BSA, 10 ng–10 μg) are run alongside the purified Sst2. The concentration of Sst2 can be sufficiently estimated by comparing its intensity to that of BSA.

Submission of Sst2 for Mass Spectral Analysis

Samples are submitted for in-gel protease digestion and mass spectral analysis in dried SDS–PAGE gels. We resolve 1 μg of Sst2 per lane using 16 cm × 20 cm × 0.75 mm 8% SDS–PAGE gels (PROTEAN II, Bio-Rad). The gels are run at a constant current of 50 mA per gel with a voltage increase from approximately 200 V to over 600 V. The outer buffer chamber is completely filled with SDS–PAGE running buffer and cool water is run through the inner chamber to help dissipate the heat produced. To obtain the best separation, the gel is run until Sst2 is approximately halfway down the separating gel, as assessed by the position of a similarly sized prestained molecular weight standards. Using these conditions the dye front is run off the bottom of the gel.

The proteins are visualized using a silver staining procedure that does not interfere with the subsequent proteolysis and mass spectral analysis (method provided by Ruedi Aebersold, University of Washington, Seattle, WA). As with all silver staining, it is essential not to touch the gel except with new latex gloves rinsed with ethanol. This particular staining produces a more yellowish color, and one that is not as dark as that produced by the Bio-Rad Silver Stain Plus kit. Use deionized water in all solutions. All of the steps are carried out with sufficient buffer to cover the gel, at room temperature, and with gentle shaking (40 rpm).

1. Fix the gel for 30 min to overnight in 50% ethanol, 10% acetic acid (v/v).
2. Discard the fixation solution and incubate in 30% ethanol (v/v) for 15 min. The solutions can be removed by aspiration since the gel is very fragile.

3. Wash 3×5 min with deionized water.
4. Sensitize the gels using 0.2 g/liter of freshly prepared sodium thiosulfate for 1.5 min. Save some of the unused sodium thiosulfate solution to add to the developing solution.
5. Wash 3×30 sec with deionized water.
6. Incubate in 2 g/liter silver nitrate for 25 min.
7. Wash 2×30 sec with excess deionized water.
8. Develop the gel by incubating in developing solution (60 g/liter sodium carbonate, 20 ml/liter sodium thiosulfate solution from the sensitization step, 500 μl/liter 37% formaldehyde, in deionized water).
9. Once the desired level of staining is achieved, stop the development by discarding the developing solution and adding 6% acetic acid (v/v) for 10 min.
10. Wash a minimum of 4×15 min with deionized water.
11. Dry the cell between cellophane sheets. Drying frames and sheets can be purchased from Owl Separation Systems (Portsmouth, NH).

Nondenaturing Purification of Sst2 from *Escherichia coli*

We have purified Sst2 from *E. coli* using either a single His tag or a combination of a glutathione *S*-transferase (GST) and His tag. The isolated protein can be used in enzymological studies such as single-turnover GAP assays.

Escherichia coli strain

We have found that the protease-deficient strain BL21 (DE3) (Stratagene, La Jolla, CA) is the most efficient in its ability to express Sst2. Other strains such as BL21 (pLys) have been tested, but the expression is attenuated relative to the BL21 (DE3) strain.

Solutions

The equilibration buffer for both of the affinity chromatography procedures contain 50 mM Tris, pH 8, 10 mM 2-mercaptoethanol, and 150 mM NaCl. The lysis buffer consists of equilibration buffer with the addition of 0.1 mM phenylmethyl sulfonyl fluoride (PMSF, Sigma). Freshly prepare the PMSF by making a 10 mM stock solution in anhydrous methanol or ethanol. Alternatively, 4-(2-aminoethyl)benzenesulfonyl fluoride hydrochloride (AEBSF, Calbiochem, La Jolla, CA; Roche Diagnostics, Mannheim, Germany), which is more stable than PMSF in aqueous solutions, can be used at a final concentration of 0.2 mM. The Tris buffer will change its pH with changing temperature so it must be titrated at 4°. This can be conveniently performed by making a 1–2 liter stock solution of

1.0 M Tris buffer and titrating to pH 8.0 while at 4° prior to making the lysis buffer. The elution buffer for the GSH-Sepharose affinity chromatography is equilibration buffer containing 25 mM reduced glutathione (Calbiochem, Sigma) and is titrated back to pH 8. A linear gradient of 5–250 mM imidazole in equilibration buffer is used to elute Sst2 in the Ni-NTA affinity chromatography. The ion-exchange buffer is composed of 50 mM HEPES, pH 8, 2 mM dithiothreitol (DTT), 20 μM GDP, and 1 mM MgCl$_2$. The isolated Sst2 is stored at −80° in 50 mM HEPES, pH 8, 2 mM DTT, 150 mM NaCl, and 20% (v/v) glycerol.

Expression of GST–His$_6$ Tagged Sst2 in Escherichia coli

1. Prepare a starter culture by inoculating 3–4 ml of LB medium containing 50 μg/ml ampicillin (American Bioanalytical, Natick, MA) with a single colony. Stocks of transformed *E. coli* can be frozen at −80°; however, protein expression is not usually as robust as newly transformed cells. Also, carbenicillin (Sigma) can be substituted for ampicillin throughout the expression process. Carbenicillin is more stable and more expensive than ampicillin. Unlike the aqueous, 100 mg/ml stock of ampicillin, a carbenicillin stock solution is prepared by dissolving in 50% ethanol (v/v).
2. Grow with shaking (250 rpm) at 37° for 6–8 hr.
3. Add 1.5 ml of the starter culture to 150 ml of LB medium containing 50 μg/ml ampicillin.
4. Grow with shaking (250 rpm) at 24° for 12–15 hr.
5. Add 10 ml of the 150 ml culture to each 6 × 1 liters of fresh medium in 4-liter Erlenmeyer flasks. Use enriched medium (2% tryptone, 1% yeast extract, 0.5% NaCl, 0.2% glycerol, 50 mM sodium phosphate, pH 7.2) containing 50 μg/ml ampicillin.
6. Grow with shaking (250 rpm) at 24° until an OD$_{600}$ of 0.5–0.6 is reached. It will take 3–5 hr to reach the correct density. For convenience, the shaker speed can be reduced to 180 rpm to increase the length of time it takes to reach the correct density.
7. Add isopropyl-β-D-thiogalactopyranoside (IPTG, American Bioanalytical) to a final concentration of 10 μM. The IPTG is prepared as an aqueous, 100 mM stock.
8. Grow with shaking (180 rpm) at 24° for 6 hr.
9. To harvest the cells, pour the cultures into 250 ml polycarbonate centrifugation bottles and chill on ice while waiting to be centrifuged. Centrifuge at 4° for 10 min at 3840g. Pour off the supernatant and add more culture on top of the cell pellet. After all the cells are harvested, remove the cell paste from the bottle with a metal spatula and place directly in liquid nitrogen. Carefully pour off all the liquid nitrogen and break the frozen pellet into small pieces using a hammer prior to storage at −80° in 50 ml conical tubes.

Breaking the cell pellet allows it to thaw and dissolve more quickly during the lysis step.

Comments

In general, degradation of Sst2 is minimized by inducing with a low concentration of IPTG (10 μM), at the lower temperature of 24–30°, and with shaking reduced to 180 rpm. Cells are allowed to reach a density of OD_{600} of 0.5–0.6 before addition of the IPTG. We monitor the expression after induction for 3–12 hr and have found that 5–6 hr generates the highest yield of solubilized Sst2 protein. Over this time period the use of large Erlenmeyer flasks (4 liters) with 1 liter of cultured LB media increases the aeration and improves expression.

Lysis of Escherichia coli

All steps are performed at 0–4°.

1. Thaw the frozen cell pellets (from 12 liters of culture) in approximately 250–350 ml of lysis buffer.
2. Stir the solution at low speed using a magnetic stir plate. Low-speed stirring reduces denaturation. Plastic containers should be used in this and all other purification steps since protein has a tendency to stick to glass.
3. Prepare a fresh stock of 10–40 mg/ml lysozyme (Roche Diagnostics) in lysis buffer. Add lysozyme to a final concentration of 0.2 mg/ml and stir until the solution becomes viscous. This step will take between 15 and 30 min depending on the protein concentration. As the buffer thickens slightly increase the stirring speed for complete and efficient lysis of the cells.
4. Add $MgCl_2$ to a final concentration of 5 mM; follow with the addition of DNase I powder (Roche Diagnostics) to a final concentration of 0.02 mg/liter. Continue to stir for approximately 10–15 min until the viscosity is dramatically decreased. Alternatively, the cells may be lysed by passing the solution through a Microfluidizer (Microfluidics Corp., Newton, MA) and collecting the supernatant.
5. Centrifuge the lysate in 250 ml polycarbonate bottles for 60 min at 20,000g and collect the supernatant

GSH-Sepharose Affinity Chromatography

All steps of the purification procedure are completed at 4°. The following steps are only applicable to the purification of GST-His$_6$-Sst2. For the purification of His$_6$-Sst2 skip to the Ni-NTA affinity chromatography procedure below.

1. Divide the supernatant into two equal aliquots of approximately 125 ml in 250 ml polycarbonate bottles. Add 5 ml of equilibrated GSH-Sepharose 4B resin (Amersham Pharmacia Biotech) to each bottle. Equilibrate the resin by washing 4 × 50 ml of equilibration buffer.
2. Allow the proteins to batch bind for 60–120 min by gently rocking the bottle. Rotate the bottle periodically to prevent settling.
3. Centrifuge at 110g for 10 min to recover the resin.
4. Remove the supernatant from each bottle, being careful not to disturb the resin.
5. Resuspend the resin in 35 ml of equilibration buffer containing 300 mM NaCl.
6. Transfer the slurry to two 50 ml conical tubes and centrifuge as described above.
7. Remove the supernatant from each tube, being careful not to disturb the resin.
8. Add 20 ml of equilibration buffer to each tube.
9. Pour the resin into an XK 16/20 column (Amersham Pharmacia Biotech) containing 5 ml of equilibration buffer. Allow the resin to settle and place the top column adapter approximately 1–2 cm from the top of the resin bed. Be careful to remove any air bubbles in the process. A peristaltic pump can be used for all steps. Alternatively, the column can be moved to an FPLC (fast protein liquid chromatography) system (Amersham Pharmacia Biotech) for the elution step to facilitate the collection of samples and the detection of absorbance.
10. Wash with 10 CV of equilibration buffer containing 300 mM NaCl at approximately 2 ml/min. Afterward lower the column adapter to the top of the resin.
11. Elute with elution buffer at 2.5 ml/min collecting 5.0 ml fractions. The fractions are analyzed by resolving the proteins on an 8% SDS–PAGE gel and visualizing using Coomassie blue.
12. Pool the peak fractions (approximately 10 ml) in a 50 ml conical tube and add CaCl$_2$ to a final concentration of 2.5 mM.
13. Cleave off the GST by adding 1 μl/ml human α thrombin (Haematologic Technologies, Essex Junction, VT; Amersham Pharmacia Biotech) and rocking slowly overnight at 4°. To verify complete GST cleavage resolve the proteins on an 8% SDS–PAGE gel and immunoblot using an antibody for GST (Santa Cruz Biotechnology, Santa Cruz, CA).

Ni-NTA Affinity Chromatography

The following procedure is used to remove the cleaved GST, human thrombin, and other contaminating proteins from the His$_6$-Sst2. Also, use the following procedure to purify C-terminally His_6-tagged Sst2 (Sst2-His). All steps are performed at $4°$.

1. Add 5–10 ml Ni-NTA agarose resin (Qiagen) that has been equilibrated with equilibration buffer. For Sst2-His, divide the supernatant into two equal aliquots of approximately 125 ml in 250 ml polycarbonate bottles and add 5 ml of resin to each bottle. For thrombin-treated His-Sst2, add 5 ml of resin to the 50 ml tube containing the cleavage reaction mixture.
2. Allow the proteins to batch bind for 60 min by gently rocking the bottle. Rotate the bottle periodically to prevent settling.
3. Centrifuge at $110g$ for 2 min to recover the resin.
4. For Sst2-His, follow steps 4–9 of the GSH-Sepharose affinity chromatography procedure to pack a XK 16/20 column (Amersham Pharmacia Biotech). For thrombin-treated His-Sst2 follow step 9 of the GSH-Sepharose purification to pack the column. An FPLC system (Amersham Pharmacia Biotech) can be used; however, any system capable of producing gradients is sufficient.
5. Wash with 10 CV of equilibration buffer containing 250 mM NaCl at 1.5 ml/min.
6. Wash with 10 CV of equilibration buffer containing 5 mM imidazole at 1.5 ml/min.
7. Elute Sst2 with 5 CV of a linear gradient of 5–250 mM imidazole in equilibration buffer at 2 ml/min. Collect 2 ml fractions. The fractions are analyzed by resolving the proteins on 8% SDS–PAGE gels and visualized by Coomassie blue staining and immunoblotting using an antibody for the His_6-tag (Qiagen).
8. Concentrate the fractions containing Sst2 approximately 10-fold using an Ultrafree 30 NMWL membrane (Millipore). Dilute the concentrated protein 10-fold with storage buffer and repeat the concentration procedure.
9. Flash freeze aliquots of the purified protein in liquid nitrogen and store at $-80°$.

Ion-Exchange Chromatography

If Sst2 degradation products are present and their removal is warranted for future experiments, purification using Mono Q ion-exchange chromatography can be performed before concentrating Sst2. All steps are performed at $4°$.

1. Dilute the Sst2-containing fractions with 3 volumes of ion exchange buffer.
2. Pack an HR 10/10 column (Amersham Pharmacia Biotech) with 5 ml of equilibrated Mono Q resin (Amersham Pharmacia Biotech) according to manufacturer's directions.

3. Load the protein on the column at 1 ml/min using a FPLC system (Amersham Pharmacia Biotech); however, any system capable of producing gradients can be used.
4. Wash with 5 CV of ion-exchange buffer at 1 ml/min.
5. Elute Sst2 with 20 CV of a linear gradient of 0–700 mM NaCl in ion exchange buffer at 1 ml/min. Collect 2 ml fractions. The fractions are analyzed by resolving the proteins on 8% SDS–PAGE gels and visualized by Coomassie blue staining and immunoblotting using an antibody for the His$_6$-tag (Qiagen).
6. Concentrate the fractions containing Sst2 approximately 10-fold using an Ultrafree 30 NMWL membrane (Millipore). Dilute the concentrated protein 10-fold with storage buffer and repeat the concentration procedure.

[45] RGS Domain: Production and Uses of Recombinant Protein

By DAVID YOWE, KAN YU, THOMAS M. WILKIE, and SERGUEI POPOV

Introduction

Regulators of G protein signaling (RGS) proteins in fungi, *Dictyostelium*, and animals are related by a conserved RGS domain that is at least 127 amino acids in length. Mammals express five subfamilies of functionally distinct RGS proteins that are distinguished by amino acid similarities both within the RGS domain and in their flanking sequences.[1] The RGS domain alone is capable of binding Gα subunits and accelerating GTP hydrolysis; truncation or internal deletions in this minimal domain abolished these activities.[2] The structure of the RGS domain is two 4-helical bundles, as defined by X-ray crystallographic and nuclear magnetic resonance (NMR) analyses of three proteins, RGS4, GAIP, and axin.[3–5] These structures revealed two sites of protein–protein interaction. The RGS domain interacts with the switch regions of Gα across an interface that includes three interhelical loops, 3/4, 5/6, and 7/8. On the opposite face, helices 4 and 5 of axin binds APC, whereas this region of RGS4 and related RGS proteins binds

[1] E. Ross and T. M. Wilkie, *Ann. Rev. Biochem.* **69**, 795 (2000).
[2] S. Popov, K. Yu, T. Kozasa, and T. M. Wilkie, *Proc. Natl. Acad. Sci. U.S.A.* **94**, 7216 (1997).
[3] J. J. Tesmer, D. M. Berman, A. G. Gilman, and S. R. Sprang, *Cell* **89**, 251 (1997).
[4] K. E. Spink and P. P. Weis, *EMBO J* **19**, 2270 (2000).
[5] E. de Alba, L. de Vries, M. G. Farquhar, and N. Tjandra, *J. Mol. Biol.* **291**, 927 (1999).

to Ca^{2+}/calmodulin. This face may provide an allosteric regulatory site because PIP_3 binding in this region inhibits the GTPase-activating protein (GAP) activity of several RGS proteins. Ca^{2+}/calmodulin binding displaces phosphatidylinositol trisphosphate (PIP_3) and relieves its inhibition of GAP activity.[6] These interactions may provide a mechanism for feedback inhibition of $G\alpha_q$-coupled Ca^{2+} oscillations in pancreatic acini and other secretory cells.[7] The RGS domain also appears to interact with active, ligand-bound receptors, although high-affinity and receptor-selective interactions are conveyed by flanking sequences.[8] Thus, the RGS domain provides an essential regulatory component of G protein–receptor complexes. In this chapter, we provide protocols for the preparation of the RGS domain of RGS4 and a single-turnover GAP assay for use with $G\alpha_i$ and $G\alpha_q$.

Preparation of Box Recombinant Protein

The RGS domains of RGS4 (4Box) and other RGS proteins are expressed in *Escherichia coli* as insoluble proteins. Preparation of recombinant 4Box that is suitable for GAP assays and *in vivo* studies necessitates protein denaturation and renaturation. The following buffers are used for preparation of 4Box protein:

Buffer A: 50 mM HEPES (pH 8.0), 1 mM dithiothreitol (DTT), 0.05% (v/v) $C_{12}E_{10}$ (Merck), 5 mM EDTA
Buffer B: 8 M Urea (pH 8.0), 0.1 M NaH_2PO_4, 10 mM Tris-HCl
Buffer C: 8 M Urea (pH 6.3), 0.1 M NaH_2PO_4, 10 mM Tris-HCl
Buffer D: 8 M Urea (pH 8.0), 20 mM Tris-HCl (pH 7.4), 500 mM NaCl, 20% (v/v) glycerol
Buffer E: 20 mM Tris (pH 7.4), 500 mM NaCl, 20% glycerol
TPB Buffer: 50 mM Tris-HCl (pH 8.0), 20 mM 2-mercaptoethanol, 0.1 mM phenylmethylsulfonyl fluoride (PMSF)

4Box (amino acids 58–177 of RGS4) is cloned into a modified pQE60 vector which places a hexahistidine (His_6) tag at the N terminus (MGHHHHHMG).[2] Recombinant 4Box is expressed either in *E. coli* JM109 or BL21. Briefly,[2] 1 liter of T7 medium containing 100 μg/ml ampicillin is inoculated with an overnight culture which originated from a single colony. 4Box protein expression was induced with 10 μM isopropyl-β-D-thiogalactoside at an OD_{600} of 0.6. Cultures are shaken overnight at room temperature, and cells were pelleted by centrifugation at 4000g for 15 min at 4°. 4Box is expressed as an insoluble protein; hence, cell pellets are

[6] S. G. Popov, U. Murali Krishna, J. R. Falck, and T. M. Wilkie, *J. Biol. Chem.* **275,** 18962 (2000).
[7] X. Luo, S. Popov, A. K. Bera, T. M. Wilkie, and S. Muallem, *Mol. Cell* **7,** 651 (2001).
[8] W. Zeng, X. Xu, S. Popov, S. Mukhopadhyay, P. Chidiac, J. Swistok, W. Danho, K. Yagaloff, S. Fisher, E. Ross, S. Muallem, and T. M. Wilkie, *J. Biol. Chem.* **273,** 34687 (1998).

gently resuspended and lysed in Buffer B (5 ml/g cell pellet) by gentle stirring with a stir bar for 1 h at room temperature. The lysate is sonicated to shear DNA and centrifuged at 22,000g for 30 min at 4°. The supernatant is applied to a 5 ml Ni-NTA column at a flow rate of 10–15 ml/h. The column is washed with 5 to 10 volumes of Buffer B and then washed with 5 volumes of Buffer C until the A_{280} of the flow through is less than 0.01. 4Box is renatured using 100 ml of a linear gradient (requires about 1.5 h) of 6 M to 1 M urea in Buffer E containing protease inhibitors (Boehringer Mannheim, Indianapolis, IN). A final wash of 20 ml TPB buffer removes residual urea. 4Box is eluted with 9 ml Buffer E, 250 mM imidazole and concentrated with an Ultrafree 15 device (Millipore, Bedford, MA) in Buffer A. The purity is 90% as assessed by SDS–PAGE analysis with Coomassie blue staining. 4Box prepared by this protocol of denaturation/renaturation is as active as full-length RGS4 in single-turnover GAP assays and has unique activities in patch-clamp and permeabilized cells.[8]

RGS-Catalyzed Gα–GTP Hydrolysis

The single-turnover GAP assay is based on the ability of Gα subunits to bind GTP, followed by GTP hydrolysis and the release of P_i. The first reaction step involves generation of the Gα–GTP complex, which in some cases can be isolated from the components of the loading reaction. The stability of Gα–GTP in the absence of a GAP is mainly determined by its intrinsic GTPase activity. The second step involves GAP acceleration and measurement of the released P_i from the Gα–GTP complex. Finally, the P_i release data is converted to kinetic parameters of GAP activity. The following reactions (Scheme 1) reflect major kinetic features of this assay.[9]

Gα–GDP + GTP → Gα–GTP + GDP	(Loading reaction)
Gα–GTP → Gα–GDP + P_i	(Intrinsic hydrolysis)
Gα–GTP + GAP → Gα–GTP–GAP	
→ Gα–GDP+ GAP + P_i	(GAP reaction)

SCHEME 1

Preparation of Gα–GTP Substrate

Recombinant Gα subunits or membrane preparations containing G proteins are commonly made in their most stable GDP bound form. In order to obtain a high loading efficiency, the intrinsic GTPase activity of Gα should be slower than the preceding reactions of GDP dissociation and GTP association. It is important to use

[9] J. Wang, J. Tu, J. Woodson, X. Song, and E. M. Ross, *J. Biol. Chem.* **272,** 5732 (1997).

Mg^{2+}-free conditions to inhibit the intrinsic GAP activity of $G\alpha$ subunits, thereby increasing the yield of loaded GTP. Even trace amounts of Mg^{2+} substantially accelerate intrinsic GTPase activity.[9] A buffer without free Mg^{2+} is required to obtain a high loading efficiency for $G\alpha_i$, $G\alpha_o$, and $G\alpha_{12}$ subfamilies. However, for $G\alpha_z$ no difference in the intrinsic GTPase activity was found over a broad range of Mg^{2+} concentrations.[9] Usually, some experimentation is required for optimizing the loading conditions and we suggest the conditions below as good starting points for assays using $G\alpha_i$, $G\alpha_q$, and $G\alpha_{12}$ specific GAPs.

For recombinant $G\alpha_{i-1}$ expressed in *E. coli*, the optimal conditions for GTP-loading were 10 mM Na–HEPES (pH 8.0), 2 mM DTT, 5 mM EDTA, and 4 μM $[\gamma\text{-}^{32}P]GTP$.[6] No detergent needs to be present, and this allows faster and more accurate pipetting of small aliquots using polypropylene tips. $G\alpha_{12}$ and $G\alpha_{13}$ can be loaded in similar conditions in the presence of 0.05% polyoxyethylene 10-lauryl ester.[10] To reduce background counting from $[\gamma\text{-}^{32}P]P_i$ present in preparations of $[\gamma\text{-}^{32}P]GTP$ and from the intrinsic GTPase activity of $G\alpha$, a rapid gel filtration through Sephadex G-50 (or other suitable medium) at 4° may be required.[9,10] Using a reaction temperature of about 30° makes it possible to obtain an acceptable loading efficiency of 30 to 50% in 20 to 30 min.

For $G\alpha_q$ subunits the relatively high rate of intrinsic GTP hydrolysis makes loading inefficient. The use of an R183C $G\alpha_q$ subunit mutant that retains its GAP sensitivity was proposed to overcome this limitation.[11] The mutant protein R183C $G\alpha_q$-GTP can be prepared by incubating 100 nM $G\alpha_q$ with 100 μM $[\gamma\text{-}^{32}P]GTP$ at 20° in 20 mM Na–HEPES (pH 7.5), 5.5 mM CHAPS, 1 mM DTT, 0.9 mM $MgSO_4$ (10 μM final free Mg^{2+}), 30 mM $(NH_4)_2SO_4$, 2% glycerol, and 0.1 mg/ml albumin. The reaction reaches steady state in about 2 h and approximately 30% of active R183C $G\alpha_q$ binds $[\gamma\text{-}^{32}P]GTP$. The reaction mix is then diluted with an equal volume of 50 mM Na–HEPES (pH 7.5), 0.3% cholate, 1 mM DTT, 1 mM EDTA, 0.9 mM $MgSO_4$, and 0.1 mg/ml albumin at 0°. GTP loading and apparent substrate specificity for different RGSs to particular GAPs are sensitive to the type of detergent that is used in the reaction. For example, the GAP activity of RGS4 is unaltered by octyl glucoside, dodecyl maltoside, cholate, or deoxycholate, whereas the GAP activity of PLC-β1 can only be observed in the presence of cholate.[11]

The specific activity of $[\gamma\text{-}^{32}P]GTP$ used by different authors in single-turnover assays varies over a broad range from 3 to 100 cpm/fmol. The concentration of the G_α subunit is usually in the range of 30 to 500 pM. If appropriate, the preparation of $G\alpha$–GTP can be diluted to the required concentration at the GAP reaction step. The following values of $t_{1/2}$ (half-life) illustrate the stability of different $G\alpha$-GTPs in the presence of Mg^{2+}: $G\alpha_i$, ~2 min at 0°; $G\alpha_q$, ~1 min at 30°; and R183C $G\alpha_q$,

[10] T. Kozasa, X. Jiang, M. J. Hart, P. M. Sternweis, W. D. Singer, A. G. Gilman, G. Bollag, and P. C. Sternweis, *Science* **280**, 2109 (1998).

[11] P. Chidiac and E. M. Ross, *J. Biol. Chem.* **274**, 19639 (1999).

140 min at 20°. Therefore, experiments with $G\alpha_i$ are often carried out at lower temperatures (0° to 15°) to allow measurement of released P_i over a convenient time. In the case of the R183C $G\alpha_q$–GTP mutant, the rate of GTP dissociation ($k_d = 0.005$ min^{-1}, $t_{1/2} = 140$ min) is also measured, and has to be taken into account for slow GAPs.

GAP Reaction

GTP hydrolysis is initiated with addition of the GAP along with other components to $G\alpha$–GTP at a chosen reaction temperature. Routinely, Mg^{2+} (about 1 mM) and unlabeled GTP (about 100 μM) are in the initiating reaction buffer. Mg^{2+} ions accelerate GAP activity. Unlabeled GTP reduces background signal. Because of [γ-^{32}P]GTP dissociation from contaminating GTP-binding proteins and rebinding to $G\alpha$, followed by a second round of GTP hydrolysis. This is especially problematic when the loading reaction mix from whole lysates is used for the assay without preliminary isolation of $G\alpha$–GTP complexes. GAP activities of different proteins may vary over a wide range, with half-lives ranging from seconds to hours, depending on their nature and the chosen experimental conditions. It is therefore a good idea to perform a time course of GAP activity to determine representative time points for further measurements.

Protocol for Performing GAP Assay Using $G\alpha_{i-1}$ Subunits

Prepare 500 nM $G\alpha$ and 4 μM [γ-^{32}P]GTP solutions in 10 mM HEPES (pH 8.0), 5 mM EDTA, and 1 mM DTT, and keep the solutions on ice. Start the GTP loading reaction by mixing equal volumes of $G\alpha$ and [γ-^{32}P]GTP solutions in 1.5 ml microfuge tubes. To determine the volume of the final mix calculate the (number of time points $+1$) $\times$ 25 μl. Add an extra 20 to 40 μl to compensate for dispensing errors. Place the tube in a water bath preheated to 30° and incubate for 20 to 30 min. While the loading reaction takes place, prepare a rack with test tubes (Falcon, NJ) for dispensing a 5% charcoal (Fisher, Philadelphia, PA) suspension in 50 mM NaH$_2$PO$_4$. Use the number of tubes equivalent to the number of time points required. When preparing the charcoal solution it is necessary to continuously stir it until an even suspension is formed, then dispense 375 μl aliquots into each test tube using a pipettor with a 1 ml pipette tip (cut the end of the tip off in order to prevent clogging with charcoal particles). Transfer the tubes into a 4° cold room. Place the required amount of 5 mM GTP/500 mM MgCl$_2$ solution onto the wall of the reaction tube, and place the tube on ice. Typically, the final free MgCl$_2$ concentration is in the 1 to 5 mM range (remember to account for the concentration of EDTA in the assay buffer). Use the assay buffer to make an appropriate dilution of the RGS protein. Place a drop of the RGS solution onto the wall of the tube next to the drop of GTP/MgCl$_2$. Do not mix the RGS and GTP/MgCl$_2$ solutions. For a negative control use the assay buffer alone instead of the RGS solution. After the loading reaction is completed, immediately place the tubes on ice for 5 min to equilibrate to 0° and transfer the ice bucket to the cold room. In a convenient position place a preset timer, a rack of charcoal tubes, and a box of prechilled pipette tips. Quickly mix 25 μl of the loading mix with the charcoal suspension and place it in the first tube. Vortex the tube for 1 to 2 sec. This tube will serve as the zero time point. Take the amount of loading mix equal to (number of time points + 1) × 25 μl and quickly inject it into the reaction tube with 4Box/GTP/MgCl$_2$. Make sure that the flow of loading mix is directed toward the drops of 4Box/GTP/MgCl$_2$ so it is mixed efficiently. Immediately start counting the reaction time. Withdraw 25-μl aliquots of the reaction solution at appropriate time points and immediately inject it into the corresponding tubes with charcoal, then vortex. After the reaction is complete (5 min for Gα_i) spin the tubes with charcoal in a tabletop microcentrifuge at 1000 to 2000g for 5 min at room temperature to the pellet charcoal. Depart the cold room. Without disturbing the charcoal pellet, carefully remove 200 μl of supernatant from each tube and place it into a scintillation vial containing 4 ml of scintillation cocktail. Count the radioactivity in the vials that will correspond to the amount of P$_i$ in 12.5 μl of reaction solution. Correct the zero time point count by taking into account that further time point counts correspond to more diluted solutions. Calculate the total reaction volume as a sum of the loading mix volume and the volume of 4Box/GTP/MgCl$_2$ added. Calculate the correction coefficient by dividing the loading mix volume by the total volume. Subtract the corrected value from each count. The resulting data will represent the time course of P$_i$ release during hydrolysis of GTP bound to Gα.

Retention of Full GAP Activity by 4Box in Single-Turnover Assay

Figure 1 shows that full-length RGS4 protein and 4Box have similar GAP activity on Gα_i–GTP in a single-turnover assay. Comparisons of the kinetic constants derived from these time course curves (Fig. 1C) were used to estimate k_1 to be 9×10^5 $M^{-1}s^{-1}$ at 0° for either 4Box or full-length RGS4 (extrapolated to 1 μM).[2]

FIG. 1. The RGS domain and full-length RGS4 have similar catalytic activities. To obtain kinetic curves of GTP hydrolysis, Gα_{i-1} (500 nM) was preloaded with [γ-^{32}P]GTP at 30° (estimated concentration of GTP–Gα_{i-1} was 215 nM), then incubated at 0° with different concentrations of (A) full-length RGS4 and (B) the RGS domain of RGS4 (4Box, amino acids 58 to 177 of RGS4) (R4). (C) A plot of the observed first-order kinetic constant verses initial concentration of RGS4 and R4 derived from (A) and (B) and additional experiments. Each data point corresponds to one time course curve, error bars indicate one standard deviation in each k_{obs} calculation. (Reproduced with permission from S. Popov, K. Yu, T. Kozasa, and T. M. Wilkie, *Proc. Natl. Acad. Sci. U.S.A.* **94,** 7216 (1997).

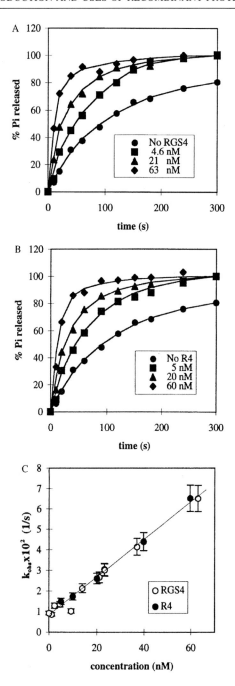

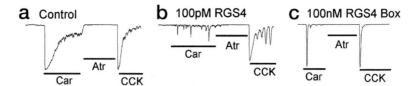

FIG. 2. 4Box is a nonselective inhibitor of Ca^{2+} signaling evoked by $G\alpha_q$-coupled agonists. Patch-clamped pancreatic acinar cells were stimulated with 100 μM carbachol (Car), inhibited with 10 μM atropine (Atr), and then stimulated with 10 nM CCK8, as indicated. (a) Cell dialyzed with buffer. (b) Addition of 100 pM RGS4 showed that carbachol-dependent signaling was about 15 ± 3 ($n = 17$)-fold more sensitive to RGS4 than CCK-dependent signaling. (c) 4Box inhibited the response to carbachol and CCK equivalently. With 100 nM 4Box, the ratio of CCK to carbachol (CCK/Car) response was 0.98 ± 0.07 ($n = 33$). (Reproduced with permission from W. Zeng *et al.*, *J. Biol. Chem.* **273**, 34687 (1998).)

Inhibition by 4Box of Agonist-Bound Receptor Complexes in Cells

Although 4Box retains GAP activity, it is much less efficient than full-length RGS4 at inhibiting receptor-catalyzed $G\alpha_q$ signaling.[8] In lipid vesicles reconstituted with ml muscarinic receptors, the rate of steady-state GTP hydrolysis on $G\alpha_q$ is 100-fold higher with RGS4 than in the presence of 4Box.[8] In patch-clamped cells dialyzed with recombinant RGS proteins, full-length RGS4 is at least 1000-fold more potent than 4Box in inhibiting $G\alpha_q$ signaling. Figure 2 shows that 100 pM RGS4 almost completely inhibits Ca^{2+} oscillations evoked by carbachol. By contrast, 100 nM 4Box has essentially no effect on the initial release of Ca^{2+} from internal stores. The RGS domain of other RGS proteins was essentially without effect in yeast[12] and tissue culture cells (our unpublished observations). However, when dialyzed into cells, 4Box obviously inhibits sustained Ca^{2+} signaling (Fig. 2c). Furthermore, 4Box appears to be recruited to active (agonist-bound) receptors, which explains why Ca^{2+} signaling is rapidly expunged following the initial Ca^{2+} spike (Fig. 2c and unpublished observations). Finally, 4Box inhibits Ca^{2+} signaling evoked by different $G\alpha_q$-coupled agonists equally well whereas full-length RGS4 preferentially inhibits signaling evoked by carbachol compared with CCK (Fig. 2b). These observations (and others) justify the approach of adding recombinant proteins to cells, rather than overexpressing RGS proteins in transfected cells, to characterize the role of RGS proteins in G protein signaling.

Evaluation of Kinetic Data

As a final point, kinetic analysis provides valuable mechanistic data for interpreting RGS function in cells. Scheme 1 showed interactions that are commonly

---

[12] S. P. Srinivasa, L. S. Berstein, K. J. Blumer, and M. E. Linder, *Proc. Natl. Acad. Sci. U.S.A.* **95**, 5584 (1998).

assumed to provide an adequate kinetic description of the single-turnover GAP assay; however, it may not always be the case. According to this scheme, interaction of $G\alpha$–GTP with a GAP leads to a formation of the intermediate tertiary $G\alpha$–GTP–GAP complex. The stability and dynamics of this interaction are unknown. Literature data is limited to K_m value estimates for several RGS proteins determined under the assumption that the kinetics of the GAP catalyzed reaction follows the traditional Michaelis–Menten equation. This is usually the case when the concentration of $G\alpha$–GTP–GAP is considered steady state, and therefore low, compared to enzyme and substrate concentrations (the rate-limiting step is the formation of the complex). For example, the GAP activity of RGS4 on $G\alpha_{i-1}$–GTP (the release of P_i) follows the first-order equation[2]: $k_{obs} = k_0 + (k_{cat}/K_m)[\text{RGS4}]_0$. In this case, intrinsic hydrolysis by $G\alpha$ was also taken into account. The plot of the observed kinetic constant vs RGS concentration is a straight line. Similar behavior was found for $G\alpha_z$–GAP.[9] However, it should always be kept in mind that any particular kinetic data can fit many alternative models. For example, the model with rapid equilibrium at the first step of the GAP reaction is $G\alpha$–GTP + RGS = $G\alpha$–GTP–RGS, also predicts a linear increase of k_{obs} with $[\text{RGS}]_0$.

When the GAP activity of a protein is low, the reaction in the $G\alpha$–GTP–GAP complex may become rate limiting. This means for experimentally reliable detection of GAP activity the concentration of a GAP should be high. In such conditions, the maximal rate of P_i release may be determined by the concentration of $G\alpha$–GTP, and not by the GAP concentration. Figure 3 illustrates the properties of an RGS4 mutant that has four amino acid substitutions: Lys to Gln at residues 99 and 100 and Lys to Glu at residues 112 and 113 (RGS4QE). Its GAP activity, measured as the rate of P_i release from 200 nM $G\alpha_i$-GTP, is low compared to wild-type RGS4, and does not increase with an increase of GAP concentrations above 200 nM (Fig. 3B).

It is also commonly assumed that the affinity of a GAP for $G\alpha$–GDP is low, so that the GAP dissociates from $G\alpha$–GTP–GAP immediately after GTP hydrolysis. However, RGS1, RGS2, and RGS16 were immunoprecipitated with GDP bound forms of $G\alpha_i$ subunits,[13-15] indicating a stable interaction which should not be ignored in an adequate model.

GTP–GDP exchange activity was previously demonstrated for the guanine nucleotide exchange factor (GEF) proteins on small G proteins. The GEF domain of p115RhoGEF, an RGS-like protein, was found to be active on Rho proteins.[16] Currently, no GEF activity has been reported for RGS domains on $G\alpha$ subunits

[13] N. Watson, M. E. Linder, K. M. Druey, J. H. Kehrl, and K. J. Blumer, *Nature* **383**, 172 (1996).

[14] C. Chen, B. Zheng, J. Han, and S. C. Lin, *J. Biol. Chem.* **272**, 8679 (1997).

[15] C. Chen and S. C. Lin, *FEBS Lett.* **422**, 359 (1998).

[16] M. J. Hart, X. Jiang, T. Kozasa, W. Roscoe, W. D. Singer, A. G. Gilman, P. C. Sternweis, and G. Bollag, *Science* **280**, 2112 (1998).

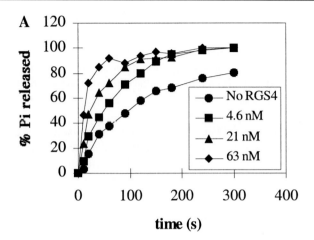

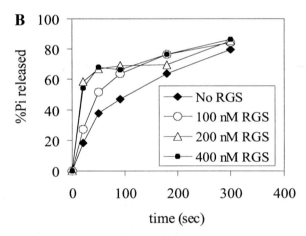

FIG. 3. Time courses for single-turnover GTP hydrolysis of $G\alpha_{i-1}$ catalyzed by wild-type and mutant RGS4. $G\alpha_{i-1}$GTP (250 nM) was held constant with increasing concentrations of RGS proteins assayed at 0°. (A) The apparent rate of GTP hydrolysis increased with RGS4. (B) The GAP activity of the mutant RGS4 protein RGS4QE is less than wild-type and the apparent rate of GTP hydrolysis did not increase above 200 nM RGS4QE.

of heterotrimeric G proteins. However, we found that a mutant RGS4 protein displayed both GEF and GAP activities (S. Popov and T. Wilkie, personal communication, 2000). Hence, it may be informative to assay for GEF activity before performing single-turnover assays.[17]

[17] D. M. Berman, T. M. Wilkie, and A. G. Gilman, *Cell* **86,** 445 (1996).

Conclusion

We have described detailed methods for the preparation of recombinant RGS4 domain (4Box) in *E. coli,* and its uses in single-turnover GTP-hydrolysis assays and in primary cells. These procedures are probably applicable to the production and characterization of the RGS domain of any RGS protein.

[46] Screening for Interacting Partners for Gα$_{i3}$ and RGS–GAIP Using the Two-Hybrid System

By LUC DE VRIES and MARILYN GIST FARQUHAR

Introduction

In the classical heterotrimeric G protein signaling pathways at the plasma membrane, the Gα subunit interacts with serpentine receptors, downstream effectors, and Gβγ dimers. Gα subunits are also localized on intracellular membranes. For example, Gα$_{i3}$ is localized in the Golgi region,[1,2] and at this site only less abundant βγ subunits have been described as possible interaction partners.[3] Isolation of proteins that interact with Gα$_{i3}$ should shed a light on the role of Gα$_{i3}$ in this location.

Among several methodological approaches used to isolate Gα$_{i3}$ interacting proteins, the two-hybrid system has the enormous advantage that it directly delivers the corresponding cDNA of the interacting proteins. However, the system also presents a drawback: it cannot be biased toward finding interacting proteins in a specific region of the cell, and localization studies of newly identified interacting proteins must subsequently be performed.

The use of wild-type Gα subunit as bait in the two-hybrid system presents the theoretical advantage of being able to isolate all interacting proteins, regardless of its GDP/GTP-dependent conformational state. This approach was used by us for Gα$_{i3}$, because we did not want to exclude any potential interacting partners in the Golgi region.[4,5]

[1] L. Ercolani, J. L. Stow, J. F. Boyle, E. J. Holtzman, H. Lin, J. R. Grove, and D. A. Ausiello, *Proc. Natl. Acad. Sci. U.S.A.* **87,** 4635 (1990).

[2] B. S. Wilson, M. Komuro, and M. G. Farquhar, *Endocrinology* **134,** 233 (1994).

[3] S. Denker, J. M. McCaffery, G. E. Palade, P. A. Insel, and M. G. Farquhar, *J. Cell Biol.* **133,** 1027 (1996).

[4] L. De Vries, M. Mousli, A. Wurmser, and M. G. Farquhar, *Proc. Natl. Acad. Sci. U.S.A.* **92,** 11916 (1995).

The two-hybrid approach should be applicable to any Gα subunit as bait, provided the screening procedure takes place with a tissue- or cell type-specific cDNA library that expresses this subunit. However, successful screenings have been published so far only for Gα_{i1},[6] Gα_{i2},[7,8] Gα_{i3},[4,5,9] Gα_z,[10,11] and Gα_o,[12,13] all of which belong structurally to the Gα_i subclass.[14] One exception is the positive result obtained screening with yeast *GPA2*, which resembles Gα_i structurally but Gα_s functionally.[15]

The two-hybrid system can be biased toward finding a specific type of interacting protein by using mutants of the Gα subunit as a bait. For example, mutants that mimic the activated conformation of the Gα subunit are expected to interact preferentially with effectors and GTPase activating proteins (GAPs). This approach has been used with success in the search for GAPs.[9,10,13]

We have also used the two-hybrid system to isolate interacting partners for RGS proteins. To date, we have obtained positive results using RGS proteins as a bait for GAIP[16] and for RGS16.[17]

We have successfully used the Matchmaker 1 and 2 systems (Clontech, Palo Alto, CA) for screening purposes with Gα_{i3} and with GAIP and will limit the description to the use of these systems. Since the original description of the two-hybrid screening system,[18] several improved systems have become commercially available. Improvements include inducible systems (LexA-based), GFP-tagged selection systems, and a system with selection after mating of bait and prey yeast strains.

The principle of all yeast two-hybrid systems is outlined in Fig. 1. Briefly, it consists of the reconstitution of the activity of a transcription factor (*GAL4*) on the promoters of two reporter genes present in the yeast nucleus: LacZ for blue/white

[5] P. Lin, H. Le-Niculescu, R. Hofmeister, J. M. McCaffery, M. Jin, H. Hennemann, T. McQuistan, L. De Vries, and M. G. Farquhar, *J. Cell Biol.* **141**, 1515 (1998).

[6] N. Mochizuki, Y. Ohba, E. Kiyokawa, T. Kurata, T. Murakami, T. Ozaki, A. Kitabatake, K. Nagashima, and M. Matsuda, *Nature* **400**, 891 (1999).

[7] N. Mochizuki, M. Hibi, Y. Kanai, and P. A. Insel, *FEBS Lett.* **373**, 155 (1995).

[8] N. Mochizuki, G. Cho, B. Wen, and P. A. Insel, *Gene* **181**, 39 (1996).

[9] T. W. Hunt, T. A. Fields, P. J. Casey, and E. G. Peralta, *Nature* **383**, 175 (1996).

[10] J. L. Glick, T. E. Meigs, A. Miron, and P. J. Casey, *J. Biol. Chem.* **273**, 26008 (1998).

[11] J. Meng, J. L. Glick, P. Polakis, and P. J. Casey, *J. Biol. Chem.* **274**, 36663 (1999).

[12] Y. Luo and B. M. Denker, *J. Biol. Chem.* **274**, 10685 (1999).

[13] J. D. Jordan, K. D. Carey, P. J. Stork, and R. Iyengar, *J. Biol. Chem.* **274**, 21507 (1999).

[14] M. I. Simon, M. P. Strathmann, and N. Gautam, *Science* **252**, 802 (1991).

[15] Y. Xue, M. Batlle, and J. P. Hirsch, *EMBO J.* **17**, 1996 (1998).

[16] L. De Vries, X. Lou, G. Zhao, B. Zheng, and M. G. Farquhar, *Proc. Natl. Acad. Sci. U.S.A.* **95**, 12340 (1998).

[17] B. Zheng, D. Chen, and M. G. Farquhar, *Proc. Natl. Acad. Sci. U.S.A.* **97**, 3999 (2000).

[18] C. T. Chien, P. L. Bartel, R. Sternglanz, and S. Fields, *Proc. Natl. Acad. Sci. U.S.A.* **88**, 9578–9582 (1991).

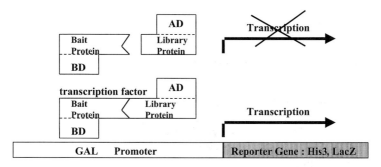

FIG. 1. Principle of the two-hybrid system. BD, DNA binding domain; AD, activation domain.

screening, and *HIS3,* a nutritional gene essential for survival on selective medium (lacking histidine). The protein sequence between the DNA binding domain (BD) and the activation domain (AD) of the transcription factor is not essential. AD and BD can complement each other to reform an active transcription factor, provided the nonessential sequence is replaced by a bait protein fused to the BD and a prey (library) protein fused to the AD, and provided bait and prey proteins interact. Bait and prey fusion proteins are directed to the nucleus because of the presence of nuclear localization signals (NLS) in the fusion proteins. Activity of the transcription factor on the reporter genes in the yeast nucleus is restored when bait and prey protein interact.

Several of the protocols that follow have been adapted from Clontech's Two-Hybrid manual (Matchmaker 1 and 2).

Materials

Plasmids

pGBT9 is used as the "bait" vector and has a moderate-strength promoter for expression in yeast to avoid eventual toxicity problems due to overexpression. It contains a multiple cloning site (MCS) at the C-terminal end of the *GAL4* DNA binding domain (BD) for expression as *GAL4* BD-fusion proteins and a tryptophan selectable marker (TRP1 nutritional gene) for maintenance of the plasmid in the yeast strain. pGADGH and pACT2 are both used as "prey" vectors (see libraries) and have strong promoters for expression in yeast, to ensure expression of normally low-expressing proteins. Library inserts are cloned into the MCS of pGADGH or pACT2 located at the C-terminal end of the *GAL4* activation domain (AD) for expression as *GAL4* AD-fusion proteins. Both prey plasmids carry the leucine selectable marker (*LEU2* nutritional gene). Maps of all plasmids are available at www.clontech.com.

Libraries

Initially we used a HeLa S3 cell line cDNA library (human) in pGADGH vector, which is commercially available from Clontech. Inserts are directionally cloned [oligo(dT) primer] into the *Eco*RI (5′) and *Xho*I (3′) sites of the MCS. The library contains ~6 × 10⁶ independent clones.

We have also used a rat pituitary cell line cDNA library (GC cells, subcloned from GH3 cells) in pACT2 vector constructed by H. Henneman using a random primed cDNA synthesis kit (Stratagene, La Jolla, CA).[5] Inserts are cloned into the *Eco*RI (5′) and *Xho*I (3′) sites of the MCS. The library contains ~2 × 10⁶ independent clones.

Other tissue-specific libraries are commercially available and have been used with success.[6–8,10,12]

Yeast and Bacterial Strains

Yeast strain HF7c is used for interaction screening and for verifying 1 to 1 interactions. Yeast strain SFY526 is used for one-to-one interactions and for the semiquantitative liquid β-Gal assays (see below). Growth of HF7c cells depends on the presence of essential amino acids, tryptophan (Trp), leucine (Leu), and histidine (His); when absent, these amino acids must be provided by expression of nutritional genes *TRP1* and *LEU2* encoded by bait and prey plasmids, respectively, as well as by the activation of the *HIS3* nutritional gene through interaction of bait and prey. Growth of SFY526 cells depends only on the presence of Trp and Leu, because this strain does not have the interaction-dependent *HIS3* nutritional gene. HF7c contains the *HIS3* reporter gene under control of the full *GAL1* promoter and the LacZ reporter under control of a minimal promoter. SFY526 contains the LacZ reporter only, under control of the full *GAL1* promoter. Genotypes of both strains are available at *www.clontech.com.*

For general and additional information on yeast manipulations, we refer the reader to the "*Guide to Yeast Genetics and Molecular Biology.*"[19]

Escherichia coli strain HB101 (containing LeuB mutation) is used to rescue pGADGH or pACT2 plasmids isolated from positively selected yeast colonies (see below).

Media

YPD (rich) medium, containing all essential amino acids, is used for maintenance of parental yeast strains. For screening purposes synthetic medium (SD) without amino acids supplemented with adenine (hemisulfate salt, 20 mg/liter final), lysine (30 mg/liter), and tyrosine (30 mg/liter) is used. When needed for

---

[19] "Guide to Yeast Genetics and Molecular Biology." *Methods Enzymol.* **194** (C. Guthrie and G. R. Fink, eds.). Academic Press, San Diego, 1991.

specific selections, Leu (100 mg/liter), Trp (20 mg/liter), or His (20 mg/liter) are added. All media contain 2% (w/v) dextrose. Although not essential, we regularly add ampicillin (50 μg/ml) to yeast media to avoid bacterial contamination.

For *E. coli* HB101, minimal M9 medium supplemented with 0.4% glucose and ampicillin (50 μg/ml) is used.

Other

Whatman (Clifton, NJ) #5 paper filters (137 and 82 mm diameter), amino acids, X-Gal stock solution (20 mg/ml in DMF), acid-washed glass beads, ONPG (*O*-nitrophenyl-β-D-galactopyranoside) and 3-aminotriazole (3-AT) are purchased from Sigma (St. Louis, MO).

All stock solutions, water, and media are filter-sterilized or autoclaved.

General Procedures for Two-Hybrid Library Screening

Performing a screening procedure with the two-hybrid system requires the following important steps:

Construct the bait or target protein in the BD vector (pGBT9) and construct or obtain the prey library in the AD vector (pGADGH, pACT2). Bait vectors are confirmed negative for transactivation of the LacZ and *HIS3* reporter genes in strain HF7c before screening to ensure minimal leakiness of the system due to expression of the bait protein alone. Transform (the term used for introducing a plasmid into yeast cells, the equivalent of transfection for mammalian cells) HF7c yeast cells with bait and prey vectors. Either perform cotransformation—i.e., the simultaneous addition of bait and library (prey) plasmids—or alternatively, sequential transformation by first making a yeast bait strain which is subsequently transformed with library plasmid. Isolate prey plasmid from positive (blue) yeast colonies after a β-Gal filter lift assay, perform interaction controls in the two-hybrid system, and sequence inserts of positive prey plasmids. To validate the positive interaction obtained in the two-hybrid system, interaction of both proteins needs to be confirmed in another *in vitro* or *in vivo* system (see below).

Construction of Bait Vector

The full-length wild-type rat Gα$_{i3}$ cDNA [originally obtained by RT-PCR (reverse transcription polymerase chain reaction) from a rat embryonic fibroblast (REF52) cell line] is subcloned from plasmid pGEX-KG (Pharmacia, Piscataway, NJ; checked for correct expression of Gα$_{i3}$ in *Escherichia coli*) into bait plasmid pGBT9 using standard molecular biology techniques.[20] Rat RGS–GAIP$_{23-216}$

---

[20] F. M. Ausubel, R. Brent, R. E. Kingston, D. D. Moore, J. G. Seidman, J. A. Smith, and K. Struhl, eds., "Current Protocols in Molecular Biology." J. Wiley and Sons, New York, 1990.

cDNA (obtained from 2-hybrid screen with $G\alpha_{i3}$) is subcloned from prey vector pACT2 into bait vector pGBT9.

Yeast Transformation Protocols

Yeast cells are transformed by the lithium acetate method.[21] This method yields an efficiency of 10^3–10^4 cotransformants/μg total plasmid for cotransformations or $\sim 10^5$ transformants/μg plasmid for single transformations. The basic yeast transformation protocol we used is as follows:

1. Inoculate 20 ml YPD medium with a single HF7c colony; grow overnight at 30° with agitation (200 rpm).
2. Dilute to $OD_{600} \sim 0.3$ into 300 ml prewarmed YPD, grow for 3–4 hr at 30° with agitation (200 rpm); OD_{600} must be less than 1.0.
3. Pellet cells (1000g, 5 min at room temperature), wash with 50 ml H_2O.
4. Pellet cells (1000g, 5 min at room temperature), resuspend in 1–2 ml lithium acetate/TE solution (100 mM lithium acetate, 10 mM Tris-HCl, 1 mM EDTA, pH 7.5).
5. Aliquot 100 μl into microcentrifuge tubes, keep at room temperature, and use competent cells within 2 hr.
6. Add 0.2 μg/plasmid + 50 μg denatured salmon sperm carrier DNA/tube, mix.
7. Add 600 μl polyethylene glycol (PEG) lithium acetate solution [40% (w/v) PEG 4000, 100 mM lithium acetate, 10 mM Tris-HCl, 1 mM EDTA, pH 7.5], mix gently, incubate 30 min at 30° (no agitation needed).
8. Add 70 μl dimethyl sulfoxide (DMSO), mix, incubate 15 min at 42° (heat shock), chill 2 min on ice.
9. Pellet cells (15 sec at 14,000 rpm), resuspend in 0.5 ml H_2O per tube, plate 100 μl on selective SD medium agar dish (100 mm).
10. Incubate for 3 days (for single plasmid transformations) at 30° or until colonies are big enough (1–2 mm) for a filter lift assay (see below).

Opting for cotransformation or sequential transformation for screening in the two-hybrid system may depend on several factors: (1) eventual toxicity of the bait protein: cotransformation is less risky; (2) time constraints; cotransformation is faster; (3) limited amount of cDNA library plasmid; sequential transformation is more efficient.

Both cotransformation with bait vector and prey library and single transformation with prey library into a preestablished HF7c $G\alpha_{i3}$ bait strain are performed in two independent screening procedures (Table I). For the RGS–GAIP bait, screening is performed by cotransformation.

[21] D. Gietz, A. St. Jean, R. A. Woods, and R. H. Schiestl, *Nucleic Acids Res.* **20**, 1425 (1992).

TABLE I
Gα$_{i3}$ AND GAIP SCREENING RESULTS[a]

Parameter	Gα$_{i3}$ cotransformation	Gα$_{i3}$ sequential transformation	GAIP cotransformation
Library	HeLa	Rat pituitary	Rat pituitary
Number of yeast colonies screened	4×10^5	10^6	6×10^5
Number of Leu, Trp, His survivors	120	400	300
Number of β-Gal positives	9	60	38
Identity of positive clones	GAIP (2)	Calnuc (13) GAIP (3) Karyopherin (3) LGN (2) AGS3 (2) Uncharacterized (3)	GIPC (3) Hsp60 (1) cytochrome-c oxidase (1) Uncharacterized (2)

[a] Only proteins that show confirmed 1-to-1 interaction with bait protein are included. Numbers in parentheses are independent positive yeast colonies isolated.

Sequential Transformation. For making the Gα$_{i3}$ bait strain by sequential transformation, use 0.2 μg of pGBT9-Gα$_{i3}$ and plate on SD lacking Trp to select for bait plasmid expressing cells. Expression of Gα$_{i3}$ in the bait strain can be verified by PCR on a yeast DNA prep (see below) or by immunoblotting a yeast lysate. For standard one-to-one interactions, 0.2 μg of each bait and prey vector is used as above, and 200 μl is plated on 100 mm dishes (SD lacking Leu and Trp).

For screening, the Gα$_{i3}$ bait strain is grown overnight in selective medium and transferred to YPD medium for preparation of competent cells as described above. A total of 50 to 100 μg prey library plasmid and denatured salmon sperm carrier DNA (50 μg/tube) is divided over microcentrifuge tubes corresponding to the number of 150 mm dishes to plate for screening. Two hundred μl of transformed cells is plated/150 mm dish (SD lacking Leu, Trp, and His). To suppress leakiness of the *HIS* reporter gene, 10 to 30 mM 3-aminotriazole can be added to the selection medium.

Cotransformation. For screening using cotransformation, bait plasmid (100 to 500 μg), prey plasmid (library, 100 to 500 μg), and denatured salmon sperm carrier DNA (50 μg/tube) are mixed, aliquotted to the tubes containing competent cells for transformation, and plated on 150 mm dishes as described above.

Always plate an aliquot of transformed cells on a 100 mm dish (SD lacking Leu and Trp only) and count surviving colonies to estimate the total number of yeast colonies screened. The quantities of plasmid DNA to use depends on the scale of the screening (number of plates), the efficiency of transformation (single- or cotransformation), and the availability of library plasmid DNA.

Screening Procedure

A first $G\alpha_{i3}$ screening is performed by cotransformation with $G\alpha_{i3}$ bait vector (100 μg total) and HeLa library vector (200 μg total). A total of 12 dishes (150 mm) lacking Leu, Trp, and His are plated with 250 μl of transformed cells per dish. A total of $\sim 4 \times 10^5$ colonies are plated. Results of this screening are summarized in Table I.

In a second screening a $G\alpha_{i3}$ bait strain is obtained first by transformation with bait plasmid pGBT9-$G\alpha_{i3}$. The bait strain is then transformed with the rat pituitary cDNA library in pACT2 (50 μg) and plated on 25 dishes (150 mm) lacking Leu, Trp, and His. A total of $\sim 10^6$ colonies are screened. Results of this screening are summarized in Table I.

Screening with GAIP as a bait is performed by cotransformation using 400 ug pGBT9-GAIP bait vector and 120 μg rat pituitary library vector. A total of $\sim 6 \times 10^5$ colonies are screened over 20 dishes (150 mm); the results are summarized in Table I. We always screen fewer colonies than the number of independent clones in the library we use, i.e., we perform "nonsaturated" screening procedures.

Typically transformed cells are grown for 6 to 7 days at $30°$ (these are harsh growth conditions) and surviving colonies are transferred onto duplicate plates: one is used for colony lift β-Gal assay, and the other is the master plate.

β-Gal/Colony Lift Assay Protocol

1. Surviving colonies are lifted onto Whatman filters by placing the filter directly on the agar. Make orientation marks with a needle and syringe filled with waterproof ink.
2. Permeabilize colonies by dipping filters for 7 sec in liquid nitrogen.
3. Place filter on another filter presoaked with 1–2 ml X-Gal/Z buffer solution (0.3 mg/ml X-Gal final in Z buffer: 100 mM sodium phosphate, pH 7, 10 mM KCl, 1 mM MgSO$_4$) with 30 mM 2-mercaptoethanol in a 150 mm dish. Cover the dish and incubate at room temperature.
4. Register appearance of blue colonies (LacZ activity) over at least 8 hr (24 hr if no background color is detected).
5. Giving a rank order is especially important when many positive colonies are found. Rate of color development and intensity of color are good indications for relative interaction strength.

Identification of Prey cDNA Insert. Once positive colonies have been identified by the β-Gal/colony lift assay, the prey plasmid that expressed the interacting protein must be isolated from the yeast colony for identification of its cDNA insert.

Prey Plasmid Isolation Protocol

1. Pick positive colonies from master plate; grow up overnight at 30° in 3 ml SD medium containing Trp and His but lacking Leu.
2. Take 500 μl for glycerol (20%, v/v) stocks of individual positive clones; store at −70°.
3. Pellet yeast cells in microcentrifuge tubes, resuspend in 200 μl lysis buffer (2% Triton X-100, 1% SDS, 0.1 M NaCl, 10 mM Tris-HCl, pH 8, 1 mM EDTA).
4. Add 200 μl phenol/chloroform/isoamyl alcohol (25 : 24 : 1) and 0.3 g acid washed glass beads, vortex for 2 min.
5. Spin (14,000 rpm, 5 min), ethanol precipitate the aqueous phase, resuspend yeast minipreparation DNA in 20 μl of 10 mM Tris-HCl, pH 8, and store at −20°.

At this point the DNA from each positive colony can be analyzed by PCR using pGADGH or pACT2 specific primers. The fact that a yeast cell can incorporate more than one prey vector can possibly complicate PCR analysis. Therefore we electroporate the yeast DNA into *E. coli* cells which take up only 1 plasmid. The *E. coli* HB101 strain contains a *leuB* mutation that can be complemented by the *Leu2* gene in the prey vectors, allowing for an enriched selection of cells transformed by prey vector. Electrocompetent HB101 cells can be prepared by standard protocols or are commercially available (Bio-Rad, Hercules, CA).

Transformation of Escherichia coli with Yeast DNA

1. Add 2 μl yeast miniprep DNA to 40 μl cold electrocompetent HB101, electroporate in 0.1 cm cuvettes (Bio-Rad) at 200 Ω, 25 μF, 1.7 kV. We use Bio-Rad's Gene Pulser.
2. Allow *E. coli* to recover for 60 min at 37° in LB medium (no antibiotics).
3. Pellet bacteria, resuspend in 100 μl M9 minimal medium, plate on M9 agar supplemented with 50 μg/ml ampicillin, 40 μg/ml proline, 1 mM thiamin hydrochloride lysine, adenine, and Trp (other amino acids can be added, but not Leu). Allow 2 days at 37° for growth of colonies.
4. Pick ∼6 colonies for minipreparations; grow overnight at 37° with agitation (250 rpm) in LB medium with ampicillin. Isolate plasmid DNA according to standard procedures.
5. Analyze minipreparation plasmid DNA by restriction analysis or by PCR with prey-specific oligonucleotides.
6. Classify inserts by size and restriction pattern; sequence from 5′ end with pGADGH- or pACT2-specific primer. Nucleotide sequence data are then fed

into classical homology search programs (BLAST analysis) to determine the identity of the interacting protein sequence or its relationship to other protein families or protein motifs.

Confirmation of Interactions between Bait and Prey Proteins

Sometimes the relevance of a positive interaction in the two-hybrid system is already revealed by the identity of the isolated cDNA and its relation to the function of the bait protein, but this is not always the case. Before claiming a relevant interaction of a newly isolated cDNA in the two-hybrid system, it is essential to confirm this interaction in the two-hybrid system (see below) and by using other techniques such as glutathione transferase (GST) pulldown, GST precipitation, or coimmunoprecipitation (see below).

Verification of Interaction in Two-Hybrid System

Verifications in the two-hybrid system include showing that the newly isolated prey protein and a bait protein unrelated to the bait protein used in the screening procedure do not activate the system on their own. To verify lack of self-activation of the system by the prey protein, the isolated prey plasmid is transformed into HF7c cells, plated on SD medium lacking Leu, and scored for negative LacZ activity using the β-Gal filter lift assay described above. Cotransformation of prey plasmid with an unrelated bait plasmid (we use lamin C in pGBT9) plated on SD medium lacking Leu and Trp should give no blue colonies in this assay. A positive interaction should be detected again by cotransforming bait and prey plasmid in HF7c and SFY526 cells (SD medium minus Leu and Trp). The use of yeast strains having different *GAL1* promoters for their reporter genes strengthens the positive interaction result.

Complementary Verifications

Complementary techniques often used to confirm true interaction are GST pulldown, GST precipitation or coimmunoprecipitation experiments, which range from strict *in vitro* to more *in vivo* interaction assays. These techniques are also very useful for showing GDP/GTP-dependent interactions of $G\alpha_{i3}$, for example, by adding GDP, GTPγS, or AIF_4^- to the interaction buffer or cell lysate. Showing colocalization of both (preferably endogenous) proteins in the cell is a strong argument for a relevant *in vivo* interaction, but one has to bear in mind that very transient interactions—detectable in the two-hybrid system—might not be visualized by immunocytochemistry. An *in vivo* or *in vitro* functional assay showing an effect of prey protein on the function of bait protein (or vice versa) is undoubtedly the strongest argument for a relevant interaction.

GST Pulldown Protocol on in Vitro Translated Products

1. Express one of the proteins (e.g., GAIP) as a GST-fusion protein in *Escherichia coli*. We use pGEX-KG vector (Pharmacia) and *E. coli* BL21 (DE3) for IPTG-induced expression following a standard protocol from Pharmacia.

2. Bind 1–5 μg GST-fusion protein to glutathione agarose beads.

3. Subclone the cDNA of the other interacting partner (e.g., Gα$_{i3}$) in a plasmid behind the T7 RNA polymerase promoter. We use pcDNA3 (Invitrogen, Carlsbad, CA) or pET28a (Novagen, Milwaukee, WI), which are useful for transfection experiments and histidine-tagged protein purifications on Ni^{2+}-beads, respectively.

4. Produce [^{35}S]methionine radiolabeled protein using the T7 polymerase-based *in vitro* transcription/translation TNT kit (Promega) according to the manufacturer's guidelines.

5. Add an aliquot (typically 5 μl of a 50 μl TNT reaction) of radiolabeled protein to the GST-fusion protein bound to the agarose beads. Interaction buffers vary; we used phosphate-buffered saline (PBS), pH 7.4, +0.1% Triton X-100 (250 μl) for Gα/GAIP and buffer A (50 mM HEPES, pH 7.2, 2 mM MgCl$_2$, 1 mM EDTA, 100 mM NaCl) for GAIP/GIPC. Incubate 1–2 hr at room temperature, wash 3 times with 5 volumes incubation buffer.

6. Boil beads 5 min in Laemmli buffer, separate proteins on 10% or 12% acrylamide gels, and perform autoradiography on dried gel.

Show specific binding of the radiolabeled protein to the GST-fusion protein and no binding to GST alone, as shown in Fig. 2. Requirements for binding can be established by changing the incubation buffer. For example, addition of 5 μM GTPγS or 2 mM NaF + 30 μM AlCl$_3$ (in this order, to form AlF$_4^-$) to the *in vitro* translated Gα$_{i3}$ enhances its binding to GST–GAIP significantly, showing GAIP binds Gα$_{i3}$ preferentially in its GTP-bound or its GTP to GDP hydrolysis transition state, respectively (Fig. 3).[22] In another example, addition of a C-terminal antibody to GAIP or a peptide corresponding to the C terminus of GAIP to the incubation buffer greatly reduced the GST-GAIP/GIPC interaction, showing that the C terminus of GAIP interacts with GIPC (Fig. 4).[16] In yet another example the Ca^{2+}/Mg^{2+} dependency of the Gα$_{i3}$/calnuc interaction could be shown.[23]

GST Precipitation Protocol. This protocol is basically the same as for a GST pulldown, but a cell lysate is added to the GST fusion protein bound to glutathione

[22] L. De Vries, E. Elenko, L. Hubler, T. L. Jones, and M. G. Farquhar, *Proc. Natl. Acad. Sci. U.S.A.* **93**, 15203 (1996).

[23] P. Lin, T. Fischer, T. Weiss, and M. G. Farquhar, *Proc. Natl. Acad. Sci. U.S.A.* **97**, 674 (2000).

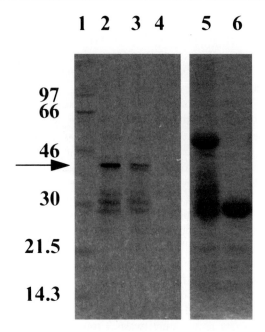

FIG. 2. Interaction of GAIP–GST with *in vitro* translated Gα_{i3}. GST-GAIP fusion protein bound to glutathione-agarose beads was incubated with ^{35}S-labeled *in vitro* translated Gα_{i3}. The bound products were separated by SDS/10% acrylamide PAGE and detected by overnight autoradiography. ^{35}S-Labeled *in vitro*-translated Gα_{i3} (arrow) binds to GST–GAIP beads (lane 3) but not to control beads with GST alone (lane 4). Lane 1, ^{14}C-labeled molecular weight markers. Lane 2, ^{35}S-labeled *in vitro* translated Gα_{i3} product. Lanes 5 and 6, Coomassie blue staining of GST–GAIP fusion protein and GST protein alone corresponding to lanes 3 and 4. (From Ref. 4.)

agarose beads. Variants include using a lysate of metabolically labeled cells or a lysate of transiently transfected cells (overexpressing the interaction partner), followed by immunoblotting of the bound fraction.

Coimmunoprecipitation Protocol

1. Subclone prey protein into a mammalian expression vector. Often no specific antiprey antiserum is available for immunoprecipitation, so include a commonly used epitope tag provided by the vector (HA, FLAG, myc, V5, . . .).
2. Transfect prey protein into a readily transfectable mammalian cell line (e.g., Cos, HeLa, HEK 293) that expresses the bait protein; harvest cells 24 hr after transfection.

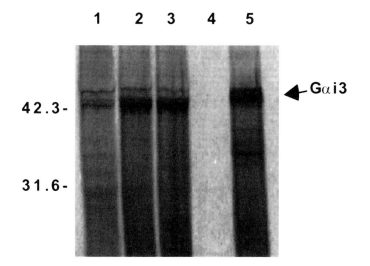

FIG. 3. GAIP interacts with the GTP bound form of $G\alpha_{i3}$ *in vitro*. GST-GAIP fusion protein bound to glutathione-agarose beads was incubated with *in vitro* translated $G\alpha_{i3}$ in the presence or absence of GTPγS and AlF$_4$⁻. The bound products were separated by 10% SDS PAGE and detected by autoradiography. ^{35}S-Labeled *in vitro* translated $G\alpha_{i3}$ binds to GST-GAIP beads 4–5 times more efficiently in the presence of GTPγS (lane 2) or AlF$_4$⁻ (lane 3) than in their absence (lane 1). Lane 4, Control beads with GST alone; Lane 5, ^{35}S-Labeled *in vitro* translated $G\alpha_{i3}$ (arrow). Molecular mass markers (kDa) are indicated on the left. (From Ref. 22.)

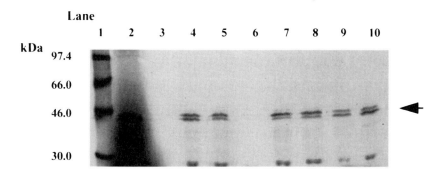

FIG. 4. GIPC interacts with the C terminus of GAIP. GST–GAIP fusion protein bound to glutathione-agarose beads was incubated with *in vitro* translated GIPC (lane 4). Addition of 30 μg (lane 5) or 300 μg (lane 6) anti-GAIP C-terminal IgG reduced the binding of GIPC to GAIP, whereas 300 μg anti-CALNUC antibody (lane 7) had minimal effect. When 10 μM (lane 8) or 1 mM (lane 9) GAIP C-terminal peptide were added, GIPC's binding to GAIP was also reduced 16% and 43%, respectively. A control peptide (1 mM, lane 10) had little effect on binding. Binding to GST alone (lane 3) was taken as background, and the signal obtained after binding of GIPC to GST-GAIP (lane 4) (~42 kDa doublet, with background substracted) was defined as 100% in arbitrary units. Lane 1, molecular weight markers. Lane 2, ^{35}S-labeled *in vitro* translated GIPC product. (From Ref. 16.)

3. Wash cells with PBS, lyse at 4° in RIPA buffer [0.1% (w/v) SDS, 1% Nonidet P-40 (NP-40), 0.5% (w/v) deoxycholate, 150 mM NaCl, 50 mM Tris, pH 8.0] (stringent) or 1% (v/v) NP-40 buffer or 1% (w/v) Triton X-100 (mild) containing protease inhibitors; solubilize by rocking for 30 min at 4°. Addition of AlF$_4^-$ or GTPγS to the lysate will enhance Gα/RGS interaction.

4. Preclear with protein A or G beads, add antitag antibody (1:100) plus protein A or G beads, wash 4× with lysis buffer.

5. Boil beads 5 min in Laemmli buffer, separate proteins on 10% acrylamide gel, and transfer to polyvinyl difluoride (PVDF) membrane for immunoblotting with antibait protein antiserum.

Although we have not used GST precipitation or coimmunoprecipitation for confirmation of the Gα_{i3}/GAIP interaction, we have successfully used this methodology for other Gα interacting proteins such as GIPC, calnuc, and RGS16.[5,16,17]

One-to-One Interactions in Two-Hybrid System

Liquid β-Gal Assay: Measuring Relative Strength of Interaction. This semiquantitative assay allows one to express numerically the relative strength of interaction of two proteins or protein domains compared to a standard defined as 100% interaction strength (see Tables II and III). This method is by no means a way

TABLE II
INTERACTION OF GAIP WITH DIFFERENT Gα SUBUNITS[a]

Bait	β-Gal Filter	β-Gal Liquid (%)
Gα_{i1}	+++	55
Gα_{i2}	+	6.3
Gα_{i3}	+++	100
Gα_o	+++	52
Gα_z	+	5.0
Gα_s	---	<1
Gα_s(Q227L)	---	<1
Gα_q	---	<1
Gα_{13}	---	<1

[a] The β-Gal filter assay was performed on (Leu$^-$, Trp$^-$) plates, and intensity of color was scored after 8 hr. ---, No color; +, weak color; +++, strong color. For the β-Gal liquid assay (Leu$^-$, Trp$^-$), the Gα_{i3}/GAIP was taken as 100%. Yeast cotransformed with void bait and prey vectors were taken as background. Baits were constructed in pGBT9, and GAIP prey vector was pACT2-rGAIP. For each experiment three colonies were picked. Values represent the mean of three independent experiments. (From Ref. 22.)

TABLE III
RGS DOMAIN OF GAIP: INTERACTION WITH $G\alpha_{i3}$

Bait	Prey	β-Gal filter[a]	β-Gal liquid[b]
GAIP 22-217	$G\alpha_{i3}$	+++	100%
GAIP 1-217	$G\alpha_{i3}$	++	57%
GAIP 1-79	$G\alpha_{i3}$	− − −	<0.5%
GAIP 79-206	$G\alpha_{i3}$	+++	92%

[a] The β-Gal filter assay was performed in HF7c strain on (Leu$^-$, TrP$^-$) plates and intensity of color was scored after 15 hr. −, No color; +, weak color; ++, moderate color; +++, strong color.

[b] Performed in SFY526 yeast strain. (From Ref. 4.)

to estimate K_d values of interaction. Strain SFY526 is better suited for this assay than HF7c because it has a stronger promoter for the LacZ reporter gene. Care must be taken to cotransform yeast cells with exactly the same amounts of bait and prey vectors for the relative comparison to be valid. Ideally, equal expression of proteins could be followed by immunoblotting the cotransformed yeast lysates. However, all bait or prey proteins of a series should contain the same epitope tag to avoid variation due to antibody affinities.

Protocol

1. Cotransform SFY526 cells with 0.2 μg bait and prey plasmid (see above), and plate on SD medium lacking Leu and Trp. Use empty vectors as negative control.
2. Grow 3 colonies overnight at 30° in 3 ml selective medium. Submit the other colonies (20–100) to a β-Gal colony lift filter assay as control.
3. Inoculate 2 ml of the overnight culture into 6 ml YPD; grow ~3 hr at 30° until $OD_{600} > 0.7$.
4. Remove triplicate 1 ml aliquots per colony; measure exact OD_{600}.
5. Pellet cells, wash with 1 ml Z buffer (see β-Gal/colony lift assay protocol for composition), resuspend in 200 μl Z buffer (concentration factor 5).
6. Freeze–thaw 100 μl cells in liquid nitrogen/37°. For strong interactions a greater dilution of cells (10–100 μl in 100 μl final) can be used, as it is important to stay within the linear range of the assay (OD_{420} readings between 0.02 and 1.0; see below). Include a blank with Z buffer only.
7. Add 0.7 ml Z buffer supplemented with 30 mM 2-mercaptoethanol; add 160 μl ONPG (at 4 mg/ml in Z buffer); start timed colorimetric reaction (yellow) by incubating at 30°, typically 3–8 hr.
8. Stop reaction by adding 400 μl 1 M Na$_2$CO$_3$; centrifuge (14,000 rpm, 2 min) at room temperature; transfer supernatant to fresh tube.

9. Read absorbance (420 nm) in spectrophotometer; calibrate against blank tube.

10. Determine β-galactosidase units according to the formula:

$$U = \frac{1000OD_{420}}{t \, V OD_{600}}$$

where t is reaction time in minutes and V is 0.1 ml × concentration factor. For 1 experiment, the mean of 9 points (3 colonies × triplicate reading) per one-to-one interaction is made to score for relative strengths of interaction.

GDP/GTP-Dependent Interactions

The two-hybrid system can also be used to determine preferential interaction of a protein with the activated (GTP bound) or inactivated (GDP bound) state of the Gα subunit. For Gα_{i3}, the G203A mutant mimics the GDP bound state because it cannot dissociate from $\beta\gamma$ subunits, whereas the Q204L mutant mimics the GTP bound state because of its abolished GTPase activity. The equivalent mutants also exist for all other Gα subunits. Both mutants are subcloned into the pGBT9 bait vector and positively interacting proteins from the screening procedure are tested in a one-to-one interaction protocol with these Gα mutants. For example GAIP, which is a GAP and thus interacts preferentially with activated Gα_i subunits, displays a net preference for interaction with the Q204L mutant vs the G203A mutant. Another Gα_{i3} interacting protein, calnuc, or nucleobindin, does not show a preference between these mutants, an observation that was confirmed by *in vitro* pulldown assays.[5]

Gα Subtype-Dependent Interactions

Interaction specificity of a prey protein for different Gα subunits can also be analyzed qualitatively and semiquantitatively, provided all Gα subunits are expressed from the same bait-type plasmid. For example, GAIP was found to interact more strongly with Gα_{i3} and Gα_{i1} than with Gα_{i2} (Table II), a result that was later confirmed and shown due to a difference in a single amino acid (D229 in Gα_{i1} and Gα_{i3}, A230 in Gα_{i2}).[24] So far all proteins isolated by interaction with a Gα_i subunit seem to interact exclusively with members inside this subfamily with one exception: Calnuc/nucleobindin interacts with Gαs.[5] It was noted, however, that Gαs-expressing yeast colonies were significantly smaller and showed slower growth kinetics on selective medium, which may explain why very few positive interactions with Gαs in the two-hybrid system have been described up to now.

---

[24] D. S. Woulfe and J. M. Stadel, *J. Biol. Chem.* **274**, 17718 (1999).

Prey Protein Subdomain-Dependent Interactions

The two-hybrid system is also very useful to determine the regions of prey proteins that interact with the Gα subunit. Regions or motifs of high homology in a protein family, for which one of the members shows up positively in a Gα two-hybrid screen, are obvious candidates for interaction domains. Finding a positive one-to-one interaction of the isolated region/motif with Gα in the two-hybrid system defines it as a Gα-interacting domain, and producing further deletions or mutations within it may help delineate the minimal requirements for interaction. Generally this delineation is first performed using the β-Gal filter assay, and a liquid β-Gal assay is usually not necessary. This is how RGS domains[4] were initially discovered (see Table III). Often the correct folding or the borders of a domain in a newly isolated protein are unknown; hence some trial and error in making two-hybrid constructs (in the prey plasmid) of the putative interaction domain should be taken into account.

The reverse approach, determining the site of interaction within the Gα subunit, has been less useful because several deletion mutants have been shown to misfold. Preliminary experiments to show proper folding of Gα deletion mutants are strongly advised before using them in two-hybrid assays.

Acknowledgments

This work was supported by the National Institutes of Health Grants CA58689 and DK17780 to M.G.F.

[47] Assay of RGS Protein Activity *in Vitro* Using Purified Components

By ANDREJS M. KRUMINS and ALFRED G. GILMAN

G protein α subunits, found as Gαβγ heterotrimers *in vivo,* cycle between active (Gα–GTP) and inactive (Gα–GDP) forms. The duration of G-protein-mediated signaling is directly related to the time that Gα remains bound to GTP. GTPase-activating proteins (GAPs) can thus act as negative regulators of these signaling systems. Measurement of the GTPase activity of the α subunits of heterotrimeric G proteins provides a fundamental method for examination of the interaction between these signaling molecules and members of the recently discovered family of regulators of G protein signaling (RGS) proteins.[1] RGS proteins act as

[1] E. R. Ross and T. M. Wilkie, *Annu. Rev. Biochem.* **69,** 795 (2000).

GAPs, augmenting the intrinsic GTPase activity of selected G protein α subunits.[2] Definition of this selectivity is critical for understanding the regulatory role played by RGS proteins.

Because the steady-state GTPase activity of an isolated G protein α subunit is limited by the rate of release of one of the reaction products, GDP, effects of RGS proteins to accelerate intrinsic GTPase catalytic activity are invisible in this setting. Instead, single-turnover assays (measurement of the rate of one round of conversion of bound substrate, GTP, to GDP and P_i) may be performed to assess catalytic capacity. However, if GTPase activity is measured in the presence of a nucleotide exchanger, such as a G-protein-coupled receptor, the rate of dissociation of product from the GTPase may be accelerated sufficiently that steady-state GTPase assays will reveal the activity of a GAP, if present.

This chapter describes simple *in vitro* methods for examination of the interactions between purified Gα subunit and RGS proteins, both in solution and in reconstituted proteoliposomes. Furthermore, the section describing the single-turnover GTPase assay also includes methods for determining Michaelis–Menten constants, fractional RGS activity, and the apparent affinity (K_I) of competitor Gα subunits toward the RGS protein in question. Methods for purification of Gα subunits are described elsewhere (G$_s\alpha$/G$_i\alpha$/G$_o\alpha$,[3] G$_q\alpha$/G$_{11}\alpha$,[4] G$_z\alpha$,[5] and G$_{12}\alpha$/G$_{13}\alpha$[6]) as are techniques for purification of some RGS proteins (RGS1,[7] RGS2,[8] RGS4,[2] RGS6,[9] RGS7,[9] RGS9,[10] RGS10,[11] RGS11,[12] RGS14,[13] RGS16,[14] GAIP,[2] G$_z$-GAP[15]). A comprehensive review of this subject has also been published by Wang *et al.*[16]

[2] D. M. Berman, T. M. Wilkie, and A. G. Gilman, *Cell* **86,** 445 (1996).

[3] E. Lee, M. E. Linder, and A. G. Gilman, *Methods Enzymol.* **237,** 146 (1994).

[4] G. H. Biddlecome, G. Berstein, and E. M. Ross, *J. Biol. Chem.* **271,** 7999 (1996).

[5] P. J. Casey, H. K. Fong, M. I. Simon, and A. G. Gilman, *J. Biol. Chem.* **265,** 2383 (1990).

[6] T. Kozasa, X. Jiang, M. J. Hart, P. M. Sternweis, W. D. Singer, A. G. Gilman, G. Bollag, and P. C. Sternweis, *Science* **280,** 2109 (1998).

[7] N. Watson, M. E. Linder, K. M. Druey, J. H. Kehrl, and K. J. Blumer, *Nature* **383,** 172 (1996).

[8] T. Ingi, A. M. Krumins, P. Chidiac, G. M. Brothers, S. Chung, B. E. Snow, C. A. Barnes, A. A. Lanahan, D. P. Siderovski, E. M. Ross, A. G. Gilman, and P. F. Worley, *J. Neurosci.* **18,** 7178 (1998).

[9] B. A. Posner, A. G. Gilman, and B. A. Harris, *J. Biol. Chem.* **274,** 31087 (1999).

[10] W. He, L. Lu, X. Zhang, H. M. El-Hodiri, C. K. Chen, K. C. Slep, M. I. Simon, M. Jamrich, and T. G. Wensel, *J. Biol. Chem.* **275,** 37093 (2000).

[11] T. W. Hunt, T. A. Fields, P. J. Casey, and E. G. Peralta, *Nature* **383,** 175 (1996).

[12] B. E. Snow, A. M. Krumins, G. M. Brothers, S. F. Lee, M. A. Wall, S. Chung, J. Mangion, S. Arya, A. G. Gilman, and D. P. Siderovski, *Proc. Natl. Acad. Sci. U.S.A.* **95,** 13307 (1998).

[13] C. Hyeseon, T. Kozasa, K. Takekoshi, J. D. Gunzburg, and J. H. Kehrl, *Mol. Pharm.* **58,** 569 (2000).

[14] X. Xu, W. Zheng, S. Popov, D. M. Berman, I. Davignon, K. Yu, D. Yowe, S. Offermanns, S. Muallem, and T. M. Wilkie, *J. Biol. Chem.* **274,** 3549 (1999).

[15] J. Wang, A. Ducret, Y. Tu, T. Kozasa, R. Aebersold, and E. M. Ross, *J. Biol. Chem.* **273,** 26014 (1998).

[16] J. Wang, Y. Tu, S. Mukhopadhyay, P. Chidiac, G. H. Biddlecome, and E. M. Ross, *in* "G Proteins Techniques of Analysis" (D. R. Manning, ed.), p. 123. CRC Press LLC, Boca Raton, FL, 1999.

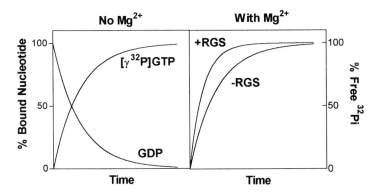

FIG. 1. Simulation of guanine nucleotide binding to a Gα subunit (left) and hypothetical RGS protein-dependent single-turnover of Gα–GTP (right). Gα subunits cycle between an inactive (Gα–GDP bound) form and an active (Gα–GTP bound) form. Inactivation of Gα–GTP occurs via Mg^{2+}-dependent hydrolysis of GTP, resulting in production of Gα-GDP and release of free P_i. Single-turnover assays exploit the Mg^{2+} dependence for GTP hydrolysis. *Left:* In the absence of Mg^{2+} and in the presence of saturating [GTP], Gα–GTP begins to accumulate in a first-order process that is dependent on the rate of GDP dissociation. *Right:* GTPase activity is initiated with the addition of Mg^{2+}. GTP hydrolysis occurs in a first-order manner, and activity is monitored via the release of free P_i. The effects of RGS protein activity are measured as increases in the rate of Gα–GTP hydrolysis. The plots were generated using single exponential equations and applying arbitrary rates for nucleotide binding and GTPase activity.

Overview of Single-Turnover Gα Protein GTPase Activity

Single-turnover GTPase assays may be conducted either spectroscopically, by monitoring nucleotide-dependent effects on the intrinsic tryptophan fluorescence of Gα subunits,[17,18] or by quantification of inorganic phosphate (P_i) released during the hydrolysis of Gα–GTP to Gα–GDP.[19] The intrinsic tryptophan fluorescence of Gα proteins is enhanced by binding of Mg^{2+}/GTP and decays when GTP is hydrolyzed to GDP. Measurement of this parameter offers high temporal resolution, but the signal-to-noise ratio is low if the proteins present in the assay contain a large number of tryptophan residues. Measurement of P_i production, usually as $[^{32}P]P_i$, represents a simple and rapid alternative.

Single-turnover assays are conducted in two steps (Fig. 1) and take advantage of the fact that Mg^{2+} is required for nucleotide hydrolysis but not for nucleotide binding. GTP is first allowed to bind to purified Gα–GDP protein in the absence of Mg^{2+}. Although dissociation of GDP from Gα may be very slow,

[17] T. Higashijima, K. M. Ferguson, P. C. Sternweis, E. M. Ross, M. D. Smigel, and A. G. Gilman, *J. Biol. Chem.* **262,** 752 (1987).

[18] K.-L. Lan, H. Zhong, M. Nanamori, and R. R. Neubig, *J. Biol. Chem.* **275,** 33497 (2000).

[19] D. Cassel and Z. Selinger, *Biochem. Biophys. Acta* **452,** 538 (1976).

Gα–GTP accumulates in the presence of a high relative concentration of GTP. Excess unbound GTP and any free P$_i$ are then removed quickly, and GTP hydrolysis is initiated by addition of Mg^{2+}. Hydrolysis occurs in a first-order fashion at a rate determined by the intrinsic GTPase activity of the Gα protein in question. Effects of RGS proteins are measured as increases in the rate of GTP hydrolysis.

Binding of [^{35}S]GTPγS to Purified Gα Subunits

Gα–GDP proteins will bind GTP under a variety of conditions, although the choice of conditions (e.g., ionic strength and detergent) may affect the rate and/or maximal extent of binding. The conditions for binding of GTP should be optimized using [^{35}S]GTPγS, a nonhydrolyzable analog of GTP. [^{35}S]GTPγS binding is also used to quantify the amount of active Gα protein in any given preparation.

Binding of [^{35}S]GTPγS (PerkinElmer Life Sciences, Boston, MA) to Gα subunits is initiated by addition of 150 μl of GTPγS loading buffer [12 μM GTPγS (5000 cpm/pmol), 50 mM Na–HEPES (pH 8.0), 10 mM MgCl$_2$, 5 mM EDTA, 1 mM dithiothreitol (DTT), 0.02 mg/ml bovine serum albumin (BSA), 5% (v/v) glycerol, and 0.05% (v/v) polyoxyethylene 10 lauryl ether (C12E10)] to 30 μl of purified Gα–GDP subunit diluted in the same buffer in the absence of GTPγS. Aliquots (10 μl) of the reaction mixture are removed at specified times and diluted into 500 μl of ice-cold wash buffer (20 mM Tris-HCl (pH 7.7), 10 mM MgCl$_2$, and 100 mM NaCl), followed by immediate vacuum filtration onto BA-85 nitrocellulose filters. Filters are washed four times with 2 ml of wash buffer, dried, placed into 5 ml of scintillation fluid, and counted. Dissociation of bound GTPγS from Gα proteins is exceedingly slow in the presence of Mg^{2+}.

For performance of GTPase assays (see below), [γ^{32}P]GTP (PerkinElmer Life Sciences, Boston, MA) replaces GTPγS, and MgCl$_2$ is omitted from the reaction mixture until hydrolysis of GTP is initiated. Since nanomolar concentrations of free Mg^{2+} are sufficient to support GTPase activity, it is critical to maintain a high concentration of EDTA.

HPLC Purification of [γ^{32}P]GTP

GTPase assays monitor the production of free [^{32}P]P$_i$, and it is thus critically important that the amount of free [^{32}P]P$_i$ added to the reaction mixture be low. Commercial preparations of [γ-^{32}P]GTP generally contain a significant ($>1\%$) amount of free [^{32}P]P$_i$; HPLC purification of [γ-^{32}P]GTP will reduce free [^{32}P]P$_i$ to $\sim$0.1%.

Purification is accomplished with a Synchropak Q-300 6.5 U strong anion-exchange column (Alltech Assoc.) Labeled GTP (1–3 mCi) is applied to the column equilibrated in 70% H_2O and 30% 1 M KPO_4, pH 7.2; the flow rate is set to 0.5 ml/min. GTP is purified by increasing the percentage of phosphate buffer in the eluate to 55% in 20 min. The absorbance (OD_{253}) of the eluate is measured, and 400-μl fractions are collected in tubes containing 10 μl of 1 M Tricine (pH 8.0). Free $[^{32}P]P_i$ elutes in the void, GMP at $\sim$4 ml, GDP at $\sim$7 ml, and GTP at $\sim$12 ml. The three peak fractions of GTP are pooled, diluted to 2–4 $\times$ 10^6 cpm/μl in 20 mM Tricine (pH 8.0), divided into aliquots, and stored at $-80°$. Recovery approximates 50%.

$[\gamma-^{32}P]$GTP Binding to $G_i\alpha$, $G_o\alpha$, $G_s\alpha$, $G_{12}\alpha$, $G_{13}\alpha$, and $G_z\alpha$

The concentration of $G\alpha$–GDP used during the GTP-binding phase of the GTPase assay may vary between 50 and 13,000 nM, depending on availability of protein and the amount of $G\alpha$–GTP substrate required to quantify RGS/$G\alpha$–GTP interactions. The specific activity of $[\gamma-^{32}P]$GTP used during the binding reaction is in turn determined by the initial concentration of $G\alpha$–GDP, the affinity of the $G\alpha$ protein for GDP, and the rate of GTP hydrolysis for a particular $G\alpha$ protein. When high concentrations of $G\alpha$ are used the concentration of GTP should be increased to maintain a $\sim$4- to 5-fold excess over $G\alpha$–GDP. The specific activity of $[\gamma-^{32}P]$GTP should be $\sim$5 cpm/fmol for proteins with relatively low affinity for GDP, such as $G_o\alpha$ and $G_s\alpha$. These proteins can be loaded efficiently with $[\gamma-^{32}P]$GTP. For $G_i\alpha$, $G_z\alpha$, $G_{12}\alpha$, and $G_{13}\alpha$ (proteins with relatively high affinity for GDP), the specific activity should be increased to $\sim$100 cpm/fmol or more. More efficient loading of $G_z\alpha$ is also achieved using 0.1% (v/v) Triton X-100 in lieu of C12E10.

The incubation times and temperatures required to achieve a level of $[\gamma-^{32}P]$GTP binding sufficient to ensure an adequate signal from GTP hydrolysis are determined by the protein's affinity for GDP and its intrinsic GTPase activity, as well as its thermal stability. Equilibrium binding of GTP is achieved in 20 min at $20°$ with $G_s\alpha$ and $G_o\alpha$. Considerably longer times ($\sim$2 hr) are required for $G_i\alpha$, $G_z\alpha$, $G_{12}\alpha$, and $G_{13}\alpha$. However, incubations are carried out for 30 min at $30°$ to prevent thermal denaturation. The total amount of label that is incorporated into active $G\alpha$ protein thus varies between 10 and 80%. Following binding of $[\gamma-^{32}P]$GTP, samples are transferred to ice to lower the rates of $[\gamma-^{32}P]$GTP dissociation and GTP hydrolysis that will occur in the presence of trace concentrations of Mg^{2+}.

$[\gamma-^{32}P]$GTP Binding to $G_q\alpha$ and $G_t\alpha$

$G_q\alpha$ and $G_t\alpha$ have very high affinity for GDP and relatively rapid GTPase activity; these properties interfere with efficient accumulation of $G\alpha$–GTP during the binding phase of the GTPase assay.[20,21] There are at least two ways to improve this situation. The first involves inclusion of ammonium sulfate in the binding buffer.[20] Ammonium sulfate facilitates dissociation of GDP from $G\alpha$ proteins.[22] At concentrations of $(NH_4)_2SO_4$ below 100 mM, the rate of GTP hydrolysis exceeds the rate of GTP binding and there is little benefit. Concentrations of $(NH_4)_2SO_4$ above 100 mM increase the rate of GDP dissociation, permitting accumulation of $G\alpha$–GTP. However, higher concentrations of $(NH_4)_2SO_4$ also promote denaturation of $G\alpha$ proteins. The second method involves use of a catalytically impaired $G\alpha$ subunit mutant (e.g., $G_q\alpha$ R183C).[8] Mutation of the catalytic arginine residue dramatically reduces the rate of intrinsic GTP hydrolysis without altering affinity for GDP.[23,24] This facilitates accumulation of $G\alpha$–GTP. RGS proteins can still stimulate the GTPase activity of this mutant.[2]

In practice, 40 mM $(NH_4)_2SO_4$ and 0.9 mM $MgSO_4$ are added to the labeled GTP binding mix to decrease GDP affinity and increase GTP affinity, respectively, when a catalytic $G\alpha$ mutant is used. Because the rate of GTP hydrolysis for the catalytic $G\alpha$ mutants is drastically reduced compared to the GTP hydrolytic rates for wild-type $G\alpha$ proteins, the small relative increases in the rate of GDP dissociation caused by 40 mM $(NH_4)_2SO_4$ permits significant catalytic mutant $G\alpha$–GTP accumulation. Moreover, while the $G\alpha$–GTP mutants are kept on ice, no significant GTP hydrolysis results in the presence of micromolar concentrations of Mg^{2+}.[16] Incubation times for $G_q\alpha$ R183C and $G_t\alpha$ R174C mutants are increased to 2–3 hr at 20°. The total amount of label that is incorporated varies from 10 to 20%.

Removal of [^{32}P]P$_i$ Released during GTP Binding

Some hydrolysis of GTP will occur during the GTP binding reaction, despite the presence of EDTA. This may increase the background of the assay to intolerable levels, particularly with $G\alpha$ proteins that release GDP slowly. We recommend removal of free [^{32}P]P$_i$ at the end of the binding reaction by the use of centrifugal gel filtration spin columns. Gel filtration spin columns (Poly-Prep chromatography columns 8 × 40 mm, Bio-Rad, Hercules, CA) are prepared using a 1-ml bed volume of Sephadex G-25 or G-50 superfine resin (Pharmacia, Piscataway, NJ) prepared and stored as a 1 : 1 slurry in H$_2$O. Columns are washed with 10 volumes of buffer [50 mM Na–HEPES (pH 8.0), 5 mM EDTA, 1 mM DTT, 0.02 mg/ml BSA, and 0.05% C12E10] at room temperature and subsequently chilled to 4°

[20] P. Chidiac, V. S. Markin, and E. M. Ross, *Biochem Pharm.* **58**, 39 (1999).

[21] L. Ramdas, R. M. Disher, and T. G. Wensel, *Biochemistry* **30**, 11637 (1991).

[22] K. M. Ferguson, T. Higashijima, M. D. Smigel, and A. G. Gilman, *J. Biol. Chem.* **261**, 7393 (1986).

[23] M. Freissmuth and A. G. Gilman, *J. Biol. Chem.* **264**, 21907 (1989).

[24] C. A. Landis, S. B. Masters, A. Spada, A. M. Pace, H. R. Bourne, and L. Vallar, *Nature* **340**, 692 (1989).

for 1 hr. Excess buffer in the column is removed by centrifugation at 200g for 3 min in a Beckman TJ6 centrifuge. The resin should appear dry but not cracked. A capless 1.5 ml Eppendorf collection tube is placed under the column. Immediately prior to gel filtration and in departure from the [^{35}S]GTPγS binding assay, 37.5 μl of column wash buffer is added to the 150-μl [γ-^{32}P]GTP binding reaction, and 175 μl of the diluted reaction mixture is applied to the center of the resin. The spin column is centrifuged for 3 min at 1000 rpm. The void volume (~150 μl) containing Gα–GTP is diluted 4-fold in ice-cold buffer in the absence of Mg^{2+} [20 mM Na–HEPES (pH 8.0), 80 mM NaCl, 5 mM EDTA, 1 mM DTT, 0.1 mM GTP, 0.01 mg/ml BSA, and 0.05% C12E10] and stored on ice for use in the GTPase reaction. When concentrations of Gα proteins or the specific activity of GTP are higher than normal, it may be necessary to increase the resin bed volume and the centrifugation time to ~3 ml and 5 min at 1000 rpm, respectively, to ensure that the P$_i$ and unbound GTP are retained on the column.

Single-Turnover GTPase Reaction

The single-turnover GTPase reaction is initiated by addition of 50 μl of diluted Gα–GTP protein to 650 μl of reaction mixture [20 mM Na–HEPES (pH 8.0), 80 mM NaCl, 5 mM EDTA, 1 mM DTT, 7.5 mM MgCl$_2$, 0.1 mM GTP, 0.01 mg/ml BSA, and 0.05% C12E10] in 12 × 75 mm polypropylene tubes in the presence or absence of a purified RGS protein. Unlabeled GTP is included to minimize a second cycle of hydrolysis of any free [γ-^{32}P]GTP that may remain. Simple qualitative effects of RGS protein activity can be assessed by its inclusion at saturating concentrations (1–10 μM). Aliquots (25 μl) of the reaction are removed at specified times and are immediately quenched in 975 μl of ice-cold 5% Norit SX 2 charcoal (Norit Americas Inc., Marshall, TX) suspended in 50 mM NaH$_2$PO$_4$. The charcoal is prepared by thorough washing in 50 mM NaH$_2$PO$_4$ to remove the fines. The acidic charcoal suspension quenches GTPase activity by denaturing Gα–GTP protein and by adsorbing nucleotides, while allowing the free P$_i$ to remain in solution. The charcoal is removed by centrifugation for 10 min at 3000 rpm in a Beckman J6 centrifuge (Beckman Coulter, Fullerton, CA). The supernatant (600 μl) is removed carefully without disturbing the charcoal pellet, transferred to scintillation vials containing 5 ml of scintillation fluid, and counted. Aliquots (~2 μl) of the GTP binding mix and diluted Gα–GTP are also counted to determine the specific activity of [γ-^{32}P]GTP and the amount of [γ-^{32}P]GTP bound to Gα subunits. Radioactivity added to the assay as bound [γ-^{32}P]GTP should be recovered as [^{32}P]P$_i$ at the conclusion of the single-turnover GTPase reaction. Zero points are taken by removing an aliquot of the diluted Gα–GTP and adding it directly into the 5% charcoal suspension prior to addition of the Mg^{2+}-containing reaction mixture.

The RGS-independent rate of GTP hydrolysis varies for each $G\alpha$ subunit. For wild-type $G_i\alpha$, $G_o\alpha$, and $G_s\alpha$, the rate varies between $\sim$2 and 4 min^{-1} at 20°.[25,26] This means that 10% of the reaction is complete in 1–2 sec and 50% of the reaction is complete in 10–20 sec. The rate of the GTPase reaction for $G_i\alpha$, $G_o\alpha$, and $G_s\alpha$ can be slowed to $\leq$1 min^{-1} when the temperature is reduced to 4° (Fig. 2). In contrast to the relatively rapid GTPase reactions catalyzed by $G_i\alpha$, $G_o\alpha$, and $G_s\alpha$, the rates of GTP hydrolysis for $G_z\alpha$ (0.05 min^{-1} at 30°),[5] $G_{12}\alpha$ (0.07 min^{-1} at 15°),[6] and $G_{13}\alpha$ (0.024 min^{-1} at 15°)[6] are relatively slow. The $G_q\alpha$ and $G_t\alpha$ proteins containing a mutation in the catalytic arginine residue display extremely slow rates of GTP hydrolysis ($\sim$0.005 min^{-1} at 20°).[8] Reactions catalyzed by these proteins will not go to completion even over the course of many hours; in these cases the amount of $G\alpha$–GTP added into a GTPase reaction can be determined by BA-85 filter binding.

RGS and $G\alpha$ Subunit Interactions

Monitoring the rate of GTP hydrolysis in the presence of excess RGS protein ($\sim$1–10 μM) can yield a simple qualitative assessment of GAP activity toward the chosen $G\alpha$–GTP protein. However, single-turnover GTPase assays can be used to generate Michaelis–Menten constants, to determine quantitatively the fraction of active protein in a preparation of RGS protein, or to examine the apparent affinity (K_I) of competing $G\alpha$ subunits toward an RGS protein.

Michaelis–Menten Constants

Michaelis–Menten constants (K_m and V_{max}) have been generated for RGS/$G\alpha$–GTP interactions by assuming that $G\alpha$–GTP represents the substrate and RGS protein the enzyme in a simple unimolecular reaction.[27,28] For $G\alpha$–GTP/RGS interactions where the K_m is low ($\sim$1–10 nM), determination of the Michaelis–Menten constants does not pose a problem because $G\alpha$–GTP substrate can readily be produced and used in the 0.1 nM–10 μM concentration range. The assay is conducted by measuring the RGS protein-dependent initial GTPase activity (k_{gap}) over a range of 0.1 nM to 10 μM $G\alpha$–GTP substrate while maintaining a constant enzyme (RGS protein) concentration. The initial GTPase rate is defined as the linear rate of hydrolysis (P_i/min) when less than 10% of the $G\alpha$–GTP substrate has been utilized, and RGS protein-dependent GTPase activity (k_{gap}) is defined as the rate of initial GTPase activity in the presence of RGS protein less the basal GTPase activity [$k_{gap} = k_{(G\alpha\ basal+RGS)} - k_{(G\alpha\ basal-RGS)}$]. A reasonable RGS

[25] M. E. Linder, D. A. Ewald, R. J. Miller, and A. G. Gilman, *J. Biol. Chem.* **265**, 8243 (1990).

[26] M. P. Graziano, M. Freissmuth, and A. G. Gilman, *J. Biol. Chem.* **264**, 409 (1989).

[27] D. M. Berman, T. Kozasa, and A. G. Gilman, *J. Biol. Chem.* **271**, 27209 (1996).

[28] B. A. Posner, S. Mukhopadhyay, J. J. Tesmer, A. G. Gilman, and E. M. Ross, *Biochemistry* **38**, 7773 (1999).

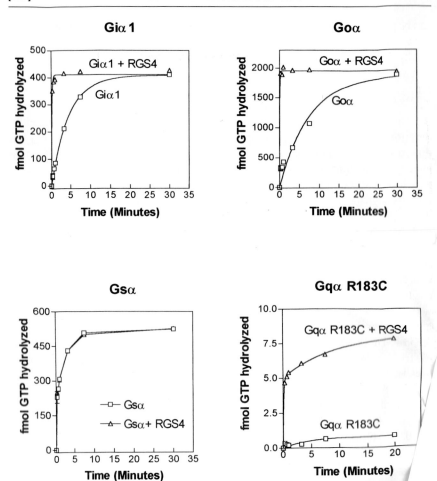

FIG. 2. Single-turnover GTPase assays. Comparison of the effects of RGS4 on the GTPase of $G_i\alpha$, $G_o\alpha$, $G_s\alpha$, and $G_q\alpha$ R183C. $G\alpha$–GTP substrates were prepared and single-turnover assays performed as described in the text. The RGS4-independent intrinsic GTPase activities (□) of $G_i\alpha$–GTP (0.2 min^{-1} at 4°), 20 nM $G_o\alpha$–GTP (0.1 min^{-1} at 4°), and 24 nM $G_q\alpha$ R183C (0.0 at 20°) were increased in the presence (△) of 100 nM RGS4. RGS4 did not accelerate the activity of 5.3 nM $G_s\alpha$–GTP (1.0 min^{-1} at 4°). Each point on the curve represents the average range of duplicate measurements.

protein concentration is one that increases the initial GTPase activity of $G\alpha$–GTP twofold. Fitting k_{gap} and [$G\alpha$–GTP] to the Michaelis–Menten will provide values for K_m and V_{max}. For RGS/$G\alpha$–GTP interactions is >100 nM, it is more difficult to determine Michaelis–Menten con of higher backgrounds and the limitations on the amount of $G\alpha$–GTP

can be produced. Nevertheless, Michaelis–Menten constants have been determined as described using k_{gap} values and a derivation of the Briggs–Haldane equation for steady-state enzyme activity.[16]

Determining Fractionally Active RGS Protein Pool

The fraction of active RGS protein in a purified preparation is determined by monitoring the effect of increasing amounts of an RGS "inhibitor" on k_{gap}. In this case the "inhibitor" is a second pool of Gα-GDP (same or different Gα) that is activated with aluminum tetrafluoride (AlF_4^-).[29] $Gα–GDP–AlF_4^-$ mimics the transition-state conformation of the protein[30] and forms a relatively stable complex with its cognate RGS protein.[27] As the concentration of the competing Gα–GDP–AlF_4^- protein increases, the effective concentration of the RGS protein and thus its GAP effect should decrease until RGS protein-dependent effects on GTPase activity are eliminated. A plot of k_{gap} vs the concentration of Gα–GDP–AlF_4^- should resemble intersecting lines with the inflection point equaling the active concentration of RGS protein.[31]

In practice, the assay resembles the single turnover assay described above with one significant variation.[27,28] The RGS protein (used at a concentration that increases basal GTPase activity twofold) is first incubated with increasing amounts of Gα–GDP in 70 μl of buffer [20 mM Na–HEPES (pH 8.0), 80 mM NaCl, 5 mM EDTA, 1 mM DTT, 0.01 mg/ml BSA, and 0.05% C12E10] containing 10 mM $MgCl_2$, 10 mM NaF, 10 μM GDP, and 20 μM $AlCl_3$ for 30 min on ice prior to initiation of the GTPase reaction. The single turnover assay is initiated by addition of 50 μl of [γ-^{32}P]GTP-bound Gα to 580 μl of reaction buffer [20 mM Na–HEPES (pH 8.0), 80 mM NaCl, 5 mM EDTA, 1 mM DTT, 0.1 mM GTP, 0.01 mg/ml BSA, and 0.05% C12E10] followed in rapid succession by 70 μl of the mixture of RGS protein plus Gα–GDP–AlF_4^-.

Determining K_I Values for Gα/RGS Interactions

Experiments similar to those described above using Gα–GDP–AlF_4^- can be designed to determine the apparent affinity (K_I) of a Gα protein (bound with GDP and AlF_4^-) for an RGS protein.[27,32] This type of assay is particularly useful with $G_q\alpha$ or $G_t\alpha$, which, as noted above, are difficult to prepare in their GTP-bound form. If the competing Gα–GDP–AlF_4^- protein interacts with the RGS

C. Sternweis and A. G. Gilman, *Proc. Natl. Acad. Sci. U.S.A.* **79**, 4888 (1982).

E. Coleman, A. M. Berghuis, L. Lee, M. E. Linder, A. G. Gilman, and S. R. Sprang, *Science* **265**, 405 (1994).

Fersht, in "Enzyme Structure and Mechanism." W. H. Freeman and Co., New York, 1977.

Tu, J. Wang, and E. M. Ross, *Science* **278**, 1132 (1997).

protein, then k_{gap} activity toward another $G\alpha$ subunit (for example, $G_i\alpha$ or $G_o\alpha$) will decrease in the presence of increasing concentrations of $G\alpha$–GDP–AlF$_4^-$. The affinity of the competing $G\alpha$–GDP–AlF$_4^-$ protein can be estimated from the IC$_{50}$ of the inhibition curve, or it can be determined more accurately by applying the Cheng–Prussof relationship[33]: $K_I = IC_{50}/([S] + K_{ms})$, where K_I is the affinity of $G\alpha$–GDP–AlF$_4^-$ for the RGS protein, IC$_{50}$ is the concentration of $G\alpha$–GDP–AlF$_4^-$ required to yield 50% inhibition of the k_{gap} activity, [S] is the concentration of $G\alpha$–GTP substrate, and K_{ms} is the Michaelis–Menten constant representing the apparent affinity of the RGS protein for the $G\alpha$–GTP substrate.

Steady-State GTPase Activity in Reconstituted Proteoliposomes

Steady-state GTPase assays using purified phospholipid vesicles containing receptors and heterotrimeric G proteins can be extremely useful for examination of RGS protein activity because the presence of an activated receptor increases the rate of GDP dissociation to the point where it is not rate limiting. The sensitivity of the assay is also increased because of multiple rounds of catalysis. Reconstitution of purified components into phospholipid vesicles also increases their local concentration many-fold, permitting observation of interactions that are not favored in solution.

The failure to detect interactions between specific RGS proteins and $G\alpha$ subunits in solution does not preclude the possibility that an RGS protein may stimulate the GTPase activity of the $G\alpha$ protein in a phospholipid bilayer or *in vivo*. For example RGS2, a protein with presumed specificity for $G_q\alpha$,[34] displays little GAP activity toward $G_i\alpha_1$ in solution. However, addition of RGS2 to reconstituted vesicles containing M2-muscarinic receptors, $G_i\alpha_1$, and $G\beta_1\gamma_2$ revealed a dramatic, muscarinic agonist-dependent increase in the GTPase activity of $G_i\alpha_1$.[8] Likewise, $G\beta_5$/RGS9 displays little or no GAP activity toward a number of $G\alpha$-GTP substrates in solution (T. K. Harden and A. G. Gilman, unpublished observations, 1999). However, addition of $G\beta_5$/RGS9 to phospholipid vesicles containing M2-muscarinic receptors, $G_o\alpha$, and $G\beta_1\gamma_2$ results in significant agonist-dependent GAP activity. Surprisingly, $G\beta_5$/RGS9 also facilitated interaction between the M2 receptors and $G_o\alpha$ in the absence of $G\beta_1\gamma_2$. Many elegant reviews have covered the methods, strategy, and rationale for producing reconstituted proteoliposomes[35–37];

[33] Y. Cheng and W. Prusoff, *Biochem. Pharm.* **22**, 3099 (1973).

[34] S. P. Heximer, S. P. Srinivasa, J. L. Bernstein, M. E. Linder, J. R. Hepler, and K. J. Blumer, *J. Biol. Chem.* **274**, 34253 (1999).

[35] E. M. Ross, in "The β-Adrenergic Receptors" (J. Perkins, ed.), p. 125. Humana Press Inc., Clifton Park, NJ, 1991.

[36] R. A. Cerione and E. M. Ross, *Methods Enzymol.* **195**, 329 (1991).

[37] R. A. Cerione, in "Receptor–Effector Coupling: A Practical Approach" (E. C. Hulme and N. J. M. Birdsall, eds.). IRL Press, Oxford, U.K., 1990.

the description herein will thus focus on the reconstitution of vesicles containing M2 receptors, $G_o\alpha$, and $G\beta_5$/RGS9 and a brief protocol for the steady-state GTPase assay.

Reconstitution of M2-Muscarinic Receptors, $G_o\alpha$, and $G\beta_5$/RGS9 in Phospholipid Vesicles

Vesicles containing receptors, $G\alpha$ subunits, and $G\beta\gamma$ subunits can be prepared as described.[38] The preparation begins with the addition of 50 μl of a 2× detergent : lipid mixture [0.55 mg/ml phosphatidylethanolamine, 0.35 mg/ml phosphatidylserine, 8.7 μg/ml cholesterol hemisuccinate, 0.4% deoxycholate, and 0.04% cholate, in 20 mM Na–HEPES (pH 8.0), 1 mM EDTA, and 100 mM NaCl] with an equal volume containing 30 pmol M2-muscarinic receptor,[39] 100 pmol myristoylated-$G_o\alpha$, and 300 pmol $G\beta_1\gamma_2$[40] in buffer A [20 mM Na–HEPES (pH 8.0), 100 mM NaCl, 1 mM EDTA, and 2 mM MgCl$_2$]. The mixture is vortexed briefly and applied immediately to a 17-ml AcA 34 column (9 × 270 mm). Fractions ($\sim$140 μl) are collected and assayed for [^{3}H]QNB binding activity. The three peak fractions are pooled and used for subsequent steady-state GTPase assays.

Preparation of vesicles containing M2-muscarinic receptors, $G_o\alpha$, and $G\beta_5$/RGS9 requires a two-step process necessitated by the sensitivity of $G\beta_5$/RGS9 to the detergents used to disperse the purified phospholipids. The first step consists of mixing 50 μl of the 2× detergent : lipid mixture (above) with an equal volume of buffer A containing 30 pmol of purified M2-muscarinic receptors. Vesicles containing the M2 receptor are formed following the application of the mixture to a 2-ml Sephadex G-50 (Fine) column (7 mm × 100 mm). Eight 250-μl fractions are collected, and those containing peak [^{3}H]QNB binding activity are pooled. Next, 200 pmol of myristoylated $G_o\alpha$ and 600 pmol of $G\beta_5$/RGS9 in buffer A are added to the vesicles, and the mixture is incubated on ice for at least 30 min to allow $G_o\alpha$ and $G\beta_5$/RGS9 to associate with the vesicles. The free proteins are removed by gel filtration (Superdex 200 10/30 column; Pharmacia) in buffer A. Fractions (0.2 ml) are collected and the five fractions with peak [^{3}H]QNB binding activity are pooled for use in steady-state GTPase assays.

Steady-State GTPase Assays

Steady-state GTPase assays using vesicles containing M2 receptors, $G_o\alpha$, and $G\beta_5$/RGS9 are conducted essentially as described for vesicles prepared with

[38] S. Mukhopahyay and E. M. Ross, *Proc. Natl. Acad. Sci. U.S.A.* **96,** 9539 (1999).

[39] E. M. Parker, K. Kameyama, T. Higashijima, and E. M. Ross, *J. Biol. Chem.* **266,** 519 (1991).

[40] T. Kozasa, in "G Proteins Techniques of Analysis" (D. R. Manning, ed.), p. 23. CRC Press LLC, Boca Raton, FL, 1999.

conventional G$\beta\gamma$ subunits.[38] Vesicles (10–20 μl) are equilibrated for 5 min at 10–20° in 100 μl of buffer A [20 mM Na–HEPES (pH 8.0), 100 mM NaCl, 1 mM EDTA, and 2 mM MgCl$_2$] supplemented with 0.1 mg/ml BSA in the presence and absence of 0.1 mM carbachol. This first incubation permits accumulation of receptor–heterotrimeric G protein complexes. Reactions are initiated with the addition of 100 μl of buffer A containing 0.4 μM GTP (specific activity 200 cpm/fmol) and 0.1 mg/ml BSA in the presence or absence of 0.1 mM carbachol. At specified time points, 25-μl aliquots of the reaction mixture are quenched in 975 μl of ice-cold 5% Norit A charcoal suspended in 50 mM NaH$_2$PO$_4$. The charcoal is centrifuged and the supernatant is counted to quantify [^{32}P]P$_i$ as described above.

Summary

Single-turnover and steady-state GTPase assays are an effective means to identify and characterize interactions between RGS and Gα proteins *in vitro*. The advantage of the single turnover GTPase assay is that it permits simple and rapid assessment of RGS protein activity toward a putative Gα–GTP substrate. Moreover, once an interaction between an RGS protein and a Gα–GTP subunit has been identified, the single-turnover assay can be used to determine Michaelis–Menten constants and/or K_I values for other competing Gα substrates. A disadvantage of the single-turnover assay is that a negative result does not preclude the possibility of an interaction between given RGS and Gα proteins *in vivo*. Inappropriate reaction conditions or the presence (or absence) of appropriate posttranslational modifications may result in small or undetectable increases in RGS protein-dependent GTPase activity. In these cases it may be tempting to examine RGS protein activity using steady-state GTPase assays in phospholipid vesicles reconstituted with receptors and heterotrimeric G proteins. The advantage to monitoring steady-state GTPase activity in reconstituted proteoliposomes is that ligand-dependent activation of the receptor facilitates GDP dissociation, such that effects of RGS proteins can be observed; multiple cycles of GTP binding and hydrolysis then amplify the GTPase signal. Additionally, the presence of the phospholipid membrane can increase the local RGS protein concentration $\sim$10^4-fold,[35] permitting observation of interactions that are weak in solution. The primary disadvantage of the reconstituted system is the requirement for receptor purification, a technically demanding undertaking in comparison to the purification of Gα, G$\beta\gamma$, and most RGS proteins.

[48] Measuring RGS Protein Interactions with $G_q\alpha$

By Peter Chidiac, Martha E. Gadd, and John R. Hepler

Introduction

Many hormones, neurotransmitters, and sensory inputs rely on heterotrimeric guanine-nucleotide regulatory proteins (G proteins) to exert their actions at target cells and tissues.[1-3] G proteins act by directly coupling cell surface receptors to the regulation of specific effector proteins at the plasma membrane. All G proteins are composed of three subunits designated as $G\alpha$ (39–46 kDa), $G\beta$ (35–39 kDa), and $G\gamma$ (8–11 kDa), and, to date, at least 21 unique $G\alpha$, 6 $G\beta$, and 12 $G\gamma$ subunits have been identified.[1-3] The identity of the individual $G\alpha$ subunit defines the G protein trimer, and $G\alpha$ subunits have been classified into four major subfamilies (G_s, G_i, G_q, and G_{12}) based on amino acid sequence identities and functional similarities.[1-3] At target cells, hormone or neurotransmitter receptor activation results in stimulation of phospholipase C, which hydrolyzes phosphatidylinositols to generate two important second messengers, inositol (1,4,5)trisphosphate [Ins(1,4,5)P$_3$] and diacylglycerol. Ins(1,4,5)P$_3$ mobilizes internal Ca^{2+} stores whereas diacylglycerol activates certain forms of protein kinase C. Members of the G_q subfamily of G proteins ($G_q\alpha$, $G_{11}\alpha$, $G_{14}\alpha$, and $G_{15/16}\alpha$) directly couple cell surface receptors to activation of the β isoforms of phospholipase C-β (PLC-β) and inositol lipid signaling.

G proteins serve as molecular switches in cell signaling. Activated receptors promote GTP binding to $G\alpha$ subunits, and the duration of a particular G-protein-directed signaling event is dictated by the lifetime of the $G\alpha$–GTP species. $G\alpha$ subunits are GTPases, and G protein signaling is terminated by $G\alpha$-catalyzed hydrolysis of bound GTP. As such, $G\alpha$–GTPase activity represents an important cellular control point for modulating hormone and neurotransmitter signaling events. Recent findings indicate that $G\alpha$ GTPase activity and linked signaling events are regulated by a newly identified class of signaling proteins, the regulators of G protein signaling (RGS proteins). RGS proteins comprise a large family (more than 30 mammalian forms) of highly diverse, multifunctional signaling proteins which share a conserved 120 amino acid domain (RGS domain).[4-9]

[1] H. E. Hamm, *J. Biol. Chem.* **273,** 669 (1998).
[2] J. R. Hepler and A. G. Gilman, *Trends Biochem. Sci.* **17,** 383 (1992).
[3] M. I. Simon, M. P. Strathman, and N. Gautam, *Science* **252,** 802 (1991).
[4] M. R. Koelle, *Curr. Opin. Cell Biol.* **9,** 143 (1997).
[5] H. G. Dohlman and J. Thorner, *J. Biol. Chem.* **272,** 3871 (1997).
[6] D. M. Berman and A. G. Gilman, *J. Biol. Chem.* **273,** 1269 (1998).
[7] J. R. Hepler, *Trends Pharmacol. Sci.* **20,** 376 (1999).
[8] L. DeVries and M. G. Farquhar, *Trends Cell Biol.* **9,** 138 (1999).
[9] D. P. Siderovski, B. Strockbine, and C. I. Behe, *Crit. Rev. Biochem. Mol. Biol.* **34,** 251 (1999).

RGS domains bind directly to activated $G\alpha$ subunits to act as GTPase-activating proteins (GAP) and/or effector antagonists to modulate hormone receptor and G-protein-directed signaling. Nearly all RGS proteins studied to date serve as GAPs for $G_i\alpha$ family members and/or $G_q\alpha$.[4-9] RGS4 was the first family member recognized to be an inhibitor of $G_q\alpha$-directed signaling.[10,11] Whereas RGS4 can act as a GAP for both $Gi\alpha$ family members and $G_q\alpha$, RGS2 was the first family member identified to be a selective inhibitor of $G_q\alpha$ function.[12-14] More recent studies have identified many other RGS proteins that also can regulate $G_q\alpha$ signaling functions (see Refs. 7, 8, and references therein), although not all block $G_q\alpha$ signaling by accelerating $G\alpha$–GTPase activity. For example, the well-studied G-protein-coupled receptor kinase GRK2 contains an RGS domain that binds tightly to $G_q\alpha$ to block its interactions with PLC-β without affecting $G_q\alpha$–GTPase activity.[15] In this regard, GRK2 and other RGS proteins[10,12,14] can act as effector antagonists.

Because of the broad importance of $G_q\alpha$-regulated inositol lipid signaling in cell physiology[16] and a new appreciation of the role for certain RGS proteins as regulators of G_q function,[4-9] considerable research interest has now focused on understanding and measuring RGS interactions with $G_q\alpha$. Numerous studies have examined RGS modulation of G_q-directed inositol lipid and Ca^{2+} signaling in various intact cell systems (see Refs. 7, 8, and references therein). In these cases, RGS/$G_q\alpha$ interactions were observed as a consequence of introducing cDNA and overexpression of recombinant RGS proteins in target cells. This approach has proven to be very useful for qualitatively confirming the negative regulatory effects of various RGS proteins on G_q signaling pathways. However, these methods do not allow for the quantitative measurement of direct RGS/$G_q\alpha$ interactions. This chapter will describe currently available methods for quantitatively measuring direct RGS interactions with $G_q\alpha$ *in vitro*. Three methods will be discussed in detail: (1) RGS stimulation of $G_q\alpha$–GTPase activity; (2) RGS inhibition of $G_q\alpha$-directed stimulation of phospholipase C-β; and (3) RGS inhibition of hormone receptor and $G_q/11\alpha$-directed stimulation of PLCβ in broken cell membrane preparations. Other available methods for measuring RGS/$G_q\alpha$ interactions that offer special advantages will also be discussed, but only briefly because of the general unavailability of required materials.

[10] J. R. Hepler, D. M. Berman, A. G. Gilman, and T. Kozasa, *Proc. Natl. Acad. Sci. U.S.A.* **94**, 428 (1997).

[11] Y. B. Yan, P. P. Chi, and H. R. Bourne, *J. Biol. Chem.* **272**, 11924 (1997).

[12] S. P. Heximer, N. Watson, M. E. Linder, K. J. Blumer, and J. R. Hepler, *Proc. Natl. Acad. Sci. U.S.A.* **94**, 14389 (1997).

[13] T. Ingi, A. M. Krummins, P. Chidiac, G. M. Brothers, S. Chung, B. E. Snow, C. A. Barnes, A. A. Lanahan, D. P. Siderovski, E. M. Ross, A. G. Gilman, and P. F. Worley, *J. Neurosci.* **18**, 7178 (1998).

[14] S. P. Heximer, S. P. Srinivasa, L. S. Bernstein, J. L. Bernhard, M. E. Linder, J. R. Hepler, and K. J. Blumer, *J. Biol. Chem.* **274**, 34253 (1999).

[15] C. V. Carman, J. L. Parent, P. W. Day, A. N. Pronin, P. M. Sternweis, P. B. Wedegaertner, A. G. Gilman, J. L. Benovic, and T. Kozasa, *J. Biol. Chem.* **274**, 34483 (1999).

Special Materials

Purified Recombinant RGS Proteins

Recombinant, hexahistidine(His$_6$)-tagged RGS4, RGS2, and RGS-GAIP are synthesized in *Escherichia coli* and purified by Ni-NTA affinity chromatography as previously described.[12,17]

Purified Recombinant Hexahistidine Tagged G$_q\alpha$

G$_q\alpha$ is not an abundant protein in native tissues and, as such, is very difficult to purify from native tissue sources such as bovine brain or liver. However, reasonable amounts (0.1–1 mg) of active recombinant G$_q\alpha$ protein can be obtained by using baculoviral expression systems and overexpression in Sf9 insect cells. In most cases, G$_q\alpha$ must be coexpressed with recombinant G$\beta\gamma$ subunits to generate properly folded, detergent-soluble Gα subunit that is active. Procedures for the purification of recombinant G$_q\alpha$ from Sf9 (*Spodoptera frugiperda* ovary) cell membranes using conventional chromatography have been described.[18] However, G$_q\alpha$ engineered to contain a hexahistidine tag at the carboxy terminus (G$_q\alpha$: CH$_6$) can be coexpressed as a functional heterotrimer with either Gβ_1 or Gβ_2 in combination with Gγ_2, and readily purified to homogeneity as free G$_q\alpha$:CH$_6$ by sequential Ni-NTA affinity and anion-exchange chromatography steps as previously described.[19] Addition of the hexahistidine tag does not affect recombinant G$_q\alpha$: CH$_6$ interactions with G$\beta\gamma$,[19] PLC-β_1,[19] and RGS proteins.[10,12]

Purified Recombinant Wild-Type G$_q\alpha$ and G$_q\alpha$R183C

Both of these proteins can be purified essentially as described originally by Biddlecome *et al.*,[20] with some minor modifications.[21] Briefly, either G$_q\alpha$ or G$_q\alpha$R183C is coexpressed in Sf9 cells with recombinant, hexahistidine tagged forms of Gβ_2 and Gγ_2 (His$_6$-Gβ_2/His$_6$-Gγ_2). Following the extraction of membranes with cholate, the solubilized trimer is diluted into lubrol and bound to Ni-NTA, followed by extensive salt washing of the resin. Subsequently, the column is brought to room temperature and washed with a cholate/lubrol buffer

[16] M. J. Berridge, *Nature* **361,** 315 (1993).

[17] D. M. Berman, T. M. Wilkie, and A. G. Gilman, *Cell* **86,** 445 (1996).

[18] J. R. Hepler, T. Kozasa, A. V. Smrcka, M. I. Simon, S. G. Rhee, P. C. Sternweis, and A. G. Gilman, *J. Biol. Chem.* **268,** 14367 (1993).

[19] J. R. Hepler, G. H. Biddlecome, C. Kleuss, L. A. Camp, S. L. Hofmann, E. M. Ross, and A. G. Gilman, *J. Biol. Chem.* **271,** 496 (1996).

[20] G. H. Biddlecome, G. Berstein, and E. M. Ross, *J. Biol. Chem.* **271,** 7999 (1996).

[21] P. Chidiac and E. M. Ross, *J. Biol. Chem.* **274,** 19639 (1999).

containing GTPγS, which elutes unwanted endogenous (insect) Gα subunits from the resin, leaving recombinant $G_q\alpha$ bound to His_6-$G\beta_2$/His_6-$G\gamma_2$. Finally, $G_q\alpha$ or $G_q\alpha$R183C is eluted with aluminum, fluoride, and magnesium ions, which frees Gα from the His_6-$G\beta2$/His_6-$G\gamma2$ bound to the Ni-NTA resin. For the final desalting of eluted protein, CHAPS is recommended over cholate, which may promote denaturation.[21] Under ideal conditions, this protocol should yield virtually pure (>95%) wild-type recombinant $G_q\alpha$. $G_q\alpha$R183C, which expresses poorly, should yield a preparation that is about 40% pure, which is adequate for single-turnover experiments but not recommended for reconstitution studies.

Baculoviruses Encoding Recombinant Wild-Type and Hexahistidine-Tagged Gβ and Gγ Subunits

Generation of baculoviruses and purification of recombinant forms of mammalian $G\beta_1$, $G\beta_2$, and $G\gamma_2$ subunits from Sf9 insect cells has been described in detail.[22] Baculovirus encoding hexahistidine-tagged forms of $G\beta_2$ and $G\gamma_2$ for use in the purification of $G_q\alpha$ has also been been described.[20]

Purified Phospholipase C-β₁

Purified $G_q\alpha$ family members activate the β isoforms of PLC (PLC-β_1, -β_2, and -β_3).[19,24,44] PLC-β_1 is abundant in brain cytosol and is readily obtained in a partially purified form using protocols that rely on sequential conventional chromatography steps.[23,25,26]

Antisera

Antisera that specifically recognize RGS4, RGS2, $G_q\alpha$, and PLC-β_1 are commercially available (Santa Cruz Biotechnology, Santa Cruz, CA; Upstate Biologicals, Lake Placid, NY). Available anti-$G_q\alpha$ sera recognize both wild-type $G_q\alpha$ and $G_q\alpha$R183C. Note that placement of a hexahistidine tag at the carboxy terminus of $G_q\alpha$ can interfere with antibody recognition at this site.

RGS Stimulation of $G_q\alpha$ GTPase Activity

General Considerations Regarding GTPase Assays

G proteins traverse a cycle of GTP binding, hydrolysis, and GDP + P_i dissociation. Agonist-bound receptors promote GDP dissociation and thus facilitate

[22] J. Iniguez-Llui, M. I. Simon, J. D. Robishaw, and A. G. Gilman, *J. Biol. Chem.* **267,** 23409 (1992).
[23] A. V. Smrcka and P. C. Sternweis, *J. Biol. Chem.* **268,** 9667 (1993).
[24] D. Wu, C. H. Lee, S. G. Rhee, and M. I. Simon, *J. Biol. Chem.* **267,** 1811 (1992).
[25] S. H. Ryu, K. S. Cho, K. Y. Lee, P. G. Suh, and S. G. Rhee, *J. Biol. Chem.* **262,** 12511 (1987).
[26] S. G. Rhee, S. H. Ryu, and K. S. Cho, *Methods Enzymol.* **197,** 502 (1991).

GTP binding and G protein activation. RGS proteins and other GAPs turn off activated G proteins by accelerating the rate of GTP hydrolysis. In the presence of activated receptor, this GAP effect serves to accelerate the steady-state turnover of GTP. However, GAPs have little or no effect on the steady-state GTPase activity of isolated Gα since GDP dissociation is already rate-limiting under such conditions.[27] Therefore, GAP assays must be performed either under pre-steady-state conditions (i.e., GTP-bound to Gα) or in the presence of activated receptor.

The direct GAP effect of an RGS protein on Gα can be detected as an increase in the rate of a single round of hydrolysis and P_i release by Gα–GTP. The primary advantage of this pre-steady-state, bimolecular method is that it allows GAP effects to be quantified as direct fold-stimulation of the basal rate of GTP hydrolysis, and also permits an approximation of the affinity between RGS and free Gα–GTP.[28] To assay GAP activity under steady-state conditions is technically simpler but, as noted, requires receptor as well as Gβγ, which themselves may influence RGS–Gα interactions.[7,21] With G_q, both single-turnover and steady-state GTPase assays pose difficulties not encountered with G_i, G_o, or G_s.

RGS-Directed Stimulation of $G_q\alpha R183C$ GTPase Activity

Special Considerations. The direct GTPase activating effect of RGS proteins on GTP-bound $G_{i/o}\alpha$ subunits is readily detectable using purified Gα subunits and RGS proteins in solution, as described elsewhere in this volume.[29] In contrast, analogous assays using wild-type $G_q\alpha$ are not practicable because of difficulties in isolating $G_q\alpha$–GTP. This stems primarily from the extremely slow dissociation of bound GDP, and probably also reflects the intrinsic instability of the nucleotide-free form of $G_q\alpha$.[30] These properties also account for the high nucleotide concentrations needed to observe G_q binding. To facilitate the isolation of the GTP-bound species, the point mutant $G_q\alpha R183C$ can be substituted for wild-type $G_q\alpha$. This mutant has a greatly decreased intrinsic rate of GTP hydrolysis,[21] as well as an increased rate of GDP dissociation. Slowly hydrolyzing G protein mutants in some cases are GAP-insensitive,[17]; however, as with $G_i\alpha R178C$, the arginine-to-cystine substitution does not prevent $G_q\alpha R183C$ from hydrolyzing GTP more rapidly in response to GAPs.[21] The rate of GTP hydrolysis by $G_q\alpha R183C$ is accelerated by phospholipase

[27] O. Saitoh, Y. Kubo, Y. Miyatani, T. Asano, and H. Nakata, *Nature* **390,** 525 (1997).

[28] J. Wang, Y. Tu, S. Mukhopadhyay, P. Chidiac, G. H. Biddlecome, and E. M. Ross, *in* "G Proteins: Techniques of Analysis" (D. R. Manning, ed.), p. 123. CRC Press, Boca Raton, FL, 1998.

[29] A. M. Krumins and A. G. Gilman, *Methods Enzymol.* **344,** [47] 2002 (this volume).

[30] P. Chidiac, V. S. Markin, and E. M. Ross, *Biochem. Pharmacol.* **58,** 39 (1999).

C-β_1,[21] RGS2,[13] RGS4,[31] RGS-GAIP, RGS1, and RGS3,[32] but not by rho-GEF,[33] GRK2,[15] RGS12,[34] RGS10, RGS11, or RGS14.[32]

Preparation for Assay. G$_q\alpha$R183C–[γ-^{32}P]GTP is prepared by incubating 100 nM G$_q\alpha$R183C with 10 μM [γ-^{32}P]GTP at 20° in 20 mM Na–HEPES (pH 7.5) containing 5.5 mM CHAPS, 1 mM dithiothreitol (DTT), 1 mM EDTA, 900 μM MgSO$_4$ (10 μM final free Mg^{2+}), 30 mM (NH$_4$)$_2$SO$_4$, 2% (v/v) glycerol, and 0.1 mg/ml albumin for 2–3 hr, at which point approximately 30% of active G$_q\alpha$R183C will be bound to [γ-^{32}P]GTP. The reaction mixture is then diluted with an equal volume of ice-cold buffer A [50 mM Na–HEPES (pH 7.5), plus either 5.5 mM CHAPS or 0.3% cholate], and after 5 min on ice, G$_q\alpha$R183C–[γ^{32}-P]GTP is recovered by centrifugal gel filtration on Sephadex G-25 equilibrated in buffer A. This procedure removes 99.9% of free nucleotide and yields 20–30% recovery of G$_q\alpha$R183C–[γ-^{32}P]GTP. A word is in order here concerning the stability and reactivity of G$_q\alpha$R183C in various detergents. We have found that Lubrol, CHAPS, Triton X-100, and dodecyl maltoside all yield approximately equal amounts of [γ-^{32}P]GTP and [^{35}S]GTPγS binding to G$_q\alpha$R183C, whereas cholate clearly has a detrimental effect on [γ-^{32}P]GTP binding. Cholate was the only detergent tested that permitted the GAP effect of PLC-β_1 on G$_q\alpha$R183C–[γ-^{32}P]GTP, whereas RGS4 GAP activity was evident to varying extents in the presence of cholate, CHAPS, Triton X-100, octylglucoside, dodecylmaltoside, deoxycholate, and Lubrol. Thus it may be necessary to optimize detergent conditions when assaying novel GAPs.

Performing Assay. To measure GTP hydrolysis, isolated G$_q\alpha$R183C–[γ-^{32}P]GTP is diluted approximately 10-fold into 20 mM HEPES buffer (pH 7.5) containing the GAP under investigation plus 80 mM NaCl, 1 mM dithiothreitol, 1 mM EDTA, 900 μM MgSO$_4$, 0.04 mg/ml albumin, and 2% glycerol plus appropriate detergent (e.g., 0.2–0.3% cholate). GTP hydrolysis is assessed by the release of free [^{32}P]P$_i$ into the buffer, which is isolated from GqαR183C–[γ-^{32}P]GTP and free [γ-^{32}P]GTP by the addition of a charcoal slurry followed by centrifugation, which leaves only free [^{32}P]P$_i$ in the supernatant.[20,21,28] Total G$_q\alpha$R183C–[γ-^{32}P]GTP should be quantified by vacuum filtration over nitrocellulose filter discs at the outset of the assay.[20,21,28] In the absence of added GAP, the rates of both GTP hydrolysis and GTP dissociation are approximately 0.05 per minute

[31] W. Zeng, X. Xu, S. Popov, S. Mukhopadhyay, P. Chidiac, J. Swistok, W. Danho, K. A. Yagaloff, S. L. Fisher, E. M. Ross, S. Muallem, and T. M. Wilkie, *J. Biol Chem.* **273,** 34687 (1998).

[32] P. Chidiac, unpublished observations, 1997.

[33] T. Kozasa, C. J. Jiang, M. J. Hart, P. M. Sternweiss, W. D. Singer, A. G. Gilman, G. Bollag, and P. C. Sternweis, *Science* **280,** 2108 (1998).

[34] B. E. Snow, R. A. Hall, A. M. Krumins, G. M. Brothers, D. Bouchard, C. A. Brothers, S. Chung, J. Mangion, A. G. Gilman, R. J. Lefkowitz, and D. P. Siderovski, *J. Biol. Chem.* **273,** 17749 (1998).

at 20°. GTP dissociation thus can be appreciable over the time required to carry out an assay, and one should monitor binding as well as hydrolysis over the course of an experiment. It is also important to note that biphasic GTP hydrolysis is generally observed in the presence of GAP, suggesting that a fraction of the total $G_q\alpha R183C–[\gamma-^{32}P]GTP$ (typically 25–40%) is GAP-insensitive. This must be taken into account when fitting or calculating hydrolysis rates, and procedures for doing so have been described in detail.[21,28]

Results. In the experiment shown in Fig. 1, RGS4 increased the rate of GTP hydrolysis by $G_q\alpha R183C–[\gamma-^{32}P]GTP$ by a factor of about 100. In this experiment, the decrease in total $G_q\alpha R183C–[\gamma-^{32}P]GTP$ (open symbols) followed a similar time course to that of $[^{32}P]P_i$ generation (closed symbols), indicating

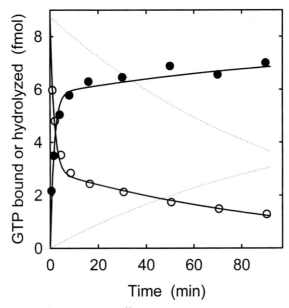

FIG. 1. Stimulation of $G_q\alpha R183C–[\gamma-^{32}P]GTP$ GTPase activity by RGS4. $G_q\alpha R183C–[\gamma-^{32}P]GTP$ was prepared and isolated as described and diluted with 9 volumes of assay buffer containing 0.2% cholate either without or with RGS4. At the times indicated, duplicate 20 μl aliquots were withdrawn and assayed for release of $[^{32}P]P_i$ or $G_q\alpha$-bound $[\gamma^{32}P]GTP$. All four sets of data were fitted simultaneously using a 2-component equation assuming a GAP-sensitive plus a GAP-insensitive fraction (see Ref. 21) to yield the fitted lines shown (dotted lines, no RGS4; solid lines, 120 nM RGS4). The solid and open symbols represent respectively the release of $[^{32}P]P_i$ and the amount of $G_q\alpha$-bound $[\gamma^{32}P]GTP$ in the presence of RGS4. Data in the absence of RGS4 have been omitted for clarity. In this experiment, 67% of the total $G_q\alpha R183C–[\gamma-^{32}P]GTP$ was sensitive to the GAP effect of RGS4. Basal rates of GTP hydrolysis and dissociation respectively were 0.0056 min^{-1} and 0.0038 min^{-1}, and the rate of hydrolysis was stimulated approximately 100-fold by RGS4, to 0.61 min^{-1}. Reproduced with permission from P. Chidiac and E. M. Ross, *J. Biol. Chem.* **274**, 19639 (1999).

that RGS4 did not increase the rate of GTP dissociation. Note that both RGS4 curves are biphasic, indicating the presence of a GAP-insensitive population of $G_q\alpha R183C-[\gamma-^{32}P]GTP$. Fitted curves corresponding to data acquired in the absence of RGS4 (dotted lines) show that loss of $G_q\alpha R183C-[\gamma-^{32}P]GTP$ (which results from both dissociation and hydrolysis) occurs at about twice the rate at which $[^{32}P_i]$ is produced, indicating that the basal rates of GTP dissociation and hydrolysis are similar.

Receptor-Promoted Steady-State GTPase Assays

Unlike studies with G_i, G_o, and G_s, attempts to assay receptor-promoted steady-state GTP turnover by G_q in broken cell preparations have been largely unsuccessful. It is possible to clearly observe agonist-stimulated $G_q\alpha$ GTPase activity and the acceleration of that activity by GAPs using m1 muscarinic receptors coreconstituted into phospholipid vesicles with heterotrimeric G_q.[20] This method requires that purified, detergent-solubilized receptor, $G_q\alpha$, and $G\beta\gamma$ be combined together, followed by gel filtration (or dialysis) to remove residual detergent. Although this technique presents a powerful means to study G_q–RGS interactions,[35] it is technically challenging in that each of the proteins used must first be purified in a nondenatured state, and typically over half of each protein is lost during the reconstitution process. For these reasons, this method is not readily approachable by most investigators.

A number of other experimental systems have been described that could be utilized for the study of receptor- and RGS-stimulated $G_q\alpha$–GTPase activity. Several investigators have developed novel assays for receptor-stimulated $[^{35}S]GTP\gamma S$ binding to G_q based on insect cell/baculovirus expression systems.[36-38] These techniques take advantage of the high receptor levels possible with baculovirus expression, combined in some instances with either the coexpression of G_q[38] or the addition of exogenous purified G_q.[37] Fusion proteins consisting of G protein coupled receptors linked directly to $G\alpha$ subunits have also been described[39,40] that allow direct study of receptor and G-protein-stimulated GTPase activity. Although none of receptors described thus far have been fused to $G_q\alpha$, there are no obvious barriers to this possibility and such a receptor/$G_q\alpha$ fusion protein could be used to study RGS interactions.[40] At least one experimental model system allows for the study of endogenous receptors linked to native $G_q/11$ and does not rely

[35] X. Xu, W. Zeng, S. Popov, D. M. Berman, I. Davignon, K. Yu, D. Yowe, S. Offermanns, S. Muallem, and T. M. Wilkie, *J. Biol. Chem.* **271**, 24684 (1999).
[36] G. Reyes-Cruz, J. Vazquez-Prado, W. Muller-Esterl, and L. Vaca, *J. Cell. Biochem.* **76**, 658 (2000).
[37] J. L. IV Hartman and J. K. Northup, *J. Biol. Chem.* **271**, 22591 (1996).
[38] A. J. Barr, L. F. Brass, and D. R. Manning, *J. Biol. Chem.* **272**, 2223 (1997).
[39] R. Seifert, K. Wenzel-Seifert, and B. K. Kobilka, *Trends Pharmacol. Sci.* **20**, 383 (1999).
[40] G. Milligan, *Trends Pharmacol. Sci.* **21**, 24 (2000).

upon recombinant proteins. Turkey erythrocyte ghosts contain native G_{11}-linked receptors (e.g., P2y-purinergic) which stimulate GTP-supported phospholipase C activity[41] and can be used to study RGS protein interactions with G_{11} and receptors.[42] While each of these experimental systems offers special advantages, they also require the use of special reagents and materials that make them generally unavailable to most laboratories. Nevertheless, it may be possible to adapt one or more of these methods to measure receptor- and RGS-stimulated $G_q\alpha$–GTPase activity in a partial reconstitution system.

RGS Inhibition of $G_q\alpha$ Signaling Functions

RGS Inhibition of $G_q\alpha$-Stimulated Phospholipase C-β_1 Activity

In addition to their defined roles as $G\alpha$–GAPs, RGS proteins also act as effector antagonists to block G protein signaling.[10,15,43] Consistent with this idea, RGS proteins form a stable complex with activated $G_q\alpha$ in *in vitro* reconstitution assays[10,12,15] and in cell membranes,[10] which prevents $G_q\alpha$/PLC-β interactions and blocks inositol lipid signaling. Since surface residues on $G\alpha$ responsible for $G\alpha$/RGS interactions are predicted to differ from those involved with $G\alpha$/effector interactions,[43] RGS capacity to block PLC binding to $G_q\alpha$ is likely due to steric hindrance. Based on these interactions, we will describe a method for detecting RGS interactions with active $G_q\alpha$ by monitoring changes in $G_q\alpha$-directed PLCβ activity that is measured as hydrolysis of exogenously supplied [³H]PIP$_2$ and formation of the water-soluble product, [³H]Ins(1,4,5)P$_3$. A technical advantage of this assay over the measurement of $G\alpha$ GTPase activity is that the wild-type recombinant $G_q\alpha$ required for these assays is more readily isolated than the $G_q\alpha$R183C mutant, which, at present, must be purified by complex formation with hexahistidine-tagged $G\beta_2/G\gamma_2$ (discussed above). An important experimental advantage of this method is that it can detect RGS proteins that bind to but do not serve as stimulators of $G_q\alpha$ GTPase activity, as is the case with GRK2.[15] The primary disadvantages of this approach are that the it only detects indirect RGS effects on $G_q\alpha$ in the form of inhibition of measured PLC-β activity, and it requires availability of PLC-β_1 and establishing protocols for measuring [³H]PIP$_2$ hydrolysis.

Special Considerations. Under defined assay conditions[44] where PLC-β and [³H]PIP$_2$ substrate are held constant (1 ng/assay PLCβ and 50 μM [³H]PIP$_2$) and substrate consumption is within a linear range (3–5 min), the half-maximal

[41] J. L. Boyer, C. P. Downes, and T. K. Harden, *J. Biol. Chem.* **264**, 884 (1989).

[42] M. Cunningham, G. Waldo, S. Hollinger, J. R. Hepler, and T. K. Harden, *J. Biol. Chem.* **276**, 5438 (2001).

[43] J. J. G. Tesmer, D. M. Berman, A. G. Gilman, and S. R. Sprang, *Cell* **89**, 251 (1997).

[44] A. V. Smrcka, J. R. Hepler, K. O. Brown, and P. C. Sternweis, *Science* **251**, 804 (1991).

concentration of $G_q\alpha$ required to activate PLC-β is approximately 30 nM, with maximal enzyme activity observed around 300 nM. Since RGS inhibition of PLC-β binding to $G_q\alpha$ is a competitive reaction, it is necessary to use submaximal amounts of $G_q\alpha$ in the assay (1–3 nM) in order to completely inhibit $G_q\alpha$ function by RGS. Under these assay conditions, uninhibited $G_q\alpha$-stimulated PIP$_2$ consumption occurs at a much lower rate and product formation is linear for up to 60 min. As such, these assays need to be carried out for longer time periods (at least 30–40 min) in order to accumulate measurable amounts of [^{3}H]Ins(1,4,5)P$_3$.

Preparation of $G_q\alpha$, RGS Proteins, and Buffers. Proteins necessary for these assays include RGS (e.g., RGS2 or RGS4), $G_q\alpha$, and PLC-β_1. If possible, relatively pure proteins should be used to allow control of the total amounts and relative ratio of each protein in the assay. Proteins should be prepared as concentrated stocks (0.1 mg/ml or higher) to promote stability and to allow for flexibility in dilution. Two buffers are used for these assays: Buffer 1 [50 mM sodium HEPES, pH 7.2, 1 mM EDTA, 3 mM EGTA, 5 mM MgCl$_2$, 2 mM dithiothreitol (DTT), 100 mM NaCl, and either 1% sodium cholate or 0.6% octylglucoside], and Buffer 2 (50 mM Na–HEPES, pH 7.2, 3 mM EGTA, 1 mM DTT, 80 mM KCl). Buffer 1 is used to prepare the $G_q\alpha$ and RGS proteins, whereas Buffer 2 is used to prepare PLC-β_1 mix, [^{3}H]PIP$_2$ phospholipid vesicles, and the Ca^{2+}mix. When all components are combined, the reactions are performed in a final total volume of 70 μl in a buffer consisting of 50 mM Na–HEPES (pH 7.2), 3 mM EGTA, 0.2 mM EDTA, 0.83 mM MgCl$_2$, 20 mM NaCl, 30 mM KCl, 1 mM dithiothreitol (DTT), 0.1 mg/ml ultrapure bovine albumin (Calbiochem, La Jolla, CA), 0.16% sodium cholate, and 1.5 mM CaCl$_2$ (to yield $\sim$150–200 nM free Ca^{2+}). Detergent is included to preserve the solubility of the lipid-modified $G_q\alpha$. Although PLC-β activity is inhibited by many detergents, activity is preserved in the presence of sodium cholate (Calbiochem, La Jolla, CA) or octyl glucoside (Calbiochem). Ca^{2+} is an essential cofactor for PLC-β activity, and Ca^{2+} mix is prepared as a 9 mM solution of CaCl$_2$ in Buffer 2 and added as 10 μl/assay.

Prior to assay, solutions containing each of the reaction components are prepared separately and stored on ice. These include the following: (1) $G_q\alpha$ mix (10 μl/assay), (2) RGS mix (10 μl/assay), and (3) Ca^{2+} mix (10 μl/assay). The remaining components, which include PLC-β and [^{3}H]PIP$_2$ containing phospholipid vesicles, are prepared separately and are described below. Since the final reaction volume is 70 μl, RGS and $G_q\alpha$ are prepared at a concentration sevenfold higher than that desired in the final assay. RGS domains will form a stable complex with $G_q\alpha$ activated with either nonhydrolyzable analogs of GTP (e.g., GTPγS) or with AlF$_4^-$, which traps Gα in the active transition state during GTP hydrolysis. $G_q\alpha$ is activated with either 1 mM GTPγS for 1 hr at 30° or with 10 mM NaF and 30 μM AlCl$_3$ for 15 min at room temperature in Buffer 1. RGS proteins are prepared as a stock concentration and then subjected to serial dilution in Buffer 1. Whereas

RGS4 is relatively stable in solution, RGS2 is sensitive to ionic strength and will form insoluble aggregates that precipitate if stored in buffers containing physiological salt concentrations.[12] As such, RGS2 is prepared and diluted in Buffer 1 containing 500 mM NaCl. The high salt conditions do not interfere with $G_q\alpha$ activation of PLCβ_1.

Preparation of PIP$_2$ Substrate Phospholipid Vesicles and PLCβ_1 Mix. In this method, RGS interactions with $G_q\alpha$ are detected as RGS capacity to inhibit $G_q\alpha$-directed PLCβ_1 activity. PLC-β activity is measured as hydrolysis of [^{3}H]PIP$_2$ and formation of the water-soluble product, [^{3}H]Ins(1,4,5)P$_3$. Activation of PLC-β is achieved by reconstituting enzyme with activated Gqα and mixed phospholipid vesicles containing lipid substrate as described previously.[18] Radiolabeled PIP$_2$ substrate is prepared as a mixture of phosphatidylinositol 4,5-bisphosphate (PIP$_2$; Sigma, St. Louis, MO) and bovine brain phosphatidylethanolamine (PE; Sigma) in a ratio of 1:10 containing 5000 to 10,000 cpm/assay of added [^{3}H]PIP$_2$ (New England Nuclear, Boston, MA) as a radiolabel trace. Phospholipids (stored at $-20°$ in chloroform under argon gas) are prepared first by drying under nitrogen gas at room temperature until organic solvent is removed and a lipid film remains at the bottom of the tube. The lipid film is then rehydrated, and vesicles are formed by brief sonication in Buffer 2 using a bath sonicator (Laboratory Supplies, Co.; Hicksville, NY). A probe sonicator may substituted for this purpose if available. The amount of PIP$_2$ necessary for the experiment is calculated based on preparing a stock solution at 175 μM added as 20 μl/assay which yields 50 μM (3000 pmol) final/reaction tube. The resulting lipid vesicles containing PIP$_2$, PE, and trace amounts of [^{3}H]PIP$_2$ (henceforth referred to as [^{3}H]PIP$_2$:PE mix) are stored on ice until ready for use. PLC-β is prepared in Buffer 2 containing 1 mg/ml bovine serum albumin (BSA), which is included to stabilize enzyme activity. PLC-β_1 purified from bovine brain or as recombinant protein can be used at a final concentration of 1 ng/20 μl/reaction which will provide a low basal activity and a robust response to $G_q\alpha$. For simplicity, the [^{3}H]PIP$_2$:PE mix and PLC-β mix are combined (20 μl each/reaction tube) and added as 40 μl/reaction tube immediately before the reaction is started (see below).

Performing Assay and Processing Samples. Round-bottom, polypropylene tubes (5 ml, 12 mm × 75 mm) are labeled and placed in a tube rack on ice. To each tube is added 10 μl Buffer 1 (for blanks and measurement of basal PLCβ activity) or Buffer 1 containing activated $G_q\alpha$. RGS proteins or an equivalent volume of Buffer 1 (10 μl/tube) are added and incubated with $G_q\alpha$ for at least 30 min on ice to allow for protein complex formation. After this time, Ca^{2+} solution (10 μl) is added to all tubes, and the reaction mix is stored on ice. The assay is started by adding the PLC-β/[^{3}H]PIP$_2$ mix (40 μl) to each tube on ice, and samples are vortexed immediately. For this addition, it is helpful to use a repeat pipetter with Combitips (Eppendorf). Reactions are initiated by transferring tubes to a 30° water bath for the desired time. Under these defined assay conditions (50 μM [^{3}H]PIP$_2$,

1 ng PLC-β_1, 1 nM G$_q$α) substrate consumption is linear for up to 50–60 min, and reactions are typically allowed to proceed for 40 min. Reactions are terminated by rapid addition of 200 μl of 10% trichloroacetic acid (TCA), followed by 100 μl of bovine serum albumin (10 mg/ml). Samples are then transferred immediately to an ice bath and vortexed.

Precipitation of the albumin in acidic conditions will coprecipitate unhydrolyzed radiolabeled PIP$_2$. To isolate released [^{3}H]Ins(1,4,5)P$_3$, samples are centrifuged at 2000g for 10 min at 4°, and supernatant is recovered. Radioactivity in the supernatant is measured by liquid scintillation counting.

Results. In the example experiment provided (Fig. 2), the relative capacities of RGS2 and RGS4 to block G$_q$α activation of PLCβ_1 were compared. The assays involved only three purified recombinant proteins (G$_q$α, RGS, PLC-β_1) and

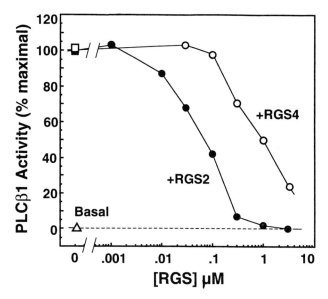

FIG. 2. RGS-directed inhibition of G$_q$α-mediated phospholipase Cβ_1 (PLCβ_1) activation. Purified recombinant G$_q$α was incubated with 1 mM GTPγS for 1 hr at 30°. Activated G$_q$α–GTPγS (1 ng) was preincubated with various concentrations of RGS2 (closed circles), RGS4 (open circles), or no RGS proteins (squares), and then mixed with purified PLCβ_1 and [^{3}H]phosphatidylinositol 4,5-bisphosphate-containing phospholipid vesicles. Synthesis of [^{3}H]inositol 1,4,5-trisphosphate ([^{3}H]InsP$_3$) was measured and basal unstimulated PLCβ_1 activity (open triangle; 170 pmol/min/ng PLCβ_1) was subtracted from each value. Blank values (i.e., [^{3}H]InsP$_3$ accumulation in the absence of PLCβ_1) were 155 pmol/min/assay. Values are expressed as a percentage of the total [^{3}H]InsP$_3$ accumulated over 20 min at 30° in the presence of G$_q$α–GTPγS (100% = 716 and 571 pmol/min/ng PLCβ_1 for RGS4 and RGS2, respectively). Reproduced with permission from S. P. Heximer, N. Watson, M. E. Linder, K. J. Blumer, and J. R. Hepler, *Proc. Natl. Acad. Sci. U.S.A.* **94,** 14389 (1997). Copyright 1997 National Academy of Sciences, U.S.A.

exogenously supplied substrate in the form of [^{3}H]PIP$_2$:PE phospholipid vesicles. RGS2 and RGS4 each bound to activated G$_q\alpha$ and blocked its capacity to stimulate PLC-β_1 activity over a broad concentration range (0.001–3 μM). Under the defined assay conditions, RGS2 inhibited the actions of 1 nM G$_q\alpha$ with a $K_{0.5}$ value of approximately 30 nM and was 10- to 30-fold more potent than RGS4 at blocking G$_q\alpha$/PLC-β_1 interactions (Fig. 2).

RGS Inhibition of Hormone Receptor and G$_q\alpha$-Stimulated Inositol Lipid Signaling in Broken Cell Preparations

The negative regulatory effects of RGS proteins on G$_q\alpha$ signaling can also be detected in broken cell or crude cell membrane preparations. The principal advantage of this approach over defined reconstitution systems is that it allows for examination of RGS effects on receptor and G$_q$-directed signaling in the partially intact native environment provided by the cell membranes. While the amount of RGS protein can be controlled quantitatively, the relative amounts of receptor, G protein, and PLC-β present are undefined and remain a function of the amount of washed membranes used in the assay. The approach is conceptually and technically similar to that described above with purified proteins, except that in this case endogenous PLC-β, G$_q\alpha$, and linked receptor are supplied by the cell membranes. Exogenous radiolabeled substrate is supplied as [^{3}H]PIP$_2$:PE phospholipid vesicles (as described above).[10,45] In theory, the method described in this section will work with membranes derived from any cell type. While G$_q$-directed inositol lipid signaling responses can be observed in many broken cell preparations, retention of linked hormone responses varies following cell lysis and should be determined empirically for each cell line. Among the broken cells that are responsive to hormone/neurotransmitter stimulation are NG-108 neuroblastoma $\times$ glioma cells which contain bradykinin receptors linked to Gq/11.[10,45] We will describe a method for measuring RGS mediated inhibition of bradykinin receptor and guanine nucleotide stimulation of PLC-β activity in broken NG-108 cell preparations.

Special Considerations. Unlike the reconstitution assays described above, there is no step for preactivating native G proteins with guanine nucleotide. Instead, the assay relies on addition of hormone or spontaneous rates of nucelotide exchange to load nonhydrolyzable guanine nucleotide (GTPγS) to endogenous G$_q\alpha$. Under these defined assay conditions, hormone plus guanine nucleotide responses occur at a much faster rate and are much more robust than those observed with guanine nucleotide alone due to receptor activation. GDPβS is included in the guanine nucleotide mix at a concentration in excess of GTPγS (10 μM and 3 μM,

---

[45] S. Gutowski, A. V. Smrcka, L. Nowak, D. Wu, M. I. Simon, and P. C. Sternweis, *J. Biol. Chem.* **266,** 20519 (1991).

respectively). This stable form of GDP binds with higher affinity than GTPγS to Gα that are uncoupled from receptors but with relatively lower affinity to Gα that are coupled to activated receptors. As such, inclusion of GDPβS helps to maximize loading of GTPγS to $G_q\alpha$ linked to activated receptors while, at the same time, minimizing nucleotide loading to uncoupled $G_q\alpha$ due to basal exchange. Ideal assay conditions would also permit GTP loading of $G_q\alpha$, which would allow measurement of RGS effects on GTPase activity. However, for unknown reasons, very few, if any, broken cell preparations of mammalian origin retain GTP-sensitive hormone activation of $G_q\alpha$ and PLC-β. An exception to this rule are specialized membrane preparations derived from turkey erythrocyte ghosts.[41] These preparations retain very robust GTP-sensitive hormone responses and, as discussed, could provide an alternative model system for studying RGS/Gα interactions.[42]

Preparation of NG-108 Cell Membranes. NG108-15 cell membranes are prepared as described.[10,45] NG108-15 cells are grown to confluency on 3 to 5 150 mm culture plates at 37° in Dulbecco's modified Eagle's medium (DMEM) supplemented with 15% fetal calf serum, 0.1 mM hypoxanthine, 0.4 mM aminopterin, 16 μM thymidine, and appropriate antibiotics. Cells are rinsed with cold phosphate-buffered saline (pH 7.2) on ice and collected by scraping in the same buffer containing protease inhibitors. Recovered cells are placed into a nitrogen cavitation bomb (Parr Instruments, Moline, IL) under pressure (400–600 psi) for 30 min at 4°. Cells are lysed by rapid decompression, and lysates are recovered. Cell nuclei are removed by centrifugation at 500 g for 10 min at 4°. Total membranes are recovered from the supernatant by centrifugation and resuspended by homogenization (10 strokes on ice) in Buffer 3 (Buffer 1 without detergents), and protein concentrations are determined. At this point, concentrated membranes (preferably greater than 5 mg protein/ml) can be stored at 4° for immediate use, or frozen by liquid N_2 and stored at −80° until further use. Receptor responses will lose some activity (30–50%) following one or more freeze–thaw cycles. In practice, crude, unwashed membrane preparations provide a more robust inositol lipid signaling response. Reasons for this are uncertain, but may be due, at least in part, to the fact that extensive washings remove PLC-β_1 from plasma membranes, which is bound only by ionic interactions.

Preparation of PIP$_2$ Substrate/Phospholipid Vesicles. [³H]PIP$_2$:PE vesicles are prepared by drying down a mixture of [³H]PIP$_2$, PIP$_2$, and PE and resuspending the lipid film in Buffer 2 by sonication as described above. The final concentration of lipid is adjusted to provide 50 μM PIP$_2$ and 500 μM PE and trace amounts of [³H]PIP$_2$ (5000–10,000 cpms) (i.e., [³H]PIP$_2$:PE mix) in the final assay volume of 60 μl. Thus, a 3-fold concentrated [³H]PIP$_2$:PE mix is prepared to add as 20 μl/reaction tube.

Preparation of RGS Proteins, Hormone and Guanine Nucleotide Mix, and Ca^{2+} Mix. RGS proteins are prepared as concentrated stocks in Buffer 2 and subjected to serial dilutions as desired in the same buffer. Bradykinin alone (10 μl/assay),

GDPβS and GTPγS as a mixture (10 μl/assay), or bradykinin plus guanine nucleotides as a mixture (10 μl/assay) are also prepared as concentrated stocks in Buffer 3. In each case, stocks are prepared at concentrations sixfold greater than that desired in the final reaction mix; each stock is added as 10 μl per reaction tube. Final desired concentrations for each component are 1 μM for bradykinin, 3 μM for GTPγS, and 10 μM GDPβS. Ca^{2+} mix is prepared as a 9 mM solution of CaCl$_2$ in Buffer 2 to be added as 10 μl per reaction tube.

Performing Assay and Processing Samples. Round-bottom, polypropylene tubes (5 ml, 12 mm × 75 mm) are labeled and placed in a tube rack on ice. To each tube is added 10 μl of membranes (5 μg/assay) and either 10 μl of Buffer 2 (for blank and basal PLC activity) or 10 μl of Buffer 2 containing RGS proteins. Membranes and RGS proteins are incubated together at 4° for 30 min. [^{3}H]PIP$_2$:PE phospholipid vesicles are added as 20 μl/assay followed by 10 μl of bradykinin and/or guanine nucleotides. Reactions are initiated by rapid addition of CaCl$_2$ mix (10 μl/tube), vortexing, and transferring the tube rack to a 30° water bath for the desired time. Under these assay conditions [^{3}H]PIP$_2$ consumption is linear for up to 50–60 min, and reactions typically proceed for 30 min. As described above, reactions are terminated by rapid addition of 200 μl of 10% trichloroacetic acid, followed by 100 μl of bovine serum albumin (10 mg/ml); samples are then transferred immediately to an ice bath and vortexed. For the start and stop addition, it is helpful to use a repeat pipetter with Combitips (Eppendorf).

As discussed above, precipitation of the albumin in acidic conditions will coprecipitate unhydrolyzed radiolabeled PIP$_2$, PE, and membranes. To isolate released [^{3}H]Ins(1,4,5)P$_3$, each of the samples is centrifuged at 2000g for 10 min, and the supernatant is recovered. Radioactivity in the supernatant is measured by liquid scintillation counting.

Results. In the example experiment provided (Fig. 3), the relative capacities of RGS4 and RGS-GAIP to block bradykinin receptor and/or GTPγS-mediated stimulation of PLC-β are compared. The assays involved one purified recombinant RGS protein, crude cell membranes containing desired G$_{q/11}$-linked hormone receptors, and exogenously supplied substrate in the form of [^{3}H]PIP$_2$:PE phospholipid vesicles. RGS4 was a fully effective inhibitor of both hormone receptor-(bradykinin plus GTPγS) and G-protein-directed (GTPγS alone)stimulation of PLC-β activity (Fig. 3A), with inhibition observed over a broad concentration range (0.001–1 μM) (Fig. 3B). Half-maximal concentration for RGS4 inhibition ($K_{0.5}$) was approximately 100 nM, with complete inhibition observed with 1μM RGS4. By contrast, RGS–GAIP was a much less potent inhibitor of G$_q\alpha$ function, consistent with the idea that it is a physiological regulator of signaling by G$_i\alpha$ family members, but not G$_q\alpha$.[10,46] This method allows for

[46] L. DeVries, M. Mousli, A. Wursmer, and M. G. Farquhar, *Proc. Natl. Acad. Sci. U.S.A.* **92,** 11916 (1995).

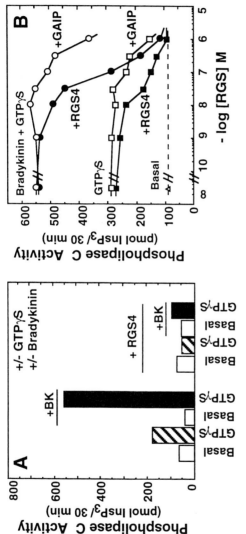

FIG. 3. RGS-directed inhibition of GTPγS and bradykinin-activated synthesis of inositol 1,4,5-trisphosphate by NG-108 cell membranes. (A) Total unwashed NG-108 membranes (5 μg protein/assay) were incubated with 10 μM GDPβS (basal) or 10 μM GDPβS and 3 μM GTPγS in the presence or absence of 1 μM bradykinin (BK) and/or 1 μM RGS4. The resulting accumulation of [^{3}H]inositol 1,4,5-trisphosphate ([^{3}H]InsP$_3$) over 30 min at 30° was measured. (B) NG-108 cell membranes were incubated as described in (A) in the presence of bradykinin (BK) and GTPγS (circles), or GTPγS (squares), and the concentration of RGS4 (filled symbols) or RGS-GAIP (open symbols) was varied as indicated. Basal PLC-β_1 activity is indicated (open square). Reproduced with permission from J. R. Hepler, D. M. Berman, A. G. Gilman, and T. Kozasa, *Proc. Natl. Acad. Sci. U.S.A.* **94**, 428 (1997). Copyright 1997 National Academy of Sciences, U.S.A.

measurement of RGS interactions with $G_q\alpha$ in a native cell membrane environment under the control of a linked receptor.

Acknowledgments

This work was supported by funds from the National Institutes of Health Grant NS37112 (to J.R.H.), the American Heart Association, GA Affiliate (to J.R.H.), the Pharmaceutical Researchers and Manufacturers Association of America (to J.R.H.), and funds from the Heart and Stroke Foundation of Ontario (to P.C.). P.C. is a Heart and Stroke Foundation of Canada Research Scholar.

[49] Assays of Complex Formation between RGS Protein Gγ Subunit-like Domains and Gβ Subunits

By DAVID P. SIDEROVSKI, BRYAN E. SNOW, STEPHEN CHUNG, GREG M. BROTHERS, JOHN SONDEK, and LAURIE BETTS

Introduction

Regulators of G-protein signaling (RGS) proteins accelerate the hydrolysis of GTP bound to heterotrimeric G protein α subunits[1,2] and are thus considered key negative regulators of G protein-coupled signaling pathways[3] by promoting formation of inactive G-protein heterotrimers (Gα–GDP/Gβ/Gγ complexes). Outside the hallmark RGS box, many RGS proteins have additional structural features that are thought to impart additional functions in signaling regulation.[4] We have identified[5] a subfamily of RGS proteins in which each member possesses both DEP (Dishevelled, EGL-10, Pleckstrin)[6] and GGL (G protein γ subunit-like)[5,7] domains N-terminal to the RGS box. This subfamily not only includes the mammalian RGS proteins RGS6, -7, -9, and -11, as depicted in Fig. 1, but also the *Caenorhabditis elegans* proteins EGL-10[8] and EAT-16[9], and the *Drosophila* RGS7 homolog.[10]

[1] D. M. Berman, T. M. Wilkie, and A. G. Gilman, *Cell* **86,** 445 (1996).
[2] B. E. Snow, R. A. Hall, A. M. Krumins, G. M. Brothers, D. Bouchard, C. A. Brothers, S. Chung, J. Mangion, A. G. Gilman, R. J. Lefkowitz, and D. P. Siderovski, *J. Biol. Chem.* **273,** 17749 (1998).
[3] D. P. Siderovski, A. Hessel, S. Chung, T. W. Mak, and M. Tyers, *Curr. Biol.* **6,** 211 (1996).
[4] D. P. Siderovski, B. Strockbine, and C. I. Behe, *Crit. Rev. Biochem. Mol. Biol.* **34,** 215 (1999).
[5] B. E. Snow, A. M. Krumins, G. M. Brothers, S.-F. Lee, M. A. Wall, S. Chung, J. Mangion, S. Arya, A. G. Gilman, and D. P. Siderovski, *Proc. Natl. Acad. Sci. U.S.A.* **95,** 13307 (1998).
[6] C. Ponting and P. Bork, *Trends Biochem. Sci.* **21,** 245 (1996).
[7] B. E. Snow, L. Betts, J. Mangion, J. Sondek, and D. P. Siderovski, *Proc. Natl. Acad. Sci. U.S.A.* **96,** 6489 (1999).
[8] M. R. Koelle and H. R. Horvitz, *Cell* **84,** 115 (1996).

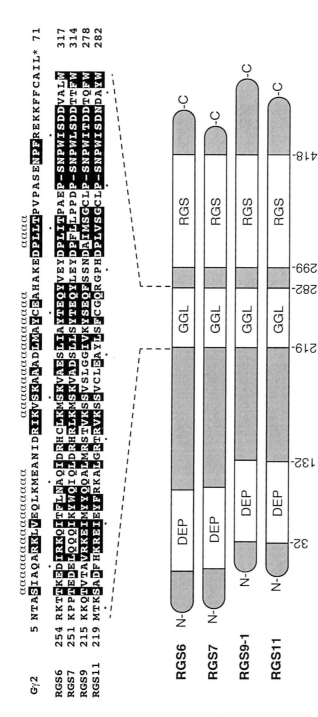

FIG. 1. Common, multidomain structure shared between RGS6, RGS7, RGS9, and RGS11 and sequence comparison of G protein γ-subunit like (GGL) domains with Gγ2 primary sequence. Conserved residues are shaded and regions within Gγ2 of α helix secondary structure, as determined by X-ray crystallography [M. A. Wall, D. E. Coleman, E. Lee, J. A. Iniguez-Lluhi, B. A. Posner, A. G. Gilman, and S. R. Sprang, *Cell* **83**, 1047 (1995)], are denoted with "α" symbols. Numerals below RGS11 denotes amino acid numbering of domain boundaries. The domain structure of RGS9 found in the retina (GenBank AAC99481), is illustrated; a striatal-specific isoform (RGS9-2; GenBank AAD20014) also exists that contains an alternative polypeptide sequence C-terminal to the RGS box.

We and others have demonstrated[5,7,11,12] that the GGL domains of RGS proteins bind selectively to the $G\beta$ subunit $G\beta_5$. However, these complexes do not appear to share the same functionality as conventional $G\beta/G\gamma$ complexes: $G\beta_5/GGL$ complexes do not appear to form heterotrimers with GDP-bound $G\alpha$ subunits *in vitro* nor modulate adenylyl cyclase activity nor activate phospholipase C-β isoforms.[5,13] The $G\beta_5/GGL$ complex can apparently inhibit $G\beta_1/G\gamma_2$-mediated activation of PLC-β_2 *in vitro*,[13] as well as potentiate acccleration of G-protein-gated inwardly rectifying K^+ (GIRK) channel kinetics[14]; however, a clearly defined biochemical function for such novel $G\beta/GGL$ complexes has yet to be ascertained.

Here, we present methods for the study of $G\beta/GGL$ domain complex assembly, including the use of *in vitro* cotranslation of $G\beta$ and RGS proteins in reticulocyte lysates, as well as the transient cotransfection of $G\beta$ and RGS protein expression vectors in mammalian cell lines. We also present a detailed protocol for the purification of $G\beta_5/RGS11$ complexes from insect cells infected with recombinant *Autographa californica* nuclear polyhedrosis viruses (AcPNV) or "baculoviruses." A purification strategy is employed whereby the $G\beta_5$ subunit is modified with a protease-cleavable N-terminal hexahistidine (His$_6$) tag. This allows the $G\beta_5/RGS11$ complex to be purified by metal chelation chromatography, after which the extraneous His$_6$ tag is easily removed by proteolysis. Heterodimers thus produced are fully soluble and behave on gel filtration as a 1 : 1 heterodimer of $G\beta_5/RGS11$. The ability to prepare purified, soluble $G\beta_5/RGS$ complexes using this protocol should facilitate atomic resolution structure determinations by X-ray crystallography as well as biochemical studies of potential effects of these novel heterodimers on G protein signaling, including receptor coupling and $G\alpha_o/G\alpha_q$ cross-talk.[9]

In Vitro Analysis of $G\beta/GGL$ Domain Association

Cloning of Expression Plasmids

Methods detailing the cloning of GGL domain-containing RGS family members have previously been described.[5,7] The open-reading frames for all six known G-protein β subunits (i.e., $G\beta_1$–$G\beta_5$ and the retinal-specific $G\beta_5$ isoform $G\beta_5L$[15]),

[9] Y. M. Hajdu-Cronin, W. J. Chen, G. Patikoglou, M. R. Koelle, and P. W. Sternberg, *Genes Dev.* **13**, 1780 (1999).

[10] T. Elmore, A. Rodriguez, and D. P. Smith, *DNA Cell Biol.* **17**, 983 (1998).

[11] E. R. Makino, J. W. Handy, T. Li, and V. Y. Arshavsky, *Proc. Natl. Acad. Sci. U.S.A.* **96**, 1947 (1999).

[12] K. Levay, J. L. Cabrera, D. K. Satpaev, and V. Z. Slepak, *Proc. Natl. Acad. Sci. U.S.A.* **96**, 2503 (1999).

[13] B. A. Posner, A. G. Gilman, and B. A. Harris, *J. Biol. Chem.* **274**, 31087 (1999).

[14] A. Kovoor, C. K. Chen, W. He, T. G. Wensel, M. I. Simon, and H. A. Lester, *J. Biol. Chem.* **275**, 3397 (2000).

[15] A. J. Watson, A. M. Aragay, V. Z. Slepak, and M. I. Simon, *J. Biol. Chem.* **271**, 28154 (1996).

as well as for three G-protein γ subunits (Gγ_1, Gγ_2, Gγ_3) are isolated by reverse transcriptase–polymerase chain reaction (RT-PCR) amplifications using primer pairs and cDNA templates outlined in Table I and employing previously described molecular biology techniques.[16] After trapping PCR products in the Invitrogen cloning vector pCR2.1-TOPO and verifying the correct sequence has been amplified [by fluorescent dideoxynucleotide sequencing (ABI/Perkin Elmer, Foster City, CA)], cDNAs are subcloned into eukaryotic expression vectors, based on pcDNA3.1 (Invitrogen, San Diego, CA), bearing in-frame N-terminal epitope tags for immunoprecipitation and immunodetection. The use of pcDNA3.1 serves a dual purpose: (a) the presence of a T7 RNA promoter allows for *in vitro* transcription of Gβ and Gγ/GGL subunits and (b) the presence of a strong cytomegalovirus promoter allows for robust transcription on plasmid transfection into mammalian cells without recloning.

In Vitro Transcription/Translation

It is important that all steps involving ^{35}S-labeled methionine be performed in a fume hood equipped and licensed to handle radioactivity, as volatile ^{35}S-containing radiolytic breakdown products should not be inhaled.

Coupled transcription/translation reactions are performed using the TnT T7 coupled reticulocyte lysate system (Promega, Madison, WI) in a final volume of 25 μl containing 12.5 μl of TnT rabbit reticulocyte lysate, 1 μl of TnT reaction buffer, 2 μl of translation-grade [^{35}S]methionine (1000 Ci/mmol, NEN Life Science Products, Boston, MA), 0.5 μl of TnT T7 RNA polymerase, 0.5 μl of 1 mM amino acid (aa) mix minus methionine, 20 units (0.5 μl) of RNase Inhibitor (Boehringer Mannheim/Roche, Nutley, NJ), 0.5 μg (0.5 μl) of template DNA, and 7.5 μl of diethyl pyrocarbonate (DEPC)-treated sterile water. Reactions are incubated for 1 hr at 30° to allow for sufficient transcription and translation of each cDNA. After this first hour, reactions are combined: 5 μl of each Gβ subunit reaction is added to a new 0.5 ml tube containing 5 μl of the appropriate tandem HA-tagged Gγ or GGL domain reaction. Combined reaction mixtures are then incubated for an additional hour at 37° to allow formation of Gβ/Gγ or Gβ/GGL heterodimers.

Coimmunoprecipitation

Cotranslated lysate (10 μl) is then transferred to a new 0.5 ml tube containing 90 μl of ice-cold buffer D (50 mM NaCl, 10 mM MgCl$_2$, 50 mM Tris pH 8.0, 1 mM EDTA, 0.05% C12E10 (polyoxyethylene-10-lauryl ether), 20% (v/v) glycerol, 10 mM 2-mercaptoethanol (added fresh), 1 Complete Mini protease inhibitor cocktail tablet (Roche; added fresh for each 10 ml of buffer), and 20 μl of a 50% (v/v) slurry (in buffer D) of protein A-Sepharose CL-4B (Sigma), for a total

---

[16] B. E. Snow, G. M. Brothers, and D. P. Siderovski, *Methods Enzymol.* **344,** [51] 2002 (this volume).

TABLE I
OLIGONUCLEOTIDE PRIMERS USED IN PCR CLONING OF Gβ AND Gγ OPEN-READING FRAMES

Subunit	Primer pair (5' to 3')	cDNA source	Subcloned into pcDNA3.1 with...		
			No tag	Myc-tag[d]	HA$_2$-tag[e]
Gβ$_1$	ACCATGAGTGAGCTTGACCAGTTACGG & GTCGACTCCACATGCTACTGGCGTTAGTTC	Human brain[a]	+	+	−
Gβ$_2$	ACCATGAGTGAGCTGGAGCAACTGAG & GTCGACGCCTGTAGTGTGGGCATGGGCAG	Human brain	+	+	−
Gβ$_3$	ACGATGGGGAGATGGAGCAACTGC & GTCGACTTCTCCAGCCTCCTCAGTTCCAGA	Human retina[b]	+	+	−
Gβ$_4$	GCACCATGAGCGAGCTGGAGCAGC & GTCGACTCAATTCCAGATTCTAAGAAAACTG	Mouse retina[c]	+	+	−
Gβ$_5$	AGCATGGCAACCGATGGGCTGCACG & GTCGACATGATTATGCCCAAACTCTTAGG	Mouse retina	+	+	−
Gβ$_5$L	ACCATGTGCGATCAGACCTTCCTGG & GTCGACATGATTATGCCCAAACTCTTAGG	Mouse retina	+	+	−
Gγ$_1$	GAATTCGCTTAAGATGCCAGTAATCAATATTGAGGAC & GTCGACCTATGAAATCACACAGCCTCCTTTGAG	Human retina	−	−	+
Gγ$_2$	GGATCCCCGATGGCCAGCAACAACACC & GTCGACTTAAAGGATGGCACAGAAAAACTTC	Human brain	−	−	+
Gγ$_3$	AAGATCTATGAAAGGTGAGACCCCGGTGAAC & AGTTGTGAGAAGGGACAGGGGAGC	Human brain	−	−	+

[a] Marathon-Ready human brain cDNA (Clontech).

[b] Marathon-Ready human retinal cDNA (Clontech).

[c] A 1:1 mixture of oligo-dT/random hexamer-primed mouse retinal cDNA prepared from C57B1/6 mouse retinal total RNA as previously described [B. E. Snow, L. Antonio, S. Suggs, and D. P. Siderovski, Gene 206, 247 (1998)].

[d] pcDNA3.1 vector with N-terminal c-myc epitope tag and hexahistidine tag: MAARGHPFEQKLISEEDLNMHTGHHHH-HHCGIRL.

[e] pcDNA3.1 vector with N-terminal double hemagglutinin (HA)-tag: MYPYDVPDYAGYPYDVPDYA.

volume of 120 μl. Tubes are rocked on a nutator at 4° for 30 min to preclear any protein that might nonspecifically bind to protein A-Sepharose. Tubes are then centrifuged at 4° for 1 min at 10,000g to pellet the protein A beads and each supernatant is transferred to a new 0.5 ml tube containing 0.4 μg of anti-HA (hemagglutinin) monoclonal antibody 12CA5 (Roche). Supernatants are then rocked at 4° for 45 min to allow formation of antibody/protein complexes. After this 45 min period, tubes are briefly centrifuged at 10,000g to reduce radioactive aerosols from the lids of the tubes and then 25 μl of 50% (v/v) protein A-Sepharose slurry is added to each tube. Tubes are then rocked at 4° for 1 hr to allow antibody/bead complexes to form and centrifuged at 4° for 1 min at 10,000g to pellet the beads.

[At this stage, we aliquot 35 μl of the supernatant into a new tube containing 35 μl of 2× Tris–glycine sodium dodecyl sulfate (SDS) sample buffer (Novex, Carlsbad, CA), containing freshly added 2-mercaptoethanol (2-ME) to a final concentration of 0.7 M. This "supernatant control" sample can be resolved by SDS–PAGE and autoradiography to detect the presence of the tandem HA-tagged Gγ or GGL protein (e.g., lower panels in Fig. 2). If there is no evidence of HA-tagged protein still present in the supernatant (i.e., all of the protein has been

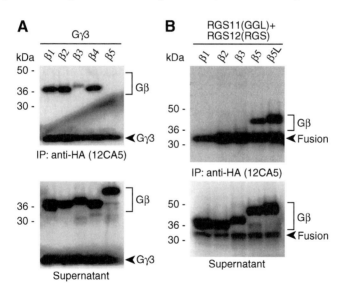

Fig. 2. Gβ subunit binding specificity of Gγ_3 and RGS11 GGL-domain fusion proteins. Indicated Gβ subunits were cotranslated *in vitro* in reticulocyte lysates with (A) tandem HA-tagged Gγ_3 or (B) a tandem HA-tagged chimeric protein ["RGS11(GGL) + RGS12(RGS)" or "Fusion"] composed of the RGS11 GGL domain (aa 219–300) fused to the rat RGS12 RGS-box (aa 716–838). Lysates were immunoprecipitated (IP) using the anti-HA antibody 12CA5 and washed in low-detergent buffer D. Immunoprecipitated proteins (upper panels) and clarified supernatants (lower panels) were separately resolved by SDS–PAGE. ^{35}S-Labeled Gβ and Gγ/GGL proteins were visualized by autoradiography. Relative positions of SeeBlue prestained molecular mass standards (Novex) are indicated (kDa).

immunoprecipitated), then the amount of 12CA5 antibody used in the immuno-precipitation must be reduced. The amount of anti-HA antibody employed in this experiment should be titrated so that only the minimum amount required to im-munoprecipitate the majority of HA-tagged protein is added to lysate samples; in our experience, too much 12CA5 antibody brings down translated proteins nonspecifically and thus results in a high background. Titration of immunopre-cipitating antibody is also important for reducing background in the cell lysate coimmunoprecipitation protocol described in the next section.]

Supernatant is removed and beads are washed using one of two buffer systems. "Low-detergent" immunoprecipitations are washed three times in 0.4 ml of buffer D (i.e., in the presence of 0.05% C12E10); these low-stringency washes are nec-essary to detect complex formation using the RGS7 GGL domain (see below). "High-detergent" immunoprecipitations are washed two times with 0.4 ml of RIPA-500 buffer [500 mM NaCl, 50 mM Tris-HCl pH 8.0, 1% Triton X-100 (v/v), 0.5% sodium deoxycholate (Sigma), 0.1% SDS, freshly added Complete protease inhibitors], followed by two washes with 0.4 ml of buffer D containing 0.1% Triton X-100 rather than 0.05% C12E10. During each wash, the beads are resuspended by repeated flicking of the base of the tube; beads are then repelleted by centrifu-gation at 10,000g for 1 min. Following the last wash, all remaining supernatant is removed and 70 μl of 2× Tris-glycine SDS sample buffer (containing 2-ME) is added prior to boiling for 5 min. Samples are then stored frozen at −80° or directly resolved by SDS–PAGE using 14% Novex Tris–glycine precast polyacrylamide gels for Gβ/Gγ coimmunoprecipitations and 10% or 12% Novex Tris–glycine gels for studies of Gβ/RGS interactions. Aliquots of clarified supernatants are sepa-rately resolved by SDS–PAGE to confirm that all input cDNAs are transcribed and translated (e.g., Fig. 2, lower panels).

Detection of Complex Formation and Detergent Considerations

After electrophoresis, SDS–PAGE gels are fixed for 30 min in 10% glacial acetic acid/25% methanol. The [^{35}S]methionine signal is amplified by the use of Enlightening Rapid Autoradiography Enhancer (NEN). The gel is then transferred to 3MM Whatman (Clifton, NJ) paper, covered with plastic wrap (avoiding wrin-kles, which can lead to cracks in the gel), and dried under vacuum for 1 hr at 80° [e.g., using a Speedvac Speed Gel SG200 gel drier (Thermo Savant, Holbrook, NY)]. After removing the plastic wrap, skim-milk powder is lightly sprinkled over the dried gel surface in order to eliminate static and stickiness. After shaking off excess milk powder, the dried gel is exposed to BIOMax film (Kodak, Rochester, NY) at −80° using a BIOMax intensifying screen until the desired exposure is obtained; an overnight exposure is usually taken first to assess signal strength.

Figure 2 shows two representative results of the *in vitro* transcription/trans-lation/coimmunoprecipitation protocol. The Gβ binding specificity of tandem

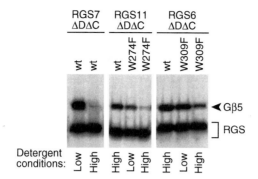

FIG. 3. Detergent sensitivity of the Gβ₅/GGL domain complex. Low- and high-detergent buffer conditions, used in washing coimmunoprecipitated Gβ₅/GGL/antibody complexes prior to gel electrophoresis, are defined in the text. The Gβ₅/RGS7ΔDΔC interaction is completely disrupted using high-detergent conditions (*left*), whereas the stability of Gβ₅/RGS11ΔDΔC (*middle*) and Gβ₅/RGS6ΔDΔC (*right*) complexes in high-detergent conditions is decreased by tryptophan-to-phenylalanine mutations (W274F for RGS11, W309F for RGS6) in the conserved, C-terminal Asn-Pro-Trp tripeptide (Fig. 1).

HA-tagged Gγ₃ is shown in Fig. 2A, revealing robust interactions with Gβ₁, Gβ₂, and Gβ₄, weak interaction with Gβ₃, and no binding with Gβ₅. Figure 2B illustrates the Gβ₅ binding specificity of a tandem HA-tagged fusion protein composed of the RGS11 GGL domain (aa 219–300 of SwissProt O94810) and the RGS12 RGS-box (aa 716–838 of SwissProt O08774). (This RGS11-GGL/RGS12-RGS-box fusion was created to obtain stable expression of the GGL fusion partner without additional polypeptide sequence from RGS11, as part of our assessment[5] of the minimal polypeptide sequence responsible for Gβ₅ association; we have observed that the RGS11 GGL domain, when expressed alone, is highly unstable and quickly degraded in this system.) Consistent with our previous findings of the fidelity of GGL domain association with Gβ subunits,[5,7] only Gβ₅ and Gβ₅L proteins are coprecipitated with the GGL/RGS-box fusion protein.

The coimmunoprecipitations displayed in Fig. 2 were washed using low-detergent conditions (0.05% C12E10) prior to SDS–PAGE. Detergent conditions are an important consideration in this protocol; it has previously been reported that the Gβ₅/Gγ₂ complex is unusually sensitive to low levels of detergent[17] compared to other Gβ/Gγ pairings. We have observed similar detergent sensitivity with the RGS7 GGL domain; washing of coimmunoprecipitated Gβ₅/RGS7/antibody complexes with high-detergent buffers (e.g., RIPA-500 buffer or buffer D + 0.1% Triton X-100) destroys the Gβ₅/GGL interaction (Fig. 3, left panel). Binding of Gβ₅ subunits to the GGL domains of RGS11 and RGS6 is less sensitive to

[17] M. B. Jones and J. C. Garrison, *Anal. Biochem.* **268**, 126 (1999).

high-detergent conditions (Fig. 3, middle and right panels). However, destabilizing the predicted $G\beta_5$/GGL interface[7] by replacing the tryptophan of the highly conserved GGL domain motif Asn-Pro-Trip with phenylalanine (the residue found in all $G\gamma$ subunits) results in increased detergent sensitivity [e.g., in Fig. 3, compare relative $G\beta_5$ coimmunoprecipitation using RGS11ΔDΔC(wt) vs RGS11ΔDΔC (W274F) and RGS6ΔDΔC(wt) vs RGS6ΔDΔC(W309F) in high-detergent conditions].

Cell Lysate Coimmunoprecipitation Analysis of $G\beta$/GGL Domain Association

Cell Cotransfection

All plasmid DNA stocks used in transient transfections are prepared using the Qiagen (Valencia, CA) QIAfilter Plasmid Maxi kit exactly according to the manufacturer's instructions and adjusted to a final concentration of 1 μg/μl in 1 mM EDTA/10 mM Tris pH 8.0. DNA transfections are performed on $\sim$80% confluent 10 cm plates of adherent monkey kidney fibroblasts (COS-7 cell-line, American Type Culture Collection, Manassas, VA) using SuperFect Transfection Reagent (Qiagen) according to the manufacturer's instructions for 10 cm-plate transfections with the following modification: for dual plasmid DNA cotransfections (i.e., a $G\beta$ subunit expression plasmid and a $G\gamma$/$G\gamma$-like subunit expression plasmid), 7 μg of each plasmid DNA is pooled for a total of 14 μg DNA. Cell monolayers are incubated for 2.5 hr at 37$°$ and 5% (v/v) CO_2 with complexed SuperFect reagent/plasmid DNA, washed 3 times with phosphate-buffered saline (PBS; 137 mM NaCl, 2.7 mM KCl, 4.3 mM NaH$_2$PO$_4$, 1.47 mM KH$_2$PO$_4$, pH 7.4), and then cultured for 48 hr in Dulbecco's modified Eagle's medium (DMEM) with 10% fetal bovine serum.

Cell Lysis

After incubation for 48 hr, cells are washed once with 10 ml of PBS. After the PBS wash is aspirated off, 1 ml of ice-cold RIPA-150 lysis buffer [150 mM NaCl, 50 mM Tris-HCl pH 7.5, 20 mM EDTA, 0.5% sodium deoxycholate, 1% Nonidet P-40 (NP-40), 0.1% SDS, freshly added Complete protease inhibitors] is added to the cell monolayer. Cells are scraped into a 1.5 ml microcentrifuge tube using a Costar cell lifter (Costar Corp., Cambridge, MA) and tubes are placed on ice; the cell lysate will resemble a "slimy blob" given the release of genomic DNA from lysed nuclei. Cell lysates are vortexed at maximum speed for 15 sec. Using a 1 ml syringe, the cell lysate is passed through an 18-gauge needle 10 times and then an additional 10 times using a 23-gauge needle to shear genomic DNA. Care is taken to thoroughly disperse the DNA blob, while avoiding foaming of the lysate when passing it through the syringe barrel. Tubes are then rocked for 30 min at 4$°$ on a Nutator [Clay Adams brand (Becton Dickinson, Franklin Lakes, NJ)] to complete the lysis.

Lysates are centrifuged at maximum speed ($\sim$16,000g) in a 4° microcentrifuge for 30 min; insoluble cellular "debris" will form a visible pellet at the tube bottom. The supernatant is transferred to a new, chilled 1.5 ml microcentrifuge tube and kept on ice. Protein content is quantitated using the detergent-compatible Dc Protein (Lowry) Assay (Bio-Rad, Hercules, CA) at 750 nm wavelength for absorbance according to the manufacturer's instructions. All lysates are adjusted to the same protein concentration by appropriate dilutions with ice-cold RIPA-150 lysis buffer. We usually obtain concentrations of 0.5–1.0 mg of protein per ml of cell lysate starting from an $\sim$80% confluent 10 cm plate of COS-7 cells. The lysates are stored frozen at −80° until the immunoprecipitation is performed.

Coimmunoprecipitation

To preclear lysates prior to immunoprecipitation, 20 μl of fresh 50% (v/v) protein A-Sepharose CL-4B slurry in RIPA-150 buffer is added to 1 ml of each lysate (>0.5 mg of total protein). Tubes are rocked for 45 min at 4° and then centrifuged for 1 min at 10,000g to pellet beads. Supernatant is transferred into a new 1.5 ml microfuge tube containing 2.8 μg of the anti-HA monoclonal antibody 12CA5. (The amount of 12CA5 antibody added at this step must be determined empirically depending on the particular lot of 12CA5. Once the appropriate amount is determined, the antibody should be aliquoted in small working volumes and stored frozen at −80°.) Tubes are rocked for 1 hr at 4° to allow for protein/antibody complex formation and briefly centrifuged at maximum speed to remove liquid from tube lids. The entire lysate mixture is transferred into a new 1.5 ml microfuge tube containing 40 μl of 50% (v/v) protein A-Sepharose slurry. Tubes are rocked again for 1 hr at 4° to allow for antibody/protein A complex formation and then centrifuged at 4° for 1 min at 10,000g to pellet beads. Supernatant is aspirated off the protein A beads using finely tapered gel-loading pipette tips and beads are washed twice by resuspension in 1 ml of ice-cold RIPA-500 buffer, followed by 3 washes in 1 ml of ice-cold RIPA-150 buffer. Following the last wash, all remaining liquid is aspirated off the protein A beads and 70 μl of 2× Tris–glycine SDS sample buffer (containing fresh 2-mercaptoethanol) is added to each tube; this volume provides enough protein sample for two immunoblotting experiments.

Electrophoresis and Immunoblotting

Protein samples are boiled for 5–10 min and resolved by SDS–PAGE: 10% and 12% Novex Tris–glycine polyacrylamide gels are suitable for Gβ/RGS protein coimmunoprecipitations, whereas 14% Novex Tris–glycine gels are used for Gβ/Gγ coimmunoprecipitations given the smaller molecular weight of Gγ subunits. Upon completion of electrophoresis, gels are soaked for 15 min in semidry transfer buffer (11.62 g Tris–base, 5.86 g glycine, 0.75 g SDS, 400 ml methanol, deionized water up to 2 liter). Proteins are then electroblotted (10 V, 20 min) onto

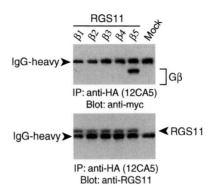

FIG. 4. Specific coimmunoprecipitation of RGS11 and Gβ$_5$. pcDNA3.1-based expression vectors for full-length, tandem HA-tagged RGS11 and *myc*-tagged Gβ subunits (β$_1$–β$_5$) were transiently cotransfected into COS-7 cells. Forty-eight hr later, cell lysates were harvested and immunoprecipitated (IP) with the anti-HA mouse monoclonal antibody 12CA5, and coimmunoprecipitated Gβ subunits were detected by immunoblotting (Blot) with anti-c-*myc* mouse monoclonal primary antibody and sheep anti-mouse Ig secondary antibody. IgG-heavy denotes heavy chain of immunoprecipitating 12CA5 antibody present in all immunoprecipitations, including from COS-7 cell lysate mock-transfected with parental pcDNA3.1 vector only (Mock). Immunoprecipitation of RGS11 is confirmed by immunoblotting with an affinity-purified rabbit antiserum raised against the RGS11 GGL domain.

semidry transfer buffer-presoaked nitrocellulose (BA85 Protran 0.45 μm pore nitrocellulose; Schleicher & Schuell, Keene, NH) using the Bio-Rad Trans-Blot SD semidry transfer cell exactly as described by the manufacturer's instructions. Nitrocellulose blots are blocked in 1× Blotto [5% nonfat dry milk and 0.1% Tween 20 in phosphate-buffered saline (PBS)] for at least 1 hr at room temperature, followed by incubation with primary antibody diluted in 1× Blotto. Following 4 washes with 0.5× Blotto (diluted in PBS), blots are incubated with horseradish peroxidase (HRP)-linked secondary antibody (if required) diluted in 1× Blotto, washed 4 times with 0.5× Blotto, and developed using enhanced chemiluminescence (ECL, Amersham Pharmacia, Piscataway, NJ) exactly as described by manufacturer's instructions. Sample results in Fig. 4 demonstrate the specific association of Gβ$_5$ with coexpressed full-length RGS11; if the chosen antibody detection system (see below) will also cross-react with the IgG heavy chain, it is important to include a mock-transfected (i.e., pcDNA3.1 parental vector DNA) control in the immunoprecipitations (see "Mock" lane in Fig. 4).

Antibodies for Western Blot Analysis

For detection of untagged Gβ subunits, blots are first incubated with a mixture of two rabbit anti-Gβ antisera (both diluted 1 : 1000) from Chemicon (Temecula, CA): anti-Gβ$_5$ and anti-Gβ common, which reacts against Gβ$_{1-4}$. Blots are then washed and incubated with a 1 : 6000 dilution of HRP-linked anti-rabbit anti-body (Amersham) prior to ECL detection. The use of N-terminally c-*myc* epitope-tagged Gβ expression vectors allows for alternative modes of detection. We have previously used 1 : 500-diluted anti-c-*myc* mouse monoclonal antibody (Invitrogen), followed by 1 : 6000-diluted HRP-linked sheep anti-mouse immunoglobulin (Ig) antibody (Amersham) (e.g., Fig. 4, upper panel), as well as 1 : 250-diluted anti-c-*myc*-HRP (Invitrogen). However, we now routinely use a 1 : 1000 dilution of HRP-linked anti-c-*myc* mouse monoclonal 9E10 (Cat#1814 150; Roche). (We have observed that C-terminally c-*myc* epitope-tagged Gβ subunits do not interact as efficiently with Gγ or GGL domain-containing proteins in cell lysate coimmunoprecipitations.)

Gγ subunits can be detected using 1 : 1000-diluted anti-Gγ$_1$, anti-Gγ$_2$, or anti-Gγ$_3$ rabbit antisera from Santa Cruz Biotechnology (Santa Cruz, CA), followed by HRP-linked anti-rabbit secondary antibody as above; however, the presence of an N-terminal tandem HA-epitope tag on each Gγ subunit allows for more facile detection using a 1 : 1000 dilution of HRP-linked anti-HA rat monoclonal antibody 3F10 (Roche). The N-terminal tandem HA-epitope tag present on RGS proteins allows their detection using anti-HA(3F10)-HRP as well; however, we have also successfully immunoblotted for RGS11 expression using affinity-purified rabbit antiserum raised against the RGS11 GGL domain[5] (e.g., Fig. 4, lower panel). We have used this anti-GGL antiserum in attempts to detect expression from a pcDNA3.1-based plasmid encoding just the RGS11 GGL domain (aa 219–281) but have not observed stable GGL polypeptide expression in transfected COS-7 cells.

Purification of Gβ$_5$/RGS11 Heterodimers from Insect Cell Expression

Generation of Baculoviral Shuttle Vectors

To create a recombinant baculovirus expressing His$_6$-tagged Gβ$_5$, the *Gβ5* open-reading frame is subcloned into the shuttle vector pFastBacHTb (Life Technologies, Gaithersburg, MD) directly 3′ of (and in-frame with) the tobacco etch virus (TEV) protease cleavage site. This strategy produces Gβ$_5$ protein with an N-terminal hexahistidine tag which, after TEV cleavage, leaves only one extra nonnative aminoacid (glycine) at the N terminus; especially for crystallization trials, it is often desirable to reduce or eliminate extraneous, nonnative amino acid extensions to purified recombinant proteins.

To subclone *Gβ5* into pFastBacHTb, the open-reading frame of mouse *Gβ5* is PCR amplified from the plasmid pcDNA3.1mGβ5 using sense primer 5′-ATGGCAACCGATGGGCTGCACGAGAACGAGACG-3′, antisense primer 5′-TTATGCCCAAACTCTTAGGGTGTGATCCCATGATCC-3′, and a proofreading thermostable DNA polymerase (e.g., *Pfu* from Stratagene, La Jolla, CA) with reaction conditions and methods as previously described.[16] PCR products are resolved by agarose gel electrophoresis and purified from gel slices using the QIAquick gel extraction system according to the manufacturer's directions (Qiagen Inc., Valencia, CA) in order to remove free nucleotides, excess primers, *Pfu* enzyme, and traces of input template DNA that could otherwise be carried into subsequent ligation reactions. Prior to ligation, purified PCR product is treated with ATP and T4 polynucleotide kinase (New England Biolabs, Beverly, MA) using manufacturer's directions to add a 5' phosphate group to both ends (given that synthetic oligonucleotide PCR primers lack 5' phosphates) and then directly repurified using the QIAquick gel extraction system. pFastBacHTb plasmid DNA is cleaved with the restriction enzyme *Ehe*I (New England Biolabs), which cuts the vector immediately 3' of the TEV protease recognition site, leaving blunt ends. EheI-cleaved vector DNA is purified by agarose gel electrophoresis and QIAquick gel extraction, then treated with calf intestinal phosphatase (New England Biolabs) using manufacturer's instructions to prevent vector religation. Blunt-end ligation is performed by mixing purified, linearized pFastBacHTb vector with *Gβ5* PCR product at a molar ratio of ~1 : 2 and incubating with T4 DNA ligase (New England Biolabs) using manufacturer's T4 DNA ligase buffer at 14° overnight. Ligation destroys the EheI site and places the N-terminal methionine for Gβ5 immediately after the TEV protease cleavage site. One μl of ligation mixtures (including a separate ligation reaction lacking Gβ5 insert to control for vector religation) is used to transform chemically competent *E. coli* strain XL-1 Blue (Stratagene) using manufacturer's protocols.

Ampicillin-resistant bacterial colonies are first screened by PCR amplification using the same *Gβ5* sense and antisense primers described above. A fraction of each colony is picked with a sterile pipette tip and homogeneously resuspended in 20 μl PCR mix containing 1× *Taq* PCR buffer (Life Technologies), 200 μ*M* dNTPs, 1.5 m*M* MgCl$_2$, 0.25 μ*M* of each primer, and 1 unit of Taq polymerase (Life Technologies). Amplification is performed with the following thermal cycling: initial denaturation at 94° for 10 min, 25 cycles of 94° for 1 min, 55° for 1 min, and 72° for 1 min, and a final extension at 72° for 10 min.

Potential positive clones are identified by the presence of a 1.1 kb PCR product (expected size of the *Gβ5* cDNA) upon agarose gel electrophoresis. Plasmid DNA is then purified from overnight liquid cultures of PCR-positive colonies by alkaline lysis and anion-exchange chromatography (QIAGEN Plasmid Mini columns; Qiagen Inc.), and digested with restriction endonucleases to check for proper insert orientation. Clones bearing correctly oriented *Gβ5* insert are sequenced using pFastBacHT-specific primers 5'-CCATCACCATCACCATCA-3' and 5'-GTTTCAGGTTCAGGGGGA-3' to confirm ligation junctions and eliminate the possibility of PCR-generated missense mutations.

To create the untagged baculoviral shuttle vector pFastBac 1hRGS11ΔD, the open-reading frame of human *RGS11* spanning the GGL domain, RGS-box, and C terminus (aa 219–467 of SwissProt O94810) is PCR amplified from the plasmid pcDNA3.1hRGS11 using sense primer 5′-<u>GAATTC</u>CGACATGACCAAGAGTG-CAGATTTCC-3′ (*Eco*RI site underlined), antisense primer 5′-*GTCGAC*TAAGC-CACCCCATCTCCACCCCCAGG-3′ (*Sal*I site in italics), and *Pfu* DNA polymerase, employing reaction conditions previously described.[16] The PCR product is subcloned into pCR2.1-TOPO vector and sequence verified as previously described.[16] The RGS11 cDNA fragment is isolated by digestion with *Eco*RI and *Sal*I and ligated into pFastBac1 vector DNA cleaved with *Eco*RI and *Xho*I. Positive clones are identified as detailed above and sequence verified using pFastBac1-specific primers 5′-CTGTTTTCGTAACAGTTTTG-3′ and 5′-GTTTCAGGTT-CAGGGGG-3′.

To create the baculoviral shuttle vector pFastBac1hRGS11ΔDΔC (i.e., with an *RGS11* open-reading frame spanning just the GGL domain and RGS-box regions; aa 219–418), site-directed mutagenesis is performed to replace the codon encoding Ala-419 with a stop-codon (5′-TGA-3′). Then, 28-nucleotide, overlapping sense (5′-CAAGGCCCTCCTG*TGA*GAGGCTGGGATC-3′) and antisense (5′-GATCCCAGCCTC*TCA*CAGGAGGGCCTTG-3′) primers are used in a PCR amplification of parental vector pFastBac1hRGS11ΔD with *Pfu* DNA polymerase, followed by *Dpn*I cleavage of input wild-type vector template, using the conditions of the QuikChange site-direction mutagenesis system (Stratagene).

[We have also created a third shuttle vector, pFastBac1hRGS11ΔDΔR, solely encoding the GGL domain (aa 219–281) as amplified by sense primer 5′-GAATTCGGCTCATATGACCAAGAGTGCAGATTTCC-3′ and antisense primer 5′-GCTAGCACTTAGTAGGCGTCATTGTCTGAGATC-3′. We have confirmed shuttle vector construction by sequencing and also verified proper transposition into the AcNPV bacmid by PCR screening (e.g., lanes E–H of Fig. 5); however, after generation of recombinant baculovirus and infection of Sf9 cells (or even coinfection with Gβ5-expressing virus), we have not been able to detect protein expression by immunoblotting with anti-RGS11 GGL-domain antiserum. This is consistent with our findings of lack of detectable, stable expression of the RGS11 GGL domain as an isolated polypeptide in COS-7 cell transfection experiments.]

Generation of Recombinant Baculoviral Genomes

Recombinant baculoviral genomes are created by transposition of pFastBac-HTbmGβ5 and pFastBac1hRGS11 shuttle vectors into the AcNPV bacmid pMON14272 using the Bac-to-Bac expression system (Life Technologies). pFast-Bac plasmid DNA is diluted to 2 ng/μl in 1 m*M* EDTA/10 m*M* Tris pH 8.0 and 1 μl (2 ng) added to 50 μl of competent DH10Bac *Escherichia coli* (containing pMON14272 and the helper plasmid pMON7124, which provides Tn7 transposition machinery *in trans*). The mixture of DNA and cells is incubated on ice for 30 min, at 42° for 45 sec, and then again on ice for 2 min. Two hundred μl of S.O.C medium is added to the mixture prior to incubation for 4 hr at 37°. Cell mixture is then spread evenly onto Luria agar plates containing 50 μg/ml kanamycin (to select for the bacmid), 7 μg/ml gentamicin (to select for the pFastBac vector), 10 μg/ml tetracycline (to select for the helper plasmid), 40 μg/ml IPTG and 100 μg/ml Bluogal (to allow for blue/white selection). After at least 24 hr growth at 37°, white colonies (along with one blue colony as a negative control) are used to inoculate 4 ml overnight cultures of LB medium supplemented with 50 μg/ml kanamycin, 7 μg/ml gentamicin, 10 μg/ml tetracycline. Bacmid DNA ($\sim$135 kb) is isolated from overnight cultures using alkaline lysis and anion-exchange chromatography (QIAGEN Plasmid Mini columns; Qiagen).

To confirm successful transposition, bacmid DNA is screened by PCR using M13 amplification primers (5′-GTTTTCCCAGTCACGAC-3′ and 5′-AACAGC-TATGACCATG-3′) directed at sequences flanking the site of transposition. PCR is performed using the Expand Long Template system (Boehringer Mannheim/Roche): A 50 μl final volume of H_2O containing 350 μM dNTPs, 0.5 μM of each primer, 10 mM Tris-HCl pH 8.3, 1.75 mM $MgCl_2$, 50 mM KCl, 0.2 μl of bacmid DNA, and 2.6 units of Long Template DNA polymerase is amplified using a "Touch-down" cycling protocol: initial denaturation of 94° for 2 min, 10 cycles of 94° for 10 sec, 58–0.4°/cycle for 30 sec, 68° for 3 min, followed by 20 cycles at 93° for 10 sec, 56° for 30 sec, 68° for 3 min + 20 sec/cycle, and a final incubation of 68° for 10 min. PCR products are then resolved by 0.8–1% agarose gel electrophoresis (e.g., Fig. 5). Bacmid DNA lacking transposition of the

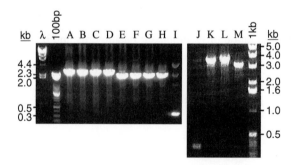

FIG. 5. Identification of recombinant AcNPV genomes by PCR amplification and agarose gel electrophoresis of PCR products. Lanes are as follows: AcNPV bacmid DNA transposed with pFastBac1hRGS11ΔD (lanes A–D), transposed with pFastBac1hRGS11ΔDΔR (lanes E–H), untransposed (lanes I–J), transposed with pFastBacHTbhGβ2 (lane K), transposed with pFastBacHTbmGβ5 (lane L), and transposed with pFastBac1hRGS11ΔDΔC (lane M). λ, *Hind*III-digested bacteriophage λ molecular weight marker; 100 bp, 100 bp ladder; 1kb, 1 kb pair ladder; ΔD, DEP domain deletion; ΔC, carboxy terminus deletion; ΔR, RGS-box deletion.

pFastBac shuttle vector yields a 300 bp PCR amplicon (lanes I, J of Fig. 5), whereas bacmid DNA transposed with empty pFastBac vectors yields a ~2300–2400 bp product. Transposition with pFastBac vectors containing inserts of RGS11 (lanes A–H and M, Fig. 5) or Gβ (lanes K–L, Fig. 5) cDNA increases the size of the expected PCR product correspondingly.

Generation of Recombinant Baculovirus and Testing for Protein Expression

To generate baculovirus, recombinant AcNPV bacmid DNA is transfected into *Spodoptera frugiperda* Sf9 cells (ATCC #CRL-1711). A stock culture of Sf9 cells is maintained in a 27° incubator in a spinner flask containing Grace's medium (Life Technologies) supplemented with 10% fetal bovine serum, 2 mM L-glutamine (Sigma), and an antibiotic/antimycotic mixture of penicillin, streptomycin, and amphotericin B (Life Technologies). Sf9 cells (9×10^5) are seeded into 35 mm wells of a 6-well plate and allowed to adhere by culturing in 2 ml of medium at 27° for 1 hr. Bacmid DNA (1 μg of Qiagen-column purified DNA) is transfected into the Sf9 cells using 6 μl of CellFECTIN lipid reagent (Life Technologies) exactly according to manufacturer's instructions. Then, 72 hr post transfection, 2 ml of cell culture medium is harvested and clarified by centrifugation for 5 min at 500g, and supernatant harboring baculovirus is transferred to a new tube for storage at 4° away from light.

After two rounds of amplification[18] of recombinant baculoviral stocks using low multiplicities of infection (MOI < 0.1 viral particle per cell), lysates from infected cells are tested by immunoblotting for protein production. Small-scale cultures of adherent Sf9 cells (1.8×10^6 cells in 1 ml Grace's medium per well of a 12-well plate) are infected with amplified baculovirus at an MOI of 3–5 (assuming a viral titer of ~1×10^8 pfu/ml) from the second round of amplification). Next, 48 hr postinfection, cells are harvested by centrifugation at room temperature for 5 min at 500g and washed once with PBS, and the cell pellet is lysed directly in 2× Tris–glycine SDS sample buffer for polyacrylamide electrophoresis and immunoblotting as previously described above. Sample results in Fig. 6 demonstrate expression of Gβ5 and RGS11ΔDΔC proteins upon infection (and coinfection) of Sf9 cells. Although both proteins are expressed alone (lanes A and B of Fig. 6), their stability on purification as uncomplexed monomeric proteins has not been tested.

Viral Stock Generation and Coinfection

Whereas *Spodoptera frugiperda* Sf9 cells are used for virus stock production, *Trichoplusia ni* cells[19] are used for protein production (available commercially from Invitrogen as the "High Five" cell line). We have found that High Five cells

[18] M. Lalumiere and C. D. Richardson, *Methods Mol. Biol.* **39,** 161 (1995).

[19] T. J. Wickham, T. Davis, R. R. Granados, M. L. Shuler, and H. A. Wood, *Biotechnol. Prog.* **8,** 391 (1992).

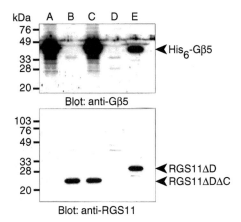

FIG. 6. Confirmation of baculoviral expression of His$_6$-Gβ_5 and RGS11ΔDΔC proteins. Western blots of whole insect cell extracts harvested 48 hr after infection with recombinant baculovirus encoding either His$_6$-Gβ_5 (lane A) or RGS11ΔDΔC (lane B), or coinfection with both viruses (lane C). Lysate from uninfected cells is present in lane D. As a positive control for antibody detection, purified Gβ_5/RGS11ΔD protein is present in lane E.

produce from slightly more to double the amount of Gβ_5/RGS complex as Sf9 cells, and High Five cells are able to be cultured in the simpler Ex-Cell 405 medium (JRH Biosciences, Lenexa, KS) supplemented only with 10 μg/ml gentamicin. High Five cells are grown in suspension in nonbaffled shaker flasks (Bellco, Vineland, NJ) at 27° at 130 rpm. High Five cultures are easily prone to aggregation if under- or overgrown; therefore, a strong effort is made to keep culture density in the range ~0.5–3 × 10^6 cells/ml. High Five cells are infected with baculovirus at a density of 1.5 × 10^6 cells/ml; infections are always performed using actively growing cells in mid-logarithmic phase, rather than on cultures freshly diluted down to 1.5 × 10^6 cells/ml from higher densities.

To produce viral stocks for High Five infections, Sf9 cell cultures (volumes from 250 ml to 600 ml) are grown at 27° to a density of 1 × 10^6 cells/ml, at which point cultures are infected with baculovirus at an MOI of ~0.1. Infection of cells at higher initial densities is found to produce much lower titer viral stocks. Infected cells are grown for 4 days and pelleted by centrifugation for 10 min at 2000g, and the virus-containing supernatant is collected; whereas 1 ml aliquots are stored in sterile cryovials in −80° for long-term storage, working stocks are stored in sterile 50 ml polypropylene conical tubes at 4° protected from light. Virus is not routinely titered for plaque-forming units (pfu)[18] but rather is assumed to be ~1 × 10^8 pfu/ml.

For each batch of new baculovirus stock, small-scale infections of High Five suspension culture are performed at MOI values between 2 and 5, while also varying the ratio of the two viruses in the coinfection; for example, Gβ5-expressing baculovirus is used at an MOI of 2 in coinfection with RGS11-expressing baculovirus at MOIs of 2, 3, and 5. The relative and total amounts of the two proteins produced in each coinfection is analyzed by small-scale Ni-NTA chromatography (see below) to identify coinfection conditions that maximize complex yield.

Once virus ratio and amounts are established, a 4 liter total volume of High Five cells in Ex-Cell 405 medium [i.e., two 2 liter cultures in 6 liter nonbaffled glass flasks (Bellco, Vineland, NJ)] is coinfected with Gβ5- and RGS11-expressing baculoviruses. Cells are cultured at 27° for 48–60 hr.

Cell Lysis

Infected High Five cells are harvested in 1 liter bottles by centrifugation at 4° for 5 min at 1000g in a swinging bucket rotor. [Just before this centrifugation, a 1 ml sample of cell culture is saved to test for protein expression (see below) before commiting to lysate preparation with the full amount of cells.] Cell pellets are washed in 0.1× volume of ice-cold PBS, and repelleted by centrifugation in 50 ml conical tubes. Supernatant is poured off and the cell pellets are flash-frozen in a dry-ice/2-propanol bath. Frozen cell pellets are stored at −80° until required.

Cells within the 1 ml test sample are pelleted by brief centrifugation at maximum speed in a microfuge, washed once with ice-cold PBS, and resuspended in 100 μl PBS. To estimate total protein content of this cell suspension, 1 μl and 2 μl aliquots are withdrawn and mixed with 800 μl H_2O and 200 μl Protein Assay Dye Reagent (Bio-Rad); after 5 min incubation, the absorbance of resultant solutions is measured at 595 nm and compared to a BSA standard curve to convert to μg/ml protein. An aliquot of the cell suspension corresponding to ∼25 μg total protein is then mixed with an equal volume of 2× Tris–glycine SDS sample buffer, boiled for 5 min, and resolved by electrophoresis through a 12% polyacrylamide Tris–glycine gel (Novex) and Coomassie blue staining. Good expression will be evidenced by the appearance of protein bands corresponding to predicted sizes (e.g., ∼48 kDa for His_6-tagged Gβ5 and ∼23 kDa for RGS11ΔDΔC) when compared to the same amount of total protein from uninfected High Five cells.

Frozen cell pellets from the 4 liter infected culture are thawed on ice and resuspended in 0.05× volume of lysis buffer (see Table II for buffers used in purification procedures). Thirty ml aliquots of resuspended cells in 50 ml polypropylene conical tubes are each sonicated by four 30-sec continuous bursts at 11–12 watts (RMS) on a Fisher Scientific (Pittsburgh, PA) Model 60 Sonic Dismembrator, with cooling on ice between 30-sec bursts. After sonication, insoluble material in the lysate is pelleted by centrifugation at 4° for 35 min at 100,000g in a Beckman ultracentrifuge in a fixed angle rotor. Although the presence of the nonionic detergent β-octylglucoside (β-OG) in the lysis buffer solubilizes a good quantity of the Gβ5/RGS11 complex, SDS–PAGE/Coomassie staining analysis of the insoluble

TABLE II

BUFFERS USED IN PURIFICATION OF $G\beta_5$/RGS11 COMPLEXES

Buffer	Tris-HCl, pH 8.0	MgCl$_2$	NaCl	Protease inhibitors[a]	DTT	β-Octylglucoside
Lysis	20 mM	1 mM	150 mM	+	—	0.5%
M	20 mM	—	20 mM ~ 500 mM		2 mM	—
S	20 mM	—	100 mM	—	2 mM	—

[a] 100 μg/ml pepstatin, 20 μg/ml phenylmethylsulfonyl fluoride (PMSF), 20 μg/ml tosyllysine chloromethyl ketone (TLCK), 30 μg/ml soybean trypsin inhibitor.

material postsonication reveals a substantial amount of insoluble complex. It is not known whether the protein remaining in the pellet is denatured or more tightly associated with the cell membrane than that fraction gently extracted by the use of β-OG lysis buffer. Detergent is not required for maintaining the solubility of the $G\beta_5$/RGS11 complex in later purification steps and is thus omitted from these subsequent steps.

Nickel Chelation Affinity Chromatography

Supernatant from the centrifuged lysate is pooled and batch-mixed with 5 ml of Ni-NTA Superflow resin (Qiagen) previously washed in lysis buffer containing 15 mM imidazole. (A stock solution of 3 M imidazole is adjusted to pH 7.5 before addition to protein purification buffers.) The lysate/Ni-NTA bead mixture is rocked gently for 1.5–2 hr at 4° and then transferred into a Bio-Rad Econo-Column at 4°. The flow-through is collected at 1 ml/min and the resin is then sequentially washed with the following buffers:

W1 : 30 ml lysis buffer containing 15 mM imidazole
W2 : 30 ml lysis buffer containing 15 mM imidazole and 0.5 M NaCl
W3 : 30 ml lysis buffer containing 30 mM imidazole

The $G\beta_5$/RGS11 complex is eluted with five applications of 5 ml lysis buffer containing 0.3 M imidazole. The major peak of eluted protein is usually found in the second fraction (e.g., lane E3 of Fig. 7).

Ion-Exchange Chromatography

Peak fractions from the metal chelation chromatography are placed in Spectra-pore dialysis tubing (Spectrum Labs, Rancho Dominguez, CA) along with 100–500 units of recombinant TEV protease (Life Technologies) and simultaneously cleaved by protease and dialyzed against 20 mM Tris-HCl, pH 8.0, 20 mM NaCl, 2 mM DTT ("low-salt" buffer M). Cleavage/dialysis is performed overnight at 4°.

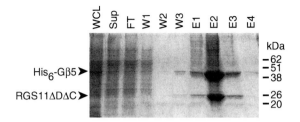

His₆-Gβ5 ►

RGS11ΔDΔC ►

FIG. 7. 12% SDS–PAGE of column fraction from Ni-NTA batch binding and elution of His₆-Gβ₅/RGS11ΔDΔC complex. Proteins are visualized by staining with Coomassie blue. WCL, whole cell lysate; Sup, supernatant after centrifugation of whole cell lysate; FT, flow-through after application of supernatant to Ni-NTA resin; W1, wash with lysis buffer + 15 mM imidazole; W2, wash with lysis buffer + 15 mM imidazole/0.5 M NaCl; W3, wash with lysis buffer + 30 mM imidazole; E1–E4, serial elutions with lysis buffer + 300 mM imidazole.

Dialysis into low-salt buffer M prepares the protein preparation for subsequent loading onto a 1 ml Mono Q strong anion-exchange column (Pharmacia Mono Q HR 5/5). Before column loading, a sample of dialyzed protein is tested for TEV cleavage of the hexahistidine tag by gel electrophoresis through a 20% PhastGel (Pharmacia) in parallel with a sample of uncleaved protein. Occasionally, some protein precipitates during the overnight dialysis; therefore, the protein sample is centrifuged at 4° for 10 min at 20,000g before it is loaded on the Mono Q column.

Sample is loaded onto the Mono Q column on a Pharmacia FPLC system at 4° at ∼0.5-1 ml/min, followed by a wash with low-salt buffer M until no protein is released from the column [as measured by in-line absorbance detection (OD_{280nm})]. This is followed by a similar OD_{280nm}-monitored wash with buffer M containing 50 mM NaCl. For elution, a 40 ml gradient of buffer M from 50 mM to 500 mM NaCl is run over the column; the Gβ₅/RGS11ΔDΔC complex elutes around 225 mM. Figure 8 illustrates the elution of the complex. Peak fractions contain not only Gβ5 and RGS11ΔDΔC, but also an unidentified contaminant (∼15 kDa), which we suspect is a TEV protease fragment, as this protein is not present in the Ni-NTA eluate prior to TEV protease addition ("−TEV" lane, Fig. 8), but is present in the starting material applied to the Mono Q column ("+TEV" and "input" lanes, Fig. 8).

Gel Filtration Chromatography

Peak fractions from the Mono Q elution (F10–12, Fig. 8) are pooled and loaded onto a 150 ml size exclusion column (Pharmacia Superdex 75 XK 16/60) pre-equilibrated in buffer S (Table II). This chromatography step serves two purposes: (1) it separates the Gβ₅/RGS11ΔDΔC complex (predicted molecular weight of 61,865) from the ∼15 kDa TEV protease artifact, and (2) it establishes that the

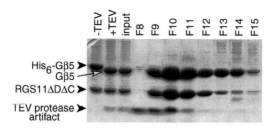

FIG. 8. TEV protease cleavage of hexahistidine tag and Mono Q anion exchange chromatography of resultant Gβ5/RGS11ΔDΔC complex. Protein samples are resolved by 12% SDS–PAGE and visualized by staining with Coomassie blue. "−TEV," eluted His6-Gβ5/RGS11ΔDΔC complex from Ni-NTA chromatography; "+TEV," complex after concomitant dialysis and cleavage with recombinant protease from tobacco etch virus; "input," complex prior to Mono Q chromatography; lanes F8–F15, column fractions during elution. Fraction F9 contains a small amount of uncleaved protein; thus, fractions F10–F12 are pooled prior to gel filtration chromatography.

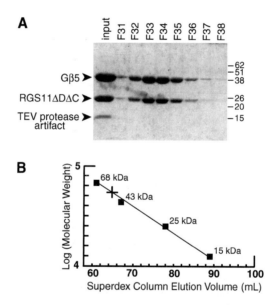

FIG. 9. Gel filtration chromatography of Gβ5/RGS11ΔDΔC complex. (A) 12% SDS–PAGE/ Coomassie stain detection of eluted protein within column fractions from Superdex 75 gel filtration. Peak fractions F10–F12 containing Gβ5/RGS11ΔDΔC protein from Mono Q column elution were applied ("input" lane). After gel filtration, the TEV protease artifact has been removed and the Gβ5/RGS11ΔDΔC complex is now homogeneous. (B) Determination of the approximate molecular weight of the Gβ5/RGS11ΔDΔC complex. Closed squares demarcate elution volume of calibrating molecular weight standards, whereas the plus symbol (+) represents elution volume of Gβ5/RGS11ΔDΔC complex. The complex elutes as a single symmetrical peak at a position corresponding to ∼58 kDa, in excellent agreement with the calculated molecular weight (62 kDa) of a monodisperse Gβ5/RGS11ΔDΔC heterodimer.

Gβ5/RGS11ΔDΔC complex behaves as a nonaggregated heterodimer. Monodispersity is important to establish before using purified Gβ5/RGS11ΔDΔC protein in biochemical assays (including Gα-directed GAP activity[5] and in crystallization trials). Figure 9A illustrates the peak fractions for the Gβ5/RGS11ΔDΔC complex as analyzed by SDS–PAGE and Coomassie staining, demonstrating the removal of larger contaminants and, especially, the TEV protease artifact, on gel filtration. Figure 9B shows the calibration curve for the Superdex 75 column and the position at which the Gβ5/RGS11ΔDΔC complex elutes—corresponding to a molecular mass of ~58 kDa.

Concentration and Storage

Peak fractions from gel filtration are pooled and concentrated by ultrafiltration using a 10-kDa molecular weight cutoff membrane (Centricon-10) following manufacturer's instructions (Millipore/Amicon, Bedford, MA). Final protein concentration is measured by diluting a small aliquot in 6 M guanidine hydrochloride and measuring absorbance (OD_{280nm}). The extinction coefficient for the complex (e.g., 9.7×10^4 M^{-1} cm^{-1} for Gβ5/RGS11ΔDΔC) is calculated using the proteomics tool "ProtParam" on the Swiss Institute for Bioinformatics ExPASy Web site: www.expasy.ch) and used to determine the protein concentration in mg/ml. The final total yield of Gβ5/RGS11ΔDΔC complex is generally ~1–2 mg per liter of infected High Five culture. Concentrated protein is divided into 50 μl aliquots at 0.5 ml microfuge tubes and flash-frozen in liquid nitrogen prior to long-term storage at −80°.

Acknowledgments

This work was supported in part by the National Institutes of Health Grant GM62338 (to D.P.S.), a University of North Carolina Faculty Research Grant (to L.B.), and a supported research agreement from Covance Biotechnology (to J.S.).

[50] RGS Function in Visual Signal Transduction

By WEI HE and THEODORE G. WENSEL

GTP Hydrolysis and Recovery of Light Responses

The biochemical machinery by which the absorption of light by photoreceptor cells of the vertebrate retina elicits a change in membrane potential represents a classic G protein pathway.[1] The first amplified step in the phototransduction cascade is catalysis by photoexcited rhodopsin (R*) of GTP binding to the transducin α subunit ($G_t\alpha$). GTP-bound $G_t\alpha$ activates the key effector enzyme cGMP phosphodiesterase (PDE), and apparently continues to do so until GTP is hydrolyzed, returning G_t to the resting state. Thus GTP hydrolysis is a prerequisite for recovery of the light response, as demonstrated by studies employing nonhydrolyzable GTP analogs.[2-4] It has been proposed that the kinetics of photoresponses are limited by a single rate-limiting step giving rise to a "dominant time constant" of phototransduction[5,6] and GTP hydrolysis is a viable candidate for that rate-limiting step.[4] In this chapter we focus on methods of analyzing the molecular mechanisms for control of $G_t\alpha$ GTP hydrolysis kinetics, with particular emphasis on the role of RGS proteins in this regulation.

Proteins Implicated in Regulating GTP Hydrolysis in Photoreceptors

A number of different proteins have been implicated in GTP hydrolysis regulation. Early studies suggested that at high membrane concentrations, including those in intact cells, GTP hydrolysis was much faster than observed when membranes were diluted or purified G_t was used.[7-10] Biochemical studies pointed to an accelerating factor in rod outer segments,[11,12] but unequivocal determination

[1] L. Stryer, *Annu. Rev. Neurosci.* **9**, 87 (1986).

[2] T. D. Lamb and H. R. Matthews, *J. Physiol.* **407**, 463 (1988).

[3] W. A. Sather and P. B. Detwiler, *Proc. Natl. Acad. Sci. U.S.A.* **84**, 9290 (1987).

[4] M. S. Sagoo and L. Lagnado, *Nature* **389**, 392 (1997).

[5] D. R. Pepperberg, M. C. Cornwall, M. Kahlert, K. P. Hofmann, J. Jin, G. J. Jones, and H. Ripps, *Vis. Neurosc.* **8**, 9 (1992).

[6] S. Nikonov and E. N. Pugh, Jr., *J. Gen. Physiol.* **111**, 7 (1998).

[7] E. A. Dratz, J. W. Lewis, L. E. Schaechter, K. R. Parker, and D. S. Kliger, *Biochem. Biophys. Res. Commun.* **146**, 379 (1987).

[8] R. Wagner and R. Uhl, *FEBS Lett.* **234**, 44 (1988).

[9] V. Y. Arshavsky, M. P. Antoch, and P. P. Philippov, *FEBS Lett.* **224**, 19 (1987).

[10] V. Y. Arshavsky, M. P. Antoch, K. A. Lukjanov, and P. P. Philippov, *FEBS Lett.* **250**, 353 (1989).

[11] J. K. Angleson and T. G. Wensel, *Neuron* **11**, 939 (1993).

[12] T. M. Vuong and M. Chabre, *Proc. Natl. Acad. Sci. U.S.A.* **88**, 9813 (1991).

of the factor's identity at the molecular level has taken much longer than the identification of GTPase accelerating proteins, or GAPs for small G proteins.[13,14] Along the way evidence accumulated that one or more subunits of the effector PDE could also influence GTP hydrolysis kinetics,[15–17] although it is now clear that the inhibitory γ subunit of PDE does not itself act as a GAP.[11,18] Rather it acts via enhancement of the activity of the GAP, now known to be a complex of the RGS (regulator of G protein signaling) protein RGS9-1 and the G protein β homolog Gβ_{5L}.[19,20]

Recognition of the existence a sizeable family of proteins[21,22] negatively regulating G protein signaling through GTPase acceleration by a conserved RGS domain led to the search for photoreceptor RGS proteins. This search led to reports that multiple RGS proteins are present in the retina, and that two initial candidates, RGS16[23] and RET-RGS1,[24] are enriched in the retina relative to other tissues. What we describe here are approaches used, with selected procedures described in detail, to find and characterize the RGS protein that mediates rapid recovery of the vertebrate light response, RGS9-1.[20,25–27]

Biochemical Assays of GTPase Acceleration

A key method for studies of RGS9-1 function is an *in vitro* single-turnover assay for GTPase acceleration. Methods for measuring transducin GTPase kinetics have been described in an earlier volume of "Methods in Enzymology."[28] Here we describe the application of this method to assaying recombinant proteins for GAP activity.

Solutions and Materials

GAPN Buffer: 10 mM Tris-HCI, pH 7.4, 100 mM NaCl, 2 mM MgCl$_2$, 1 mM dithiothreitol (DTT), solid PMSF

[13] M. Trahey and F. McCormick, *Science* **238,** 542 (1987).
[14] M. S. Boguski and F. McCormick, *Nature* **366,** 643 (1993).
[15] V. Y. Arshavsky and M. D. Bownds, *Nature* **357,** 416 (1992).
[16] F. Pages, P. Deterre, and C. Pfister, *J. Biol. Chem.* **267,** 22018 (1992).
[17] F. Pages, P. Deterre, and C. Pfister, *J. Biol. Chem.* **268,** 26358 (1993).
[18] J. K. Angleson and T. G. Wensel, *J. Biol. Chem.* **269,** 16290 (1994).
[19] E. R. Makino, J. W. Handy, T. Li, and V. Y. Arshavsky, *Proc. Natl. Acad. Sci. U.S.A.* **96,** 1947 (1999).
[20] C. K. Chen, M. E. Burns, W. He, T. G. Wensel, D. A. Baylor, and M. I. Simon, *Nature* **403,** 557 (2000).
[21] M. R. Koelle and H. R. Horvitz, *Cell* **84,** 115 (1996).
[22] D. M. Berman, T. M. Wilkie, and A. G. Gilman, *Cell* **86,** 445 (1996).
[23] C. K. Chen, T. Wieland, and M. I. Simon, *Proc. Natl. Acad. Sci. U.S.A.* **93,** 12885 (1996).
[24] E. Faurobert and J. B. Hurley, *Proc. Natl. Acad. Sci. U.S.A.* **94,** 2945 (1997).
[25] W. He, C. W. Cowan, and T. G. Wensel, *Neuron* **20,** 95 (1998).
[26] C. W. Cowan, R. N. Fariss, I. Sokal, K. Palczewski, and T. G. Wensel, *Proc. Natl. Acad. Sci. U.S.A.* **95,** 5351 (1998).

Charcoal slurry: 0.05 g activated charcoal/ml potassium phosphate, 50 mM
   P$_i$, pH 7.5
[γ-^{32}P]GTP (typically 1–2 × 10^4 dpm/μl, 150 nM total GTP, frequently sup-
   plemented with AMP-PNP (5'-adenylyl imidodiphosphate), 1–2 mM
Trichloroacetic acid (5%; 6% perchloric acid may be substituted)
Urea-stripped ROS membranes (UROS) reconstituted with purified G$_t\alpha\beta\gamma$[28]
Tape recorder, vortexer, stopwatch

Procedure

The assay begins with the G$_t$/UROS mixture at concentrations of 15 μM
rhodopsin, 1 μM G$_t$, in GAPN buffer. Just before the assay begins, the sample
is exposed to room light, the tested proteins are added to the desired concentration,
and the total volume is filled to 14 μl by GAPN buffer. At time zero, with the
tape recorder running, 7 μl of the GTP stock (50 nM final GTP concentration) is
added while reaction is being vortexed, and verbally marked on the tape recorder.
A few seconds later, the reaction is quenched by addition of 100 μl TCA and
another verbal note recorded. Released [^{32}P]P$_i$ is assayed by charcoal binding and
scintillation counting, with appropriate controls, as described.[28] The stopwatch
and taperecorder are used to determine the elapsed reaction time for each sample.

Candidate Gene Approach

The approach that turned out to be most successful for understanding the role
of various proteins in GTPase regulation in photoreceptors was based on homology to other G-protein-coupled signaling systems and a hunt for candidate RGS
proteins.[25] The RGS domain is conserved sufficiently for PCR with degenerate
primers to be an effective approach to identifying a range of RGS proteins expressed in a particular tissue.[21] To enhance the chances of finding RGS proteins
selective for photoreceptors or for neuronal signaling, use was made of the identification of human RGS7 and RGS4, and of *Caenorhabditis elegans* EGL-10 as
RGS proteins important for neuronal signaling. Primer pairs were designed with
sufficient degeneracy to allow for amplification of the corresponding cDNAs in
the hope that the photoreceptor GAP would share greater structural similarity with
these RGS proteins than with those, for example, implicated in B cell function. In
addition, primer pairs used in successful amplification of RGS domain-encoding
fragments from rat brain cDNA[21] were used. To further enhance chances of finding
the desired photoreceptor RGS, several mammalian retinal cDNA libraries were
screened.

---

[27] K. Zhang, K. A. Howes, W. He, J. D. Bronson, M. J. Pettenati, C. Chen, K. Palczewski, T. G. Wensel,
   and W. Baehr, *Gene* **240**, 23 (1999).
[28] C. W. Cowan, T. G. Wensel, and V. Y. Arshavsky, *Methods Enzymol.* **315**, 524 (2000).

Sequencing of the products revealed the presence of multiple RGS cDNAs in the mammalian retinal libraries: RGS3, RGS4, RGS6, RGS7, RGS9, RGS11, and RGS16 from the screen described here, and RGS16, RGS2, RGS8, and RET-RGS1 from screens in other laboratories.[23,24] Three of these, RGS9-1, RGS16, and RET-RGS1, were subsequently found to be enriched in the retina relative to most other tissues, and two, RGS7 and RGS4, were already known to be largely confined to the central nervous system. Several of the PCR products were used to screen the retinal cDNA libraries for clones, and ultimately these were obtained: full-length bovine RGS4, RGS7, RGS9, RGS16, and RET-RGS1, as well as full-length murine RGS7, RGS9, and RGS16, and partial murine RGS6 and RGS11.

mRNA Expression Analysis

One criterion applied for initial screening of the candidate RGS members was whether their expression was enriched in the photoreceptor layer of the retina. Phototransduction components are in general overrepresented in total retinal mRNA. The reasons are (a) the protein components of phototransduction are highly concentrated in the disks of the photoreceptor outer segments, and (b) there is a daily turnover of disk membranes due to the disk shedding process, requiring continual resynthesis of the phototransduction components. For identifying messages present in and enriched in the retina, we used Northern analysis of RNA (Fig. 1) isolated from retina and other tissues, as well as commercially available Multiple Tissue Northern blots. The procedure described below is for isolation and analysis of bovine retinal RNA and is modified from a standard procedure.[29]

Solutions and Materials

> GITC buffer: 4 M guanidine isothiocyanate, 30 mM sodium acetate, pH 7;
>   add 1% 2-mercaptoethanol before use
> CsCl buffer: 5.7 M CsCl, 30 mM sodium acetate, pH 7, 10 mM EDTA
> Salt-saturated phenol
> Chloroform/isoamyl alcohol (24:1)
> Absolute ethanol
> Diethyl pyrocarbonate (DEPC) treated water
> Beckman ultracentrifuge with SW 28 rotor
> Tissue blender (Waring, Philadelphia, PA)
> High shear tissue homogenizer (Tekmar-Dohrmann, Mason, OH)
> RNase inhibitor spray for treating labware and surfaces
> RNase-free pipette tips

---

[29] L. Davis, M. Kuehl, and J. Battey, "Basic Methods in Molecular Biology," p. 322. Appleton & Lange, Norwalk, CO, 1994.

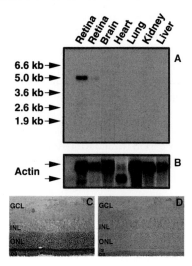

FIG. 1. Photoreceptor-specific expression of bovine RGS9-1 mRNA. Reproduced by permission from (Ref. 25). (A) Northern blot analysis of total RNA (from left to right): retina (10 μg), retina (3 μg), brain (10 μg), heart (6 μg), lung (10 μg), kidney (10 μg), liver (10 μg). (B) Control blot using human β-actin probe. (C, D) Localization of RGS9-1 mRNA to photoreceptor cell layer of the retina. *In situ* hybridization of murine eye sections using mouse RGS9 antisense (C) or sense (D) riboprobe, labeled with digoxigenin (C and D). GCL, ganglion cell layer; INL, inner nuclear layer; ONL, outer nuclear layer (contains cell bodies and nuclei of rod and cone photoreceptors); is, photoreceptor inner segment layer; os, photoreceptor outer segment layer. Reproduced with permission from W. He, C. W. Cowan, and T. G. Wensel, *Neuron* **20,** 95 (1998). Copyright 1998 Cell Press.

Sterile culture tubes and RNase-free microfuge tubes
Centrifugal vacuum evaporator

Procedure

Twenty-five frozen bovine retinas (from Lawson, ~25 g) are placed in pre-chilled 200 ml GITC buffer in tissue blender and blended immediately until no big tissue chunks can be observed. These retina lysates are then aliquotted into 50 ml conical tubes and further homogenized on ice four times for 15 s by Polytron homogenizer. The retina lysates can be stored at −80° for extended periods without significant RNA degradation.

For RNA preparation using an SW 28 rotor, 9.5 ml of 5.7 *M* CsCl solution is poured into the bottom of the centrifuge tubes at room temperature, then 29 ml retina lysates is carefully layered on the top of the CsCl solution. The mixture is spun at 20°, 25,000 rpm for 24 hr. Most of supernatants are removed by suction using a disposable 25 ml pipet. The centrifuge tubes are turned upside-down and the bottom (~1 cm) of the tubes is cut away by razor blade. The retinal RNA at the bottom of the tubes is resuspended in 400 μl DEPC water, then the bottom of the tubes is rinsed twice with addition of 400 μl water each time. For 4 SW28 tubes, the total volume is 4.8 ml. The pooled RNA is supplemented with 480 μl 3 M sodium acetate, then extracted sequentially with 4.8 ml of phenol/chloroform (1 : 1) and 4.8 ml chloroform. The extracted RNA is then aliquotted into 12 microcentrifuge tubes (400 μl in each) and precipitated by 1 ml ethanol per tube. After ethanol removal, the samples are dried under vacuum and stored dry at $-80°$, under which conditions they are stable indefinitely. One tube is sacrificed for RNA quantification by UV absorbance after RNA is dissolved in DEPC-treated water. It should be possible to obtain at least 200 μg per retina. Up to 15 μg can be loaded onto a gel for Northern blotting without band distortion due to overloading, and up to 30 μg if some distortion can be tolerated. The message for RGS9-1, a protein present at about 1/1600 of the level of rhodopsin, can be easily detected in 3 μg total RNA. For more rare messages, poly(A)$^+$ RNA can be easily prepared from total retinal RNA.[29]

To localize the cell layers of the retina in which RGS mRNAs are expressed, we used digoxigenin-based *in situ* hybridization (DIG-ISH)[25] starting with paraffin-embedded sections, or frozen sections. The following is a DIG-ISH protocol for paraffin-embedded sections modified from the ^{35}S *in situ* protocol.[30]

Solutions and Materials

    Xylene
    Ethanol
    Phosphate-buffered saline (PBS) buffer (GIBCO, Grand Island, NY)
    Proteinase K (GIBCO)
    TE: 10 mM Tris, pH 7.6, 1 mM EDTA
    PFA buffer: 4% Paraformaldehyde in PBS
    NTE: 100 mM Tris, pH 8.0, 0.5 M NaCl, 1 mM EDTA
    Acetylation buffer: 240 μl acetic anhydride in 100 ml triethanolamine, pH 8.0
    SSC: 30 mM Sodium citrate, pH 7.0, 300 mM NaCl
    Prehybridization solution:

Formamide	5.0 ml
SSC (20×)	2.0 ml
Denhardt's reagent (50×)	0.2 ml
Herring sperm DNA(10 mg/ml)	0.5 ml
Yeast tRNA (10 mg/ml)	0.25 ml
Dextran sulfate (50%)	2.0 ml

[30] U. Albrecht, G. Ichele, J. A. Helms, and H.-C. Lu, *in* "Molecular and Cellular Methods in Developmental Toxicology" (G. P. Daston, ed.), p. 23. CRC Press, Boca Raton, FL, 1997.

Buffer 1: 100 mM Tris, pH 7.5, 150 mM NaCl
Buffer 2: 0.3% Triton X-100 in buffer 1
Buffer 3: 100 mM Tris, pH 9.5, 100 mM NaCl, 50 mM MgCl$_2$
NBT/BCIP: 50 μl NBT [100 mg/ml in 70% dimethylformamide (DMF),
    Roche] and 37.5 μl BCIP (50 mg/ml in DMF) and 1 mM levamisole (Sigma,
    St. Louis, MO) in buffer 3
Coplin jars
Hot shaker (Bellco)
37° water bath
42° incubator
Wet chamber

Procedure

The paraffin-embedded sections are first dewaxed in xylene for 2 × 10 min,
then rehydrated sequentially in 100% ethanol, 2 × 2 min; 95%, 80%, 70%, 50%
ethanol, 20 sec each; 30% ethanol/0.9% NaCl, 20 sec; 0.9% NaCl, 5 min; PBS,
5 min. To get good penetration of RNA probes and anti-DIG antibodies, the sections are then permeabilized sequentially by 0.2 N hydrochloric acid for 20 min;
2× SSC (start at 65°, then let it cool down with sections) for 15 min; and 20 μl/ml
proteinase K (in 5× TE) at 37° for 10–30 min. Then the sections are washed by
PBS for 5 min, postfixed by PFA buffer for 5 min, and washed again by PBS for
5 min. Nonspecific binding sites on the sections are blocked in acetylation buffer
for 2× 5 min. The sections are equilibrated in 2× SSC for 10 min and used for
hybridization.

For hybridization, 200 μl prehybridization solution, which has been heated
at 95° for 5 min, cooled down on ice, and supplemented with dithiothreitol (50 μl
1 M DTT/1 ml prehybridization solution), is added on each slide. Then each slide
is covered with a piece of Parafilm and kept in a wet chamber at 42°. After 2 hr of
incubation, the prehybridization solution is removed, and 30 μl prehybridization
solution with the probe, which has been heated at 95°, 5 min, is added on each
section. The sections are then covered with a piece of Parafilm and kept in a wet
chamber at 42° overnight.

For washes, the slides are first dipped into 5× SSC at 62–66° for 20 min to
remove the Parafilm covers, then sequentially washed by 50% formamide/2× SSC
at 62–66° for 30 min; NTE at 37° for 3× 10 min; NTE with 20μg/ml RNase A
at 37° for 30 min; NTE at 37° for 10 min; 50% formamide/2× SSC at 62–66° for
30 min; 2× SSC at room temperature for 15 min; 0.1× SSC at room temperature
for 15 min.

For immunological detection, the slides are equilibrated in buffer 1 for 10 min,
then blocked with 2% sheep serum (Sigma) in buffer 2 for 1 hr. Anti-DIG antibody
is diluted 1 : 500 in buffer 2 with 1% sheep serum, then placed on the slides. After

5 hr of incubation, the antibody is washed by buffer 1 for 5×10 min. Then the slides are equilibrated in buffer 3 for 10 min and stained by NBT/BCIP solution for 1–12 hr in the dark in a wet chamber.

Protein Localization by Antibodies

In order to play a role in phototransduction, the RGS protein must be not only expressed in photoreceptors, but localized at the subcellular level to the rod or cone outer segments, the organelles containing the phototransduction machinery. Immunoblots are applied during rod outer segment purification to test RGS9-1 for copurification with rod outer segments as monitored by the marker protein rhodopsin (Fig. 2).

Solutions and Materials

> Light-tight darkroom with safelights or infrared imaging equipment
> Beckman ultracentrifuge with SW 28 rotor
> Peristaltic pump
> Auto Densi-Flow II Gradient puller (Labconco, Kansas City, MO)
> Spectrophotometer (in darkroom)
> Western electrophoretic transfer apparatus
> Supported introcellulose (Osmonics, Westborough, MA)
> Transfer buffer: 25 mM Tris, pH 8.4, 192 mM glycine, 0.1% sodium dodecyl sulfate (SDS), 20% methanol
> TBS: 50 mM Tris, pH 7.5, 100 mM NaCl
> TBST: 0.1% Tween 20 in TBS buffer
> Blotto 1 : 5% Nonfat dry milk in TBST
> Blotto 2 : 0.5% Nonfat dry milk in TBST
> RGS9 specific antiserum (polyclonal, rabbit)
> HRP-conjugated goat anti-rabbit IgG (Promega, Madison, WI)
> ECL (enhanced chemiluminescence) reagents (Amersham Pharmacia), Piscataway, NJ

Procedure

ROS are first purified from 25 bovine retinas by a discontinuous sucrose density gradient in the dark.[31] One ml fractions are carefully collected from the top of the sucrose gradient using an automatic gradient puller and peristaltic pump. A 100 μl aliquot of each fraction is analyzed by change of absorbance at 500 nm before and after bleach in 1.5% LDAO (N,N-dimethyldodecylamine N-oxide) detergent in order to obtain the concentration of rhodopsin in each fraction. The major

---

[31] D. S. Papermaster and W. J. Dreyer, *Biochemistry* **13**, 2438 (1974).

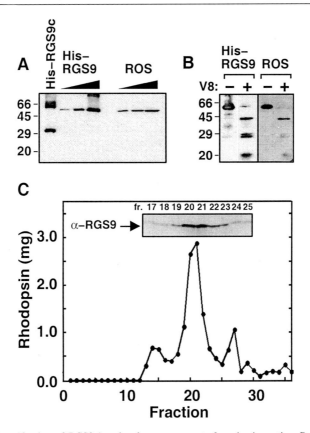

FIG. 2. Copurification of RGS9-1 and rod outer segments from bovine retina. Reproduced by permission from Ref. 25. (A) Immunoblots of recombinant (lane 1) His-RGS9c fragment (residues 226–484) used as immunizing antigen for rabbit antisera, (lanes 2–4) full-length His-RGS9, and (lanes 6–8) purified ROS. Amounts of recombinant protein loaded were 3 ng (lane 2), 30 ng (lane 3), or 300 ng (lane 4), and amounts of rhodopsin loaded were 3.4 μg (lane 6), 10 μg (lane 7), or 34 μg (lane 8). Preimmune serum did not detect any of these bands. (B) Immunoblot following treatment with 50 ng V8 protease of His-RGS9 (left) or immunoreactive protein extracted from ROS gel (right). (C) RGS9 col`calization with rhodopsin in ROS purification from bovine retinas. ROS were purified from bovine retinas by a discontinuous sucrose density gradient,[31] and fractions were analyzed by change in rhodopsin absorbance at 500 nm after bleach (closed circles). Fractions were also analyzed by immunoblot using anti-RGS9c rabbit polyclonal serum (*inset*). Reproduced with permission from W. He, C. W. Cowan, and T. G. Wensel, *Neuron* **20**, 95 (1998). Copyright 1998 Cell Press.

peak of rhodopsin in the sucrose gradient localizes to the layer between 26% and 30% sucrose. Five μl of each fraction is mixed with 5 μl of sample application buffer, resolved by standard SDS–PAGE[32] using 0.5 mm thick minigels, 12% acrylamide, and then transferred to supported nitrocellulose for 1 hr at 350 mA. The blots are blocked with Blotto 1, then incubated overnight at 4° with anti-RGS9 rabbit polyclonal serum (1 : 1000 dilution in Blotto 2). After washing with TBST, the specific bands are visualized using the chemiluminescence system according to the manufacturer's suggested protocol. RGS9 colocalizes with rhodopsin on the sucrose gradient. This method can be used as a general one to determine whether a protein is actually localized to ROS, or simply a contaminant of ROS preparations.[33]

An important control experiment to verify the identity of an immunoreactive protein is to extract the protein from the polyacrylamide gel following SDS–PAGE, and then analyze the proteolytic fragmentation pattern by Western blotting. Identical proteins should give identical patterns in this procedure. A detailed protocol is provided in Ref. 25.

Further information on subcellular localization has been obtained using antibodies directed to RGS9-1 for immunofluorescence staining of retinal sections. In these experiments it has been valuable to use different antibody preparations. One type[25] is polyclonal sera raised in rabbits against a highly immunogenic fragment of RGS9-1 (residues 226–484, "RGS9-c"). Another[28] is a mouse monoclonal antibody directed against an epitope including a small part of the RGS domain and adjacent portions of the C-terminal domain. The third is polyclonal rabbit sera raised against a C-terminal peptide of RGS9-1.[27] Each has its own patterns of specificity and cross-reactivity, which of course are likely to be different in tissue staining vs immunoblotting. We have used either agarose-embedded vibratome sections[26,27] or frozen sections, and the following method describes the procedure for frozen sections from mouse retina.

Solutions and Materials

   PBS (GIBCO)
   PFA buffer: 4% Paraformaldehyde in PBS
   Sucrose buffer: 30% sucrose in PBS
   PBST: 0.1% Triton X-100 in PBS
   Blocking buffer: 10% goat serum in PBST
   Cryo-microtome
   OCT (Tissue-Tek Compound, Sakura Finetek, Torrance, CA)
   Superfrost slides (Fisher)

Procedure

Mouse eyes are removed and immediately immersed in prechilled PFA buffer overnight and then soaked in sucrose buffer for 6 hr. The eyes are embedded in OCT and frozen in liquid nitrogen. For sectioning, the embedded mouse eyes are warmed up to $-20°$ and cut into sections of 20 μm thickness. If lens staining is not of interest, it is useful either to remove the lens before beginning to section,

[32] U. K. Laemmli, *Nature* **227**, 6809 (1970).
[33] Z. Yang and T. G. Wensel, *J. Biol. Chem.* **267**, 24634 (1992).

or to orient the eye so that sectioning begins at the optic nerve and moves forward toward the lens.

For immunostaining, the slides are fixed 4 min at $-20°$ in methanol, then 4 min in acetone. Slides are washed 2×10 min in PBST, then placed in blocking buffer for 1.5 hr at room temperature. Then anti-RGS9 antibodies are added to the slides at various dilutions (1 : 100–1 : 500) in PBST and incubated for 3 hr. The slides are washed 2×10 min again in PBST. The goat anti-rabbit FITC secondary antibody (Vector) is applied to the slides for 1 hr at 1 : 300 dilution in PBST. Finally, the slides are washed 2×5 min in PBS and mounted in Vectashield (Vector Laboratories, Burlingame, CA) mounting medium.

Immunodepletion

Biochemical studies with purified or recombinant proteins can demonstrate that a particular protein is capable of providing the GAP activity needed to accelerate $G_t\alpha$ GTPase into the physiological range. However, such studies are incapable of demonstrating that such a protein is required for the physiological acceleration of interest, or even of indicating what portion of the endogenous activity the protein accounts for. For such experiments it is necessary to remove the candidate protein, and to do so selectively. Perhaps the most elegant approach to accomplishing this is a genetic one, as described below, but a very useful and much faster alternative is to use immunodepletion. Quantitative removal of a specific protein from a cell homogenate or extract using antibodies can be used to assess the contribution of the protein recognized by the antibodies to the total activity. The following procedure has been used to deplete RGS9-1 from a detergent extract of ROS membranes (see Fig. 3). It should be kept in mind in designing such experiments that the goals and requirements are rather different from those of standard "IP" experiments, which typically precipitate only a small fraction of the total antigen.

Solutions and Materials

> ROS solubilization buffer: 10 mM Tris, pH 7.4, 100 mM NaCl, 2 mM MgCl$_2$, 40 mM octylglucoside
> Ni-NTA agarose (Qiagen)
> Protein A-Sepharose CL-4B (Amersham Pharmacia)
> CNBr-activated Sepharose 4B (Amersham Pharmacia)

Procedure

Rabbit IgG is purified from preimmune serum or immune serum against RGS9 by protein A-Sepharose CL-4B using standard techniques.[34] Purified RGS9

---

[34] E. Harlow and D. Lane, "Antibodies: A Laboratory Manual," p. 309. Cold Spring Harbor Laboratory, Cold Spring Harbor, NY, 1988.

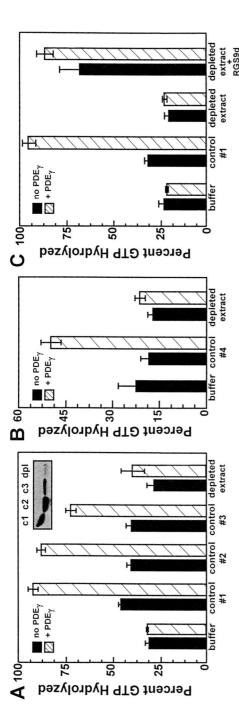

FIG. 3. Removal of ROS GAP activity by immunodepletion of RGS9-1 from bovine rod outer segment membrane extracts. Reproduced by permission from (Ref. 26). (A) RGS9 antibodies were immobilized on protein A-Sepharose and incubated with octylglucoside-solubilized ROS as described in the text. After removal by centrifugation, the Ros supernatants were analyzed for ROS GAP activity, PDEγ-enhanced ROS GAP activity, and RGS9 protein remaining in the supernatant (immunoblot, *inset*). Control samples are: 1, solubilized ROS extracts; 2, solubilized extract after incubation with immobilized preimmune rabbit antibodies; and 3, solubilized extract after incubation with RGS9 antibodies, which had been preincubated with recombinant RGS9 bound to Ni²⁺-NTA agarose prior to immobilization on protein A-Sepharose. Depleted extract, solubilized ROS extract after incubation with RGS9 antibodies, which had been preincubated with Ni²⁺-NTA agarose only prior to immobilization on protein A-Sepharose. For assays containing PDEγ (hatched boxes), a final concentration of 1.84 μM was used. (B) Immunodepletion by directly coupled antibodies. Either CNBr-activated Sepharose 4B with covalently coupled purified RGS9 antibodies or rabbit IgG-agarose was incubated with OG-solubilized ROS (control 4), and the supernatants were assayed for PDEγ-enhanced ROS GAP activity. PDEγ (hatched boxes) was present at 1.17 μM. (C) Extract was immunodepleted as in A, without preincubation of IgG with Ni²⁺-NTA agarose, and recombinant RGS9d was added (10 μM) with or without PDEγ (0.3 μM). Control #1 as in A.

immune immunoglobulin G (IgG) (0.75 mg) is incubated with either 100 μl of Ni^{2+}-NTA agarose or 100 μl of Ni^{2+}-NTA agarose preincubated with 1.5 mg of recombinant His_6-RGS9 (to deplete RGS9-specific antibodies), for 2 hr at 4° with constant mixing. Then the supernatants (with remaining IgG) after low-speed centrifugation are transferred and incubated with 50 μl of protein A-Sepharose CL-4B beads for 1 hr at 4° with constant mixing. The protein A beads are then washed three times with ROS solubilization buffer. ROS are solubilized in ROS solubilization buffer under dim-red light at final concentration of 60 μM rhodopsin, and centrifuged at $73,000g$ for 30 min at 4°. The supernatant is incubated overnight at 4° with protein A-Sepharose CL-4B with bound preimmune IgG, RGS9 immune IgG, or RGS9 immune IgG depleted of RGS9-specific antibodies.

Immunodepletion can also be performed by covalently coupling purified RGS9 immune IgG to CNBr-activated Sepharose 4B (Pharmacia) according to the manufacturer's recommended protocol. Five mg of purified IgG is applied to 1 ml of resin. Rabbit whole IgG-agarose (Sigma) is used as a control resin. Five hundred μl of solubilized ROS (60 μM rhodopsin) is incubated with 50 μl of RGS9 IgG-Sepharose or rabbit IgG-agarose overnight at 4°.

The beads are then pelleted by low-speed centrifugation, and the supernatants after low-speed centrifugation are assayed for GAP activity and PDEγ-enhanced GAP activity in single-turnover GTPase assays as well as for RGS9 protein by immunoblotting. In these experiments we use octylglucoside, but active RGS9-1/Gβ_{5L} can also be solubilized with lauryl sucrose.[19]

Gene Inactivation

Perhaps the most definitive test for the function of RGS9 in vision has been provided by experiments in which the RGS9 gene is deleted in mice.[20] The results demonstrate clearly that RGS9 is required for normal recovery kinetics, and also for effects of PDEγ on GTP hydrolysis kinetics (see Fig. 4). They also provide evidence that RGS9 is required for stable expression of GMβ_{5L}. Methods for gene inactivation in mice have been described extensively. In the case of RGS9, conventional methods were used.[20,35] The phenotype was assessed by standard histology, revealing no morphological changes, by immunoblots, by electrophysiological techniques, and by biochemical assays. The following is the procedure for isolation of rod outer segments from a small number of mouse retinas, using a modified version of a previously described[36] method, and for assaying GTP hydrolysis kinetics under single-turnover conditions.

[35] R. Ramirez-Solis, A. C. Davis, and A. Bradley, *Methods Enzymol.* **225**, 855 (1993).
[36] S. H. Tsang, M. E. Burns, P. D. Calvert, P. Gouras, D. A. Baylor, S. P. Goff, and V. Y. Arshavsky, *Science* **282**, 117 (1998).

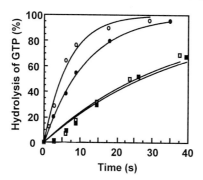

FIG. 4. GTPase assays using rod outer segments from wild-type and RGS9-knockout mice. Reproduced by permission from Ref. 20. Single-turnover GTP hydrolysis by $G_t\alpha$ in ROS from RGS9$^{+/+}$ (circles) and RGS9$^{-/-}$ (squares) mice with (open) or without (filled) exogenous PDEγ (1.33 μM). Rate constants were determined by fitting the results with a single exponential function; $k_{inact} = $: filled circles, 0.087 ± 0.002 s^{-1}; filled squares, 0.026 ± 0.002 s^{-1}; open circles, 0.148 ± 0.011 s^{-1}; open squares, 0.027 ± 0.002 s^{-1}. Reproduced with permission from C-K. Chen, M. E. Burns, W. He, T. G. Wensel, D. A. Baylor, and M. I. Simon, *Nature* **403**, 557 (2000).

Solutions and Materials

ROS buffer: 10 mM MOPS, pH 7.4, 30 mM NaCl, 60 mM KCl, 2 mM MgCl$_2$, 1 mM DTT, solid PMSF
GAPN buffer (described above)
Beckman ultracentrifuge with SW 40 rotor
Beckman TL 100 ultracentrifuge with TL 100.3 rotor

Procedure

Mice used for mouse ROS preparation are dark-adapted for 12 hr. Mouse retinas are removed from the eyecups under dim red light, frozen immediately in liquid nitrogen, and stored in $-80°$ before use. Mouse ROS are isolated on ice under dim red light in ROS buffer. Ten retinas are thawed and placed in 200 μl of 24% sucrose in ROS buffer and vortexed for 40 s. The tube is then spun at 4° and 250g for 40 s. The supernatant containing the ROS is gently removed. Vortexing and sedimentation are repeated five more times. The pooled supernatant is loaded on a discontinuous gradient of 24%, 26%, 30%, and 34% sucrose in ROS buffer (2.5 ml each). The tube is centrifuged (SW-40) for 1 h at 30,000g, and ROS are collected from the interface between 30% and 34% sucrose. Note that at this interface, primarily broken ROS are obtained. A second band of more intact ROS is generally found at the 26%/30% interface. The ROS are then diluted threefold with GAPN buffer and pelleted down at 80,000g for 30 min. The pellet is washed by 1 ml GAPN. Finally, the ROS pellet is hypotonically shocked in 90 μl H$_2$O by intense mixing for 10 s, then adjusted to the right salt concentration by adding 10 μl 10× GAPN buffer. About 4–6 μg rhodopsin can be obtained from one mouse retina. Single-turnover assays are carried out at a final concentration of 20 μM rhodopsin, with 50 μM AMP-PNP, and addition of 0 or 1.33 μM PDEγ. Initial GTP concentration is 50 nM.

Heterologous Expression Systems

In order to understand the role of RGS9-1 and its complex with Gβ_{5L} in phototransduction, it is essential to characterize their biochemical properties. Knowledge of kinetic and equilibrium constants governing their interactions with G$_t\alpha$–GTP, membranes, and other outer segment proteins, together with estimates of their intracellular concentrations and those of their interacting partners, should provide an accurate and complete picture of their roles in photoreceptor physiology. Purified protein is required for these constants to be determined, and we have yet to develop methods for preparing highly purified RGS9-1 or RGS9-1Gβ_{5L} complex in active form from mammalian retina. Therefore heterologous expression has been necessary to obtain purified proteins. In addition to providing material with which to conduct measurements of catalytic and binding constants, the expressed protein has allowed the production of antibodies that have been extremely useful in analyzing protein localization and intracellular concentrations, and in developing purification schemes for the endogenous proteins (e.g., Ref. 19). Heterologous expression also allows the introduction of site-specific mutations for structure/function studies.[37] Unpublished work from this laboratory indicates that baculovirus expression in insect cells is perhaps the most useful system for preparing large amounts of the RGS9-1/Gβ_{5L} complex. So far, published descriptions of active RGS9-1 have included among others, two kinds of expression systems: one is cells from *Xenopus laevis,* either melanocytes[38] or oocytes,[39] in which purified protein is not obtained, but in which cellular responses to stimuli can be measured. The other is *Escherichia coli,* in which catalytically active fragments of RGS9-1 as well as inactive but immunologically useful full-length protein have been prepared. Below are described methods for preparing full-length RGS9-1 with an N-terminal His$_6$ tag, in denatured form for immunization and immunoblot calibration, and for preparing a catalytically active fragment of RGS9-1 with either a His$_6$ tag or GST tag at the amino terminus (see Fig. 5).

Solutions and Materials

Lysis buffer: 50 mM Sodium phosphate, pH 7.4, 300 mM NaCl, 1 mM DTT
Guanidine buffer: 6 M Guanidinium hydrochloride, 100 mM sodium phosphate, 10 mM Tris, pH 8.0

[37] R. L. McEntaffer, M. Natochin, and N. O. Artemyev, *Biochemistry* **38,** 4931 (1999).
[38] Z. Rahman, S. J. Gold, M. N. Potenza, C. W. Cowan, Y. G. Ni, W. He, T. G. Wensel, and E. J. Nestler, *J. Neurosci.* **19,** 2016 (1999).
[39] A. Kovoor, C. K. Chen, W. He, T. G. Wensel, M. I. Simon, and H. A. Lester, *J. Biol. Chem.* **275,** 3397 (2000).

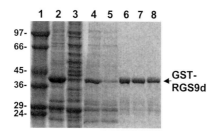

FIG. 5. Recombinant GST-tagged RGS9d prepared by extraction of proteins from inclusion bodies under denaturing conditions, renaturation by dialysis, and affinity purification. Lanes: 1, size standards; 2, postsonication pellet (inclusion bodies); 3, postsonication supernatant (soluble proteins); 4, guanidine-solubilized protein after renaturation by dialysis; 5, unbound fraction after incubation with GST-beads; 6, first GSH elution fraction; 7, second GSH elution fraction; 8, third GSH elution fraction.

Urea buffer: 8 M urea, 100 mM sodium phosphate, 10 mM Tris, pH 8.0
Renaturing buffer: 50 mM sodium phosphate, pH 7.4, 300 mM NaCl, 10%
    glycerol, 0.1% 2-mercaptoethanol
Ni^{2+}-NTA agarose (Qiagen)
GSH Sepharose 4B (Amersham Pharmacia)
Dialysis tubing (Spectrum)

Procedure

His-RGS9 (corresponding to the full-length protein), His-RGS9d (residues 291–418), and His-RGS9dc (residues 291–484) are expressed using pET14b (Novagen), which incorporates an N-terminal His$_6$ tag. GST-RGS9d and GST-RGS9dc are expressed using pGEX-2TK, which incorporates an N-terminal GST-tag. RGS9 recombinant proteins are expressed in BL21 (DE3) pLysS cells, by addition of 1 mM (for pET14b) or 0.1 mM (for pGEX-2TK) IPTG at a cell density of A_{600} of 0.6–0.8, and grown for 3 hr more at 37°. The cells are harvested by centrifugation at 4000g (20 min, 4°), and pellets sonicated and washed twice using lysis buffer (30 ml per 1 liter culture). The second pellet contains most of the recombinant protein in inclusion bodies.

The recombinant His$_6$-tagged proteins are extracted from inclusion bodies using guanidine buffer (30–35 ml per 1 liter culture, rotated 1–2 hr at room temperature), and purified with Ni^{2+}-NTA agarose under denaturing conditions using the manufacturer's recommended protocol. His-RGS9d and His-RGS9dc are also extracted using guanidine buffer, purified under denaturing conditions, and then renatured by dialysis at 4° against a 10,000-fold greater volume of renaturing buffer. Insoluble proteins are removed after dialysis by two rounds of centrifugation at 20,000g and 4° for 30 min each. After dialysis, renatured recombinant proteins are then concentrated by centrifugal ultrafiltration (Centricon, Millipore, Bedford, MA). In order to obtain good yields of renatured proteins from dialysis, we routinely dilute our starting material to 0.1 mg/ml with urea buffer before dialysis.

Glutathione transferase (GST)-tagged recombinant proteins cannot be purified directly from *Escherichia coli* inclusion bodies under denaturing conditions. Therefore, we first renature proteins in solubilized inclusion bodies by dialysis as described above, then purify GST-tagged proteins using a GSH-affinity column by the manufacturer's recommended protocol.[40] The proteins prepared by either procedure are exchanged into GAPN buffer using centrifugal ultrafiltration, and then glycerol is added to 40% (v/v) for storage at $-20°$.

---

[40] M. E. Sowa, W. He, T. G. Wensel, and O. Lichtarge, *Proc. Natl. Acad. Sci. U.S.A.* **97**, 1483 (2000).

[51] Molecular Cloning of Regulators of G-Protein Signaling Family Members and Characterization of Binding Specificity of RGS12 PDZ Domain

By BRYAN E. SNOW, GREG M. BROTHERS, and DAVID P. SIDEROVSKI

Introduction

Heterotrimeric G proteins, composed of $G\alpha$, $G\beta$, and $G\gamma$ subunits, translate cell-surface receptor/ligand interactions into intracellular signaling cascades via receptor-catalyzed exchange of GDP for GTP on the $G\alpha$ subunit.[1] It was originally thought that the duration of heterotrimeric G-protein signaling (i.e., the lifetime of $G\alpha$ in a GTP-bound form) could be modulated by only two factors: the intrinsic guanosine triphosphatase rate of the G protein α ($G\alpha$) subunit and acceleration of that rate by some $G\alpha$ effectors (e.g., phospholipase C-β[2]). We and others have since discovered a new family of regulatory proteins[3–6] that accelerate $G\alpha$ GTPase activity[7–9]: the "regulator of G-protein signaling" or RGS proteins. Each RGS protein contains a conserved $\sim$130 amino acid "RGS box" composed of three discrete primary sequence motifs[3] (GH1, GH2, GH3), which are presumed to fold

[1] A. G. Gilman, *Annu. Rev. Biochem.* **56**, 615 (1987).
[2] G. Berstein, J. L. Blank, D.-Y. Jhon, J. H. Exton, S. G. Rhee, and E. M. Ross, *Cell* **70**, 411 (1992).
[3] D. P. Siderovski, A. Hessel, S. Chung, T. W. Mak, and M. Tyers, *Curr. Biol.* **6**, 211 (1996).
[4] L. De Vries, M. Mousli, A. Wurmser, and M. G. Farquhar, *Proc. Natl. Acad. Sci. U.S.A.* **92**, 11916 (1995).
[5] M. R. Koelle and H. R. Horvitz, *Cell* **84**, 115 (1996).
[6] K. M. Druey, K. J. Blumer, V. H. Kang, and J. H. Kehrl, *Nature* **379**, 742 (1996).
[7] D. M. Berman, T. M. Wilkie, and A. G. Gilman, *Cell* **86**, 445 (1996).
[8] N. Watson, M. E. Linder, K. M. Druey, J. H. Kehrl, and K. J. Blumer, *Nature* **383**, 172 (1996).
[9] T. W. Hunt, T. A. Fields, P. J. Casey, and E. G. Peralta, *Nature* **383**, 175 (1996).

into a nine α helix bundle that contacts the Gα switch regions and stabilizes the transition state for GTP hydrolysis.[10,11] In this article, we describe the cloning of RGS box-containing cDNAs using degenerate PCR and characterization of the PDZ domain of one of these RGS box-containing proteins, RGS12, using surface plasmon resonance biosensors and yeast two-hybrid techniques.

Molecular Cloning of Novel RGS Family Members

Design of Degenerate Primers

Highly conserved central portions of the RGS-box GH1 and GH3 motifs are used to design degenerate 5' sense and 3' antisense oligonucleotide primer pools for polymerase chain reaction (PCR) amplification of new RGS family members (Fig. 1). The 5' sense primers are designed to end with the single codon-encoded (5'-TGG-3') amino acid tryptophan (i.e., 106-CEENIEF<u>W</u>-113 in G0S8/RGS2[12]); the 3' antisense primers end with the single codon-encoded (5'-ATG-3') amino acid methionine (i.e., 181-<i>M</i>ENNSYPRF-189 in G0S8/RGS2[12]). For RGS family members only identified by the internal GH2 motif, as derived from degenerate primer PCR cloning,[5] contiguous sequences spanning the GH1 through GH3 motifs are assembled using expressed sequence tags (ESTs) from the GenBank dbEST database (National Center for Biotechnology Information; http://www.ncbi.nlm.nih.gov).

Starting from the invariant codon position (TGG or ATG), each nucleotide position in turn is examined for common differences from the consensus; thus the motif sequences are subgrouped into clades (e.g., S-1 to S-3 for GH1 motif primers and AS-1 to AS-3 for GH3 motif primers; Fig. 1) to which degenerate primers are designed. For the group of <i>RGS</i> sequences depicted in Fig. 1, the following set of three 24-mer sense (S) and three 27-mer antisense (AS) oligodeoxyribonucleotide primer pools was synthesized (fold degeneracy in brackets):

S-1: 5'-AG(CT)GA(AG)GA(AG)AA(CT)(ACT)TIGAITTCTGG-3'   (768-fold)
S-2: 5'-AG(CT)GA(AG)GA(AG)AA(CT)(AG)TI(CT)TITT(CT)TGG-3'   (2048-fold)
S-3: 5'-AG(CT)TCIGA(AG)AA(CT)(CT)TIAG(AG)TTCTGG-3'   (512-fold)

AS-1: 5'-(AG)AA(AG)CGIGG(AG)TAIGA(AG)TCITT(CT)TCCAT-3'   (2048-fold)
AS-2: 5'-(AG)AA(AG)CGIGI(AG)TA(AG)CT(AG)TC(AG)CT(CT)TTCAT-3'   (2048-fold)
AS-3: 5'-(AG)AA(AG)CGIGI(AG)TA(AG)CT(AG)TC(AG)(AGT)A(CT)TTCAT-3'   (6144-fold)

The base inosine (I) is used to allow promiscuous base pairing at those positions that are fully degenerate. With three sense primer pools and three antisense primer

[10] J. J. Tesmer, D. M. Berman, A. G. Gilman, and S. R. Sprang, <i>Cell</i> 89, 251 (1997).
[11] D. M. Berman, T. Kozasa, and A. G. Gilman, <i>J. Biol. Chem.</i> 271, 27209 (1996).
[12] H. K. Wu, H. H. Heng, X. M. Shi, D. R. Forsdyke, L. C. Tsui, T. W. Mak, M. D. Minden, and D. P. Siderovski, <i>Leukemia</i> 9, 1291 (1995).

```
                          5' sense region                    S    ATG start region                            AS

Consensus           AGT.GAG.GAA.AAT.ATT.GAG.TTC.TGG          ATG.GAG.AAG.GAC.TCT.TAT.CCT.GCG.TTC      F
                     S   E   E   N   I   E   F   W             M   E   K   D   S   Y   P   R   F            AS-1

Human RGS1          AGT.GAG.GAG.AAT.ATT.GAG.TTC.TGG   S-1    ATG.GAA.AAG.GAC.TCT.TAT.CCC.aGg.TTC      F
[sp|Q07918]          S   E   E   N   I   E   F   W             M   E   K   D   S   Y   P   R   F            AS-1

Human G0S8/RGS2     TGT.GAA.GAA.AAT.ATT.GAA.TTC.TGG   S-1    ATG.GAG.AAC.aAC.TCT.TAT.CCT.CGT.TTC      F
[sp|P41220]          C   E   E   N   I   E   F   W             M   E   N   N   S   Y   P   R   F            AS-1

Human RGS3          AGT.GAG.GAG.AAT.CTG.GAG.TTC.TGG   S-1    ATG.GAA.AAG.GAC.TCG.TAC.CCT.CGC.TTT      F
[gp|U27655]          S   E   E   N   L   E   F   W             M   E   K   D   S   Y   P   R   F            AS-1

Human RGS4          AGT.GAG.GAG.AAT.CTG.GAC.TTC.TGG   S-1    ATG.GAG.AAG.GAT.TCC.TAC.CgC.CGC.TTC      F
[gp|U27768]          S   E   E   N   L   D   F   W             M   E   K   D   S   Y   R   R   F            AS-1

Human RGS5          AGT.GAG.GAA.AAC.CTT.GAG.TTC.TGG   S-1    ATG.GAA.AAG.GAT.TCT.ctG.CCT.CGC.TTT      F
[ESTs]               S   E   E   N   L   E   F   W             M   E   K   D   S   L   P   R   F            AS-1

Human S194/RGS6     AGT.TCA.GAA.AAC.CTC.AGG.TTC.TGG   S-3    ATG.AAG.AGT.GAC.AGC.TAT.GCC.CGC.TTC      F
[ESTs]               S   S   E   N   L   R   F   W             M   K   S   D   S   Y   A   R   F            AS-2

Human RGS7          AGC.TCG.GAA.AAT.TTA.AGA.TTC.TGG   S-3    ATG.AAA.AGT.GAT.tca.TAC.CCA.CGT.TTT      F
[gp|U32439]          S   S   E   N   L   R   F   W             M   K   S   D   S   Y   P   R   F            AS-2

Human RGS10         AGT.GAA.GAA.AAT.GTT.TTG.TTT.TGG   S-2    ATG.AAG.TAC.GAC.AGC.TAC.AgC.CGC.TTC      F
[ESTs]               S   E   E   N   V   L   F   W             M   K   Y   D   S   Y   S   R   F            AS-3

Human RGS12         AGT.GAA.GAA.AAC.ATT.TTA.TTC.TGG   S-2    ATG.AAG.TTT.GAT.AGC.TAC.ACT.CGC.TTT      F
[ESTs]               S   E   E   N   I   L   F   W             M   K   F   D   S   Y   T   R   F            AS-3

Human RGS13         AGT.GAc.GAG.AAT.ATT.CaA.TTC.TGG   S-2    n/a
[ESTs]               S   D   E   N   I   Q   F   W

Human RGS14         n/a                                      ATG.AAG.TTC.GAC.AGC.TAT.GCG.CGC.TTC      F
[ESTs]                                                         M   K   F   D   S   Y   A   R   F            AS-3

Human GAIP          AGC.GAG.GAG.AAC.ATG.CTC.TTC.TGG   S-2    ATG.cAc.cGg.GAC.TCC.TAC.CCC.CGC.TTC      F
[gp|X91809]          S   E   E   N   M   L   F   W             M   H   R   D   S   Y   P   R   F            AS-1

Primer S-1          AGY.GAR.GAR.AAY.HTN.GAN.TTC.TGG          ATG.GAR.AAN.GAY.TCN.TAY.CCN.CGY.TTY      Primer AS-1
                     S   E   E   N   LI  DE  F   W             M   E   KN  D   S   Y   P   R   F            F

Primer S-2          AGY.GAR.GAY.AAY.RTN.YTN.TTY.TGG          ATG.AAR.AGY.GAY.TAY.NCN.CGY.TTY          Primer AS-2
                     S   E   E   N   IMV L   F   W             M   K   S   D   S   Y   SPTA R   F           F

Primer S-3          AGY.TCN.GAR.AAY.YTN.AGR.TTC.TGG          ATG.AAR.THY.GAY.AGY.TAY.NCN.CGY.TTY      Primer AS-3
                     S   S   E   N   L   R   F   W             M   K   FYS D   S   Y   SPTA R   F           F
```

pools, a total of nine sense/antisense pairings are possible. Synthesized primer pools are individually diluted to 25 μM in sterile H_2O and stored in aliquots at $-20°$.

RNA Isolation

Eight 10-cm dishes of rat C6 glioma cells ($\sim2.5 \times 10^6$ cells/plate) are each lysed by repeated pipetting in 8 ml of TRIzol reagent (Gibco-BRL, Burlington, Ontario) and each plate is transferred into a 12×75 mm round-bottom polypropylene tube (Falcon #2059). Lysate is incubated for 5 min at room temperature prior to addition of 1.6 ml ($0.2\times$ volume) of chloroform and vigorous shaking for 15 sec. After 3 min standing at room temperature, lysate is centrifuged at 9000 rpm for 15 min in a $4°$ prechilled Beckman JA-17 rotor (Beckman Coulter, Fullerton, CA) to separate the aqueous and organic phases. The top aqueous phase (~4.8 ml) is transferred into a new tube and mixed with 4 ml of 2-propanol (invert five times to mix together). After 10 min standing at room temperature, RNA is precipitated by 11 min of centrifugation at 9000 rpm in a $4°$ JA-17 rotor. Supernatant is carefully poured off and the RNA pellet is washed by a brief vortexing in 8 ml of ice-cold 75% ethanol. After repelleting RNA by centrifugation at 7000 rpm for 7 min in a $4°$ JA-17 rotor, the ethanol wash is decanted and the RNA pellet is air-dried for 10 min. On addition of 25 μl sterile water [pretreated with diethyl pyrocarbonate (DEPC) and containing 1 μl (40 units) of RNase inhibitor (Cat #799017; Boehringer Mannheim/Roche, Nutley, NJ) per 1 ml DEPC-treated H_2O], the RNA pellet is resuspended by gentle pipetting and incubated for 20 to 30 min at $55°$. RNA samples are then pooled and quantitated by measuring the OD_{260nm} of diluted aliquots (1.0 OD_{260nm} units $= 40$ μg/ml RNA). Our typical yield using this protocol is 400–500 μg total RNA. Stock solutions are diluted to 2 μg/μl in DEPC-treated H_2O and frozen in 75 μl aliquots at $-80°$.

Reverse Transcription/Polymerase Chain Reaction (RT-PCR) Amplification

First strand cDNA is prepared by the reverse transcription of polyadenylated mRNA within the total RNA pool using oligo (dT) primers and the Superscript-II first strand cDNA synthesis system (Gibco BRL) exactly as directed by the

---

FIG. 1. Design of degenerate primer pools for PCR amplification of RGS box-containing proteins. All nucleotide sequences are displayed 5′ to 3′ in codon triplets separated by periods; lowercase indicates nucleotides found in native RGS mRNA sequences but not accounted for in the degeneracy of the designed primer series. Translated amino acid sequences are displayed in italics. Swiss-Prot (sp) and GenBank (gp) accession numbers are denoted in square brackets. Clusters of expressed sequence tags (ESTs) were used to generate contiguous sequence encoding RGS-box regions of RGS5 (GenBank accession H85251, D31257, R35272, N34362, N34348), RGS6 (L40394, H09621), RGS10 (H87415, N95393, N31637, N38837, N57092), RGS12 (N31659, R35472, R77405, T57943), RGS13 (H70046, T94013), and RGS14 (H06460, R11933, F13091); "n/a" denotes not applicable (i.e., no ESTs were available from this part of the RGS-box). Note that the antisense primers AS-1, AS-2, and AS-3, are synthesized as the *reverse complement* of the sequence illustrated in the figure, as described in the text. Definitions for degenerate nucleotide positions: H = A,C,T; N = A,C,G,T; R (purine) = A,G; Y (pyrimidine) = C,T.

manufacturer's instructions. To control for amplification of potential genomic DNA contaminants within the total RNA pool, a duplicate first strand cDNA synthesis reaction is performed without input of Superscript-II reverse transcriptase. PCR amplifications are performed in a 50 μl total volume in thin-walled 0.2 ml tubes (Robbins Scientific) containing 10 mM Tris-HCl (pH 8.3), 50 mM KCl, 1.5 mM MgCl$_2$, 200 μM dNTPs, 2 μM sense primer pool (S-1, S-2, or S-3), 2 μM antisense primer pool (AS-1, AS-2, or AS-3), 2 ng of first-strand cDNA, and 5 units of *Taq* DNA polymerase (Boehringer Mannheim/Roche).

The PCR reactions are carried out in a PTC-100 programmable thermal cycler (MJ Research Inc.) using a "Touch-down/Touch-up" protocol in which the annealing temperature (T_{an}) and extension time (t_{ex}) are varied over the course of 35 total cycles of amplification. The basic amplification cycle consists of a denaturation phase at 94°, an annealing phase at a temperature near the optimal annealing temperature for the primer/target duplex, and an extension phase at 72° (Fig. 2A). The optimal annealing temperature for sense/antisense primer pairs is calculated by the program Oligo (Hitachi Genetic Systems; www.hitachi-soft.com): for the S-1,2,3/AS-1,2,3 primer pairs, an optimal $T_{an} \approx 45°$ is predicted. In the Touch-down/Touch-up protocol, the annealing temperature is initially set at 5° above the predicted optimum and decreased 2° per cycle over the following four cycles (Fig. 2B) to minimize mispriming to non-*RGS* templates in the cDNA pool.[13] Over the next five cycles (cycles 6–10), the annealing temperature is increased 1° per cycle, reaching a final T_{an} that is 2° above optimum, an annealing temperature used throughout the remaining amplication cycles (cycles 11–35). The time of incubation (t_{ex}) at 72° for *Taq* polymerase extension of new amplicons is set at 20 sec for the first cycle and increased 1 sec per cycle to allow longer extension times at later reaction cycles when reagents may become limiting (Fig. 2C). Hence, the program to run the Touch-down/Touch-up protocol on an MJ Research thermal cycler is as follows:

Step 1	95.0°	00:25 sec	(initial denaturation phase)
Step 2	94.0°	00:20 sec	
Step 3	50.0°	00:20 sec − 2.0°/cycle	(using "increment" option)
Step 4	72.0°	00:20 sec + 1 sec/cycle	(using "extend" option)
Step 5	4 Times to Step 2		(using GOTO command)
Step 6	94.0°	00:20 sec	
Step 7	43.0°	00:20 sec + 1.0°/cycle	
Step 8	72.0°	00:25 sec + 1 sec/cycle	
Step 9	4 Times to Step 6		
Step 10	94.0°	00:20 sec	
Step 11	47.0°	00:20 sec	
Step 12	72.0°	00:30 sec + 1 sec/cycle	
Step 13	24 Times to Step 10		
Step 14	72.0°	02:00 sec	(final extension phase)
Step 15	4.0°	00:00 sec	(i.e., continuous incubation)
Step 16	END		

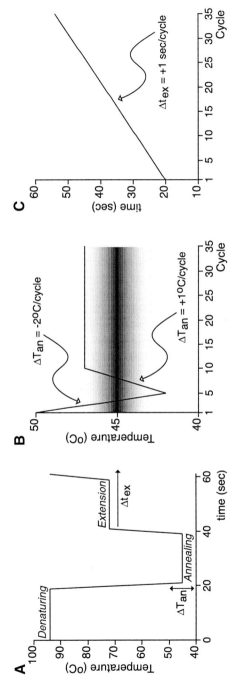

FIG. 2. Touch-down/Touch-up protocol for PCR amplification using degenerate RGS primer pools. (A) Single amplification cycle of denaturation, primer annealing, and primer extension. Over the course of 35 cycles, the annealing temperature (T_{an}) and extension time (t_{ex}) are varied. (B) Variation of the annealing temperature over the first 10 cycles. The temperature range that is centered about the predicted optimal annealing temperature of ~45° is illustrated by gradient shading. (C) Increase in extension time over the course of 35 amplification cycles.

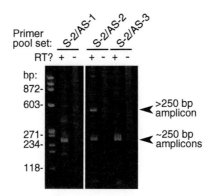

FIG. 3. Analysis of degenerate PCR products by 6% polyacrylamide gel electrophoresis. Primer pool sets S-2/AS-1, S-2/AS-2, and S-2/AS-3 (as defined in Fig. 1) were used in RT-PCR amplification of rat C6 glioma mRNA samples treated (RT⁺) or not treated (RT⁻) with reverse transcriptase during first-strand cDNA synthesis.

Analysis and Cloning of PCR Products

Fifteen μl of reaction product from each reaction is initially resolved by electrophoresis through precast 6% polyacrylamide gels buffered with Tris/borate/EDTA (6% TBE gels; Novex, San Diego, CA) and visualized by ethidium bromide staining (Fig. 3); the rest of the reaction product is stored at $-80°$ to preserve the single $3'$ deoxyadenosine tails on PCR amplicons as added by Taq polymerase terminal transferase activity. To isolate the $\sim$250 bp amplicons for cloning, 35 μl of remaining product is resolved by electrophoresis through 1.9% agarose (buffered with 40 mM Tris-acetate, 1 mM EDTA) and DNA purified from gel slices using chaotropic salt solubilization and binding to silica-gel spin-column membranes (QIAquick gel extraction system, Cat #28704; Qiagen Inc., Valencia, CA).

Four μl of DNA (eluted in total volume of 30 μl sterile 10 mM Tris pH 8.0, 1 mM EDTA) is added to 1 μl of topoisomerase I-activated, linearized, $3'$ deoxythymidine-tailed vector (pCR2.1-TOPO; Invitrogen, San Diego, CA) and incubated for 5 min at room temperature; ligation product is then transformed into *Escherichia coli* TOP10F′ cells and clones with inserts identified by blue/white selection exactly according to the manufacturer's instructions. Plasmid DNA from white colonies is purified by alkaline lysis of overnight liquid cultures and anion-exchange chromatography (QIAGEN Plasmid Mini columns; Qiagen Inc.). Correct sized inserts ($\sim$250 bp) are confirmed by plasmid DNA digestion with *Eco*RI and 6% TBE polyacrylamide gel electrophoresis, prior to fluorescent dideoxynucleotide sequencing (ABI/Perkin Elmer, Foster City, CA) using T7

---

[13] R. H. Don, P. T. Cox, B. J. Wainwright, K. Baker, and J. S. Mattick, *Nucleic Acids Res.* **19,** 4008 (1991).

(5'-TAATACGACTCACTATAGGG-3') and M13 reverse (5'-CAGGAAACAGCT-ATGACC-3') primers.

Applying the S-1,2,3/AS-1,2,3 primer pool set to rat C6 glioma cDNA, we have isolated 22 clones from RGS box-containing mRNA: *Rgs2* (one), *Rgs3* (six), *Rgs8* (one), *Rgs10* (four), *Rgs12* (five), and *Rgs14* (five). Three non-RGS clones were also isolated: two clones from the rat ortholog of human 100 kDa coactivator (GenBank U22055) and one clone from the rat ortholog of human hnRNP A2 (GenBank M29065). This same degenerate primer set and protocol has also been used to identify *RGS6* expression in human T lymphocytes (Dr. Eleanor Fish, University of Toronto; personal communication, 2000) and *RGS-GAIP*, −4, −6, −7, −8, −9, −10, −12, −16, and −Z1 in embryonic chick dorsal root ganglia neurons (Dr. María Diversé-Pierluissi, Mount Sinai, NY; personal communication, 2000). As the definition of an RGS-box has now expanded[14] to include "outliers" lacking conserved tryptophan and methionine codons necessary to anchor S-1,2,3 and AS-1,2,3 primers (e.g., RGS boxes of p115RhoGEF[15], PDZ-RhoGEF[16], GRK2[3,17]), newly designed degenerate primer pools will need to be employed to isolate family members related to these RGS outliers. In addition, isolation and sequencing of PCR amplicons greater than 250 bp [e.g., Fig. 3, lane S-2/AS-2(RT+)] may also facilitate identifying new RGS family members, as it is clear from the fungal RGS proteins (e.g., *Saccharomyces cerevisiae* Sst2[18] and *Emericella nidulans* FlbA[19]) that large intervening polypeptide sequences can be accomodated within the RGS-box between the GH1, GH2, and GH3 motifs.[3]

Isolation of Full-Length RGS cDNA Clones

The full-length sequences of novel *RGS* genes initially trapped by degenerate PCR can be obtained using a combination of cDNA library screening and PCR-based rapid amplification of cDNA ends (RACE) using well-documented protocols[20,21] from this series. To isolate full-length rat *Rgs12* and *Rgs14* cDNA,[22] ~1 × 10^6 phage from an adult rat brain (corpus striatum) Lambda ZAP-II phagemid library (Stratagene, La Jolla, CA) are screened by filter-hybridization

[14] D. P. Siderovski, B. Strockbine, and C. I. Behe, *Crit. Rev. Biochem. Mol. Biol.* **34,** 215 (1999).
[15] T. Kozasa, X. Jiang, M. J. Hart, P. M. Sternweis, W. D. Singer, A. G. Gilman, G. Bollag, and P. C. Sternweis, *Science* **280,** 2109 (1998).
[16] S. Fukuhara, C. Murga, M. Zohar, T. Igishi, and J. S. Gutkind, *J. Biol. Chem.* **274,** 5868 (1999).
[17] C. V. Carman, J. L. Parent, P. W. Day, A. N. Pronin, P. M. Sternweis, P. B. Wedegaertner, A. G. Gilman, J. L. Benovic, and T. Kozasa, *J. Biol. Chem.* **274,** 34483 (1999).
[18] D. M. Apanovitch, K. C. Slep, P. B. Sigler, and H. G. Dohlman, *Biochemistry* **37,** 4815 (1998).
[19] B. N. Lee and T. H. Adams, *Mol. Microbiol.* **14,** 323 (1994).
[20] J. M. Short and J. A. Sorge, *Methods Enzymol.* **216,** 495 (1992).
[21] M. A. Frohman, *Methods Enzymol.* **218,** 340 (1993).
[22] B. E. Snow, L. Antonio, S. Suggs, H. B. Gutstein, and D. P. Siderovski, *Biochem. Biophys. Res. Commun.* **233,** 770 (1997).

to radiolabeled ~250 bp probes derived from the degenerate PCR cloning. *Rgs12* and *Rgs14* DNA probes are excised from the pCR2.1-TOPO vector by *Eco*RI digestion and gel purified prior to labeling with [α-^{32}P]dCTP using the Multiprime DNA Labeling System (Amersham Pharmacia, Piscataway, NJ) exactly according to the manufacturer's instructions. Twenty filter lifts (~50,000 plaques per filter) are made on Hybond-N$^+$ membrane (Amersham Pharmacia) and incubated at 42° for 8 hr in prehybridization buffer: 5× SSC (75 m*M* sodium citrate, pH 7.0; 750 m*M* NaCl), 1× Denhardt's solution,[23] 20 m*M* sodium phosphate, 1% SDS, 50% formamide, and 270 μg/ml sheared, denatured salmon sperm DNA. Filters are then hybridized overnight at 42° in a volume of ~100 ml hybridization buffer, identical to the prehybridization buffer except also containing 10% (w/v) dextran sulfate and 0.75–1.5 × 10^6 cpm/ml of ^{32}P-labeled RGS DNA probe. Filters are then washed twice at room temperature with 2× SSC, 0.2% SDS for 15 min and twice at 65° for 30 min before exposure to film. Positive phage picks are reselected in a secondary hybridization screening prior to the generation of pBluescript phagemid by *in vivo* excision using the ExAssist/SOLR system[20] (Stratagene). DNA from independent phagemid clones is prepared and fragmented by *Hin*dIII/*Xba*I double digests and *Bgl*II/*Bam*HI/*Eco*RI triple digests in order to assess insert size and facilitate the grouping of clones into classes of similar restriction map patterns; the longest clones are sequenced by primer walks, starting with the T3 and T7 promoter primers flanking the pBluescript polylinker.[24]

Combined sequence from overlapping *Rgs12* phagemid clones was found to encode a 2853 bp (950 aa) open-reading frame and the longest *Rgs14* phagemid clone encoded a 1635 bp (544 aa) open-reading frame.[22] To identify the complete 5' and 3' ends of rat *Rgs12* and *Rgs14* mRNA transcripts, we initially tried "rapid amplification of cDNA ends" or RACE reactions[21] using the Gibco-BRL Life Technologies 5' RACE and 3' RACE systems with rat C6 glioma total RNA as the mRNA source. We then switched to the use of commercially available, adaptor-ligated cDNA (i.e., Marathon-Ready rat brain cDNA, Clontech, Palo Alto, CA) and followed the nested PCR protocol exactly as suggested by the manufacturer's instructions. Two nested, 28-mer or 24-mer gene-specific primers (GSPs) were designed for each cloning strategy (nucleotide numbering corresponds to GenBank records U92280 for *Rgs12* and U92279 for *Rgs14*):

3' *Rgs12* GSP1:	5'-TATATCAAGTCTGGATGGACAGCGGGTC-3'	(nt 3369–3396)
3' *Rgs12* GSP2:	5'-ATCCCCGGCTTTCAAAGAGAGAAGAATC-3'	(nt 3581–3608)
5' *Rgs12* GSP1:	5'-GCAGTCCTTCAGGTCAGCTCCCGTCAAC-3'	(nt 2148–2121)
5' *Rgs12* GSP2:	5'-CCTGAATGCTTTGTTAAGTCCCAGAACC-3'	(nt 1607–1580)
5' *Rgs14* GSP1:	5'-TGGTAGATGTTGTGGGCCTCCTGA-3'	(nt 632–609)
5' *Rgs14* GSP2:	5'-CTAGCTGTTTGGTGTCGCTGGCTG-3'	(nt 607–584)

[23] D. T. Denhardt, *Biochem. Biophys. Res. Commun.* **23,** 641 (1966).
[24] M. A. Alting-Mees, J. A. Sorge, and J. M. Short, *Methods Enzymol.* **216,** 483 (1992).

PCR products from multiple independent RACE reactions are sequenced, in order to build a consensus and thus eliminate any PCR-generated frameshift or point mutations. 5′ and 3′ RACE product sequence analysis extended the rat *Rgs12* open-reading frame to a total of 4164 bp (1387 aa), whereas the *Rgs14* open-reading frame was not extended; instead, 5′ untranslated sequence from *Rgs14* RACE clones revealed in-frame stop-codons 5′ of the predicted initiator methionine codon. BLAST[25] searches of the GenBank nonredundant nucleotide sequence database using the full-length rat *Rgs12* cDNA sequence identified human genomic sequence from several cosmid clones.[22] These nucleotide sequences were aligned and intronic sequences removed following identification of exon/intron boundaries using Macintosh GeneWorks software (Oxford Molecular Group, Oxford, UK); this resulted in a predicted full-length human *RGS12* cDNA encoding a 1376 amino acid protein with 85% identity to rat RGS12.

To clone human *RGS12*, oligonucleotide primers flanking the predicted open-reading frame [sense primer 5′-ATATGGCTCCAAGGGAACAATGAGACG-3′, nts 14533–14507 of cosmid L129H7 (GenBank Z68274); antisense primer 5′-TA-CGGGGGCCAAGGTGGAGGGATCAG-3′, nts 4784–4760 of cosmid L60G9B (GenBank Z69363)] are used to PCR amplify Marathon-Ready human brain cDNA (Clontech). PCR is performed using the Expand Long Template system (Boehringer Mannheim/Roche) essentially as previously employed for the cloning of human RGS16[26]: A 50 μl final volume of H_2O containing 350 μM dNTPs, 0.5 μM of each primer, 10 mM Tris-HCl pH 8.3, 1.75 mM $MgCl_2$, 50 mM KCl, 0.5 ng of human brain cDNA, and 2.6 units of Long Template DNA polymerase is amplified using a "Touch-down" cycling protocol consisting of an initial denaturation of 95° for 2 min, 10 cycles of 94° for 10 sec, 67° −0.4°/cycle for 30 sec, 68° for 3 min, followed by 20 cycles at 93° for 10 sec, 63° for 30 sec, 68° for 3 min + 20 sec/cycle, and a final incubation of 68° for 7 min. The resultant 4.3 kb PCR product is cloned into pCR2.1-TOPO vector and sequenced as described above.

In Silico Identification of Domain Structures within Novel RGS Members

Once the entire open-reading frame of a novel *RGS* gene has been established, computational tools are used to predict the presence of established and/or novel domain structures in regions N- and/or C-terminal of the RGS-box. For detection of established domain structures, polypeptide sequences are analyzed using the Simple Modular Architecture Research Tool[27] (http://SMART.embl-heidelberg.de). SMART contains an extensively annotated collection of cytoplasmic signaling domain alignments and enables sensitive detection of additional

[25] S. F. Altschul, T. L. Madden, A. A. Schaffer, J. Zhang, Z. Zhang, W. Miller, and D. J. Lipman, *Nucleic Acids Res.* **25,** 3389 (1997).
[26] B. E. Snow, L. Antonio, S. Suggs, and D. P. Siderovski, *Gene* **206,** 247 (1998).
[27] J. Schultz, F. Milpetz, P. Bork, and C. P. Ponting, *Proc. Natl. Acad. Sci. U.S.A.* **95,** 5857 (1998).

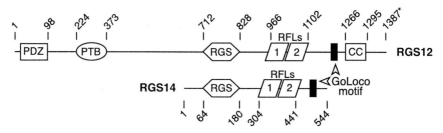

FIG. 4. Schematic illustration of the multidomain structures of rat RGS12 and RGS14 proteins. The numbers above and below the horizontal lines represent amino-acid numbering of domain boundaries, corresponding to GenBank records U92280 and U92279, respectively. The GoLoco motif is present at amino acids 1188–1206 in rat RGS12 and amino acids 498–516 in rat RGS14. Domain structures are defined in the text. PDZ, PSD-95/Dlg/ZO-1 domain; PTB, phosphotyrosine binding domain; RGS, regulator of G-protein signaling box; RFL, c-Raf1-like domain; GoLoco, novel $G\alpha_i/_o$ interaction motif first found in *Drosophila* Loco; CC, coiled-coil region. Asterisk denotes the existence within human RGS12 isoforms [B. E. Snow, R. A. Hall, A. M. Krumins, G. M. Brothers, D. Bouchard, C. A. Brothers, S. Chung, J. Mangion, A. G. Gilman, R. J. Lefkowitz, and D. P. Siderovski, *J. Biol. Chem.* **259**, 17749 (1998)] of an alternative C-terminal region that, in surface plasmon resonance biosensor binding assays, has been shown to be a binding target for the N-terminal PDZ domain.

domain homologs using profile hidden Markov models (HMMs).[28] Alternatives include HMM searching using the PFAM[29] database of protein domain family alignments (http://www.sanger.ac.uk/Pfam or http://pfam.wustl.edu) and BLAST[30] pairwise similarity searches of collections of domain sequences containing all identifiable members (e.g., the Schnipsel database at http://SMART.embl-heidel-berg.de); the latter is valuable to perform as domain outliers can be missed by profile HMM searches yet detected by their pairwise similarity to established member(s) of a domain family.[28]

SMART analysis of the rat RGS12 polypeptide sequence[14,31] identifies two well-characterized protein–protein interaction domains within the first 400 N-terminal amino acids (Fig. 4): PDZ (PSD-95/Dlg/ZO-1) and phosphotyrosine-binding (PTB) domains. PDZ domains are capable of binding internal regions[32] or, more often, the extreme carboxy termini[33] of target proteins and are found singly,

[28] J. Schultz, R. R. Copley, T. Doerks, C. P. Ponting, and P. Bork, *Nucleic Acids Res.* **28**, 231 (2000).
[29] E. L. Sonnhammer, S. R. Eddy, E. Birney, A. Bateman, and R. Durbin, *Nucleic Acids Res.* **26**, 320 (1998).
[30] S. F. Altschul, W. Gish, W. Miller, E. W. Myers, and D. J. Lipman, *J. Mol. Biol.* **215**, 403 (1990).
[31] B. E. Snow, R. A. Hall, A. M. Krumins, G. M. Brothers, D. Bouchard, C. A. Brothers, S. Chung, J. Mangion, A. G. Gilman, R. J. Lefkowitz, and D. P. Siderovski, *J. Biol. Chem.* **273**, 17749 (1998).
[32] B. J. Hillier, K. S. Christopherson, K. E. Prehoda, D. S. Bredt, and W. A. Lim, *Science* **284**, 812 (1999).
[33] Z. Songyang, A. S. Fanning, C. Fu, J. Xu, S. M. Marfatia, A. H. Chishti, A. Crompton, A. C. Chan, J. M. Anderson, and L. C. Cantley, *Science* **275**, 73 (1997).

or in tandem repeats, in a variety of scaffold proteins that localize enzymatic activities and other protein–protein interaction modules to specific submembraneous regions.[34] Methods for characterizing PDZ domain binding specificity are discussed in the remaining sections of this chapter. PTB domains were first assumed[35] to be strictly targeted to phosphotyrosine in the context of the tetrapeptide motif Asn-Pro-X-Tyr(p); however, it is now clear that PTB domains can also have nonphosphorylated polypeptide[36] or acidic phospholipid[37] binding targets. We have identified an initial binding target for the RGS12PTB domain in embryonic chick dorsal root ganglion neurons: the N-type Ca^{2+} channel (D.P.S., M.L. Schiff, G.M.B., B.E.S., M.A. Diversé-Pierluissi, manuscript submitted).

For detection of novel domain structures, polypeptide sequences present between recognizable domain structures identified by SMART are analyzed using a Position-Specific Iterated BLAST or "PSI-BLAST" search.[25] We have previously detailed[38] the use of PSI-BLAST in our discovery of a novel $G\alpha$-interacting module, the GoLoco motif, within RGS12 and RGS14 (Fig. 4); PSI-BLAST searches using the polypeptide region between the RGS domain and GoLoco motif (our unpublished observations and those published by Ponting[39]) have also revealed the presence of a tandemly-repeated c-Raf1-like (RFL) region (Fig. 4). The PSI-BLAST algorithm automatically constructs an *ad hoc* profile HMM based on the original query sequence and its significant pairwise similarity matches ("iteration 0") and replaces the original query with this HMM in a subsequent search ("iteration 1") of the nonredundant protein database. Newly identified sequence alignments with statistical significance above a set threshold are incorporated into a new version of the HMM for subsequent search rounds. The interactive version of PSI-BLAST (http://www.ncbi.nlm.nih.gov/blast/psiblast.cgi) allows the user at each iteration to override the automated inclusion or exclusion of specific alignments. One important consideration for enhancing predictive value is exploiting this interactivity: alignments are selected for inclusion into the next iteration if any additional information is available connecting the protein possessing the aligned sequence similarity to intracellular signaling machinery, irrespective of whether they fall above or below the default expect-value threshold of $E = 0.001$. Specifically, at each iteration of the PSI-BLAST search, a critical appraisal is made of all available literature on each reported similar sequence; this appraisal is greatly facilitated by using the HTML hot-links within the sequence

[34] C. P. Ponting, C. Phillips, K. E. Davies, and D. J. Blake, *Bioessays* **19**, 469 (1997).

[35] G. Wolf, T. Trub, E. Ottinger, L. Groninga, A. Lynch, M. F. White, M. Miyazaki, J. Lee, and S. E. Shoelson, *J. Biol. Chem.* **270**, 27407 (1995).

[36] S. E. Dho, S. Jacob, C. D. Wolting, M. B. French, L. R. Rohrschneider, and C. J. McGlade, *J. Biol. Chem.* **273**, 9179 (1998).

[37] S. E. Dho, M. B. French, S. A. Woods, and C. J. McGlade, *J. Biol. Chem.* **274**, 33097 (1999).

[38] D. P. Siderovski, M. A. Diversé-Pierluissi, and L. De Vries, *Trends Biochem. Sci.* **24**, 340 (1999).

[39] C. P. Ponting, *J. Mol. Med.* **77**, 695 (1999).

descriptor, provided by the interactive PSI-BLAST implementation, that allows immediate cross-referencing to the PubMed literature database and other NCBI Entrez services (http://www.ncbi.nlm.nih.gov/entrez). For example, in the case of our GoLoco motif discovery,[38] several proteins with "below-threshold" alignments (LGN; Pcp2, Rap1 GAP) were found to be described in the literature as interacting with Gα subunits or monomeric, Ras-superfamily G proteins and thus were included in subsequent search iterations. For a PSI-BLAST pairwise alignment derived from a protein sequence without associated literature (e.g., anonymous open-reading frames), a SMART search is performed on the entire protein sequence (not just the portion in the alignment) to ascertain whether established domain structures can be identified that suggest a functional role for this protein in intracellular signal transduction; if so, the alignment from this protein is included in subsequent search iterations.

Characterization of RGS12 PDZ Domain Binding

Subcloning of RGS12 PDZ Domains

The isolated PDZ domains of rat RGS12 (amino acids 1–94 of SwissProt O08774) and human RGS12 (amino acids 1–110 of SwissProt O14924) are expressed in *E. coli* as fusions with glutathione S transferase (GST). To construct the appropriate *E. coli* expression vectors, cDNA fragments corresponding to these polypeptide regions are first PCR amplified from the appropriate rat or human *RGS12* full-length cDNA plasmid using a proof-reading thermostable DNA polymerase (i.e., *Pfu* from Stratagene) and one of these sets of PDZ-domain flanking primers:

Rat *Rgs12* PDZ sense primer	5′-CATATGATGTACAGGGCTGGGGAGC-3′
Rat *Rgs12* PDZ antisense primer	5′-GTCGACGATCACCATGCGCAGCAC-3′
Human *RGS12* PDZ sense primer	5′-GTCGACAGAATGTTTAGAGCTGGGGAGG-3′
Human *RGS12* PDZ antisense primer	5′-AAGCTTTTCTTCATCACTGGAACAGGATTCG-3′

A 50 μl final volume containing 1× Pfu reaction buffer [20 mM Tris-HCl pH 8.8, 2 mM MgSO$_4$, 10 mM KCl, 10 mM (NH$_4$)$_2$SO$_4$, 0.1% Triton X-100, 100 μg/ml BSA], 200 μM dNTPs, 0.5 μM of each primer (sense and antisense), 200 ng of plasmid template, and 2.5 units of cloned *Pfu* DNA polymerase is amplified using the Touch-down/Touch-up cycling protocol (Fig. 2) with the following parameter modifications: optimal $T_{an} = 63°$, initial $t_{ex} = 20$ sec, total number of cycles = 12. In setting up the amplification reaction, care is taken to add the DNA polymerase last and then immediately begin thermocycling to minimize primer degradation by the 3′ to 5′-exonuclease activity of the proofreading polymerase. After thermocycling is complete, 1 unit of *Taq* polymerase (Roche) is added and the reaction mixture incubated for 10 min at 72°, in order to add 3′-adenine overhangs to PCR amplicons and thereby allow for ligation into the pCR2.1-TOPO vector as described above. Gel purification of PCR amplicons is imperative prior to ligation into pCR2.1-TOPO and *E. coli* transformation, given the large quantity of input supercoiled plasmid template (200 ng) in the reaction mixture which, if not removed, will also transform competent *E. coli*.

Sequencing is performed on positive pCR2.1-rRGS12PDZ and pCR2.1-hRGS12PDZ plasmid clones to eliminate the possibility of PCR-generated mutations, as well as to identify the reading frame of the EcoRI site (within the pCR2.1-TOPO polylinker) that is 5' of the PDZ coding sequence. The PDZ coding sequence is then subcloned into the appropriate version of pGEX (Amersham Pharmacia) for in-frame fusion with the GST coding sequence: the human RGS12 PDZ insert was ligated into the pGEX4T2 *Eco*RI site [after treatment of *Eco*RI-cut pGEX4T2 DNA with calf intestinal phosphatase (New England Biolabs, Beverly, MA)], whereas the rat RGS12 PDZ insert was forced-directionally cloned into the *Eco*RI and *Xho*I sites of pGEX4T1 (using the *Sal*I site encoded within the rat *Rgs12* PDZ antisense primer, as underlined above).

Purification of RGS12 PDZ Domains

The resultant pGEX/PDZ expression plasmids are transformed into the protease-deficient *Escherichia coli* strain BL21 (Stratagene) and plated onto LB-agar (Life Technologies, Rockville, MD) containing 50 μg/ml carbenicillin to select for stable transformants (overnight growth at 37°). A single colony is selected and grown overnight at 37° shaking at 250 rpm in 30 ml of LB broth (Life Technologies) supplemented with carbenicillin. The overnight culture is used to seed three 1 liter volumes of 2× YT broth (Life Technologies) in baffled 6 liter flasks, which are incubated at 37° with shaking at 250 rpm until reaching a density of $OD_{600nm} = 0.7$–0.8. Expression of GST-fusion protein is induced by the addition of isopropyl-β-D-thiogalactopyranoside (IPTG) to a 0.8–1.0 mM final concentration and further shaking incubation at 37° for 4–5 hours. Cells are pelleted by centrifugation at 4° for 15 min at 6000g, washed once with phosphate-buffered saline (PBS: 137 mM NaCl, 2.7 mM KCl, 4.3 mM NaH$_2$PO$_4$, 1.47 mM KH$_2$PO$_4$, pH 7.4) containing Complete protease inhibitor cocktail (Roche; used as per manufacturer's instructions), flash-frozen in a dry-ice/2-propanol bath, and stored frozen at −80° prior to the lysis procedure below.

Frozen cell pellets are thawed in 10 volumes (10 ml per gram wet weight; 3 liters of culture weighs ∼11 g) of lysis buffer [PBS containing 1% Nonidet P-40 (NP-40), 5 mM 2-mercaptoethanol, Complete protease inhibitor cocktail, 2 μg/ml pepstatin A]. The cell pellet is resuspended with a glass Dounce homogenizer and lysed by sonication. Lysate is then clarified by centrifugation at 27,000g for 30 min at 4°. Cleared lysate is passed over a 2 ml bed volume of glutathione-Sepharose 4B column (Amersham Pharmacia) preequilibrated with lysis buffer (gravity flow). The column is washed with 40 ml of lysis buffer prior to elution of GST-PDZ protein

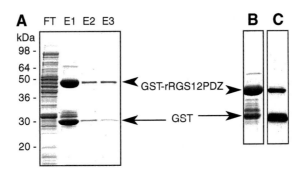

FIG. 5. Purification of GST-rRGS12PDZ(1-94) fusion protein by glutathione-Sepharose 4B chromatography. (A) Samples from sequential elutions (lanes E1–E3) of GST-rRGS12PDZ(1-94) fusion protein bound to glutathione-Sepharose 4B were resolved by SDS–PAGE on a 10–20% polyacrylamide gradient gel and visualized by Coomassie blue staining. FT, sample of flowthrough material after the application of *E. coli* lysate to glutathione-Sepharose 4B column. (B) and (C) represent Coomassie blue and anti-GST immunoblot detection, respectively, of the final concentrated GST-rRGS12PDZ(1-94) preparation. A breakdown product of ~30 kDa present as a minor contaminant in the fusion protein preparation is identified as GST by immunoblotting (panel C).

by incubation for 1 hr with 4 ml of elution buffer (10 mM glutathione/50 mM Tris-HCl, pH 7.8). Eluant is collected and the resin is washed with an additional 6 ml elution buffer. Eluted protein (10 ml pooled volume) is dialyzed against storage buffer (50 mM Tris-HCl, pH 7.6, 150 mM NaCl, 1 mM EDTA, 5 mM DTT, 10% glycerol), concentrated by ultrafiltration using an Amicon Series 8000 stirred-cell pressure concentrator with a YM-10 filter (Millipore, Bedford, MA), and stored in small aliquots at −80°. The final yield per liter of cell culture is ~2 mg/liter for both proteins. (Protein aliquots to be used in biosensor assays are thawed once and never refrozen.) Using this procedure, GST-hRGS12PDZ(1-110) protein (predicted molecular weight of 40.2 kDa) was obtained at a final concentration of 1.9 mg/ml, and GST-rRGS12PDZ(1-94) protein (predicted molecular weight of 38.4 kDa) was obtained at a final concentration of 0.88 mg/ml. An example of purification of GST-rRGS12PDZ(1-94) using this procedure is illustrated in Fig. 5.

Surface Plasmon Resonance Biosensor Analysis of PDZ Binding Specificity

Methods for identifying candidate C-terminal polypeptide PDZ-docking sites have been described that employ screening of chemically synthesized peptide libraries[33] or Lac repressor/C-terminal peptide fusion libraries.[40] In the case of the RGS12 PDZ domain, we noticed sequence similarity[31] to the first PDZ domain

[40] N. L. Stricker, P. Schatz, and M. Li, *Methods Enzymol.* **303,** 451 (1999).

of NHERF, a regulator of the Na^+/H^+-exchanger type 3 (NHE3) that associates with the carboxy terminus of the β_2-adrenergic receptor.[41] Assuming that such sequence similarity might also reflect functional similarity in binding G protein-coupled receptor (GPCR) C termini, we devised a rapid method of screening chemically synthesized, biotinylated peptides corresponding to GPCR C termini using streptavidin-coated surface plasmon resonance biosensors. Our choice of receptor C termini to test for PDZ binding[31] was aided by an excellent Internet-based resource for accessing GPCR primary amino-acid sequences (http://www.expasy.ch/cgi-bin/lists?7tmrlist.txt), curated by Dr. Amos Bairoch at the Swiss Institute of Bioinformatics.

Peptides corresponding to the final 12 to 20 amino acids of candidate C-terminal PDZ-docking sites are chemically synthesized by a commercial vendor or in-house peptide synthesis facility. These peptides are modified by the addition of biotin to the primary amine group at the N terminus; we have routinely requested N-terminal biotinylation using succinimidyl 6-[biotinamido]hexanoate (Sigma) which incorporates a "long-chain" aminocaproyl spacer between the biotin group and the peptide N terminus. (If possible, it is valuable to request a split synthesis of biotinylated and N-terminally unmodified peptide; the latter is useful in competition binding studies.) Direct amine coupling of peptides to carboxymethylated dextran biosensor surfaces via surface matrix activation with N-hydroxysuccinimide and N-ethyl-N'-(dimethylaminopropyl)carbodiimide[42] has been described for use in PDZ binding studies[33]; however, the use of N-terminal biotinylated peptides and a streptavidin capture surface allows for the uniform orientation of peptide and avoids the potential for covalent coupling to amine-bearing amino acid side chains critical for PDZ binding.

One to 5 mg samples of lyophilized peptide stocks are weighed out directly in 50 ml conical polypropylene tubes; as static is a considerable problem (some peptides have a tendency to "fly away" during transfer), avoid wearing gloves and use an antistatic gun (e.g., Zerostat3; Sigma) to neutralize static surrounding the analytical balance. Peptide samples are initially dissolved in H_2O or, if necessary for solubilization, dimethyl sulfoxide (DMSO) to a stock concentration of 10 mg/ml. Subaliquots are diluted 10-fold in H_2O to a 1 mg/ml concentration, and subaliquots of this solution are serially diluted in BIA running buffer [10 mM HEPES, pH 7.4, 150 mM NaCl, 3 mM EDTA, 0.005% (v/v) surfactant P20] to a final concentration of 0.1 μg/ml; all stock and subdiluted peptide solutions are stored long-term at $-80°$.

Four independent biosensor surfaces (flow cells 1-4 or Fc1-4) are created and analyzed for PDZ binding affinity using the BIAcore 2000 (Biacore Inc.,

[41] R. A. Hall, R. T. Premont, C. W. Chow, J. T. Blitzer, J. A. Pitcher, A. Claing, R. H. Stoffel, L. S. Barak, S. Shenolikar, E. J. Weinman, S. Grinstein, and R. J. Lefkowitz, *Nature* **392,** 626 (1998).
[42] B. Johnsson, S. Lofas, and G. Lindquist, *Anal. Biochem.* **198,** 268 (1991).

Piscataway, NJ). A streptavidin-coated sensor chip (Sensor Chip SA, Code No. BR-1000-32; Biacore) is mounted into the BIAcore 2000 using the DOCK command. Freshly filtered (0.22 μm) and degassed BIA running buffer ($\sim$200 ml) is added to the intake lines and the pumps are flushed using the PRIME command. A new sensorgram is initiated using the RUN command, specifying an initial flow rate of 20 μl/min, multichannel detection (Fc1-4), and multichannel flow path (Fc1-4). Prior to peptide injections, the sensor surface is pretreated with three 20 μl pulses of regeneration buffer (1 M NaCl/50 mM NaOH) using the INJECT command. Flow rate is then decreased to 5 μl/min prior to loading four different peptides to the four independent flow cells. At least one flow cell should be a negative control surface or "reference cell" to control for nonspecific binding; for RGS12 PDZ binding studies, we have used peptides derived from the last 12 amino acids of the mouse Notch1 protein (biotin-PSQITHIPEAFK-cooh) or the rat β_1-adrenergic receptor ("rβ1AR"; biotin-PGRQGFSSESKV-cooh).

Switch the flow path to the desired single flow cell using the FLOWPATH command. Using the MANUAL INJECT command, load the autosampler loop with 20–30 μl of dilute peptide solution (0.1 μg/ml) and begin injecting in manual mode. Inject peptide until the response units (RUs) for that flow cell plateaus at a new maximum. For a 12 amino acid, biotinylated peptide (1500–1600 Da), expect an injection of $\sim$10 μl to yield a fully loaded surface of $\sim$300 RU. (For the purposes of these rapid relative binding measurements, saturated streptavidin–peptide surfaces are employed; for accurate kinetic rate and binding affinity measurements,[43, 44] varied surface conjugation levels are required to minimize mass transport effects.[45]) We routinely use dilute peptide solutions (0.1–2 μg/ml) during surface loading to minimize cross-channel "carryover" observable on switching flow cells after injections of highly concentrated peptide solutions.

After repeating the peptide loading procedure for each of the four flow cells in turn, the flow path is returned to all four flow cells in series (Fc1-4) for two 10 μl pulses of regeneration buffer across the entire biosensor at a flowrate of 20 μl/min. All subsequent protein injections are performed at a 10 μl/min flow rate, with intervening surface regenerations performed by single 10 μl injections of regeneration buffer at 20 μl/min flow rate. (It is important to use the EXTRA-CLEAN option on regeneration pulse injections in order to avoid carryover of regeneration buffer into subsequent protein injections.)

GST-RGS12PDZ protein is diluted to 10 μM in BIA running buffer. A 200 μl aliquot of diluted protein and a $\sim$2 ml aliquot of BIA running buffer (along with 6 empty plastic vials) are added to the autosampler rack. Using multiple DILUTE

[43] D. J. O'Shannessy, M. Brigham-Burke, K. K. Soneson, P. Hensley, and I. Brooks, *Methods Enzymol.* **240,** 323 (1994).

[44] T. A. Morton and D. G. Myszka, *Methods Enzymol.* **295,** 268 (1998).

[45] A. A. Kortt, L. C. Gruen, and G. W. Oddie, *J. Mol. Recognit.* **10,** 148 (1997).

commands (with default parameters that specify a 50% mixing of 100 μl protein with 100 μl BIA running buffer), a 2-fold dilution series of the 10 μM GST-RGS12PDZ stock is generated by the autosampler (down to 1/64th or 0.16 μM). Each protein dilution is separately injected over all four flow cells (with intervening regeneration pulses) using the KINJECT command (parameters: injected volume $= 50 \mu$l $= 300$ sec at a 10 μl/min flow rate, dissociation time of 300 sec). Use of the KINJECT command allows for a standard time of dissociation to be specified for each protein injection; alternatively, the COINJECT command can be used to inject free (unbiotinylated) peptide immediately after GST-RGS12PDZ injection (rather than the default BIA running buffer) to minimize any potential for rebinding events during dissociation. [Purified GST protein, diluted to 10 μM in BIA running buffer and injected over peptide–streptavidin surfaces, has routinely given ($>$30 peptides tested[31]) very low background responses ($<$25 RU).]

After all injections are complete, the sensorgram is stopped and saved to disk. The sensorgram file is then imported into BIAevaluation 3.0 (Biacore). Areas in the lengthy sensorgram tracing containing binding curves from individual protein injections are then isolated and saved as separate files; these curves are clearly demarcated by the large bulk shifts in response units arising from regeneration buffer injections. For each set of binding curves arising from a single protein injection, two normalizations are performed: (a) the y axis is normalized so that all curves begin at a relative $RU = 0$ prior to the GST–PDZ injection time and (b) the x axis of each curve is separately adjusted to set the beginning of GST–PDZ injection at timepoint $= 0$ sec. The relative increases in RU values for each flow cell with test peptides ("test") at timepoint $= 286$ sec (RU_{286s}; an arbitrarily chosen timepoint just before injection cessation) is used to calculate a relative binding ratio [Eq. (1)] normalized to a positive control (wt) and negative control (ctrl) peptide surfaces. We routinely normalize binding data to that observed between RGS12-PDZ and the C-terminal tail of the interleukin (IL)-8 receptor CXCR2.

$$\text{Relative binding} = \frac{RU_{286s}^{test} - RU_{286s}^{ctrl}}{RU_{286s}^{wt} - RU_{286s}^{ctrl}} \tag{1}$$

Sample results are depicted in Fig. 6 in using the biosensor binding assay to assess relative binding avidity of GST-hRGS12PDZ(1-110) to a peptide derived from the alternative C terminus of human RGS12 protein ("hRGS12Tail2"; biotin-KPKTSAHHATFV-cooh[31]). As normalized to binding observed to the positive control peptide, hCXCR2 (biotin-VGSSSGHTSTTL-cooh), the relative binding of hRGS12-PDZ protein to the alternative RGS12 C-terminal peptide is (428 − 74.9)/(1149 − 74.9) = 0.33. This metric is useful for ranking the relative binding avidity of the PDZ domain to different GPCR C-terminal peptides, as well as ranking avidity for modified peptide sequences vs wild-type (e.g., alanine/serine-scanning mutagenesis, C-terminal amidation). We have used both forms of modification to the wild-type CXCR2 peptide to demonstrate that the PDZ domains of

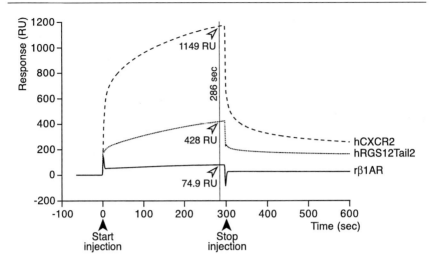

FIG. 6. Normalized binding curves from surface plasmon resonance biosensor analysis of PDZ domain binding to peptide/streptavidin surfaces. Time 0 to 300 sec indicates the association phase during which 50 μl of 10 μM GST-hRGS12PDZ(1-110) fusion protein is injected over three independent peptide surfaces, namely, the C termini of the human interleukin-8 receptor CXCR2, the alternative human RGS12 isoform, and the rat β_1-adrenergic receptor (see text for peptide sequences). Figure adapted with permission from D. P. Siderovski, B. Strockbine, and C. I. Behe, *Crit. Rev. Biochem. Mol. Biol.* **34**, 215 (1999). © CRC Press, Boca Raton, Florida.

rat and human RGS12 bind selectively to C-terminal (A/S)-T-X-(L/V) motifs,[31] a binding specificity characteristic of "Group I" PDZ domains.[33]

Yeast Interaction Trap Assay for PDZ Binding Specificity

PDZ domain binding specificity can also be assessed using the yeast two-hybrid system. As the fundamentals of this technique are well-documented in this series,[46,47] only those details specific to testing the binding specificity of the RGS12 PDZ domain[31] are discussed below. Although biosensor measurements of PDZ/peptide interactions have an advantage in allowing the potential for (a) incorporation of nonnatural amino acids and carboxy-terminal modifications within PDZ-docking targets and (b) quantitation of binding kinetics, the yeast two-hybrid system is clearly a less expensive and more accessible approach. We have employed the yeast two-hybrid system to confirm our biosensor results regarding the sequence requirements for RGS12 PDZ domain binding to the C-terminal tail of the IL-8 receptor CXCR2.

[46] P. L. Bartel and S. Fields, *Methods Enzymol.* **254**, 241 (1995).
[47] C. Bai and S. J. Elledge, *Methods Enzymol.* **283**, 141 (1997).

To express a "bait" composed of the rat RGS12 PDZ domain (aa 1–94) fused to the DNA-binding domain (DBD) of the yeast Ga14p transcription factor (aa 1–147), the *Eco*RI/*Sal*I fragment of pCR2.1-rRGS12PDZ (as described above) is forced-directionally cloned into the *Eco*RI and *Sal*I sites of the pAS1 plasmid.[48] To construct a "prey" composed of the rat CXCR2 C terminus (aa 322–359 of SwissProt P35407) fused to the Ga14p activation domain (AD), the corresponding *CXCR2* cDNA fragment is first PCR amplified using 0.5 ng of Marathon-Ready rat brain cDNA (Clontech) and the Expand High Fidelity PCR system (as per manufacturer's instructions; Cat #1732641, Roche) with the following primers:

Rat *CXCR2* sense primer	5′-*GAATTC*ATGGACTTCTCAAGATCATGGCT-3′
Rat *CXCR2* antisense primer	5′-CTCGAGAAGGACAAGAAGGGAACCCAGAGG-3′

A separate PCR reaction is performed using the above sense primer with an alternative antisense primer (5′-CTCGAGctaGGTGTTCGCTGAAGAAGAGCCAAC-3′); this primer encodes a premature stop-codon (in lowercase) which eliminates the final pentapeptide sequence (Thr355-Ser-Thr-Thr-Leu359) of the CXCR2 C terminus ("Δ355-359"). PCR products are gel purified, cloned into pCR2.1-TOPO, and sequence verified as described above. Both *CXCR2* cDNA fragments are then forced-directionally subcloned into *Eco*RI and *Xho*I sites of the Ga14p-AD expression vector pACTII,[48] using the *Eco*RI and *Xho*I sites designed into the primer sequences (italics and underlined, respectively, above). Point mutations to the CXCR2 C-terminal three amino acids are generated within the pCR2.1-rCXCR2 vector, prior to subcloning into pACTII, using the QuikChange site-directed mutagenesis system (Stratagene) according to manufacturer's instructions and the following, overlapping primer pairs (mutated nucleotides in italics):

Leu-359>Ala sense	5′-GAACACCTCCACTACC*GCT*TAAGACTGTTTAC-3′
Leu-359>Ala antisense	5′-GTAAACAGTCTTA*AGC*GGTAGTGGAGGTGTTC-3′
Thr-357>Ser sense	5′-CTTCAGCGAACACCTCG*TCG*ACCCTCTAAGACTG-3′
Thr-357>Ser antisense	5′-CAGTCTTAGAGGGTCGACGAGGTGTTCGCTGAAG-3′
Thr-358 sense	5′-CAGCGAACACCTCCACT(*CGT*)(*ACGT*)CCTCTAGGACTGTTTAC-3′
Thr-358 antisense	5′-GTAAACAGTCTTAGAGG(*ACGT*)(*ACG*)AGTGGAGGTGTTCGCTG-3′

The last primer set (Thr-358 sense/antisense) is degenerate in the first two nucleotides of the Thr-358 codon, allowing for 12 different amino acids to be encoded at this position (Ala, Arg, Asp, Cys, Gly, His, Leu, Phe, Pro, Ser, Tyr, Val) without the possibility for a stop-codon to be generated given the cytosine in the third codon position (underlined above).

Prototrophic growth of the *S. cerevisiae* strain PJ69-4A[49] on histidine- and adenine-deficient medium is used to test for interaction between Ga14p-DBD-rRGS12PDZ and Ga14p-AD-rCXCR2 fusion proteins. A single, isolated colony

[48] T. Durfee, K. Becherer, P. L. Chen, S. H. Yeh, Y. Yang, A. E. Kilburn, W. H. Lee, and S. J. Elledge, *Genes Dev.* **7**, 555 (1993).
[49] P. James, J. Halladay, and E. A. Craig, *Genetics* **144**, 1425 (1996).

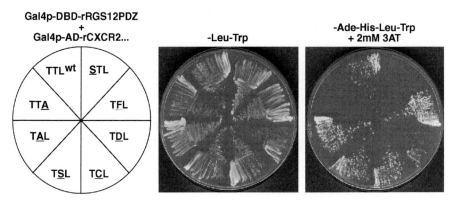

FIG. 7. Yeast two-hybrid analysis of rat RGS12 PDZ domain interaction with the wild-type (wt) or single residue-substituted rat CXCR2 C termini. Yeast strain PJ69-4A is cotransformed with bait/prey plasmid pairs expressing Gal4p-DBD-rRGS12PDZ and Gal4p-AD-rCXCR2 fusion proteins. *Left:* Relative position of Gal4p-AD-rCXCR2 prey mutants and the corresponding sequence of their C-terminal three amino acids. Growth in all eight sections of the middle panel (-Leu-Trp) demonstrates that all transformed yeast colonies are competent for growth on medium lacking leucine and tryptophan, indicating incorporation of bait and prey plasmids. Lack of growth (over 4 days) of yeast transformed with Gal4p-AD-rCXCR2-(TT$\underline{A}$) and -($\underline{S}$TL) mutant prey expression plasmids (*right:* -Ade-His-Leu-Trp + 2mM 3AT) indicates loss of PDZ/CXCR2 interaction, consistent with the known importance of the −2 and terminal residue positions to RGS12/CXCR2 binding as observed in biosensor studies [B. E. Snow, R. A. Hall, A. M. Krumins, G. M. Brothers, D. Bouchard, C. A. Brothers, S. Chung, J. Mangion, A. G. Gilman, R. J. Lefkowitz, and D. P. Siderovski, *J. Biol. Chem.* **259,** 17749 (1998)].

of PJ69-4A yeast is used to inoculate a 100 ml volume of rich YPD broth (BIO 101, Vista, CA) in a baffled 1000 ml glass flask. This culture is grown for 12 hr at 30° with vigorous aeration (~250 rpm) in a shaking incubator and culture density measured by absorbance for confirmation of mid-log growth phase status ($OD_{600nm} = 0.4$–0.6). Yeast cells are then treated with lithium acetate[50] for cotransformation of pAS1-rRGS12PDZ and pACTII-rCXCR2 plasmid pairs using the alkali–cation transformation system (BIO 101) exactly according to manufacturer's instructions. Negative control plasmid pairs (i.e., pAS1 parental vector + pACTII-rCXCR2; pAS1-rRGS12PDZ + pACTII parental vector) are also cotransformed at this time. Cotransformed yeast are then plated onto synthetic dropout medium plates (SDA-Leu-Trp; BIO 101) lacking leucine (to select for incorporation of the pACTII plasmid bearing the *LEU2* gene) and tryptophan (to select the pAS1 plasmid bearing the *TRP1* gene) and incubated for at least 2 days at 30°.

To test for interaction of PDZ bait and CXCR2 prey, six independent colonies from each SDA-Leu-Trp plate are streaked onto synthetic dropout medium plates

---

[50] R. D. Gietz, R. H. Schiestl, A. R. Willems, and R. A. Woods, *Yeast* **11,** 355 (1995).

(SDA-Ade-His-Leu-Trp; BIO 101 Cat #4842-925) lacking adenine and histidine. [For the strain PJ69-4A, this medium is routinely supplemented with 2 mM 3-amino-1,2,4-triazole (3-AT; BIO 101), an inhibitor of the histidine synthesis pathway; a stock solution of 1 M 3-AT is prepared in sterile water, filter-sterilized (0.22 μm), and stored away from light at$-20°$.] Plates are incubated for 3–4 days at $30°$ to allow for growth. Sample results testing the relative importance of amino acids 357 to 359 ("TTL") in the rat CXCR2 C terminus to RGS12 PDZ domain binding are illustrated in Fig. 7; mutation of the -2 and terminal residues of the C-terminus (Thr-357 to serine and Leu-359 to alanine, respectively) abolishes PDZ/CXCR2 interaction, whereas several different amino acids at the -1 position (Thr-358 to alanine, aspartate, cysteine, phenylalanine, or serine) do not disrupt the interaction. These results are entirely consistent with the known binding specificity of Group I PDZ domains,[33] as well as human RGS12 PDZ/human CXCR2 interaction data generated by biosensor studies.[31]

Concluding Remarks

Novel RGS proteins can be cloned using this degenerate PCR strategy (e.g., RGS12 and RGS14). These novel RGS proteins, with their multidomain structures, suggest higher-order functionality (beyond Gα-directed GAP activity) in the coordination of signal transduction pathways transacted by heterotrimeric G proteins, monomeric G proteins, and/or tyrosine kinases. The methods herein described for assessing PDZ domain binding specificity should prove useful in the analysis of other PDZ domain proteins including PDZ–RhoGEF, the second RGS protein found to contain an N-terminal PDZ domain.[16]

Acknowledgments

We thank Praag Arya, Sudha Arya, Denis Bouchard, Carol Anne Brothers, Stephen Chung, and Joan Da Costa for their contributions, and Cindy Behe for critical appraisal of the manuscript. D.P.S. is supported by the NIH Grant GM62338.

Author Index

Subject Index

A

Activator of G protein signaling, AGS3–G
  protein interactions using glutathione
  S-transferase fusion proteins, 533–535
Adenovirus, $G_s\alpha$ constitutively active mutant
  expression
  replication-defective vector construction
    containing $G_s\alpha$ Q227L mutant
    advantages of system, 328
    Cre-mediated site-specific recombination,
      overview, 328–329
    cyclic AMP assay for functional
      characterization, 341–343
    FLAG epitope tagging, 331
    materials, 329–331
    plasmid rescue, 334, 336
    plasmid sequencing, 332–334
    purification of recombinant virus, 338–339
    shuttle vector insertion of gene, 331–332
    titering, 339
    two-plasmid method, 328
    viral clone isolation, 336–338
    Western blotting, 340–341
  transduction in mouse hippocampus
    immunohistochemical localization, 349
    long-term spatial memory studies, 346–348
    protein kinase A induction, 345
    rationale, 345, 349
Adenylyl cyclase
  $G\beta$, interactions with adenylyl cyclase 2
    peptide
    advantages and limitations of molecular
      modeling, 585–586
    cross-linking studies, 577–578
    docking modeling, 581–584
    experimental verification of interaction
      sites, 584–585
    residue solvent accessibility calculation,
      578–579
    secondary structure prediction for peptide,
      579–580
    surface electrostatic potential visualization,
      580–581

$G\beta\gamma$ mutation effects, assay, 429–430
G_s inhibitor assay, 476–477
  receptor-mediated activation of G_s, assay,
    271–273
ADP-ribosylation factor, *N*-myristoyltransferase
  coexpression in *Escherichia coli* for
  purification
  acylation efficiency
    factors affecting, 186–187, 189
    studies, 190–193
  methionine aminopeptidase role in system,
    186, 191–193
  rationale, 187
  vectors, 188–190
α_2-Adrenergic receptor
  signal restoration assay
    applications, 144–145
    brain G protein preparation, 142
    incubation conditions, 142–143
    materials, 142
    membrane preparations, 142
    overview, 140–141
    properties of assay system, 143–144
    receptor density determination, 143
  solution phase assay for G protein regulators
    applications, 149–150, 152
    incubation conditions, 148
    membrane extracts, 147–148
    overview, 145, 147
    properties of assay system, 148–149
β-Adrenergic receptor kinase,
  see G protein-coupled receptor kinase
Affinity shift assay, G protein for receptors
  advantages, 79
  G protein concentration, 79–80
  incubation conditions, 80–81
  normalized affinity shift activity analysis, 81
  principles, 78–79
  receptor density, 80
  reconstitution, 80
AGS, *see* Activator of G protein signaling
Antisense oligodeoxynucleotide,
  $G\alpha$ suppression

ISBN 0-12-182245-1

90051